Nanotechnology

Volume 3: Information Technology I
Edited by Rainer Waser

Related Titles

Nanotechnologies for the Life Sciences

Challa S. S. R. Kumar (ed.)

Volume 1: Biofunctionalization of Nanomaterials
2005
978-3-527-31381-5

Volume 2: Biological and Pharmaceutical Nanomaterials
2005
978-3-427-31382

Volume 3: Nanosystem Characterization Tools in the Life Sciences
2005
978-3-527-31383-9

Volume 4: Nanodevices for the Life Sciences
2006
978-3-527-31384-6

Volume 5: Nanomaterials - Toxicity, Health and Environmental Issues
2006
978-3-527-31385-3

Volume 6: Nanomaterials for Cancer Therapy
2006
978-3-527-31386-0

Volume 7: Nanomaterials for Cancer Diagnosis
2006
978-3-527-31387-7

Volume 8: Nanomaterials for Biosensors
2006
978-3-527-31388-4

Volume 9: Tissue, Cell and Organ Engineering
2006
978-3-527-31389-1

Volume 10: Nanomaterials for Medical Diagnosis and Therapy
2007
978-3-527-31390-7

Nanotechnology

Günter Schmid (ed.)

Volume 1: Principles and Fundamentals
2008
978-3-527-31732-5

Harald Krug (ed.)

Volume 2: Environmental Aspects
2008
978-3-527-31735-6

Rainer Waser (ed.)

Volume 3: Information Technology I
2008
978-3-527-31738-7

Rainer Waser (ed.)

Volume 3: Information Technology II
2008
978-3-527-31737-0

Viola Vogel (ed.)

Volume 5: Nanomedicine and Nanobiotechnology
2009
978-3-527-31736-3

Harald Fuchs (ed.)

Volume 6: Nanoprobes
2009
978-3-527-31733-2

Michael Grätzel, Kuppuswamy Kalyanasundaram (eds.)

Volume 7: Light and Energy
2009
978-3-527-31734-9

Lifeng Chi (ed.)

Volume 8: Nanostructured Surfaces
2009
978-3-527-31739-4

www.wiley.com/go/nanotechnology

G. Schmid, H. Krug, R. Waser, V. Vogel, H. Fuchs,
M. Grätzel, K. Kalyanasundaram, L. Chi (Eds.)

Nanotechnology

Volume 3: Information Technology I

Edited by Rainer Waser

WILEY-VCH Verlag GmbH & Co. KGaA

The Editor

Prof. Dr.-Ing. Rainer Waser
RWTH Aachen
Institut für Elektrotechnik II
Sommerfeldstr. 24
52074 Aachen

Cover: Nanocar reproduced with kind permission of Y. Shirai/Rice University

All books published by Wiley-VCH are carefully produced. Nevertheless, authors, editors, and publisher do not warrant the information contained in these books, including this book, to be free of errors. Readers are advised to keep in mind that statements, data, illustrations, procedural details or other items may inadvertently be inaccurate.

Library of Congress Card No.: applied for

British Library Cataloguing-in-Publication Data
A catalogue record for this book is available from the British Library.

Bibliographic information published by the Deutsche Nationalbibliothek
Die Deutsche Nationalbibliothek lists this publication in the Deutsche Nationalbibliografie; detailed bibliographic data are available in the Internet at http://dnb.d-nb.de.

© 2008 WILEY-VCH Verlag GmbH & Co. KGaA, Weinheim

All rights reserved (including those of translation into other languages). No part of this book may be reproduced in any form – by photoprinting, microfilm, or any other means – nor transmitted or translated into a machine language without written permission from the publishers. Registered names, trademarks, etc. used in this book, even when not specifically marked as such, are not to be considered unprotected by law.

Typesetting Thomson Digital, Noida, India
Printing betz-druck GmbH, Darmstadt
Binding Litges & Dopf Buchbinderei GmbH, Heppenheim

Printed in the Federal Republic of Germany
Printed on acid-free paper

ISBN: 978-3-527-31738-7

Contents

Preface *XV*
List of Contributors *XIX*

I   **Basic Principles and Theory** *1*

1   **Phase-Coherent Transport** *3*
Thomas Schäpers
1.1   Introduction *3*
1.2   Characteristic Length Scales *4*
1.2.1   Elastic Mean Free Path *4*
1.2.2   Inelastic Mean Free Path *5*
1.2.3   Phase-Coherence Length *5*
1.2.4   Transport Regimes *6*
1.3   Ballistic Transport *7*
1.3.1   Landauer–Büttiker Formalism *7*
1.3.2   Split-Gate Point Contact *9*
1.4   Weak Localization *13*
1.4.1   Basic Principles *14*
1.4.2   Weak Localization in One and Two Dimensions *15*
1.4.3   Weak Localization in a Magnetic Field *16*
1.5   Spin-Effects: Weak Antilocalization *19*
1.6   Al'tshuler–Aronov–Spivak Oscillations *21*
1.7   The Aharonov–Bohm Effect *23*
1.8   Universal Conductance Fluctuations *27*
1.8.1   Basic Principles *27*
1.8.2   Detailed Analysis *29*
1.8.3   Fluctuations in Long Wires *31*
1.8.4   Energy and Temperature Dependence *32*
1.9   Concluding Remarks *33*
    References *34*

2	**Charge Transport and Single-Electron Effects in Nanoscale Systems** 37
	Joseph M. Thijssen and Herre S.J. van der Zant
2.1	Introduction: Three-Terminal Devices and Quantization 37
2.2	Description of Transport 40
2.2.1	Structure of Nanoscale Devices 40
2.2.1.1	The Reservoirs 40
2.2.1.2	The Leads 41
2.2.1.3	The Island 41
2.2.2	Transport 42
2.2.2.1	Coherent-Incoherent Transport 43
2.2.2.2	Elastic–Inelastic Transport 44
2.2.2.3	Resonant–Off-Resonant Transport 44
2.2.2.4	First-Order versus Higher-Order Processes 44
2.2.2.5	Direct Tunneling 45
2.3	Resonant Transport 45
2.4	Constant Interaction Model 49
2.5	Charge Transport Measurements as a Spectroscopic Tool 53
2.5.1	Electronic Excitations 55
2.5.2	Including Vibrational States 57
2.6	Second-Order Processes 59
2.6.1	The Kondo Effect in a Quantum Dot with an Unpaired Electron 60
2.6.2	Inelastic Co-Tunneling 61
	References 63

3	**Spin Injection–Extraction Processes in Metallic and Semiconductor Heterostructures** 65
	Alexander M. Bratkovsky
3.1	Introduction 65
3.2	Main Spintronic Effects and Devices 67
3.2.1	TMR 67
3.2.2	GMR 68
3.2.3	(Pseudo)Spin-Torque Domain Wall Switching in Nanomagnets 70
3.3	Spin-Orbital Coupling and Electron Interference Semiconductor Devices 72
3.3.1	Spin-Hall Effect (SHE) and Magnetoresistance due to Edge Spin Accumulation 74
3.3.2	Interacting Spin Logic Circuits 76
3.4	Tunnel Magnetoresistance 77
3.4.1	Impurity Suppression of TMR 80
3.4.2	Negative Resonant TMR? 81
3.4.3	Tunneling in Half-Metallic Ferromagnetic Junctions 82
3.4.4	Surface States Assisted TMR 84
3.4.5	Inelastic Effects in TMR 84
3.5	Spin Injection/Extraction into (from) Semiconductors 86
3.5.1	Spin Tunneling through Modified (Delta-Doped) Schottky Barrier 89

3.5.2	Conditions for Efficient Spin Injection and Extraction	95
3.5.3	High-Frequency Spin-Valve Effect	97
3.5.4	Spin-Injection Devices	99
3.5.5	Spin Source of Polarized Radiation	101
3.6	Conclusions	104
	References	104

4 Physics of Computational Elements 109
Victor V. Zhirnov and Ralph K. Cavin

4.1	The Binary Switch as a Basic Information-Processing Element	109
4.1.1	Information and Information Processing	109
4.1.2	Properties of an Abstract Binary Information-Processing System	110
4.2	Binary State Variables	111
4.2.1	Essential Operations of an Abstract Binary Switch	111
4.2.2	The Use of Particles to Represent Binary Information	111
4.3	Energy Barriers in Binary Switches	113
4.3.1	Operation of Binary Switches in the Presence of Thermal Noise	113
4.3.2	Quantum Errors	114
4.3.3	A Combined Effect of Classical and Quantum Errors	116
4.4	Energy Barrier Framework for the Operating Limits of Binary Switches	116
4.4.1	Limits on Energy	116
4.4.2	Limits on Size	117
4.4.3	Limits on Speed	118
4.4.4	Energy Dissipation by Computation	119
4.5	Physics of Energy Barriers	119
4.5.1	Energy Barrier in Charge-Based Binary Switch	120
4.5.2	Energy Barrier in Spin-Based Binary Switch	124
4.5.3	Energy Barriers for Multiple-Spin Systems	128
4.5.4	Energy Barriers for the Optical Binary Switch	130
4.6	Conclusions	131
	References	132

II Nanofabrication Methods 135

5 Charged-Particle Lithography 137
Lothar Berger, Johannes Kretz, Dirk Beyer, and Anatol Schwersenz

5.1	Survey	137
5.2	Electron Beam Lithography	141
5.2.1	Introduction	141
5.2.1.1	Electron Sources	141
5.2.1.2	Electron Optics	142
5.2.1.3	Gaussian Beam Lithography	147
5.2.1.4	Shaped Beam Lithography	148
5.2.1.5	Patterning	151

5.2.2	Resists	*153*
5.2.3	Applications	*158*
5.2.3.1	Photolithography Masks	*158*
5.2.3.2	Direct-Write Lithography	*162*
5.2.3.3	Maskless Lithography	*164*
5.2.3.4	Imprint Templates	*170*
5.2.4	Integration	*170*
5.3	Ion Beam Lithography	*172*
5.3.1	Introduction	*172*
5.3.1.1	Ion Sources	*173*
5.3.1.2	Ion Optics	*173*
5.3.1.3	Patterning	*173*
5.3.2	Applications	*173*
5.3.2.1	Direct-Structuring Lithography	*174*
5.3.2.2	Imprint Templates	*176*
5.4	Conclusions	*176*
	References	*177*

6	**Extreme Ultraviolet Lithography**	*181*
	Klaus Bergmann, Larissa Juschkin, and Reinhart Poprawe	
6.1	Introduction	*181*
6.1.1	General Aspects	*181*
6.1.2	System Architecture	*182*
6.2	The Components of EUV Lithography	*185*
6.2.1	Light Sources	*185*
6.2.1.1	Plasmas as EUV Radiators	*186*
6.2.1.2	Laser-Induced Plasmas	*187*
6.2.1.3	Gas Discharge Plasmas	*188*
6.2.1.4	Source Concepts and Current Status	*189*
6.2.2	Collectors and Debris Mitigation	*191*
6.2.3	Multilayer Optics	*194*
6.2.4	Masks	*198*
6.2.5	Resist	*199*
6.3	Outlook	*203*
	References	*204*

7	**Non-Optical Lithography**	*209*
	Clivia M. Sotomayor Torres and Jouni Ahopelto	
7.1	Introduction	*209*
7.2	Nanoimprint Lithography	*210*
7.2.1	The Nanoimprint Process	*210*
7.2.2	Polymers for Nanoimprint Lithography	*211*
7.2.3	Variations of NIL Methods	*215*
7.2.3.1	Single-Step NIL	*215*

7.2.3.2	Step-and-Stamp Imprint Lithography	216
7.2.3.3	Step-and-Flash Imprint Lithography	216
7.2.3.4	Roll-to-Roll Printing	217
7.2.4	Stamps	217
7.2.5	Residual Layer and Critical Dimensions	222
7.2.6	Towards 3-D Nanoimprinting	227
7.2.7	The State of the Art	230
7.3	Discussion	230
7.4	Conclusions	234
	References	235

8	**Nanomanipulation with the Atomic Force Microscope**	**239**
	Ari Requicha	
8.1	Introduction	239
8.2	Principles of Operation of the AFM	242
8.2.1	The Instrument and its Modes of Operation	242
8.2.2	Spatial Uncertainties	247
8.3	Nanomanipulation: Principles and Approaches	250
8.3.1	LMR Nanomanipulation by Pushing	250
8.3.2	Other Approaches	253
8.3.3	Manipulation and Assembly of Nanostructures	256
8.4	Manipulation Systems	260
8.4.1	Interactive Systems	260
8.4.2	Automated Systems	261
8.5	Conclusion and Outlook	265
	References	267

9	**Harnessing Molecular Biology to the Self-Assembly of Molecular-Scale Electronics**	**275**
	Uri Sivan	
9.1	Introduction	275
9.2	DNA-Templated Electronics	278
9.2.1	Scaffolds and Metallization	278
9.2.2	Sequence-Specific Molecular Lithography	281
9.2.3	Self-Assembly of a DNA-Templated Carbon Nanotube Field-Effect Transistor	284
9.3	Recognition of Electronic Surfaces by Antibodies	288
9.4	Molecular Shift-Registers and their Use as Autonomous DNA Synthesizers [11]	293
9.4.1	Molecular Shift-Registers	293
9.4.2	Error Suppression and Analogy Between Synthesis and Communication Theory	298
9.5	Future Perspectives	300
	References	301

10	**Formation of Nanostructures by Self-Assembly** 305
	Melanie Homberger, Silvia Karthäuser, Ulrich Simon, and Bert Voigtländer
10.1	Introduction 305
10.2	Self-Assembly by Epitaxial Growth 306
10.2.1	Physical Principles of Self-Organized Epitaxial Growth 306
10.2.1.1	Epitaxial Growth Techniques 306
10.2.1.2	Kinetically Limited Growth in Homoepitaxy 307
10.2.1.3	Thermodynamically Stable Nanostructures 309
10.2.1.4	Nanostructure Formation in Heteroepitaxial Growth 311
10.2.2	Semiconductor Nanoislands and Nanowires 313
10.2.2.1	Stranski–Krastanov Growth of Nanoislands 313
10.2.2.2	Lateral Positioning of Nanoislands by Growth on Templates 314
10.2.2.3	Silicide Nanowires 315
10.2.2.4	Monolayer-Thick Wires at Step Edges 315
10.2.3	Hybrid Methods: The Combination of Lithography and Self-Organized Growth 317
10.2.4	Inorganic Nanostructures as Templates for Molecular Layers 318
10.3	Molecular Self-Assembly 320
10.3.1	Attaching Molecules to Surfaces 321
10.3.1.1	Preparation of Substrates 322
10.3.1.2	Preparation of Self-Assembled Monolayers 322
10.3.1.3	Preparation of Mixed Self-Assembled Monolayers 323
10.3.2	Structure of Self-Assembled Monolayers 324
10.3.2.1	Organothiols on Metals 325
10.3.2.2	Carboxylates on Copper 326
10.3.3	Supramolecular Nanostructures 327
10.3.4	Applications of Self-Assembled Monolayers 330
10.3.4.1	Surface Modifications 330
10.3.4.2	Adsorption of Nanocomponents 330
10.3.4.3	Steps to Nanoelectronic Devices 331
10.4	Preparation and Self-Assembly of Metal Nanoparticles 334
10.4.1	Preparation of Metal Nanoparticles 334
10.4.2	Assembly of Metal Nanoparticles 337
10.4.2.1	Three-Dimensional Assemblies 337
10.4.2.2	Two-Dimensional Assemblies: The Formation of Monolayers 339
10.4.2.3	One-Dimensional Assemblies 341
10.5	Conclusions 344
	References 344
III	**High-Density Memories** 349
11	**Flash-Type Memories** 351
	Thomas Mikolajick
11.1	Introduction 351
11.2	Basics of Flash Memories 353

11.2.1	Programming and Erase Mechanisms	*353*
11.2.1.1	Hot Carrier Injection	*354*
11.2.1.2	Fowler–NordheimTunneling	*357*
11.2.1.3	Array Architecture	*358*
11.3	Floating-Gate Flash Concepts	*359*
11.3.1	The Floating-Gate Transistor	*359*
11.3.2	NOR Flash	*361*
11.3.3	NAND Flash	*363*
11.3.4	Reliability Aspects of Floating-Gate Flash	*365*
11.3.5	Scaling of Floating-Gate Flash	*366*
11.4	Charge-Trapping Flash	*370*
11.4.1	SONOS	*370*
11.4.2	Multi-Bit Charge Trapping	*372*
11.4.3	Scaling of Charge-Trapping Flash	*375*
11.5	Nanocrystal Flash Memories	*376*
11.6	Summary and Outlook	*378*
	References	*379*

12 Dynamic Random Access Memory *383*
Fumio Horiguchi

12.1	DRAM Basic Operation	*383*
12.2	Advanced DRAM Technology Requirements	*384*
12.3	Capacitor Technologies	*385*
12.4	Array Transistor Technologies	*389*
12.5	Capacitorless DRAM (Floating Body Cell)	*393*
12.6	Summary	*395*
	References	*395*

13 Ferroelectric Random Access Memory *397*
Soon Oh Park, Byoung Jae Bae, Dong Chul Yoo, and U-In Chung

13.1	An Introduction to FRAM	*397*
13.1.1	1T1C and 2T2C-Type FRAM	*398*
13.1.2	Cell Operation and Sensing Scheme of Capacitor-Type FRAM	*399*
13.2	Ferroelectric Capacitors	*401*
13.2.1	Ferroelectric Oxides	*401*
13.2.2	Fatigue	*402*
13.2.3	Retention	*403*
13.2.3.1	Crystallinity of PZT Film	*406*
13.2.3.2	The MOCVD Deposition Process	*406*
13.2.3.3	Perovskite Oxide Electrode	*407*
13.3	Cell Structures	*408*
13.3.1	CUB Structure	*408*
13.3.2	COB Structure	*410*
13.4	High-Density FRAM	*410*
13.4.1	Area Scaling	*410*

13.4.2	Voltage Scaling 412
13.4.3	3-D Capacitor Structure 413
13.4.3.1	Limitation of Planar Capacitor 413
13.4.3.2	Demonstration of a 3-D Capacitor 413
13.5	Summary and Conclusions 417
	References 417

14	**Magnetoresistive Random Access Memory** *419*
	Michael C. Gaidis
14.1	Magnetoresistive Random Access Memory (MRAM) 419
14.2	Basic MRAM 420
14.3	MTJ MRAM 422
14.3.1	Antiferromagnet 426
14.3.2	Reference Layer 427
14.3.3	Tunnel Barrier 427
14.3.4	Free Layer 428
14.3.5	Substrate 428
14.3.6	Seed Layer 428
14.3.7	Cap Layer 429
14.3.8	Hard Mask 429
14.4	MRAM Cell Structure and Circuit Design 429
14.4.1	Writing the Bits 429
14.4.2	Reading the Bits 433
14.4.3	MRAM Processing Technology and Integration 436
14.4.3.1	Process Steps 437
14.5	MRAM Reliability 439
14.5.1	Electromigration 439
14.5.2	Tunnel Barrier Dielectrics 440
14.5.3	BEOL Thermal Budget 440
14.5.4	Film Adhesion 441
14.6	The Future of MRAM 441
	References 443

15	**Phase-Change Memories** *447*
	Andrea L. Lacaita and Dirk J. Wouters
15.1	Introduction 447
15.1.1	The Non-Volatile Memory Market, Flash Memory Scaling, and the Need for New Memories 447
15.1.2	PCM Memories 448
15.2	Basic Operation of the Phase-Change Memory Cell 449
15.2.1	Memory Element and Basic Switching Characteristics 449
15.2.2	SET and RESET Programming Characteristics 452
15.3	Phase-Change Memory Materials 453
15.3.1	The Chalcogenide Phase-Change Materials: General Characteristics 453

15.3.1.1	The Pseudo-Binary GeTe-Sb$_2$Te$_3$ Compositions	454
15.3.1.2	Compositions Based on the Sb$_{70}$Te$_{30}$ "Eutectic" Compound	454
15.3.1.3	Other Material Compositions	455
15.3.1.4	N- or O-Doped GST	455
15.3.2	Material Structure	455
15.3.2.1	Long-Range Order: Crystalline State in GST and Doped Sb-Te	455
15.3.2.2	Short-Range Order in Crystalline versus Amorphous State	455
15.3.3	Specific Properties Relevant to PCM	457
15.4	Physics and Modeling of PCM	458
15.4.1	Amorphization and Crystallization Processes	458
15.4.2	Band-Structure and Transport Model	459
15.4.3	Modeling of the SET and RESET Switching Phenomena	462
15.4.4	Transient Behavior	463
15.5	PCM Integration and Cell Structures	464
15.5.1	PCM Cell Components	464
15.5.2	Integration Aspects	466
15.5.3	PCM Cell Optimization	467
15.5.3.1	Concentrating the Volume of Joule Heating	467
15.5.3.2	Improving the Thermal Resistance	467
15.6	Reliability	469
15.6.1	Introduction	469
15.6.2	Retention for PCM: Thermal Stability	469
15.6.3	Cycling and Failure Modes	470
15.6.4	Read and Program Disturbs	472
15.7	Scaling of Phase-Change Memories	472
15.7.1	Temperature Profile Distributions	472
15.7.2	Scaling of the Dissipated Power and Reset Current	473
15.7.3	Voltage Scaling	475
15.7.4	Cell Size Scaling	476
15.7.5	Scaling and Cell Performance: Figure of Merit for PCM	478
15.7.6	Physical Limits of Scaling	478
15.8	Conclusions	479
	References	480
16	**Memory Devices Based on Mass Transport in Solid Electrolytes**	**485**
	Michael N. Kozicki and Maria Mitkova	
16.1	Introduction	485
16.2	Solid Electrolytes	486
16.2.1	Transport in Solid Electrolytes	486
16.2.2	Major Inorganic Solid Electrolytes	488
16.2.3	Chalcogenide Glasses as Electrolytes	490
16.2.4	The Nanostructure of Ternary Electrolytes	492
16.3	Electrochemistry and Mass Transport	494
16.3.1	Electrochemical Cells for Mass Transport	494
16.3.2	Electrodeposit Morphology	497

16.3.3	Growth Rate	*500*
16.3.4	Charge, Mass, Volume, and Resistance	*501*
16.4	Memory Devices	*504*
16.4.1	Device Layout and Operation	*504*
16.4.2	Device Examples	*506*
16.4.3	Technological Challenges and Future Directions	*511*
16.5	Conclusions	*513*
	References	*513*

Index *517*

Preface

Beyond any doubt, Information Technology constitutes the area in which nanotechnology is most advanced. Since its origination during the 1960s, semiconductor technology as the driving force of information technology has advanced and continues to advance at an exponential pace. Today, semiconductor-based information technology penetrates almost all areas of contemporary society – and we are still only at the beginning of a new era. Within the coming decades completely new applications may emerge such as personal real-time translation systems, fully automatic navigation systems for cars, intelligent software agents for the internet, and autonomous robots to assist in our daily lives.

The main ingredient of the tremendous evolution of semiconductor circuitry has been the technological opportunity of ever-shrinking the minimum feature size in the fabrication of semiconductor chips. This led to a corresponding increase in the component density on the chips, decreasing energy consumption of the individual logic and memory cells, as well as higher clock frequencies and the development of multi-core architectures. All this added up to a doubling of the computer performance of chips approximately every 18 months, known as Moore's law. During the first decades of development, semiconductor technology was referred to as *microelectronics*, but this was changed to *nanoelectronics* a few years ago when the minimum feature size was reduced to below 100 nm. At about the same time, the component density has surpassed the one billion per chip mark and continues to progress at an unrestrained pace. Research areas related to nanoelectronics, however, comprise much more than simply the extension of current semiconductor technology to still smaller structures. More importantly, they cover the entire physics of nanosized objects with manifold properties that are unmatched in the macroscopic world and which might one day be exploited to store, to transmit, and to process information. In addition, they deal with technological approaches which are completely different to the *top-down* concept based on lithographical methods. The alternative *bottom-up* concept starts with the chemistry of, for example, organic molecules, nanocrystals, nanotubes, or nanowires, and strives for the self-organization of structures which can themselves act as assemblies of functional devices. Furthermore, nanoelectronics research investigates completely new computational concepts and architectures.

This text on *Information Technology* within the series *Nanotechnology*, is divided into two volumes and covers the concepts of potential future advances of the semiconductor technology right up to their physical limits, as well as alternative concepts which might one day augment the semiconductor technology, or even replace it in designated areas. Some readers may be familiar with the book *Nanoelectronics and Information Technology* (Wiley-VCH, 2nd edition, 2005) which I have edited. Although the topic of the present book is quite similar, the target is somewhat different. While the first volume represents an advanced text book, the present two volumes emphasize encyclopedic reviews in-line with the concept of the series. Yet, wherever possible, I have strived for a complementarity of the topics covered in the two texts.

This volume covers three parts:

Part One – *Basic Principles and Theory* – includes chapters on the mesoscopic transport of electrons, single electron effects and processes dominated by the electron spin. Furthermore, the fundamental physics of computational elements and its limits are covered.

Part Two – *Nanofabrication Methods* – starts with the prospects of various optical and non-optical lithography techniques, describes the manipulation of nanosized objects by probe methods, and closes with chemistry- and biology-based bottom-up concepts.

Part Three – *High-Density Memories* – begins with an outlook at the future potential of current memories such as Flash and DRAM, attributes magnetoresistive and ferroelectric RAM, and reports about the perspectives of resistive RAM such as phase-change RAM and electrochemical metallization RAM.

Nanotechnology Volume 4 will cover the following topics:

– *Logic Devices and Concepts* – ranges from advanced and non-conventional CMOS devices and semiconductor nanowire device, via various spin-controlled logic devices, and concepts involving carbon nanotubes, organic thin films, as well as single organic molecules, to the visionary idea of intramolecular computation.

– *Architectures and Computational Concepts* – covers biologically inspired structures, and quantum cellular automata, and finalizes by summarizing the main principles and current approaches to coherent solid-state-based quantum computation.

There are many people to whom I owe acknowledgments. First of all, I would like to express my sincere thanks to the authors of the chapters, for their dedication, their patience, and their willingness whenever I requested modifications.

Next, I must pay tribute to the following colleagues (in alphabetical order) for critically reviewing the concept of the text and for their advice on topic and author selection: George Bourianoff (Intel Corp.), Ralph Cavin (Semiconductor Research Corp.), U-In Chung (Samsung Electronics), James Hutchby (Semiconductor Research Corp.), Christoph Koch (Caltech), Phil Kuekes (Hewlett Packard Research Laboratories), Heinrich Kurz (RWTH Aachen University), Rich Liu (Macronix Intl.

Ltd.), Hans Lüth (FZ Jülich), Siegfried Mantl (FZ Jülich), Tobias Noll (RWTH Aachen University), Stanley Williams (Hewlett Packard Research Laboratories), and Victor Zhirnov (Semiconductor Research Corp.).

Heartfelt thanks are due to Günther Schmid, editor of the series *Nanotechnology*, who invited and motivated me, and the staff of Wiley-VCH, in particular Gudrun Walter and Steffen Pauly, who supported me in every possible way.

Last – but certainly not least – I was greatly assisted by Dagmar Leisten, who redrew most of the original figures in order to improve their graphical quality, by Thomas Pössinger for his layout work and design ideas aiming at a more consistent appearance of the book, and by Maria Garcia for all the organizational work around such a project and for her sustained support.

Aachen, January 2008 *Rainer Waser*

List of Contributors

Jouni Ahopelto
VTT Micro and Nanoelectronics
Micro and Nanoelectronics
P.O. Box 1000
02044 VTT
Finland

Byoung Jae Bae
Process Development Team,
Semiconductor R&D Division
Samsung Electronics Co., Ltd.
San #24, Nongseo-Dong
Giheung-Gu, Yongin-City
Gyeonggi-Do 449-711
Korea

Lothar Berger
Fraunhofer Center for Nanoelectronic
Technologies
Königsbrücker Straße 180
01099 Dresden
Germany
and
Zuken Electronic Design Automation
Europe GmbH
Airport Business Centre
85399 Hallbergmoos
Germany

Klaus Bergmann
Fraunhofer Institut für Lasertechnik
Steinbachstr. 15
52074 Aachen
Germany

Dirk Beyer
Vistec Electron Beam GmbH
Goeschwitzer Strasse 25
07745 Jena
Germany

Alex M. Bratkovsky
Hewlett-Packard Laboratories
1501 Page Mill Road
MS 1123
Palo Alto, CA 94304
USA

Ralph K. Cavin
Semiconductor Research Corporation
1101 Slater Road
Durham, NC 27703
USA

List of Contributors

U-In Chung
Process Development Team,
Semiconductor R&D Division
Samsung Electronics Co., Ltd.
San #24, Nongseo-Dong
Giheung-Gu, Yongin-City
Gyeonggi-Do 449-711
Korea

Michael C. Gaidis
IBM T.J. Watson Research Center
1101 Kitchawan Road
Route 134
P.O. Box 218
Yorktown Heights, NY 10598
USA

Melanie Homberger
Forschungszentrum Jülich
CNI - Center of Nanoelectronic Systems
for Information Technology
Institute for Solid State Research (IFF)
52425 Jülich
Germany

Fumio Horiguchi
Toyo University
Department of Computational Science
and Engineering
2100 Kujirai
Kawagoe
Saitama 350-8585
Japan

Larissa Juschkin
RWTH Aachen
Lehrstuhl Technologie optischer
Systeme
Steinbachstr. 15
52074 Aachen
Germany

Silvia Karthäuser
RWTH Aachen University
Institute for Inorganic Chemistry (IAC)
Landoltweg 1
52074 Aachen
Germany

Michael N. Kozicki
Arizona State University
Center for Applied Nanoionics
Tempe, AZ 85287-6206
USA

Johannes Kretz
Qimonda Dresden GmbH & Co. OHG
Electron Beam Lithography
Competence Center
Koenigsbruecker Strasse 180
01099 Dresden
Germany

Andrea L. Lacaita
Politecnico di Milano
Dipartimento di Elettronice
e Informazione
Piazza L. da Vinci, 32
20133 Milan
Italy

Thomas Mikolajick
University of Technology and Mining
Institute of Electronic- and Sensor
Materials
Gustav-Zeuner-Strasse 3
09596 Freiberg
Germany

Maria Mitkova
Boise State University
Department of Electrical and Computer
Engineering
Boise, ID 83725-2075
USA

Soon Oh Park
Process Development Team,
Semiconductor R&D Division
Samsung Electronics Co., Ltd.
San #24, Nongseo-Dong
Giheung-Gu, Yongin-City
Gyeonggi-Do 449-711
Korea

Reinhart Poprawe
RWTH Aachen
Lehrstuhl für Lasertechnik
Steinbachstr. 15
52074 Aachen
Germany

Ari Requicha
University of Southern California
Computer Science Department
Laboratory for Molecular Robotics
941 Bloom Walk
Los Angeles, CA 90089-0781
USA

Thomas Schäpers
Forschungszentrum Jülich
Institut für Bio- and Nanosysteme
(IBN-1)
52425 Jülich
Germany

Anatol Schwersenz
Wuerth Electronics GmbH & Co. KG
Salzstrasse 21
74676 Niedernhall
Germany

Ulrich Simon
RWTH Aachen University
Institute for Inorganic Chemistry (IAC)
Landoltweg 1
52074 Aachen
Germany

Uri Sivan
Technion - Israel Institute of Technology
Department of Physics and the Russell
Berrie Nanotechnology Institute
Haifa 32000
Israel

Clivia M. Sotomayor Torres
University College Cork
Tyndall National Institute
Photonics Nanostructures Group
Lee Maltings
Cork
Ireland
and
Catalan Institute of Nanotechnology
Phononic and Photonic Nanostructures
Group
Campus Bellaterra - Edifici CM7
08193-Bellaterra
Barcelona
Spain

Joseph M. Thijssen
Delft University of Technology
Kavli Institute of Nanoscience
Lorentzweg 1
2628 CJ Delft
The Netherlands

Bert Voigtländer
Forschungszentrum Jülich
CNI - Center of Nanoelectronic Systems
for Information Technology
Institute for Bio- and Nanosystems
(IBN)
52425 Jülich
Germany

Dirk J. Wouters
IMEC
Memory Group, Division RDO/PT/
CMOSDR
Kapeldreef 75
3001 Leuven
Belgium

Dong Chul Yoo
Process Development Team,
Semiconductor R&D Division
Samsung Electronics Co., Ltd.
San #24, Nongseo-Dong
Giheung-Gu, Yongin-City
Gyeonggi-Do 449-711
Korea

Herre S. J. van der Zant
Delft University of Technology
Kavli Institute of Nanoscience
Lorentzweg 1
2628 CJ Delft
The Netherlands

Victor V. Zhirnov
Semiconductor Research Corporation
1101 Slater Road
Durham, NC 27703
USA

I
Basic Principles and Theory

1
Phase-Coherent Transport

Thomas Schäpers

1.1
Introduction

From elementary quantum mechanics it is known that electrons possess wave properties in addition to their appearance as a particle. Often, these wave properties are difficult to observe directly, the main reason being that in many cases the electron wavelength is quite small – that is, in metals the wavelength of the electrons at the Fermi energy is only of the order of a few nanometers. Therefore, one possible approach to observing the phenomena related to the wave properties of the electrons is to reduce the sample size to dimensions close to the electron wavelength, as performed in a quantum point contact. Nevertheless, the wave nature of the electrons is sometimes revealed under much more relaxed conditions. An essential perquisite here is that the coherent wave propagation is maintained over sufficiently long distances, so that interference effects can occur. In most cases this condition is only fulfilled at low temperatures in the Kelvin range, where inelastic scattering is suppressed to a large extent.

In diffusive conductors, one possible way to achieve electron interference is if the diffusive motion allows electrons to propagate coherently in closed loops. This so-called "weak localization effect" can even be observed in macroscopic structures. The electron interference can be significantly modified if spin precession (i.e., due to spin-orbit coupling) comes into play. Well-controlled electron interference can be achieved if the wave propagation is guided by the shape of the conductor, and an excellent example in this respect is the Aharonov–Bohm effect, which is observed in ring-shaped conductors.

This discussion of phase-coherent transport in nanostructures begins by introducing the relevant length scales and the different transport regimes in Section 1.2. Subsequently, in Section 1.3 the Landauer–Büttiker formalism and ballistic transport through a split-gate point contact are discussed. Section 1.4 provides an explanation for the weak localization effect, which leads to an enhanced resistance, whilst in Section 1.5 it is shown that spin precession can result in the reversal of the weak localization effect. Phase-coherent transport in ring-shaped structures is discussed in Sections 1.4 and 1.5, while in Section 1.6 it is shown that the finite number of

Nanotechnology. Volume 3: Information Technology I. Edited by Rainer Waser
Copyright © 2008 WILEY-VCH Verlag GmbH & Co. KGaA, Weinheim
ISBN: 978-3-527-31738-7

scattering centers in very small structures can result in pronounced fluctuations in conductance. Although, within this chapter, transport phenomena in two- and one-dimensional structures are outlined, zero-dimensional structures – namely quantum dots – are discussed in detail in Chapter 2.

1.2
Characteristic Length Scales

Transport in nanoelectronic systems can be classified by relating its size to some specific characteristic length scales [1, 2] which determine how the carriers propagate through the sample. In the following sections, the elastic and inelastic mean free path are introduced, which quantify the degree of elastic and inelastic scattering occurring in the structure, respectively. A length scale, which provides information about loss of the phase memory is termed the phase-coherence length.

1.2.1
Elastic Mean Free Path

The elastic mean free path l_e is a measure of the distance between subsequent elastic scattering events. Such events occur due to the fact that the conductor is not ideal but rather contains irregularities in the lattice, such as impurities or dislocations. The scattering can be considered as *elastic*, which means that the electron energy is conserved. A typical example is the scattering of an electron at a charged impurity. If we assume a stationary scattering center, then effectively no energy is transferred during the scattering event, whereas the direction of the electron momentum can change greatly.

In order to determine the elastic mean free path l_e within the Drude model, one must first calculate the average time between elastic scattering events, τ_e. Its value can be extracted from the electron mobility μ_e, given by

$$\mu_e = \frac{e\tau_e}{m^*} \tag{1.1}$$

The quantities m^* and e are the effective electron mass and the elementary charge, respectively. The electron mobility is a measure of the increase of the drift velocity v_{drift} in a conductor with increasing electric field E: $v_{\text{drift}} = -\mu_e E$. In practice, the electron mobility is determined from the electron concentration n_e and the Drude conductivity σ_0 by

$$\mu_e = \frac{\sigma_0}{en_e} \tag{1.2}$$

Experimentally, the electron concentration n_e is obtained from Hall measurements, while the conductivity σ_0 is deduced from resistance measurements at zero magnetic field.

Effectively, only electrons at the Fermi energy E_F contribute to the electron transport. Therefore, the elastic mean free path l_e is given by the length an electron with the Fermi velocity v_F propagates until it is elastically scattered after the elastic scattering time τ_e:

$$l_e = \tau_e v_F \tag{1.3}$$

As an example, for a typical two-dimensional (2-D) electron gas in an AlGaAs/GaAs heterostructure (see Section 1.3), low-temperature mobilities of around $10^6\,\text{cm}^2\,(\text{Vs})^{-1}$ at $n_e = 3\times 10^{11}\,\text{cm}^{-2}$ are achieved. For a 2-D system the Fermi velocity is given by $v_F = \hbar\sqrt{2\pi n_e}/m^*$. With $m^* = 0.067 m_e$ and using Equation 1.3, the length of the elastic mean free path is $9\,\mu\text{m}$.

1.2.2
Inelastic Mean Free Path

In addition to the elastic scattering discussed above, electron scattering can also be connected to an energy transfer. A typical example is the effect of lattice vibrations on electron transport. An electron moving within a crystal will be scattered by these lattice vibrations and either lose or gain energy, depending on whether it excites the lattice vibrations or is excited by them. As an energy transfer occurs, these scattering processes are considered to be *inelastic*. Similar to the previous discussion, one can define an inelastic scattering length l_{in} as a measure for the length between inelastic scattering events. Besides electron–phonon scattering, electron–electron scattering is another possible process, where a considerable amount of energy can be exchanged between both scattering partners [3].

1.2.3
Phase-Coherence Length

The phase-coherence length l_φ is the relevant length scale, which determines if phase-coherent transport can be observed in nanolectronic systems [2]. It is a measure of the distance that the electron propagates phase coherently before its phase is randomized. At low temperatures, the phase-coherence length can be larger than the elastic mean free path l_e. Thus, a number of elastic scattering events occur before the phase information is finally lost. During an elastic scattering event (i.e., at an impurity), the phase of an electron is not randomized; it is only shifted by well-defined amount. If the electron propagates along the identical path a second time, the phase accumulation will be exactly the same. This is in strong contrast to inelastic scattering events (e.g., electron–phonon scattering), where the scattering target changes with time. Consequently, the phase shift that the electron would acquire is different each time. However, care must be taken to identify l_φ right away with the inelastic mean free path l_{in}, as they are not identical in all cases; that is, spin-flip scattering is considered to be phase-breaking and thus contributing to l_φ whilst it may be elastic at the same time. In addition, small-energy-transfer electron–electron scattering, which is due to the fluctuation of the electric field produced by the electrons (Nyquist contribution), can contribute to a large extent to l_φ [4]. As mentioned above, at low temperatures a number of elastic scattering events occur

Table 1.1 Comparison of the different transport regimes.

Diffusive	Classical	$\lambda_F, l_e \ll L, l_\varphi < l_e$
	Quantum	$\lambda_F, l_e \ll L, l_\varphi > l_e$
Ballistic	Classical	$\lambda_F \ll L < l_e, l_\varphi$
	Quantum	$\lambda_F \approx L < l_e, l_\varphi$

until the phase is broken, implying that the characteristic phase-breaking time τ_φ is larger than the elastic scattering time τ_e. Owing to the diffusive motion during the time τ_φ, the phase-coherence length l_φ must be expressed by

$$l_\varphi = \sqrt{D\tau_\varphi} \tag{1.4}$$

Here, D is the diffusion constant defined as

$$D = \frac{1}{d} v_F^2 \tau_e \tag{1.5}$$

with d the dimensionality of the system. Typical values for the phase-coherence length of an AlGaAs/GaAs 2-D electron gas below 1 K are of the order of several micrometers [5].

1.2.4
Transport Regimes

By comparing l_e and l_φ with the dimension L of the sample and the Fermi wavelength λ_F, different transport regimes can be classified, and these are summarized in Table 1.1. For the case where the elastic mean free path l_e is smaller than the dimensions of the sample, many elastic scattering events occur while the electrons propagate through the structure. The carriers are traveling randomly (*diffusive*) through the crystal, as illustrated in Figure 1.1a. If the phase-coherence

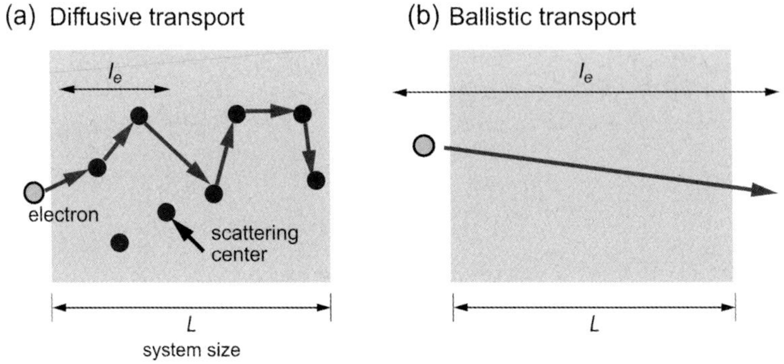

Figure 1.1 Illustration of (a) a diffusive conductor, and (b) a ballistic conductor. In the diffusive transport regime many elastic scattering events occur, while the electron crosses the sample. In the ballistic regime, the electron crosses the sample without any elastic scattering event.

length l_φ is shorter than the elastic mean free path l_e, the transport is considered as classical. In contrast, if $l_\varphi > l_e$, then quantum effects owing to the wave nature of the electrons can be expected. This diffusive regime is thus called the *quantum regime*. As illustrated in Figure 1.1b, in the case that l_e is larger than the dimensions of the sample, the electrons can transverse the system without any scattering; this regime is called *ballistic*. Depending on the magnitude of the Fermi wavelength λ_F in comparison to the dimension of the sample, the transport can either be regarded as classical ballistic or quantum ballistic. In the following section, ballistic transport will first be discussed, and later the transport phenomena in the diffusive regime.

1.3 Ballistic Transport

In this section transport in the ballistic transport regime will be discussed; that is, where the elastic mean free path exceeds the dimensions of the sample. First, the Landauer–Büttiker formalism is explained, where the resistance of a sample is described in terms of transmission and reflection probabilities, which is a very convenient scheme to analyze the transport in the ballistic regime. Subsequently, the quantized conductance of a split-gate point contact will be discussed, making use of the Landauer–Büttiker formalism.

1.3.1 Landauer–Büttiker Formalism

In order to analyze the electronic transport properties of a sample, usually a current is allowed to flow between two contacts while the response of the system is measured by two voltage probes. The latter are not necessarily different from the current contacts. The ratio between the voltage drop U and the current I can be defined as a *macroscopic resistance*. (The expression *macroscopic* is used here as only the global properties of the sample are measured.)

A very intuitive interpretation of the macroscopic resistance, R, of a sample can be obtained if the so-called Landauer–Büttiker formalism is used [6–9]. In this model, the resistance

$$\mathcal{R}_{mn,kl} = \frac{U_{kl}}{I_{mn}} \tag{1.6}$$

is defined by the voltage measured between contacts k and l and the current flowing between contacts n and m.

In order to keep things simple, the discussion is restricted to a conductor connected via ideal one-dimensional (1-D) ballistic leads to four corresponding reservoirs. The geometry of the sample is depicted in Figure 1.2. The ballistic wires should consist of only a single 1-D. The reservoirs with the corresponding chemical potentials μ_i ($i = 1, \ldots, 4$) serve as source and drain for carriers flowing in and out of

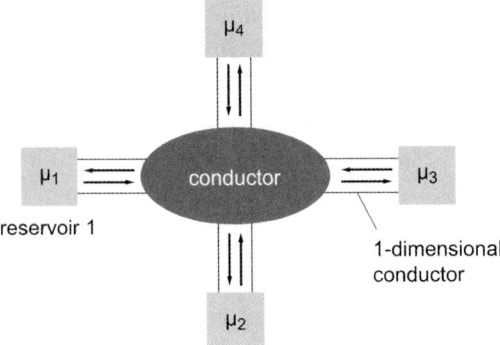

Figure 1.2 Schematic illustration of a four-terminal resistance measurement set-up. The conductor is connected by ideal one-dimensional leads to four corresponding reservoirs.

the conductor. At zero temperature, the i-th reservoir can supply electrons to the conductor up to a maximum energy of μ_i. Each carrier from the lead, which reaches the reservoir is absorbed by the reservoir, irrespective of the phase and energy of the carriers. As discussed above, inelastic scattering is forbidden within the leads, so that electrons once injected into the conductor maintain their energy until they reach one of the reservoirs.

As an example, we will study the current contributions in the 1-D lead 1, which results in the net current I_1. The current injected from reservoir 1 is given by:

$$I_{inj} = e \int_0^{\mu_1} D_{1D}(E) v(E) dE \tag{1.7}$$

where $v(E)$ is the velocity of the electrons. As the wire is 1-D, the density of states of a 1-D system must be inserted, which is given by

$$D_{1D}(E) = \frac{2}{hv(E)} \tag{1.8}$$

So far, only the states propagating from reservoir 1 are considered, and the density of states used here is half of the commonly known value because there is only one direction of propagation [1]. It can be seen directly that the product of the 1-D density of states $D_{1D}(E)$ and the velocity $v(E)$ is constant, and therefore the current leaving reservoir 1 has the following simple form:

$$I_{inj} = \frac{2e}{h} \mu_1 \tag{1.9}$$

Part of the current supplied by reservoir 1 will be reflected back into the conductor. If R_{ii} is defined as the reflection probability for a reflection of carriers from lead i back into lead i, then the current reflected into lead 1 can be written as

$$I_R = -\frac{2e}{h} R_{ii} \mu_i \tag{1.10}$$

In addition, electrons are transmitted from the other three leads into lead 1. By defining the transmission probability from lead j into lead i ($i \leftarrow j$) as T_{ij}, we arrive at the following expression for the current transmitted into lead 1:

$$I_T = -\frac{2e}{h}\sum_{j=2}^{4} T_{1j}\mu_j \qquad (1.11)$$

By summing all of these contributions it can be seen that the net current flowing in lead 1 is finally given by:

$$\begin{aligned}I_1 &= I_{inj} + I_R + I_T \\ &= \frac{2e}{h}\left[(1-R_{11})\mu_1 - \sum_{j=2}^{4} T_{1j}\mu_j\right],\end{aligned} \qquad (1.12)$$

or, more generally, the current in lead i is given by

$$I_i = \frac{2e}{h}\left[(1-R_{ii})\mu_i - \sum_{j\neq i} T_{ij}\mu_j\right] \qquad (1.13)$$

By using Equation 1.13, the above-defined resistance $\mathcal{R}_{mn,kl}$ can be determined for given reflection and transmission probabilities of the sample. According to the initial definition of $\mathcal{R}_{mn,kl}$, as given in Equation 1.6, the net current I_{mn} flows between contacts n and m. The leads k and l do not carry a net current in case of an ideal voltage measurement. The voltage drop U_{kl} is given by the difference of the electrochemical potentials divided by e: $(\mu_k - \mu_l)/e$. In the following section, Equation 1.13 will serve as a basis to describe the transport properties of a split-gate quantum point contact.

1.3.2
Split-Gate Point Contact

In split-gate quantum point contacts the transport is limited to only one dimension. This is obtained by first restricting the propagation of the electrons to a plane. In these so-called "two-dimensional electron gases" (2DEGs), the carriers are confined at an interface of two different semiconductor layers. A typical example of a 2DEG realized in an AlGaAs/GaAs layer system is depicted in Figure 1.3. Here, the carriers are located at the AlGaAs/GaAs interface and, owing to the conduction band offset between AlGaAs and GaAs, a triangular quantum well is formed at the interface. The electrons in the quantum well are supplied by an n-type δ-doped (very thin) layer. In order to prevent ionized impurity scattering, the electrons in the quantum well are separated from the δ-doped layer by an undoped AlGaAs spacer layer. Using this scheme, very large electron mobilities and thus very long elastic mean free paths of the order of several micrometers can be achieved.

A further restriction of the electron propagation to only one dimension can be realized by using split-gate point contacts [10, 11]. As illustrated in Figure 1.4, two opposite gate fingers are separated by a distance of a few hundreds of nanometers. Split-gate electrodes are usually prepared by using electron beam lithography. Since

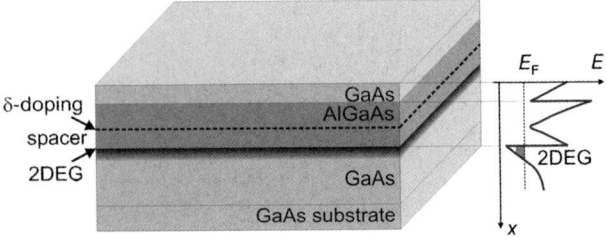

Figure 1.3 Layer sequence of an AlGaAs/GaAs heterostructure containing a two-dimensional electron gas at the AlGaAs/GaAs interface. A schematic illustration of the conduction band profile is shown on the right-hand side.

the Fermi wavelength λ_F of a 2-D electron gas is typically a few tenths of a nanometer, the separation of the split-gates is comparable with λ_F. The length of the channel formed by the gate electrodes is usually smaller than 1 μm, and thus smaller than the elastic mean free path l_e. According to the classification introduced in Section 1.2, the transport can be considered as ballistic.

By applying a sufficiently large negative voltage to the gate fingers, the underlying 2-D electron gas is depleted underneath the gate fingers (see Figure 1.4a). Only a small opening between the gate fingers remains for the electrons to propagate from one side to the opposite side; however, by varying the gate voltage it is possible to control the effective width of the opening. An increase of the negative bias voltage enlarges the depletion area and thus reduces the opening width. At sufficiently large negative bias voltages the opening can even be closed completely (pinch-off).

Owing to the depletion area underneath the split-gate electrodes, it can be assumed that the electrons in the 2DEG are confined in a potential well along the y-axis, while the free propagation takes place along the x-axis. If the potential profile in the plane of the 2DEG induced by the split-gate electrodes is expressed by $V(x, y)$, the Hamilton

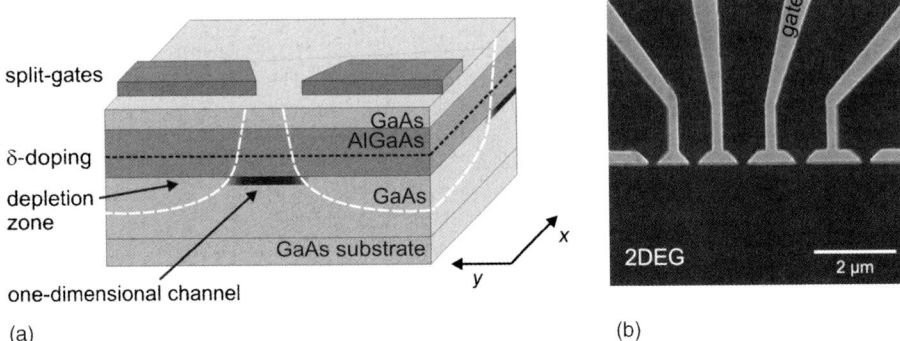

Figure 1.4 (a) Schematic illustration of a split-gate point contact on an AlGaAs/GaAs heterostructures. By applying a negative gate voltage to the split-gate electrodes, the electron gas underneath is depleted. The electrons can only pass the small opening. (b) An electron beam micrograph of split-gate point contacts.

operator has the following form:

$$H = \frac{\hbar^2}{2m^*}\left(\frac{\partial^2}{\partial x^2} + \frac{\partial^2}{\partial y^2}\right) + V(x,y) \tag{1.14}$$

In order to determine the precise shape of the potential $V(x, y)$ as a function of the gate voltage, elaborated self-consistent simulations are required [12]. However, for most applications it is sufficient to assume an approximated potential profile. For low gate voltages an appropriate approximation is a rectangular potential profile, while for higher negative gate voltages the potential well can be approximated by a parabolic potential. As an example, we will consider here the latter potential shape. Due to the short length of the channel formed by the split-gates, the 2-D potential profile will be saddle-shaped. However, if the potential shaped along the constriction is smooth (adiabatic limit), it is sufficient to consider only the narrowest point of the channel, which can be expressed by

$$V(y) = \frac{1}{2}m^*\omega_0^2 y^2 + V_0 \tag{1.15}$$

Here, ω_0 is the characteristic frequency of the parabolic potential, while V_0 represents the height of the inflection point of the saddle-shaped potential. For the energy dispersion of the 1-D subbands in the point contact, we obtain

$$E_n(k_x) = E_n^0 + V_0 + \frac{\hbar^2 k_x^2}{2m^*}, \quad n = 1, 2, 3, \ldots, \tag{1.16}$$

with

$$E_n^0 = (n - 1/2)\hbar\omega_0, \tag{1.17}$$

the energy eigenvalues of the harmonic oscillator. By changing the gate voltage at the split-gate electrodes, the effective width of the opening can be adjusted. In the parabolic approximation ω_0 is increased if a more negative gate voltage is applied, and this leads to an increased separation of the energy eigenvalues. As a consequence, lesser levels are occupied up to the Fermi energy (see Figure 1.5a and b).

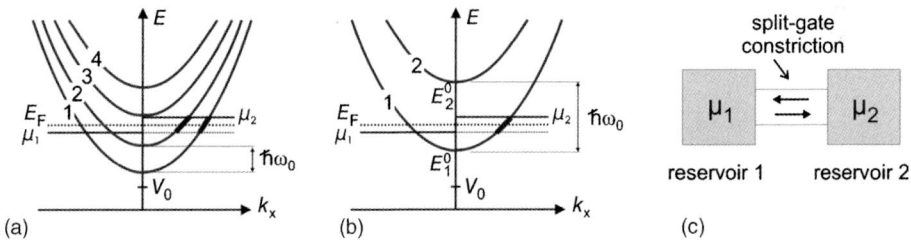

Figure 1.5 (a) Energy dispersion of a one-dimensional channel with the two lowest levels lying below the Fermi energy E_F. (b) Corresponding situation with only one subband occupied. The energy separation between the levels given by $\hbar\omega_0$ is larger compared to the situation shown in (b). (c) A one-dimensional conductor; that is, the channel formed by the split-gate electrodes, connected by two reservoirs with the electrochemical potential μ_1 and μ_2, respectively.

Before examining the experimental outcome of measurement of the split-gate point contact resistance, the conduction of a 1-D conductor by using the Landauer–Büttiker formalism will be briefly discussed. It must first be assumed that the conductor is connected on both terminals to reservoirs with the electrochemical potentials μ_1 and μ_2, respectively (i.e., the 2DEG on both sides of the split-gates), as shown in Figure 1.5c.

For a set-up with only two reservoirs, and where only the lowest subband is occupied, the following expression is obtained according to the Landauer–Büttiker formalism [cf. Equation 1.13]:

$$(h/2e)I = (1 - R_{11})\mu_1 - T_{12}\mu_2 \tag{1.18}$$

$$-(h/2e)I = (1 - R_{22})\mu_2 - T_{21}\mu_1 \tag{1.19}$$

At zero magnetic field ($B=0$), the transport is time-inversion invariant so that the following relationships hold:

$$T_{12} = T_{21} = T = 1 - R_{11} = 1 - R_{22} \tag{1.20}$$

Thus, finally we arrive at the expression for the conductance of the constriction:

$$G = \frac{I}{U} = \frac{Ie}{\mu_1 - \mu_2} = \frac{2e^2}{h} T \tag{1.21}$$

As illustrated in Figure 1.5b, only carriers with energy between μ_1 and μ_2 contribute to the conductance. If backscattering is neglected ($T=1$), the conduction through a constriction is simply given by:

$$G = \frac{2e^2}{h}. \tag{1.22}$$

It should be stressed that the constant conductance is a result of the cancellation of the energy dependence of the density of states and the velocity for the 1-D case [cf. Equation 1.7], which is not the case for 2-D or three-dimensional (3-D) systems. In analogy, the conductance can be calculated if N subbands are occupied. The occupied subbands taking part in the transport are usually called channels; the situation for two channels ($N=2$) is illustrated in Figure 1.5a. If N one-dimensional channels are assumed, then the total transmission probability from reservoir j to reservoir i ($i \leftarrow j$) can be expressed as

$$T_{ij} = \sum_{mn}^{N} T_{ij,mn} \tag{1.23}$$

where $T_{ij,mn}$ denotes the transmission probability from the n-th subband of lead j into the m-th subband of lead i. If ideal transmission and no intersubband scattering is assumed, then the total transmission probability of a 1-D channel with N subbands is given by $T = N$. Thus, each subband contributes with $2e^2/h$ to the conductance so that the total conductance of a constriction with N subbands occupied is given by

$$G = \frac{2e^2}{h} N \tag{1.24}$$

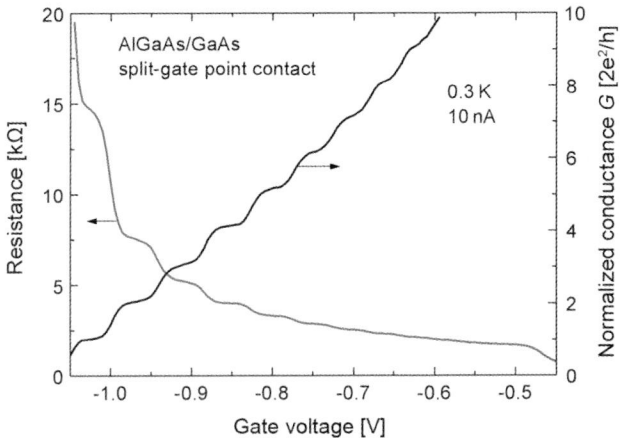

Figure 1.6 Resistance and conductance of an AlGaAs/GaAs split-gate point contact as a function of the gate voltage. The conductance is plotted in units of $2e^2/h$.

This remarkable result indicates that the conductance of a 1-D constriction changes in steps equal to $2e^2/h$, if the number of channels is altered by adjusting the widths of the constriction. The latter can be achieved by applying an appropriate voltage to the split-gate electrodes.

An experimental result of the resistance and conductance of quantum point contact based on a 2DEG in an AlGaAs/GaAs heterostructure is shown in Figure 1.6. With a more negative gate voltage, the resistance of the point contact increases, as the width of the constriction becomes increasingly narrower. As can be seen in Figure 1.6, if the conductance G is plotted, it can clearly be seen that G decreases stepwise by multiples of $2e^2/h$ with increasing negative gate voltage.

The experimentally observed curves can deviate in many aspects from the ideal curves. The calculations given above were restricted to zero temperature, but at finite temperatures the broadening of the Fermi distribution function results in a broadening of the steps owing to the partial occupation and emptying of the 1-D channels. The geometrical shape of the point contact opening also affects the transmission through the point contact. For example, sharp edges of the point contact opening can result in reflections of the incoming and transmitted electrons waves at the inlet and outlet of the 1-D channel. As a result, oscillations are expected in the plateaus of the steps [13, 14].

1.4
Weak Localization

Interference effects of electron waves due to phase coherent transport can be seen even in large samples, where the phase coherence length is much smaller than the dimensions of the sample. This effect, called weak localization, results in an increased resistance compared to the classically expected value [15, 16]. Weak localization is observed if the temperature is sufficiently low so that the phase coherence length l_φ is

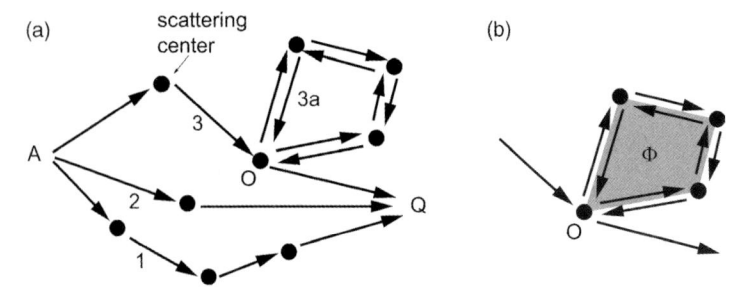

Figure 1.7 (a) Possible trajectories of electrons propagating from point A to Q. The trajectory 3a represents a closed loop. (b) Detail of a closed loop with a magnetic flux Φ penetrating this loop.

larger than the elastic scattering length l_e. As we will see below, the effect of weak localization depends strongly on the dimensionality of the system. The lower the dimension of the system is, the stronger the effect of weak localization is, that is in quasi one-dimensional wire structures weak localization is most pronounced. In order to illustrate the general mechanisms leading to weak localization, we will first introduce a simple model. Later on more quantitative expressions for the conductivity corrections will be given.

1.4.1
Basic Principles

Let us consider a diffusive conductor, in which an electrons starting at point A propagate to point Q. Some typical trajectories of an electron are sketched in Figure 1.7, illustrating that there are many possibilities for an electron to propagate from A to Q.

It is assumed that the elastic mean free path l_e is smaller than the distance between A and Q. Thus, an electron undergoes many elastic scattering events on its way. However, during elastic scattering the electron does not lose its phase memory. If it is assumed that the phase coherence length is longer than the distance between A and Q, the phase information is not lost. By following Feynman, each path j can be described from the initial state A to the final state Q by a complex probability amplitude C_j given by [17, 18]:

$$C_j = c_j \exp(i\varphi_j) \tag{1.25}$$

Here, φ_j is the phase shift that the electron acquires on its way from A to Q while propagating along path j. Often, there are many possible paths for an electron to propagate between A and Q. For example, for free electron propagation the phase accumulation along the path j can be calculated from the action S_j by

$$\varphi_j = \frac{S_j}{\hbar} \tag{1.26}$$

The non-relativistic action S_j is defined by

$$S_j = \int_{t_A}^{t_Q} dt L(\mathbf{r}, \dot{\mathbf{r}}, t) \tag{1.27}$$

with

$$L(\mathbf{r}, \dot{\mathbf{r}}, t) = \frac{m}{2}\dot{r}^2 \tag{1.28}$$

the Lagrangian function of a free propagating electron. Here, t_A is the time when the electron starts at A, and t_Q the time when it arrives at Q. The quantities $\mathbf{r}$ and $\dot{\mathbf{r}}$ are the position and velocity of the particle, respectively. However, the electron acquires not only a phase shift during free propagation but also well-defined phase shifts by the elastic scattering events, so that the total phase accumulated along the path is the sum of both contributions. The total amplitude for the propagation from A to Q is given by the sum of the amplitudes C_j of all undistinguished paths. Finally, the total probability P_{AQ} for an electron to be transported from A to Q is determined by the square of the total amplitude

$$P_{AQ} = \left| \sum_j c_j e^{i\varphi_j} \right|^2 \tag{1.29}$$

In systems with a large number of possible paths, the phases φ_j are usually randomly distributed, and therefore the wave nature should have no effect on the electron transport due to averaging. Nevertheless, the fact that an increase of the resistance is observed, compared to the classical transport, is a result of closed loops (see Figure 1.7a, trajectory 3a). Along these loops, an electron can propagate in two opposite orientations with the corresponding complex amplitudes $C_{1,2} = c_{1,2}\exp(i\varphi_{1,2})$. The current contribution of the current returning to the starting point of the loop (O) is given by

$$P_{OO} = |C_1 + C_2|^2 = |C_1|^2 + |C_1|^2 + 2\mathrm{Re}(C_1^* C_2) \tag{1.30}$$

Since, for time-reversed paths $c_1 = c_2$ and $\varphi_1 = \varphi_2$, we obtain

$$|C_1 + C_2|^2 = 4|C_1|^2 \tag{1.31}$$

For classical non-phase-coherent transport, the probability would simply be $|C_1|^2 + |C_2|^2$, which is a factor of 2 smaller than for the phase-coherent case. A larger probability to return to the origin implies that the net current through the sample is reduced. Hence, the carriers are *localized* within the loop. Such localization does not depend on the size of the loop as long as its length is smaller than the phase-coherence length. It is important to note here that constructive interference occurs for *all* possible closed loops in the conductor, and is therefore not averaged out. As a result, the total resistance is increased compared to the classical case.

1.4.2
Weak Localization in One and Two Dimensions

In the following section, it is briefly sketched how a value for the correction of the conductance due to weak localization can be obtained quantitatively [18]. For the weak localization effect we are interested only in those processes where the electrons return to their starting points. The discussion will first be restricted to a 2-D system, for example a 2-D electron gas in an AlGaAs/GaAs heterostructure. A larger number of scattering centers increase the probability for backscattering of the electrons. The larger the number of scattering centers is, the smaller is the diffusion constant; as a consequence one obtains for the return probability due to diffusive motion: $1/(4\pi Dt)$. For the total return probability, it must be ensured that the phase of the electrons is preserved up to time τ_φ, which provides a pre-factor $\exp(-t/\tau_\varphi)$. Furthermore, it is required that the electron is at least once elastically scattered; thus, a pre-factor $[1 - \exp(-t/\tau_e)]$ must be included. In total, the correction to the conductance can be expressed as [19, 20]:

$$\Delta\sigma_{2D} = -\frac{2\hbar}{m^*}\sigma_0 \int_0^\infty dt \frac{1}{4\pi Dt}(1-e^{-t/\tau_e})e^{-t/\tau_\varphi}$$

$$= -\frac{e^2}{2\pi^2\hbar}\ln\left(1+\frac{\tau_\varphi}{\tau_e}\right) \qquad (1.32)$$

Here, σ_0 is the classical Drude conductivity of a 2-D system. The localization vanishes, if the phase-breaking time τ_φ is much smaller than τ_e, since then the logarithmic factor tends towards zero. The ratio of the correction due to weak localization to the Drude conductivity $\Delta\sigma_{2D}/\sigma_0$ is usually small and of the order of $1/k_F l_e$. Here, $k_F = m^* V_F/\hbar$ is the Fermi wavenumber. For a typical 2-D electron gas with $\mu_e = 10^6$ cm^2 V s^{-1} at $n_e = 3 \times 10^{11}$ cm^{-2}, a correction of less than 0.1% would be expected.

For a quasi 1-D structure of width W with $l_\varphi \gg W$, the diffusion is effectively reduced to one dimension, so that the return probability can now be expressed by $W^{-1}(4\pi Dt)^{-1/2}$. The conductivity correction in this case is given by [20]:

$$\Delta\sigma_{1D} = -\frac{e^2}{\pi\hbar}\frac{l_\varphi}{W}\left[1-\left(1+\frac{\tau_\varphi}{\tau_e}\right)^{-1/2}\right] \qquad (1.33)$$

A comparison of the 1-D and 2-D cases reveals that the weak localization correction to the conductivity is much larger for the 1-D case. In the latter case, the ratio $\Delta\sigma_{1D}/\sigma_0$ is of the order $(l_\varphi/W)(1/k_F l_e)$. If a phase-breaking time of $\tau_\varphi = 10^{-10}$ s and a width of $W = 200$ nm are assumed, the result is a ratio $\Delta\sigma_{1D}/\sigma_0$ of 6%, for a wire based on the 2-D electron gas as specified above. Clearly, this value is much larger than the corresponding value for a 2-D system.

1.4.3
Weak Localization in a Magnetic Field

If the sample is penetrated by a magnetic field **B**, the phase accumulation along a certain trajectory is modified, since the Lagrangian function L [cf. Equation 1.28] of an electron with charge $-e$ contains an additional term

$$L(r,\dot{r},t) = \frac{m}{2}\dot{r}^2 - e[\dot{r}A(r,t)] \qquad (1.34)$$

Here, **A** is the vector potential defined by **B** = rot **A**. In the presence of a vector potential, the probability amplitude C_1 of a closed loop propagated in clockwise orientation acquires an additional phase factor

$$C_1 \to C_1 \exp\left(-i\frac{e}{\hbar}\oint \mathbf{A}d\mathbf{l}\right) = C_1 \exp\left(i\frac{2\pi\Phi}{\Phi_0}\right) \qquad (1.35)$$

Here, $\Phi = BS$ is the magnetic flux penetrating the enclosed area S of the loop, with $\Phi_0 = h/e$ the magnetic flux quantum. For the propagation in the opposite orientation one obtains

$$C_2 \to C_2 \exp\left(-i\frac{2\pi\Phi}{\Phi_0}\right) \qquad (1.36)$$

The phase difference accumulated between both time-reversed paths is therefore

$$\Delta\varphi = 4\pi\frac{\Phi}{\Phi_0} \qquad (1.37)$$

Thus, if a magnetic field is applied, the property that constructive interference occurs for *all* loops in case of $B = 0$ is lost. Generally, many loops enclosing different areas are found in a diffusive conductor and, depending on the size of the loops, different phase shifts $\Delta\varphi$ develop. Thus, for a particular magnetic field the localization is lifted to a different extent depending on the loop size. If the magnetic field is increased starting from zero, the constructive interference is destroyed first for the largest loops. Finally, if the magnetic field is sufficiently large, the phase difference will be randomly distributed between the ensemble of loops. On average, the degree of localization decreases with increasing magnetic field, resulting in a continuous decrease of the resistance.

For a quantitative approach one must take into account that, in addition to the usual phase breaking at zero magnetic field, the phase is also broken effectively by a magnetic field. Similar to l_φ a length l_m is defined, which is characterized by the condition that the area l_m^2 corresponds to the case that the penetrating flux is equal to Φ_0. Thus, l_m is defined by $\sqrt{\hbar/eB}$. As outlined above, for a flux Φ_0 the phase difference between time-reversed paths is already significant. The characteristic magnetic relaxation time τ_B related to l_m can be estimated from the relationship $l_m \sim \sqrt{D\tau_B}$, in analogy to Equation 1.4 defining l_φ. The expression that quantitatively describes the increase of the conductivity with increasing magnetic field is given by [21, 22]:

$$\Delta\sigma_{2D}(B) - \Delta\sigma_{2D}(0) = \frac{e^2}{2\pi^2\hbar}\left[\Psi\left(\frac{1}{2} + \frac{\tau_B}{2\tau_\varphi}\right) - \Psi\left(\frac{1}{2} + \frac{\tau_B}{2\tau_e}\right) + \ln\left(\frac{\tau_\varphi}{\tau_e}\right)\right] \qquad (1.38)$$

where $\Psi(x)$ is the digamma function. The exact expression for τ_B, which must be inserted into Equation 1.38, is given by $\tau_B = l_m^2/2D$. At zero magnetic fields the relevant maximum size of the loops at which the phase coherence is broken is given by l_φ^2. In a finite magnetic field, weak localization is suppressed if a noticeable phase shift between time-reversed loops is accumulated. This is the case for loops with the area of about l_m^2. By comparing both relationships, it is clear that the magnetic field has a significant effect on the conductance for $l_\varphi^2 \approx l_m^2$. This relationship defines a critical

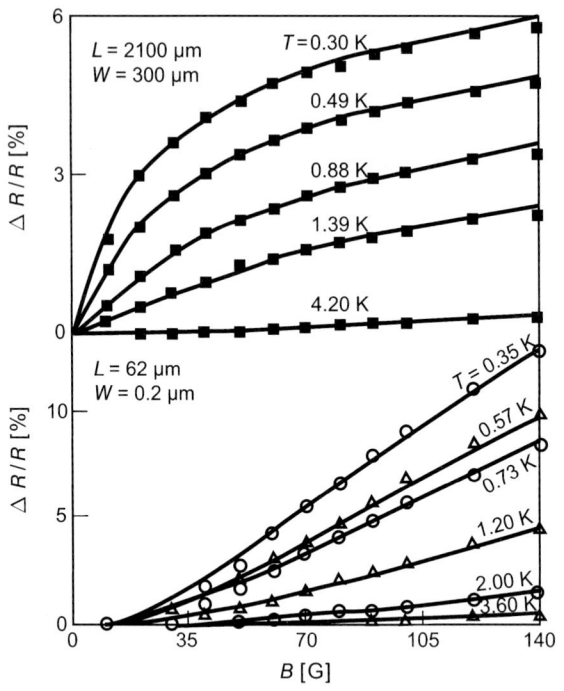

Figure 1.8 Comparison of the weak localization effect in a two-dimensional (upper graph) and a one-dimensional electron gas (lower graph) in AlGaAs/GaAs. For the one-dimensional structures a much higher magnetic field is required to suppress the weak localization effect. (Reprinted with permission from [23]. Copyright (1987) by the American Physical Society.)

magnetic field B_c, which is given by

$$B_c = \frac{\hbar}{2el_\varphi^2} \tag{1.39}$$

Thus, at the characteristic field of B_c one expects a suppression of weak localization. For semiconductor structures, l_φ may be of the order of 1 μm, and result in a critical field of about 1 mT. In the case of a 2-D electron gas, weak localization is suppressed at relatively low magnetic fields (see Figure 1.8).

In 1-D systems in the dirty metal limit, defined as $l_e \ll W \ll l_\varphi$, the closed trajectories contributing to weak localization are quenched in one direction, with a typical enclosed area of the loop given by $W\sqrt{D\tau_B}$ (see Figure 1.9a). For a unit phase shift this area corresponds to l_m^2, resulting in a magnetic relaxation time of $\tau_B \sim l_m^4/DW^2$ and a critical field of $B_c \sim \hbar/eWl_\varphi$. The full expression for the weak localization correction of one-dimensional systems in the dirty limit is given by [24]

$$\Delta\sigma_{1D}(B) = \frac{e^2}{\pi\hbar} \frac{\sqrt{D}}{W} \left[\left(\frac{1}{\tau_\varphi} + \frac{1}{\tau_B} \right)^{-1/2} - \left(\frac{1}{\tau_\varphi} + \frac{1}{\tau_e} + \frac{1}{\tau_B} \right)^{-1/2} \right] \tag{1.40}$$

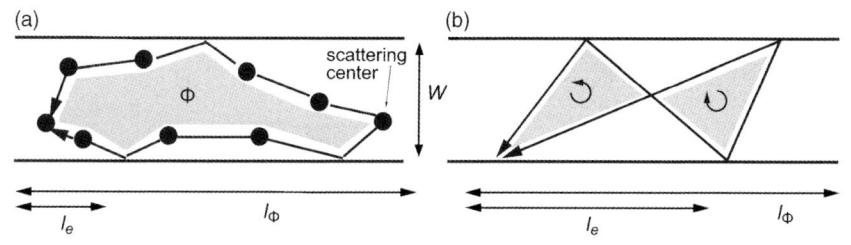

Figure 1.9 (a) Typical closed trajectory in a dirty metal one-dimensional conductor ($l_e \ll W \ll l_\phi$). (b) Typical closed trajectory in a narrow one-dimensional structure with $W \ll l_e$. Here, diffusive boundary scattering results in loops which self-interact. The net flux is cancelled in this configuration.

with magnetic relaxation time in this case given by $\tau_B = 3 l_m^4 / W D$. It should be noted that, at zero magnetic fields, Equation 1.32 is recovered. Furthermore, a closer inspection of B_c reveals, that if the width of the wire is reduced, the critical field is increased, ensuring that the weak localization effect is preserved up to much higher magnetic fields compared to the 2-D case. This is confirmed by the measurements shown in Figure 1.8, where the magnetoresistance peak is wider in the 1-D case. In wire structures based on high-mobility, 2-D electron gases, the elastic mean free path l_e may be larger than the width of the wire: $W \ll l_e$. In this ballistic regime, the electrons propagate without any scattering between the wire boundaries. As illustrated in Figure 1.9b, owing to diffusive boundary scattering the typical closed loops will self-interact. As both parts of the loop area are traversed in opposite orientation, the net flux is basically cancelled [20]. Clearly, the flux cancellation results in a further increase of the critical field.

1.5
Spin-Effects: Weak Antilocalization

So far, the effect of spin on the electron interference has been neglected, and this approach is valid as long as the spin orientation is conserved. However, in many materials the spin changes its orientation while the electron propagates along the closed loops, resulting in the weak localization effect.

It can be assumed that $|s\rangle$ is the initial spin state, this generally being a superposition of the spin up $|\uparrow\rangle$ and spin down $|\downarrow\rangle$ states. In principle, there are two possibilities of how the spin orientation can be changed:

- The Elliot–Yafet mechanism. Here, the potential profile of the scattering centers can lead to spin-orbit coupling; this results in a spin rotation, while the electron is scattered at the impurities (see Figure 1.10a).
- The so-called D'yakonov–Perel mechanism, where the spin precesses while the electron propagates *between* the scattering centers (see Figure 1.10b). The origin of the spin precession may either be a lack of inversion symmetry (i.e., in zinc blende

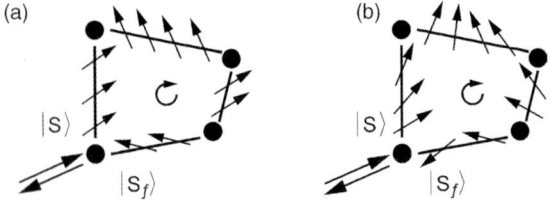

Figure 1.10 (a) Typical closed trajectory in forward direction with spin scattering at the impurities. The initial spin state $|s\rangle$ is transformed to the final spin state $|s_f\rangle$. The spin orientation is preserved while propagation between the scattering centers. (b) The situation where a spin precession occurs while the electron propagates between the scattering centers.

crystals; the Dresselhaus effect [25]), or an asymmetric potential shape of the quantum well forming a 2-D electron gas (the Rashba effect) [26].

Further details on spin precession are provided in Chapter 3 of this volume and Chapter 5 of volume 4 of this series (Bandyopadhyay, S., Monolithic and Hybrid Spintronics. In: Schmid, G. (ed), Nanotechnology, Vol 4, Chapter 5).

Regardless of the underlying mechanism, if an electron propagates along a closed loop, its spin orientation is changed. The modification of the spin orientation can be expressed by a rotation matrix **U** [27]. For the propagation along the loop in forward (f) direction the final state $|s_f\rangle$ can be expressed by

$$|s_f\rangle = \mathbf{U}|s\rangle \tag{1.41}$$

where **U** is the corresponding rotation matrix. For propagation along the loop in a backwards directions (b), the final spin state is given by

$$|s_b\rangle = \mathbf{U}^{-1}|s\rangle \tag{1.42}$$

Here, use is made of the fact that the rotation matrix of the counter-clockwise propagation is simply the inverse of **U**. For interference between the clockwise and counter-clockwise electron waves, not only the spatial component is relevant but also the interference of the spin component:

$$\begin{aligned}\langle s_b|s_f\rangle &= \langle \mathbf{U}^{-1}s|\mathbf{U}s\rangle \\ &= \langle s|\mathbf{U}^\dagger \mathbf{U}|s\rangle \\ &= \langle s|\mathbf{U}^2|s\rangle.\end{aligned} \tag{1.43}$$

The final expression was obtained by making use of the fact that **U** is a unitary matrix: $\mathbf{U}^{-1} = \mathbf{U}^\dagger$, with $\mathbf{U}^\dagger$ the adjoint (complex conjugated and transposed) matrix of **U**. Weak localization – and thus constructive interference – is recovered if the spin orientation is conserved in the case that **U** is the unit matrix **1**.

However, if the spin is rotated during electron propagation along a loop, in general no constructive interference can be expected. Moreover, for each loop a different interference will be expected. Interestingly, averaging over all possible trajectories even leads to a reversal of the weak localization effect such that, instead of an increase

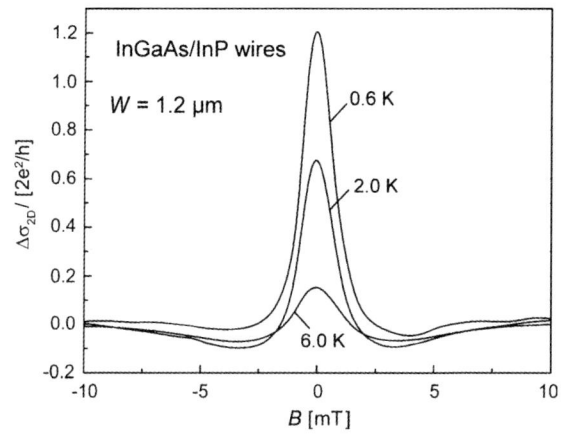

Figure 1.11 Magneto-conductivity measured on a set of 160 InGaAs/InP wires at various temperatures [29]. The Rashba spin-orbit coupling, present in this type of quantum well, results in weak antilocalization, an enhanced conductivity at zero magnetic field. The wires had a geometrical width of 1.2 μm.

in the resistance, a decrease occurs [22, 27, 28]. As the sign of the quantum mechanical correction to the conductivity is reversed, this effect is referred to as "weak antilocalization".

The weak antilocalization measurements of a set of InGaAs/InP wires are shown in Figure 1.11. In contrast to the weak localization effect, an enhanced conductivity is found at $B = 0$. However, if a magnetic field is applied then the weak antilocalization effect is gradually suppressed. Notably, important parameters characterizing the spin scattering and spin precession can be extracted from weak antilocalization measurements. In fact, detailed information of the Rashba and Dresselhaus contributions in a particular material can be obtained by fitting the experimental curves to the appropriate theoretical model. It should be noted that both contributions are important for the spin field effect transistor, as introduced in Chapter 3 of this volume and Chapter 5 of volume 4 of this series (Bandyopadhyay, S., Monolithic and Hybrid Spintronics. In: Schmid, G. (ed), Nanotechnology, Vol 4, Chapter 5).

1.6
Al'tshuler–Aronov–Spivak Oscillations

The fact that, in a metallic conductor, closed loops are responsible for the reduction of the resistance if a magnetic field is applied raises the question: Is it possible to observe resistance oscillations due to interference if the shape of the closed loops are restricted by a fixed, well-defined geometry? In the following sections, it will be shown that indeed these oscillations – the so-called Al'tshuler–Aronov–Spivak oscillations – can be observed in ring-shaped conductors, if the rings are penetrated by a magnetic flux. A series of interconnected ring-shaped conductors is shown schematically in

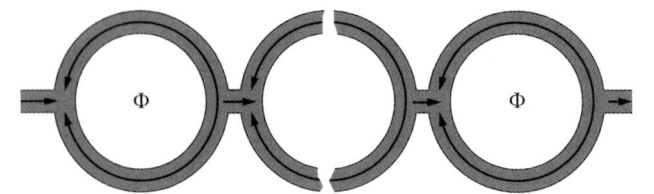

Figure 1.12 A series of interconnected ring-shaped conductors. Each ring is penetrated by a magnetic flux Φ. The interference of the time-reversed trajectories leads to the Al'tshuler–Aronov–Spivak oscillations as a function of a magnetic field.

Figure 1.12, where the enclosed magnetic flux Φ is the same for each ring. Thus, the phase shift $\Delta\varphi$ between time-reversed paths, as given by Equation 1.37, is approximately the same in all rings. By using Equation 1.30, the total amplitude in a loop is given by [30, 31]:

$$P = \left| C_1 \exp\left(i2\pi \frac{\Phi}{\Phi_0}\right) + C_2 \exp\left(-i2\pi \frac{\Phi}{\Phi_0}\right) \right|^2 \quad (1.44)$$
$$= 2|C_1|^2 [1 + \cos(4\pi\Phi/\Phi_0)].$$

From the equation given above it can be concluded that the resistance in this type of structure should oscillate with a period of $\Phi_0/2$.

The first demonstration of this type of weak localization resistance oscillations was provided by Sharvin and Sharvin [31], who evaporated a thin Mg film onto the surface of a quartz filament. The magnetic field was applied in axial orientation with respect to the filament while the current was flowing through the Mg film along the filament. A comparison with the cross-section of the filament confirmed, that the resistance oscillations indeed had a period of $\Phi_0/2$.

Beside cylindrical samples, Al'tshuler–Aronov–Spivak oscillations can also be observed in planar quantum wire networks, similar to the structure shown in Figure 1.12 [33]. The closed trajectories are realized by squares connected to a chain or to a mesh, as shown in Figure 1.13 (inset).

The relative resistance difference $\Delta R/R_0$ as a function of a perpendicular magnetic field is shown in Figure 1.13. Pronounced oscillations are found in the chain as well as in the mesh structure. For smaller square elements, the oscillation period is larger as a larger magnetic field is required to generate a magnetic flux of $\Phi_0/2$. In order to observe Al'tshuler–Aronov–Spivak oscillations, the phase-coherence length must be larger than the circumference of the squares. For the chain structure, the total resistance is given by adding the contribution of each single square. Depending on the type of material, each ring produces either a maximum or minimum at $B = 0$, depending on the absence or presence of spin scattering. This ensures that, after summation of the contribution of each element of the chain, the Al'tshuler–Aronov–Spivak oscillations are not averaged out.

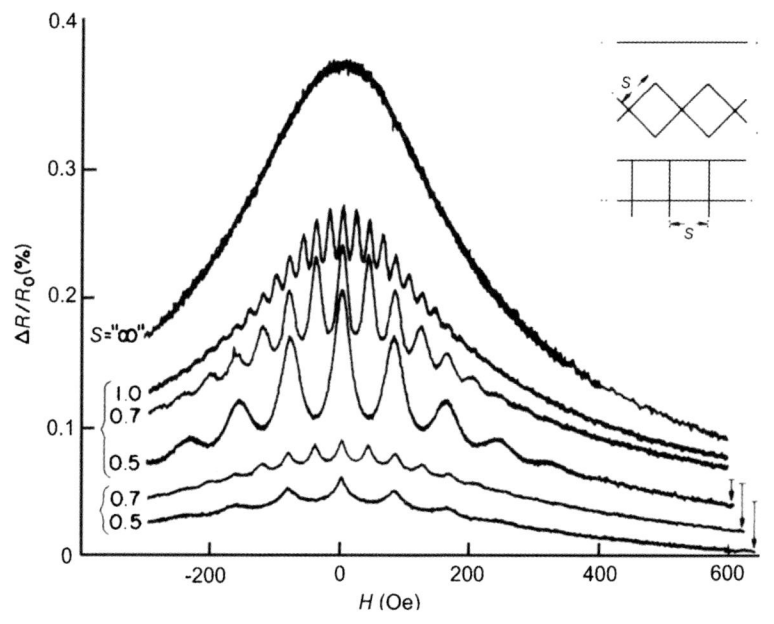

Figure 1.13 Magnetoresistance of 21 nm-thick and 55 nm-wide lithium wires of different geometry measured at 0.13 K. The upper curve shows the measurement of a single wire. Here, a resistance maximum at $B=0$ due to weak localization is found. The following three curves show the resistance for a chain of squares. The size of the elements decreases for lower curves, respectively. The lowest two curves show the measurement on a mesh structures. The size S of the unit cell (in micrometers) is indicated next to each curve. (Reprinted with permission from [33]. Copyright (1986) by the American Physical Society.)

1.7
The Aharonov–Bohm Effect

In contrast to the Al'tshuler–Aronov–Spivak oscillations, which originate from the interference of electrons propagating along time-reversed paths interfering at the point of departure, the Aharonov–Bohm effect is based on electron waves propagating along two different branches of a ring structure and interfering at the opposite side of the ring. This situation is illustrated schematically in Figure 1.14a.

The Aharonov–Bohm effect was predicted, from a theoretical point of view, in 1959 [34]. The essence of this effect is that the vector potential A affects the interference of the electron waves, even in the case when the magnetic field B in the conductor is zero. In order to clarify this point, the experimental set-up shown in Figure 1.14a will be discussed. Here, the magnetic field B is restricted to an area within the ring structure (the gray-shaded area in Figure 1.14a), and is zero in the ring-shaped conductor. Classically no effect is expected since, at the location of the electrons, no magnetic field is present. However, as seen in Section 1.6, the vector potential A, which is non-zero in the conductor, will induce a phase shift of the electron wave and thus affect the electron transport.

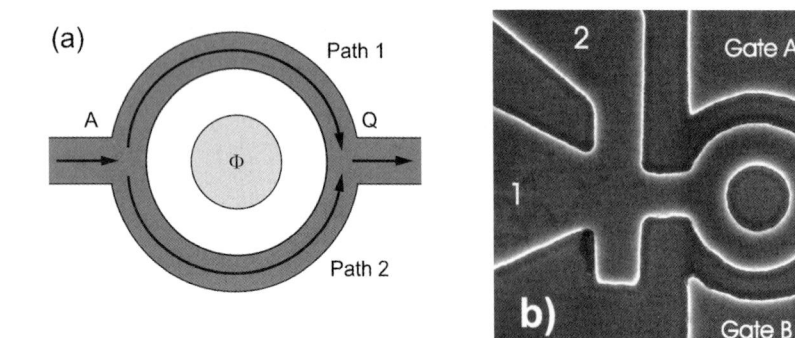

Figure 1.14 (a) Electron trajectories in a ring-shaped conductor. For the Aharonov–Bohm effect the ring is penetrated by a magnetic field within the inner diameter of the conductor. No magnetic field is applied within the conductor. The magnetic flux within the ring is Φ. (b) Electron beam micrograph of an AlGaAs/GaAs ring structure with two in-plane gates (A and B).

The phase difference $\Delta\varphi$ of two electron waves propagating along the upper and the lower branches of the ring (paths 1 and 2 in Figure 1.14a) and interfering at the end point Q of the ring is given by

$$\Delta\varphi = \chi_1 - \chi_2 - \frac{e}{\hbar} \int_{path1} A dl + \frac{e}{\hbar} \int_{path2} A dl$$

$$= \Delta\chi + \frac{e}{\hbar} \oint A dl. \quad (1.45)$$

Here, χ_1 and χ_2 are the phases that the electron waves acquire during their propagation along path 1 and path 2 at zero magnetic field in the interior of the ring. In contrast to weak localization, the paths are different and therefore not time-reversed. Since the impurity configurations on both branches usually differ, the accumulated phases are different in both branches.

We will now return to the Aharonov–Bohm effect itself. By making use of $rot\, A = B$, Equation 1.45 results in

$$\Delta\varphi = \Delta\chi + \frac{e}{\hbar} \int B df$$

$$= \Delta\chi + 2\pi \frac{\Phi}{\Phi_0}. \quad (1.46)$$

The surface integral over B corresponds to the magnetic flux Φ penetrating the ring. As illustrated in Figure 1.14a, the area penetrated by the magnetic field does not need to be as large as the opening of the ring. As can be inferred from Equation 1.46, a phase shift of 2π is acquired if the magnetic flux is changed by a magnetic flux quantum Φ_0. Thus, the period is twice as large as the period of the Al'tshuler–Aronov–Spivak oscillations discussed above.

1.7 The Aharonov–Bohm Effect

For the first experiments demonstrating the Aharonov–Bohm effect, a set-up was used where the electrons were not exposed to a magnetic field [36, 37]. These experiments were performed with an electron beam in a vacuum and a shielded magnet coil. In solid-state the Aharonov–Bohm effect was first demonstrated in Au rings, with the diameter of the ring structure being less than 1 μm and a wire width of a few tens of nanometers [38]. In metallic ring structures, the magnetic field cannot usually be prevented from penetrating the wire itself. Nevertheless, the vector potential *A* is still responsible for the effect on the electron interference pattern. A typical ring structure defined in an AlGaAs/GaAs semiconductor heterostructures is shown in Figure 1.14b [35]. One important difference between the Aharonov–Bohm experiments on nanoscaled rings and experiments using electron beams in vacuum, is that in the former case the electrons are usually scattered many times within the conductor before reaching the opposite side of ring. Thus, the elastic mean free path is most often smaller than the ring size. In addition, in metallic or semiconducting ring structures the phase-coherence length is in the order of the ring diameter at low temperatures, and consequently many electrons lose their phase memory while propagating through the ring. This is the reason why the oscillation amplitude is considerably smaller than the total resistance of the structure. As can be seen in Figure 1.15, pronounced Aharonov–Bohm oscillations were observed in ring structures based on 2-D electron gases in an $In_{0.77}Ga_{0.23}As$/InP heterostructure [39]. A comparison of the enclosed area of the ring confirmed that the oscillation period corresponded to a magnetic flux quantum Φ_0. Owing to the low effective electron mass and to the high mobility in these heterostructures, the phase-coherence length can exceed 1 μm at temperatures below 1 K, and consequently large oscillation amplitudes are achieved. Previously, resistance modulations of up to 12% have been observed in this type of structure.

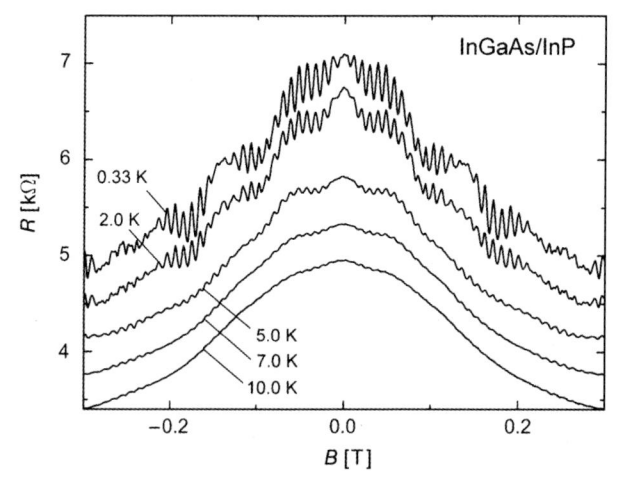

Figure 1.15 Magnetoresistance of an $In_{0.77}Ga_{0.23}As$/InP ring structure measured at 0.3 K. The ring had a diameter of 820 nm, with a width of the wires forming the ring of about 85 nm. (Reprinted with permission from [39]. Copyright (1995) by the American Physical Society.)

When the Al'tshuler–Aronov–Spivak oscillations were discussed, it was found that at $B = 0$ a maximum is observed in the resistance, owing to the constructive interference between time-reversed loops (see Figure 1.13). For the Aharonov–Bohm effect, the interference at the branching point is determined by the two different paths 1 and 2 along the two branches of the ring, as illustrated in Figure 1.14a. The phase difference $\Delta\chi$ at $B = 0$ between the two paths depends to a large extent on the distribution of scattering centers in the ring; that is, for different rings – regardless of whether they have the same geometry – a different phase shift $\Delta\chi$ is accumulated. As a result, no clear maximum or minimum, as for the Al'tshuler–Aronov–Spivak oscillations, is expected at $B = 0$. In fact, it could be shown that the amplitude of the Aharonov–Bohm oscillations is decreased if the signal of many rings is averaged [40]. This is due to the fact that the contributions of interferences of the different rings with the same oscillation period but with statistically distributed phase shifts $\Delta\chi$ are averaged out.

Besides the magnetic control of the interference pattern of an Aharonov–Bohm ring structure, the oscillation pattern can also be changed electrostatically by means of a gate electrode. A typical AlGaAs/GaAs sample with two in-plane gates is shown in Figure 1.14b. By applying a voltage to one of the gates, the electron concentration in the corresponding branch of the ring is altered. A change of the carrier concentration goes along with a change of the Fermi wavelength and, as a result, the phase accumulated in this branch of the ring is changed. Clearly this will immediately affect the interference pattern, as can be seen in Figure 1.16.

It is interesting to observe that the oscillation pattern of the sample is symmetric with respect to the magnetic field. This can be seen in Figure 1.16, where the resistance oscillations are shown as a grayscale plot as a function of magnetic field

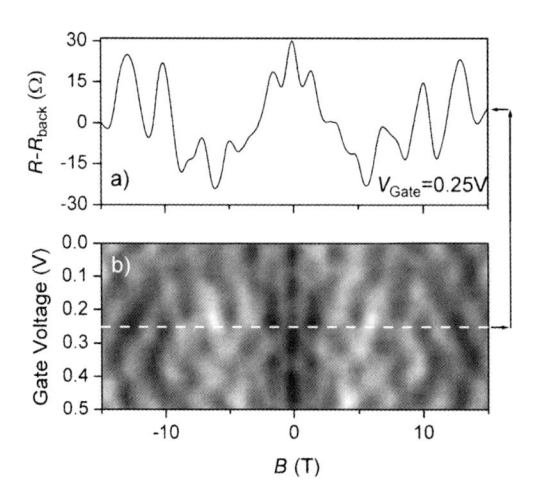

Figure 1.16 (a) Magnetoresistance of the AlGaAs/GaAs ring structure shown in Figure 1.14b for 0.25 V applied to gate B. The voltage at gate A was set to zero. The background resistance R_{back} was subtracted from the total resistance. (b) Grayscale plot of the magnetoresistance R-R_{back} for different voltages applied to gate A and B. The dark and light regions correspond to large and low resistance values, respectively. The dashed line indicated the measurement shown in (a).

and gate voltage. The symmetric pattern is due to the fact that, although four terminals are used during the measurement, the measurement is effectively a two-terminal measurement. For such a measurement it can be shown in general that the resistance is symmetric under reversal of the magnetic field: $R(-B) = R(B)$ [9]. The reason for the two-terminal nature of the measurement is the coupling of the two contacts on each side, although these are not independent, as would be required for a pure four-terminal measurement.

1.8
Universal Conductance Fluctuations

A closer inspection of the Aharonov–Bohm effect measurements reveals that irregular fluctuations are often superimposed on the regular oscillations. This can be seen clearly in the measurements shown in Figure 1.15 where a long-wavelength underground is observed superimposed on the oscillations. The fluctuations are reproducible if the measurements are repeated for the same sample [42, 43]; however, if different samples with the same geometry and fabricated using the same material are compared, it is found that a different fluctuation pattern belongs to each sample. This is the reason, why the individual fluctuation pattern of a sample is sometimes referred to as a *fingerprint*. Conductance fluctuations are observed in semiconducting structures as well as in metallic samples [44, 45]. However, as the size of semiconductor devices becomes smaller, the statistical distribution of the remaining few dopant atoms will result in a spread of the device characteristics. Thus, resistance fluctuations are not only important in the phase-coherent regime but are also becoming much more of an issue in device applications.

1.8.1
Basic Principles

The reason, why each sample shows a different conductance fluctuation pattern becomes clear when it is realized that very few scattering centers (i.e., impurities) are present in very small structures. In fact, if the sample possesses very few impurities then it is the impurity configuration that governs the transport properties. Moreover, if only a finite number of scattering centers is present, an ensemble average cannot be applied for the theoretical description, as this does not take into consideration the particular spatial distribution of the scattering centers. Of course fluctuations may be observed not only in ring structures but also in a single quantum wire, as they originate from the spatial configuration of the scattering centers in the wire. The conductance fluctuations of a single Au wire are shown in Figure 1.17.

The fluctuations shown in Figure 1.17 cover an interval of approximately $\pm e^2/h$, the fluctuation amplitude of which, as proven by a detailed theoretical analysis, is universal [47, 48]. For a qualitative explanation of the physical origin of the conductance fluctuations, the reader is referred to the above-mentioned Aharonov–Bohm effect. As illustrated in Figure 1.18a, the electron is able to propagate along a certain number of paths in order to cross the wire.

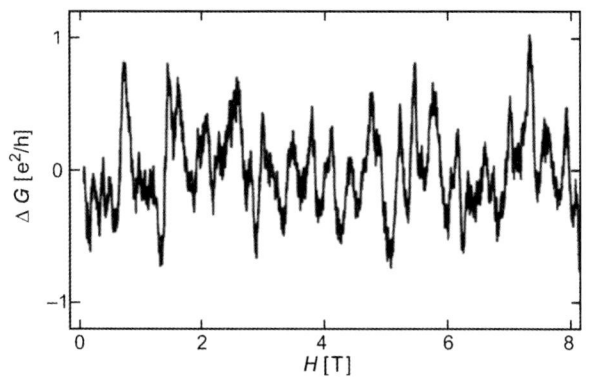

Figure 1.17 Conductance fluctuations as a function of magnetic field of a 310 nm-long and 25 nm-wide Au wire measured at a temperature of 10 mK [46]. (© Taylor &Francis Ltd.).

The total transmission probability results from the squared amplitude of all possible trajectories. Among these trajectories, a limited number of paths may be found which meet again after a certain distance. If a magnetic field is applied, the paths become penetrated by a magnetic flux Φ; subsequently, if the magnetic field is varied, superpositioning of the electron waves of two paths (which cross twice) leads to a variation in the transmission probability due to the Aharonov–Bohm effect. Then, in contrast to a well-defined ring structure, the encircled areas differ among the various locations of the wire, such that a different Aharonov–Bohm period is developed for each area. Superposition of the different quasi Aharonov–Bohm rings then produces an irregular conductance pattern [43]. It is important that only a limited number of trajectories exists, so that an effective averaging out of the

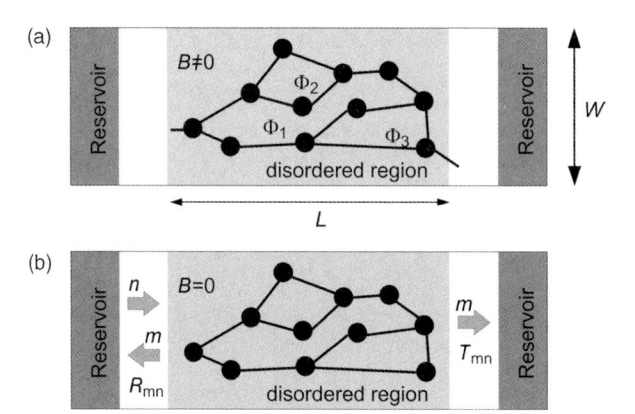

Figure 1.18 (a) Electron trajectories in a quantum wire. If a magnetic field is applied, loops are penetrated by a magnetic flux Φ_1, Φ_2, Φ_3, ... (b) Sample configuration considered for the calculation of the conductance fluctuations using the Landauer–Büttiker formalism. The wire consists of a disordered region connected by ballistic areas to the phase-randomizing reservoirs. Here, n denotes the incoming channel, while m is the outgoing channel (transmitted or reflected).

oscillations is prevented. In addition to varying the magnetic field, it is also possible to observe conductance fluctuations by increasing the applied voltage, but this will lead to a change in the Fermi wavelength of the electrons.

1.8.2
Detailed Analysis

A detailed theoretical description of conductance fluctuations is based on the particular scattering center configuration [47, 48]. Hence, by using this theoretical approach it was possible to calculate the average oscillation amplitude of sample-specific conductance fluctuations. One important point of the theoretical model is that a variation in the magnetic field or the Fermi energy induces the same type of fluctuation as would an ensemble average (quasi-ergodic hypothesis). This allows the measurement of only a single sample rather than measuring numerous different wires at zero magnetic field and constant Fermi energy.

In the following section it will first be shown that the magnitude of the conductance fluctuations at zero temperature is of the order e^2/h, independent of the sample size. The only requirement is that the transport takes place within the diffusive regime. For the derivation, an ensemble of different wires with different impurity configurations at zero magnetic field is considered. An explanation of the universality of conductance fluctuations follows the approach of Lee [49], although for the following considerations use will be made of the Landauer–Büttiker formalism (see Section 1.3.1), where conductance is expressed by the transmission probabilities of the different quantum channels. A scheme of the conductor is shown in Figure 1.18b, where the current is flowing from the left to the right reservoir. The disordered region of the wire is connected by ballistic areas to the phase-randomizing reservoirs.

The first step of the process is to express the Drude conductance of a diffusive conductor by the Landauer–Büttiker formula. Hence, the Drude conductance for a single spin direction can be written as

$$G = \frac{e^2}{h} \sum_{m,n}^{N} T_{mn} = \frac{e^2}{h} \left(N - \sum_{m,n}^{N} R_{mn} \right) \qquad (1.47)$$

where $N = k_F W/\pi$ is the number of quantum channels of a 1-D conductor of width W, and k_F is the Fermi wavenumber. The quantities T_{mn} and R_{mn} denote the transmission and reflection probabilities from channel n into channel m, respectively [cf. Equation 1.23 and 1.20, the indices i, j for the channels are omitted, here]. Interest has been shown in the variations of the conductance for different impurity configurations, and the quantity related to this is the variance of the conductance, which is defined by:

$$\text{var}(G) = \langle \Delta G^2 \rangle = \langle (G - \langle G \rangle)^2 \rangle \qquad (1.48)$$

Here, $\langle \cdots \rangle$ denotes the average over different impurity configurations. The square root of the variance, $\delta G \equiv \sqrt{\text{var}(G)}$, is a measure of the magnitude of the conductance fluctuations, the quantity, which is determined experimentally. With the expression for the conductance, given by Equation 1.47, one obtains for the variance:

$$\text{var}(G) = \left(\frac{e^2}{h}\right)^2 \text{var}\left(N - \sum_{m,n}^{N} R_{mn}\right)$$

$$= \left(\frac{e^2}{h}\right)^2 \text{var}\left(\sum_{m,n}^{N} R_{mn}\right) \qquad (1.49)$$

$$= \left(\frac{e^2}{h}\right)^2 N^2 \text{var}(R_{mn}),$$

and it is assumed that $\text{var}(R_{mn})$ is independent of m and n. The question might be asked as to why the variance is not calculated directly by using the transmission probabilities T_{mn}. Following Lee [49], this causes a problem, since for transmission processes in the diffusive transport regime with many impurity collisions a sequence of scattering events is shared by different channels of the conductor. As a consequence, the different channels are not completely uncorrelated, so that problems are encountered in the proceeding averaging procedure. The situation is different, however, if reflection processes are considered, where it may be assumed that only a few scattering events are responsible for the back-reflection. This is also the reason, why it can be assumed that the reflections in different channels are uncorrelated.

In order to calculate the variance of R_{mn}, use is made of the concept of Feynman paths. By analogy to the probability to propagate between to points A and Q, as expressed by Equation 1.29, the probability for a reflection from channel n into m by the square of the total amplitude of all possible paths j which propagate from the incoming channel n into the outgoing channel m can be expressed as:

$$R_{mn} = \left|\sum_{j} C_j\right|^2 \qquad (1.50)$$

According to Equation 1.49, the variance of R_{mn} must first be calculated in order to obtain the variance of G. The variance of R_{mn} is given by

$$\text{var}(R_{mn}) = \langle R_{mn}^2 \rangle - \langle R_{mn} \rangle^2 \qquad (1.51)$$

The last term is the square of the average reflection probability $\langle R_{mn} \rangle$, which can be expressed by

$$\langle R_{mn} \rangle = \sum_{ij} \langle C_i C_j^* \rangle \qquad (1.52)$$

If uncorrelated paths are assumed, as discussed above, so that

$$\langle C_i C_j^* \rangle = 0 \qquad (1.53)$$

one finally obtains

$$\langle R_{mn} \rangle = \sum_{i} \langle C_i C_i^* \rangle = \sum_{i} \langle |C_i|^2 \rangle \qquad (1.54)$$

For the first term in Equation 1.51, the following can be written:

$$\langle R_{mn}^2 \rangle = \sum_{ijkl} \langle C_i C_j C_k^* C_l^* \rangle$$
$$= \sum_{ijkl} \{\langle |C_i|^2 \rangle \langle |C_j|^2 \rangle \delta_{ik}\delta_{jl} + \langle |C_i|^2 \rangle \langle |C_j|^2 \rangle \delta_{il}\delta_{jk}\} \quad (1.55)$$
$$= 2\sum_{ij} \langle |C_i|^2 \rangle \langle |C_j|^2 \rangle = 2\langle R_{mn} \rangle^2.$$

Thus, the variance of R_{mn} is given simply by $\langle R_{mn} \rangle^2$.

For small transmission probabilities $T_{mn} \to 0$, which is the case for a sufficient number of scattering centers in the wire ($l_e \ll L$), the average reflection probability may be approximated by [49]

$$\langle R_{mn} \rangle \approx \frac{1}{N} \quad (1.56)$$

so that finally one obtains

$$\text{var}(G) = \left(\frac{e^2}{h}\right)^2 N^2 \text{var}(R_{mn}) \approx \left(\frac{e^2}{h}\right)^2 \quad (1.57)$$

for the variance of the conductance in the diffusive limit. As can be seen here, the conductance fluctuations are of the order of e^2/h. The universal magnitude of the conductance fluctuations is found for example in the measurement shown in Figure 1.17.

1.8.3
Fluctuations in Long Wires

Until now, it has been assumed that the wire length is smaller than the phase-coherence length, but the question must also be addressed as to what happens if the length L of the wire exceeds the phase-coherence length l_φ. In this situation, the wire may be cut into $N = L/l_\varphi$ phase-coherent pieces connected in series (see Figure 1.19).

Each of these segments produces resistance fluctuations, δR_0, so that the total resistance fluctuations are given by

$$\delta R = \sqrt{N} \delta R_0 \quad (1.58)$$

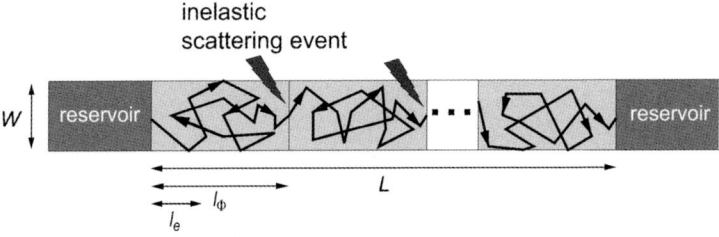

Figure 1.19 The conductance fluctuations are determined by cutting the wire into $N = L/l_\varphi$ coherent pieces.

By using the total resistance $R = NR_0$, where R_0 is the resistance of a single segment, the total conductance fluctuations can be calculated:

$$\delta G = \left| -\frac{\delta R}{R^2} \right| = \frac{e^2}{h} \frac{\sqrt{N}}{N^2} = \frac{e^2}{h} N^{-3/2} \tag{1.59}$$

If N is substituted by the ratio between total length and phase coherence length, the following is finally obtained:

$$\delta G = \frac{e^2}{h} \left(\frac{l_\varphi}{L} \right)^{3/2} \tag{1.60}$$

It is important to note that no exponential decrease of the fluctuations with respect to length is expected. In contrast, only a relatively weak decrease of δG with increasing length is predicted, and this was indeed confirmed experimentally [50].

1.8.4
Energy and Temperature Dependence

In semiconductor structures the electron concentration – and thus the Fermi energy – can be controlled by means of a gate electrode. An impression of how the Fermi energy affects the conductance fluctuations can be gained by comparing the change of E_F with the characteristic correlation energy E_{Th} (Thouless energy).

First, an insight must be obtained into the nature of the correlation energy E_{Th}. For the sake of simplicity, an ideal system can be assumed where any scattering is neglected. Along the length L, the phase develops as

$$\varphi = kL \tag{1.61}$$

If a state is now considered with a slightly different wavevector $k \to k' = k + \Delta k$, the phase difference between both waves is

$$\Delta \varphi = \Delta k L \tag{1.62}$$

The energy difference between both states can be quantified as

$$\Delta E = \frac{dE}{dk} \Delta k = \hbar v_F \frac{\Delta \varphi}{L} = \hbar \Delta \varphi \frac{1}{\tau_{Th}} \tag{1.63}$$

Here, τ_{Th} is the time that the wave requires to propagate along length L. The Thouless energy, E_{Th}, is defined as the energy difference where both states are uncorrelated, at which point if the phase difference $\Delta \varphi$ is equal to 1, the following is obtained:

$$E_{Th} = \frac{\hbar v_F}{L} = \frac{\hbar}{\tau_{Th}} \tag{1.64}$$

Thus, the Thouless energy is connected to the time $\tau_L = L/v_F$ that an electron wave requires to cover the dimension L of the system. Up to now, only a ballistic system has been considered, but in the diffusive regime the corresponding characteristic time is given by $\tau_{Th} = L^2/D$. Only if the Fermi energy is changed by a value comparable to E_{Th} is the next energy level reached and the conductance changed to a value uncorrelated

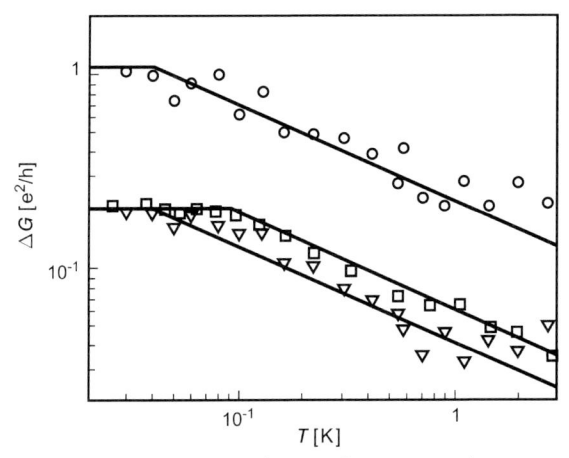

Figure 1.20 Decrease in conductance fluctuations with temperature for a ring structure with a diameter of 820 nm (○). In addition, the temperature dependence of the Aharonov–Bohm oscillations of a 820 nm-wide ring (▽) and for a 325 nm-wide ring (□) are shown [46]. (© Taylor &Francis Ltd.).

to the conductance value of the previous energy value. The correlation energy for small quantum wires has a relative large value, which would not be expected if a large number of uncorrelated trajectories were to be averaged.

If the temperature is increased, then the Fermi distribution becomes smeared out. However, while the temperature remains sufficiently low that the width of the Fermi distribution is smaller than E_{Th}, the maximum fluctuation amplitude is maintained. When the smearing of the Fermi distribution exceeds E_{Th}, a number of approximately $N = (k_B T)/E_{Th}$ segments contribute. As these N segments are uncorrelated, the fluctuation amplitude decreases with $1/\sqrt{N}$, thus $1/\sqrt{T}$. This behavior was confirmed experimentally, as shown in Figure 1.20 [46]. At low temperature the fluctuation amplitude is found initially to be constant, but if the temperature is increased above a critical value a continuous decrease following $1/\sqrt{T}$ is observed. From the starting point of the decrease at $T = 0.1$ K, the correlation energy can be estimated. Typically, a value for E_{Th} of the order of 10 μeV would be obtained for this sample.

1.9
Concluding Remarks

It is clear that many interesting phenomena related to phase-coherence transport can be observed in semiconducting or metallic nanostructures. Although, many of these effects are quite well understood and the theoretical models well established, in other cases open questions remain. For example, does the phase coherence time always saturate if the temperature is sufficiently low [2]? Very recently, the issue of phase coherence in nanostructures has attracted much attention in connection with solid-state quantum computation, where the maintenance of phase coherence is a critical issue. Furthermore, spin-related phenomena are a subject of current interest, as phase-coherent spin manipulation is regarded as an interesting option for future electronic devices.

Clearly, this chapter can provide only a brief overview of the most important phenomena connected with phase-coherent transport in nanostructures. However, for further information, the reader is referred to various textbooks [1, 51, 52] and reviews [2, 18, 46].

References

1 Datta, S. (1995) *Electron Transport in Mesoscopic Systems*, Cambridge University Press, Cambridge.
2 Lin, J.J. and Bird, J.P. (2002) *Journal of Physics: Condensed Matter*, **14**, R501.
3 Giuliani, G.F. and Quinn, J.J. (1982) *Physical Review B*, **26**, 4421.
4 Al'tshuler, B.L., Aronov, A.G. and, Khmelnitsky, D.E. (1981) *Solid State Communications*, **39**, 619.
5 Choi, K.K., Tsui, D.C. and Alavi, K. (1987) *Physical Review B*, **36**, 7751.
6 Landauer, R. (1957) *IBM Journal of Research and Development*, **21**, 223.
7 Landauer, R. (1987) *Zeitschrift für Physik B*, **68**, 217.
8 Büttiker, M., Imry, Y., Landauer, R. and Pinhas, S. (1985) *Physical Review B*, **31**, 6207.
9 Büttiker, M. (1986) *Physical Review Letters*, **57**, 1764.
10 van Wees, B.J., van Houten, H., Beenakker, C.W.J., Willamson, J.G., Kouwenhoven, L.P., van der Marel, D. and Foxon, C.T. (1988) *Physical Review Letters*, **60**, 848.
11 Wharam, D.A., Thornton, T.J., Newbury, R., Pepper, M., Ahmed, H., Frost, J.E.F., Hasko, D.G., Peacock, D.C., Ritchie, D.A. and Jones, G.A.C. (1988) *Journal of Physics C: Solid State Physics*, **21**, L209.
12 Laux, S.E., Frank, D.J. and Stern, F. (1988) *Surface Science*, **196**, 101.
13 Szafer, A. and Stone, A.D. (1989) *Physical Review Letters*, **62**, 300.

14 Kirczenov, G. (1989) *Physical Review B*, **39**, 10452.
15 Abrahams, E., Anderson, P.W., Licciardello, D.C. and Ramakrishnan, T.V. (1979) *Physical Review Letters*, **42**, 673.
16 (a) Gorkov, L.P., Larkin, A.I. and Khmel'nitskii, D.E. (1979) *Pis'ma Zhurnal Eksperimentalnoi I Teoreticheskoi Fiziki*, **30**, 248; (b) Gorkov, L.P., Larkin, A.I. and Khmel'nitskii, D.E. (1979) *JETP Letters*, **30**, 228.
17 Feynman, R.P. and Hibbs, A.R. (1965) *Quantum Mechanics and Path Intergrals*, McGraw-Hill, New York.
18 Beenakker, C.W.J. and van Houten, H. (1991) (eds. H. Ehrenreich and D. Turnbull), *Semiconductor Heterostructures and Nanostructures in Solid State Physics*, Volume 44, Academic Press, New York. p. 1.
19 Chakravarty and Schmid, A. (1986) *Physics Reports*, **140**, 193.
20 Beenakker, C.W.J. and van Houten, H. (1988) *Physical Review B*, **38**, 3232.
21 Al'tshuler, B.L., Khmelnitskii, D., Larkin, A.I. and Lee, P.A. (1980) *Physical Review B*, **22**, 5142.
22 Hikami, S., Larkin, A.I. and Nagoka, Y. (1980) *Progress of Theoretical Physics*, **63**, 707.
23 Choi, K.K., Tsui, D.C. and Alavi, K. (1987) *Physical Review B*, **36**, 7751.
24 (a) Al'tshuler, B.L. and Aronov, A.G. (1981) *Pis'ma Zhurnal Eksperimentalnoi I Teoreticheskoi Fiziki*, **33**, 515; (b) Al'tshuler,

B.L. and Aronov, A.G. (1981) *JETP Letters*, **33**, 499.
25 Dresselhaus, G. (1955) *Physical Review*, **100**, 580.
26 Bychkov, Yu.A. and Rashba, E.I. (1984) *Journal of Physical Chemistry (Solid State Physics)*, **17**, 6039.
27 Bergmann, G. (1982) *Solid State Communications*, **42**, 815.
28 Iordanskii, S.V., Lyanda-Geller, Yu.B. and Pikus, G.E. (1994) *JETP Letters*, **60**, 206.
29 Guzenko, V.A., Schäpers, Th., Indlekofer, K.M. and Knobbe, J. (2006) *Physica E*, **32**, 333.
30 (a) Al'tshuler, B.L., Aronov, A.G. and Spivak, B.Z. (1981) *Pis'ma Zhurnal Eksperimentalnoi I Teoreticheskoi Fiziki*, **33**, 101; (b) Al'tshuler, B.L., Aronov, A.G. and Spivak, B.Z. (1981) *JETP Letters*, **33**, 94.
31 Aronov, A.G. and Sharvin, Yu.V. (1987) *Reviews of Modern Physics*, **59**, 755.
32 (a) Sharvin, Yu.D. and Sharvin, Yu.V. (1981) *Pis'ma Zhurnal Eksperimentalnoi I Teoreticheskoi Fiziki*, **34**, 285; (b) Sharvin, D.Yu. and Sharvin, Yu.V. (1981) *JETP Letters*, **34**, 272.
33 Dolan, G.J., Licini, J.C. and Bishop, D.J. (1986) *Physical Review Letters*, **56**, 1493.
34 Aharonov, Y. and Bohm, D. (1959) *Physical Review*, **115**, 485.
35 Krafft, B., Förster, A., van der Hart, A. and Schäpers, Th. (2001) *Physica E*, **9**, 635.
36 Chambers, R.G. (1960) *Physical Review Letters*, **5**, 3.
37 Tonomura, A., Matsuda, T., Suzuki, R., Fukuhara, A., Osakabe, N., Umezaki, H., Endo, J., Shinagawa, K., Sugita, Y. and Fujiwara, H. (1982) *Physical Review Letters*, **48**, 1443.
38 Webb, R.A., Washburn, S., Umbach, C.P. and Laibovitz, R.B. (1985) *Physical Review Letters*, **54**, 2696.
39 Appenzeller, J., Schäpers, Th., Hardtdegen, H., Lengeler, B. and Lüth, H. (1995) *Physical Review B*, **51**, 4336.
40 Murat, M., Gefen, Y. and Imry, Y. (1984) *Physical Review B*, **34**, 659.
41 Büttiker, M. (1986) *Physical Review Letters*, **57**, 1761.
42 Umbach, C.P., Washburn, S., Laibowitz, R.B. and Webb, R.A. (1984) *Physical Review B*, **30**, 4048.
43 Stone, A.D. (1985) *Physical Review Letters*, **54**, 2692.
44 Kaplan, S.B. and Hartstein, A. (1986) *Physical Review Letters*, **56**, 2403.
45 Licini, J.C., Bishop, D.J., Kastner, M.A. and Melngailis, J. (1985) *Physical Review Letters*, **55**, 2987.
46 Washburn, S. and Webb, R.A. (1986) *Advances in Physics*, **35**, 375 (http://www.informaworld.com).
47 (a) Al'tshuler, B.L. (1985) *Pis'ma Zhurnal Eksperimentalnoi I Teoreticheskoi Fiziki*, **41**, 530; (b) Al'tshuler, B.L. (1985) *JETP Letters*, **41**, 648.
48 Lee, P.A. and Stone, A.D. (1985) *Physical Review Letters*, **55**, 1622.
49 Lee, P.A. (1986) *Physica A*, **140A**, 169.
50 Umbach, C.P., van Haesendonk, C., Laibowitz, R.B., Washburn, S. and Webb, R.A. (1986) *Physical Review Letters*, **56**, 386.
51 Ferry, D.K. and Goodnick, S.M. (2005) *Transport in Nanostructures*, Cambridge University Press, Cambridge.
52 Heinzel, T. (2003) *Mesoscopic Electronics in Solid State Nanostructures*, Wiley-VCH, Weinheim.

2
Charge Transport and Single-Electron Effects in Nanoscale Systems
Joseph M. Thijssen and Herre S.J. van der Zant

2.1
Introduction: Three-Terminal Devices and Quantization

In electronics, charges are manipulated by sending them through devices which have a few terminals: a *source* which injects the charge; and a *drain* which removes the charge from the device. Occasionally, a third terminal, called a *gate*, is present, and this is used to manipulate the charge flow through the device. The gate does neither inject charge into, nor removes it from the device. Three-terminal devices are standard elements of electronic circuits, where they act as switches or as amplifying elements. Semiconductor-based three-terminal switches are responsible for the tremendous increase in computer speed achieved over the past few decades.

Feynman, in his famous lecture [1], pointed out that the possible scale reduction from the standards of that period was still enormous, and he also suggested that quantum mechanical behavior may result in a different way of operation of the devices, which may open new horizons for applications. Indeed, as we now know, two aspects become important when the size of the device is reduced. The first aspect is indeed the quantum mechanical behavior, and the second is the quantization of the charges flowing into and out of the devices. It is interesting to analyze how the energy scales at which the two effects become noticeable, depend on the device size.

The charge quantization is subtle in view of quantum mechanics: in principle, the charge carried by an electron is distributed in space. In quantum mechanics, a single charge may be distributed according to $|\psi(r)|^2$, where $\psi(r)$ is the quantum mechanical wave function, and this leaves open the possibility of having a fractional charge inside the device. Therefore, the discrete nature of charge does not seem to play a role in the charge transport. However, if the device were to be uncoupled from its surroundings, we would only find integer charges residing on it. This puzzle is solved by realizing that the expectation value of the electrostatic energy, which must be included into the Hamiltonian governing the electron behavior, is dominated by the charge distribution which occurs most of the time. It can be shown that the charge within a device that is *weakly* coupled to its surroundings, is always very close to an integer. Therefore, in order to observe Coulomb effects resulting from the discreteness of the electron charge, it is necessary to consider devices that are weakly coupled to the surroundings.

For the charge quantization, the energy scale associated with the discreteness of the electron charge is given by

$$E_C = \frac{e^2}{2C},$$

where C is the capacitance of the device. This is the energy needed to add a unit charge to the device – it is called the "charging energy". Taking as an estimate the capacitance of a sphere with radius R, we have

$$E_C = \frac{e^2}{8\pi\varepsilon_0 R} = \frac{1}{2R} E_H, \tag{2.1}$$

where, in the rightmost expression, R is given in atomic units (Bohr radii), as is the energy (E_H is the atomic unit of energy – it is called the Hartree and it is given by 27.212 eV). In Section 2.4, we shall present a more detailed analysis for the case where the device is (weakly) coupled to a source, drain and gate.

The energy scale for quantum effects is given by the distance between the energy levels of an isolated device. As a rough estimate, we consider the particle in the (cubic) box problem with energy levels separated by a level splitting Δ given by

$$\Delta = \text{const} \times \frac{\hbar^2}{mL^2} = \text{const} \times \frac{1}{L^2} E_H, \tag{2.2}$$

where m is the electron mass and L is the box size (which must be given in atomic units in the rightmost expression). The first multiplicative constant is of order 1; it depends on the geometry and on the details of the potential.

In the case of carbon nanotubes, the device is much smaller in the lateral direction than along the tube axis. In such cases it is useful to distinguish between the two sizes. The lateral size leads to a large energy splitting and the longitudinal splitting may become vanishingly small. For a metallic nanotube, the level spacing associated with the tube length L is

$$\Delta = \frac{\hbar v_F}{2L},$$

where v_F is the Fermi velocity $v_F = \hbar k_F/m$ with $v_F \approx 8 \times 10^5 \, \text{m s}^{-1}$.

Equations 2.1 and 2.2 tell us how the typical Coulomb and quantum energies scale with the device size (R or L). In Figure 2.1, several experimental realizations are shown of small gated devices that may be weakly coupled to source, drain. Most of these devices have the layout shown in Figure 2.2. An order of magnitude estimate for the charging energy and level splitting for some typical three-terminal devices is provided in Table 2.1. Semiconducting and nanotube quantum dots have been studied in great detail, and their behavior is fairly well understood; however, at the time of writing, the properties of molecular quantum dots are still much less established mainly because it is difficult to fabricate them in a reliable manner.

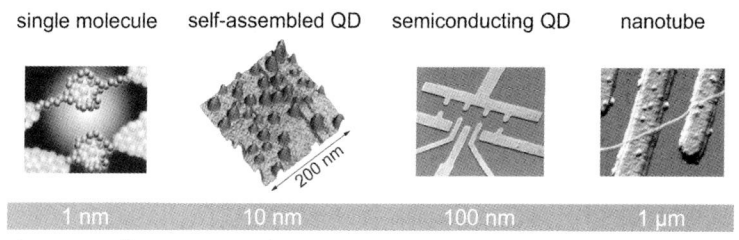

2.1 Introduction: Three-Terminal Devices and Quantization

Figure 2.1 Different quantum dot systems.

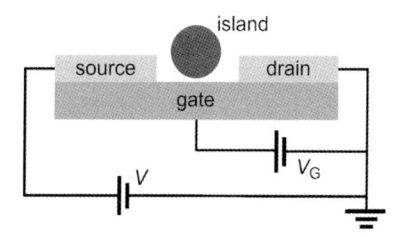

Figure 2.2 A schematic diagram of the three-terminal device layout.

When studying transport through a small island, weakly coupled to a source and a drain, information can be obtained about the quantum level splitting Δ and the charging energy E_C if the energy of the particles flowing through the device can be controlled with precision high enough to resolve these energy splittings. Pauli's principle states that electrons can only flow from an occupied state in the source to an empty state in the drain. The separation between empty and occupied states in the leads is only sharp enough when the temperature is sufficiently low. It can be seen that a low operation temperature is essential for observing the quantum and charge quantization effects. The energy scale associated with the temperature is given by $k_B T$, so we must have

$$k_B T \leq \Delta, E_C.$$

Note that for molecular devices, with their relatively large values of Δ and E_C, quantum and charge quantization effects should still be observable at room temperature. In a typical metallic island, $\Delta \ll k_B T$, and the Coulomb blockade dominates the level separation. In this case we speak of a *classical dot* (see also Chapter 21 of this volume).

In the present chapter we explain the different aspects of charge transport, with emphasis on those devices in which the level spacing and the charging energy plays an essential role in the transport properties. This is the case in quantum dots and in many molecular devices.

Table 2.1 Typical charging energies and level spacings for various three-terminal devices.

	Ga As quantum dot	Carbon nanotube[a]	Molecular transistor
E_C	0.2–2 meV	3 meV	>0.1 eV
Δ	0.02–0.2 meV	3 meV	>0.1 eV

[a] Metallic nanotube, 500 nm in length.

2.2
Description of Transport

In this section, we present a qualitative discussion of the different transport mechanisms, after which attention will be focused on the weak-coupling case.

The major question here is what picture should be used to describe transport through small devices. In solids, electrons are usually thought of in terms of the independent particle model, in which the wave function of the many-electron system is written in the form of a Slater determinant built from one-electron orbitals. This is an exact solution for a Hamiltonian, which is a sum of one-electron Hamiltonians:

$$H = \sum_i h_i. \qquad (2.3)$$

The electrostatic repulsion between the electrons:

$$V_{ES} = \frac{1}{4}\pi\varepsilon_0 \frac{e^2}{|\vec{r}_i - \vec{r}_j|}.$$

does not satisfy this requirement. Also, the electrons couple electrostatically to the motion of the nuclei, which interact among themselves via a similar Coulomb interaction. Several schemes exist for building a Hamiltonian, such as Equation 2.3, in which the interaction between the electrons is somehow moved into a (possibly non-local) average electrostatic potential. The best known such schemes are the Hartree–Fock (HF) and the density functional theory (DFT). The question is now whether the independent electron picture can survive in the study of transport through small devices. The answer is that single-electron orbitals still form a useful basis for understanding this transport, but that the Coulomb and electron–nucleus interaction must be included quite explicitly into the description in order to understand single-electron effects.

2.2.1
Structure of Nanoscale Devices

Although it often cannot be used in the transport itself, the single particle picture is still suitable for the bulk-like systems to which the device is coupled, and for the narrow leads which may be present between the island and the bulk reservoirs. These elements are described in Chapter 1, and their properties will be recalled only briefly here, with emphasis on the issues needed in the context of the present chapter.

2.2.1.1 The Reservoirs
The reservoirs are bulk-like regions where the electrons are in equilibrium. These regions are maintained at a specified temperature, and the number of electrons is variable as they are connected to the voltage source and the leads to the device (see below). The electrons in these reservoirs are therefore distributed according to Fermi functions with a given temperature T and a chemical potential μ:

$$f_{FD}(E) = \frac{1}{\exp[(E-\mu)/k_B T]+1}.$$

This function falls off from 1 at low energy to 0 at high energy. For $(\mu - E_0) \ll k_B T$, where E_0 is the ground state energy, this reduces to a sharp step down from 1 to 0 at $E = \mu$, and μ can in that case be identified with the Fermi energy (the highest occupied single-particle energy level).

In order to have a current running through the device and the leads, the source and drain reservoirs are connected to a voltage source. A bias voltage causes the two leads to have different chemical potentials.

2.2.1.2 The Leads

Sometimes it is useful to consider the leads as a separate part of the system, in particular for convenience of the theoretical analysis. The leads are channels, which may be considered to be homogeneous. They form the connection between the reservoirs and the island (see below). They are quite narrow and relatively long. Electrons in the leads can still be described by single-particle orbitals. If the leads have a discrete or continuous translational symmetry, the states inside them are Bloch waves. By separation of variables, we can write the states as

$$e^{ik_z z} u_T(x, y)$$

with energy

$$E = E_T + \frac{\hbar^2 k_z^2}{2m}.$$

It is seen that the states can be written as a transverse state $u_T(x, y)$ which contributes an amount E_T to the total energy, times a plane wave along z. The quantum numbers of the transverse wave function $u_T(x, y)$ are used to identify a channel.

In this chapter, usually no distinction is made between reservoirs and leads: rather, they are both simply described as baths in equilibrium with a particular temperature and chemical potential (which may be different for the source and drain lead). However, for a theoretical description of transport, it is often convenient to study the scattering of the incoming states into outgoing states – in that case, the simple and well-defined states of the leads facilitate the description.

2.2.1.3 The Island

This is the part of the system which is small in all directions (although in a nanotube, the transverse dimensions are much smaller than the longitudinal); hence, this is the part where the Coulomb interaction plays an important role. To understand the device, it is useful to take as a reference the isolated island. In that case we have a set of quantum states with discrete energies (levels). The density of states of the device consists of a series of delta-functions corresponding to the bound state energies.

Now imagine there is a knob by which we can tune the coupling to the leads. This is given in terms of the rate $\Gamma/\hbar$ at which electrons cross the tunnel barriers separating the island from the leads. The transport through the barriers is a tunneling process which is fast and, in most cases, it can be considered as elastic: the energy is conserved in the tunneling process. Generally speaking, when the island is coupled to the leads (or directly to the reservoirs), the level broadens as a result of the continuous density of states in the leads (or reservoirs), and it may shift due to charge transfer from the leads to the island. Two limits can be considered. For weak coupling, $\Gamma \ll E_C, \Delta$, the density of states should be close to that of the isolated device: it consists of a series of peaks, the width of which is proportional to Γ. Sometimes, we wish to distinguish between the coupling to the source and drain lead, and use Γ_S and Γ_D, respectively. For strong coupling, that is, $\Gamma \gg E_C, \Delta$, the density of states is strongly influenced by that of the leads, and the structure of the spectrum of the island device is much more difficult to recognize in the density of states of the coupled island.

If we keep the number of electrons within the island fixed, we still have the freedom of distributing the electrons over the energy spectrum. The only constraint is the fact that not more than one electron can occupy a quantum state as a consequence of Pauli's principle. The change in total energy of the device is then mainly determined by the level splitting which is characterized by the energy scale, Δ. If we wish to *add* or *remove* an electron to or from the device, we must pay or we gain in addition a charging energy respectively.

It should be noted that, in principle, Γ may depend on the particular charge state on the island. This is expected to be the case in molecules: the charge distribution usually strongly differs for the different orbitals and this will certainly influence the degree in which that orbital couples to the lead states.

At this stage, an important point should be emphasized. From statistical mechanics, it is known that a particle current is driven by a chemical potential difference. Therefore, the chemical potential of the island is the relevant quantity driving the current to and from the leads. However, in an independent particle picture, a single particle energy is identical to the chemical potential (which is defined as the difference in *total* energy between a system with $N+1$ and N particles). Therefore, if we speak of a single-particle energy of the island, this should often be read as "chemical potential".

2.2.2
Transport

For an extensive discussion of the issues discussed in this paragraph, the reader is referred to the monograph by Datta [2].

As seen above, in the device we can often distinguish discrete states as (Lorentzian) peaks with finite width in the density of states. A convenient representation of transport is then given in Figure 2.3, which shows that the effect of the gate is to shift the levels of the device up and down, while leaving the chemical potentials μ_S and μ_D of the leads unchanged (for small devices, the gate field is inhomogeneous due to the

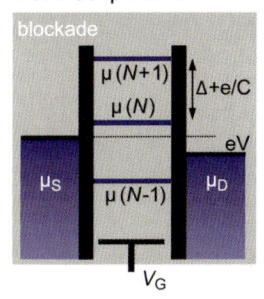

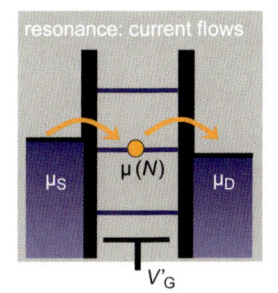

first order processes

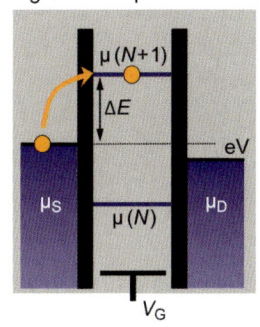

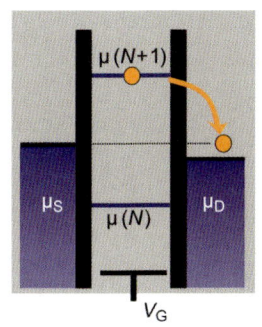

Figure 2.3 Schematic representation of the electrochemical potentials of an island connected to two reservoirs, across which a small (negative) bias voltage V is applied. A voltage on the gate electrode can be used to shift the electrostatic potential of the energy level. Top: Resonant transport becomes possible when the gate voltage pushes one of the levels within the bias window eV. The $\mu(N)$ level is aligned with μ_S and the number of electrons on the dot alternates between N and $N-1$ (sequential tunneling). Bottom: The levels are not aligned. The Coulomb blockade fixes the number of electrons on the dot to N. Transport, however, is possible through a virtual co-tunnel process in which an unoccupied level is briefly occupied. A similar process exists for the occupied level, $\mu(N)$, which may be briefly unoccupied. In contrast to resonant transport, the level is empty (full) most of the time. For all panels it should be noted that, in reality, the levels are not sharp lines but rather have a finite width, Γ. Similarly, the edge between the occupied (blue) and unoccupied states is blurred by temperature via the Fermi–Dirac function.

effect of the leads; moreover, the electrostatic potential in the surface region of the leads will be slightly affected by the gate voltage).

The transport through the device can take place in many different ways. A few classifications will now be provided which may help in understanding the transport characteristics of a particular transport process.

2.2.2.1 Coherent-Incoherent Transport

First, the transport may be either *coherent* or *incoherent*, a notion which pertains to an independent particle description of the electrons where the electrons occupy one-particle orbitals. In the case of coherent transport, the phase of the orbitals evolves deterministically. In the case of incoherent processes, the phase changes in an unpredictable manner due to interactions which are not contained in the independent particle Hamiltonian. Such interactions can be either electron–electron or electron–phonon interactions, or between the electrons and an electromagnetic field.

If the electrons spend a long time on the island – which occurs when the couplings to the leads are weak – then the decoherence will be complete. Only for short traversal times the phase will be well preserved.

2.2.2.2 Elastic–Inelastic Transport

Another distinction is that between elastic and inelastic transport. In the latter case, interactions may cause energy loss or gain of the electrons flowing through the device. This energy change may be caused by the same interactions as those causing decoherence (electron–electron, electron–phonon, electron–photon). It should be noted, however, that decoherent transport can still be elastic.

2.2.2.3 Resonant–Off-Resonant Transport

This classification is relevant for elastic tunneling in combination with weak coupling to the leads. In resonant transport, electrons are injected at an energy corresponding to a resonance of the island. Such a resonance corresponds to a discrete energy level of the isolated device. The transport resonance energy corresponds to the center of the shifted peak; this is seen as a peak in the transport current for that energy or, more specifically, an increase of the current as soon as a resonance enters the bias window. The fact that the coupling to the leads is weak causes the time that an electron resides in the device to be rather long. If this time is longer than the time taken for the electron orbital to lose its coherence, we speak of *sequential tunneling*, as the transport process may then be viewed as electrons hopping from the lead to the island where they stay a while before hopping off to the drain. In off-resonant transport, the electrons are injected at energies (far) off the resonance.

2.2.2.4 First-Order versus Higher-Order Processes

The standard technique for calculating the current arising from coherent processes is time-dependent perturbation theory. In this theory, the transition from one particular state to another is calculated in terms of transitions between the initial, intermediate and final states. The first-order process (Figure 2.3, top) corresponds to a direct transition from the initial to the final state and, for this process, the current is proportional to the couplings Γ between device and leads. In off-resonant first-order processes, the current decays rapidly with the energy difference between the closest discrete level on the island and the Fermi energies of the leads (ΔE in Figure 2.3).

Second-order transport processes, often called *co-tunneling*, take place via an intermediate state, as illustrated in the lower panel of Figure 2.3. In these processes, the current is proportional to higher powers of the couplings, but they are less strongly suppressed with increasing distance (in energy) between the states in the leads and on the island. Therefore, they may sometimes compete with – or even supersede – first-order processes, provided that the intermediate state is sufficiently far in energy (chemical potential) from those in the leads. Currents due to second-order processes vary quadratically with the coupling strengths.

Molecules can often be viewed as chains of weakly coupled sites. If the Fermi energy of the source lead (i.e., the injection energy) is at some distance below the on-site energies of the molecule, the dominant transport mechanism is through higher order processes, which in electron transfer theory are known as *superexchange* processes. This term also includes hole transport through levels below the Fermi energy of the leads.

2.2.2.5 Direct Tunneling
It should be noted that if the device is very small (e.g., a molecule), there is a possibility of having direct tunneling from the source to the drain, in which the resonant states of the device are not used for the transport.

2.3 Resonant Transport

We start this section by studying resonant transport qualitatively [2]. Suppose we have one or more sharp resonant levels which can be used in the transport process from source to drain. We neglect inelastic processes inside the device during tunneling from the leads to the device, or vice versa. In order to send an electron into the device at the resonant energy, we need occupied states in the source lead. This means that the density of states in that lead must be non-zero for the resonant energy (otherwise there is no lead state at that energy), and that the Fermi–Dirac distribution must allow for that energy level to be occupied. Furthermore, for the electron to end up in the drain, the states in the drain at the resonant energy should be empty according to Pauli's principle. We conclude that for the transport to be possible, the resonance should be *inside the bias-window*. This window is defined as the range of energies between the Fermi energies of the source and the drain.

The process is depicted in Figure 2.3 (top), and from this picture we can infer the behavior of the current as a function of the bias voltage. It can be seen that no current is possible (left panel) for a small bias voltage as a result of a finite difference in energy ΔE between the energy of the resonant state on the island and the nearest of the two chemical potentials of the leads. The current sets off as soon as the bias window encloses the resonance energy (right panel). Any further increase of the bias voltage does not change the current, until another resonance is included. The mechanism described here gives rise to current–voltage characteristics shown in Figure 2.4.

At this point, two remarks are in order. First, the image sketched here supposes weak coupling and a low temperature. Increasing the temperature blurs the sharp edge in the spectrum between the occupied and empty states, and this will cause the sharp steps seen in the I/V curve to become rounded. Second, the differential conductance, dI/dV as a function of the bias voltage V shows a peak at the positions where the current steps up.

In the previous section it was noted that the coupling $\Gamma = \Gamma_S + \Gamma_D$ between leads and device can be given in terms of the rate at which electrons hop from the lead onto the device. From this, an heuristic argument leads via the time–energy uncertainty

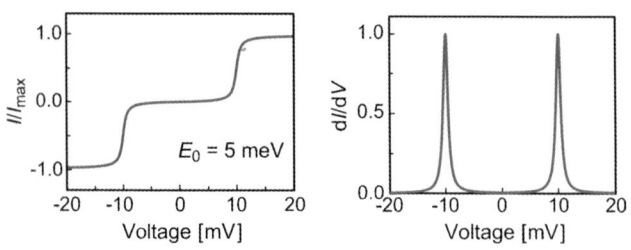

Figure 2.4 Left: Current–voltage characteristic calculated with Equation 2.6 for a level that is located 5 meV from the nearest Fermi energy of one of the electrodes. A symmetric coupling to the leads is assumed with a total broadening of 0.5 meV. Right: Corresponding differential conductance with a peak height equal to the conductance quantum. Note that the peak width is of the order of the total broadening.

relation to the conclusion that Γ gives us the extent to which an energy level[1] E_0 on the island is broadened. Simple models for leads and device yield a Lorentzian density of states on the device [2]:

$$D(E) = \frac{1}{2\pi} \frac{\Gamma}{(E-E_0)^2 + (\Gamma/2)^2}.$$

Further analysis, which is based on a balance between ingoing and outgoing electrons [2] gives the following expression for the current:

$$I(E) = -\int \frac{e}{\hbar} D(E) \frac{\Gamma_S \Gamma_D}{\Gamma_S + \Gamma_D} \left[f_{FD}(E-\mu_S) - f_{FD}(E-\mu_D) \right] dE. \quad (2.4)$$

It must be remembered that the bias voltage (the potential difference between source and drain) is related to the chemical potentials μ_D and μ_S as

$$-eV = \mu_S - \mu_D;$$

where $e > 0$ is unit charge. A positive bias voltage drives the electrons from right to left, such that the current is then from left to right; this is defined as the positive direction of the current.

If the density of states has a single sharp peak, then current is only possible when this peak lies inside the bias window. Indeed, replacing $D(E)$ by a delta-function centered at E_0 directly gives

$$I = \frac{-e}{\hbar} \frac{\Gamma_S \Gamma_D}{\Gamma_S + \Gamma_D} \left[f_{FD}(E_0 - \mu_S) - f_{FD}(E_0 - \mu_D) \right].$$

At low temperature, the factor in square brackets is 1 when E_0 lies inside the bias window and 0 otherwise. It can be seen that the maximum value of the current is found as

---

1) Note that the energy should be identified with the chemical potential of the island; see the comment in the previous section.

$$|I_{max}| = \frac{e}{\hbar}\frac{\Gamma_S\Gamma_D}{\Gamma_S+\Gamma_D}. \tag{2.5}$$

For low temperature, the Fermi functions in Equation 2.4 become sharp steps, and the integral of the Lorentzian can be carried out analytically, yielding

$$I = \frac{e}{\pi\hbar}\frac{\Gamma_S\Gamma_D}{\Gamma_S+\Gamma_D}\left[\arctan\left(2\frac{\mu_S-E_0}{\Gamma}\right)-\arctan\left(2\frac{\mu_D-E_0}{\Gamma}\right)\right]. \tag{2.6}$$

Equation 2.4 is valid in the limit where we can describe the transport in terms of the independent particle model. It has the form of the Landauer formula:

$$I = \frac{e}{\hbar}\int T(E)[f_{FD}(E-\mu_D)-f_{FD}(E-\mu_S)]dE,$$

which is discussed extensively in Chapter 1 of this volume. In that chapter it is shown that the transmission per channel (which corresponds to the eigenvalues of the matrix $T(E)$) has a maximum value of 1, so that the current assumes for low temperatures a maximum value of

$$I_{max} = \frac{e^2}{\hbar}nV, \tag{2.7}$$

where n is the number of channels inside the bias window. It should be noted that this maximum occurs only for *reflectionless* contacts, for which a wave incident from the leads onto the device, is completely transmitted. This usually occurs when the device and the leads are made from the same material. The strong-coupling result in Equation 2.7 has been given in order to emphasize that the two Equations 2.5 and 2.7 hold in quite opposite regimes.

Often, in experiments the differential conductance dI/dV is measured, and this can be calculated from the expression in Equation 2.4:

$$\frac{dI}{dV} = -\frac{e^2}{\hbar}\frac{\Gamma_S\Gamma_D}{\Gamma_S+\Gamma_D}$$
$$\int dE D(E)\{\eta f'_{FD}(E-\bar{\mu}+\eta eV)-(1-\eta)f'_{FD}[E-\bar{\mu}-(1-\eta)eV]\}, \tag{2.8}$$

where f'_{FD} denotes the first derivative of the Fermi–Dirac distribution with respect to its argument, and $\bar{\mu} = (\mu_S+\mu_D)/2$. The parameter η specifies how the bias voltage is distributed over the source and drain contact; for $\eta=1/2$, this distribution is symmetric. For $T=0$, the Fermi–Dirac distribution function reduces to a step function, and its derivative is then a delta-function. For low bias ($V\approx 0$), the integral picks up a contribution from both delta functions occurring in the integral in Equation 2.8. The result is

$$\frac{dI}{dV} = 4\frac{e^2}{\hbar}\frac{\Gamma_S\Gamma_D}{\Gamma_S+\Gamma_D}D(\mu),$$

where the energy E in Equation 2.8 is taken at the Fermi energy of either the source or the drain. As the maximum value of $D(E)$ is given as

$$D(E)_{max} = \frac{2}{\pi}\frac{1}{\Gamma_S+\Gamma_D},$$

it follows that the maximum of the differential conductance occurs when $\Gamma_S = \Gamma_D$ and is then given by e^2/h. Note that this holds even when the current is much smaller than the quantum conductance limit [see Equation 2.7] which follows from the Landauer formula.

At *finite* temperature, for $k_B T \gg \Gamma$ and zero bias, working out the derivative with respect to bias of Equation 2.8 gives:

$$\frac{dI}{dV} = \frac{e^2 \Gamma_S \Gamma_D}{4 k_B T (\Gamma_S + \Gamma_D)} \left[\cosh \frac{e\alpha(V_G - V_0)}{2 k_B T} \right]^{-2}. \tag{2.9}$$

This line shape (see Figure 2.5) is characterized by a maximum value $e^2 \Gamma_S \Gamma_D / 4 k_B T (\Gamma_S + \Gamma_D)$, attained when the gate voltage reaches the resonance $V_0 = E_0/e$. The full-width half maximum (FWHM) of this peak is $3.525 k_B T/e\alpha$. The parameter α is the gate coupling parameter: the potential on the island varies linearly with the gate voltage, $\Delta V_I = \alpha \Delta V_G$. These features are often used as a signature for true quantum resonant behavior as opposed to classical dots, where the small value of Δ renders the spectrum of levels accessible to an electron continuous. For a classical dot, the peak height is independent of temperature and the FWHM is predicted to increase by a factor 1.25 [3,4]. It should be noted that, in a quantum dot, Γ sets a lower bound for the temperature dependence of the peak shape: for $\Gamma > k_B T$ the peak height and shape are independent of temperature (not visible in Figure 2.5 due to the small value for Γ chosen there).

Interestingly, the finite width of the density of states, which is given by $\Gamma_S + \Gamma_D$, can in principle be measured experimentally from the resonance line widths at low temperature. It should be noted that the expressions for the current and differential conductance depend only on the combinations $\Gamma_S + \Gamma_D$ and $\Gamma_S \Gamma_D / (\Gamma_S + \Gamma_D)$. If both are extracted from experimental data, the values of Γ_S and Γ_D can be determined (although the symmetry between exchange of source and drain prevents us from identifying which value belongs to the source).

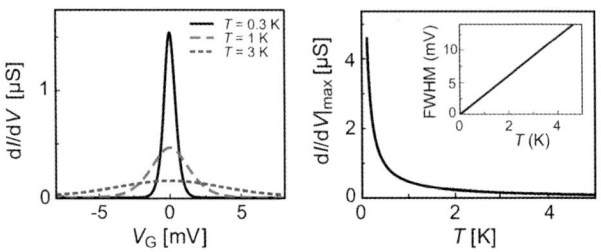

Figure 2.5 Left: Temperature-dependence of the Coulomb peak height [Equation 2.9] in the resonant transport model, showing the characteristic increase as the temperature is lowered. Right: Peak height as a function of temperature. The inset shows the full-width half maximum (FWHM) of the Coulomb peak as a function of temperature (see text). Calculations are performed with $\Gamma = 10^9 \, s^{-1}$ and a gate coupling of 0.1 in the regime $\Gamma < k_B T$.

2.4
Constant Interaction Model

In Section 2.1.1 it was seen that in the weak-coupling regime, energy levels can be discrete for two reasons: quantum confinement (the fact that the state must "fit" into a small island), and charge quantization effects. The scale for the second type of splitting is the charging or Coulomb energy, E_C. It is important to realize that this energy will only be noticeable when the coupling to the leads is small in comparison with E_C; this situation is referred to as the *Coulomb blockade* regime. In this situation, a clear distinction should be made a between one or two electrons occupying a level, as their Coulomb interaction contributes significantly to the total energy. The transport process may be analyzed using the so-called *constant interaction model* [3], which is based on the set-up shown in Figure 2.6. Elementary electrostatics provides the following relation between the different potentials and the charge Q on the island:

$$CV_I - C_S V_S - C_D V_D - C_G V_G = Q,$$

where $C = C_S + C_D + C_G$. Note that this equation can be written in the form:

$$V_I = V_{ext} + \frac{Q}{C},$$

with

$$V_{ext} = (C_S V_S + C_D V_D + C_G V_G)/C.$$

It is seen that the potential on the dot is determined by the charge residing on it and by the induced potential V_{ext} of the source, drain and gate.

We take as a reference configuration the one for which all voltages and the charge are zero. For total energies, U rather than E is used in order to avoid confusion with the single-particle energies E_n resulting from solving the single-particle Schrödinger equation. The electrostatic energy $U_{ES}(N)$ with respect to this reference configuration after changing the source, drain and gate potentials and putting N electrons (of charge $-e$) on the island is then identified as the work needed to place this extra charge on the island, and the energy cost involved in changing the external potential when a charge Q is present:

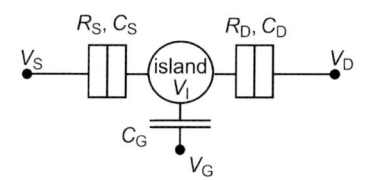

Figure 2.6 The capacitance model. A schematic drawing of an island connected to source and drain electrodes with tunnel junctions; the gate electrode shifts the electrostatic potential of the island.

$$U_{ES}(N) = \int_{Q=0, V_{ext}=0}^{-Ne, V_{ext}} (V_I dQ + Q dV_{ext}) = \frac{(Ne)^2}{2C} - NeV_{ext}.$$

The integral is over a path in Q, V_{ext} space; it is independent of the path – that is, of how the charge and external potential are changed in time.

The result for the total energy, including the "quantum energy" due to the orbital energies is

$$U(N) = \frac{(Ne)^2}{2C} - NeV_{ext} + \sum_{n=1}^{N} E_n.$$

The energy levels E_n correspond to states which can be occupied by the electrons in the device, provided that their total number does not change, as changing this number would change the Coulomb energy, which is accounted for by the first term. This expression for the total energy is essentially the constant interaction model.

From non-equilibrium thermodynamics, it is known that a current is driven by a chemical potential difference; hence, we should compare the chemical potential on the device,

$$\mu(N) = U(N) - U(N-1) = (N - 1/2)\frac{e^2}{C} - eV_{ext} + E_N, \tag{2.10}$$

with that of the source and drain in order to see whether a current is flowing through the device. From the definition of V_{ext} we see that the effective change in the chemical potential due to a change of the gate voltage (while keeping source and drain voltage constant), carries a factor C_G/C; this is precisely the gate coupling, which is called the α factor (this was referred to at the end of Section 2.3).

It is important to be aware of the conditions for which the constant interaction model provides a reliable description of the device. The first condition is weak coupling to the leads; the second condition is that the size of the device should be sufficiently large to make a description with single values for the capacitances possible. Finally, the single-particle levels E_n must be independent of the charge N. The constant-interaction model works well for weakly coupled quantum dots, for which it is very often used. However, for molecular devices the presence of a source and drain both of which are large chunks of conducting material separated by very narrow gaps, reduces the gate field to be barely noticeable close to the leads and far from the gate. This inhomogeneity of the gate field may lead to a dependence of the gate capacitance C_G with N due to the difference in structure of subsequent molecular orbitals, and the chemical potential on the molecule will vary non-linearly with the gate potential.

As will be seen below, the distance between the different chemical potential levels can be inferred from three-terminal measurements of the (differential) conductance. From Equation 2.10, this distance is given by

$$\mu(N+1) - \mu(N) = \frac{e^2}{C} + E_{N+1} - E_N.$$

2.4 Constant Interaction Model

It should be noted that the difference in energy levels occurring in this expression ($E_{N+1} - E_N$) is nothing but the splitting Δ mentioned at the very start of this chapter. However, for typical metallic and semiconductor quantum dots, this splitting is usually significantly smaller than the charging energy, so that this quantity determines the distance between the energy levels:

$$\mu(N+1) - \mu(N) = \frac{e^2}{C}.$$

Note that this *addition energy* is twice the energy of a charge on the dot (as the addition energy is the second derivative of the energy with respect to the charge).

Now, the current can be studied as a function of bias and gate voltage. In Section 2.2.2 it was seen that, in the weak coupling regime and at low temperature, the current is suppressed when all chemical potential levels lie outside of the bias window. As the location of these levels can be tuned by using the gate voltage, it is interesting to study the current and differential conductance of the device as a function of the bias *and* of the gate voltage.

We can calculate the line in the V, V_G plane which separates a region of suppressed current from a region with finite current; this line is determined by the condition that the chemical potential of the source (or drain) is aligned with that of a level on the island. Again, it is assumed that the drain is grounded (as in Figure 2.2). From the expression in Equation 2.10 for the chemical potential, and using the definition for V_{ext}, we find the following condition for the chemical potential to be aligned to the source (keeping the dot's charge constant):

$$V = \beta(V_G - V_C),$$

where $\beta = C_G/(C_G + C_D)$ and $V_C = (N - 1/2)\frac{e}{C_G} + \frac{C}{C_G}\frac{E_N}{e}$; that is, the voltage corresponding to the chemical potential on the dot in the absence of an external potential. If the chemical potential is aligned with the drain, we have

$$V = \gamma(V_C - V_G)$$

with $\gamma = C_G/C_S$. The expressions given here are specific for a grounded drain electrode. It is easily verified that, irrespective of the grounding, it holds that

$$\frac{C}{C_G} = \frac{1}{\alpha} = \frac{1}{\beta} + \frac{1}{\gamma}.$$

Each resonance generates two straight lines separating regions of suppressed current from those with finite current. For a sequence of resonances, the arrangement shown in Figure 2.7a is obtained. The diamond-shaped regions are traditionally called "Coulomb diamonds", as they were very often studied in the context of metallic dots, where the chemical potential difference of the levels is mainly made up of the Coulomb energy. The name is also used in molecular transport, although this is – strictly speaking – not justified there as Δ may be of the same order as the Coulomb interaction.

From the Coulomb diamond picture we can infer the values of some important quantities. First, we consider two successive states on the molecule with chemical potentials $\Delta\mu_{(N)}$ and $\Delta\mu_{(N+1)}$. If we suppose that both states have the

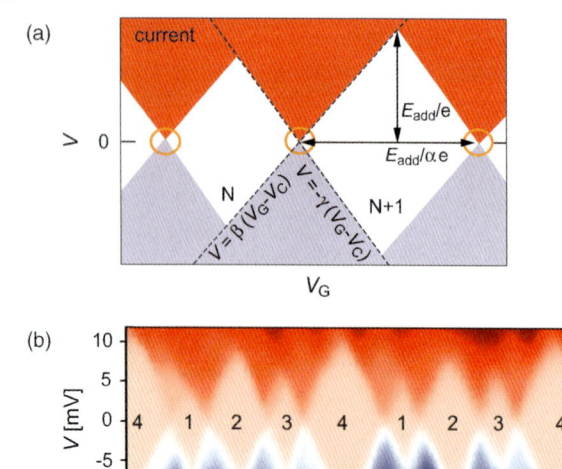

Figure 2.7 Linear transport. (a) Two-dimensional plot of the current as a function of bias and gate voltage (stability diagram). For a small bias, current flows only in the three points corresponding to the situation shown in Figure 2.3 (top right). These points are known as the "degeneracy points". Red = positive currents; blue = negative currents; white = blockade, no current. (b) Measured stability diagram of a metallic, single-walled carbon nanotube, showing the expected fourfold shell filling. The blockade regime is shown in pink. (Data from Ref. [5].).

same gate-coupling parameter α, it can be seen that the upper and lower vertices of the diamond are both at a distance

$$\Delta V = \frac{|\mu(N) - \mu(N+1)|}{e} = \frac{E_{add}}{e}$$

from the zero-bias line. This difference in chemical potentials is the electron addition energy, E_{add}. If the addition energy is dominated by the charging energy, then the total capacitance can be determined. Combining this with the slopes of the sides of the diamond, which provide the relative values of C_G, C_S and C_D, all of these capacitances we can be determined explicitly.

One interesting consequence of the previous analysis is that, if the capacitances do not depend on the particular state being examined, then the height of successive Coulomb diamonds is constant. If, in addition to the Coulomb energy, a level splitting is present, this homogeneity will be destroyed, as can be seen in Figure 2.7b, which shows the diamonds for a carbon nanotube (CNT) [5]. The alternation of a large diamond with three smaller ones can be nicely explained with a model Hamiltonian [6]. In the case of transport through molecules there is no obvious underlying structure in the diamonds.

The electron addition energy is sometimes connected to the so-called HOMO–LUMO gap. [These acronyms represent the Highest Occupied (Lowest Unoccupied)

Molecular Orbital, and denote orbitals within an independent particle scheme.] If the Coulomb interaction is significant, the HOMO–LUMO gap can be related to the excitation energy for an optical absorption process in which an electron is promoted from the ground state to the first excited state, without leaving the system. In that case, the change in Coulomb energy is modest, and the energy difference is mostly made up of the quantum splitting Δ. It should be noted, however, that the HOMO and LUMO are usually calculated using a computational scheme whereby the orbitals are calculated for the ground-state configuration – that is, without explicitly taking into account the fact that all orbitals change when, for example, an electron is excited to a higher level.

The addition energies are partly determined by quantum confinement effects and partly by Coulomb effects. A difficulty here is that these energies will be different for a molecular junction, in which a molecule is either physisorbed or chemisorbed to conducting leads, than for a molecule in the gas phase. There are several effects responsible for this difference. First, if there is a chemical bond present, then the electronic orbitals extend over a larger space, which reduces the confinement splitting. Second, a chemical bond may cause a charge transfer from lead to molecule, which in turn causes the potential on the molecule to change. Third, the charge distribution on the molecule will polarize the surface charge on the leads, which can be represented as an *image charge*. Such charges have the effect of reducing the Coulomb part of the addition energy. In experiments with molecular junctions, much smaller addition energies are often observed than in gas-phase molecules. At the time of writing, there is no quantitative understanding of the addition energy in molecular three-terminal junctions, although the effects mentioned here are commonly held responsible for the observed gaps.

2.5
Charge Transport Measurements as a Spectroscopic Tool

A stability diagram can be used not only for finding addition energies, but also to form a spectroscopic tool for revealing subtle excitations that arise on top of the ground state configurations of an island with a particular number of electrons on it. These excitations appear as lines running parallel to the Coulomb diamond edges. An example taken from Ref. [7] is shown in Figure 2.8a, where the white arrows indicate the excitation lines. At such a line, a new state (electronic or vibrational) enters the bias window, thus creating an additional transport channel. The result is a stepwise increase in the current, and a corresponding peak in the differential conductance. The energy of an excitation can be determined by reading off the bias voltage of the intersection point between the excitation line and the Coulomb diamond edge through the same argument used for finding addition energies. The excitations correspond to the charge state of the Coulomb diamond at which they ultimately end (see Figure 2.8c). The width of the lines in the dI/dV plot (or, equivalently, the voltage range over which the stepwise increase in current occurs) is determined by the larger of the energies k_BT and Γ. In practice, this means that sharp lines – and thus accurate information on spectroscopic features – are obtained at low temperatures and

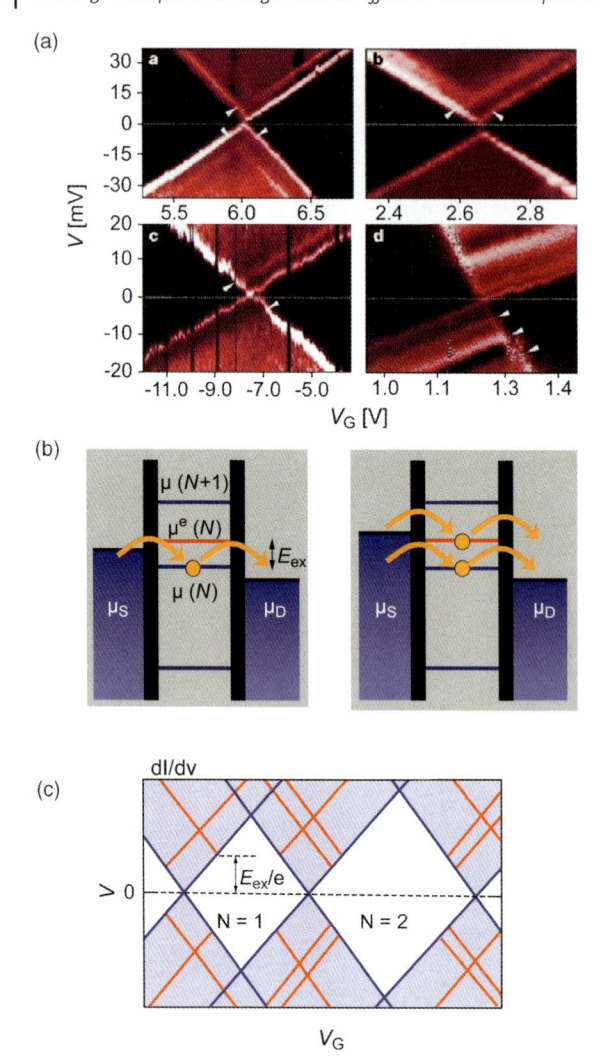

Figure 2.8 Non-linear transport and excited states. (a) Four different conductance maps (stability diagrams) of C_{60} molecules trapped between two electrodes [7]. Excitations lines are indicated by arrowheads. These run parallel to the diamond edges, and are due to vibrational modes of the C_{60} molecule. (b) Electrochemical potential plot of a dot with three electronic energy levels and one excited state (red). Transport through an excited level becomes possible as soon as the red level enters the bias window. (c) Schematic representation of a differential conductance map. The red lines show the positions at which excited states enter the bias window. The associated stepwise increases in current appear as lines running parallel to the edges of the diamond-shaped regions. Blue: dI/dV is zero but the current is not (sequential-tunneling regime). White: current blockade.

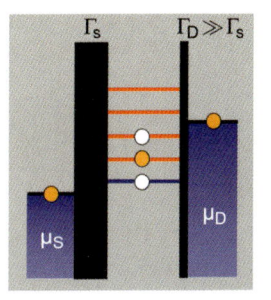

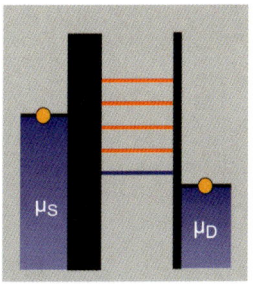

Figure 2.9 Asymmetric coupling to the electrodes leads to an almost full occupancy if tunneling out of the level is limited by the thick barrier (left: low tunnel rate) and almost zero occupancy of the level if tunneling out is determined by the thin barrier (right: high tunnel rate). Note, that of the three levels in the bias window, only one of them can be occupied at the same time. An increase of the bias voltage such that another excited level enters the bias window yields a very small current increase in the case (as depicted on the left-hand side) because the thick barrier remains the limiting factor for the current. In contrast, on the right-hand side a new transport channel becomes available and the current shows a clear stepwise increase.

for weak coupling to the leads. However, it should be noted that the current is proportional to Γ [Equations 2.4 and 2.5], so that Γ should not be too small; in fact, a Γ-value in the order of 0.1–1 meV seems typical in experiments that allow for spectroscopy.

An important experimental issue here is that, for a particular charge state, the lines are often visible on only one side of the Coulomb diamond (see Figure 2.8a, lower right panel). This is due to an asymmetry in the coupling – that is, for $\Gamma_D \gg \Gamma_S$ (or $\Gamma_S \gg \Gamma_D$). The situation at the two "main" diamond edges is illustrated in Figure 2.9. A thick and a thin barrier between the island and the source/drain represent these anti-symmetric couplings. It is clear that if the chemical potential in the lead connected through the thin barrier is the higher one, then the island will have one of its transport channels filled. The limiting step for transport is the thick barrier, and only the occupied orbital will contribute to the current. When an extra transport orbital becomes available, this will have only a minor effect on the total current, but if the chemical potential of the lead beyond the thick barrier is high, then the transport levels on the island will all be empty. The lead electrons which must tunnel through the thick barrier have as many possible channels at their disposal as there are possible empty states: the more orbitals, the more channels there are, and therefore a stepwise increase occurs each time a new excitation becomes available.

2.5.1
Electronic Excitations

In order to study how detailed information on the electronic structure of the island can be obtained from conduction measurements, we consider a system consisting of levels that are separated in energy by the Δ_i (see Figure 2.10). It should be noted that this level splitting does not include a charging energy: the levels can be occupied in charge-neutral excitations. For one extra electron on the island, $N = 1$, the ground

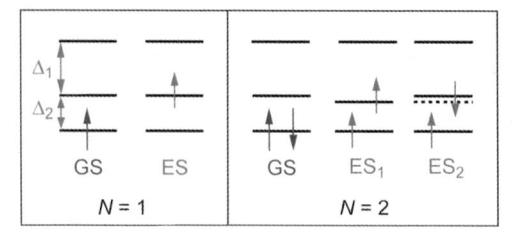

Figure 2.10 Schematic drawing of the ground state (GS) filling and the excited states (ES). Left: The island contains one electron and the first excited state involves a transition to the nearest unoccupied level. (In a zero magnetic field there is an equal probability to find a down spin on the dot.) Right: Two electrons with opposite spin occupy the lowest level. The first excited state involves the promotion of one of the spins to the nearest unoccupied level. A ferromagnetic interaction favors a spin flip. The antiparallel configuration (ES$_2$) has a higher energy (see text).

state is the one in which it occupies the lowest level. As discussed above, as soon as this level is inside the bias window, the current begins to flow, thereby defining the edges of the Coulomb diamonds. When the bias increases further, transport through the excited level becomes possible. This leads to a step-wise increase of the current as there are now two states available for resonant transport, and this increases the probability for electrons to pass through the island. It should be noted that both levels cannot be occupied at the same time, as this requires a charging energy in addition to the level splitting. The resulting peak in the dI/dV forms a line (red) inside the conducting region (blue), ending up at the "$N=1$" diamond (white), as shown in Figure 2.8c. ($E_{ex} = \Delta_1$ in this case). A second excitation is found at $\Delta_1 + \Delta_2$; subsequent excitations intersect the diamond edge at bias voltages $\sum_i \Delta_i$, but they are only visible if $\sum_i \Delta_i < e^2/C$.

Now, we consider the case where two electrons are added to the neutral island ($N=2$). When two electrons occupy the lower orbital, the Fermi principle requires their spin to be opposite. The first excited state is the one in which one of the electrons is transferred to the higher orbital, which costs an energy of Δ_1. A ferromagnetic exchange coupling favors a triplet state with a parallel alignment. If we take only exchange interactions between different orbitals into account, this results in an energy gain of J with respect to the situation with opposite spins. Thus, the first excitation is expected to be at $\Delta_1 - J$, and the second one (corresponding to opposite spins) at Δ_1. The energy difference between the two excitations in Figure 2.8c provides a direct measure of J. In some systems, J may be negative (antiferromagnetic case) and the antiparallel configuration has a lower energy.

The simple analysis presented here captures some of the basic features of few-electron semiconducting quantum dots [8] in which the charge states to which the levels belong, can be identified. The complete electron spectrum has also been determined in metallic CNT quantum dots [5,9]. Although, for a nanotube, many densely spaced excitations occur, level spectroscopy is possible as the regularly spaced levels are well separated from each other with $E_C \approx \Delta$. Careful inspection of the excitation and addition spectra of CNTs shows that the exchange coupling J is ferromagnetic and that it is small, of the order of a few meV, or less. Further identification of the states can be performed in a magnetic field, using the Zeeman effect as a diagnostic tool. Doublet states are expected to split into two levels, and triplet states into three.

One final remark concerns the $N=0$ diamond. In systems such as semiconducting quantum dots, where there is a gap separating the ground state from the first excited state, Δ_1 may be of the order of hundreds of meV, and in that case no electronic excitations are expected to end up in this diamond.

2.5.2
Including Vibrational States

An interesting phenomenon in molecular transport occurs when the molecular vibrations couple to the electrons, giving rise to excitations available for transport (as mentioned above). This phenomenon has been studied quite extensively in recent years, and the basics will briefly be discussed at this point (for further details, see Refs. [10,11]).

Molecules are rather "floppy" in nature, and from classical mechanics it is known that small deformations of a molecule with respect to its lowest energy conformation can be described in terms of *normal modes*. These are excitations in which all nuclei oscillate with the *same* frequency ω (although some nuclei may stand still). In particular, these excitations have the form

$$R_{i,\alpha}^{(l)}(t) = X_{i,\alpha}^{(l)} \exp(i\omega^{(l)}t),$$

where $R_{i\alpha}^{(l)}$ is the Cartesian coordinate $\alpha = x, y, z$ of nucleus i; l labels the normal mode; $X_{i,\alpha}^{(l)}$ is a fixed vector which determines the amplitudes of the oscillation for the degree of freedom labeled by i,α. The vibrations are described by a harmonic oscillator, which has a spectrum with energy levels separated by an amount $\hbar\omega^{(l)}$:

$$E_\nu^{(l)} = \hbar\omega^{(l)}(\nu + 1/2), \quad \nu = 0, 1, 2, \ldots.$$

For molecular systems, the normal modes are often referred to as *vibrons* (in analogy with phonons in a periodic solid). These modes couple with the electrons as the electrons feel a change in the electrostatic potential when the nuclei move in a normal mode. The coupling is determined by the electron–vibron coupling constant, called λ.

The presence of vibrational excitations can be detected in transport measurements. However, it should be noted that for this to happen, the vibrational modes must be excited, which can occur for two reasons: (i) the thermal fluctuations excite these modes; or (ii) they can be excited through the electron–vibron coupling.

In order to study the effect of electron–vibron coupling on transport, for simplicity the discussion is restricted to a single vibrational mode and a single electronic level. The nuclear part of the Hamiltonian is

$$H = \frac{P^2}{2M} + \frac{1}{2}M\omega^2 X^2$$

where P, X and M represent the momentum, position and mass of the oscillator.

It twins out [11] that the electron-vibron coupling has the form:

$$H_{e-v} = \lambda \hbar \omega \hat{n} X / u_0,$$

where $\hat{n}$ is the number operator, which takes on the values 0 or 1 depending on whether there is an electron in the orbital under consideration; $u_0 = \sqrt{\hbar/(2M\omega)}$ is the zero-point fluctuation associated with the ground state of the harmonic oscillator. The electron–vibron coupling λ is given as (φ is the electronic orbital):

$$\lambda = \frac{1}{\hbar \omega} \sqrt{\frac{\hbar}{2\omega}} \frac{1}{\sqrt{M}} \left\langle \varphi \left| \frac{\partial H_{el}}{\partial X} \right| \varphi \right\rangle.$$

When the charge in the state φ increases from 0 to 1, the equilibrium position of the harmonic oscillator (i.e., the minimum of the potential energy) is shifted over a distance $-2\lambda u_0$ along X, and it is also shifted down in energy (see Figure 2.11a). Fermi's "golden rule" states that the transition rate for going from the neutral island in the conformational ground state to a charged island in some

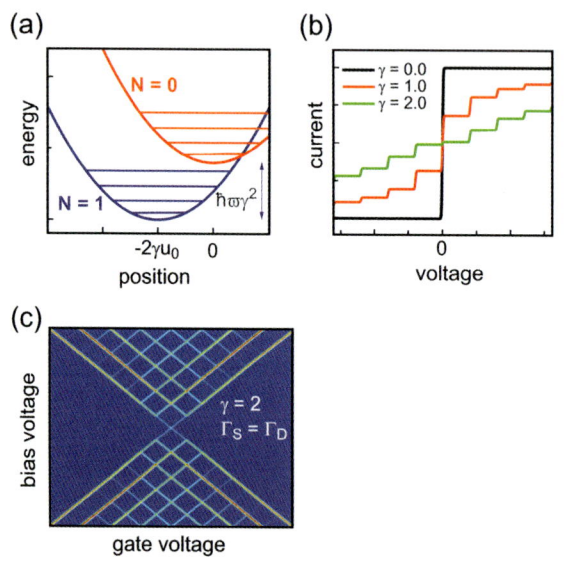

Figure 2.11 (a) Potential of the harmonic oscillator for the empty (red) and occupied state (blue). When an electron tunnels onto the island, the position of the potential minimum is shifted in space and energy. (b) Current–voltage characteristics calculated for three different values of the electron–phonon coupling constant. For non-zero coupling steps appear, which are equally spaced in the voltage (harmonic spectrum). (c) Differential conductance plotted in a stability diagram for an island coupled to a single vibrational mode. Lines running parallel to the diamond edges correspond to the steps, forming a lozenge pattern of excitation lines in (b). Around zero bias the current is suppressed for this rather large electron–phonon coupling (phonon blockade).

excited vibrational state is proportional to the square of the overlap between the initial and final states. Hence, this rate is proportional to the overlap of the ground state of the harmonic oscillator corresponding to the higher parabola and the excited state of the oscillator corresponding to the shifted parabola (to be multiplied by the coupling between lead and island). This overlap is known as the Frank–Condon factor. It is clear that for large displacements, this overlap may be larger for passing to a vibrationally excited state than for passing to the vibrational ground state of the shifted oscillator. The Franck–Condon factors may be calculated analytically (see for example Ref. [10]).

The sequential tunneling regime, which corresponds to weak coupling, can be described in terms of a rate equation: the *master equation*. This describes the time evolution of the probability densities for the possible states on the molecular island. The master equation can be used for any sequential tunneling process and is particularly convenient when vibrational excitations play a role. The details of formulating and solving master equations are beyond the scope of this chapter, but the interested reader is referred to Refs. [11,12] for further details.

Figure 2.11b and c were prepared using such a master equation analysis. For sufficient electron–phonon coupling, steps appear in the current–voltage characteristics (Figure 2.11b), which for $\Gamma_D = \Gamma_S$ leads to the lozenge pattern in a stability diagram, as illustrated in Figure 2.11c. It should be noted that, if the vibrational modes are excited, they may in turn lose their energy through coupling to the leads or other parts of the device. This can be represented by an effective damping term for the nuclear degrees of freedom. For actual molecules, solving the master equations by using Frank–Condon factors obtained from quantum chemical calculations may be used to compare theory with experiment. This is especially useful because the observed vibrational frequencies can be used as a "fingerprint" of the molecule under study [7,13–15] (see also Figure 2.8a).

2.6
Second-Order Processes

In the analysis presented so far, sequential tunneling events do not contribute to the current inside Coulomb diamonds as they are blocked in these regions. However, it should be realized that elastic co-tunneling processes (as depicted in the upper part of Figure 2.3) always take place, albeit that the current levels are generally very small. For second-order processes, the current is proportional to $\Gamma_S \Gamma_D$ instead of showing a linear dependence (on $\Gamma_S \Gamma_D/(\Gamma_S + \Gamma_D)$) as for first-order processes. Consequently, co-tunneling becomes more important for larger Γ-values. In some cases, higher order coherent processes involving virtual states give rise to observable features inside Coulomb diamonds. In the following section, two examples are discussed; namely, the Kondo effect in quantum dots, which is an elastic co-tunneling process conserving the dot energy; and inelastic co-tunneling, which leaves the dot in an excited state.

2.6.1
The Kondo Effect in a Quantum Dot with an Unpaired Electron

The Kondo effect has long been known to cause a resistance increase at low temperatures in metals with magnetic impurities [16]. However, in recent years Kondo physics has also been observed in semiconducting [17], nanotube [18] and single-molecule quantum dots [19]. It arises when a localized unpaired spin interacts by antiferromagnetic exchange with the spin of the surrounding electrons in the leads (see Figure 2.12a). The Heisenberg uncertainty principle allows the electron to tunnel out for only a short time of about $\hbar/\Delta E$, where ΔE is the energy of the electron relative to the Fermi energy and is taken as positive. During this time, another electron from the Fermi level at the opposite lead can tunnel onto the dot, thus conserving the total energy of the system (elastic co-tunneling). The exchange interaction causes the majority spin in the leads to be opposite to the original spin of the dot. Therefore, the new electron entering from these leads is more likely to have the opposite spin. This higher-order process gives rise to a so-called *Kondo resonance* centered around the Fermi level. The width of this resonance is proportional to the characteristic energy scale for Kondo physics, T_K. For $\Delta E \gg \Gamma$, T_K is given by:

$$k_B T_K = \frac{\sqrt{\Gamma U}}{2} \exp\left[\frac{\pi \Delta E (\Delta E + U)}{\Gamma U}\right]. \tag{2.11}$$

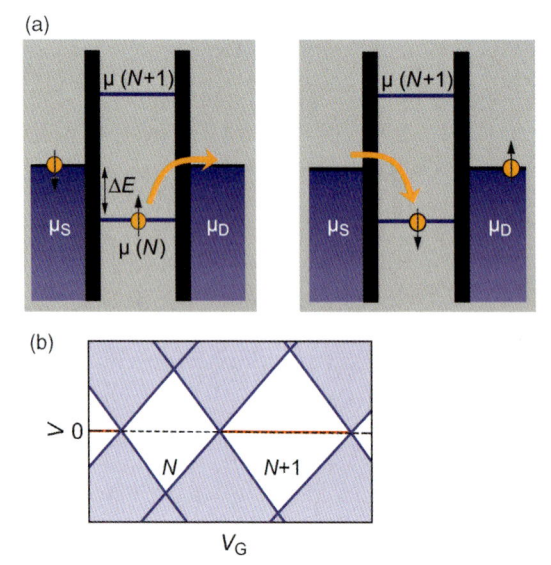

Figure 2.12 (a) A schematic drawing of the two-step Kondo process which occurs for odd occupancy of the island. (b) The Kondo effect leads to a zero bias conductance peak (red lines) in the differential conductance plots. N is even in this case.

Typical values for T_K are 1 K for semiconducting quantum dots, 10 K for CNTs, and 50 K for molecular junctions. This increase of T_K with decreasing dot size can be understood from the prefactor, which contains the charging energy ($U = e^2/C$).

In contrast to bulk systems, the Kondo effect in quantum dots leads to an *increase* of the conductance, as exchange makes it easier for the spin states belonging to the two electrodes to mix with the state (of opposite spin) on the dot, thereby facilitating transport through the dot. The conductance increase occurs only for small bias voltages, and the characteristic feature is a peak in the trace of the differential conductance versus bias voltage (see Figure 2.10b, red lines). The peak occurs at zero bias inside the diamond corresponding to an odd number of electrons. (For zero spin, no Kondo is expected; for $S = 1$ a Kondo resonance may be possible, but the Kondo temperature is expected to be much smaller.) The full width at half maximum (FWHM) of this peak is proportional to T_K: FWHM $\approx 2k_B T_K/e$. Equation 2.11 indicates that T_K is gate-dependent because ΔE can be tuned by the gate voltage. Consequently, the width of the resonance is the smallest in the middle of the Coulomb blockade valley and increases towards the degeneracy point on either side.

Another characteristic feature of the Kondo resonance is the logarithmic decrease in peak height with temperature. In experiments, this logarithmic dependence of the conductance maximum is often used for diagnostic means, and in the middle of the Coulomb blockade valley it is given by:

$$G(T) = \frac{G_C}{[1 + (2^{1/S} - 1)(T/T_K)^2]^S}, \qquad (2.12)$$

where $S = 0.22$ for spin-1/2 impurities and $G_C = 2e^2/h$ for symmetric barriers. For asymmetric barriers, G_C is lower than the conductance quantum. Equation 2.12 shows that for low temperatures, the maximum conductance of the Kondo peak saturates at G_C while at the Kondo temperature it reaches a value of $G_C/2$.

2.6.2
Inelastic Co-Tunneling

The inelastic co-tunneling mechanism becomes active above a certain bias voltage, which is independent of the gate voltage. At this point, the current increases stepwise because an additional transport channel opens up. In the stability diagram, this results in a horizontal line inside the Coulomb-blockaded regime. This conductance feature appears symmetrically around zero at a source-drain bias of $\pm\Delta/e$ for an excited level that lies at an energy Δ above the ground state. Co-tunneling spectroscopy therefore offers a sensitive measure of excited-state energies, which may be either electronic or vibrational. Often, in combination with Kondo peaks, inelastic co-tunneling lines are commonly observed in semiconducting, nanotube and molecular quantum dots. An example of inelastic co-tunnel lines (dashed horizontal lines) for a metallic nanotube quantum dot is shown in Figure 2.13a.

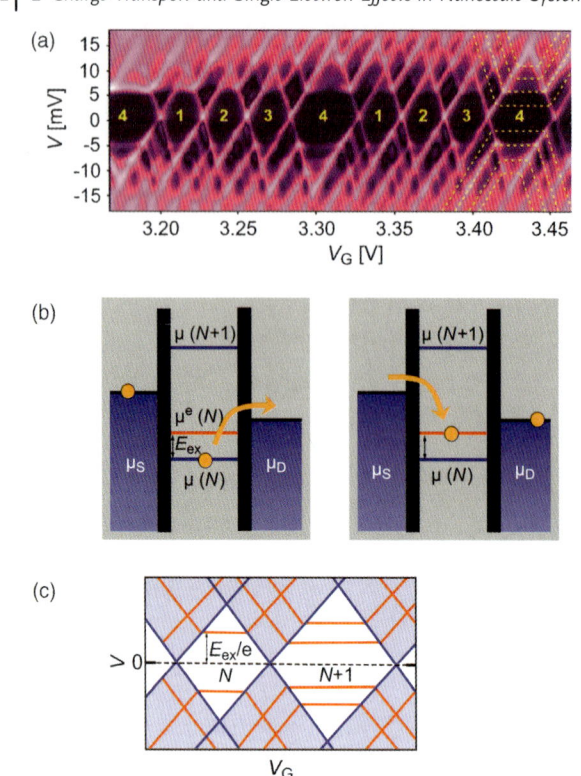

Figure 2.13 Inelastic co-tunneling. (a) A measured stability diagram of a metallic, single-walled carbon nanotube (from Ref. [5]). The dashed horizontal lines indicate the presence of inelastic co-tunnel lines. (b) Schematic drawing of this two-step tunneling process, leaving the dot in an excited state. (c) Inelastic co-tunneling gives rise to horizontal lines in the blocked current region. The energy of the excitation can directly be determined from these plots, as indicated in the figure.

Figure 2.13b shows the mechanism of inelastic co-tunneling. An occupied state lies below the Fermi level, and this can only virtually escape for a small time, as governed by the Heisenberg uncertainty relation. If, in the meantime, an electron from the left lead tunnels onto the dot in the excited level (red), then effectively one electron has been transported from left to right. The dot is left in an excited level, and the energy difference E_{ex} must be paid by the bias voltage; hence, this two-step process is only possible for $|V| > E_{ex}/e$. Relaxation inside the dot may then lead to the dot to decay into the ground state.

Acknowledgments

The authors thank Menno Poot for his critical reading of the manuscript.

References

1. Feynman, R.P. (1961) There is plenty of room at the bottom, in *Miniaturization*, (ed. H.D. Hilbert), Reinhold, New York, pp. 282–286.
2. Datta, S. (1995) *Electronic transport in mesoscopic systems*, Cambridge University Press, Cambridge.
3. Beenakker, C.W.J. (1991) Theory of Coulomb-blockade oscillations in the conductance of a quantum dot. *Physical Review B-Condensed Matter*, **44**, 1646–1656.
4. Foxman, E.B. *et al.* (1994) Crossover from single-level to multilevel transport in artificial atoms. *Physical Review B-Condensed Matter*, **50**, 14193–14199.
5. Sapmaz, S., Jarillo-Herrero, P., Kong, J., Dekker, C., Kouwenhoven, L.P. and van der Zant, H.S.J. (2005) Electronic excitation spectrum of metallic nanotubes. *Physical Review B-Condensed Matter*, **71**, 153402.
6. Oreg, Y., Byczuk, K. and Halperin, B.I. (2000) Spin configurations of a carbon nanotube in a nonuniform external potential. *Physical Review Letters*, **85**, 365–368.
7. Park, H., Park, J., Lim, A.K.L., Anderson, E.H., Alivisatos, A.P. and McEuen, P.L. (2000) Nanomechanical oscillations in a single-C_{60} transistor. *Nature*, **407**, 57–60.
8. Kouwenhoven, L.P., Austing, D.G. and Tarucha, S. (2001) Few-electron quantum dots. *Reports on Progress in Physics*, **64**, 701–736.
9. Liang, W., Bockrath, M. and Park, H. (2002) Shell filling and exchange coupling in metallic single-walled carbon nanotubes. *Physical Review Letters*, **88**, 126801.
10. Flensberg, K. and Braig, S. (2003) Incoherent dynamics of vibrating single-molecule transistors. *Physical Review B-Condensed Matter*, **67**, 245415.
11. Mitra, A., Aleiner, I. and Millis, A.J. (2004) Phonon effects in molecular transistors: Quantum and classical treatment. *Physical Review B-Condensed Matter*, **69**, 245302.
12. Koch, J. and von Oppen, F. (2005) Franck–Condon blockade and giant fano factors in transport through single molecules. *Physical Review Letters*, **94**, 206804.
13. Djukic, D., Thygesen, K.S., Untiedt, C., Smit, R.H.M., Jacobsen, K.W. and van Ruitenbeek, J.M. (2005) Stretching dependence of the vibration modes of a single-molecule Pt-H_2-Pt bridge. *Physical Review B-Condensed Matter*, **71**, 161402.
14. Pasupathy, A.N., Park, J., Chang, C., Soldatov, A.V., Lebedkin, S., Bialczak, R.C., Grose, J.E., Donev, L.A.K., Sethna, J.P., Ralph, D.C. and McEuen, P.L. (2005) Vibration-assisted electron tunneling in C140 single-molecule transistors. *Nano Letters*, **5**, 203–207.
15. Osorio, E.A., O'Neill, K., Stuhr-Hansen, N., Faurskov Nielsen, O., Bjørnholm, T. and van der Zant, H.S.J. (2007) Addition energies and vibrational fine structure measured in electromigrated single-molecule junctions based on an oligophenylenevinylene derivative. *Advanced Materials*, **19**, 281–285.
16. Kondo, J. (1964) Resistance minimum in dilute magnetic alloys. *Progress of Theoretical Physics*, **32**, 37.
17. (a) Goldhaber-Gordon, D., Shtrikman, H., Mahalu, D., Abusch-Magder, D., Meirav, U. and Kastner, M.A. (1998) Kondo effect in a single-electron transistor. *Nature*, **391**, 157–159.(b) Cronenwett, S.M., Oosterkamp, T.H. and Kouwenhoven, L.P. (1998) A tunable Kondo effect in quantum dots. *Science*, **281**, 540–544.
18. Nygård, J., Cobden, D.H. and Lindelof, P.E. (2000) Kondo physics in carbon nanotubes. *Nature*, **408**, 342–346.
19. (a) Park, J. *et al.* (2002) Coulomb blockade and the Kondo effect in single-atom transistors. *Nature*, **417**, 722–725. (b) Liang, W., Shores, M.P., Bockrath, M.,

Long, J.R. and Park, H. (2002) Kondo resonance in a single-molecule transistor. *Nature*, **417**, 725–729. (c) Yu, L.H., Keane, Z.K., Ciszek, J.W., Cheng, L., Tour, J.M., Baruah, T., Pederson, M.R. and Natelson, D. (2005) Kondo resonances and anomalous gate dependence in the electrical conductivity of single-molecule transistors. *Physical Review Letters*, **95**, 256803.

3
Spin Injection–Extraction Processes in Metallic and Semiconductor Heterostructures
Alexander M. Bratkovsky

3.1
Introduction

Spin transport in metal-, metal-insulator, and semiconductor nanostructures holds promise for the next generation of high-speed, low-power electronic devices [1–10]. Amongst important spintronic effects already used in practice are included a giant magnetoresistance (MR) in magnetic multilayers [11] and tunnel MR (TMR) in ferromagnet-insulator-ferromagnet (FM-I-FM) structures [12–19]. The injection of spin-polarized electrons into semiconductors is of particular interest because of relatively large spin relaxation time (~1 ns in semiconductors, ~1 ms in organics) [2] during which the electron can travel over macroscopic distances without losing polarization, or stay in a quantum dot/well. This also opens up possibilities, albeit speculative ones, for quantum information processing using spins in semiconductors.

The potential of spintronic devices is illustrated most easily with a simple spin-dependent transport process, which is a tunneling magnetoresistance in FM-I-FM structure (see the next section). The effect is a simple consequence of the golden rule that dictates a dependence of the tunnel current on the density of initial and final states for tunneling electron. Most of the results for tunnel spin junctions may be reused later in describing the spin injection from ferromagnets into semiconductors (or vice versa) in the later sections.

It is worth noting from the outset that there are two major characteristics of the spin transport processes that will define the outcome of a particular measurement, namely *spin polarization* and *spin injection efficiency*. These may be very different from each other, and this may lead (and frequently does) to a confusion among researchers. The spin polarization measures the imbalance in the *density* of electrons with opposite spins (spin accumulation/depletion),

$$P = \frac{n_\uparrow - n_\downarrow}{n_\uparrow + n_\downarrow}, \tag{3.1}$$

while the injection *efficiency* is the *polarization* of injected *current J*

$$\Gamma = \frac{J_\uparrow - J_\downarrow}{J_\uparrow + J_\downarrow}, \qquad (3.2)$$

where $\uparrow(\downarrow)$ refers to the electron spin projection on a quantization axis. In case of ferromagnetic materials, the axis is antiparallel to the magnetization moment $\vec{M}$. Generally, $P \ne \Gamma$, but in some cases they can be close. Since in the ferromagnets the spin density is constant, a reasonable assumption can be made that the current is carried independently by two electron fluids with opposite spins (Mott's two-fluid model [20]). Then, in the FM bulk the injection efficiency parameter is

$$\Gamma = \Gamma_F = \frac{(\sigma_\uparrow - \sigma_\downarrow)}{\sigma}, \qquad (3.3)$$

where $\sigma_{\uparrow(\downarrow)}$ are the conductivities of up-, (down)spin electrons in ferromagnet, $\sigma = \sigma_\uparrow + \sigma_\downarrow$.

In the case of spin tunneling, it is found that Γ characterizes the value of MR in magnetic *tunnel* junctions, which is quite obvious as there one measures the difference between *currents* in two configurations: with parallel (P) and antiparallel (AP) moments on electrodes. The tunnel current is small; hence the injected spin density is minute compared to metallic carrier densities. At the same time, in experiments where one injects spin (creates a non-equilibrium spin population) in a quantum well, this results in the emission of polarized light (spinLED), the measured intensity of which is, obviously, proportional to the spin polarization P (see below).

We will outline the major spin-transport effects here in the Introduction, starting with an analysis of tunnel magnetoresistance (TMR), followed by giant magnetoresistance (GMR) and spin-torque (ST) switching in magnetic nanopillars. We will then outline the spin-orbit effects in three-dimensional (3-D) and two-dimensional (2-D) semiconductor systems (Dresselhaus and Vasko-Rashba effects) and use this for further discussion of Datta–Das interference device and the Spin-Hall effect. A brief assessment will then be made of spin logic devices, showing why they are impractical because of difficulty in precise manipulation of individual spins, in addition to practically gapless excitation of the spin waves that easily destroy a particular spin configuration of the multispin system. Spin ensemble-based quantum computing – that is, coherent manipulation of spin ensembles over a number of steps – is, even less likely than using classical spin logic.

We then turn to spin injection/extraction effects in ferromagnetic metal-semiconductor heterojunctions. It is shown that efficient spin injection is possible with modified Schottky junctions, and is a strongly non-linear effect. A brief discussion is then provided of a few possible spin-injection effects and devices, some of which are likely to be demonstrated in the near future. The final section is devoted to a complementary topic of spin injection in degenerate semiconductors. These processes are significantly different from the case of non-degenerate semiconductors so as to warant a separate discussion.

3.2
Main Spintronic Effects and Devices

3.2.1
TMR

Tunnel magnetoresistance is observed in metal–insulator–metal magnetic tunnel junctions (MTJ), usually with Ni–Fe, Co–Fe electrodes and (amorphous) Al_2O_3 tunnel barrier where one routinely observes upward of 40–50% change in conductance as a result of the changing relative orientation of magnetic moments on electrodes. A considerably larger effect, about 200% TMR, is found in Fe/MgO/Fe junctions with an epitaxial barrier, that may be related to surface states and/or peculiarities of the band structure of the materials. As will be seen shortly, TMR is basically a simple effect of an asymmetry between densities of spin-up and -down (initial and final tunneling) states. TMR is intimately related to giant magnetoresistance [11]; that is, a giant change in conductance of magnetic multilayers with relative orientation of magnetic moments in the stack.

We can estimate the TMR using the golden rule which states that the tunnel current at small bias voltage V is $J_\sigma = G_\sigma V$, $G_\sigma \propto |M|^2 g_{i\sigma} g_{f\sigma}$, where $g_{i(f)\sigma}$ is the density of initial (final) tunneling states with a spin projection σ, and M is the tunneling matrix element. Consider the case of electrodes made from the same material. It is clear from the band schematic shown in Figure 3.1 that the total rates of tunneling in parallel (a) and antiparallel (b) configurations of moments on electrodes are different.

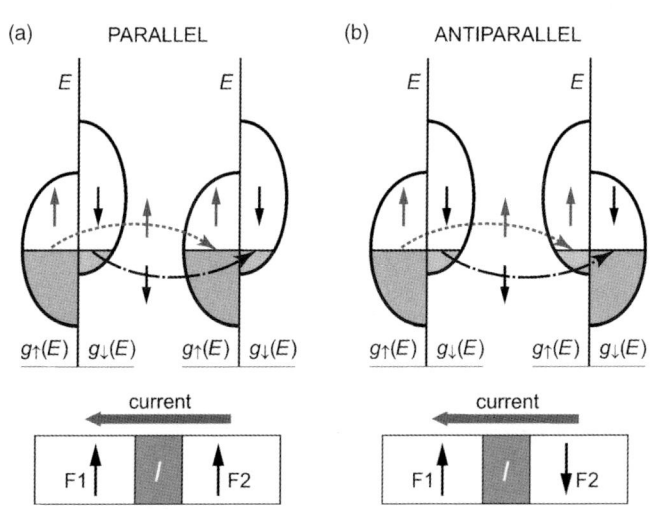

Figure 3.1 Schematic illustration of spin tunneling in FM-I-FM junction for (a) parallel (P) and (b) antiparallel (AP) configuration of moments on ferromagnetic electrodes. Top panels: band diagram; bottom panels: schematic of the corresponding magnetic configuration in the junction with regards to current direction.

Indeed, denoting $D=g_\uparrow$ and $d=g_\downarrow$ as the partial densities of states (DOS), we can write down the following golden rule expression for parallel and antiparallel moments on the electrodes:

$$G_P \propto D^2 + d^2, \quad G_{AP} \propto 2Dd, \tag{3.4}$$

and arrive at the expression for TMR first derived by Jullieres [12]

$$\text{TMR} \equiv \frac{G_P - G_{AP}}{G_{AP}} = \frac{(D-d)^2}{2Dd} = \frac{2P^2}{1-P^2}, \tag{3.5}$$

where we have introduced a polarization P, which fairly approximates the polarization introduced in Equation 3.1, at least for narrow interval of energies:

$$P = \frac{D-d}{D+d} \equiv \frac{g_\uparrow - g_\downarrow}{g_\uparrow + g_\downarrow}. \tag{3.6}$$

Below, we shall see that the "polarization" entering expression for a particular process, depends on particular physics and also on the nature of the electronic states involved. It should be noted that, for instance, the DOS entering the above expression for TMR, is not the *total* DOS but rather the one for states that contribute to tunneling current. Thus, Equation 3.6 may lead one to believe that the *tunnel* polarization in elemental Ni should be negative, as there is a sharp peak in the minority carrier density of states at the Fermi level. The data, however, suggest unambiguously that the tunnel polarization in Ni is positive [14], $P > 0$. This finds a simple explanation in a model by M.B. Stearns, who highlighted the presence in elemental 3d metals parts of Fermi surface with almost 100% d-character and a small effective mass close to one of a free electron [21]. A detailed discussion of TMR effects is provided below.

3.2.2
GMR

There are important differences between TMR and GMR processes. Indeed, GMR is most reminiscent of TMR for current-perpendicular-to-planes (CPP) geometry in FM-N-FM-... stacks, where N stands for normal metal spacer (Figure 3.2a). In the CPP geometry, the spins cross the nanometer-thin normal spacer layer (N) without spin flip, similarly to tunneling through the oxide barrier, but the elastic mean free path is comparable or smaller than the N thickness, so that a drift-diffusive electrons transport takes place in metallic GMR stacks. In commonly used current-in-plane geometry (CIP), the electrons bounce between different ferromagnetic layers, (Figure 3.2b) effectively showing the same motif in transport across the layers as in the CPP geometry. Comparing with TMR, the latter (and spin injection efficiency) depends on the difference between the densities of states g_σ, spin $\sigma = \uparrow(\downarrow)$ at the Fermi level, while GMR depends on relative conductivity

$$\sigma_\sigma = e^2 \langle g_\sigma v_\sigma^2 \tau_\sigma \rangle_F,$$

where the angular brackets indicate an average over the Fermi surface that involves the Fermi velocity v_σ and the momentum relaxation time τ_σ. One can still use the

Mott's two independent spin fluid picture, but as one is dealing with metallic heterostructure, the continuity (or Boltzmann) equations must be solved for a periodic FM-N-FM-... stack to find the ramp of an electrochemical potential that defines the total current. Neglecting any slight imbalance of electrochemical potentials for two spins in the N regions (spin accumulation), one may construct an equivalent circuit model for CPP stack in the spirit of Mott's model, and thus qualitatively explain the GMR. The parallel "spin" layer resistances would be $R_{\uparrow(\downarrow)} \sigma_{\uparrow(\downarrow)} L_F$ for the FM layers, and $r = \sigma_N^{-1} L_N$ in the normal N regions with thicknesses $L_F(L_N)$, respectively. For conductances in two configurations of the moments, we then obtain (Figure 3.2c):

$$G_P = \frac{1}{R_P} = \frac{1}{2R_\uparrow + r} + \frac{1}{2R_\downarrow + r}, \qquad (3.7)$$

$$G_{AP} = \frac{1}{R_{AP}} = \frac{2}{R_\uparrow + R_\downarrow + r}, \qquad (3.8)$$

and the GMR simply becomes

$$\mathrm{GMR} = \frac{G_P - G_{AP}}{G_{AP}} = \frac{(R_\downarrow - R_\uparrow)^2}{(2R_\uparrow + r)(2R_\downarrow + r)} \qquad (3.9)$$

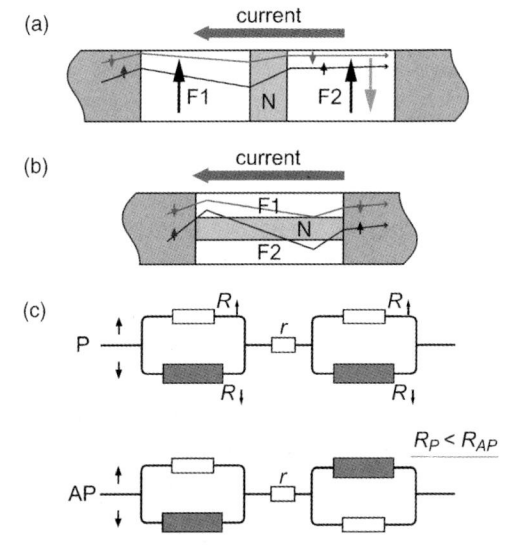

Figure 3.2 Schematic of giant magnetoresistance (GMR) in (a) current perpendicular to plane (CPP) and (b) current in-plane (CIP) geometries. Electron scattering in the GMR valve depends on the configuration of moments and spin of traversing electrons in both configurations. (c) Equivalent circuit model for GMR. Two-fluid spin model that is valid in the absence of spin relaxation, leads to higher resistivity for antiparallel arrangement of magnetic moments in the spin valve, $R_{AP} > R_P$.

which we can rewrite as

$$\text{GMR} = \frac{\Gamma_F^2}{(1+r/R)^2 - \Gamma_F^2} \approx \frac{\Gamma_F^2}{1-\Gamma_F^2}, \quad (3.10)$$

where $R = R_\uparrow + R_\downarrow$. The latter is very similar to the expression TMR [Equation 3.5], but there is an absence of a factor two in the numerator. It can be seen that the polarization entering GMR is different from that entering TMR. Even in more common CIP geometry, the electrons certainly do scatter across the interfaces; hence, this equation can also be used for semi-quantitative estimates of the GMR effect in CIP geometry (Figure 3.2c). Obviously, the effective circuit model [Equations 3.7 and 3.8] remains exactly the same because it simply reflects the two-fluid approximation for contributions of both spins. However, all effective resistances depend in rather nontrivial manner on the geometry (CPP or CIP) and electronic structure of the metals involved that is particularly complicated in magnetic transition metals.

In terms of applications, the TMR effect is used in non-volatile magnetic random access memory (MRAM) devices and as a field sensor, while GMR is widely used in magnetic read heads as field sensors. The MRAM devices are in a tight competition with semiconductor memory cards (FLASH), since MRAM is technologically more involved (and hence more expensive) than standard silicon technology. It is problematic to use those effects in building three-terminal devices with gain that would show any advantage over standard CMOS transistors.

3.2.3
(Pseudo)Spin-Torque Domain Wall Switching in Nanomagnets

Magnetic memory based on TMR is non-volatile, may be rather fast ($\sim$1 ns), and can be scaled down considerably toward paramagnetic limits observed in nanomagnets. Switching, however, requires the MTJ to be placed at a crosspoint of the bit and word wires carrying current that produces a sufficient magnetic field for switching the domain orientation in "free" (unpinned) MTJ ferromagnetic electrodes. The undesirable side effect is a crosstalk between cells, a rather complex layout, and a power budget. Alternatively, one may take a GMR multilayer in a nanopillar form with antiparallel orientation of magnetic moments and run a current through it (this would obviously correspond to a CPP geometry) (Figure 3.3). In this case, there will be a spin accumulation in the drain layer; that is, the accumulation of *minority* spins in the drain electrode for *antiparallel* configuration of moments on electrodes (for the electrodes made out the same material. This formally means a transfer of spin (angular) moment across the space layer. Injection of angular momentum means that there is a change in the spin momentum on the drain electrode, the spin projection on the quantization axis then evolves with time simply because of an influx of electrons with a different projection of polarization on the (drain electrode) quantization axis, $d\mathcal{P}_{Rz}/dt \neq 0$, $\mathcal{P}_{Rz} = n_+ - n_-$, where $\pm$ marks along (against) the quantization axis on the right electrode with $n_{+(-)}$ the densities of majority (minority) electrons. One may call this a (pseudo) torque effectively acting on a moment in the

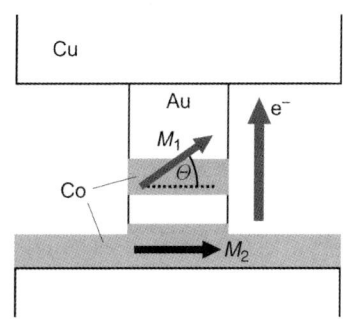

Figure 3.3 Schematic of a nanopillar device for spin-torque measurements. Current through the device results in change in angle θ between magnetic moments in the free layer in the middle and the bottom ferromagnetic electrode.

drain electrode, although this is obviously not a good term. This simple effect was predicted in Refs. [22, 23], and observed experimentally in nanopillars multilayer stacks by Tsoi et al. [24], and also in other studies.

The spin *accumulation* is proportional to density J of the spin-polarized current through the drain magnet or magnetic particle, that carries $(-\hbar J/2q)$ spin moment with it per second. The change in the longitudinal component of spin polarization $\mathcal{P}_{Rz}$ in the right electrode next to the interface is then simply

$$\begin{aligned} d\mathcal{P}_{Rz} &= -\frac{\hbar(J_+ - J_-)}{2q} = -\frac{\hbar V(\mathcal{G}_+ - \mathcal{G}_-)}{2q} \\ &= -\frac{qV(T_+ - T_-)}{4} \\ &= -\frac{\hbar V}{4q}(G_{+\uparrow} + G_{+\downarrow} - G_{-\uparrow} - G_{-\downarrow}), \end{aligned} \quad (3.11)$$

in a linear regime, where $\mathcal{G}_\pm = (q^2/\hbar)T_\pm$ are the conductances for spin-up and -down channels, expressed through the transmission probabilities $T_\pm$ and partial spin conductances G. There is an angle θ between the quantization axes on the source and drain electrode, $\cos\theta = \vec{\mathcal{P}}_L \cdot \vec{\mathcal{P}}_R/(\mathcal{P}_L \mathcal{P}_R)$. The partial spin conductances G are in turn given by the standard Landauer expression through the transmission coefficients $t_{\sigma\sigma'}$ as $T_{\sigma=\pm} = \sum_{\sigma'=\uparrow,\downarrow}|t_{\sigma\sigma'}|^2$. The above expression can be expressed through the partial conductances for arbitrary configuration of spins on the electrodes $G_{+\uparrow} = (q^2/\hbar)|t_{+\uparrow}|^2$, $G_{-\uparrow} = (q^2/\hbar)|t_{-\uparrow}|^2, \ldots$, where the orientation along (against) the spin on the right electrode, $\mathcal{P}_R$, is marked by the subscript $+(-)$. Assuming that there is no spin-flip in the oxide (non-magnetic metal) spacer, we can express the transmission amplitudes through those calculated for parallel/antiparallel configuration of spins on the electrodes with the use of a standard rule for spin wave function in a rotated frame.

Finally, the rate of change in polarization on the drain electrode due to influx of polarized current is simply

$$\frac{d\mathcal{P}_{Rz}}{dt} = -\frac{\hbar V}{4q}\left[(G_{\uparrow\uparrow} - G_{\downarrow\downarrow})\cos^2\frac{\theta}{2} + (G_{\uparrow\downarrow} - G_{\downarrow\uparrow})\sin^2\frac{\theta}{2}\right], \tag{3.12}$$

and the term on the right-hand side should be added to the driving force in the Landau–Lifshitz (LL) equation on the right-hand side. This expression should be better suited for MTJs, as in metallic spin valves one must consider spin accumulation in a metallic spacer. Note that Slonczewski obtains $d\mathcal{P}_{Rz}/dt \propto \sin\theta$ for MTJs [25], which may be inaccurate. Indeed, consider the antiparallel configuration ($\theta = \pi$) of *unlike* electrodes. Then, $d\mathcal{P}_{Rz}/dt \propto G_{\uparrow\downarrow} - G_{\downarrow\uparrow} \neq 0$, since for the unlike electrodes $G_{\uparrow\downarrow} \neq G_{\downarrow\uparrow}$, and there obviously will be a change in the spin density in the right electrode because the influx of spins into majority states would not be equal to the influx into minority states. To handle the resulting spin dynamics properly, one needs to write down the continuity equation for the spin, similar to Equation 3.23 below, with Equation 3.12 as the boundary condition at the interface.

Time-resolved measurements of current-induced reversal of a free magnetic layer in permalloy/Cu/permalloy elliptical nanopillars at temperatures from 4.2 to 160 K can be found in Ref. [26]. There is considerable device-to-device variation in the ST attributed to presence of an antiferromagnetic oxide layer around the perimeter of the Permalloy free layer (and some ambiguity in an expression used for the torque itself). Obviously, controlling this layer would be very important for the viability of the whole approach for memory applications, and so on. There are reports about the activation character of switching that may be related to pinning of the domain walls at the side walls of the pillar. The injected DC polarized current may also induce a magnetic vortex oscillation, when vortex may be formed in a magnetic island in, for example, a pillar-like spin valve. These induced oscillations have recently been found [27]. It is worth noting that the agreement between theory and experiment may be fortuitous: thus, in permalloy nanowires the speed of domain wall has substantially exceeded the rate of spin angular momentum transfer rate [28].

3.3
Spin-Orbital Coupling and Electron Interference Semiconductor Devices

In most cases of interest, such as direct band semiconductors near high-symmetry points, and a two-dimensional electron gas [29, 30], the spin-orbital (SO) coupling effects can be treated fairly well within the Kane's or Luttinger–Kohn's models in so-called kp method (see e.g. Refs. [31, 32]). The SO interaction is given by the term in electron Hamiltonian

$$H_{SO} = \frac{\hbar}{4m_0^2 c^2}\left[\vec{\nabla} U \times \vec{p}\right]\vec{\sigma}, \tag{3.13}$$

where $\vec{p} = -i\hbar\nabla$ the electron momentum operator.

SO interactions lead in some cases, as for semiconductor 2-D channels, to various effects that can be used to build electron interference devices, at least as a matter of principle. As an interesting (yet unrealized at the time of writing) example of such a

3.3 Spin-Orbital Coupling and Electron Interference Semiconductor Devices

three-terminal spintronic device, it is worth describing Datta–Das ballistic spintronic modulator/switch [3]. This is a quantum interference device with FET-like layout where a 2-D electron gas (2DEG) has a ferromagnetic source and drain. The asymmetric confinement potential induces precession of spins injected into the 2DEG channel due to specific low-symmetry SO effect (Vasko–Rashba spin splitting) [29, 30]. The resulting angle may become large, $\sim\pi$ in channels $\gtrsim \mu$m long and made of narrow-gap semiconductors with strong so coupling. Since the ferromagnetic drain works as a spin filter, one hopes that changing the gate voltage would change the shape of the confinement potential and the Vasko–Rashba coupling constant α. As a result, one may be able to change the precession angle of ballistic electrons and the current through the structure (yet to be observed). To appreciate the situation, we need to describe the SO effects in a simple kp-model for semiconductors. As we shall see, the SO effects are expectedly weak, being of relativistic nature, and so in general are the effects. It is difficult to expect that SO-based devices can outperform any of the conventional electronics devices in standard applications.

In MOSFET structures the confining potential is asymmetric, so there appears an inversion asymmetry term derived by Vasko [28] (Vasko–Rashba or simply Rashba term, see also Ref. [29]). The only contribution coming from the confinement field in SO Hamiltonian is $\propto \langle \nabla_z V \rangle$ giving the Vasko–Rashba term,

$$H_R = \alpha(k_y \sigma_x - k_x \sigma_y). \tag{3.14}$$

The magnitude of the coupling constant α depends on the confining potential, and this can in principle be modified by gating. It also defines the spin-precession wave vector $k_\alpha = \alpha m/\hbar^2$. Such a term, H_R, is also present in cubic systems with strain [35]. The Vasko–Rashba Hamiltonian for heavy holes is cubic in k, and, generally, very small.

Electric fields due to impurities (and external field) lead to extrinsic contributions of the spin-orbit coupling in the standard form

$$H_{\text{ext}} = \lambda [\vec{k} \times \nabla U] \, \vec{\sigma}, \tag{3.15}$$

where U is the potential due to impurities and an externally applied field, with the coupling constant λ derived from 8×8 Kane Hamiltonian in third-order perturbation theory [33, 34]

$$\lambda = \frac{P^2}{3}\left[\frac{1}{E_g^2} - \frac{1}{(E_g + \Delta)^2}\right], \tag{3.16}$$

where P is the matrix element of momentum found from $\langle S|p_x|X\rangle = \langle S|p_y|Y\rangle = \langle S|p_z|Z\rangle = iPm_0/\hbar$. This is the same analytical form as the vacuum spin-orbit coupling but, for $\Delta > 0$ the coupling has the *opposite sign*. Numerically, $\lambda = 5.3$ Å^2 for GaAs and 120 Å^2 for InAs, that is, spin-orbit coupling in n-GaAs is by *six orders* of magnitude larger than in vacuum [31]. This helps to generate the relatively large extrinsic spin currents observed in the spin-Hall effect (see below). In 2-D, $H_{\text{ext}} = \lambda(\vec{k} \times \nabla U)_z \cdot \sigma_z$.

Now, we can analyze the behavior of polarized electrons injected into the 2DED channel. A free-electron VR Hamiltonian H_{VR} has two eigenstates with the momenta $k_{\pm}(\epsilon)$ for opposite spins for each energy ϵ, with $k_{-} - k_{+} = 2m\alpha/\hbar^2$. Datta and Das [3] have noted that the conductivity of the device depends on the phase difference $\Delta\theta = (k_{-} - k_{+})L = 2m\alpha L/\hbar^2$ between electron carriers after crossing a ballistic channel of a length L and oscillates with a period defined by the interference condition $(k_{-} - k_{+})L = 2\pi n$, with n as integer. An equivalent description of the same phenomenon is the precession of an electron spin in an effective magnetic field $\vec{B}_{so} = (2\alpha/g\mu_B)(\vec{k} \times \hat{z})$, with μ_B the Bohr magneton, g the gyromagnetic ratio. The device is supposed to be controlled by the gate voltage V_g that modulates the SO coupling constant α, $\alpha = \alpha(V)_g$. This pioneering report generated much attention, yet to date the device appears not to have been demonstrated. Using typical parameters from Refs. [36, 37], $\hbar\alpha \approx 1 \times 10^{-11}$ eV·m and $m = 0.1\,m_0$ for the effective carrier mass, current modulation would be observable in channels with relatively large length $L \gtrsim 1$ μm. Given the above, the observation of the effect would need: (i) efficient spin injection into channel from the FM source, which is tricky and requires a modified Schottky barrier (see below); (ii) the splitting should well exceed the bulk inversion asymmetry effect; (iii) the inhomogeneous broadening of α due to impurities, that mask the Vasko–Rashba splitting, should be small; and (iv) one should be able to gate control α. All these represent great challenges for building a room-temperature interference device, where one needs to use narrow-gap semiconductors and structures with *ballistic* channels. The device is not efficient is a diffusive regime. The gate control of α has been demonstrated (see Refs. [36, 38] and references therein).

3.3.1
Spin-Hall Effect (SHE) and Magnetoresistance due to Edge Spin Accumulation

Recently, there has been a resurgence of interest in the spin-Hall effect (SHE), which is another general consequence of the spin-orbital interaction, predicted by Dyakonov and Perel in 1971 [40]. These authors found that because of the spin-orbital interaction the electric and spin currents are intertwined: an electrical current produces a transverse spin current, and *vice versa*. In the case when impurity scattering dominates, which is quite often, the transverse is caused by the Mott skew scattering of spin-polarized carriers due to the SO interaction, [see Eq. (3.13)]. Since the current drags along the polarization of the carriers, the *spin accumulation at the edges* occurs, a so-called spin-Hall effect (SHE). In ferromagnets, the appearing Hall current is termed *anomalous*, and is always accompanied by the SHE. Importantly, the edge spin accumulation results in a slight *decrease* in sample resistance. External magnetic field would destroy the spin accumulation (Hanle effect) and lead to a *positive magnetoresistance*, recently identified by Dyakonov [41].

We present here a simple phenomenological description of spin-Hall effects (direct and inverse) and Dyakonov's magnetoresistance [42]. To this end, we introduce the electron charge flux $\vec{f}$ related to the current density as $\vec{J} = -q\vec{f}$, where q is the elementary charge. For parts not related to the SO interaction, we have the usual drift-diffusion expression:

$$\vec{f}^{(0)} = -\mu n \vec{E} - D\vec{\nabla} n, \tag{3.17}$$

where μ and D are the usual electron mobility and diffusion coefficient, connected by the Einstein relation, $\vec{E}$ the electric field, and n is the electron density. The spin polarization flux t_{ij} is a tensor characterizing the flow of the jth component of the polarization *density* $\mathcal{P}_j = n_{j\uparrow} - n_{j\downarrow}$ in the direction i (spin density is $s_j(\mathbf{r}) = \frac{1}{2}\mathcal{P}_j$). It is non-zero even in the absence of spin–orbit interaction, simply because the spins are carried by electron flux, and we mark the corresponding quantity $t_{ij}^{(0)}$. Then, we have

$$t_{ij}^{(0)} = -\mu E_i \mathcal{P}_j - D\partial_i \mathcal{P}_j, \tag{3.18}$$

where $\partial_i = \partial/\partial x_i$, and one can add other sources of current, such as temperature gradient, in Equations 3.17 and 3.18. Spin–orbit interaction couples the charge and spin currents. For a material with an inversion symmetry, we have [42]:

$$f_i = f_i^{(0)} + \gamma \varepsilon_{ijk} t_{jk}^{(0)}, \tag{3.19}$$

$$t_{ij} = t_{ij}^{(0)} - \gamma \varepsilon_{ijk} f_k^{(0)}, \tag{3.20}$$

where ε_{ijk} is the unit antisymmetric tensor and $\gamma \ll 1$ is a dimensionless coupling constant proportional to the spin–orbit interaction λ [Equation 3.16] (typically, $\gamma \sim 10^{-2} - 10^{-3}$). The difference in signs in Equations 3.19 and 3.20 is consistent with the Onsager relations, and is due to the different properties of $\vec{f}$ and t_{ij} with respect to time inversion. Explicit phenomenological expressions for the two currents follow from Equations 3.17–3.20 [40, 41]:

$$\vec{J}/q = \mu n \vec{E} + D\vec{\nabla} n + \beta \vec{E} \times \vec{\mathcal{P}} + \delta \vec{\nabla} \times \vec{\mathcal{P}}, \tag{3.21}$$

$$t_{ij} = -\mu E_i \mathcal{P}_j - D\partial_i \mathcal{P}_j + \varepsilon_{ijk}(\beta n E_k + \delta \partial_k n), \tag{3.22}$$

where the parameters $\beta = \gamma \mu$, $\delta = \gamma D$, satisfy the same Einstein relation, as do μ and D. The spin polarization vector evolves with time in accordance with the continuity equation [41, 42]:

$$\partial_t \mathcal{P}_j + \partial_i t_{ij} + |\vec{\Omega} \times \vec{\mathcal{P}}|_j + \mathcal{P}_j/\tau_s = 0, \tag{3.23}$$

where the vector $\Omega = g\mu_B H/\hbar$ is the spin precession frequency in the applied magnetic field $\vec{H}$, and τ_s the spin relaxation time. The term $\beta \vec{E} \times \vec{P}$ describes the *anomalous Hall effect*, where the spin polarization plays the role of the magnetic field. We ignore the action of magnetic field on the particle dynamics, which is justified if $\omega_c \tau \ll 1$, where ω_c is the cyclotron frequency and τ is the momentum relaxation time. Since normally $\tau_s \gg \tau$, it is possible to have both $\Omega\tau_s \gg 1$ and $\omega_c \tau \ll 1$ in a certain range of magnetic fields. It is also assumed that the equilibrium spin polarization in the applied magnetic field is negligible. The fluxes [Equations 3.21 and 3.22] need to be modified for an inhomogeneous magnetic field by adding a counter-term proportional to $\partial B_j/\partial x_i$, which takes care of the force acting on the electron with a given spin in an inhomogeneous magnetic field $\vec{H}(r)$.

Equations 3.21–3.23 derived in Ref. [41] fully describe all physical consequences of spin–charge current coupling. For instance, the term $\delta \vec{\nabla} \times \vec{\mathcal{P}}$ describes an electrical current induced by an inhomogeneous spin density (so-called Inverse spin-Hall Effect) found experimentally for the first time by Bakun *et al.* [42] under the conditions of optical spin orientation. The term $\beta n \epsilon_{ijk} E_k$ (and its diffusive counterpart $\delta \epsilon_{ijk} \partial n/\partial x_k$) in Equation 3.22, describes what is now called the spin-Hall effect: an electrical current induces a transverse spin current, resulting in spin accumulation near the sample boundaries [41]. Recently, a spin-Hall effect was detected optically in thin films of n-doped GaAs and In GaAs [44] (with bulk electrons) and 2-D holes [45]. All of these phenomena are closely related and have their common origin in the coupling between spin and charge currents given by Equations 3.21 and 3.22. Any mechanism that produces the anomalous Hall effect will also lead to the spin-Hall effect, and vice versa. Remarkably, there is a single dimensionless parameter, γ, that governs the resulting physics.

It was found recently by Dyakonov that the spin-Hall effect is accompanied by a *positive* magnetoresistance due to spin accumulation near the sample boundaries [41]. The spin accumulation occurs over the spin diffusion length $L_s = \sqrt{D\tau_s}$. Therefore, it depends on the ratio L_s to the sample width L and vanishes in wide samples with $L_s/L \ll 1$. In a stripe sample with the width L (in xy plane), the z-component of $\mathcal{P}$ varies across the stripe (y-axis), $\vec{\nabla} \times \vec{\mathcal{P}} \neq 0$, this creates a positive correction to a current compared to a hypothetical case of an absent spin–orbit coupling, and the sample resistivity goes down. By applying the magnetic field in xy plane, one may destroy the spin polarization (the Hanle effect) and observe the *positive* (Dyakonov's) magnetoresistance in magnetic fields at the scale $\Omega \tau_s \sim 1$.

The data for 3-D [44] and 2-D [46] GaAs suggest the estimate $\gamma \sim 10^{-2}$, for platinum at room temperature [48] one finds $\gamma = 3.7 \times 10^{-3}$, so in these cases a magnetoresistance due to spin accumulation is on the order of 10^{-4} and 10^{-5}, respectively. It should be possible to find this MR due to its characteristic dependence on the field and the width of the sample, when it becomes comparable to the spin diffusion length. Because of the high sensitivity of electrical measurements, magnetoresistance might provide a useful tool for studying the spin–charge interplay in semiconductors and metals.

3.3.2
Interacting Spin Logic Circuits

There are various suggestions of more exotic devices based on, for example, arrays of spin-polarized quantum dots with exchange coupled single spins with typical exchange coupling energy $\delta \sim 1$ meV [47], or magnetic quantum cellular automata [48]. It is assumed that one can apply a local field or short magnetic π-pulse to flip the "input" spin that would result in nearest-neighbor spins to flip in accordance with the new ground state (of the antiferromagnetically coupled circuit of quantum dots). The idea is that those spin arrays (no quantum coherence is required) can be used to perform classical logic on bits represented by spins pointing along or against the quantizations axis, $|+z\rangle \to 1$, $|-z\rangle \to 0$. However, there are problems with using those schemes. Indeed, the standard Zeeman splitting for electron in the field of 1 T is only 0.5 K *in vacuo*, so that one needs the field of *at least* ~150 T to flip the spin (or use materials with unusually large gyromagnetic factors), or one can apply ~1 T transversal *B*-field for some 30 ps to do the same. The practicalities of building such a control system at a nanoscale is a major challenge, and would require a steep power budget. The other challenge is that instead of *nearest-neighbor* spins falling into a new shallow ground state with the directions of all other spins fixed, the initial flip would trigger spin wave(s) in the circuit, thus destroying the initial set-up. Indeed, the spin wave spectrum in large coupled arrays of *N* spins is almost gapless, with the excited state just $\sim\delta/N$ above the ground state (see e.g., Ref. [49]). Additionally, the spins are subject to a fluctuating external (effective) magnetic field that tends to excite the spin waves and destroy the direction of the spins along set quantization axis $\pm z$. For the same reason, keeping the spins in a *coherent* superposition state is unlikely, so quantum computing with coupled spins is even less practical [50].

It is clear from the above discussion, however, that it is unlikely that the Datta–Das or any other interference devices can offer any advantages over standard MOSFETs, especially as they do not have any *gain*, should operate in a ballistic regime (i.e., at low temperatures in clean systems), and require new fabrication technology.

3.4
Tunnel Magnetoresistance

Here, we describe some important aspects of TMR on the basis of simple microscopic model for elastic, impurity-assisted, surface state-assisted and inelastic contributions. Most of these results are generic, and some will be useful later to analyze room-temperature spin injection into semiconductor through a modified Schottky barrier. A model for spin tunneling has been formulated by Julliere [12], and further developed in Refs. [17, 18, 21]. It is expected to work rather well for iron-, cobalt-, and nickel-based metals, according to theoretical analysis [21] and experiments [15]. However, it disregards important points such as an impurity scattering and a reduced effective mass of carriers inside the barrier. Both issues have important implications for magnetoresistance and will be considered here, along with proposed novel half-metallic systems which should, in principle, show the ultimate performance. Enhanced performance is also found in MTJ with MgO epitaxial oxide barrier, which may be a combination of band-structure and surface effects [16, 51]. In particular, Zhang and Butler [51] predicted a very large TMR in Fe/MgO/Fe, bcc Co/MgO/Co, and FeCo/MgO/FeCo tunnel junctions, having to do with peculiar band matching for majority spin states in a metal with that in MgO tunnel barrier.

We shall describe electrons in ferromagnet-insulating barrier-ferromagnet (FM–I–FM) systems by the Schrödinger equation [17] $[-(\hbar^2/2m_i)\nabla^2 + U_i - \frac{1}{2}\vec{\Delta}_{xc}\vec{\sigma}]\psi = E\psi$ with $U(\mathbf{r})$ the potential (barrier) energy, $\vec{\Delta}_{xc}(\mathbf{r})$ the exchange splitting of, for example, d-states in 3d ferromagnets (=0 inside the barrier), $\vec{\sigma}$ stands for the Pauli matrices; index $i = 1(3)$ for left (right) ferromagnetic electrode FM1(2) and $i = 2$ for tunneling barrier (quantities for the tunnel barrier also marked t), respectively.

We start with the expression for a direct tunnel current density of spin σ from FM1 to FM2 [52]

$$J_{\sigma 0} = \frac{q}{h} \int dE [f(E - F_{\sigma 0}^{FM2}) - f(E - F_{\sigma 0}^{FM1})] \int \frac{d^2 k_\parallel}{(2\pi)^2} T_\sigma(E, \vec{k}_\parallel), \quad (3.24)$$

where $f(x)$ is the Fermi–Dirac distribution function with local Fermi level $F_{\sigma 0}^{FM1(2)}$ for ferromagnetic electrode FM1(2), $T_\sigma = \sum_{\sigma'} T_{\sigma\sigma'}$ the transmission probability from majority (minority) spin subband in FM1 $\sigma = \uparrow$ or $\downarrow$ into majority (minority) spin subband in FM2, $\sigma' = \uparrow (\downarrow)$. It has a particularly simple form for a square barrier and *collinear* [parallel (P) or antiparallel (AP)] moments on electrodes:

$$T_{\sigma\sigma'} = \frac{16 m_1 m_3 m_2^2 k_1 k_3 \kappa^2}{(m_2^2 k_1^2 + m_1^2 \kappa^2)(m_2^2 k_3^2 + m_3^2 \kappa^2)} e^{-2\kappa w}, \quad (3.25)$$

where κ is the attenuation constant for the wavefunction in the barrier $k_1 \equiv k_{1\sigma}$, $k_2 = i\kappa$, $k_3 \equiv k_{3\sigma'}$ are the momenta normal to the barrier for the corresponding spin subbands, w is the barrier width, and we have used a limit of T at $\kappa w \gg 1$ [18]. With the use of Equations 3.17 and 3.18, and accounting for the misalignment of magnetic moments in ferromagnetic terminals (given by the mutual angle θ), we obtain following expression for the junction conductance per unit area, assuming $m_1 = m_3$,

$$G = G_0 (1 + P_1 P_2 \cos\theta), \quad (3.26)$$

$$P_{1(2)} = \frac{k_\uparrow - k_\downarrow}{k_\uparrow + k_\downarrow} \frac{\kappa^2 - m_2^2 k_\uparrow k_\downarrow}{\kappa^2 + m_2^2 k_\uparrow k_\downarrow} \\
= \frac{(v_\uparrow - v_\downarrow)(v_t^2 - v_\uparrow v_\downarrow)}{(v_\uparrow + v_\downarrow)(v_t^2 + v_\uparrow v_\downarrow)}, \quad (3.27)$$

where $P_{1(2)}$ is the effective polarization of the FM1(2) electrode, $\kappa = [2m_2(U_0 - E)/\hbar^2]^{1/2}$, and U_0 is the top of the barrier. Equation 3.26 correct an expression derived earlier [17] for the effective mass of the carriers in the barrier. To obtain the last simple expression for tunnel current polarization Equation 3.27, which has exactly the same form also for FM-semiconductor modified Schottky junctions (below), we have introduced the carrier band velocities $v_{\uparrow(\downarrow)} = \hbar k_{\uparrow(\downarrow)}/m$, that are generally different for FM1(2), and "tunneling" velocity $v_t = \hbar\kappa/m_2$. These relations between the velocity an momentum are equivalent to an *effective mass* approximation.

By taking a typical value of $G_0 = 4$–$5\,\Omega^{-1}\,\text{cm}^{-2}$ (Ref. [15] $k_\uparrow = 1.09\,\text{Å}^{-1}$, $k_\downarrow = 0.42\,\text{Å}^{-1}$, $m_1 \approx 1$ (for itinerant d electrons in Fe) [21] and a typical barrier height for Al_2O_3 (measured from the Fermi level μ) $\phi = U_0 - \mu \approx 3\,\text{eV}$, and the thickness $w \approx 20\,\text{Å}$, one arrives at the following estimate for the effective mass in the barrier: $m_2 \approx 0.4$ [53]. These values give $P_{Fe} = 0.28$, which is noticeably smaller than the experimental value of 0.4–0.5 (note that neglect of the mass correction, $m_2 < 1$, as in Ref. [17], would give a negative value of the effective polarization). Below, we shall see that tunneling

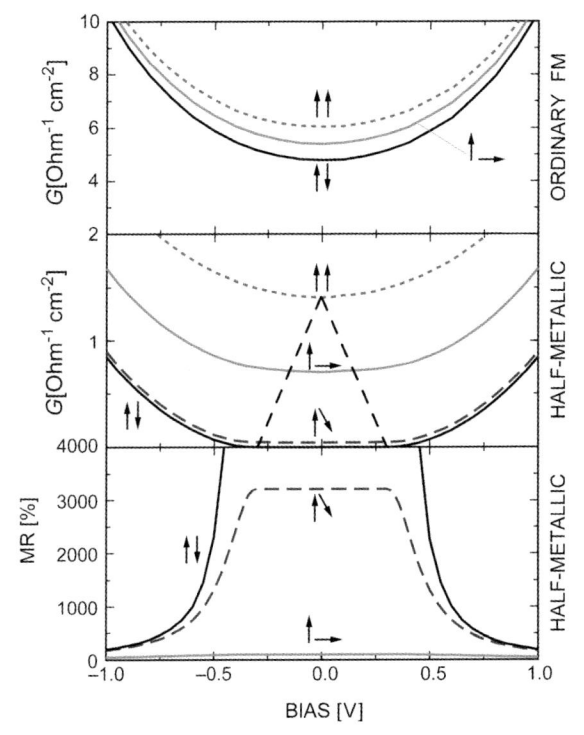

Figure 3.4 Conductance and magnetoresistance of tunnel junctions versus bias. Top panel: conventional (Fe-based) tunnel junction (for parameters, see text). Middle panel: half-metallic electrodes. Bottom panel: magnetoresistance for the half-metallic electrodes. The dashed line shows schematically a region where a gap in the minority spin states is controlling the transport. Even for imperfect antiparallel alignment ($\theta = 160°$, marked ↑↘), the magnetoresistance for half-metallics (bottom panel) exceeds 3000% at biases below the threshold V_c. All calculations have been performed at 300 K, with inclusion of multiple image potential and exact transmission coefficients. Parameters are described in the text.

assisted by polarized surface states may lead to much larger TMR, this may relevant to observed large values of TMR [19].

The most striking feature of Equation 3.26 is that MR tends to infinity for vanishing $k_\downarrow$; that is when the electrodes are made of a 100% spin-polarized material ($P = P' = 1$) because of a gap in the density of states (DOS) for minority carriers up to their conduction band minimum $E_{CB\downarrow}$. Then G^{AP} vanishes together with the transmission probability Equation 3.25, as there is a zero DOS at $E = \mu$ for both spin directions. Such a half-metallic behavior is rare, but some materials possess this amazing property, most interestingly the oxides CrO_2 and Fe_3O_4 (e.g., see recent discussion in Ref. [2]). These oxides are very interesting for future applications in combination with matching materials, as will be seen below.

Remarkably, for $|V| < V_c$ in the AP geometry one has $MR = \infty$. From the middle and the bottom panels in Figure 3.4 we see that even at 20° deviation from the AP configuration, the value of MR exceeds 3000% in the interval $|V| < V_c$, and this is indeed a very large value.

3.4.1
Impurity Suppression of TMR

An important aspect of spin-tunneling is the effect of tunneling through the defect states in the (amorphous) oxide barrier. Dangling bonds and random trap states may play the role of defects in an amorphous barrier. Since the contacts under consideration are typically short, their current–voltage (I–V) curve and MR should be very sensitive to defect resonant states in the barrier with energies close to the Fermi level, forming "channels" with the nearly periodic positions of impurities. Generally, channels with one impurity (most likely to dominate in thin barriers) would result in a monotonous behavior of the I–V curve, whereas channels with two or more impurities would produce intervals with negative differential conductance. Impurity-assisted spin tunneling at zero temperature [the general case of non-zero temperature would require integration with the Fermi factors as in Equation 3.24] can be written in the standard form [54]:

$$G_\sigma = \frac{2e^2}{\pi \hbar} \sum_i \frac{\Gamma_{L\sigma}\Gamma_{R\sigma}}{(E_i - \mu)^2 + \Gamma^2}, \tag{3.28}$$

where $\Gamma_\sigma = \Gamma_{L\sigma} + \Gamma_{R\sigma}$ is the total width of a resonance given by a sum of the partial widths $\Gamma_{L(R)}$ corresponding to electron tunneling from the impurity state at the energy E_i to the left (right) terminal. It is easiest to analyze the case of parallel (P) and antiparallel (AP) mutual orientation of magnetic moments M_1 and M_2 on electrodes with the angle θ between them. In this case, one looks at tunneling of majority (maj) and minority (min) carriers from the left electrode $L_\sigma = (L_{maj}, L_{min})$ into states $R_\sigma = (R_{maj}, R_{min})$ for parallel orientation ($\theta = 0$) or $R_\sigma = (R_{min}, R_{maj})$ in antiparallel orientation ($\theta = \pi$), respectively. The general case is then easily obtained from standard spinor algebra for spin projections. The tunnel widths can be evaluated analytically for a rectangular barrier, $\Gamma_{L\sigma} \sim g_{L\sigma}\Omega \exp[-\kappa(w + 2z_i)]$, where z_i is the coordinate of the impurity with respect to the center of the barrier, $g_{L\sigma}$ the density of states in the (left) electrode, Ω [18].

The resonant conductance Equation 3.28 has a sharp maximum $[= e^2/(2\pi\hbar)]$ when $\mu = E_i$ and $\Gamma_L = \Gamma_R$, that is for the symmetric position of the impurity in the barrier for parallel configuration. For antiparallel configuration, most effective impurities will be positioned somewhat off-center since the DOS for the majority and minority spins may be quite different. An asymmetric position of effective impurities in the AP orientation immediately suggests smaller conductance G_{AP} than G_P and *positive* ("normal") impurity TMR > 0. This result is confirmed by direct calculation. Indeed, if we assume that we have ν defect/localized levels in a unit volume and unit energy interval in a barrier, then, replacing the sum by an integral in Equation 3.28, and considering a general configuration of the magnetic moments on terminals, we obtain the following formula for impurity-assisted conductance per unit area in leading order in $\exp(-\kappa w)$:

$$G_1 = g_1(1 + \Pi_L \Pi_R \cos\theta), \tag{3.29}$$

where we have introduced the quantities

$$g_1 = \frac{e^2}{\pi\hbar} N_1, \quad N_1 = \pi^2 \nu \Gamma_1 / \kappa,$$

$$\Pi_{L(R)} = \frac{(r_\uparrow - r_\downarrow)}{(r_\uparrow + r_\downarrow)}, \quad (3.30)$$

N_1 being the effective number of one-impurity channels per unit area, and one may call Π_F a "polarization" of the impurity channels, defined by the factor $r_\sigma = [m_2 \kappa k_\sigma / (\kappa^2 + m_2^2 k_\sigma^2)]^{1/2}$ with momenta k_σ for left (right) [L (R)] electrode.

Comparing direct and impurity-assisted contributions to conductance, we see that the latter dominates when the density of localized states $\nu \gtrsim (\kappa/\pi)^3 \epsilon_i^{-1} \exp(-\kappa w)$, and in our example a crossover takes place at the density of localized states $\nu \gtrsim 10^{17}$ eV^{-1} cm^{-3}. When resonant transmission dominates, the magnetoresistance will be given by

$$MR_1 = \frac{2\Pi_L \Pi_R}{(1 - \Pi\Pi')}, \quad (3.31)$$

which is only 4% in the case of Fe. We see that indeed MR_1 is suppressed yet remains positive (unless the polarization of tunnel carriers is opposite to the magnetization direction on one of the electrodes, in this case MR is obviously inverted for trivial reasons). There are speculations about a possibility of negative MR_1, which is analyzed below in the following subsection.

We have estimated the above critical DOS for localized states for the case of Al$_2$O$_3$ barrier, in systems such as amorphous Si the density of localized states is higher because of considerably smaller band gap, and estimated as 8×10^{18} eV^{-1} cm^{-3}, mainly due to dangling bonds and band edge smearing because of disorder [55]. One can appreciate that in junctions with thin Al$_2$O$_3$ amorphous barriers (<20–25 Å) of practical interest the impurity-assisted tunneling is not the major effect, so the above consideration of elastic tunneling applies. In this seminal work, Beasley have studied a-Si barriers with wide variety of thicknesses $w = 30$–1000 Å and obtained detailed data on crossover from direct tunneling to directed inelastic hopping along statistically rare, yet highly conductive, chains of localized states. The crossover thicknesses depend heavily on the materials parameters of the barrier. The above-described suppression of TMR by impurities was confirmed experimentally for magnetic tunnel junctions in Ref. [56].

3.4.2
Negative Resonant TMR?

It should be noted that the MR becomes suppressed yet remains positive. Indeed, the conductance is dominated, but can it change sign, or become *inverted* in the case of impurity assisted tunneling? It was shown above that the asymetry of polarized DOS in contacts gives *positive* resonant tunneling TMR. Negative (inverse MR_1) can only appear if the dominating impurity levels were lined up with the Fermi level [Equation 3.28] and positioned in asymmetric positions in the barrier, with for example, the "right" width of the resonance much larger than the "left" width, $\Gamma_{R\sigma} \equiv \Gamma_\sigma \gg \Gamma_{L\sigma} \equiv \gamma_\sigma$

$$G_\sigma \approx \frac{e^2}{\pi\hbar} \frac{\gamma_\sigma}{\Gamma_\sigma},$$

$$\text{TMR} \sim -P_1 P_2, \, (?) \tag{3.32}$$

the latter was noted in Ref. [57]. However, the required coincidence is statistically very unlikely in tunnel junctions, where the number of impurity states involved is $\gg 1$, as in all usual situations with a possible exception of very small area tunnel junctions. Indeed, the data in Ref. [57] suggest that, in a tiny percentage of small area junctions, the TMR is negative. The attempts to simulate the amorphous barrier that may produce such a result in resonant tunneling regime showed, however, that one needs an unphysically large amount of disorder in the barrier to obtain traces of negative TMR. Indeed, for the barrier with height $U = 1.5$ eV, an unphysical amount of onsite disorder $\gamma = 4U = 6$ eV should be assumed. It must be concluded that the speculations about negative resonant TMR in Ref. [57] have nothing to do with most observations of inverse TMR. Averaging over disorder suppresses TMR, as predicted in Ref. [18] and observed in for example Ref. [56]. It is noted, however, that is the case when the impurity states are located at a particular interface in the barrier, perhaps as in tunnel junctions with composite barriers MgO/NiO [58], there may be a suppression and a slight inversion of TMR in a certain window of bias voltages, given by the energy interval occupied by the interfacial states, as described elsewhere [59].

3.4.3
Tunneling in Half-Metallic Ferromagnetic Junctions

Now we shall discuss a couple of systems with half-metallic behavior, CrO_2/TiO_2 and CrO_2/RuO_2 (Figure 3.5). These are based on half-metallic CrO_2, and all species have the rutile structure type with almost perfect lattice matching, which should yield a good interface and should help in keeping the system at the desired stoichiometry. TiO_2 and RuO_2 are used as the barrier/spacer oxides. The electronic structure of CrO_2/TiO_2 is truly stunning in that it has a half-metallic gap which is 2.6 eV wide and extends on both sides of the Fermi level, where there is a gap either in the minority *or* majority spin band. Thus, a huge magnetoresistance should, in principle, be seen not only for electrons at the Fermi level biased up to 0.5 eV, but also for *hot* electrons starting at about 0.5 eV above the Fermi level. We note that states at the Fermi level are a mixture of $Cr(d)$ and $O(2p)$ states, so that p–d interaction within the first coordination shell produces a strong hybridization gap, and the Stoner spin-splitting moves the Fermi level right into the gap for minority carriers (Figure 3.5). It is worth noting that CrO_2 and RuO_2 are very similar in terms of a paramagnetic band structure, but the difference in the number of conduction electrons and exchange splitting results in a usual metallic behavior of RuO_2 as compared to the half-metallic ferromagnet CrO_2.

An important difference between two spacer oxides is that TiO_2 is an insulator whereas RuO_2 is a good metallic conductor. Thus, the former system can be used in a

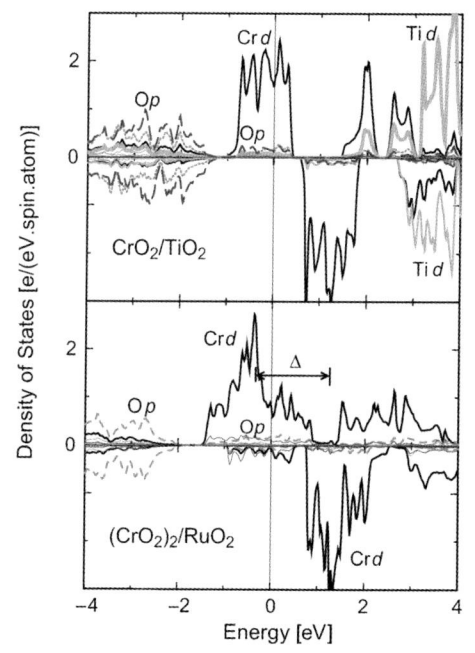

Figure 3.5 Density of states of CrO_2/TiO_2 (top panel) and $(CrO_2)_2/RuO_2$ (bottom panel) half-metallic multilayers calculated with the use of the LMTO method. The partial contributions are indicated by letters. The zero of energy corresponds to the Fermi level. Δ indicates a spin-splitting of the Cr d band near E_F (schematic). Note a strong hybridization of Cr d with O $2p$ states at E_F and below the hybridization gap. Growth direction is [001].

tunnel junction, whereas the latter will form a metallic multilayer. In the latter case the physics of conduction is different from tunnelling, but the effect of vanishing phase volume for transmitted states still works when current is passed through such a system *perpendicular to planes*. One interesting possibility is to form three-terminal devices with these systems, like a spin-valve transistor [60], and check the effect in a hot-electron region. CrO_2/TiO_2 seems to a be a natural candidate to check the present predictions about half-metallic behavior and for a possible record tunnel magnetoresistance. One important advantage of these systems is an almost perfect lattice match at the oxide interfaces. The absence of such a match of the conventional Al_2O_3 barrier with Heussler half-metallics (NiMnSb and PtMnSb) may have been among other reasons for their unimpressive performance [2]. The main concerns for achieving a very large value of magnetoresistance will be spin-flip centers, magnon-assisted events, and imperfect alignment of moments. As for conventional tunnel junctions, the present results show that presence of defect states in the barrier, or a resonant state, as in a resonant tunnel diode-type of structure, reduces their magnetoresistance several fold but may dramatically increase the current through the structure.

3.4.4
Surface States Assisted TMR

Direct tunneling, as we have seen, gives a TMR of about 30%, whereas in recent experiments TMR is well above this value, approaching 40–50% in systems with Al_2O_3 amorphous barrier, and 200% in systems with epitaxial MgO barriers [16, 52]. It will become clear below, that this enhancement is unlikely to come from the inelastic processes. Until now, we have disregarded the possibility of localized states at metal–oxide interfaces. Bearing in mind that the usual barrier AlO_x is amorphous, the density of such surface states may be high, and we must take into account tunneling into/from those states. The results for Tamm states that may exist at clean interfaces, are similar. The corresponding tunneling conductance per unit area is [19]:

$$G_s(\theta) = \frac{e^2}{\pi \hbar} B \bar{D}_s (1 + P_F P_s \cos \theta),$$

$$P_s = \frac{D_{s\uparrow} - D_{s\downarrow}}{D_{s\uparrow} + D_{s\downarrow}}, \quad \bar{D}_s = \frac{1}{2}(D_{s\uparrow} + D_{s\downarrow}), \tag{3.33}$$

where P_s is the polarization and $\bar{D}s$ is the average density of surface states, and θ is the mutual angle between moments on electrodes. The parameter $B \sim [2\pi\hbar^2 m\kappa/(m_2^2 w)]\exp(-2\kappa w)$, where w is the barrier width, κ is the absolute value of electron momentum under the barrier, m and m_2 are the free electron mass and the effective mass in the barrier, respectively. The corresponding magnetoresistance would be $MR_s = 2P_F P_s/(1 - P_F P_s)$. It is easy to show that the bulk-to-surface conductance exceeds the bulk-to-bulk one at densities of surface states $D_s > D_{sc} \sim 10^{13} \, cm^{-2} \, eV^{-1}$ per spin, comparable to those found at some metal–semiconductor interfaces. Since this result was obtained, various groups confirmed this with *ab-initio* calculations, usually without mentioning the original result obtained in Ref. [19].

If on both sides of the barrier the density of surface states is above the critical value D_{sc}, the magnetoresistance would be due to surface-to-surface tunneling with a value given by $MR_{ss} = 2P_{s1}P_{s2}/(1 - P_{s1}P_{s2})$. If the polarization of surface states is larger than that of the bulk, as is often the case even for imperfect surfaces [61], then it would result in enhanced TMR.

3.4.5
Inelastic Effects in TMR

Inelastic processes with excitation of magnon or phonon modes introduce new energy scales into the problem (30–100 meV) which correspond to a region where unusual I–V tunnel characteristics are seen (Figure 3.6). One can describe their effect on TMR fairly well within the tunnel Hamiltonian approach [19]. We obtain for magnon-assisted inelastic tunneling current at $T = 0$ with the use of tunnel Hamiltonian formalism [62]:

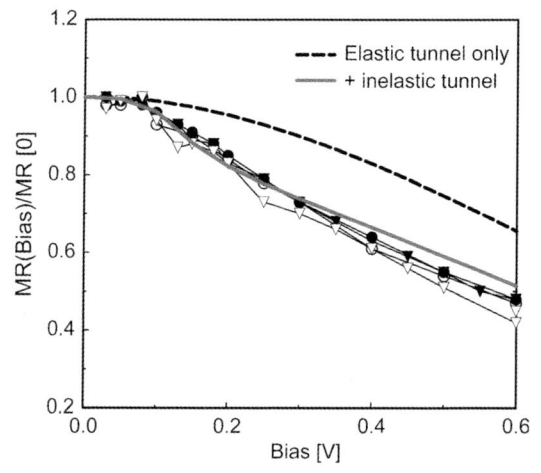

Figure 3.6 Fit to experimental data for the magnetoresistance of CoFe/Al$_2$O$_3$/NiFe tunnel junctions [9] with inclusion of elastic and inelastic (magnons and phonons) tunneling. The fit gives for magnon DOS $\propto \omega^{0.65}$, which is close to a standard bulk spectrum $\propto \omega^{1/2}$.

$$I_P^x = \frac{2\pi e}{\hbar} \sum_\alpha X^\alpha g_\downarrow^L g_\uparrow^R \int d\omega \rho_\alpha^{mag}(\omega)(eV-\omega)\theta(eV-\omega),$$

$$I_{AP}^x = \frac{2\pi e}{\hbar} \Big[X^R g_\uparrow^L g_\uparrow^R \int d\omega \rho_R^{mag}(\omega)(eV-\omega)\theta(eV-\omega) \\ + X^L g_\downarrow^L g_\downarrow^R \int d\omega \rho_L^{mag}(\omega)(eV-\omega)\theta(eV-\omega) \Big],$$

(3.34)

where X is the incoherent tunnel exchange vertex, $\rho_\alpha^{mag}(\omega)$ is the magnon density of states that has a general form $\rho_\alpha^{mag}(\omega) = (\nu+1)\omega^\nu/\omega_0^{\nu+1}$, the exponent ν depends on a type of spectrum, ω_0 is the maximum magnon frequency, $g_{L\,(R)}$ marks the corresponding electron density of states on left (right) electrode, $\theta(x)$ is the step function, $\alpha = L, R$. The analogous expressions can be written down for phonon contribution with the important distinction that electron–phonon interaction does *not* affect spin (if one ignores any small magnetoelastic contribution). The elastic and inelastic contributions together will define the total junction conductance $G = G(V,T)$ as a function of the bias V and temperature T. We find that the inelastic contributions from magnons Equation 3.34 and phonons grow as $G^x(V,0) \propto (|eV|/\omega_0)^{\nu+1}$ and $G^{ph}(V,0) \propto (eV/\omega_D)^4$ at low biases. These contributions saturate at higher biases: $G^x(V,0) \propto 1 - \frac{i+1}{\nu+2}\frac{\omega_0}{|eV|}$ at $|eV| > \omega_0$; $G^{ph}(V,0) \propto 1 - \frac{4}{5}\frac{\omega_D}{|eV|}$ at $|eV| > \omega_D$. This behavior would lead to sharp features in the I–V curves on a scale of 30–100 mV (Figure 3.6).

It is important to highlight the opposite effects of phonons and magnons on the TMR. If we take the case of the same electrode materials and denote $D = g_\uparrow$ and $d = g_\downarrow$, then we see that $G_P^x(V,0) - G_{AP}^x(V,0) \propto -(D-d)^2(|eV|/\omega_0)^{\nu+1} < 0$, whereas $G_P^{ph}(V,0) - G_{AP}^{ph}(V,0) \propto +(D-d)^2(eV/\omega_D)^4 > 0$; that is, spin-mixing due to magnons *kills* the TMR, whereas the phonons tend to reduce the negative effect of magnon emission [63]. Different bias and temperature dependence can make possible a separation of these two contributions, which are of opposite sign. At finite temperatures we obtain the contributions of the same respective sign as above. For magnons: $G_P^x(0,T) - G_{AP}^x(0,T) \propto -(D-d)^2(-TdM/dT) < 0$, where $M = M(T)$ is the magnetic moment of the electrode at a given temperature T. The phonon contribution is given by a standard Debye integral with the following results: $G_P^{ph}(0,T) - G_{AP}^{ph}(0,T) \propto +(D-d)^2(T/\omega_D)^4 > 0$ at $T \ll \omega_D$, and $G_P^{ph}(0,T) - G_{AP}^{ph}(0,T) \propto +(D-d)^2(T/\omega_D)$ at $T \gtrsim \omega_D$. It is worth mentioning that the magnon excitations are usually cut off, for example, by the anisotropy energy K_{an} at some ω_c. Therefore, at low temperatures the conductance at small biases will be almost constant. We conclude that the inelastic processes are responsible for TMR diminishing with bias voltage, the unusual shape of the I–V curves at low biases, and their temperature behavior, which is also affected by impurity-assisted tunneling. The surface states-assisted tunneling may lead to enhanced TMR, if their polarization is higher than that of the bulk. This could open up ways to improving performance of ferromagnetic tunnel junctions.

3.5
Spin Injection/Extraction into (from) Semiconductors

Much attention has been devoted recently to exploring the possibility of a three-terminal spin injection device where spin is injected into semiconductor from either metallic ferromagnetic electrode [67, 68], or from magnetic semiconductor electrode, as demonstrated in Ref. [64]. However, the magnetization in FMS usually vanishes or is too small at room temperature. Relatively high spin injection from ferromagnets (FM) into non-magnetic semiconductors (S) has recently been demonstrated at low temperatures [65], and the attempts to achieve an efficient room-temperature spin injection have faced substantial difficulties [66]. Theoretical studies of the spin injection from ferromagnetic metals, as initiated in Refs. [68, 69], have been subject of extensive research in Refs. [5–10, 69–77] that has gained much insight into the problem of spin injection/accumulation in semiconductors. As a consequence, some suggestions for *spin transistors* and other spintronic devices have appeared that are experimentally realizable, can work at room temperatures, and exceed the parameters of standard semiconductor analogues [5, 6].

As an important distinction with spin transport in magnetic tunnel junctions, one would like to create non-equilibrium spin polarization and manipulate it with external fields in semiconductors, with a possible advantage of long spin relaxation time in comparison with mean collision time. In order to be interesting for applications, there should be a straightforward method of creating substantial non-equilibrium spin polarization *density* in semiconductor. This is different from

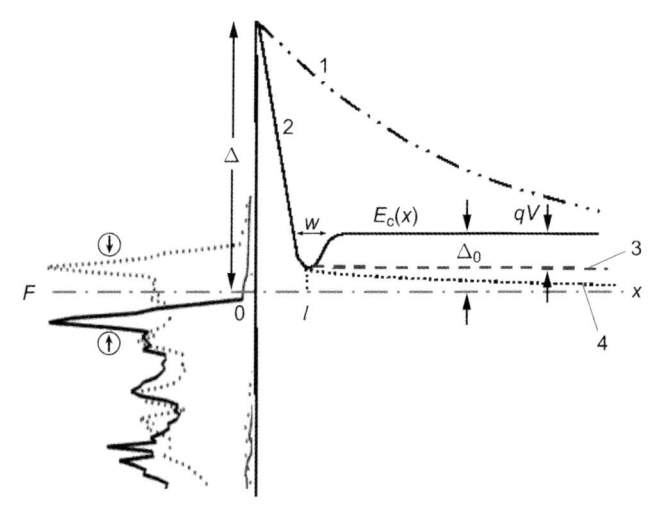

Figure 3.7 Energy diagrams of ferromagnet-semiconductor heterostructure with δ-doped layer (F is the Fermi level; Δ the height and l the thickness of an interface potential barrier; Δ_0 the height of the thermionic barrier in n-semiconductor). The standard Schottky barrier (curve 1); $E_c(x)$ the bottom of conduction band in n-semiconductor in equilibrium (curve 2), under small (curve 3), and large (curve 4) bias voltage. The spin-polarized density of states in Ni is shown at $x < 0$.

tunnel junctions, where one is interested in large spin injection efficiency; that is, in large resistance change with respect to magnetic configuration of the electrodes. Obviously, a spin imbalance in the drain ferromagnetic electrode is created in MTJ, proportional to the current density, but relatively minute, given a huge density of carriers in a metal. The principal difficulty of *spin injection* in semiconductor from ferromagnet is that the materials in the FM-S junctions usually have very different electron affinity and, therefore, high Schottky barrier forms at the interface [78] (Figure 3.7, curve 1). Thus, in GaAs and Si the barrier height $\Delta \simeq 0.5$–0.8 eV with practically all metals, including Fe, Ni, and Co [65, 78], and the barrier width is large, $l \gtrsim 100$ nm for doping concentration $N_d \lesssim 10^{17}$ cm^{-3}. The spin-injection corresponds to a reverse current in the Schottky contact, which is saturated and usually negligible due to such large l and Δ [78]. Therefore, a thin heavily doped $n^+ - S$ layer between FM metal and S is used to increase the reverse current [78] determining the spin-injection [6, 8, 65, 72]. This layer sharply reduces the thickness of the barrier, and increases its tunneling transparency [6, 78]. Thus, a substantial spin injection has been observed in FM-S junctions with a thin n^+–layer [65].

One usually overlooked formal paradox of spin injection is that a current through Schottky junctions (as derived in textbooks) depends solely on parameters of a semiconductor [78], and cannot formally be spin-polarized. Some authors even emphasize that in Schottky junctions "spin-dependent effects do not occur" [70]. In earlier reports [67–76], spin transport through FM-S junction, its spin-selective properties, and non-linear I–V characteristics have not been actually calculated. They were described by various, often contradictory, boundary conditions at the FM-S interface. For example, Aronov and Pikus assumed that *spin injection efficiency* of a FM-S interface is a constant equal to that in the ferromagnet FM, $\Gamma = \Gamma_F$, [Equation 3.3], and then studied non-linear spin accumulation in S considering spin diffusion and drift in an electric field [67]. The authors of Refs. [68–72] assumed a continuity of both the currents and the electrochemical potentials for both spins, and found that a spin polarization of injected electrons depends on a ratio of conductivities of a FM and S (the so-called "conductivity mismatch" problem). At the same time, it has been asserted in Refs. [73–76] that the spin injection becomes appreciable when the electrochemical potentials have a substantial discontinuity (produced e.g., by a tunnel barrier [74]). The effect, however, was described by the unknown spin-selective interface conductances $G_{i\sigma}$, which cannot be found within those theories.

We have developed a microscopic theory of the spin transport through ferromagnet-semiconductor junctions, which include an ultrathin heavily doped semiconductor layer (δ-doped layer) between FM and S [6, 8]. We have studied the non-linear effects of spin accumulation in S near reverse-biased modified FM-S junctions with the δ-doped layer [6] and spin extraction from S near the modified forward-biased FM-S junctions [8]. We found conditions for the most efficient spin injection, which are *opposite* to the results of previous phenomenological theories. We show, in particular, that: (i) the current of the FM-S junction does depend on spin parameters of the ferromagnetic metal but *not* its conductivity, so, contrary to the results [68–72, 74–76], the "conductivity mismatch" problem *does not exist* for the Schottky FM-S junctions. We find also that: (ii) a spin injection efficiency Γ of the FM-S junction depends strongly on the current, contrary to the assumptions in [67–72, 74–76]; and (iii) the highest spin polarization of both the injected electrons P and spin injection efficiency can be realized at room temperatures and relatively small currents in *high-resistance* semiconductors, *contrary* to claims in Ref. [71], which are of most interest for spin injection devices [3, 4, 6]. We also show that: (iv) tunneling resistance of the FM-S junction must be relatively small, which is *opposite* to the condition obtained in linear approximation in Ref. [74]; and that (v) the spin-selective interface conductances $G_{i\sigma}$ are not constants, as was assumed in Ref. [73–76], but vary with a current J in a strongly non-linear fashion. We have suggested a new class of spin devices on the basis of the present theory.

Below, we describe a general theory of spin current, spin injection and extraction in Section 3.5.1, followed by the discussion of the conditions of an efficient spin injection and extraction in Section 3.5.2. Further, we turn to the discussion of high-frequency spin valve effect in a system with two δ-doped Schottky junctions in Section 3.5.3 A new class of spin devices is detailed in Section 3.5.3 field detector, spin transistor, and square-law detector. The efficient spin injection and extraction may be a basis for efficient sources of (modulated) polarized radiation, as mentioned in Section 3.5.4.

3.5.1
Spin Tunneling through Modified (Delta-Doped) Schottky Barrier

The modified FM-S junction with transparent Schottky barrier is produced by δ-doping the interface by sequential donor and acceptor doping. The Schottky barrier is made very thin by using large donor doping N_d^+ in a thin layer of thickness l. For reasons which will become clear shortly, we would like to have a narrow spike followed by the narrow potential well with in the width w and the depth $\sim rT$, where T is the temperature in units of $k_B = 1$ and $r \sim 2$–3, produced by an acceptor doping N_a^+ of the layer w (Figure 3.7). Here $l \lesssim l_0$, where $l_0 = \sqrt{\hbar^2/[2m_*(\Delta - \Delta_0)]}$ ($l_0 \lesssim 2$ nm), the remaining low (and wide) barrier will have the height $\Delta_0 = (E_{c0} - F) > 0$, where E_{c0} is the bottom of the conduction band in S in equilibrium, q the elementary charge, and ϵ (ϵ_0) the dielectric permittivity of S (vacuum). A value of Δ_0 can be set by choosing a donor concentration in S,

$$N_d = N_c \exp\left[\frac{(F^S - E_{c0})}{T}\right] = N_c \exp\left(\frac{-\Delta_0}{T}\right) = n, \qquad (3.35)$$

where F^S is the Fermi level in the semiconductor bulk, $N_c = 2M_c(2\pi m_* T)^{3/2} h^{-3}$ the effective density of states and M_c the number of effective minima of the semiconductor conduction band; n and m_* the concentration and effective mass of electrons in S [79]. Owing to small barrier thickness l, the electrons can rather easily tunnel through the δ-spike, but only those with an energy $E \geq E_c$ can overcome the wide barrier Δ_0 due to thermionic emission, where $E_c = E_{c0} + qV$. We assume here the standard convention that the bias voltage $V < 0$ and current $J < 0$ in the reverse-biased FM-S junction and $V > 0$ ($J > 0$) in the forward-biased FM-S junction [78]. At positive bias voltage $V > 0$, we assume that the bottom of conduction band shifts upwards to $E_c = E_{c0} + qV$ with respect to the Fermi level of the metal. Presence of the mini-well allows to keep the thickness of the δ-spike barrier equal to $l \lesssim l_0$ and its transparency high at voltages $qV \lesssim rT$ (see below).

The calculation of current through the modified barrier is rather similar to what has been done in the case of tunnel junctions above, with a distinction that in the present case the barrier is triangular, (Figure 3.7). We again assume elastic coherent tunneling, so that the energy E, spin σ, and $\vec{k}_{\|}$ (the component of the wave vector $\vec{k}$ parallel to the interface) are conserved. The exact current density of electrons with spin $\sigma = \uparrow, \downarrow$ through the FM-S junction containing the δ-doped layer (at the point $x = l$; Figure 3.7) is written similarly to Equaion 3.24 as:

$$J_{\sigma 0} = \frac{q}{h}\int dE[f(E - F_{\sigma 0}^S) - f(E - F_{\sigma 0}^{FM})]\int \frac{d^2 k_{\|}}{(2\pi)^2} T_\sigma, \qquad (3.36)$$

where $F_{\sigma 0}^S$ ($F_{\sigma 0}^{FM}$) are the spin quasi-Fermi levels in the semiconductor (ferromagnet) near the FM-S interface, and the integration includes a summation with respect to a band index. Note that here we study a strong *spin accumulation* in the semiconductor. Therefore, we use *nonequilibrium* Fermi levels, $F_{\sigma 0}^{FM}$ and $F_{\sigma 0}^S$, describing distributions of electrons with spin $\sigma = \uparrow, \downarrow$ in the FM and the S, respectively, which is especially important for the semiconductor. In reality, due to very high electron density in FM metal in comparison with electron density in S, $F_{\sigma 0}^{FM}$ differs negligibly from the equilibrium Fermi level F for currents under consideration; therefore, we can assume that $F_{\sigma 0}^{FM} = F$, as in Refs. [18, 52] (see discussion below).

The current Equation 3.36 should generally be evaluated numerically for a complex band structure $E_{k\sigma}$ [79]. The analytical expressions for $T_\sigma(E, k_\parallel)$ can be obtained in an effective mass approximation, $\hbar k_\sigma = m_\sigma v_\sigma$, where $v_\sigma = |\nabla E_{k\sigma}|/\hbar$ is the band velocity in the metal. The present Schottky barrier has a "pedestal" with a height $(E_c - F) = \Delta_0 + qV$, which is opaque at energies $E < E_c$. For $E > E_c$ we approximate the δ-barrier by a triangular shape and one can use an analytical expression for $T_\sigma(E, k_\parallel)$ [5] and find the spin current at the bias $0 < -qV \lesssim rT$, including at room temperature,

$$J_{\sigma 0} = j_0 d_\sigma \left[\frac{2n_{\sigma 0}(V)}{n} - \exp\left(-\frac{qV}{T}\right) \right], \tag{3.37}$$

$$j_0 = \alpha_0 nq v_T \exp(-\eta \kappa_0 l). \tag{3.38}$$

with the most important spin factor

$$d_\sigma = \frac{v_T v_{\sigma 0}}{v_{t0}^2 + v_{\sigma 0}^2}. \tag{3.39}$$

Here $\alpha_0 = 1.2(\kappa_0 l)^{1/3}$, $\kappa_0 \equiv 1/l_0 = (2m_*/\hbar^2)^{1/2}(\Delta - \Delta_0 - qV)^{1/2}$, $v_{t0} = \sqrt{2(\Delta - \Delta_0 - qV)/m_*}$ is the characteristic "tunnel" velocity, $v_\sigma = v_\sigma(E_c)$ the velocity of polarized electrons in FM with energy $E = E_c$, $v_T = \sqrt{3T/m_*}$ the thermal velocity. At larger reverse bias the miniwell on the right from the spike in Figure 3.7 disappears and the current practically saturates. Quite clearly, the tunneling electrons incident almost normally at the interface and contribute most of the current (a more careful sampling can be done numerically [79]).

One can see from Equation 3.30 that the total current $J = J_{\uparrow 0} + J_{\downarrow 0}$ and its spin components $J_{\sigma 0}$ depend on a conductivity of a semiconductor but *not* a ferromagnet, as in usual Schottky junction theories [79]. On the other hand, $J_{\sigma 0}$ is proportional to the spin factor d_σ and the coefficient $j_0 d_\sigma \propto v_T^2 \propto T$, but not the usual Richardson's factor T^2 [78]. Equation 3.37, for current in the FM-S structure, is valid for any sign of the bias voltage V. Note that at $V > 0$ (forward bias) it determines the spin current from S into FM. Hence, it describes *spin extraction* from S [8].

Following the pioneering studies of Aronov and Pikus [67], one customarily assumes a boundary condition $J_{\uparrow 0} = (1+\Gamma_F)J/2$. Since there is a spin accumulation in S near the FM-S boundary, the density of electrons with spin σ in the semiconductor is $n_{\sigma 0} = n/2 + \delta n_{\sigma 0}$, where $\delta n_{\sigma 0}$ is a non-linear function of the current J, and $\delta n_{\sigma 0} \propto J$ at small current [67] (see also below). Therefore, the larger J the higher the $\delta n_{\sigma 0}$ and the smaller the current $J_{\sigma 0}$ [see Equation 3.37]. In other words, a type of negative feedback is realized, which decreases the spin injection efficiency Γ and

3.5 Spin Injection/Extraction into (from) Semiconductors

makes it a non-linear function of J (see below). We show that the spin injection efficiency, Γ_0, and the polarization, $P_0 = [n_\uparrow(0) - n_\downarrow(0)]/n$ in the semiconductor near FM-S junctions essentially differ, and that both are small at small bias voltage V (and current J) but *increase* with the current up to P_F. Moreover, P_F can essentially differs from Γ_F, and may ideally approach 100%.

The current in a spin channel σ is given by the standard drift-diffusion approximation [67, 76]:

$$J_\sigma = q\mu n_\sigma E + qD\nabla n_\sigma, \tag{3.40}$$

where E is the electric field; and D and μ are the diffusion constant and mobility of the electrons respectively. D and μ do not depend on the electron spin σ in the non-degenerate semiconductors. From current continuity and electroneutrality conditions

$$J(x) = \sum_\sigma J_\sigma = \text{const}, \quad n(x) = \sum_\sigma n_\sigma = \text{const}, \tag{3.41}$$

we find

$$E(x) = J/q\mu n = \text{const}, \quad \delta n_\downarrow(x) = -\delta n_\uparrow(x). \tag{3.42}$$

Since the *injection* of spin-polarized electrons from FM into S corresponds to a reverse current in the Schottky FM-S junction, one has $J < 0$, and $E < 0$ Figure 3.7. The spatial distribution of density of electrons with spin σ in the semiconductor is determined by the continuity equation [67, 71]

$$\nabla J_\sigma = \frac{q\delta n_\sigma}{\tau_s}, \tag{3.43}$$

where in the present one-dimensional case $\nabla = d/dx$. With the use of Equations 3.40 and 3.42, we obtain the equation for $\delta n_\uparrow(x) = -\delta n_\downarrow(x)$ [67, 76]. Its solution satisfying a boundary condition $\delta n_\uparrow \to 0$ at $x \to \infty$, is

$$\delta n_\uparrow(x) = C\frac{n}{2}\exp\left(-\frac{x}{L}\right) \equiv P_{n0}\exp\left(-\frac{x}{L}\right), \tag{3.44}$$

$$L_{\text{inject (extract)}} = \frac{1}{2}\left[\sqrt{L_E^2 + 4L_s^2} + (-)L_E\right] = \frac{L_s}{2}\left(\sqrt{\frac{J^2}{J_S^2} + 4} - \frac{J}{J_S}\right), \tag{3.45}$$

where the plus (minus) sign refers to forward (reverse) bias on the junction, $L_s = \sqrt{D\tau_s}$ is the usual spin-diffusion length, $L_E = \mu|E|\tau_s = L_s|J|/J_s$ the spin-drift length. Here we have introduced the characteristic current density

$$J_S \equiv qDn/L_s, \tag{3.46}$$

and the plus and minus signs in the expression for the spin penetration depth L Equation 3.45 refer to the spin *injection* at a reverse bias voltage, $J < 0$, and spin *extraction* at a forward bias voltage, $J > 0$, respectively. Note that $L_{\text{inject}} > L_{\text{extract}}$, and the spin penetration depth for injection increases with current, at large currents, $|J| \gg J_S$, $L_{\text{inject}} = L_s|J|/J_s \gg L_s$, whereas $L_{\text{extract}} = L_sJ_s/J \ll L_s$.

The degree of spin polarization of non-equilibrium electrons (i.e., a spin *accumulation* in the semiconductor near the interface) is given simply by the parameter C in Equation 3.37:

$$C = \frac{n_\uparrow(0) - n_\downarrow(0)}{n} = P(0) \equiv P_0. \qquad (3.47)$$

By substituting Equation 3.44 into Equations 3.40 and 3.37, we find

$$J_{\uparrow 0} = \frac{J}{2}\left(1 + P_0 \frac{L}{L_E}\right) = \frac{J}{2}\frac{(1+P_F)(\gamma - P_0)}{\gamma - P_0 P_F}, \qquad (3.48)$$

where $\gamma = \exp(-qV/T) - 1$. From Equation 3.41, one obtains a quadratic equation for $P_n(0)$ with a physical solution that can be written fairly accurately as

$$P_0 = \frac{P_F \gamma L_E}{\gamma L + L_E}. \qquad (3.49)$$

By substituting Equation 3.49 into Equation 3.37, we find for the total current $J = J_{\uparrow 0} + J_{\downarrow 0}$:

$$J = -J_m \gamma = -J_m(e^{-qV/T} - 1), \qquad (3.50)$$

$$J_m = \alpha_0 n q v_T (1 - P_F^2)(d_{\uparrow 0} + d_{\downarrow 0}) e^{-\eta \kappa_0 l}, \qquad (3.51)$$

for the bias range $|qV| \lesssim rT$. The sign of the Boltzmann exponent is unusual because we consider the tunneling thermoemission current in a modified barrier. Obviously, we have $J > 0$ (<0) when $V > 0$ (<0) for forward (reverse) bias.

We notice that at a reverse bias voltage $-qV \simeq rT$ the shallow potential mini-well vanishes, and $E_c(x)$ takes the shape shown in Figure 3.7 (curve 3). For $-qV > rT$, a wide potential barrier at $x > l$ (in S behind the spike) remains flat (characteristic length scale $\gtrsim 100$ nm at $N_d \lesssim 10^{17}$ cm^{-3}), as in usual Schottky contacts [78]. Therefore, the current becomes weakly dependent on V, since the barrier is opaque for electrons with energies $E < E_c - rT$ (curve 4). Thus, Equation 3.50 is valid only at $-qV \lesssim rT$ and the reverse current at $-qV \gtrsim rT$ practically saturates at the value

$$J_{\text{sat}} = qn\alpha_0 v_T(d_{\uparrow 0} + d_{\downarrow 0})(1 - P_F^2)\exp(r - \eta\kappa_0 l). \qquad (3.52)$$

With the use of Equations 3.50 and 3.45, we obtain from Equation 3.42 the spin polarization of electrons near FM-S interface,

$$P_0 = -P_F \frac{2J}{2J_m + \sqrt{J^2 + 4J_S^2} - J}. \qquad (3.53)$$

The spin injection efficiency at FM-S interface is, using Equations 3.30, 3.41, 3.38 and 3.46,

$$\Gamma_0 \equiv \frac{J_{\uparrow 0} - J_{\downarrow 0}}{J_{\uparrow 0} + J_{\downarrow 0}} = P_0 \frac{L}{L_E} = -P_F \frac{\sqrt{4J_S^2 + J^2} - J}{2J_m + \sqrt{J^2 + 4J_S^2} - J}. \qquad (3.54)$$

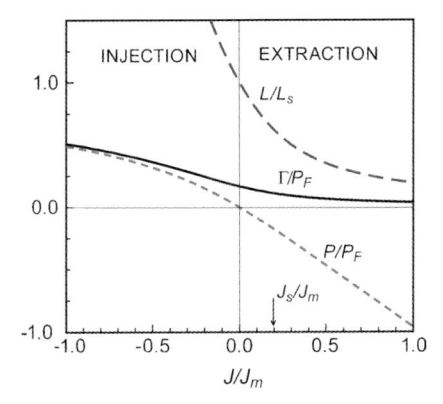

Figure 3.8 The spin accumulation $P = (n_\uparrow - n_\downarrow)/n$, the spin polarization of a current $\Gamma = (J_\uparrow - J_\downarrow)/J$, and the relative spin penetration depth L/L_s (broken line) in the semiconductor as the functions of the relative current density J/J_s for spin injection ($J < 0$) and spin extraction ($J > 0$) regimes. P_F is the spin polarization in the ferromagnet, the ratio $J_s/J_m = 0.2$, L_s is the usual spin diffusion depth. The spin penetration depth considerably exceeds L_s for the injection and smaller than L_s for the extraction.

One can see that Γ_0 strongly differs from P_0 at small currents. As expected, $P_n \approx P_F |J|/J_m \to 0$ vanishes with the current (Figure 3.8), and the prefactor differs from those obtained in Refs. [67, 71, 73, 75, 76].

These expressions should be compared with the results for the case of a degenerate semiconductor, Ref. [10], for the polarization

$$P_0 = -\frac{6 P_F J}{3\left(\sqrt{J^2 + 4J_S^2} - J\right) + 10 J_m}, \qquad (3.55)$$

and the spin injection efficiency

$$\Gamma_0 = -P_F \frac{\sqrt{4J_S^2 + J^2} - J}{2J_m + \sqrt{J^2 + 4J_S^2} - J}. \qquad (3.56)$$

In spite of very different statistics of carriers in a degenerate and non-degenerate semiconductor, the accumulated polarization as a function of current behaves similarly in both cases. The important difference comes from an obvious fact that the efficient spin accumulation in degenerate semiconductors may proceed at and below room temperature, whereas in present design an efficient spin accumulation in FM-S junctions with non-degenerate S can be achieved at around room temperature only.

In the reverse-biased FM-S junctions the current $J < 0$ and, according to Equations 3.46 and 3.47, sign $(\delta n_{\uparrow 0})$ = sign (P_F). In some realistic situations, like elemental Ni, the polarization at energies $E \approx F + \Delta_0$ would be negative, $P_F < 0$ and, therefore, electrons with spin $\sigma = \downarrow$ will be accumulated near the interface. For large currents $|J| \gg J_S$, the spin penetration depth L in Equation 3.45 increases with

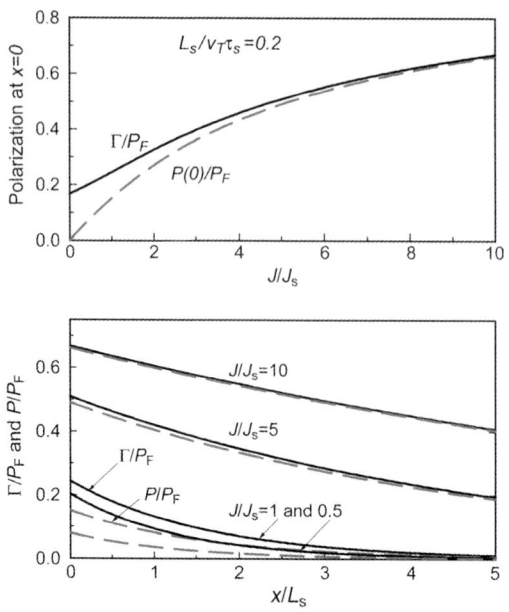

Figure 3.9 Spin polarization of a current $\Gamma = (J_\uparrow - J_\downarrow)/J$ (solid line) and spin accumulation $P = (n_\uparrow - n_\downarrow)/n$ (broken line) in the semiconductor as the functions of the relative current density J/J_s (top panel) and their spatial distribution for different densities of total current J/J_s (bottom panel) at $L_s/v_T\tau_s = 0.2$ where $J_s = qnL_s/\tau_s$, P_F is the spin polarization in the ferromagnet (see text).

current J and the spin polarization (of electron density) approaches the maximum value P_F. Unlike the spin accumulation P_0, the spin injection efficiency (polarization of current) Γ_0 does not vanish at small currents, but approaches the value $\Gamma_0^0 = P_F J_S/(J_S + J_m) \ll P_F$ in the present system with transparent tunnel δ-barrier. There is an important difference with the magnetic tunnel junctions, where the tunnel barrier is relatively opaque and the injection efficiency (polarization of current) is high, $\Gamma \approx P_F$ [18]. However, the polarization of carriers P_0, measured in, for example, spin-LED devices [66], would be minute (see below). Both P_0 and Γ_0 approach the maximum P_F only when $|J| \gg J_S$, (Figure 3.9). The condition $|J| \gg J_S$ is fulfilled at $qV \simeq rT \gtrsim 2T$, when $J_m \gtrsim J_S$.

Another situation is realized *in the forward-biased FM-S junctions* when $J > 0$. Indeed, according to Equations 3.53 and 3.54 at $J > 0$ the electron density distribution is such that sign $(\delta n_{\uparrow 0}) = -\text{sign}(P_F)$. If a system like elemental Ni is considered (Figure 3.7), then $P_F(F + \Delta_0) < 0$ and $\delta n_{\uparrow 0} > 0$; that is, the electrons with spin $\sigma = \uparrow$ would be accumulated in a non-magnetic semiconductor (NS), whereas electrons with spin $\sigma = \downarrow$ would be extracted from NS (the opposite situation would take place for $P_F(F + \Delta_0) > 0$). One can see from Equation 3.46 that $|P_0|$ one can reach a maximum P_F only when $J \gg J_s$. According to Equation 3.50, the condition $J \gg J_s$ can only be fulfilled when $J_m \gg J_s$. In this case Equation 3.46 reduces to

$$P_0 = \frac{-P_F J}{J_m} = -P_F(1 - e^{-qV/T}). \tag{3.57}$$

Therefore, the absolute magnitude of a spin polarization approaches its maximal value $|P_0| \simeq P_F$ at $qV \gtrsim 2T$ linearly with current (Figure 3.8). The maximum is reached when J approaches the value J_m, which depends weakly on bias V (see below). In this case $\delta n_{\uparrow(\downarrow)}(0) \approx \mp P_F n/2$ at $P_F > (d_\uparrow > d_\downarrow)$, so that the electrons with spin $\sigma = \uparrow$ are *extracted*, $n_\uparrow(0) \approx (1 - P_F)n/2$, from a semiconductor, while the electrons with spin $\sigma = \downarrow$ are *accumulated* in a semiconductor, $n_\uparrow(0) \approx (1 + P_F)n/2$, near the FM-S interface. The penetration length of the accumulated spin Equation 3.45 at $J \gg J_s$ is

$$L = \frac{L_s^2}{L_E} = \frac{L_s J_s}{J \ll L_s} \ll L_s \quad \text{at} \quad J \gg J_s, \tag{3.58}$$

that is, it decreases as $L \propto 1/J$ (Figure 3.8). We see from Equation 3.54 that at $J \gg J_s$

$$\Gamma_0 = \frac{P_F J_s^2}{J_m J} \to 0. \tag{3.59}$$

Hence, the behavior of the spin injection efficiency at forward bias (extraction) is very different from a spin injection regime, which occurs at a reverse bias voltage: here, the spin injection efficiency Γ_0 remains $\ll P_F$ and vanishes at large currents as $\Gamma_0 \propto J_s/J$. Therefore, we come to an unexpected conclusion that *the spin polarization of electrons*, accumulated in a non-magnetic semiconductor near forward biased FM-S junction can be relatively large for the parameters of the structure when the spin injection efficiency is actually very *small* [8]. Similar, albeit much weaker phenomena are possible in systems with *wide opaque* Schottky barriers [80] and have been probably observed [81]. Spin extraction may also be observed at low temperature in FMS-S contacts [82]. A proximity effect leading to polarization accumulation in FM-S contacts [83] may be related to the same mechanism.

3.5.2
Conditions for Efficient Spin Injection and Extraction

According to Equations 3.51 and 3.46, the condition for maximal polarization of electrons P_n can be written as

$$J_m \gtrsim J_s, \tag{3.60}$$

or, equivalently, as a condition

$$\beta \equiv \frac{\alpha_0 v_T (d_{\uparrow 0} + d_{\downarrow 0})(1 - P_F^2) e^{-\eta l/l_0} \tau_s}{L_s} \gtrsim 1. \tag{3.61}$$

Note that when $l \lesssim l_0$, the spin injection efficiency at small current is small $\Gamma_0^0 = P_F/(1+\beta) \gg P_F$, since in this case the value $\beta \simeq (d_{\uparrow 0} + d_{\downarrow 0})\alpha_0 v_T \tau_s/L_s \gg 1$ for real semiconductor parameters. The condition $\beta \gg 1$ can be simplified and rewritten as a requirement for the spin-relaxation time

$$\tau_s \gg D \left(\frac{\Delta - \Delta_0}{2\alpha_0 v_{\sigma 0}^2 T} \right)^2 \exp \frac{2\eta l}{l_0}. \tag{3.62}$$

It can be met only when the δ-doped layer is very thin, $l \lesssim l_0 \equiv \kappa_0^{-1}$. With typical semiconductor parameters at $T \simeq 300\,\text{K}$ ($D \approx 25\,\text{cm}^2\,\text{s}^{-1}$, $(\Delta - \Delta_0) \simeq 0.5\,\text{eV}$, $v_{\sigma 0} \simeq 10^8\,\text{cm}^2\,\text{s}^{-1}$ [78]) the condition in Equation 3.62 is satisfied at $l \lesssim l_0$ when the spin-coherence time $\tau_s \gg 10^{-12}\,\text{s}$. It is worth noting that it can certainly be met: for instance, τ_s can be as large as $\sim 1\,\text{ns}$ even at $T \simeq 300\,\text{K}$ (e.g., in ZnSe [84]).

Note that the higher the semiconductor conductivity, $\sigma_s = q\mu n \propto n$, the larger the threshold current $J > J_m \propto n$ [Equation 3.51] for achieving the maximal spin injection. In other words, the polarization P_0 reaches the maximum value P_F at *smaller* current in *high-resistance* lightly doped semiconductors compared to heavily doped semiconductors. Therefore, the "conductivity mismatch" [70, 74, 75] is actually irrelevant for achieving an efficient spin injection.

The necessary condition $|J| \gg J_s$ can be rewritten at small voltages, $|qV| \ll T$, as

$$r_c \ll \frac{L_s}{\sigma_s}, \tag{3.63}$$

where $r_c = (dJ/dV)^{-1}$ is the tunneling contact resistance. Here, we have used the Einstein relation $D/\mu = T/q$ for non-degenerate semiconductors. We emphasize that Equation 3.63 is *opposite* to the condition found by Rashba in Ref. [74] for small currents.

We also emphasize that the spin injection in structures considered in the literature [4, 65–76] has been dominated by electrons at the Fermi level and, according to calculation [85], $g_\downarrow(F)$ and $g_\uparrow(F)$ are such that $P_F \lesssim 40\%$. We also notice that the condition in Equation 3.61 for parameters of the Fe/AlGaAs heterostructure studied in Refs. [65] ($l \simeq 3\,\text{nm}$, $l_0 \simeq 1\,\text{nm}$ and $\Delta_0 = 0.46\,\text{eV}$) is satisfied when $\tau_s \gtrsim 5 \times 10^{-10}\,\text{s}$ and can be fulfilled only at low temperatures. Moreover, for the concentration $n = 10^{19}\,\text{cm}^{-3}$ E_c lies below F, so that the electrons with energies $E \simeq F$ are involved in tunneling, but for these states the polarization is $P_F \lesssim 40\%$. Therefore, the authors of Ref. [65] were indeed able to estimate the observed spin polarization as being $\approx 32\%$ at low temperatures.

Better control of the injection can be realized in heterostructures where a δ-layer between the ferromagnet and the *n*-semiconductor layer is made of very thin heavily doped n^+-semiconductor with larger electron affinity than the *n*-semiconductor. For instance, FM–n^+-GaAs–n-Ga$_{1-x}$Al$_x$As, FM–n^+-Ge$_x$Si$_{1-x}$–n-Si or FM–n^+-Zn$_{1-x}$Cd$_x$Se–n-ZnSe heterostructures can be used for this purpose. The GaAs, Ge$_x$Si$_{1-x}$ or Zn$_{1-x}$Cd$_x$Sen^+–layer must have the width $l < 1\,\text{nm}$ and the donor concentration $N_d^+ > 10^{20}\,\text{cm}^{-3}$. In this case, the ultrathin barrier forming near the ferromagnet-semiconductor interface is transparent for electron tunneling. The barrier height Δ_0 at Ge$_x$Si$_{1-x}$–Si, GaAs–Ga$_{1-x}$Al$_x$As or Zn$_{1-x}$Cd$_x$Se–ZnSe interface is controlled by the composition x, and can be selected as $\Delta_0 = 0.05$–$0.15\,\text{eV}$. When the donor concentration in Si, Ga$_{1-x}$Al$_x$As, or ZnSe layer is $N_d < 10^{17}\,\text{cm}^{-3}$, the injected electrons cannot penetrate relatively low and wide barrier Δ_0 when its width $l_0 > 10\,\text{nm}$.

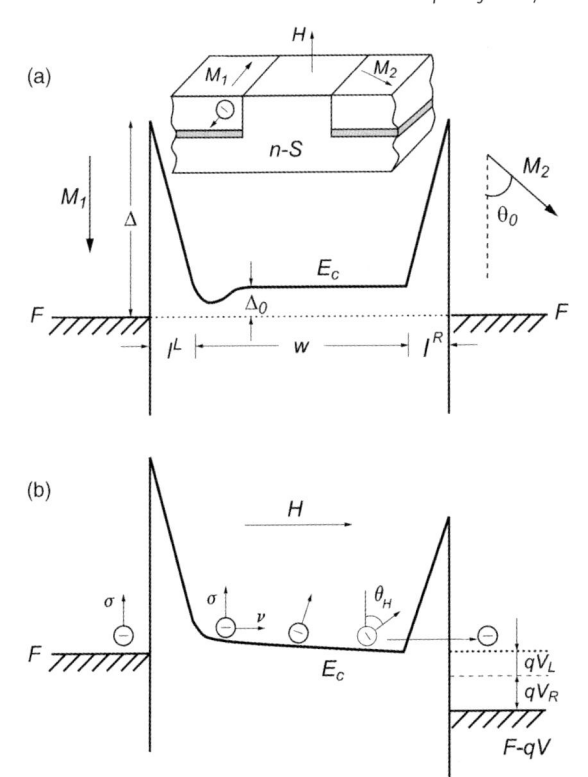

Figure 3.10 Energy diagram of a the FM-S-FM heterostructure with δ-doped layers in equilibrium (a) and at a bias voltage V (b), with V_L (V_R) the fraction of the total drop across the left (right) δ-layer. F marks the Fermi level, Δ the height, $l^{L(R)}$ the thickness of the left (right) δ-doped layer, Δ_0 the height of the barrier in the n-type semiconductor (n-S), E_c the bottom of conduction band in the n-S, w the width of the n-S part. The magnetic moments on the FM electrodes $\vec{M}_1$ and $\vec{M}_2$ are at some angle θ_0 with respect to each other. The spins, injected from the left, drift in the semiconductor layer and rotate by the angle θ_H in the external magnetic field H. Inset: schematic of the device, with an oxide layer separating the ferromagnetic films from the bottom semiconductor layer.

3.5.3
High-Frequency Spin-Valve Effect

Here we describe a new high-frequency spin valve effect that can be observed in a FM-S-FM device with two back-to-back modified Schottky contacts (Figure 3.10). We find the dependence of current on a magnetic configuration in FM electrodes and an external magnetic field. The spatial distribution of spin-polarized electrons is determined by the continuity [Equation 3.43] and the current in spin channel σ is given by Equation 3.40. Note that $J < 0$, thus $E < 0$ in a spin injection regime. With the use of the kinetic equation and Equation 3.40, we obtain the equation for $\delta n_\uparrow$, Equation 3.43 [68]. Its general solution is

$$\delta n_\uparrow(x) = \frac{n}{2}(c_1 e^{-x/L_1} + c_2 e^{-(w-x)/L_2}), \tag{3.64}$$

where $L_{1(2)} = (1/2)[\sqrt{L_E^2 + 4L_s^2} + (-)L_E]$ is the same as found earlier in Equation 3.45. Consider the case when $w \ll L_1$ and the transit time $t_{tr} \simeq w^2/(D + \mu|E|w)$ of the electrons through the n-semiconductor layer is shorter than τ_s. In this case, a spin ballistic transport takes place; that is, the spin of the electrons injected from the FM$_1$ layer is conserved in the semiconductor layer, $\sigma' = \sigma$. Probabilities of the electron spin $\sigma = \uparrow$ to have the projections along $\pm \vec{M}_2$ are $\cos^2(\theta/2)$ and $\sin^2(\theta/2)$, respectively, where θ is the angle between vectors $\sigma = \uparrow$ and $\vec{M}_2$. Accounting for this, we find that the resulting current through the structure saturates at bias voltage $-qV > T$ at the value

$$J = J_0 \frac{1 - P_R^2 \cos^2\theta}{1 - P_L P_R \cos\theta}, \tag{3.65}$$

where J_0 is the prefactor similar to Equation 3.38. For the *opposite* bias the total current J is given by Equation 3.65 with the replacement $P_L \leftrightarrow P_R$. The current J is minimal

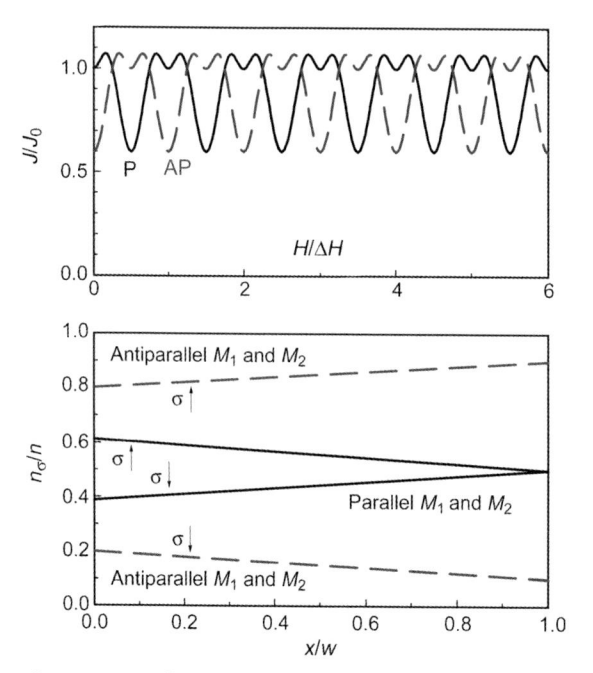

Figure 3.11 Oscillatory dependence of the current J through the structure on the magnetic field H (top panel) for parallel (P) and antiparallel (AP) moments M_1 and M_2 on the electrodes, Figure 3.10, and $P_L = P_R = 0.5$. Spatial distribution of the spin polarized electrons $n_{\uparrow(\downarrow)}/n$ in the structure for different configurations of the magnetic moments M_1 and M_2 in the limit of saturated current density J, $w = 60$ nm, $L_2 = 100$ nm (bottom panel).

for antiparallel (AP) moments $\vec{M}_1$ and $\vec{M}_2$ in the electrodes when $\theta = \pi$ and near maximal for parallel (P) magnetic moments $\vec{M}_1$ and $\vec{M}_2$.

The present heterostructure has an additional degree of freedom, compared to tunneling FM-I-FM structures that can be used for *magnetic sensing*. Indeed, spins of the injected electrons can precess in an external magnetic field H during the transit time t_{tr} of the electrons through the semiconductor layer ($t_{tr} < \tau_s$). The angle between the electron spin and the magnetization $\vec{M}_2$ in the FM$_2$ layer in Equation 3.65 is in general $\theta = \theta_0 + \theta_H$ where θ_0 is the angle between the magnetizations M_1 and M_2, and $\theta_H = \gamma_0 g H t_{tr}(m_0/m_*)$ is the spin rotation angle. Here, H is the magnetic field normal to the spin direction, $\gamma = qg/(m_* c)$ is the gyromagnetic ratio, g is the g-factor. According to Equation 3.65, with increasing H the current *oscillates* with an amplitude $(1 + P_L P_R)/(1 - P_L P_R)$ and period $\Delta H = (2\pi m_*)(\gamma_0 g m_0 t_{tr})^{-1}$ (Figure 3.11, top panel). The maximum operating speed of the field sensor is very high, since redistribution of non-equilibrium-injected electrons in the semiconductor layer occurs over the transit time $t_{tr} = w/\mu|E| = J_s w \tau_s/(JL_s)$, $t_{tr} \lesssim 10^{-11}$ s for $w \lesssim 200_*$ nm, $\tau_s \sim 3 \times 10^{-10}$ s, and $J/J_s \gtrsim 10$ ($D \approx 25$ cm^2 s^{-1}) at $T \simeq 300$ K [78]). Therefore, the operating frequency $f = 1/t_{tr} \gtrsim 100$ GHz may be achievable at room temperature. We see that: (i) the present heterostructure can be used as a sensor for an ultrafast nanoscale reading of an inhomogeneous magnetic field profile; (ii) it includes two FM-S junctions and can be used for measuring the spin polarizations of these junctions; and (iii) it is a *multifunctional* device where current depends on the mutual orientation of the magnetizations in the ferromagnetic layers, an external magnetic field, and a (small) bias voltage. Thus, it can be used as a logic element, a magnetic memory cell, or an ultrafast read head.

3.5.4
Spin-Injection Devices

The high-frequency spin-valve effect, described above, can be used for designing a new class of ultrafast spin-injection devices such as an amplifier, a frequency multiplier, and a square-law detector [6]. Their operation is based on the injection of spin-polarized electrons from one ferromagnet to another through a semiconductor layer, and spin precession of the electrons in the semiconductor layer in a magnetic field induced by a (base) current in an adjacent nanowire. The base current can control the emitter current between the magnetic layers with frequencies up to several 100 GHz. Here, we shall describe a spintronic mechanism of ultrafast amplification and frequency conversion, which can be realized in heterostructures comprising a metallic ferromagnetic nanowire surrounded by a semiconductor (S) and a ferromagnetic (FM) thin shells (Figure 3.12a). Practical devices may have various layouts, with two examples shown in Figure 3.12b and c.

Let us consider the principle of operation of the spintronic devices shown in Figure 3.12a. When the thickness w of the n-type semiconductor layer is not very small ($w \gtrsim 30$ nm), tunneling through this layer would be negligible. The base voltage V_b is applied between the ends of the nanowire. The base current J_b, flowing through the nanowire, induces a cylindrically symmetric magnetic field $H_b = J_b/2\pi\rho$ in the S layer, where ρ is the distance from the center of nanowire. When the emitter

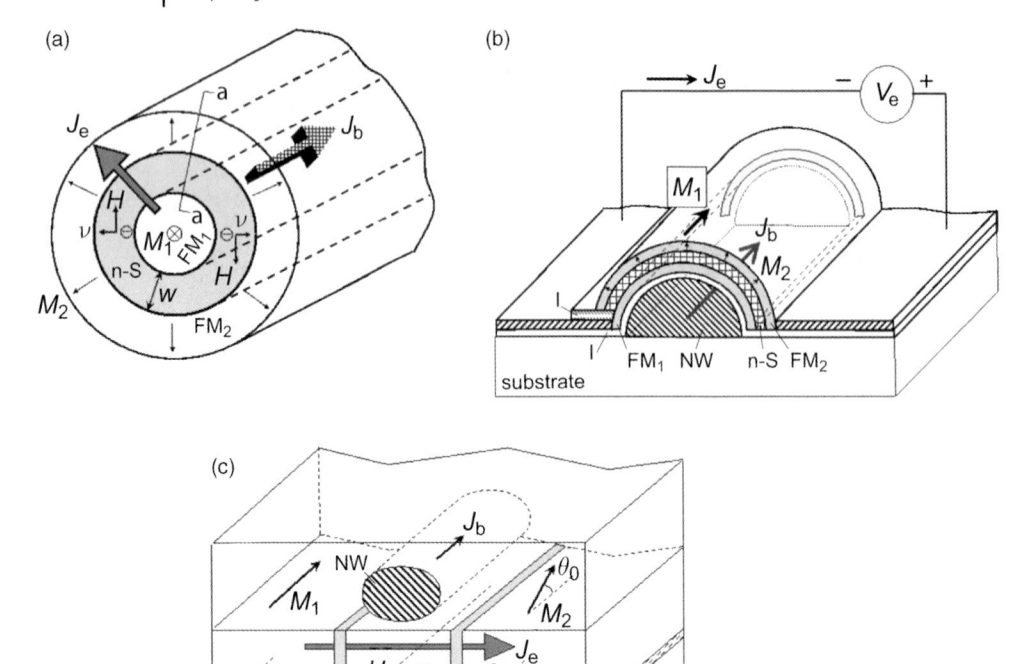

Figure 3.12 Schematic of the spin injection-precession devices having (a) cylindrical, (b) semi-cylindrical, and (c) planar shape. Here, FM$_1$ and FM$_2$ are the ferromagnetic layers; n-S the n-type semiconductors layer; w the thickness of the n-S layer; δ the δ-doped layers; NW the highly conductive nanowires; I the insulating layers. The directions of the magnetizations $\vec{M}_1$ and $\vec{M}_2$ in the FM$_1$ and FM$_2$ layers, as well as the electron spin σ, the magnetic field H_b, and the angle of spin rotation θ in S are also shown.

voltage V_e is applied between FM layers, the spin-polarized electrons are injected from the first layer (nanowire FM$_1$) through the semiconductor layer into the second (exterior) ferromagnetic shell, FM$_2$. The FM$_1$-S and FM$_2$-S junctions are characterized by the spin injection efficiencies P_1 and P_2, respectively. We assume that the transit time t_{tr} of the electrons through the S layer is less than the spin relaxation time, τ_s (i.e., we consider the case of a spin ballistic transport). The exact calculation gives a current J_e, through the structure as a function of the angle θ between the magnetization vectors $\vec{M}_1$ and $\vec{M}_2$ in the ferromagnetic layers. At small angles θ or $P_1 = P_L$ or $P_2 = P_R$

$$J_e = J_{0e}(1 + P_L P_R \cos\theta), \tag{3.66}$$

where $\theta = \theta_0 + \theta_H$, θ_0 is the angle between $\vec{M}_1$ and $\vec{M}_2$, and θ_H is the angle of the spin precesses with the frequency $\Omega = \gamma H_\perp$, where $H_\perp$ is the magnetic field component normal to the spin and γ is the gyromagnetic ratio. One can see from Figure 3.12a that $H_\perp = H_b = J_b/(2\pi\rho)$. Thus, the angle of the spin rotation is equal to $\theta_H = \gamma H_b t_{tr} = t_{tr} J_b / 2\pi \rho_s$, where ρ_s is the characteristic radius of the S layer. Then, according to Equation 3.57,

$$J_e = J_{e0}[1 + P_1 P_2 \cos(\theta_0 + k_j J_b)], \tag{3.67}$$

where $k_j = \gamma t_{tr}/2\pi\rho_s = \gamma/\omega\rho_s$ and $\omega = 2\pi/t_{tr}$ is the frequency of a variation of the base current, $J_b = J_s \cos(\omega t)$.

Equation 3.67 shows that, when the magnetization M_1 is perpendicular to M_2, $\theta_0 = \pi/2$, and $\theta_H \ll \pi$,

$$J_e = J_{e0}(1 + k_j P_1 P_2 J_b), \; G = dJ_e/dJ_b = J_{e0} k_j P_1 P_2. \tag{3.68}$$

Hence, the *amplification* of the base current occurs with the gain G, which can be relatively high even for $\omega \gtrsim 100\,\text{GHz}$. Indeed, $\gamma = q/(m_* c) \approx 2.2(m_0/m_*) 10^5 \, m/(A \cdot s)$, where m_0 is the free electron mass, m_* the effective mass of electrons in the semiconductor, and c the velocity of light. Thus, the factor $k_j \simeq 10^3 \, A^{-1}$ when $\rho_s \simeq 30\,\text{nm}$, $m_0/m_* = 14$ (GaAs) and $\omega = 100\,\text{GHz}$, so that $G > 1$ at $J_{e0} > 0.1\,\text{mA}/(P_1 P_2)$.

When M_1 is collinear with M_2 ($\theta_0 = 0, \pi$) and $\theta_H \ll \pi$, then, according to Equation 3.67, the emitter current is

$$J_e = J_{e0}(1 \pm P_1 P_2) \mp \frac{1}{2} J_{e0} P_1 P_2 k_j^2 J_b^2. \tag{3.69}$$

Therefore, the time-dependent component of the emitter current $\delta J_e(t) \propto J_b^2(t)$, and the device operates as a square-law detector. When $J_b(t) = J_{b0} \cos(\omega_0 t)$, the emitter current has a component $\delta J_e(t) \propto \cos(2\omega_0 t)$, and the device operates as a *frequency multiplier*. When $J_b(t) = J_h \cos(\omega_h t) + J_s \cos(\omega_s t)$, the emitter current has the components proportional to $\cos(\omega_h \pm \omega_s) t$; that is, the device can operate as a high-frequency *heterodyne detector* with the conversion coefficient $K = J_{e0} J_h P_1 P_2 k_j^2 / 4$. For $k_j = 10^3 A^{-1}$, one obtains $K > 1$ when $J_{e0} J_h > 4 \, (\text{mA})^2/(P_1 P_2)$.

3.5.5
Spin Source of Polarized Radiation

The spin extraction effect can be used for making an efficient source of (modulated) polarized radiation. Consider a structure containing a FM-S junction with the δ-doped layer and a double p–n'–n heterostructure where the n'-region is made from narrower gap semiconductor (Figure 3.13). We show that the following effects can be realized in the structure when both FM-S junction and the heterostructure are biased in the forward direction, and the electrons are injected from n-semiconductor region into FM and p-region. Due to a spin selection property of FM-S junction [7], spin-polarized electrons appear in n-region with a spatial extent $L \lesssim L_s$ near the FM-S interface, where L_s is the spin diffusion length in NS. When the thickness of the n-region w is smaller than L, the spin-polarized electron from the n-region and

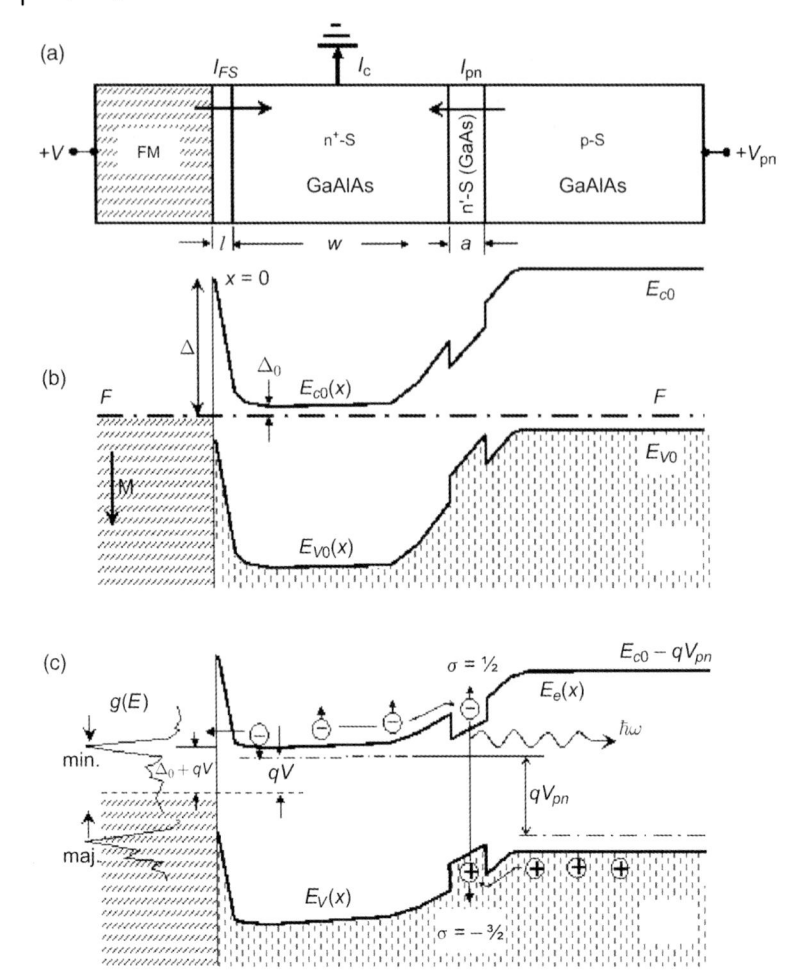

Figure 3.13 Schematic (a) of structure and the band diagram of polarized photon source containing FM-S junction with δ-doped layer and a double n^+–n'–p heterostructure without (b) and under the bias voltage V. Minority spin electrons are extracted from n^+–S semiconductor layer and the remaining (majority) electrons are recombined in n'–S quantum well. F is the Fermi level, Δ the height and l the thickness of the δ-doped layer, Δ_0 the height of a barrier in the n-type semiconductor, $E_c(x)$ the bottom of conduction band in the semiconductor. The spin density of states is shown at $x<0$ with a high peak in minority states at $E=F+\Delta_0$, typical of elemental Ni, as an example.

holes from p-region are injected and accumulated in a thin narrow-gap n'-region (quantum well) where they recombine and emit polarized photons.

The conditions for maximal polarization are obtained as follows. When the thickness of n-region is $w<L$, we can assume that $\delta n_\uparrow(x)\simeq \delta n_{\uparrow 0}$ and $P_n\simeq P_0$. In this case, integrating Equation 3.43 over the volume of the n-semiconductor region (with area S and thickness w), we obtain

$$I_{\uparrow FS} + I_{\uparrow pn} - I_{c\uparrow} = q\delta n_{\uparrow 0} wS/\tau_s = P_0 I_s w/2L_s. \quad (3.70)$$

Here, $I_s = J_s S$; $I_{\uparrow FS} = J_{\uparrow 0} S$ and $I_{\uparrow pn}$ are the electron currents with spin $\sigma = \uparrow$ flowing into the n-region from FM and the p-region, respectively; $I_{\uparrow c}$ is the spin current out of the n-region in a contact (Figure 3.13a). The current $I_{\uparrow pn}$ is determined by injection of electrons with $\sigma = \uparrow$ from the n-region into the p-region through the crossectional area S, equal to $I_{\uparrow pn} = I_{pn} n_{\uparrow 0}/n = I_{pn}(1 + P_0)/2$, where I_{pn} is the total current in the p–n junction. The current of metal contact I_c is not spin-polarized; hence $I_{c\uparrow} = (I_{pn} + I_{FS})/2$, where I_{FS} is the total current in the FM-S junction. The current in the FM-S junction I_{FS} approaches a maximal value $I_m = J_m S$ at rather small bias, $qV_{FS} > 2T$. When $I_{pn} \ll I_{FS} \simeq I_m$ and $I_m \gg I_s w/L_s$, we get $P_0 \simeq -P_F$. The way to maximize polarization is by adjusting V_{FS}. The maximal $|P_0|$ can be achieved for the process of electron tunneling through the δ-doped layer when the bottom of conduction band in a semiconductor $E_c = F + \Delta_0 + qV_{FS}$ is close to a peak in the density of states of minority carriers in the elemental ferromagnet (Figure 3.13c, curve g).

The rate of polarized radiation recombination is $R_\sigma = q n'_\sigma d/\tau_R$ and the polarization of radiation is $p = (R_\uparrow - R_\downarrow)/(R_\uparrow + R_\downarrow) = 2\delta n'_\uparrow / n'$. Since $I_{pn} = q n' d/\tau n$, we find $2\delta n'_\uparrow / n' = P_0 \tau'_s (\tau'_s + \tau_n)^{-1}$, so that $p = P_0 \tau'_s (\tau'_s + \tau_n)^{-1}$. Thus, the radiation polarization p can approach maximum $p \simeq |P_F|$ at large current $I \simeq I_m$ when $\tau < \tau'_s$. The latter condition can be met at high concentration n' when the time of radiation recombination $\tau_R \simeq \tau_n < \tau'_s$. For example, in GaAs $\tau_R \simeq 3 \times 10^{-10}$ s at $n \gtrsim 5 \times 10^{17}$ cm^{-3} [86] and τ'_s can be larger than τ_R [84]. We emphasize that spin injection efficiency near a forward-biased FM-S junction is very small.

Practical structures may have various layouts, with one example shown in Figure 3.14. It is clear that the distribution of $\delta n_\uparrow(\vec{r})$ in such a 2-D structure is characterized by the length $L \lesssim L_s$ in the direction x where the electrical field E can be strong, and by the diffusion length L_s in the (y, z) plane where the field is weak. Therefore, the spin density near FM and p–n junctions will be close to $\delta n_{\uparrow 0}$ when the size of the p-region is $d < L_s$. Thus, the above results for one-dimensional structure

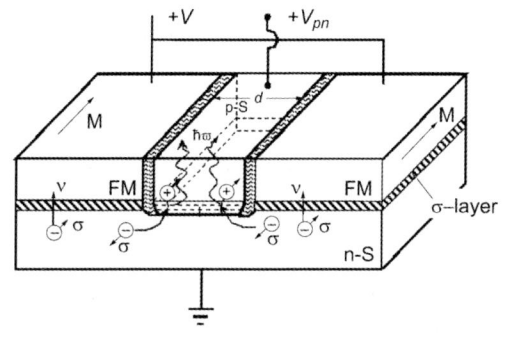

Figure 3.14 Layout of a structure from Figure 3.13, including FM layers and semiconductor n- and p-regions. Here, n' made from narrower gap semiconductor, δ-doped layers are between the FM layers and the n-semiconductor. FM layers are separated by thin dielectric layers from the p-region.

(Figure 3.13) are also valid for more complex geometry shown in Figure 3.14. The predicted effect should also exist for a reverse-biased FM-S junction where the radiation polarization p can approach $+P_F$.

3.6
Conclusions

In this chapter we have described a variety of heterostructures where the spin degree of freedom can be used to efficiently control the current: magnetic tunnel junctions, metallic magnetic multilayers exhibiting giant magnetoresistance, spin-torque effects in magnetic nanopillars. We also described a method of facilitating an efficient spin injection/accumulation in semiconductors from standard ferromagnetic metals at room temperature. The main idea is to engineer the band structure near the ferromagnet-semiconductor interface by fabricating a delta-doped layer there, thus making the Schottky barrier very thin and transparent for tunneling. A long spin lifetime in a semiconductor allows the suggestion of some interesting new devices such as field detectors, spin transistors, square-law detectors, and sources of the polarized light described in the present text. This development opens up new opportunities in potentially important novel spin injection devices. We also discussed a body of various spin-orbit effects and systems of interacting spins. In particular, Spin Hall effects result is positive magnetoresistance due to spin accumulation that may be used to extract the coeefficints for spin-orbit transport. We notice, however, that Datta–Das spinFET would have inferior characteristics compared to MOSFET. We also discussed the severe challenges facing *single-spin logic and, especially, spin-based quantum computers.*

References

1 (a) Wolf, S.A., Awschalom, D.D., Buhrman, R.A., Daughton, J.M., von Molnar, S., Roukes, M.L., Chtchelkanova, A.Y. and Treger, D.M. (2001) *Science*, **294**, 1488; (b) Awschalom D.D., Loss D. and Samarth N. (eds) (2002) *Semiconductor Spintronics and Quantum Computation*, Springer, Berlin.

2 Žutić I., Fabian, J. and Das Sarma, S. (2004) *Reviews of Modern Physics*, **76**, 323.

3 (a) Datta, S. and Das, B. (1990) *Applied Physics Letters*, **56**, 665. (b) Gardelis, S., Smith, C.G., Barnes, C.H.W., Linfield, E.H. and Ritchie, D.A. (1999) *Physical Review B-Condensed Matter*, **60**, 7764.

4 (a) Sato, R. and Mizushima, K. (2001) *Applied Physics Letters*, **79**, 1157; (b) Jiang, X., Wang, R., van Dijken, S., Shelby, R., Macfarlane, R., Solomon, G.S., Harris, J. and Parkin, S.S.P. (2003) *Physical Review Letters*, **90**, 256603.

5 Bratkovsky, A.M. and Osipov, V.V. (2004) *Physical Review Letters*, **92**, 098302.

6 Osipov, V.V. and Bratkovsky, A.M. (2004) *Applied Physics Letters*, **84**, 2118.

7 Osipov, V.V. and Bratkovsky, A.M. (2004) *Physical Review B-Condensed Matter*, **70**, 235302.

8 Bratkovsky, A.M. and Osipov, V.V. (2004) *Journal of Applied Physics*, **96**, 4525.

9 Bratkovsky, A.M. and Osipov, V.V. (2005) *Applied Physics Letters*, **86**, 071120.
10 Osipov, V.V. and Bratkovsky, A.M. (2005) *Physical Review B-Condensed Matter*, **72**, 115322.
11 (a) Baibich, M.N., Broto, J.M., Fert, A., Nguyen Van Dau, F., Petroff, F., Etienne, P., Creuzet, G., Friederich, A. and Chazelas, J. (1988) *Physical Review Letters*, **61**, 2472; (b) Berkowitz, A.E., Mitchell, J.R., Carey, M.J., Young, A.P., Zhang, S., Spada, F.E., Parker, F.T., Hutten, A. and Thomas, G. (1992) *Physical Review Letters*, **68**, 3745.
12 Julliere, M. (1975) *Physics Letters*, **54A**, 225.
13 Maekawa, S. and Gäfvert, U. (1982) *IEEE Transactions on Magnetics*, **18**, 707.
14 Meservey, R. and Tedrow, P.M. (1994) *Physics Reports*, **238**, 173.
15 Moodera, J.S., Kinder, L.R., Wong, T.M. and Meservey, R. (1995) *Physical Review Letters*, **74**, 3273.
16 (a) Yuasa, S., Nagahama, T., Fukushima, A., Suzuki, Y. and Ando, K. (2004) *Nature Materials*, **3**, 858; (b) Parkin, S.S.P., Kaiser, C., Panchula, A., Rice, P.M., Hughes, B., Samant, M. and Yang, S.-H. (2004) *Nature Materials*, **3**, 862.
17 Slonczewski, J.C. (1989) *Physical Review B-Condensed Matter*, **39**, 6995.
18 Bratkovsky, A.M. (1997) *Physical Review B-Condensed Matter*, **56**, 2344.
19 Bratkovsky, A.M. (1998) *Applied Physics Letters*, **72**, 2334.
20 Mott, N.F. (1936) *Proceedings of the Royal Society of London. Series A*, **153**, 699.
21 Stearns, M.B. (1977) *Journal of Magnetism and Magnetic Materials*, **5**, 167.
22 Berger, L. (1996) *Physical Review B-Condensed Matter*, **54**, 9353.
23 Slonczewski, J.C.(1996) *Journal of Magnetism and Magnetic Materials*, **159**, L1.
24 Tsoi, M.V. et al. (1998) *Physical Review Letters*, **80**, 4281.
25 (a) Slonczewski, J.C. (2002) *Journal of Magnetism and Magnetic Materials*, **247**, 324; (b) Slonczewski, J.C. (2005) *Physical Review B-Condensed Matter*, **71**, 024411; (c) Slonczewski, J.C., Sun, J.Z. (2007) *J. Magn. Magn. Mater.*, **310**, 169.
26 Emley, N.C. et al. (2006) *Physical Review Letters*, **96**, 247204. They have added a phenomenological spin-torque (ST) factor $\eta = A/(1 + B \cos\theta)$ to fit the data for the torque.
27 Pribiag, V.S. et al. (2007) *Nature Physics*, **3**, 498.
28 Hayashi, M., Thomas, L., Rettner, C., Moriya, R., Bazaliy, Ya.B. and Parkin, S.S.P. (2007) *Physical Review Letters*, **98**, 037204.
29 (a) Vasko, F.T. (1979) *Pisma Zh Eksp Teor Fiz*, **30**, 574; (b) Vasko, F.T. (1979) *JETP Letters*, **30**, 541.
30 (a) Yu. A., Bychkov and Rashba, E.I. (1984) *Pisma Zh Eksp Teor Fiz*, **39**, 66; (1984) *JETP Letters*, **39**, 78; (b) Yu. A., Bychkov and Rashba, E.I. (1984) *Journal of Physics C*, **17**, 6039.
31 Anselm, A. (1981) *Introduction to Semiconductor Theory*, Prentice-Hall, New Jersey.
32 Engel, H.A., Rashba, E.I. and Halperin, B.I. (2006) cond-mat/0603306.
33 Winkler, R. (2003) *Spin-Orbit Coupling Effects in Two-Dimensional Electron and Hole System*, Springer, Berlin.
34 Nozieres, P. and Lewiner, C. (1973) *Journal of Physics (Paris)*, **34**, 901.
35 Winkler, R. (2000) *Physical Review B-Condensed Matter*, **62**, 4245.
36 Pikus, G.E. and Titkov, A.N. (1984) in *Optical Orientation*, (eds F. Meier and B.P. Zakharchenya), North Holland, Amsterdam, p. 73.
37 Nitta, J., Akazaki, T., Takayanagi, H. and Enoki, T. (1997) *Physical Review Letters*, **78**, 13351338.
38 Koga, T., Nitta, J., Akazaki, T. and Takayanagi, H. (2002) *Physical Review Letters*, **89**, 046801.
39 Grundler, D. (2000) *Physical Review Letters*, **84**, 60746077.
40 (a) Dyakonov, M.I. and Perel, V.I. (1971) *JETP Letters*, **13**, 467; (b) Dyakonov, M.I. and Perel, V.I. (1971) *Physics Letters A*, **35**, 459.
41 Dyakonov, M.I. (2007) *Physical Review Letters*, **99**, 126601.

42 Bakun, A.A., Zakharchenya, B.P., Rogachev, A.A., Tkachuk, M.N. and Fleisher, V.G. (1984) *Pisma Zh Eksp Teor Fiz*, **40**, 464.

43 Kato, Y.K. et al. (2004) *Science*, **306**, 1910.

44 Wunderlich, J. et al. (2005) *Physical Review Letters*, **94**, 047204.

45 Liu, B., Shi, J., Wang, W., Zhao, H., Li, D., Zhang, S., Xue Q. and Chen, D. (2006) arXiv:cond-mat/0610150.

46 Kimura, T., Otani, Y., Sato, T., Takahashi, S. and Maekawa, S. (2007) *Phy Rev. Lett.*, **98**, 156601.

47 Bandyopadhay, S., Das, B. and Miller, A.E. (1994) *Nanotechnology*, **5**, 113.

48 Cowburn, R.P. and Welland, M.E. (2000) *Science*, **287**, 1466.

49 Rakhmanova, S. and Mills, D.L. (1996) *Physical Review B-Condensed Matter*, **54**, 9225.

50 Dyakonov, M.I. quant-ph/0610117.

51 Zhang, X.-G. and Butler, W.H. (2004) *Physical Review B-Condensed Matter*, **70**, 172407.

52 Duke, C.B. (1969) *Tunneling in Solids*, Academic Press, New York.

53 Ma, W.G. (1992) *Appl. Phys. Lett.*, **61**, 2542. Even smaller $m_2 = 0.2$ has been used by Q. Q. Shu and for Al-Al$_2$O$_3$-metal junctions.

54 Larkin, A.I. and Matveev, K.A. (1987) *Zh Eksp Teor Fiz*, **93**, 1030. (1987) *Soviet Physics JETP*, **66**, 580.

55 Xu, Y., Ephron, D. and Beasley, M.R. (1995) *Physical Review B-Condensed Matter*, **52**, 2843.

56 (a) Jansen, R. and Moodera, J.S. (2000) *Physical Review B-Condensed Matter*, **61**, 9047; (b) Jansen, R. and Moodera, J.S. (1998) *Journal of Applied Physics*, **83**, 6682.

57 Tsymbal, E.Y., Sokolov, A., Sabirianov, I.F. and Doudin, B. (2003) *Physical Review Letters*, **90**, 186602.

58 Parkin, S.S.P. private communication.

59 Bratkovsky, A.M. to be published.

60 Monsma, D.J. et al. (1995) *Physical Review Letters*, **74**, 5260.

61 Smirnov, A.V. and Bratkovsky, A.M. (1997) *Physical Review B-Condensed Matter*, **55**, 14434.

62 Mahan, G.D. (1990) *Many-Particle Physics*, 2nd edn, Plenum Press, New York, Chapter 9.

63 Zhang, S. et al. (1997) *Physical Review Letters*, **79**, 3744. These authors have assumed that surface magnons are excited, and did not consider phonons and bias dependence of direct tunneling.

64 (a) Osipov, V.V., Viglin, N.A. and Samokhvalov, A.A. (1998) *Physics Letters A*, **247**, 353; (b) Ohno, Y., Young, D.K., Beschoten, B., Matsukura, F., Ohno, H. and Awschalom, D.D. (1999) *Nature*, **402**, 790; (c) Fiederling, R., Keim, M., Reuscher, G., Ossau, W., Schmidt, G., Waag, A. and Molenkamp, L.W. (1999) *Nature*, **402**, 787.

65 (a) Hanbicki, A.T. Jonker, B.T. Itskos, G. Kioseoglou, G. and Petrou, A. (2002) *Applied Physics Letters*, **80**, 1240; (b) Hanbicki, A.T., van't Erve, O.M.J., Magno, R., Kioseoglou, G., Li, C.H. and Jonker, B.T. (2003) *Applied Physics Letters*, **82**, 4092; (c) Adelmann, C., Lou, X., Strand, J., Palmstrøm, C.J. and Crowell, P.A. (2005) *Physical Review B-Condensed Matter*, **71**, 121301.

66 (a) Hammar, P.R., Bennett, B.R., Yang, M.J. and Johnson, M. (1999) *Physical Review Letters*, **83**, 203; (b) Zhu, H.J., Ramsteiner, M., Kostial, H., Wassermeier, M., Schönherr, H.-P. and Ploog, K.H. (2001) *Physical Review Letters*, **87**, 016601; (c) Lee, W.Y., Gardelis, S., Choi, B.-C., Xu, Y.B., Smith, C.G., Barnes, C.H.W., Ritchie, D.A., Linfield, E.H. and Bland, J.A.C. (1999) *Journal of Applied Physics*, **85**, 6682; (d) Manago T. and Akinaga, H. (2002) *Applied Physics Letters*, **81**, 694. (e) Motsnyi, A.F., De Boeck, J., Das, J., Van Roy, W., Borghs, G., Goovaerts, E. and Safarov, V.I. (2002) *Applied Physics Letters*, **81**, 265; (f) Ohno, H., Yoh, K., Sueoka, K., Mukasa, K., Kawaharazuka, A. and Ramsteiner, M.E. (2003) *Japanese Journal of Applied Physics*, **42**, L1.

67 Aronov A.G. and Pikus, G.E. (1976) *Fiz Tekh Poluprovodn*, **10**, 1177. (1976) *Soviet Physics Semiconductors-USSR*, **10**, 698.

68 (a) Johnson M. and Silsbee, R.H. (1987) *Physical Review B-Condensed Matter*, **35**, 4959; (b) Johnson M. and Byers, J. (2003) *Physical Review B-Condensed Matter*, **67**, 125112.

69 (a) van Son, P.C., van Kempen, H. and Wyder, P. (1987) *Physical Review Letters*, **58**, 2271; (b) Schmidt, G., Richter, G., Grabs, P., Gould, C., Ferrand, D. and Molenkamp, L.W. (2001) *Physical Review Letters*, **87**, 227203.

70 Schmidt, G., Ferrand, D., Molenkamp, L.W., Filip A.T. and van Wees, B.J. (2000) *Physical Review B-Condensed Matter*, **62**, R4790.

71 Yu Z.G. and Flatte, M.E. (2002) *Physical Review B-Condensed Matter*, **66**, R201202.

72 Albrecht J.D. and Smith, D.L. (2002) *Physical Review B-Condensed Matter*, **66**, 113303.

73 Hershfield, S. and Zhao, H.L. (1997) *Physical Review B-Condensed Matter*, **56**, 3296.

74 Rashba, E.I. (2000) *Physical Review B-Condensed Matter*, **62**, R16267.

75 Fert A. and Jaffres, H. (2001) *Physical Review B-Condensed Matter*, **64**, 184420.

76 Yu Z.G. and Flatte, M.E. (2002) *Physical Review B-Condensed Matter*, **66**, 235302.

77 (a) Shen, M., Saikin, S. and Cheng, M.-C. (2005) *IEEE Transactions on Nanotechnology*, **4**, 40; (b) Shen, M., Saikin, S. and Cheng, M.-C. (2004) *Journal of Applied Physics*, **96**, 4319.

78 (a) Sze, S.M. (1981) *Physics of Semiconductor Devices*, Wiley, New York; (b) Monch, W. (1995) *Semiconductor Surfaces and Interfaces*, Springer, Berlin; (c) Tung, R.T. (1992) *Physical Review B-Condensed Matter*, **45**, 13509.

79 (a) Sanvito, S., Lambert, C.J., Jefferson, J.H. and Bratkovsky, A.M. (1999) *Physical Review B-Condensed Matter*, **59**, 11936; (b) Wunnicke, O., Mavropoulos, Ph., Zeller, R., Dederichs, P.H. and Grundler, D. (2002) *Physical Review B-Condensed Matter*, **65**, 241306.

80 (a) Ciuti, C., McGuire, J.P. and Sham, L.J. (2002) *Applied Physics Letters*, **81**, 4781; (b) Ciuti, C., McGuire, J.P. and Sham, L.J. (2002) *Physical Review Letters*, **89**, 156601.

81 Stephens, J., Berezovsky, J., Kawakami, R.K., Gossard, A.C. and Awschalom, D.D. (2004) cond-mat/0404244.

82 Žutić, I., Fabian, J. and Das Sarma S. (2002) *Physical Review Letters*, **88**, 066603.

83 Epstein, R.J., Malajovich, I., Kawakami, R.K., Chye, Y., Hanson, M., Petroff, P.M., Gossard, A.C. and Awschalom, D.D. (2002) *Physical Review B-Condensed Matter*, **65**, 121202.

84 (a) Kikkawa, J.M., Smorchkova, I.P., Samarth, N. and Awschalom, D.D. (1997) *Science*, **277**, 1284; (b) Kikkawa J.M. and Awschalom, D.D. (1999) *Nature*, **397**, 139; (c) Malajovich, I., Berry, J.J., Samarth, N. and Awschalom, D.D. (2001) *Nature*, **411**, 770; (d) Hagele, D., Oestreich, M., Rühle, W.W., Nestle, N. and Eberl, K. (1998) *Applied Physics Letters*, **73**, 1580.

85 (a) Mazin, I.I. (1999) *Physical Review Letters*, **83**, 1427; (b) Moruzzi, V.L., Janak, J.F. and Williams, A.R. (1978) *Calculated Electronic Properties of Metals*, Pergamon, New York.

86 Levanyuk A.P. and Osipov, V.V. (1981) *Soviet Physics Uspekhi*, **24**, 3. Usp Fiz Nauk, (1981) **133**, 427.

4
Physics of Computational Elements

Victor V. Zhirnov and Ralph K. Cavin

4.1
The Binary Switch as a Basic Information-Processing Element

4.1.1
Information and Information Processing

Information can be defined as a technically quantitative measure of distinguishability of a physical subsystem from its environment [1]. One way to create distinguishable states is by the *presence* or *absence* of material particles (information carrier) in a given location. For example, one can envision the representation of information as an arrangement of particles at specified physical locations as for instance, the depiction of the acronym "IBM" by atoms placed at discrete locations on the material surface (Figure 4.1).

Information of an arbitrary type and amount – such as letters, numbers, colors, or graphics specific sequences and patterns – can be represented by a combination of just two distinguishable states [1–3]. The two states – which are known as binary states – are usually marked as state 0 and state 1. The maximum amount of information, which can be conveyed by a system with just two states is used as a unit of information known as 1 bit (abbreviated from "binary digit"). A system with two distinguishable states forms a basis for *binary switch*.

The binary switch is a fundamental computational element in information-processing systems (Figure 4.2) which, in its most fundamental form, consists of:

- two states 0 and 1, which are equally attainable and distinguishable
- a means to control the change of the state (a gate)
- a means to read the state
- a means to communicate with other binary switches.

Nanotechnology. Volume 3: Information Technology I. Edited by Rainer Waser
Copyright © 2008 WILEY-VCH Verlag GmbH & Co. KGaA, Weinheim
ISBN: 978-3-527-31738-7

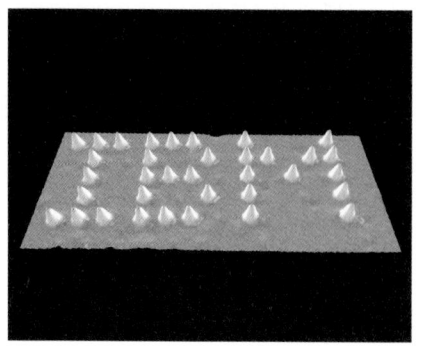

Figure 4.1 "IBM" representation via atoms. (Courtesy of IBM: http://www.almaden.ibm.com/vis/stm/atomo.html).

4.1.2
Properties of an Abstract Binary Information-Processing System

An elementary binary information-processing system consists of N binary switches connected in a certain fashion to implement a function. Each binary switch is characterized by a dimension L and a switching time t_{sw}. A related dimensional characteristic is the packing density (the number of binary switches per unit area). In order to increase the packing density, n_{bit}, the characteristic dimension, L, of the binary switch must decrease:

$$n_{bit} \sim \frac{1}{L^2} \tag{4.1}$$

Another fundamental characteristic of a binary switch is the *switching energy*, E_{sw}.

One indicator of the ultimate performance of an information processor, realized as an interconnected system of binary switches, is the *maximum binary throughput* (BIT); that is, the number of binary transitions per unit time per unit area.

$$\text{BIT} = \frac{n_{bit}}{t_{sw}} \tag{4.2}$$

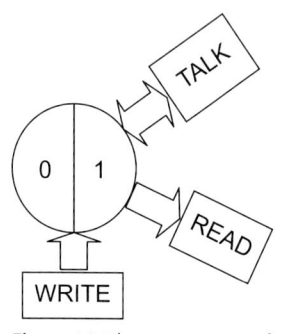

Figure 4.2 The constituents of an abstract binary switch.

One can increase the binary throughput by increasing the number of binary switches per unit area, n_{bit}, and/or decreasing the switching time – that is, the time to transition from one state to the other, t_{sw}.

It should be noted that, as each binary switching transition requires energy E_{sw}, the total power dissipation growth is in proportion to the information throughput:

$$P = \frac{n_{bit}}{t_{sw}} \cdot E_{sw} = \text{BIT} \cdot E_{sw} \quad (4.3)$$

The above analysis does not make any assumptions on the material system or the physics of switch operation. In the following sections we investigate the fundamental relations for n_{bit}, t_{sw}, E_{sw} and the corresponding implications for the computing systems.

4.2 Binary State Variables

4.2.1 Essential Operations of an Abstract Binary Switch

The three essential properties of a binary switch are *Distinguishability*, *Controllability* and *Communicativity*. It is said that a binary switch is *Distinguishable* if – and only if – the binary state (0 or 1) can be determined with an acceptable degree of certainty by a measurement (READ operation). The binary switch is *Controllable* if an external stimulus can reliably change the state of the system from 0 to 1 or from 1 to 0 (WRITE operation). The binary switch is *communicative* if it is capable of transferring its state to other binary switches (TALK operation).

4.2.2 The Use of Particles to Represent Binary Information

Information-processing systems represent system states in terms of physical variables. One way to create physically distinguishable states is by the *presence* or *absence* of material particles or fields in a given location. Figure 4.3a illustrates an abstract model for a binary switch the state of which is represented by different positions of a material particle. In principle, the particle can possess arbitrary mass, charge, and so on. The only two requirements for the implementation of a particle-based binary switch are the ability to detect the presence/absence of the particle in, for example the location x_1, and the ability to move the particle from x_0 to x_1 and from x_1 to x_0.

$\Pi_{correct}$ is defined as the probability that the binary switch is in the correct state at an arbitrary time after the command to achieve that state is given. (Alternatively, one can use the probability of error $\Pi_{err} = 1 - \Pi_{correct}$). A necessary condition for the distinguishability of a binary switch is

$$\Pi_{correct} > \Pi_{err} \quad (4.4)$$

4 Physics of Computational Elements

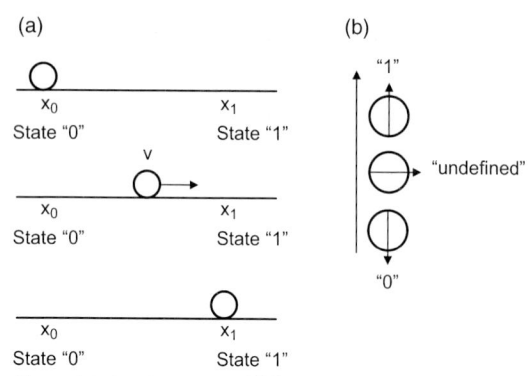

Figure 4.3 An abstract model for the operation of a binary switch formed (a) by different locations of material particles and (b) by opposite direction of the electron spin magnetic moment.

Or equivalently:

$$\Pi_{err} < 0.5 \tag{4.5}$$

As will be discussed below, in the physical realizations of binary switches, there always is some error probability ($\Pi_{err} > 0$) in the operation of the switch. As the error probability cannot exceed 0.5, in the following analysis we will use the condition in Equation 2.5 to estimate the parameters of a binary switch in the limiting case.

An elementary switching operation of a binary switch consists of three distinct steps. For example, consider the switch in Figure 4.3a switching from "0" to "1". The three steps are: (i) the initial STORE "0" mode; (ii) the transition CHANGE "0–1" mode; and (iii) the final STORE "1" mode. All three modes have characteristic times and can be described by the coordinate and velocity of the information carrier/material particle.

In STORE "0" the particle must be located in position $x = x_0$ and remain there for the time T_{state}. In CHANGE "0–1" mode, the state is *undefined*, as the particle is in the transition from x_0 to x_1 with a velocity $v_{01} > 0$ (for simplicity, velocity can be taken as the linear dimension of the switch divided by transition time). In STORE "1" ("0") the particle must be located in position $x = x_1$ (x_0) and have lifetime T_{state}. The switching time t_{sw} in this case is given by $t_{sw} = L/v_{01}$, where $L = x_1 - x_0$ is the linear size of the binary switch.

The question is, what are the requirements for T_{state} and t_{sw} in a binary switch for information processing? As binary logic operates with *two* logic states "0" and "1", while the binary switch has *three* physical states "0", "1" and UNDEFINED (i.e., CHANGE), if the READ operation of binary switch occurs in the UNDEFINED state, an error will result.

The conditions for maximum distinguishability of an ideal binary switch are:

- Unlimited lifetime of each state in the absence of control signal: $T_{state}^{max} \to \infty$ in STORE mode. More specifically, for example, synchronous circuits with clock, $T_{state} \in [T_{clock}, \infty[$, where T_{clock} is the clock period.

- Fast transition between binary states at the presence of control signal: $t_{sw} \to 0$ in CHANGE mode (say a negligible fraction of the clock period).

As $T_{state}^{max} \to \infty$ in STORE mode, the particle velocity in both 0 and 1 states must be zero, as the particle must be at rest, that is, $v_0 = v_1 = 0$; that is, the kinetic energy $E = \frac{mv^2}{2}$ of the particle should ideally be zero in both STORE modes. In the CHANGE mode, the average particle velocity is $\langle v_{01} \rangle > 0$, and $E > 0$. The switching time can then be estimated as

$$t_{sw} = \frac{L}{\langle v_{01} \rangle} = L\sqrt{\frac{m}{2E}} \qquad (4.6)$$

Equation 2.6 sets a limit for the switching speed in the non-zero distance case (non-relativistic approximation). Note that in binary switch operation, an amount of energy E must be supplied to the particle before the CHANGE operation begins, and taken out of the particle after the CHANGE.

If energy remains in the system, the information-bearing particle will oscillate between the two states with the period $2t_{sw}$. In an oscillating system, if friction is neglected, then energy is preserved (no dissipation) but the information state is not: the lifetime of each binary state $T_{state} \to 0$. The conditions (i) and (ii) of maximum distinguishability will be violated and such a system cannot act as a binary switch.

Alternatively, if one wishes to preserve the binary state (i.e., $T_{state} > 0$), the energy must be rapidly taken out from the system. The time of energy removal t_{out} must be less then half of the switching time: $t_{out} < \frac{t_{sw}}{2}$, otherwise an unintended transition to another state may occur. One rapid way to remove energy is by *thermal dissipation to the environment*. If instead, the aim is to remove the energy in a controllable manner, for example for a possible re-use, a faster switch will be needed which, according to Equation 4.6 will require a greater energy for its operation. It is concluded that non-zero energy dissipation is a necessary attribute of binary switch operations.

The above analysis considers binary switches, with states represented by the *presence* or *absence* of material particles (the information-defining particle or information carrier) in a given locations, for example the utilization of electrons as information carriers. As mentioned above, the electromagnetic field also can, in principle, be used to represent information. For example, a popular candidate is a binary switch that uses the electron spin magnetic moment, when the two opposing directions of the magnetic field represent "0" and "1" (Figure 4.3b).

4.3
Energy Barriers in Binary Switches

4.3.1
Operation of Binary Switches in the Presence of Thermal Noise

Consider again, a binary switch where the binary state is represented by particle location (see Figure 4.3a). Until now, it has been assumed that the information-

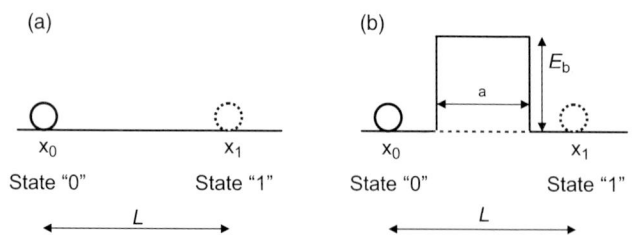

Figure 4.4 Illustration of an energy barrier to preserve the binary states in the presence of noise.

defining particle in the binary switch has zero velocity/kinetic energy, prior to a WRITE command. However, each material particle *at equilibrium* with the environment possesses kinetic energy of $1/2\, k_B T$ per degree of freedom due to thermal interactions, where k_B is Boltzmann's constant and T is temperature. The permanent supply of thermal energy to the system occurs via the mechanical vibrations of atoms (phonons), and via the thermal electromagnetic field of photons (background radiation).

The existence of random mechanical and electromagnetic stimuli means that the information carrier/material particle located in x_0 (Figure 4.4a) has a non-zero velocity in a non-zero-T environment, and that it will spontaneously move from its intended location. According to Equation 4.6, the state lifetime in this case will be

$$T_{state} \sim L\sqrt{\frac{m}{k_B T}} \tag{4.7}$$

For an electron-based switch $(m = m_e = 9.11 \times 10^{-31}\,\text{kg})$ of length $L = 1\,\mu\text{m}$ at $T = 300\,\text{K}$. Equation 4.8 gives T_{state} as ~ 15 ps, and hence the time before the system would lose distinguishability would be very small.

In order to prevent the state from changing randomly, it is possible to construct energy barriers that limit particle movements. The energy barrier, separating the two states in a binary switch is characterized by its height E_b and width a (Figure 4.4b).

The barrier height, E_b, must be large enough to prevent spontaneous transitions (errors). Two types of unintended transition can occur: "classical" and "quantum". The "classical" error occurs when the particle jumps over barrier, and this can happen if the kinetic energy of the particle E is larger than E_b. The corresponding "classic" error probability, Π_C, is obtained from the Boltzmann distribution as:

$$\Pi_C = \exp\left(-\frac{E_b}{k_B T}\right) \tag{4.8}$$

The presence of energy barrier of width a sets the minimum device size to be $L_{min} \geq a$.

4.3.2
Quantum Errors

Another class of errors, termed "quantum errors", occur due to quantum mechanical tunneling through the barrier of finite width a. If the barrier is too narrow, then spontaneous tunneling through the barrier will destroy the binary information. The conditions for significant tunneling can be estimated using the Heisenberg uncertainty principle, as is often carried out in texts on the theory of tunneling [4]:

$$\Delta x \Delta p \geq \frac{\hbar}{2} \tag{4.9}$$

The uncertainty relationship of Equation 4.9 can be used to estimate the limits of distinguishability. Consider again a "two-well" bit in Figure 4.4b. As is known from quantum mechanics, a particle can pass (tunnel) through a barrier of finite width, even if the particle energy is less than the barrier height, E_b. An estimate of how thin the barrier must be to observe tunneling can be made from Equation 4.9; for a particle at the bottom of the well, the uncertainty in momentum is $\sqrt{2mE_b}$, which gives:

$$\sqrt{2mE_b}\,\Delta x \approx \frac{\hbar}{2} \tag{4.10}$$

Equation 4.10 states that by initially setting the particle on one side of the barrier, it is possible to locate the particle on either side, with high probability, if Δx is of the order of the barrier width a. That is, the condition for losing distinguishability is $\Delta x \langle a$, and the minimum barrier width is:

$$a_{min} = a_H \approx \frac{\hbar}{2\sqrt{2mE_b}}, \tag{4.11}$$

where a_H is the *Heisenberg distinguishability length* for "classic to quantum transition".

For $a < a_H$, the tunneling probability is significant, and therefore particle localization is not possible. In order to estimate the probability of tunneling, Equation 4.10 can be re-written, taking into account the tunneling condition $a \leq \Delta x$:

$$\sqrt{2m}(a\sqrt{E_b}) \leq \frac{\hbar}{2} \tag{4.12}$$

From Equation 4.12, the "tunneling condition" can also be written in the form

$$1 - \frac{2\sqrt{2m}}{\hbar} a\sqrt{E_b} \geq 0, \tag{4.13}$$

Since for small x, $e^{-x} \sim 1 - x$, the tunneling condition then becomes

$$\exp\left(-\frac{2\sqrt{2m}}{\hbar} \cdot a \cdot \sqrt{E_b}\right) \geq 0 \tag{4.14}$$

The left-hand side of Equation 4.14 has the properties of probability. Indeed, it represents the tunneling probability through a rectangular barrier given by the Wentzel–Kramers–Brillouin (WKB) approximation [5]:

$$\Pi_{WKB} \sim \exp\left(-\frac{2\sqrt{2m}}{\hbar} \cdot a \cdot \sqrt{E_b}\right) \tag{4.15a}$$

This equation also emphasizes the parameters controlling the tunneling process, which include the barrier height E_b and barrier width a, as well as the mass m of the information-bearing particle. If separation between two wells is less than a, the structure of Figure 4.4b would allow significant tunneling. In fact, it is instructive to examine the physical meaning of Equation 4.11, which we marked as the condition of significant tunneling or "classic to quantum transition". Substituting Equation 4.11 into Equation 4.15a provides an estimate for tunneling probability through a rectangular barrier of width a_H:

$$\Pi_{WKB} \sim \exp\left(-\frac{2\sqrt{2m}}{\hbar} \cdot a_H \cdot \sqrt{E_b}\right) = \exp(-1) \approx 0.37 \tag{4.15b}$$

Thus, the Heisenberg distinguishability length a_H from Equation 4.11 corresponds to a tunneling probability of approximately 37%.

4.3.3
A Combined Effect of Classical and Quantum Errors

As discussed in the previous sections, there are two mechanisms of spontaneous transitions (errors) in binary switching: the over-barrier transition ("classic" error); and through-barrier tunneling ("quantum" error). The probabilities of the classic and quantum errors are given by Equations 4.8 and 4.15a, respectively. The joint error probability of the two mechanisms is [3]:

$$\Pi_{err} = \Pi_C + \Pi_Q - \Pi_C \cdot \Pi_Q \tag{4.16}$$

Or, from Equations 4.8 and 4.15a, we obtain:

$$\Pi_{err} = \exp\left(-\frac{E_b}{k_B T}\right) + \exp\left(\frac{2\sqrt{2m}}{\hbar} \cdot a\sqrt{E_b}\right) - \exp\left(-\frac{\hbar E_b + 2ak_B T\sqrt{2mE_b}}{\hbar k_B T}\right) \tag{4.17}$$

4.4
Energy Barrier Framework for the Operating Limits of Binary Switches

4.4.1
Limits on Energy

The minimum energy of binary transition is determined by the energy barrier. The work required to suppress the barrier is equal or larger than E_b; thus, the minimum energy of binary transition is given by the minimum barrier height in binary switch. The minimum barrier height can be found from the distinguishability condition [Equation 4.5], which requires that the probability of errors $\Pi_{err} < 0.5$. First, the case is considered when only "classic" (i.e., thermal) errors can occur. In this case, according to Equation 4.8:

$$\Pi_{err} = \Pi_C = \exp\left(-\frac{E_b}{k_B T}\right) \tag{4.18}$$

4.4 Energy Barrier Framework for the Operating Limits of Binary Switches

These classic transitions represent the thermal (Nyquist–Johnson) noise. By solving Equation 4.18 for $\Pi_{err} = 0.5$, we obtain the Boltzmann's limit for the minimum barrier height, E_{bB}

$$E_{bB} = k_B T \ln 2 \approx 0.7 k_B T \qquad (4.19)$$

Equation 4.19 corresponds to the minimum barrier height, the point at which distinguishability of states is completely lost due to thermal over-barrier transitions. In deriving Equation 4.19, tunneling was ignored – that is, the barrier width is assumed to be very large, $a \gg a_H$.

Next, we consider the case where only quantum (i.e., tunneling) errors can occur. In this case, according to Equation 4.15a

$$\Pi_{err} = \Pi_Q \sim \exp\left(-\frac{2\sqrt{2m}}{\hbar} \cdot a \cdot \sqrt{E_b}\right) \qquad (4.20)$$

By solving Equation 4.20 for $\Pi_{err} = 0.5$, we obtain the Heisenberg's limit for the minimum barrier height, E_{bH}

$$E_{bH} = \frac{\hbar^2}{8ma^2} (\ln 2)^2 \qquad (4.21)$$

Equation 4.21 corresponds to a narrow barrier, $a \sim a_H$, the point at which distinguishability of states is lost due to tunneling transitions. In deriving Equation 4.21, over-barrier thermal transitions were ignored – that is, the temperature was assumed to be close to absolute zero, $T \to 0$.

Now, we consider the case when both thermal and tunneling transitions contribute to the errors in a binary switch. In this case, the total error probability is given by Equation 4.17. An approximate solution of Equation 4.17 for $\Pi_{err} = 0.5$ is

$$E_{b\,min} = k_B T \ln 2 + \frac{\hbar^2}{8ma^2} (\ln 2)^2 \qquad (4.22)$$

Equation 4.22 provides a generalized value for minimum energy per switch operation at the limits of distinguishability, that takes into account both classic and quantum transport phenomena. The graph in Figure 4.5 shows the numerical solution of Equation 4.17 and its approximate analytical solution given by Equation 4.22 for $\Pi_{err} = 0.5$. Is it clearly seen that for $a > 5$ nm, the Boltzmann's limit, $E_{bB} = k_B T \ln 2$, is a valid representation of minimum energy per switch operation, while for $a < 5$ nm, the minimum switching energy can be considerably larger.

4.4.2
Limits on Size

The minimum size of a binary switch L cannot be smaller then the distinguishability length a_H. From both Equations 4.11 and 4.19, one can estimate the Heisenberg's length for the binary switch operation at the Boltzmann's limit of energy:

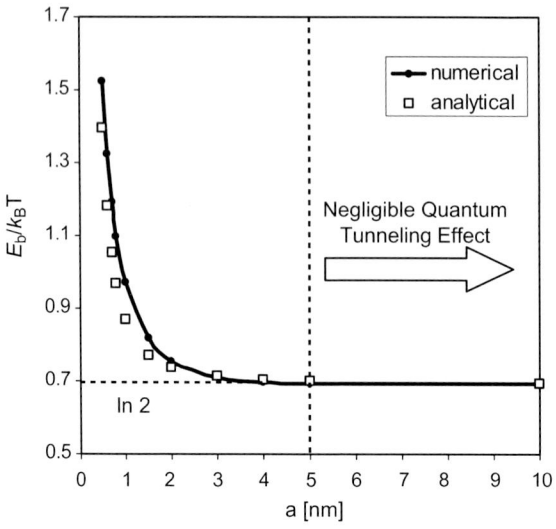

Figure 4.5 Minimum energy per switch operation as a function of minimum switch size.

$$a_{HB} = \frac{\hbar}{2\sqrt{2mk_B T \ln 2}} \qquad (4.23)$$

For electrons ($m = m_e$) at $T = 300$ K, we obtain $a_{HB} \sim 1$ nm.

The distinguishability length a_{HB} defines both the minimum size of the switch and the separation between the two neighboring switches. Thus, the maximum density of binary switches is:

$$n_{max} \leq \frac{1}{(2a_{HB})^2} \qquad (4.24)$$

For electron-based binary switches at 300 K ($a_{HB} \sim 1$ nm) and $n_{max} \sim 10^{13}$ cm^{-2}.

4.4.3
Limits on Speed

The next pertinent question is the minimum switching time, τ_{min}, which can be derived from the Heisenberg relationship for time and energy:

$$\Delta E \Delta t \geq \frac{\hbar}{2} \qquad (4.25a)$$

or

$$\tau_{min} \cong \frac{\hbar}{2\Delta E} \qquad (4.25b)$$

Equation 4.25b is an estimate for the maximum speed of dynamic evolution [6] or the maximum passage time [7]. It represents the zero-length approximation for the switching speed, in contrast with Equation 4.6, which is distance-dependent. If $L \sim a_H$, then Equation 4.6 converges to Equation 4.25b.

For the Boltzmann's limit, $E = E_{bB}$ [Equation 4.19], we obtain

$$\tau_{\min B} \cong \frac{\hbar}{2k_B T \ln 2} \approx 2 \cdot 10^{-14} \text{ s} \tag{4.26}$$

It should be noted that Equation 4.26 is applicable to all types of device, and no specific assumptions were made about any physical device structure.

4.4.4
Energy Dissipation by Computation

Using Equations 4.1–4.3 allows one to estimate power dissipation by a chip containing the smallest binary switches, $L \sim a_{HB}$ [Equation 4.23], packed to maximum density [Equation 4.24] and operating at the lowest possible energy per bit [Equation 4.19].

The power dissipation per unit area of this limit technology is given by:

$$P = \frac{n_{\max} E_{\min B}}{\tau_{\min B}} \sim \frac{10^{13} \text{ cm}^{-2} \cdot 3 \cdot 10^{-21} \text{ J}}{2 \cdot 10^{-14} \text{ s}} \sim 2 \cdot 10^6 \frac{\text{W}}{\text{cm}^2} \tag{4.27}$$

The energy density bound in the range of MW cm^{-2} obtained by invoking $k_B T \ln 2$ as the lower bound for the device energy barrier height is an astronomical number. If known cooling methods are employed, it appears that that heat-removal capacity of several hundred W cm^{-2} represents a practically achievable limit. The practical usefulness of alternative electron-transport devices may be derived from lower fabrication costs or from specific functional behavior; however, the heat removal challenge will remain.

4.5
Physics of Energy Barriers

The first requirement for the physical realization of any arbitrary switch is the creation of distinguishable states within a system of such material particles. The second requirement is the capability for a conditional change of state. The properties of distinguishability and conditional change of state are two fundamental and essential properties of a material subsystem that represents binary information. These properties are obtained by creating and control energy barriers in a material system.

The physical implementation of an energy barrier depends on the choice of the state variable used by the information processing system. The energy barrier creates a local change of the potential energy of a particle from a value U_1 at the generalized coordinate q_1 to a larger value U_2 at the generalized coordinate q_2. The difference $\Delta U = U_2 - U_1$ is the barrier height. In a system with an energy barrier, the force exerted on a particle by the barrier is of the form $F = \frac{\partial U}{\partial q}$. A simple illustration of a one dimensional barrier in linear spatial coordinates, x, is shown in Figure 4.6a. It should

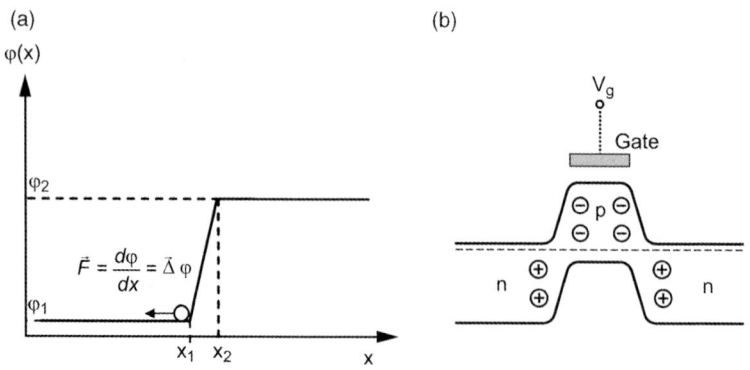

Figure 4.6 An illustration to the energy barrier in a material system. (a) Abstraction; (b) physical implementation by doping of a semiconductor.

be noted, that the spatial energy changes in potential energy require a finite spatial extension ($\Delta x = x_2 - x_1$) of the barrier (Figure 4.6a and b). This spatial extension defines a minimum dimension of energy barrier, a_{min}: $a_{min} > 2\Delta x$. In this section, we consider the physics of barriers for electron charge, electron spin, and optical binary switches.

4.5.1
Energy Barrier in Charge-Based Binary Switch

For electrons, the basic equation for potential energy is the Poisson equation

$$\nabla^2 \varphi = -\frac{\rho}{\varepsilon_0}, \tag{4.28}$$

where ρ is the charge density, $\varepsilon_0 = 8.85 \times 10^{-12}\,\mathrm{F\,m^{-1}}$ is the permittivity of free space, and φ is the potential: $\varphi = U/e$. According to Equation 4.28, the presence of an energy barrier is associated with changes in charge density in the barrier region. The barrier-forming charge is introduced in a material, for example by the doping of semiconductors. This is illustrated in Figure 4.6b, for a silicon n-p-n structure where the barrier is formed by ionized impurity atoms such as P^+ in the n-region and B^- in the p-region. The barrier height E_{b0} depends on the concentration of the ionized impurity atoms [8]:

$$E_{b0} \approx k_B T \ln \frac{N_A^- N_B^+}{n_i}, \tag{4.29}$$

where, N_A^-, N_B^+, and n_i are the concentration of negatively charged impurities (acceptors), positively charged impurities (donors), and the intrinsic carrier concentration in a semiconductor, respectively.

The minimum barrier extension is given by the Debye length [8]:

$$L_D \approx \sqrt{\frac{\varepsilon_0 \varepsilon k_B T}{e^2 N_D}}, \tag{4.30}$$

where ε is the relative dielectric permittivity of a semiconductor, and $N_D = N_A^- = N_B^+$ (abrupt p-n junction approximation). The maximum concentration of electrically active dopants N_{max} is close to the density of states in the conduction, N_c, and valence bands, N_v of the semiconductor. For silicon, $N_{max} \sim 10^{19}$ cm^{-3} [8], and it follows that $L_{Dmin} \sim 1.3$ nm. The minimum barrier length therefore is $a_{min} = 2L_{Dmin} \sim 2.6$ nm.

The energy diagram of Figure 4.6b is typical of many semiconductor devices, for example, field effect transistors (FET). The barrier region corresponds to the FET channel, while the wells correspond to the source and drain. In order to enable electron movement between the source and drain, the barrier height must be decreased (ideally suppressed to zero). To do this, the amount of charge in the barrier region needs to be changed, according to Equation 4.28. A well-known relationship connects the electrical potential difference $\Delta\varphi = V$ and charge, Δq, through capacitance

$$C = \frac{\Delta q}{\Delta \varphi} \quad (4.31)$$

In field-effect devices, in order to change the charge distribution in the barrier region – and hence lower the barrier – a voltage is applied to an external electrode (gate), which forms a capacitor with the barrier region (in bipolar devices, external charge is injected in the barrier region to control the barrier height). When voltage V_g is applied to the barrier region, the barrier will change from it initial height E_{b0} (determined by impurity concentration):

$$E_b = E_{b0} - eV_g \quad (4.32)$$

The voltage needed to suppress the barrier from E_{b0} to zero (the threshold voltage) is:

$$V_t = \frac{E_{b0}}{e} \quad (4.33)$$

Thus, the operation of all charge transport devices involves charging and discharging capacitances to change barrier height, thereby controlling charge transport in the device. When a capacitor C is charged from a constant voltage power supply V_g, the energy E_{dis} is dissipated, that is, it is converted into heat [9]:

$$E_{dis} = \frac{CV_g^2}{2} \quad (4.34)$$

The minimum energy needed to suppress the barrier (by charging the gate capacitor) is equal to the barrier height E_b. Restoration of the barrier (by discharging gate capacitance) also requires a minimum energy expenditure of E_b. Thus, the minimum energy required for a full switching cycle is at least $2E_b$.

It should be noted that in the solid-state implementation of binary switch, the number of electrons in both wells is large. This is different from an abstract system having only one electron (see above). In a multi-electron system, the electrons strike the barrier from both sides, and the binary transitions are determined by the net electron flow, as shown in Figure 4.7.

4 Physics of Computational Elements

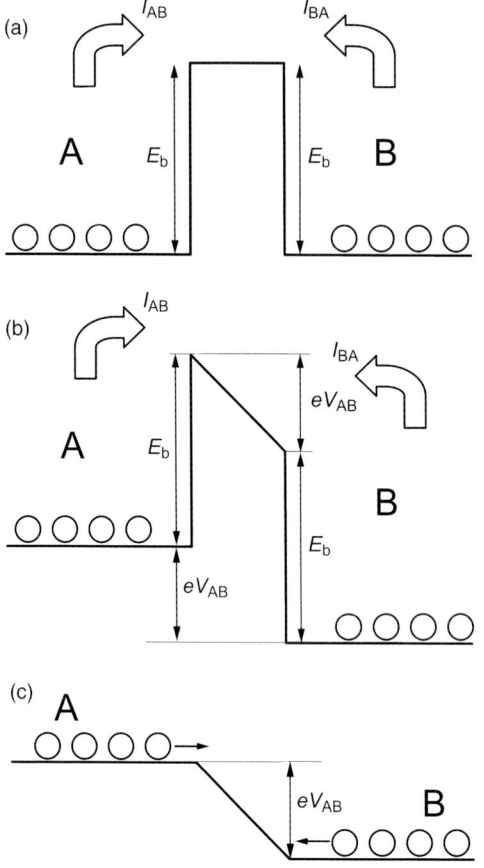

Figure 4.7 The fundamental operation of multi-electron binary switch. (a) No energy difference between wells **A** and **B**, resulting in a symmetric energy diagram. (b) Energy asymmetry is created due to energy difference eV_{AB} between the wells **A** and **B**. (c) State CHANGE operation: the barrier height E_b is suppressed by applying gate potential V_g, while energy difference eV_{AB} between the wells **A** and **B**.

Let N_0 be the number of electrons that strike the barrier per unit time. Thus, the number of electrons N_A that transition over the barrier from well **A** per unit time is

$$N_A = N_0 \exp\left(-\frac{E_b}{k_B T}\right) \tag{4.35}$$

The corresponding current I_{AB} is

$$I_{AB} = e \cdot N_A = eN_0 \exp\left(-\frac{E_b}{k_B T}\right) \tag{4.36}$$

Electrons in well **B** also can strike the barrier and therefore contribute to the over-barrier transitions with current I_{BA}. Thus, the net over-barrier current is

$$I = I_{AB} - I_{BA} \tag{4.37}$$

The energy diagram of Figure 4.7a is symmetric, hence $I_{AB} = I_{BA}$, and $I = 0$. Therefore, no binary transitions occur in the case of symmetric barrier. In order to enable the rapid and reliable transition of an electron from well **A** to well **B**, an energy asymmetry between two wells must be created. This is achieved by energy difference eV_{AB} between the wells **A** and **B** (Figure 4.7b).

For such an asymmetric diagram, the barrier height for electrons in the well **A** is E_b, and for electrons in the well **B** is $(E_b + eV_{AB})$. Correspondingly, from Equations 4.36 and 4.37 the net current is

$$\begin{aligned}I &= eN_0 \exp\left(-\frac{E_b}{k_B T}\right) - eN_0 \exp\left(-\frac{E_b - eV_{AB}}{k_B T}\right) \\ &= eN_0 \exp\left(-\frac{E_b}{k_B T}\right)\left[1 - \exp\left(-\frac{eV_{AB}}{k_B T}\right)\right]\end{aligned} \tag{4.38a}$$

By substituting Equation 4.32 for E_b, we obtain:

$$I = eN_0 \exp\left(-\frac{E_{b_0} - eV_g}{k_B T}\right)\left[1 - \exp\left(-\frac{eV_{AB}}{k_B T}\right)\right] \tag{4.38b}$$

By expressing E_{b0} as eV_t from Equation 4.33 and using the conventional notations $I = I_{ds}$ (source-drain current) and $V_{AB} = V_{ds}$ (source-drain voltage), we obtain the equation for the subthreshold *I–V* characteristics of FET [10]:

$$I_{ds} = I_0 \exp\left(\frac{e(V_g - V_t)}{k_B T}\right)\left[1 - \exp\left(-\frac{eV_{ds}}{k_B T}\right)\right] \tag{4.38c}$$

An example plot of Equation 4.38c is shown in Figure 4.8.

The minimum energy difference between the wells, eV_{ABmin}, can be estimated based on the distinguishability arguments for CHANGE operation, when is E_b is

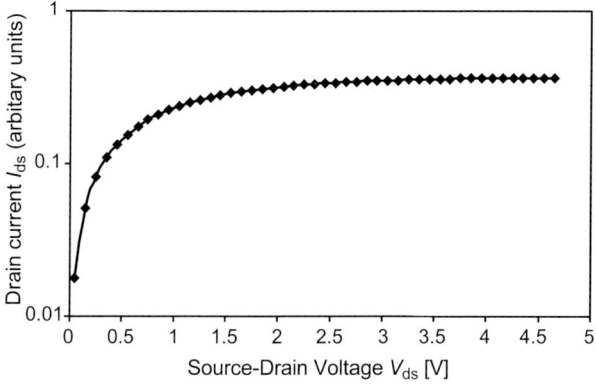

Figure 4.8 A source-drain *I-V* characteristic derived from the energy barrier model for a charge-based binary switch.

suppressed by applying the gate voltage, for example $E_b = 0$ (Figure 4.7c). For a successful change operation, the probability that each electron flowing from well **A** to well **B** is not counterbalanced by another electron moving from well **B** to well **A** should be less than 0.5 in the limiting case. The energy difference eV_{AB} forms a barrier for electrons in well **B**, but not for electrons in well **A**, and therefore, from Equation 4.18 we obtain

$$eV_{ABmin} = E_{bmin} = k_B T \ln 2 \tag{4.39}$$

If N is the number of electrons involved in the switching transition between two wells, the total minimum switching energy is

$$E_{SWmin} = 2E_b + NeV_{AB} = (N+2)k_B T \ln 2 \tag{4.40a}$$

If $N = 1$, then

$$E_{SWmin} = 3k_B T \ln 2 \approx 10^{-20} \text{ J} \tag{4.40b}$$

4.5.2
Energy Barrier in Spin-Based Binary Switch

In addition to charge, e, electrons possess intrinsic angular momentum (spin). As result, they also possess a permanent magnetic moment [11]:

$$\mu_s = \pm \frac{1}{2} g \cdot \mu_B \tag{4.41}$$

where μ_B is the Bohr magneton, $\mu_B = \frac{e\hbar}{2m_e}$, and g is the coupling constant known as the Landé gyromagnetic factor or g-factor. For free electrons and electrons in isolated atoms $g_0 = 2.00$. In solids, consisting of a large number of atoms, the effective g-factor can differ from g_0.

The energy of interaction, $E_{\mu-B}$, between a magnetic moment $\vec{\mu}$ and a magnetic field $\vec{B}$ is:

$$E_{\mu-B} = -\vec{\mu} \cdot \vec{B} \tag{4.42}$$

For the electron spin magnetic moment in a magnetic field applied in the z direction, the energy of interaction takes two values, depending on whether the electron spin magnetic moment is aligned or anti-aligned with the magnetic field. From Equations 4.41 and 4.42 one can write, assuming $g = 2$

$$E_{\uparrow\uparrow} = -\frac{e\hbar}{2m_e} \cdot B_z$$

$$E_{\uparrow\downarrow} = +\frac{e\hbar}{2m_e} \cdot B_z \tag{4.43}$$

The energy difference between the aligned and anti-aligned states represents the energy barrier in the spin binary switch and is

$$E_b = E_{\uparrow\downarrow} - E_{\uparrow\uparrow} = 2\mu_B B_z \tag{4.44}$$

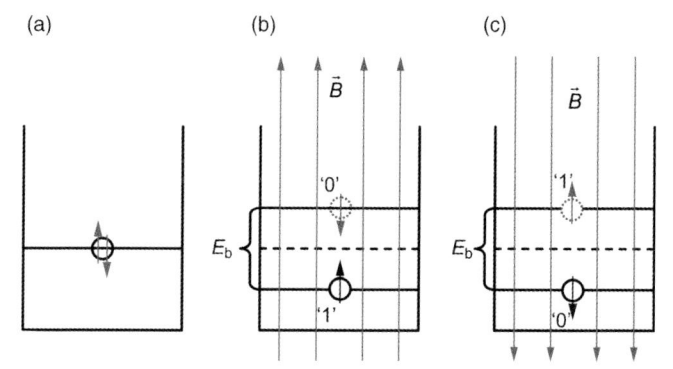

Figure 4.9 An abstract model of a single spin binary switch. (a) $B = 0$, the two states are indistinguishable. (b, c) $B \neq 0$, the two binary states are separated by the energy gap E_b.

Equations 4.43 and 4.44 represent a physical phenomenon known as Zeeman splitting [11]. The operation of a single spin binary switch is illustrated in Figure 4.9. In the absence of an external magnetic field, there is equal probability that the electron has magnetic moment $+\mu_B$ or $-\mu_B$; that is, the two states are indistinguishable (Figure 4.9a). When an external magnetic field is applied (Figure 4.9b and c), the two states are separated in energy. The lower energy state has higher probability of population and this represents the binary state "1" (Figure 4.9b) or "0" (Figure 4.9c) in this system. Binary switching occurs when the external magnetic field changes direction, as shown in Figure 4.9b and c.

This abstraction, while very simple, applies to all types of spin devices, at equilibrium with the thermal environment, including for example proposed spin transport devices [12–14] and coupled spin-polarized quantum dots [15].

The barrier-forming magnetic field B can be either a built-in field formed by a material layer with a permanent magnetization, or created by an external source (e.g., an electromagnet). In both cases, the change in the direction of the external magnetic field required for binary switching is produced by an electric current pulse in one of two opposite directions. The need for two opposite directions of electrical current flow requires additional electrical binary switches to control current flow in the electromagnet.

Some generic electrical circuits to manipulate the barrier height are a spintronic binary switch are shown in Figure 4.10. These require four electrical binary switches in the case of single power supply (Figure 4.10a), or two binary switches in the case of two power supplies (Figure 4.10b).

Thus, each spin-based binary switch requires two or four "servant" charge-based binary switches. This will result in larger area per device and also a larger energy consumption per operation, as compared to the charge-based switches. As the minimum switching energy of one charge-based device is $\sim 3k_B T$ [according to Equation 4.40b], the minimum switching energy of a spin device E_{spin} for single power supply scheme is

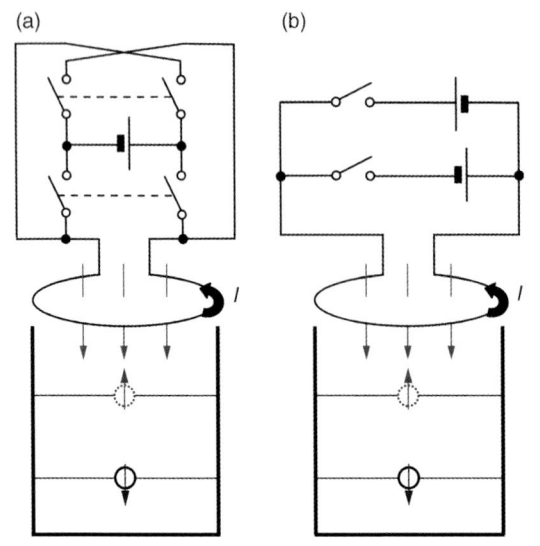

Figure 4.10 A generic electrical circuit to control spin binary switch. (a) Single power supply scheme; (b) double power supply scheme.

$$E_{spin} = E_M + 12k_B T \tag{4.45}$$

where E_M is the energy required to generate the magnetic field to change the spin state.

One possible way to address the "electrical" challenge for spin devices is to change the spin control paradigm. The paradigm described above uses a system of binary switches, each of which can be controlled independently by an external stimulus, and each switch can, in principle control any other switch in the system. However, it is not clear whether a spin state-based device can be used to control the state of subsequent spin devices without going through an electrical switching mechanism as discussed above. Although it is possible that local interactions may be used to advantage for this purpose [15, 16], feasibility assessments of these proposals in general information processing applications are clearly required.

Let us now consider a hypothetical single spin binary switch that, ideally, might have atom-scale dimensions. At thermal equilibrium there is a probability of spontaneous transition between spin states "1" and "0" in accordance with Equation 4.18. Correspondingly, for $\Pi_{err} < 0.5$, according to Equation 4.19 the energy separation between to state should be larger than $k_B T \ln 2$

$$2\mu_B B_{min} = k_B T \ln 2 \tag{4.46}$$

From Equations 4.19 and 4.44 we can obtain the minimum value of B for a switch operation

$$B_{min} = \frac{k_B T \ln 2}{2\mu_B} = \frac{m_e}{e\hbar} k_B T \ln 2 \tag{4.47}$$

Table 4.1 Maximum magnetic fields and limiting factors (from Ref. [17]).

Magnet	B_{max} [T]	Limiting factor
Conventional magnetic-core electromagnet	~2	Permeability saturation of the of magnetic core
Steady-field air-core NbTi and Nb$_3$Sn superconducting electromagnet	~20	Critical magnetic field
Steady-field air-core water-cooled electromagnet	~30	Joule heating
Pulsed-field hybrid electromagnet	~50	Maxwell stress

At $T = 300$ K, Equation 4.47 results in $B_{min} \approx 155$ Tesla (T), which is much larger than can be practically achieved (a summary of the technologies used to generate high magnetic fields is provided in Table 4.1).

One of the most difficult problems of very high magnetic field is the excessive power consumption and Joule heating in electromagnets. The relationship between power consumption, P, and magnetic field B, in an electromagnet is [17]:

$$P \sim B^2 \tag{4.48}$$

Moreover, as the production of magnetic fields with $B > 10$ T requires many megawatts of power, these magnetic field-production systems have large dimensions and a mass of many tons (see Table 4.2).

Thus, it is concluded that single electron spin devices operating at room temperature would require local magnetic fields greater than have been achieved to date with large-volume apparatus.

In multi-spin systems, it is possible to increase the magnetic moment μ and therefore, to decrease the magnitude of the external magnetic field B required for binary switch operation. The increase of μ may be due to an increase in number of co-aligned spins, which results in collective effects such as paramagnetism and ferromagnetism.

Table 4.2 Examples of practical implementations of the sources of magnetism.

Magnet	B [T]	P	Mass	Comments
Small bar magnet	~0.01	—	~g	—
Small neodymium–iron–boron magnet	~0.2	—	~g	—
Big Magnetic-core electromagnet	~2	~100 W	~kg	—
Steady-field superconducting electromagnet [19]	~16	~MW	~tons	Cryogenic temperatures
Current status of Pulse Magnet Program at National High Magnetic Field Laboratory [18]	60–65	~MW	~tons	Cryogenic temperatures; a 30-min cooling time between shots Lifetime ~400 cycles

4.5.3
Energy Barriers for Multiple-Spin Systems

In a system of N spins in an external magnetic field, there are $N_{\uparrow\uparrow}$ spin magnetic moments parallel to the external magnetic field, and the resulting magnetic moment is

$$\mu = \mu_B \cdot N_{\uparrow\uparrow} = \mu_B N(1 - \Pi_{err}) = \mu_B N\left(1 - \exp\left(-\frac{\mu_B B}{k_B T}\right)\right) \quad (4.49)$$

As seen above, for all practical cases $\mu_B B \ll k_B T$, and since $(1 - e^x) \approx x$, for $x \to 0$, there results

$$\mu \approx \frac{N \mu_B^2 B}{k_B T} \quad (4.50)$$

Equation 4.50 is known as the Curie law for paramagnetism [11]. From Equations 4.50 and 4.44 one can calculate the minimum number of electron spins required for spin binary switch operating at realistic magnitudes of the magnetic field:

$$N_{min} \approx \frac{\ln 2}{2}\left(\frac{k_B T}{\mu_B B}\right) \quad (4.51)$$

For example, for $B = 0.1$ T (a small neodymium–iron–boron magnet; see Table 4.1), $N_{min} \sim 7 \times 10^6$. If the number of electrons with unpaired spins per atom is f (f varies between 1 and 7 for different atoms), then the number of atoms needed is N_{min}/f. Correspondingly, one can estimate the minimum critical dimension a_{min} of the binary switch as:

$$a_{min} \sim \left(\frac{N_{min}}{f \cdot n_V}\right)^{\frac{1}{3}}, \quad (4.52)$$

where n_V is the density of atoms in the material structure. Assuming an atomic density close to the largest known in solids, $n_V = 1.76 \times 10^{23}$ cm^{-3} (the atomic density of diamond) and $B \sim 0.1$ T, we obtain $a_{min} \sim 41$ nm for $f = 1$ and $a_{min} \sim 22$ nm for $f = 7$. Thus, it is concluded that for reliable operation at moderate magnetic fields, the physical size of a multi-spin-based binary switch is larger than the ultimate charge-based devices. The effect of collective spin behavior is currently used, for example, in magnetic random access memory (MRAM) [14] and in electron spin resonance (ESR) technologies.

As an example, it is estimated that the energy needed to operate a spin-based binary switch for the case when the magnetic field is produced by electrical current I in a circular loop of wire surrounding the switch, as shown in Figure 4.10. The magnetic field in the center of the loop is [20]:

$$B = \frac{\mu_0 I}{2r}, \quad (4.53)$$

where μ_0 is the magnetic permeability of free space.

4.5 Physics of Energy Barriers

In order to switch the external magnetic field, for example from zero to $+\vec{B}$ or from $+\vec{B}$ to $-\vec{B}$, work must be done. If the magnetic field is formed by electrical current I produced by an external voltage source, then the energy dissipated by one half of switching cycle (e.g., the rise from zero to $+\vec{B}$):

$$E_{dis} = \frac{LI^2}{2} \qquad (4.54)$$

where L is electrical inductance. Equation 4.54 is analogous to the energy of $CV^2/2$ dissipated in charge-based devices [Equation 4.34].

If magnetic field needs to be sustained for the time period t, then additional energy will be dissipated due to resistance R, and thus the total energy dissipation is:

$$E_{dis} = \frac{LI^2}{2} + I^2 R \cdot t \qquad (4.55)$$

If the magnetic field does not need to be sustained (e.g., in ferromagnetic devices), after switching the current must be reduced to zero, in which case another amount of energy of Equation 4.54 is dissipated. Thus, the energy expenditure needed to generate the magnetic field E_M [see Equation 4.55] in a full switching cycle is

$$E_M = LI^2 \qquad (4.56)$$

By definition, electrical inductance L is a proportionality factor connecting magnetic flux Φ and the electric current I that produces the magnetic field [20]:

$$\Phi = L \cdot I \qquad (4.57)$$

The magnetic flux in turn is defined as

$$\Phi = B \cdot A \qquad (4.58)$$

where A is the area, $A = \pi r^2$, for the case of circular loop.

From Equations 4.53, 4.57 and 4.58, the inductance of a circular loop of radius r is approximately:

$$L \approx \frac{\pi \mu_0 r}{2} \qquad (4.59)$$

Equation 4.64 is an approximation assuming constant magnetic field inside the loop and ignoring the effects of wire thickness.[1]

By combining Equations 4.53, 4.56 and 4.59, we obtain:

$$E_M = \frac{2\pi}{\mu_0} r^3 B^2 \qquad (4.60)$$

For the minimum device size given by Equation 4.52, and noting that $r \geq a_{min}$ as obtained from Equations 4.19 and 4.60:

---

[1] An accurate result for the inductance of the circular loop is $L = \mu_0 r[\ln(8r/b) - 7/4]$, where b is the radius of the wire [21]. This differs from Equation 4.59 by a factor of 1.6–3 for a realistic range of r/b ratios of 10–100.

$$E_{M\,min} = \frac{\pi \ln 2}{\mu_0 n_V} \left(\frac{k_B T}{\mu_B}\right)^2 \qquad (4.61)$$

For the largest atomic density of solids, $n_V = 1.76 \times 10^{-23}\,\text{cm}^{-3}$ (diamond), we obtain:

$$E_{M\,min} \approx 2 \cdot 10^{-18}\,\text{J} \approx 480 k_B T \qquad (4.62)$$

4.5.4
Energy Barriers for the Optical Binary Switch

Optical digital computing was – and still is – considered by some as a viable option for massive information processing [22]. Sometimes, it is referred to as "computing at speed of light" [23] and, indeed, photons do move at the speed of light. At the same time, a photon cannot have a speed other then the speed of light, c, and therefore it cannot be confined within a binary switch of finite spatial dimensions.

The minimum dimension of optical switch is given by the wavelength of light, λ. If $a_{min} < \lambda$, there is high probability that the light will not "sense" the state – that is, the error probability will increase. For visible light, $a_{min} \sim 400\,\text{nm}$.

Binary state control in the optical switch is accomplished by local changes in optical properties of the medium, such as the refraction index, reflectivity or absorption, while photons are used to read the state. In many cases, the changes in optical properties are related to a rearrangement of atoms under the influence of electrical, optical, or thermal energy. The energy barrier in this case is therefore related to either inter-atomic or inter-molecular bonds. Examples of such optical switches are liquid crystal spatial light modulators [22] and non-linear interference filters [22]. Another example is a change in refractive index as the result of a crystalline-to amorphous phase change, which is used for example in the rewritable CD. The minimum switching energy of this class of optical switches is related to the number of atoms, N, and therefore to the size L. In the limiting case, $L \sim a_{min} \sim \lambda$. In order for the atoms of an optical switch to have a distinguishable change of their position, the energy supply to each atom should be larger than $k_B T$. The total switching energy is therefore:

$$E \sim N \cdot k_B T \qquad (4.63)$$

For a minimum energy estimate, we must consider the smallest possible N, which corresponds to an single-atom plane of size λ. If optical switch materials have an atomic density n, then one obtains:

$$E \sim n^{\frac{2}{3}} \cdot \lambda^2 \cdot k_B T \qquad (4.64)$$

For most solids, $n = 10^{22}\text{–}10^{23}\,\text{cm}^{-3}$. Taking $n = 5 \times 10^{22}\,\text{cm}^{-3}$, $T = 300\,\text{K}$, and $a_{min} \sim 400\,\text{nm}$, we obtain $E \sim 10^{14}\,\text{J}$, which is in agreement with estimates of the physical limit of switching energy of optical digital switches, as reported in the literature [22].

Optical switches may also be based on electroabsorption. In these devices, the absorption changes by the application of an external electric field that deforms the energy band structure. One example that has attracted considerable interest for practical application is the Quantum-Confined Stark Effect [24]. If an electrical field is applied to a semiconductor quantum well, the shape of the well is changed, perhaps from rectangular to triangular. As result, the position of energy levels also changes, and this affects the optical absorption. As the formation of an electric field requires changes in charge distribution [Equation 4.28], the analysis of electroabsorption optical switch is analogous to a charge-based switch, where the energetics is determined by charging and discharging of a capacitor [Equations 4.31 and 4.32]. It should be noted that the capacitance of an optical switch is considerably larger then the capacitance of electron switch, because of larger capacitance area of the optical switch ($\sim \lambda^2$). By using the estimated minimum size of an electron switch $a_{emin} \sim 1$ nm (as estimated in Section 4.4.2), and taking into account Equation 4.40b, we obtain an estimate for the switching energy of an electroabsorption device:

$$E \sim 3k_B T \frac{\lambda^2}{a_{e\ min}^2} \approx 1.2 \cdot 10^{-20} \text{ J} \cdot \frac{(400 \text{ nm})}{(1 \text{ nm})} \approx 10^{-15} \text{ J} \qquad (4.65)$$

The result from Equation 4.65 is in an agreement with estimates of physical limit of electro-absorption optical switch, as detailed in the literature [22, 25].

This energy barrier – and therefore the switching energy for an optical binary switch – is relatively high, with estimates for the theoretical limit for optical device switching energy varying between 10^{-14} and 10^{-15} J, for different types of optical switch [22]. It should be noted that the switching speed is the speed of re-arrangement for atoms or for charge in the material, and is not related to the speed of light.

4.6
Conclusions

Based on the idea that information is represented by the state of a physical system – for example, the location of a particle – we have shown that energy barriers play a fundamental role in evaluating the operating limits of information-processing systems. In order for the barrier to be useful in information-processing applications, it must prevent changes in the state of the processing element with high probability, and it also must support rapid changes of state when an external CHANGE command is given. If one examines the limit of tolerable operation – that is, the point at which the state of the information-processing element loses its ability to sustain a given state – it is possible to advance estimates of the limits of performance for various types of information-processing element. In these limit analyses, the Heisenberg uncertainty relationship can serve as a basis for estimating performance using algebraic manipulations only.

It was shown that charge-based devices in the limit could offer extraordinary performance and scaling into the range of a few nanometers, albeit at the cost of enormous and unsustainable power densities. Nonetheless, it appears that there is considerable room for technological advances in charge-based technologies. One could consider using electron spin as a basis for computation, as the binary system state can be defined in terms of spin orientation. However, an energy barrier analysis, based on the equilibrium room-temperature operation of a digital spin-flipping switch, has revealed that extraordinarily large external magnetic fields are required to sustain the system state, and hence that a high energy consumption would result. (Although it has been proposed that if devices can be operated out-of-equilibrium with the thermal environment, then perhaps computational state variables can be chosen to improve on the switching energy characteristic of spin-based devices [26].) The need for very large magnetic fields can be eased by utilizing multiple electron spins to represent the state of the processing element. Unfortunately, the number of electrons that must be utilized is such that the size of the processing elements would be about an order of magnitude larger than that of a charge-based device. Finally, we briefly examined the different physical realizations for optical binary elements, and found that the inability to localize a photon, although an advantage for communication systems, works against the implementation of binary optical switches. As a general rule, optical binary switches are significantly larger than charge-based switches.

Although it appears that it will be difficult to supplant charge as a mainstream information-processing state variable, there may be important application areas where the use of spin or optics could be used to advantage. While the present chapter has focused on the properties of the processing elements themselves, information-processing systems are clearly comprised of interconnected systems of these elements, and it is the system consideration that must remain paramount in any application. Nonetheless charge-based systems, by using the movement of charge to effect element-to-element communication, avoid changing any state variable to communicate, and this is a decided advantage.

References

1 Ayres, R.U. (1994) *Information, Entropy and Progress*, AIP Press, New York.
2 Brillouin, L. (1962) *Science and Information Theory*, Academic Press, New York.
3 Yaglom, A.M. and Yaglom, I.M. (1983) *Probability and Information*, D. Reidel, Boston.
4 Gomer, R. (1961) *Field Emission and Field Ionization*, Harvard University Press.
5 French, A.P. and Taylor, E.F. (1978) *An Introduction to Quantum Physics*, W.W. Norton & Co, Inc.
6 Margolus, N. and Levitin, L.B. (1998) The maximum speed of dynamical evolution. *Physica D*, **120**, 1881.
7 Brody, D.C. (2003) Elementary derivation for passage times. *Journal of Physics A-Mathematical and General*, **36**, 5587.
8 Sze, S.M. (1981) *Physics of Semiconductor Devices*, John Wiley & Sons.
9 Cavin, R.K., Zhirnov, V.V., Hutchby, J.A. and Bourianoff, G.I. (2005) Energy barriers, demons, and minimum energy operation of electronic devices. *Fluctuation and Noise Letters*, **5**, C29.
10 Taur, Y. and Ning, T.H. (1998) *Fundamentals of Modern VLSI Devices*, Cambridge University Press.
11 Singh, J. (1997) *Quantum Mechanics – Fundamentals and Applications to Technology*, John Wiley & Sons.

12 Pearton, S.J., Norton, D.P., Frazier, R., Han, S.Y., Abernathy, C.R. and Zavada, J.M. (2005) Spintronics device concepts. *IEE Proceedings-Circuits Devices and Systems*, **152**, 312.

13 Jansen, R. (2003) The spin-valve transistor: a review and outlook. *Journal of Physics D-Applied Physics*, **36**, R289.

14 Daughton, J.M. (1997) Magnetic tunneling applied to memory. *Journal of Applied Physics*, **81**, 3758.

15 Bandyopadhay, S., Das, B. and Miller, A.E. (1994) Supercomputing with spin-polarized single electrons in a quantum coupled architecture. *Nanotechnology*, **5**, 113.

16 Cowburn, R.P. and Welland, M.E. (2000) Room temperature magnetic quantum cellular automata *Science*, **287**, 1466.

17 Motokawa, M. (2004) Physics in high magnetic fields. *Reports on Progress in Physics*, **67**, 1995.

18 (a) Marshall, W.S., Swenson, C.A., Gavrilin, A. and Schneider-Muntau, H.J. (2004) Development of "Fast Cool" pulse magnet coil technology at NHMFL. *Physica B*, **346**, 594. (b) National High Magnetic Field Laboratory website at: http://www.magnet.fsu.edu/magtech/core.

19 Lietzke, A.F., Bartlett, S.E., Bish, P., Caspi, S., Dietrich, D., Ferracin, P., Gourlay, S.A., Hafalia, A.R., Hannaford, C.R., Higley, H., Lau, W., Liggins, N., Mattafirri, S., Nyman, M., Sabbi, G., Scanlan, R. and Swanson, J. (2005) Test results of HD1b, and upgraded 16 Tesla Nb3Sn Dipole Magnet. *IEEE Transactions on Applied Superconductivity*, **15**, 1123.

20 Corson, D.R. and Lorrain, P. (1962) *Introduction to Electromagnetic Fields and Waves*, W.H. Freeman and Co., San Francisco and London.

21 Jackson, J.D. (1998) *Classical Electrodynamics*, 3rd edn., John Wiley & Sons, New York.

22 Wherrett, B.S. (1996) *Synthetic Metals*, **76**, 3.

23 Higgins, T.V. (1995) *Laser Focus World*, **31**, 72.

24 Miller, D.A.B., Chelma, D.S., Damen, T.C., Gossard, A.C., Wiegmann, W., Wood, T.H. and Burrus, C.A. (1984) *Physical Review Letters*, **53**, 2173.

25 Miller, D.A.B., Chelma, D.S., Damen, T.C., Gossard, A.C., Wiegmann, W., Wood, T.H. and Burrus, C.A. (1984) *Applied Physics Letters*, **45**, 13.

26 Nikonov, D.E., Bourianoff, G.I. and Gargini, P.A. (2006) Power dissipation in spintronic devices out of thermodynamic equilibrium. *The Journal of Superconductivity and Novel Magnetism*, **19**, 497.

II
Nanofabrication Methods

5
Charged-Particle Lithography

Lothar Berger, Johannes Kretz, Dirk Beyer, and Anatol Schwersenz

5.1
Survey

The extensive functional range of modern microelectronics is being driven by the ability to pack large numbers of transistors into a small piece of silicon as an integrated circuit. Today, the method used to pattern almost all integrated circuits is *photolithography* (also referred to as *optical lithography*), where circuit patterns from master images, the transmission photomasks, are transferred to silicon wafers by projection optics. In more detail, the wafer is coated with a photoresist, which is exposed with the desired circuit pattern (see Figure 5.1). The resulting resist pattern is transferred to the wafer by subsequent process steps. A detailed introduction to photolithography can be found in Ref. [1], while a comprehensive study is presented in Ref. [2].

The pivotal device of the integrated circuits of today is the metal oxide semiconductor field effect transistor (MOSFET), for which a sample photolithographic process, producing the electrical connections, is shown in Figure 5.2. Here, the bulk transistor has already been fabricated on the wafer, and consists of the doped areas of source, drain, and gate. The wafer is then coated with a positive photoresist, and exposed to form the contact areas for the transistor (Figure 5.2a). The exposed areas of the photoresist are removed by a developer chemical (b), after which the insulating layer, now open at source, drain, and gate, is etched away (c). After metallization, the remaining unexposed photoresist is removed by a stripping chemical (d).

In 2006, the gate length of the MOSFET in the most advanced integrated circuits is typically 65 nm. The gate length, being the smallest feature required, is also known as the *critical dimension* (CD) of the pattern. A CD of 65 nm is close to the resolution limit of the current photolithography. The smallest feature is determined by:

$$CD = k_1 \frac{\lambda}{NA} \tag{5.1}$$

where k_1 is a factor determined by the projection optics and process flow, λ is the wavelength of the photons, and NA is the numerical aperture between the objective optical lens and the resist plane.

Nanotechnology. Volume 3: Information Technology I. Edited by Rainer Waser
Copyright © 2008 WILEY-VCH Verlag GmbH & Co. KGaA, Weinheim
ISBN: 978-3-527-31738-7

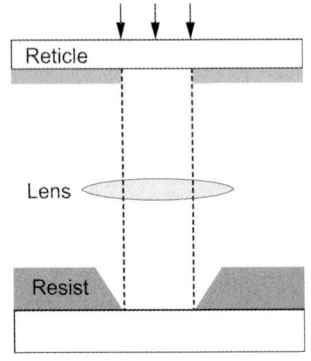

Figure 5.1 The principle of photolithography.

While a number of techniques have been developed to increase the resolution by reducing k_1 (referred to as *resolution enhancement techniques*; RET), the smallest feature that can be prepared by photolithography is ultimately dependent on the wavelength λ of the photons. Therefore, there is a history of reducing the wavelength, which is 193 nm for the current state-of-the-art *deep ultraviolet lithography* (DUVL). With $k_1 \approx 0.3$ for current projection optics and process flow, and NA ≈ 0.95 for a technically feasible projection in air, CD ≈ 61 nm represents the smallest feature.

Attempts to reduce the wavelength further have been investigated extensively, but encountered problems. For example, at $\lambda = 157$ nm no fully suitable material has been found to fabricate the transmission photomask and the lenses of the projection optics. Therefore, the current 193-nm photolithography is now scheduled to be extended into 193 nm immersion photolithography, which increases the NA by

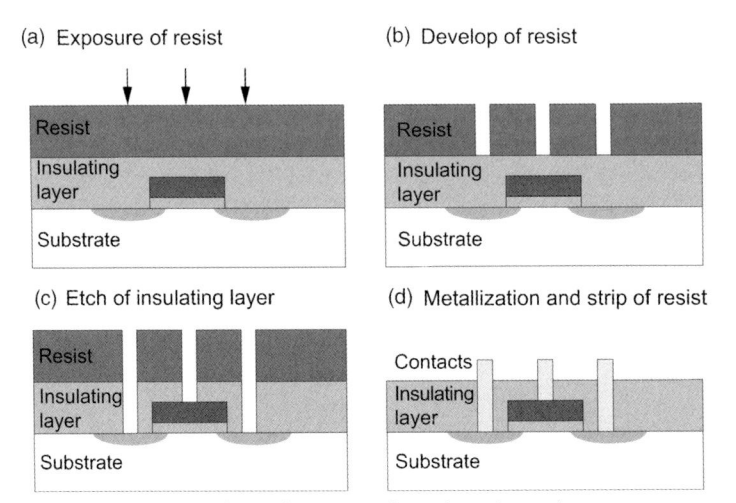

Figure 5.2 The photolithographic process for making electrical connections to a transistor.

conducting the exposure not in air, but in a liquid. With NA ≈ 1.4 for a technically feasible projection in water, CD ≈ 42 nm can be achieved. Much progress has been made with immersion photolithography, and the technique is currently at the stage of pilot production.

Another means of reducing the wavelength is to forego the use of transmission photomasks and projection optics with lenses, and to utilize reflective masks and mirror projection optics. Lithography involving reflection is no longer considered as classical photolithography. The wavelength of the photons where mirrors can be applied most effectively is 13.5 nm, and lithography of 13.5 nm involving masks with multilayer Bragg reflectors is referred to as *extended ultraviolet lithography* (EUVL), for which considerable research effort is currently being expended. At this point, it should be mentioned that EUVL differs greatly from DUVL in that it requires the redevelopment of almost all the exposure equipment and lithography processes currently in use. A comprehensive study of the process is presented in Ref. [3].

An alternative to any lithography involving photons is *charged-particle lithography*, where charged particles (electrons, ions) are used for patterning. While certain charged particle lithography techniques are already used for special applications, such as fabricating masks for photolithography, or prototyping, promising new charged-particle lithography techniques for preparing integrated circuits are currently under development, and may in time complement or even replace photolithography.

In order to illustrate the relationship between charged-particle lithography and photolithography, it is of help to examine the International Technology Roadmap for Semiconductors (ITRS), which demonstrates the status of lithographic techniques currently in use and under development within the microelectronics industry. The ITRS for the year 2006 is shown in Figure 5.3 [4]. According to the current ITRS, in 2007 the most likely successors of DUVL include EUVL, a multiple-electron-beam lithography technique called *maskless lithography* (ML2), and *nanoimprint lithography* (NIL). Charged-particle lithography techniques known as electron projection lithography (EPL) and ion projection lithography (IPL), both of which use a transmission mask and projection optics with electromagnetic lenses to direct electrons and ions, were removed from the ITRS in 2004. Proximity electron lithography (PEL), which uses a 1:1 transmission mask, was removed in 2005, although its re-emergence cannot be ruled out completely.

In this chapter we discuss the physical concepts and the principal advantages and limitations of charged-particle lithography techniques. A brief insight is also provided into the charged-particle lithography techniques currently in use and under development, while a strong focus is placed on ML2 techniques. The chapter comprises four sections: Section 5.1 includes a survey of the field, while Section 5.2 incorporates discussions of electron beam lithography and electron resists, and their major applications:

- the fabrication of transmission masks for DUVL and reflective masks for EUVL

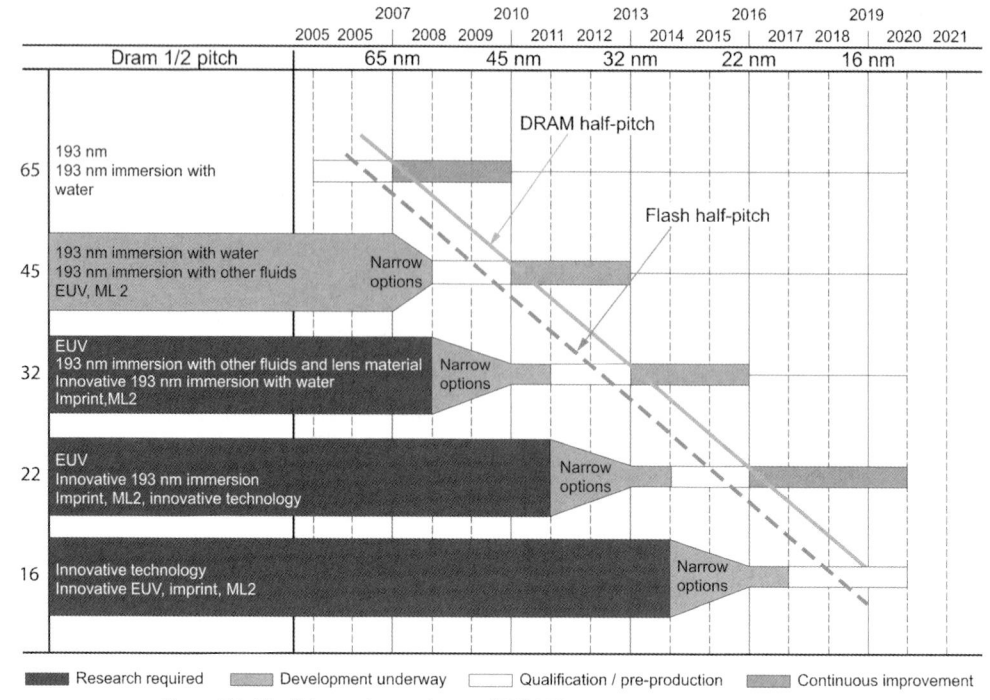

Figure 5.3 The lithography roadmap of 2006 [4].

- the direct-writing of patterns onto wafers with single beams for prototyping, low-volume production, and mix-and-match with photolithography
- the direct-writing of patterns on wafers with multiple beams for volume production: ML2
- the fabrication of imprint templates for NIL.

Section 5.2 concludes with a discussion of the special requirements for mix-and-match, namely the integration of electron beam lithography (EBL) with photolithography. Section 5.3 presents details of ion beam lithography (IBL), for which the major applications include:

- the direct-structuring of patterns on wafers without resist processing, for prototyping, low-volume production, and special applications
- the fabrication of imprint templates for NIL, with direct-structuring of patterns without resist processing.

A graphical overview of the charged-particle lithography techniques discussed in this chapter is provided in Figure 5.4. Finally, Section 5.4 provides a conclusion and outlook on charged-particle lithography techniques.

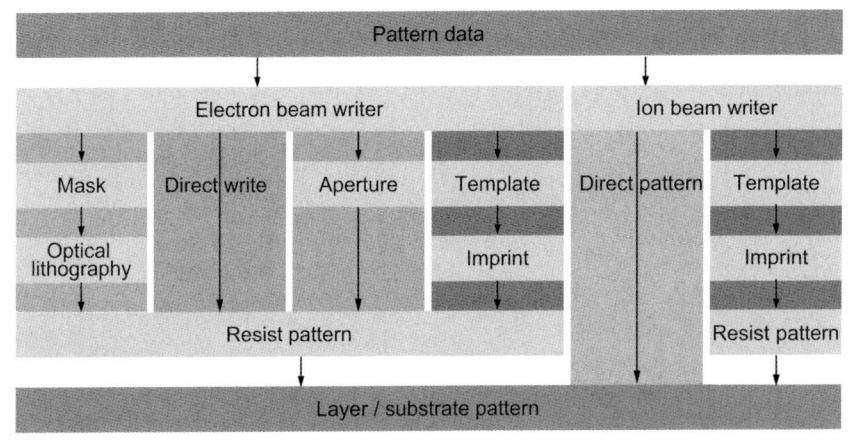

Figure 5.4 Charged-particle lithography techniques: an overview.

5.2
Electron Beam Lithography

5.2.1
Introduction

Electron beam lithography involves the use of electrons to induce a chemical reaction in an electron resist for pattern formation (the properties of electron resists are discussed in detail in Section 5.2.2). Because of the extremely short wavelength of accelerated electrons, EBL is capable of very high resolution, as $\lambda = h/(mv)$, $E_{kin} = (mv^2)/2$, and $E = eU$ gives:

$$\lambda = \frac{h}{\sqrt{2meU}} \quad (5.2)$$

Therefore, the typical acceleration voltage of $U = 100\,\text{kV}$ results in $\lambda = 0.004\,\text{nm}$, which is well below the atomic scale. Even with simple EBL systems, 10-nm patterns have been demonstrated [5], which is well beyond any other lithographic technique. Further, EBL is unaffected by the major issues of optical lithography – diffraction and reflection – and much less affected by the depth-of-focus (DOF) limit [1, 2].

The first EBL tools where based on the scanning electron microscope, and first developed during the 1960s [6]. A schematic representation of a simple EBL system is shown in Figure 5.5. The column consists of an electron source, and the electron optics. The substrate is mounted on a precision stage below the column, where its position is controlled by a laser interferometer. As the electron source and electron optics require a high vacuum, a load–unload system with an air-lock is also fitted.

5.2.1.1 Electron Sources
While early EBL systems based on scanning electron microscopy (SEM) used field emission sources, where the electrons are extracted out of a sharp-tip cathode, modern

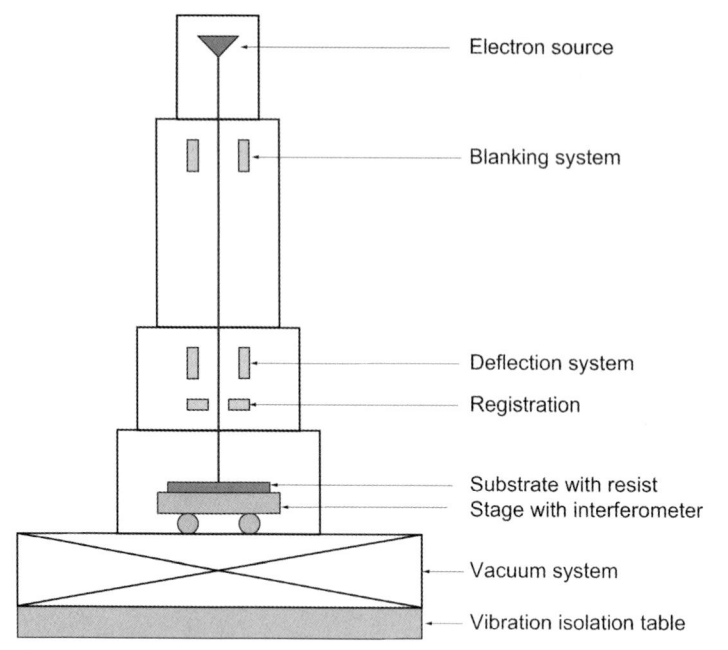

Figure 5.5 Schematic representation of a simple electron beam lithography system.

systems employ thermal sources. Such a thermal source consists of an emission region, where the cathode is heated, and an extraction region, where an electric field extracts the electrons and accelerates them to form a beam. The maximum current density j which can be obtained from this type of source for an acceleration voltage V and an acceleration distance d is limited by space-charge effects [6]:

$$j = \frac{1}{9\pi d^2} \sqrt{\frac{2q}{m}} V^{\frac{3}{2}} \tag{5.3}$$

The properties of common field emission and thermal sources are listed in Table 5.1 [6]. Because of its lower operating temperature and energy spread, LaB_6 is the source of choice for most current tools.

5.2.1.2 Electron Optics

Electron optics is based on the fact that electrons can be deflected by electromagnetic fields. The electric field between two grid electrodes, which causes bending of the trajectory of an electron, is shown in Figure 5.6.

Table 5.1 A comparison of the properties of common field emission and thermal sources [6].

Parameter	Field emission: tungsten	Thermal: tungsten	Thermal: LaB_6
Operating temperature T [K]	300	2700	1700
Energy spread [eV]	0.3	3	1.5

5.2 Electron Beam Lithography

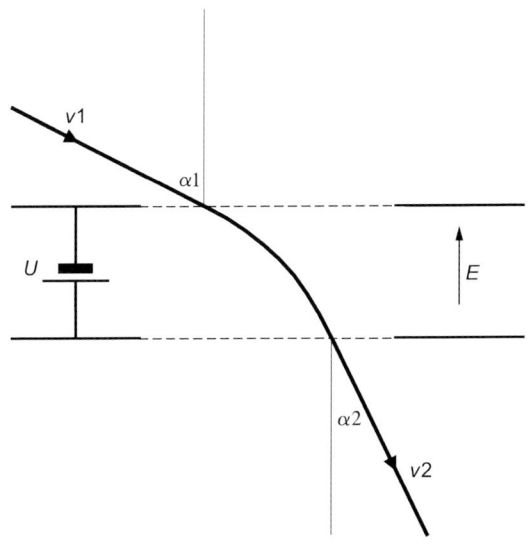

Figure 5.6 Electron–optical refraction.

For this configuration, the conservation of energy for an electron yields:

$$E_{kin,1} + eU = E_{kin,2} \tag{5.4}$$

where $E_{kin,1} = (mv_1^2)/2 = eU_0$ and $E_{kin,2} = (mv_2^2)/2$, and Equation 5.4 can be expressed as an electron–optical refraction law:

$$\frac{\sin \alpha_1}{\sin \alpha_2} = \frac{v_2}{v_1} = \sqrt{1 + \frac{U}{U_0}} \tag{5.5}$$

Electron Lenses A simple electrostatic lens (Einzel lens) consists of three rings, where the outlying rings have the same electrostatic potential (Figure 5.7).

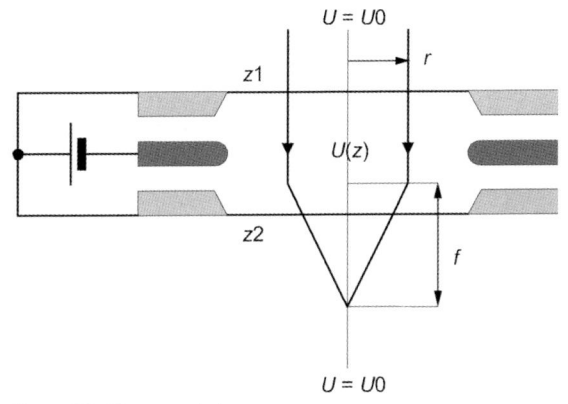

Figure 5.7 Electrostatic lens.

For this configuration, if the charge effects are negligible, then a simple equation for the paraxial trajectories of electrons can be obtained [6]:

$$\frac{d^2r}{dz^2} + \frac{1}{2U}\frac{dU}{dz}\frac{dr}{dz} + \frac{1}{4U}\frac{d^2U}{dz^2}r = 0 \qquad (5.6)$$

It should be noted that as Equation 5.6 is invariant towards the scaling of the voltage U, voltage instabilities in general do not cause a jitter of trajectories through electrostatic lenses. With the geometrical relation of the trajectory r and the focal length f, $dr(z_1)/dz = -r_1/f$, and $r \approx r_1$, Equation 5.6 yields [6]:

$$\frac{1}{f} \approx \frac{1}{8\sqrt{U_0}} \int_{z_1}^{z_2} \left(\frac{dU}{dz}\right)^2 U^{-\frac{3}{2}} dz \qquad (5.7)$$

Magnetic lenses utilize the force on an electron in a magnetic field:

$$\vec{F} = e\vec{v} \times \vec{B} \qquad (5.8)$$

A simple magnetic lens is a solenoid ring (Figure 5.8).

For this configuration, if the radial field components are negligible, then a simple equation for the paraxial trajectories of electrons can be obtained [6]:

$$\frac{d^2r}{dz^2} = -\frac{e}{m}\frac{B_z^2}{8U_0}r \qquad (5.9)$$

With the geometrical relation of the trajectory r and the focal length f, $dr(z_2)/dz = -r_1/f$, and $r \approx r_1$, Equation 5.9 yields [6]:

$$\frac{1}{f} \approx \frac{e}{8mU_0} \int_{z_1}^{z_2} B_z^2 \, dz \qquad (5.10)$$

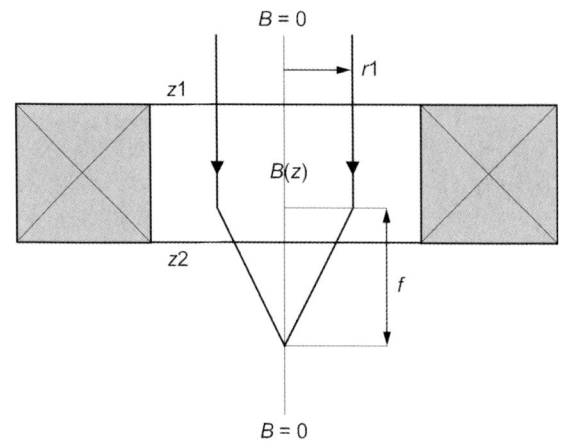

Figure 5.8 Magnetic lens.

Electron Optical Columns For the design of electron optical columns, the previous simple considerations are not adequate. It is required to derive the trajectories of electrons in a general form, which is valid also for non-rotational-symmetric lenses. A straightforward calculation can be based on the general equation of motion of electrons in an electron optical column:

$$\frac{d}{dt}(m\vec{v}) = e\left[\vec{E}(\vec{r},t) + \vec{v} \times \vec{B}(\vec{r},t)\right] \quad (5.11)$$

It is convenient to substitute with the arc element of the trajectory, $ds = |d\vec{r}| = vdt$:

$$m\frac{d\vec{v}}{ds} = e\left[\frac{\vec{E}}{v} + \frac{d\vec{v}}{ds} \times \vec{B}\right] \quad (5.12)$$

The trajectory equation can be derived in Cartesian coordinates, and with the introduction of the abbreviations $\eta = \sqrt{\frac{e}{2m_0}}$, $\varepsilon = \frac{e}{2m_0 c^2}$, $\Phi_0 = \frac{E_0}{e}$, $\hat{\Phi} = (\Phi_0 + \Phi)$ $[1 + \varepsilon(\Phi_0 + \Phi)]$, $\rho = |\vec{r}\,'| = \sqrt{1 + (x')^2 + (y')^2}$, where E_0 is the initial kinetic energy of an electron at the source, the trajectory equation results as follows [7]:

$$x'' = \frac{\rho^2}{2\hat{\Phi}}\left(\frac{\partial \hat{\Phi}}{\partial x} - x'\frac{\partial \hat{\Phi}}{\partial z}\right) + \frac{\eta \rho^2}{\sqrt{\hat{\Phi}}}\left(\rho B_y - y' B_t\right)$$

$$y'' = \frac{\rho^2}{2\hat{\Phi}}\left(\frac{\partial \hat{\Phi}}{\partial y} - y'\frac{\partial \hat{\Phi}}{\partial z}\right) + \frac{\eta \rho^2}{\sqrt{\hat{\Phi}}}\left(-\rho B_x + x' B_t\right) \quad (5.13)$$

It should be noted that this trajectory equation is valid only if all the trajectories are continuous – that is, if $x'(z)$, $y'(z)$ are finite. This is not the case with electron mirrors, although these are not used in current electron optical columns.

Aside of the trajectory representation of Equation 5.13, more elaborate mathematical methods have been applied to the analytic investigation of electron optical columns concepts and designs, focusing on the prediction of projection imperfections, called *aberrations*. These methods include the classical mechanics approach of Lagrange or Hamilton formalism [7], with the latter leading towards the Hamilton–Jacobi theory of electron optics, which is capable of treating whole sets of trajectories, and therefore is a standard tool for the design of electron optical columns with minimal aberrations [8].

Recently, with the support of computer algebra tools, a Lie algebraic electron optical aberration theory has been derived, which makes accessible high-order canonical aberration formulas, and therefore may open up new possibilities in the design of high-performance electron beam projection systems [9].

Electron Optical Aberrations Electron optical columns suffer from projection deviations termed aberrations, which are caused either by non-ideal electron optical elements, or by the physical limits of electron optics. An ideal electron optical lens should project an electron beam crossing the entrance plane at the point (x_0, y_0) onto the exit plane at the point $(x_1, y_1) = m(x_0, y_0)$, where m is a scalar called the *magnification*. Unfortunately, however, a real lens suffers from imperfections. For example, a real lens has the same focal length only for paraxial electron beams, while off-axis beams are slightly deflected. This is called *spherical aberration*, and is expressed by a coefficient C_s, relative to the beam position in the aperture plane (x_a, y_a):

$$(x_0, y_0) \rightarrow m\left(x_0 + C_s x_a (x_a^2 + y_a^2), y_0 + C_s y_a (x_a^2 + y_a^2)\right) \qquad (5.14)$$

With $C_s > 0$, an electron beam is blurred into a finite disk. Electron optical aberration theory shows that C_s cannot be made to vanish completely for rotational symmetric lenses [7], which in turn led to the introduction of non-rotational symmetric elements such as quadrupoles or octopoles. An overview of lens aberrations occurring in an electron optical column is provided in Ref. [7]. Additionally, the non-ideality of the electron source must be considered: a real electron source has a small spread in electron energy and, as the focal length of electron lenses depends on the electron energy, a non-monochromatic electron beam is blurred into a finite disk. This is referred to as *chromatic aberration*. In addition to these lens aberrations, other electron optical elements, such as beam deflectors, also introduce aberrations.

For a high-resolution EBL system, optimal projection is crucial, and therefore the aberrations discussed above must either be minimized by the design of the electron optical column, or compensated by additional electron optical elements. For example, since it is not possible to prepare a column with perfect rotational symmetry, an electron beam always suffers from astigmatism and misalignment. In order to compensate for astigmatism, a non-rotational symmetric element is required, a *stigmator*. This may be an electrostatic quadrupole, which is a circular arrangement of four electrodes. However, stigmators are usually designed as electrostatic octopoles which, in addition to quadrupole fields, can also generate dipole fields at any angle to correct beam misalignment [10].

While major efforts have been made to apply aberration theory to the design of an electron column with minimal aberrations, only powerful numerical optimization tools can be utilized for any systematic approach [11]. Such a tool generally consists of the following packages:

- an electromagnetic field computation package
- an electron beam ray tracing package
- an electron exposure spot plotting package
- an optimizer for minimizing aberrations.

An aberrations optimizer typically implements the damped least squares (DLS) method for iteratively minimizing overall aberrations. The individual aberrations $f_i(x_1, \ldots, x_n)$, depending on the column parameters x_j, (lens positions, sizes, strengths), are weighted and summed into a deviation function:

$$\psi = \sum_i (w_i f_i)^2 \tag{5.15}$$

which is then minimized. With this method, existing electron column designs can be improved, and new designs optimized automatically.

In addition to the design of electron optical columns with minimal aberrations, it has been proposed to implement adaptive aberration correction by introducing novel electron optical elements: for example, an electrostatic dodecapole, with time-dependent voltage control for the poles [12]. The learning process could be based on exposure pattern images.

5.2.1.3 Gaussian Beam Lithography

Patterning in EBL can be accomplished by focusing the electrons into beams with very narrow diameters. Because the electrons are created by a thermal source, they have a spread in energy. The trajectories of the electrons therefore vary slightly, resulting in electron beams with near-Gaussian intensity distribution after traversing the electron optics [13].

The basic principle of Gaussian electron beam exposure is *raster scanning*. Similar to a television picture tube, the electron beam is moved in two dimensions across the scanning area on the electron resist, which typically is about 1 mm². Within that area, which is termed the *deflection field*, the electron beam can be moved very rapidly by the electron optics. In order to change the position of this area on the substrate, a mechanical movement of the precision stage is required. Patterns which stretch over more than 1 mm² must be stitched together from separate deflection fields; in order to avoid discontinuities in the patterns at the boundaries of the deflection fields (these are known as *butting errors*, and potentially are caused by mechanical movement of the stage), the positions of these boundaries are calibrated and corrected in sophisticated manner.

In order to create the pattern shown in Figure 5.9a, the electron beam is moved across the area where the desired pattern is located, and is blanked on spots not intended for exposure. Raster scanning certainly has the major problem that the electron beam must target the whole scanning area, which takes a considerable time. Therefore, a great improvement is to target only the area of the desired pattern; this is termed *vector scanning* (Figure 5.9b). However, another major limitation remains in that, if a pattern consists of large and small features, the diameter of the electron

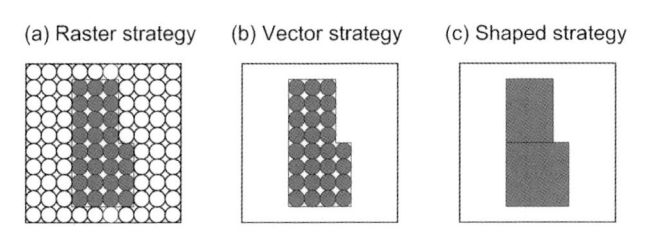

Figure 5.9 Writing a pattern with: (a) raster-, (b) vector-, and (c) vector shaped & beam strategies.

beam must be adapted to the smallest feature, thereby greatly increasing the exposure time of the large features.

5.2.1.4 Shaped Beam Lithography

In order to overcome the limitations of both raster and vector scanning, EBL tools have been developed which can shape electron beams. A shaped electron beam is created by special aperture plates and, in contrast to the near-Gaussian intensity distribution of standard EBL tools, shaped electron beam tools can apply rectangular, or even triangular, intensity distributions [14]. At present the fastest technique available is vector scanning using shaped electron beams; when using this technique the pattern shown in Figure 5.9c requires only two exposures.

The principal function of a variable-shaped beam (VSB) column is illustrated in Figure 5.10. The electron source illuminates a first shaping aperture, after which a first condenser lens projects the shaped beam onto a second shaping aperture. The beam position on the second aperture is controlled by an electrostatic deflector. A second condenser lens projects the shaped beam onto the demagnification system, consisting of two lenses, and the final aperture in between. After demagnification, the

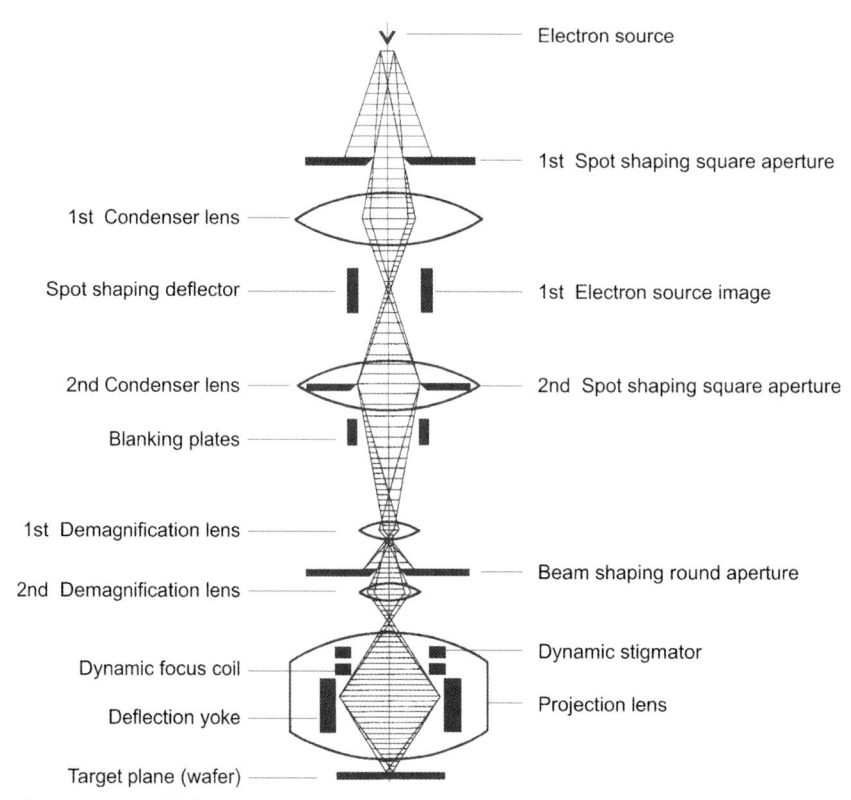

Figure 5.10 Variable beam-shaping column [15]. (© 1979, IEEE.)

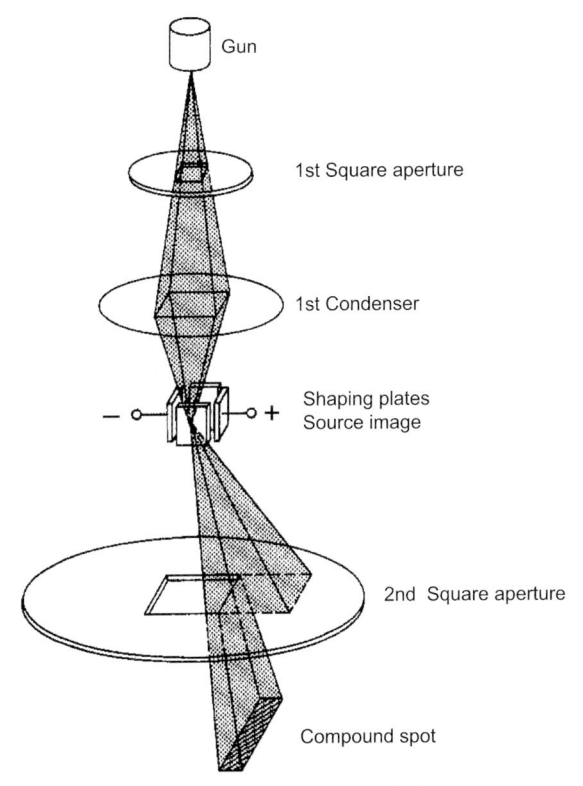

Figure 5.11 Variable beam-shaping method [15]. (© 1979, IEEE.)

shaped beam is projected onto the substrate by the final projection system, which consists of a stigmator and a projection lens with an integrated deflector.

The formation of shaped beams is illustrated for the example of rectangular spots in Figure 5.11. Two square apertures shape the spot; the image of a first square aperture, which appears in the plane of a second square aperture, can be shifted laterally with respect to the second aperture. This results in a rectangular spot which then is demagnified and projected onto the substrate.

Modern shaped electron beam tools can apply both rectangular and triangular spots. For example, the Vistec SB3050 [16] employs a LaB_6 thermal source, and utilizes a vector scan exposure strategy, a continuously moving stage, and the variable-shaped beam principle. The maximum shot area is $1.6 \times 1.6\,\mu m$, and rectangular shapes with 0° and 45° orientation, as well as triangles, can be exposed in a single shot. A detailed view of the two shaping aperture plates is shown in Figure 5.12.

The architecture and motion principle of the stage is decisive for pattern placement accuracy, so as to avoid butting errors. Position control by interferometer with a resolution <1 nm, and the use of a beam tracking system, allow write-on-the-fly exposures with stage speeds of up to $75\,mm\,s^{-1}$. The driving range of the stage is

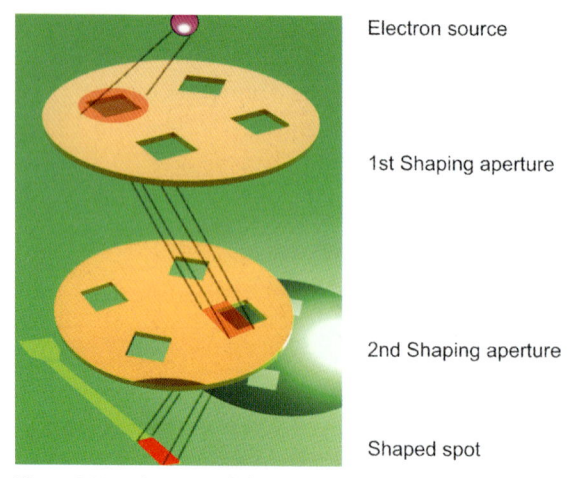

Figure 5.12 Schematic of electron beam shaping by double-aperture for rectangular and triangular spots [16].

310 × 310 mm, thus enabling the exposure of 6 in and 9 in masks, as well as 300 mm wafers (see Figure 5.13).

However, even such a precision stage cannot eliminate butting errors completely, and therefore the vector-shaped beam strategy involves overlapping exposure shapes, resulting in features being exposed at least twice, a process known as *multi-pass writing*.

A production-worthy EBL system is highly automated, with no human intervention required for operation, except for an operator issuing a command for the system to start loading the substrate and writing the pattern. The pattern is encoded in a digital data file, and stored in a computer memory or a mass storage device. Prior to writing, the original design data must be converted to a format which is usable by the writing tool. This data fracturing is accomplished using separate computer hardware, usually

Figure 5.13 Electron beam lithography precision stage [16].

(a)   (b)   (c)

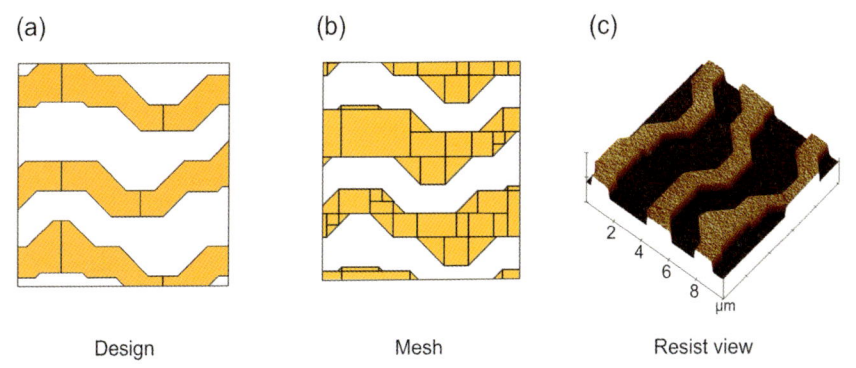

Design   Mesh   Resist view

Figure 5.14 Writing a complex pattern with a vector-shaped beam strategy. (a) Design; (b) data fracturing into shapes; (c) the final atomic force microscopy resist picture.

a high-performance cluster, operating a real-time multi-threaded operating system, and data preparation software. An example of the writing of a complex pattern with vector shaped-beam strategy is shown in Figure 5.14.

EBL tools are usually tailored specifically for either photomask or wafer applications (this is also referred to as *direct-write lithography*), and the principal advantages and limitations of these processes are presented in more detail in Sections 5.2.3.1 and 5.2.3.2. Evaluations of the current leading-edge electron beam writing systems are available in Refs. [16–18].

5.2.1.5 Patterning

As highlighted at the start of Section 5.1, EBL – unlike optical lithography – is unaffected by major issues of optical lithography, such as diffraction and reflection, and much less affected by the DOF limit. However, it also encounters specific patterning issues, which must be addressed.

The most critical issue results from the scattering of the electrons passing through matter. While this scattering in part is indeed required for transferring energy to the electron resist for exposure, many electrons scatter into different directions from their original trajectories. In Figure 5.15, the shaded areas of the resist are intended for exposure. However, the electrons show two scattering modes, from resist intended for exposure into resist not intended for exposure – this is *forward scattering* (a); and through resist intended for exposure into the substrate and back into resist

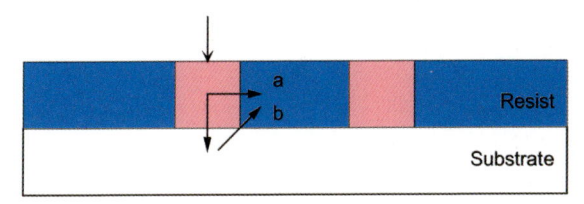

Figure 5.15 Electron-scattering modes. (a) Forward scattering; (b) back-scattering.

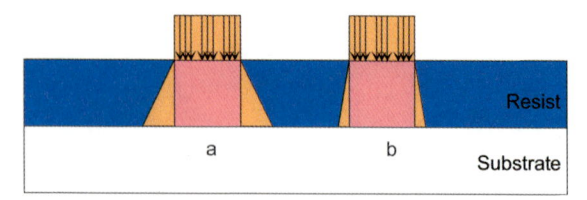

Figure 5.16 Forward scattering at: (a) low electron energy and (b) high electron energy.

not intended for exposure – this is *back-scattering* (b). Both scattering modes lead to unintended exposure, and therefore to degraded resolution and distortion of patterns. This issue is known as the *proximity effect* [19], and should be minimized as much as possible.

Forward scattering into the resist is addressed by increasing the acceleration voltage of the electrons, as shown in Figure 5.16. Although the first electron lithography tools used $U = 10\,\text{kV}$, and current tools employ 50 kV, more recently 100 kV tools have been introduced. This increase in acceleration voltage leads also to an improved projection performance, because the imperfections in the electron optics are less pronounced.

However, back-scattering intensifies with greater electron energy. With the currently required feature size being less than the range of back-scattered electrons, the features are broadened by back-scattering, thus offsetting the resolution improvement by reduced forward scattering.

For optimal resolution and minimal distortion, correction methods are therefore applied to overcome the proximity effect. A simple compensation technique, which accounts for back-scattering only, is to use a second exposure equaling the background exposure of the first one, with a reverse additional energy distribution [20]. However, as a second exposure is undesirable due to the additional writing time, a proximity correction by dose variation is introduced during data processing for converting the circuit designs [21]. Such a correction must be based on suitable models for proximity effect predictions. Here, a useful tool is the Monte-Carlo-based simulation of scattering from electron impact. Simulated trajectories for 100 electrons impacting into one point are shown in Figure 5.17, where both forward scattering and back-scattering appear distinctively.

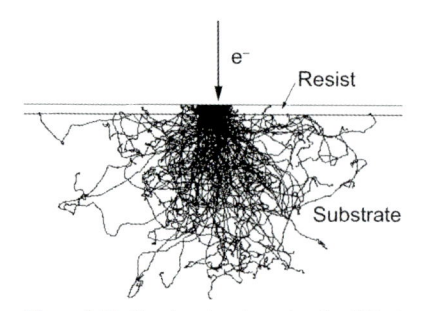

Figure 5.17 Simulated trajectories for 100 electrons impacting into one point.

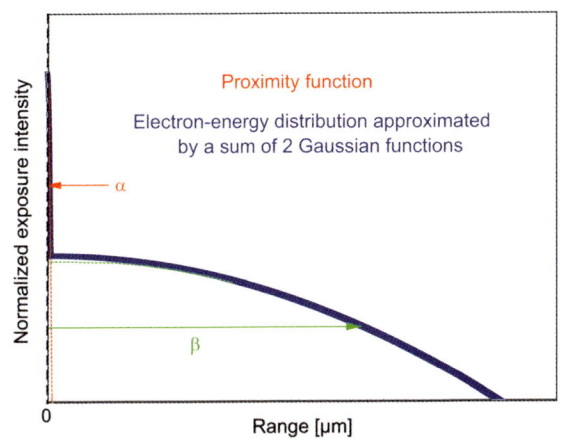

[Figure: Proximity function plot showing normalized exposure intensity vs Range [μm], with α indicating the narrow forward-scattering peak and β indicating the broader back-scattering distribution. Labeled "Proximity function — Electron-energy distribution approximated by a sum of 2 Gaussian functions".]

Figure 5.18 Proximity function (point exposure distribution fitted within two Gaussian functions).

In addition to model-based analysis, interpretations of generic pattern distortions of non-corrected patterns, as well as successive back-simulations, are utilized for the reconstruction of proximity effects [22], therefore enabling the best possible correction.

Current compensation techniques rely on either shot-by-shot modulation of the exposure dose, modification of the pattern geometry, or a combination of both methods. The proximity function $f(r)$ is usually described as a sum of two or more normalized two-dimensional Gaussian functions:

$$f(r) = \frac{1}{\pi(1+\eta)} \left[\frac{1}{\alpha^2} e^{-\frac{r^2}{\alpha^2}} + \frac{\eta}{\beta^2} e^{-\frac{r^2}{\beta^2}} \right] \tag{5.16}$$

The term α characterizes the forward scattering, and the term β the back-scattering, of the electrons. The parameter η is the deposited energy ratio of the back-scattering component towards the forward-scattering component, and r is the distance from the point of electron impact. The behavior of this function is shown by the diagram in Figure 5.18.

The concept of proximity correction by shot-to-shot dose modulation is illustrated in Figure 5.19: two features in close distance should be exposed. When applying a uniform exposure dose (Figure 5.19a) the intended resist exposure by forward scattering is compromised by back-scattering, leading to a distorted overall exposure, and to broadened features. When applying a suitably modulated exposure dose (Figure 5.19b), the resulting resist exposure by forward scattering complements the back-scattering, leading to the intended resist exposure.

5.2.2
Resists

Electron beam patterning requires specially designed electron-sensitive resists. As with photoresists, electron resists are available either as positive resists (which are insoluble in the developer chemical and become soluble when exposed), or as

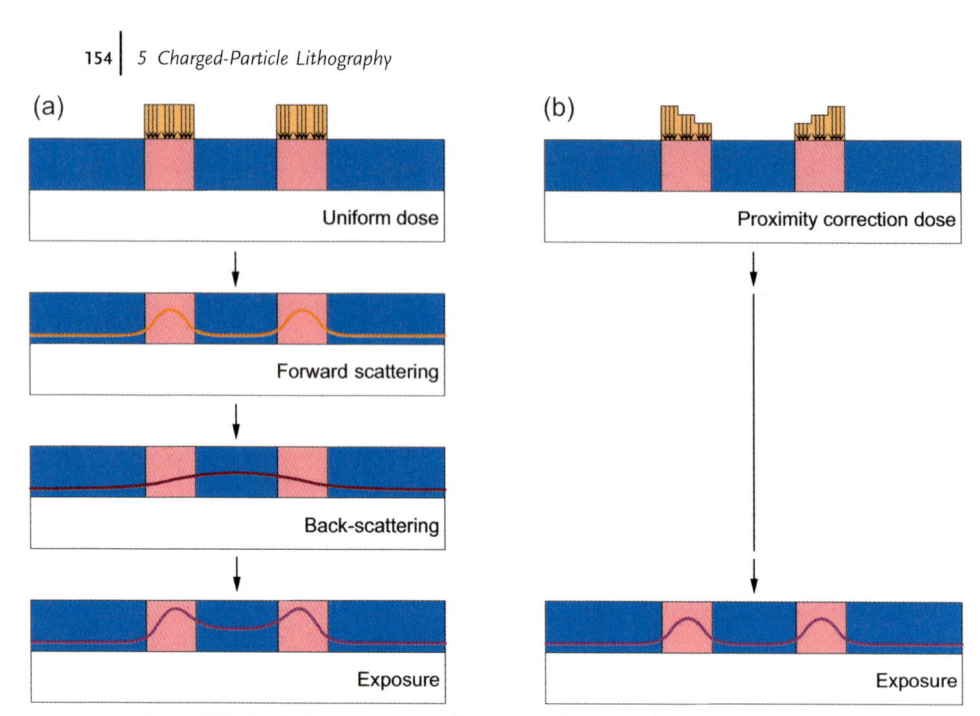

Figure 5.19 Proximity correction by shot-to-shot dose modulation.

negative resists (which are soluble from the start, and become insoluble when exposed). While the chemistry of electron resists is usually considerably different from that of photoresists, a parameter termed *contrast* can be defined to characterize the resolution of the resists.

In order to determine the contrast of an electron resist, the developing rate depending on the exposure dose is plotted; this is also called the characteristics of the electron resist. For a low-exposure dose, the resist still behaves like an unexposed resist, whereas for a high dose it is fully activated. The idealized characteristics of positive and negative electron resists are shown in Figure 5.20.

Linearization yields two parameters that define the characteristics: the resist sensitivity D_0, where the resist activation starts, and the resist activation D_v, after which the resist is fully activated. The contrast is then defined as:

$$\gamma = \frac{1}{\log \frac{D_V}{D_0}} \tag{5.17}$$

A high steepness of the transition in the characteristics therefore results in a high contrast.

Early electron resists employed just a single component for obtaining the latent image. In a positive resist, the image was created by electron-induced chain scission, with high-molecular-weight polymers with long chains being fragmented into smaller chains. The contrast of the resist was engineered by maximizing the difference in molecular weight before and after exposure. In a negative resist, the fragmentation into smaller chains generates radicals, which induce a crosslinking.

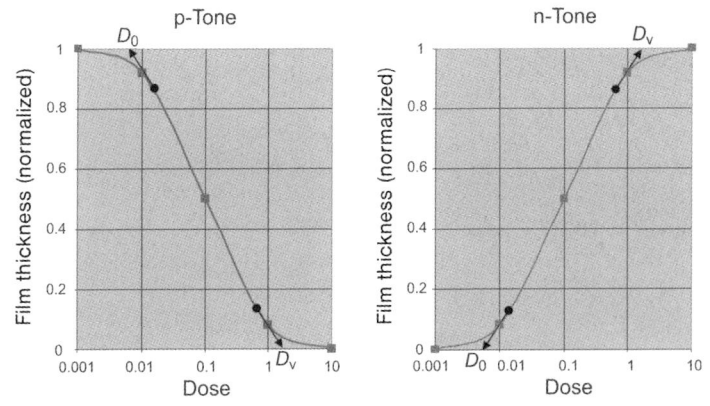

Figure 5.20 Idealized characteristics of positive and negative electron resists.

One of the first resists to be developed for EBL was polymethyl methacrylate (PMMA) [23]. Electron beam exposure breaks the polymer into fragments that are dissolved by a solvent-based developer (see Figure 5.21). Because of its very high resolution capability of <10 nm, PMMA is still used for certain R&D applications and electron beam writer resolution tests. However, it is not suitable for commercial lithography, mainly because of its poor resistance to dry etching.

Another group of early electron resists is based on a copolymer of chloromethacrylate and methylstyrene. One commercial resist for direct-write applications is ZEP-520 [24], which has considerable advantages compared to PMMA. While providing a comparable resolution of <10 nm, the sensitivity towards electrons is 10-fold higher, and the resistance to dry etching is 2.5-fold higher. Although these resists are still used for R&D applications, they have fallen out of favor for commercial lithography, because they require solvent-based developers that, because of their rapid evaporation rate in air, introduce temperature gradients on wafers, and therefore uniformity problems. A 45 nm 1 : 1 dense-lines pattern of a ZEP-520 resist is shown in Figure 5.22 [25].

For some time, UV photoresists have been used as electron resists. This group of resists is based on diazonaphthoquinone (DNQ), and has the advantage of using an aqueous developer. Their resistance to dry etching is also threefold higher. However, with resolutions <250 nm, and despite still being used extensively in optical lithography for micro-electromechanical systems (MEMS) fabrication, their time in

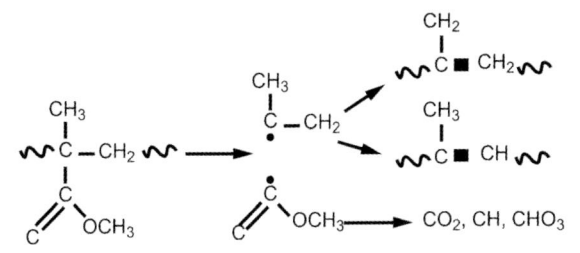

Figure 5.21 The reaction chemistry of polymethyl methacrylate (PMMA).

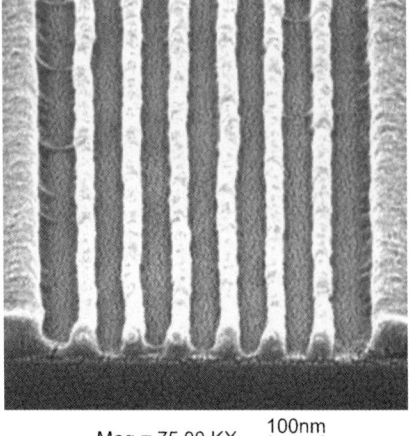

Figure 5.22 ZEP-520 : 45 nm 1 : 1 dense-lines test pattern [25].

EBL has passed. A detailed presentation of the chemistry of DNQ photoresists can be found in Ref. [26].

The impetus for the development of the electron resists used today was derived from the need for a new type of photoresist required for the introduction of DUV optical lithography. Because of the limitations of DNQ-based resists in resolution and sensitivity, *chemically amplified* (CA) resists have been introduced [27]. These are based on a multi-component scheme, where a sensitizer chemical causes dissolution modification within the exposed areas of the polymer matrix. The latent image is obtained from energy transfer to the sensitizer chemical molecules, causing a degradation into their ionic pairs or neutral species, which can catalyze the reaction events needed for solubility distinction. Commonly, photo acid generators (PAG) or photo base generators (PBG) are utilized as sensitizers in CA resists.

In a positive CA resist the PAG, upon exposure, releases an acid. During heating of the substrate after exposure (the post-exposure bake; PEB), this acid reacts with the resin, which in turn becomes soluble towards an aqueous developer. In addition, further acid is produced. With this multiplication reaction an exposed PAG molecule can trigger up to 1000 reactions. It is also acknowledged that a CA resist shows a high quantum yield compared to a DNQ-based resist.

Following the establishment of CA photoresists, specialized electron CA resists have now been developed. The reaction mechanism of a positive electron CA resist is shown in Figure 5.23. The CA electron resists that are used mainly in current commercial EBL are the positive-tone FEP-171 [28], and the negative-tone NEB-22 [29]. These resists both have resolutions <100 nm and show excellent process performance, especially in photomask fabrication [30, 31].

However, with the need for <50 nm resolution – especially for direct-write applications – the development and evaluation of more advanced positive and negative CA resists is currently the subject of intense investigation [32]. For example, the 50 nm dense lines and 70 nm dots with high contrast, obtained with a recently developed and evaluated positive CA resist [33], are shown in Figure 5.24.

Figure 5.23 Reaction chemistry of a positive electron chemically amplified resist.

The major challenges in the development of CA resists for <50 nm resolution are to: (i) reduce the diffusion length during PEB; (ii) improve etch stability; and (iii) reduce the line edge roughness. For very small features, the molecular structure of the resist contributes to the roughness of the lines, which can be a significant fraction of the linewidth. A measure of line edge roughness is the standard deviation σ of the actual line edge relative to the average line edge. The reduction of line edge roughness is pursued by the application of resins with shorter molecules.

All of the electron resists discussed so far have been based on organic polymers. Although, in principle, it has been shown that such resists can achieve a resolution close to 10 nm, before applicable in manufacturing additional points must be taken into consideration, including the above-mentioned line edge roughness. As polymers are relatively large molecules, they cannot easily form smooth edges close to the atomic scale. Hence, in parallel to the improvement of CA resists, inorganic electron beam resists, such as hydrogen silsesquioxane (HSQ), are being pursued [34]. Initially, HSQ was used as a low-dielectric (low-k) material, with a k-factor of 2.5–3.0. In addition, HSQ demonstrates good spin-coating properties, such as good gap-fill, global planarization, and crack-free adhesion. It also shows excellent proces-

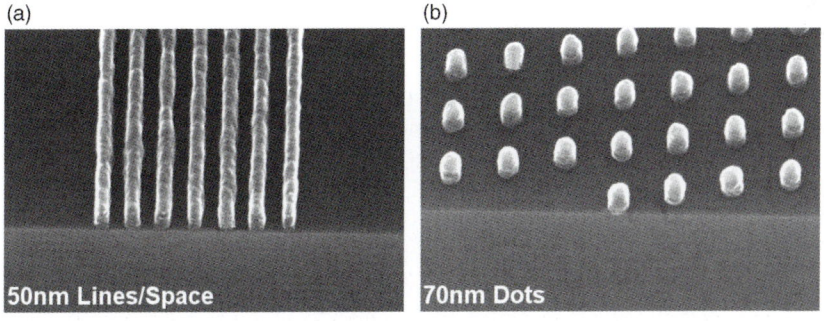

Figure 5.24 An advanced positive chemically amplified resist. (a) The 50 nm dense lines test pattern; (b) the 70 nm dots test pattern [33].

sing properties, notably a high thermal stability. HSQ is an oligomer composed of caged silsesquioxane within a linear Si–O network. A thermal curing is carried out to convert the caged species into a highly crosslinked network through the hydrolysis and condensation of the reactive Si–H functionalities.

Following the discovery that electron beam irradiation also initiates this curing reaction, HSQ was proposed as an inorganic electron resist. Our current understanding is that the Si–H bonds are broken during electron beam irradiation and are, in the presence of absorbed moisture, converted into silanol (Si–OH) groups. These silanol groups are unstable, and therefore condense, causing the caged molecule to break into a linear network. This transition drastically decreases the dissolution rate of the matrix within an aqueous base, thus enabling the use of HSQ as negative-tone electron resist. Furthermore, due to its high etch resistance HSQ can be utilized in a bi-layer resist process (BLR), where the patterns are transferred through a planarizing layer using reactive ion etching (RIE). As, with the rise in popularity of maskless lithography, BLR may become much more important, it is described in more detail in Section 5.2.3.3. With the implementation of a sensitizer chemical into the functional matrix (similar to the PAG in a polymeric CA resist), HSQ can, at least potentially, be made production-applicable.

5.2.3
Applications

Although EBL has a wide range of applications, the major focus is currently on the fabrication of transmission masks for DUVL and reflective masks for EUVL. Another important application is the direct-writing of patterns on wafers with single beams (which is also referred to as direct-write EBL; EBDWL). EBDWL is mainly used for fabricating device prototypes, and with recent improvements in shaped electron beam lithography tools, is also suitable for the low-volume production of application-specific integrated circuits (ASIC) and other specialized devices, for example hard disk heads. Multiple EBL such as maskless lithography (ML2), where a single electron beam is split into multiple beams to enable massively parallel EBDWL, shows the potential to complement or even replace optical lithography. The fabrication of imprint templates for NIL by using EBL techniques similar to photomask making is steadily gaining in importance. Finally, EBL is also combined with optical lithography in volume production, where it is used to fabricate critical structures such as gates. This approach – termed *mix-and-match lithography* production, or *hybrid lithography* if a single resist is utilized as both electron resist and photoresist – has special requirements.

5.2.3.1 Photolithography Masks

In optical lithography, the patterns on wafers are reproductions of those on a photomask. As the photomask is used for thousands of chip exposures, the quality of photomasks is critical for optical lithography. Photomasks are fabricated with techniques similar to those used in wafer processing (see Section 5.1). A photomask blank, a glass substrate with a deposited opaque film (usually chromium) is coated

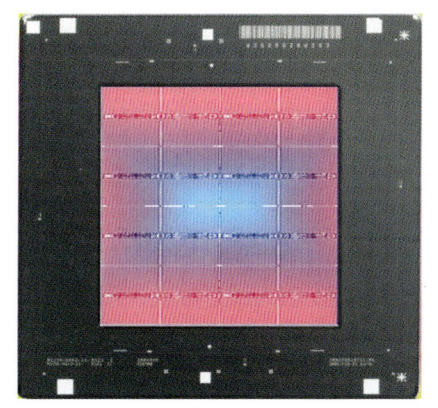

Figure 5.25 Photomask for deep ultraviolet optical lithography [35].

with a resist, and the latter is exposed with the pattern, developed, and the opaque film is etched. A processed photomask for DUV optical lithography is shown in Figure 5.25 [35].

Three types of photomask are currently in use. The simplest type is the binary mask (Figure 5.26a) [36], which employs only clear areas in the opaque chromium film to project the pattern to the wafer. Unfortunately, binary masks have the problem that, because of diffraction, the edges of the resist lines do not become straight. To rectify this problem, masks with molybdenum silicide films, which function as phase shifters, are used in phase-shift masks (PSM) (Figure 5.26b) [36]. Recently, so-called chromeless phase lithography (CPL) masks have been introduced for a resolution-enhancement technique (RET) called *off-axis illumination*, as shown in Figure 5.26c [36].

The pattern of a CPL mask featuring 125 nm lines applicable for 32 nm optical lithography is shown in Figure 5.27 [37]. An introduction to the functional principles of the different photomask types and their application in optical lithography is provided in Ref. [1].

Typically, photomask patterning is carried out with beam writers. Whilst for low- and medium-resolution masks, optical beam writers can be used, high-resolution masks are prepared using electron beam writers [38]. The general EBL techniques were described in Section 5.2.1; today, raster scanning has been replaced by vector scanning, and both Gaussian-beam and shaped-beam writers are currently employed for mask-making.

Gaussian-beam lithography is usually applied in a different way for photomasks than as described in Section 5.2.1.3. In addition to the IC patterns, photomasks also contain smaller features than the CD for optical proximity correction (OPC). When using Gaussian-beam writing, creation of the pattern in Figure 5.28 requires a beam size equivalent to the smallest feature (here, the right upper edge), but this leads to long writing times. However, it is possible to expose this pattern with a beam-size which is twice the size of the smallest feature. Because of the Gaussian intensity distribution, the 2σ circle touches the edges of the writing grid, whereupon the outline of the pattern can be made by multiple exposures of the spots; this is referred

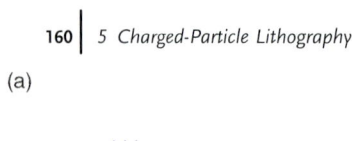

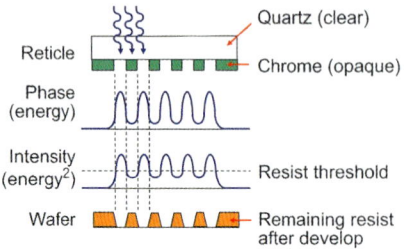

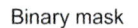

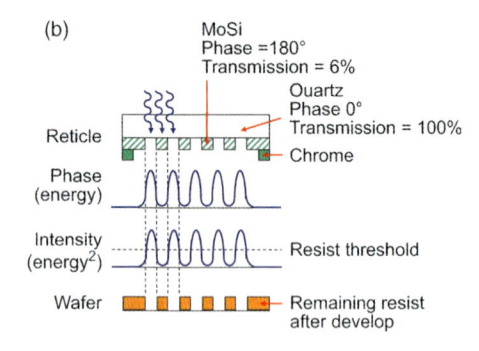

Figure 5.26 Mask types for deep ultraviolet optical lithography. (a) Binary; (b) phase-shift masks (PSM); (c) chromeless phase lithography (CPL) [36].

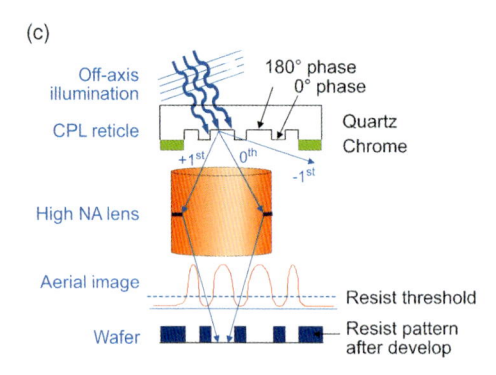

to as *multi-pass gray writing* [39]. The straight outlines are exposed four times, while the outlines of the pattern can be moved locally by exposing spots only three, two, or one times. Because of the overlap between spots and the multiple passes, this method has the additional effect of smoothing the exposure.

Although Gaussian-beam strategy is still used today, high-end mask-making has become a domain of 50 kV variable-shaped beam writers, which have a significantly higher throughput than Gaussian-beam writers. The shaped-beam strategy is applied as discussed in Section 5.2.1; however, mask-making encounters several specific issues, each of which must be addressed.

In addition to the proximity effect (see Section 5.2.1.5), which is a short-range phenomenon, long-range effects also appear which may stretch over large portions of the masks. The re-scattering of incident electrons at the objective electron lens can lead to a long-range background exposure, called a *fogging effect*. Further, an optimal exposure result may be compromised by succeeding processing steps, such as developing or etching. During developing, the concentration of the developer chemical decreases faster in areas with dense patterns, than in areas with sparse patterns. Similar effects appear in both wet and dry etching, and this may lead to long-

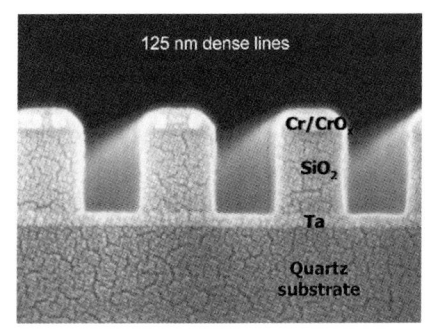

Figure 5.27 Photomask for deep ultraviolet optical lithography: 125 nm lines on mask result in 32 nm lines on the wafer [37].

range distortions of the final patterns, known as the *loading effect*. Whilst both fogging and loading effects can, in principle, be corrected within post-exposure process steps such as the PEB, the proximity effect correction methods of electron beam writers have been successfully augmented for handling both fogging and loading effects.

As shown in Section 5.2.1.5, forward scattering into the resist can be reduced by increasing the acceleration voltage of the electrons. However, this approach causes a significant increase in scattering within the substrate, and as a result the substrate is heated up. For silicon wafers, this problem is less pronounced, because silicon has a high thermal conductivity, and distortions of the wafer can be rectified by electrostatic chucks with wafer backside cooling. However, photomasks have a low thermal conductivity, and because of their large thermal capacity, the local temperature increases significantly during exposure. Therefore, mask writing tools currently do not exceed 50 kV acceleration voltage, and there is a reluctance to employ 100 kV tools.

The fabrication of high-end photomasks with EBL requires specialized post-exposure processing equipment. The PEB is a critical process step for CA resists, requiring a temperature uniformity of <0.1 K within the resist plane of the mask. Because of the large thermal capacity of photomasks, and their non-radial shape, such temperature uniformity is difficult to achieve. The PEB equipment is preferably connected directly to the electron beam writer, thus enabling the PEB and development to be conducted immediately after exposure. With such a direct connection, a

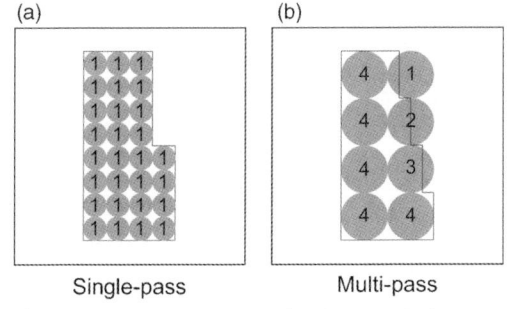

Figure 5.28 Writing a pattern with: (a) vector single-pass; and (b) with vector multi-pass strategies.

specially tailored post-exposure processing is applicable to compensate writing errors such as fogging and loading, and to improve overall pattern uniformity. The PEB is especially suitable for such compensation [40].

Besides photomask making, electron beam techniques have recently been introduced for the repair of photomasks [41] which, to date, has been the domain of ion beam structuring (see Section 5.3.2.1). If, due to a problem in the manufacturing process, a part of the opaque film is stuck when it should have been removed, it can be selectively removed by an electron beam-induced etching process; moreover, if part of the opaque film is damaged, it can be partially redeposited.

5.2.3.2 Direct-Write Lithography

Electron beam writers may also be used to create patterns on wafers directly, a process referred to as EBDWL. As shown in Section 5.2.1, EBDWL has the potential for fabricating features close to the atomic scale, and also provides very large depth-of-focus compared with optical lithography. As an example, a transistor demonstrator employing a gate with linewidth of 13 nm, and which has been fabricated by EBDWL utilizing vector Gaussian beam strategy [33], is shown in Figure 5.29.

EBDWL is especially suitable for fabricating device prototypes, as the device and the driving integrated circuit can be made directly from the computer-aided design (CAD) file. The production of device prototypes by optical lithography is not feasible, as a high-end photomask would first have to be made. Even if this effort were to be undertaken, it would not be possible to make any quick design changes. Notable applications of EBDWL in prototyping currently include research into nanoscale planar transistors (Figure 5.30) [42], and the next-generation transistor designs such as FinFET (Figure 5.31) [43, 44], or carbon nanotube FETs (CNTFET) [45].

In additional to the fabrication of device prototypes, EBDWL, with its recent improvements in repeatability and throughput of shaped EBL tools, has also been recognized as a feasible method for the low-volume production of ASICs and other special devices, for example hard disk heads.

As described in Section 5.2.1.5, forward scattering into the resist can be reduced by increasing the acceleration voltage of the electrons. While this approach causes a

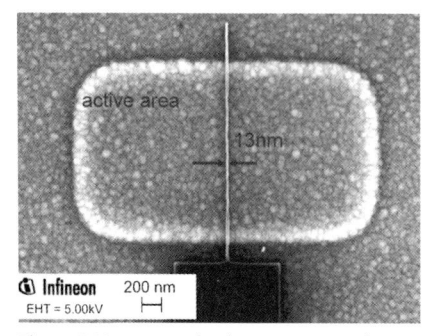

Figure 5.29 Nanoscale planar MOSFET demonstrator. A gate with a 13 nm linewidth prepared using direct-write electron beam lithography (EBDWL) [33].

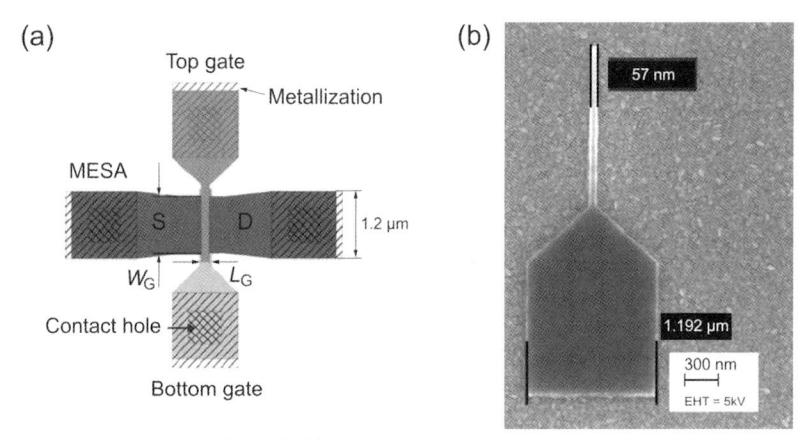

Figure 5.30 Nanoscale planar double gate transistor. (a) Top view of the design. (b) Scanning electron microscopy image of the first fabricated structure: bottom gate (produced by EBDWL) [42].

significant increase of the scattering within the substrate, and also heats up the substrate, in the case of silicon wafers this problem is less pronounced due to silicon's high thermal conductivity. As distortions of the wafer can be rectified by electrostatic chucks with wafer backside cooling, modern direct-writing tools employ an acceleration voltage of up to 100 kV.

Aside from the electron beam writer, specialized post-exposure processing equipment is required for reliable EBDWL. Similar to photomask fabrication, CA resists are employed, for which the PEB is a critical process step, requiring a temperature uniformity of <0.1 K within the resist plane of the wafer. However, because of the high thermal conductivity of silicon wafers, such uniformity is less difficult to achieve than with photomasks. As with photomask fabrication, the post-exposure processing equipment is preferably connected directly to the electron beam writer, enabling the PEB and development to be conducted immediately after exposure.

The fabrication of integrated circuits, either completely with EBL or with mix-and-match lithography, poses some significant challenges in integrating the required

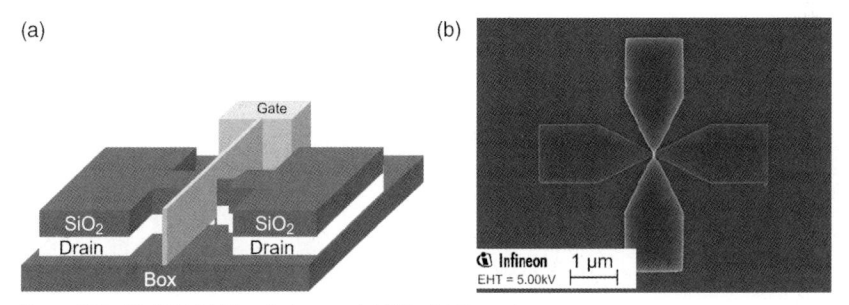

Figure 5.31 FinFET. (a) The design model [43]. (b) Scanning electron microscopy image of the actual device: with 50 nm gate (produced by EBDWL) [44].

process steps, and especially when aligning the different layers of patterned functional films. These challenges – and the methods used to overcome them – are discussed in Section 5.2.4.

5.2.3.3 Maskless Lithography

Considerable effort has been made towards making EBL available for volume production. The initial approach had been to implement either separate multiple electron optical columns [46] or separate multiple electron beams [47], to achieve massively parallel EBDWL. However, as the adequate calibration of either multiple columns or beams is a very challenging task, the suggestion was made to mimic optical lithography by introducing a transmission mask and projection optics with electromagnetic lenses to direct the electrons, which led in time to the development of electron projection lithography (EPL). In parallel, the use of a 1:1 transmission mask was also investigated, which led to the development of proximity electron lithography (PEL). As mentioned in Section 5.1, EPL was removed from the ITRS in 2004 due to significant difficulties with fabrication and application of the EPL transmission masks. PEL was subsequently removed in 2005, although its re-emergence cannot be ruled out completely. Whilst EPL and PEL are currently dormant, the advances made in electron optics have been significant, and hence the idea was conceived to devise electron projection and proximity techniques without the use of a mask – hence the term maskless lithography (ML2). As the current ML2 techniques are, loosely, derivatives of EPL and PEL, the details of both processes are explained in the following sections.

EPL: Principles and Limitations The transmission mask for EPL is a 4:1 stencil mask, a thin membrane through which holes are etched for the transmission of electrons. The stencil masks are themselves prepared using EBL, utilizing similar processes as for photomasks (see Section 5.2.3.1). Due to the instability of the membrane, fabrication has proven very challenging. In addition, a stencil mask absorbs electrons where there are no holes, thus causing the mask to undergo considerable heating, which then leads to distortions. Two concepts were devised to overcome this problem:

- SCALPEL (SCattering with Angular Limitation Projection Electron-beam Lithography) employed a scattering mask made from an extremely thin membrane (<150 nm) of low-atomic-number material (e.g., silicon nitride), through which the electrons can pass. The development of SCALPEL has been stopped, mainly because even the smallest deviation in mask membrane thickness resulted in intolerable intensity variations on the wafer. The main principles of SCALPEL are detailed in Ref. [48].
- PREVAIL (PRojection Exposure with Variable Axis Immersion Lenses) employed a stencil mask with a thick membrane (1–2 µm), thereby scattering the electrons to unexposed spots. This concept is quite similar to the vector-shaped beam strategy presented in Section 5.2.1.4, but instead of a simple shape a quadratic portion of the stencil mask was printed onto the wafer. The development of PREVAIL has also been stopped, mainly because even today it is not possible to make a 4:1 stencil

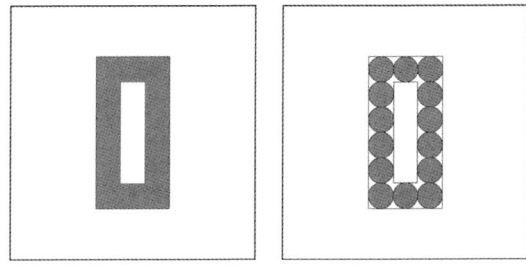

Figure 5.32 Electron projection lithography: the "donut problem".

mask covering an exposure field equal to that of an optical stepper with 26×26 mm, and therefore several masks must be stitched together to expose a single chip. Moreover, a stencil mask encounters the so-called "donut problem": the pattern shape shown in Figure 5.32 can only be made by two exposures with two complementary stencil masks, since on a stencil mask for a single exposure the center would fall out. For IPL, single stencil masks for fourfold exposure have been developed to circumvent this problem, as detailed in Section 5.3.2.1. The principle of PREVAIL is described in Ref. [49].

The electron optical columns for EPL experience a peculiar aberration in addition to the electron optical imperfections described in Section 5.2.1.2, namely the charge effects resulting from the electric charges of the electrons. Charge effects can be separated into a global charge effect, which influences any individual electron when traveling within an electron beam; this can be viewed as a continuous negative charge, with a charge density ρ. The single electron is within an electrostatic potential as of:

$$\nabla^2 \Phi = -\frac{\rho}{\varepsilon_0} \tag{5.18}$$

This electrostatic potential acts as an extended diverging lens. The global charge effect can, in principle, be compensated for by a suitable electron lens [50].

This is not the case for the stochastic charge effect, which is especially pronounced in a crossover, for example at demagnification, where all electrons interact within a small space. The stochastic charge effect leads not only to a beam blur but also to a beam energy spread, which in turn leads to chromatic aberrations. Global and stochastic charge effects are illustrated in Figure 5.33.

PEL: Principles and Limitations The transmission mask for PEL is a 1:1 stencil mask, and electrons are used for the proximity printing of this mask to the wafer. Initially, PEL employed 10 keV electrons, but currently low-energy electron-beam proximity lithography (LEEPL) [51], which utilizes low-energy electrons of 2 keV, is the PEL technique of choice.

The principle of LEEPL is shown in Figure 5.34. A single electron beam is generated in an electron beam column, and the mask is scanned. The low-energy electrons minimize the proximity effect, but forward scattering degrades the resolution.

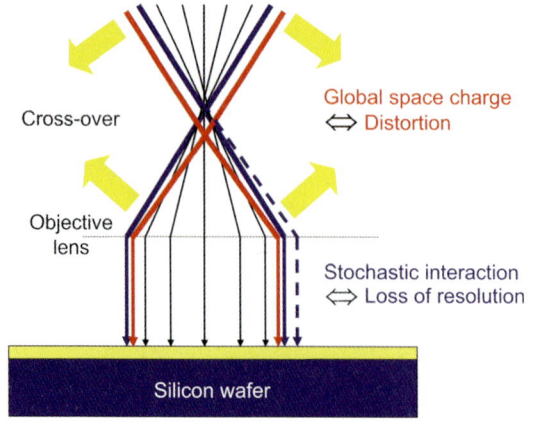

Figure 5.33 Electron projection space charge effects.

As the range of 2 keV electrons in the resist is <150 nm, the resist thickness for LEEPL is limited to 100 nm. In order to achieve the required aspect ratios, BLR [52] processes, which initially were developed for extending DUV lithography towards smaller features, must be applied (see Figure 5.35). Currently, LEEPL is not included in the ITRS as it has encountered several problems. For example, as this is a proximity technique, the distance between the stencil mask and the wafer is very small, usually <50 μm. Therefore, any distortion of the mask or wafer, or the presence of particles, would severely compromise the exposure. Furthermore, the implementation of bi-layer electron resist processes is not yet satisfactory. One positive development

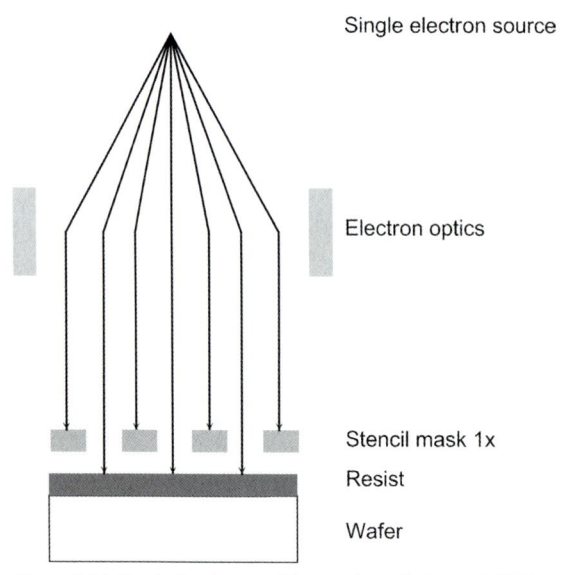

Figure 5.34 Proximity electron lithography techniques: LEEPL.

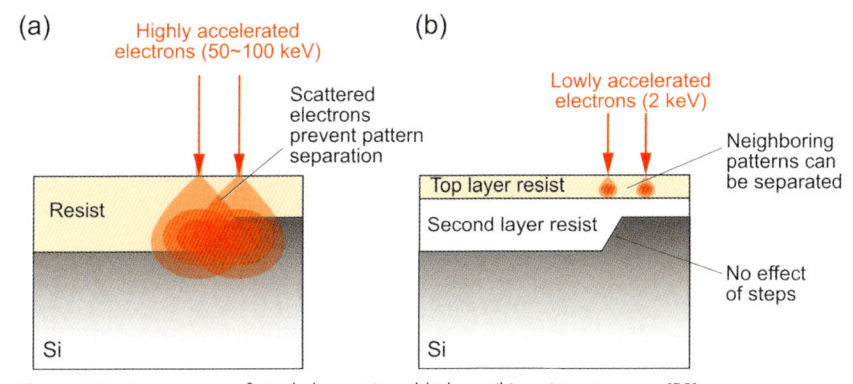

Figure 5.35 Comparison of single-layer (a) and bi-layer (b) resist processes [51].

however has been that, recently, stencil masks with a 26 × 26 mm exposure field could be prepared.

Projection Maskless Lithography Projection ML2 (PML2) [53] can be seen as the ML2 equivalent of EPL. Here, a single electron beam is split into multiple beams, with imaging being accomplished by a programmable aperture plate system (APS) [54, 55]. A range of innovative technologies was introduced to overcome the specific problems of both electron beam direct write and multiple beam application. The column of the demonstration system is shown in Figure 5.36 [53]. This employs a single electron source, therefore avoiding the control problems with multiple sources. The primary electron beam has a low acceleration voltage of $U = 5\,kV$, and is widened by condenser optics to fully cover the APS. Because of the low energy of the electrons, the APS cover plate experiences no significant thermal expansion problems. Subsequently, hundreds of thousands of separate electron beams emerge from the APS,

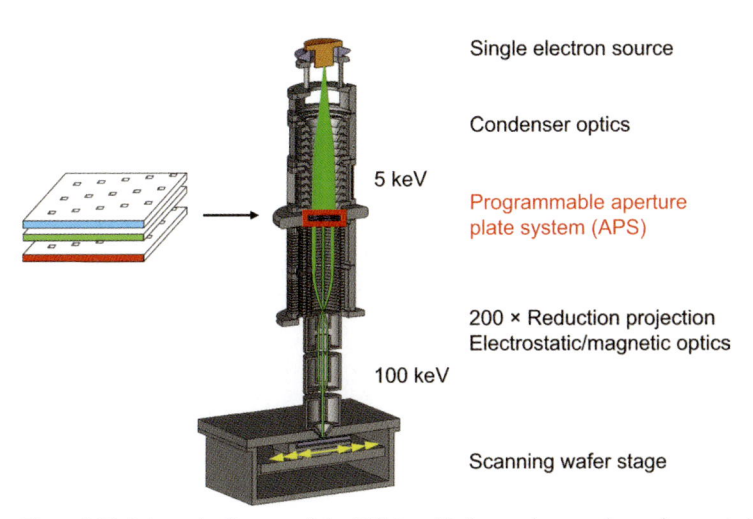

Figure 5.36 Schematic diagram of the PML2 multi-electron-beam column demonstrator [53, 54].

168 | *5 Charged-Particle Lithography*

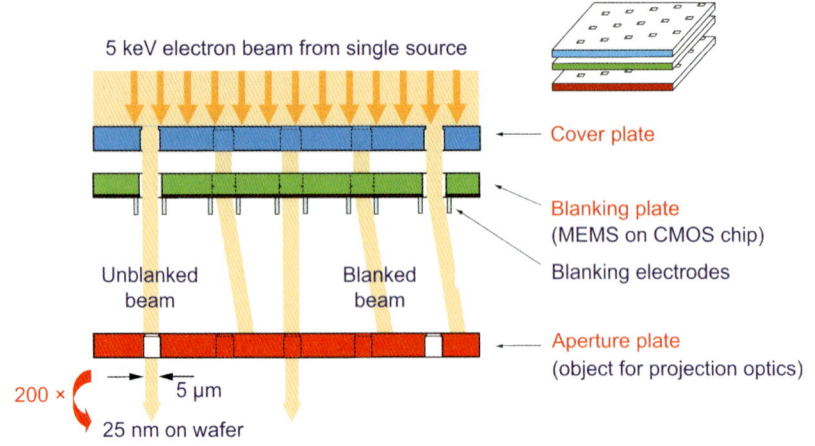

Figure 5.37 Schematic diagram of the multi-electron-beam modulator (APS) [53–55].

and are accelerated by $U = 100\,\text{kV}$, which results in a very high contrast when imaging on the wafer.

The APS, which is shown in detail in Figure 5.37 [53–55], consists of a cover, blanking, and aperture plate. The blanking plate employs MEMS-based structured electrodes for each of the transmission holes, which deflect the electron beam to strike the aperture plate and provide opaque features on the wafer. The diameter of each transmission hole is 5 µm. By using a two-stage electron optics, a reduction of ×200 is achieved, leading to a beam size of 25 nm on the wafer. Currently, the exposure area of the APS is $100 \times 100\,\mu\text{m}$.

With this exposure area, patterns on the wafer are written in stripes as shown in Figure 5.38 [53]. As discussed in Section 5.2.1, this may potentially lead to butting errors but, due to the small stripe size of <300 µm compared to current vector shaped-beam tools, no difficulties are expected.

The major challenge for PML2 is certainly to provide the required data transmission rate to the APS control electronics. A proof-of-concept (POC) tool was intended

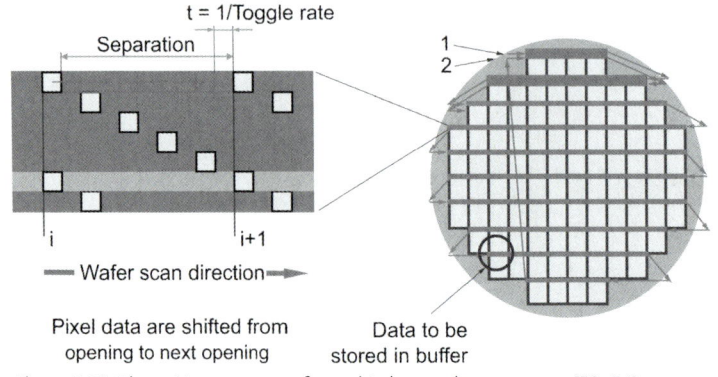

Figure 5.38 The writing strategy of a multi-electron beam system [53, 54].

within the MEDEA+ "CMOS logic 0.1 μm" project [56], for which a data transmission rate of 36 Gbit s^{-1} was demonstrated, scalable by channel count. A throughput of approximately 0.1 of a 300-mm wafer per hour was intended with this POC tool, although commercial tools should expose up to five 300-mm wafers per hour. However, this is still a small throughput compared to optical lithography steppers, with typical exposure throughputs of >100 wafers per hour. Nonetheless, this would be a great leap from EBDWL, which accomplishes much less than 0.1 wafer per hour in 65-nm patterning. Another issue is that the ×200 reduction requires all electrons to cross in a single region, thus leading to global and stochastic space charge effects, which potentially limit the throughput for the 22-nm node to less than one wafer per hour. To address this problem, an innovative PML2 scheme, with a throughput potential of up to 20 wafers per hour for the 32 and 22-nm nodes, is currently being investigated within the Radical Innovation MAskless NAnolithography (RIMANA) project [57].

Proximity Maskless Lithography Proximity ML2 (Mapper) [58] can be seen as the ML2 equivalent of LEEPL. Within Mapper, low-energy electrons of 5 keV are used, and the multiple electron beams are generated by splitting a single electron beam that originates from a single electron source. The multiple beams are then separately focused within an electrostatic lens array. The electron beams are arranged in such a way that they form a rectangular slit with a width of 26 mm, the same width as a field in current optical steppers. During exposure, the beams are deflected over 2 μm perpendicular to the wafer stage movement. With one scan of the wafer a full field of 26 × 33 mm can be exposed. During simultaneous scanning of the wafer, and deflection of the electron beams, these beams are switched on and off by light signals, one for each beam. The light signals are generated in a data system that contains the chip patterns in a bitmap format. The column of the proof-of-lithography (POL) tool, implementing 110 electron beams, is shown in Figure 5.39 [58]. A commercial tool would most likely implement 13 000 electron beams, so the bitmap would be divided over 13 000 data channels and streamed to the electron beams at up to 10 GHz, thus enabling a throughput of ten 300-mm wafers per hour.

As with LEEPL, the major challenges for Mapper are the problems arising with the proximity of the exposure, and the resist process. Additionally, as Mapper imple-

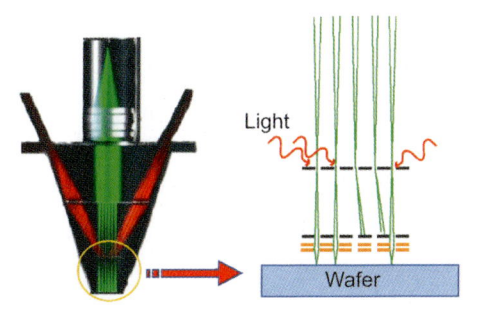

Figure 5.39 Schematic of the Mapper multi-electron-column demonstrator [58].

ments an electrostatic lens array within a <50 μm distance to the wafer, it is still unclear how the cleanliness of this array can be maintained during wafer throughput.

Overall, as ML2 techniques may provide the performance and throughput advantages of EPL and PEL, whilst avoiding the problems arising from mask fabrication and application, they should have the potential to rival optical lithography [59].

5.2.3.4 Imprint Templates

Nanoimprint lithography (NIL), considered to be a lithography method with the potential to rival optical lithography, is a technique where a patterned template is pressed onto a substrate coated with resist [60]. Currently, photoactivated NIL (PNIL), which uses a monomer resist with low viscosity, is considered to show the highest potential for volume production. The template, which must be constructed from a transparent material such as fused silica, is pressed onto the sample, after which a polymerization reaction is induced in the resist by applying UV light (thus, the technique is also called UV-NIL), and the template is removed.

Whilst NIL in itself does not involve exposure with photons or charged particles, the patterned template must be fabricated first, similar to a photomask for optical lithography. The templates are prepared by electron beam lithography, using similar processes as for photomasks (see Section 5.2.3.1). As the template is reproduced without demagnification, and therefore requires the same feature size as the pattern on the wafer, both fabrication and application still pose certain challenges. Currently, efforts are under way to utilize photomask fabrication methods for making PNIL templates [61]. As mentioned above, the templates must be transparent to UV light, and therefore fused silica photomask blanks may be used for template making. An overview of the template process flow is shown in Figure 5.40: in a first lithography step, the template is structured by EBL (first write). In a second lithography step, the pedestal required for imprint is made (second level write). Further details of the template process flow are presented in Refs. [62, 63].

When using photomask fabrication methods, currently four imprint templates are structured on a single photomask blank, as shown in Figure 5.41a and b. The photomask blank is then diced into separate templates (Figure 5.41c). As the dicing introduces contamination and mechanical strain, a modified fabrication approach must be developed before NIL can be employed in volume production. The size standard for templates resembles the exposure field of current optical lithography steppers (Figure 5.41d).

Complete details of the NIL imprint processes are described in Chapter 7 of this volume.

5.2.4
Integration

Although EBL is widely used in research because of its flexibility and high resolution, its low throughput and complex maintenance requirements of electron beam writing tools have limited the use of EBDWL in volume production. However, continuous improvements have led to the development of reliable tools with shaped beam writing

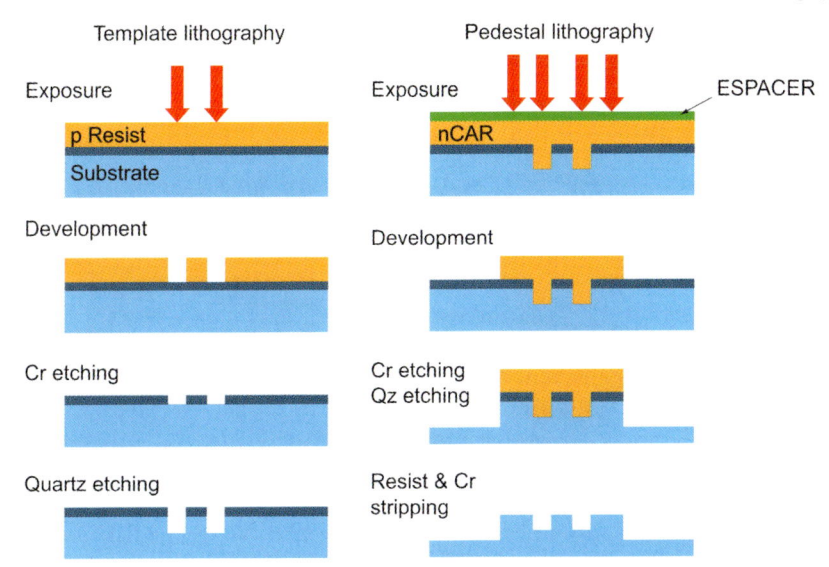

Figure 5.40 An overview of the fabrication process for two-dimensional templates [62, 63].

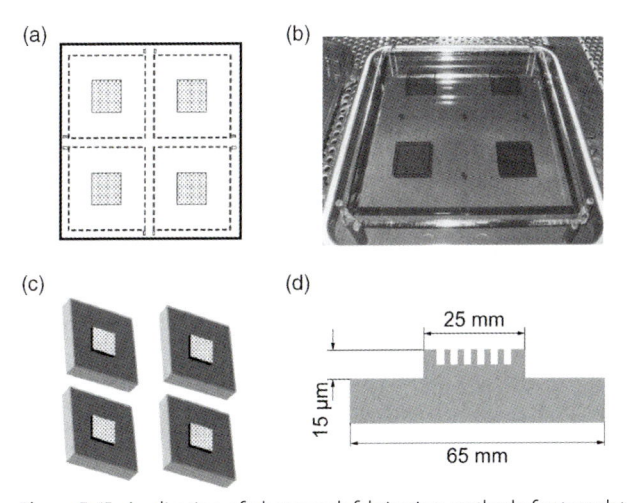

Figure 5.41 Application of photomask fabrication methods for template making [64].

strategy, which fulfill the requirements of the current 65 nm node fabrication which, according to the ITRS, are 65 nm dense lines, 45 nm isolated lines, 90 nm contact holes, and an overlay accuracy of <25 nm [16]. Yet, while the required resolution and repeatability has been achieved through developments in tools and CA resists, the overlay accuracy has long posed a major challenge.

Integrated circuits consist of several functional layers, for example metallizations and barriers, which must be fabricated sequentially. In optical lithography, each layer is patterned by a suitably made photomask. When using EBL, two scenarios can occur:

- Either all layers are made by EBL, which is the case for making device prototypes [65]
- Just one critical layer, for example the gates, are made by EBL, and all other layers are made by optical lithography.

The use of mixed EBL and optical lithography in volume production is referred to as "mix-and-match" lithography production [66], or hybrid lithography [67] if a single resist is utilized as both electron resist and photoresist.

In order to ensure that all subsequent layers are exactly matched, aligning techniques are used in optical lithography. One widely used alignment method in wafer steppers is through-the-lens alignment, where an alignment mark on the wafer is projected onto an alignment mark on the photomask, and a comparison is made. However, this approach is not possible with EBL tools; rather, two types of EBL alignment mark are currently used. The first option is to employ marks made from a film of high-atomic-weight material. This type of mark can be detected by secondary electron emission, but the method may lead to contamination issues and it is, therefore, mostly used only for back-end processing [68]. The second option is to create trenches as marks (see Figure 5.42), which are then scanned. This EBL alignment strategy has been used successfully in creating 65 nm node integrated circuits with hybrid lithography [69].

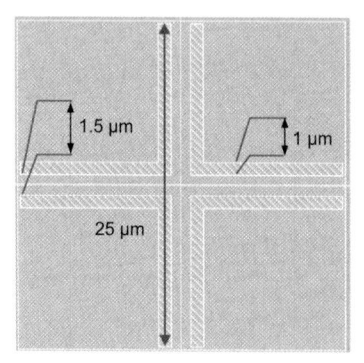

Figure 5.42 Trench alignment mark for EBDWL of a 25 × 25 μm integrated circuit [69].

5.3
Ion Beam Lithography

5.3.1
Introduction

Ion beam lithography (IBL) either utilizes ions to induce a chemical reaction in an ion resist for pattern formation, or can directly structure a functional film such as a metallization or barrier layer. When using ion resists, the wavelength of accelerated ions is even smaller than that of electrons, because of their higher mass. For example, the mass of H^+ is 2000-times the mass of an electron, and therefore a calculation analogous to that in Section 5.2.1 yields:

$$\lambda = \frac{h}{\sqrt{2mQU}} \qquad (5.19)$$

which, with an acceleration voltage of $U = 100$ kV, gives $\lambda = 0.0001$ nm.

Simple IBL tools use ion optics to focus ions from a source into a beam with a Gaussian energy distribution; therefore, they are referred to as focused ion beam (FIB) tools. The functional principle is analogous to the Gaussian EBL tools introduced in Section 5.2.1. A significant difference is that, because of the much higher mass of ions, a deflection is more difficult to achieve.

5.3.1.1 Ion Sources

IBL tools utilize volume ion sources [70], which consist of an ionizing region, where a plasma is formed, and an extraction region, where an electric field extracts the ions and accelerates them to form a beam. As with the thermal electron sources discussed in Section 5.2.1.1, the maximum current density j which can be obtained for an acceleration voltage V and an acceleration distance d is:

$$j = \frac{1}{9\pi d^2}\sqrt{\frac{2Q}{m}}V^{\frac{3}{2}} \qquad (5.20)$$

5.3.1.2 Ion Optics

Ion optics function in a very similar manner to electron optics (see Section 5.2.1.2). The general trajectory representation as Equation 5.13 remains valid, when the electron charge e is replaced by the appropriate ion charge Q. However, there is an important difference with the design of ion optical columns: while magnetic lenses are used in electron optics, because of their lower aberrations compared to electrostatic lenses [70], in ion optics only electrostatic lenses can be used because, unlike magnetic fields, electric fields focus independently of the charge to mass ratio (see Section 5.2.1.2).

5.3.1.3 Patterning

The initial idea behind using IBL instead of EBL for fabricating integrated circuits was that ions scatter very little in solids. Unlike with EBL, there is no significant proximity effect, and therefore IBL can deliver very high resolution and contrast. Advances in EBL technology (especially proximity effect corrections), together with the fact that because of their high mass, ions are likely to damage functional films or doped areas on the substrate, EBL is the currently established technology for direct-write methods.

5.3.2
Applications

Although IBL is not used for integrated circuit fabrication, it is being applied and continuously improved for the direct-structuring of functional films in the fabrication of special devices, such as nano-electromechanical systems (NEMS), nano-photonics, nanomagnetics, and molecular nanotechnology devices. Direct-structuring is also currently being investigated for the fabrication of imprint templates for NIL.

5.3.2.1 Direct-Structuring Lithography

Focused ion beam (FIB) tools can be used to create patterns in functional films, such as metallizations, or barriers on wafers directly, which is referred to as ion direct-structuring (IDS) lithography. The ions, when striking the functional film, cause the material to sputter, such that IDS is also known as *ion milling*. Another possibility is the local deposition of a functional film, with ions inducing the decomposition of a process gas at the surface of the wafer.

One major application of FIB tools is the repair of photomasks for optical lithography. As noted above (see Section 5.2.3.1), photomask quality is of utmost importance for yield. However, if due to a problem in the manufacturing process a part of the opaque film is sticking where it should have been removed, it can be sputtered away (see Figure 5.43) [71] or, if part of the opaque film is damaged, then it can be partially redeposited (see Figure 5.44) [71].

For some applications it is also reasonable to employ techniques initially developed for IPL, which introduced a transmission mask and projection optics with electromagnetic lenses to direct ions. This derivative of IPL is called ion projection direct structuring (IPDS). As mentioned in Section 5.1, IPL was removed from the ITRS in 2004 due to significant problems with fabrication and the application of the transmission masks.

IPL requires the use of stencil masks (see Section 5.2.3.3), because ions cannot pass through membrane masks, even with the thinnest imaginable membrane. However, a stencil mask absorbs the ions where there are no holes, and the resultant heating of the mask leads to its distortion. Further, with stencil masks it is not possible to make all required patterns with a single exposure, and so sets of complementary masks are required (see Section 5.2.3.3). Although initially these issues appear to make IPL impractical, studies to rectify the situation are ongoing. For example, thermal radiation cooling could be utilized to solve the heating problem,

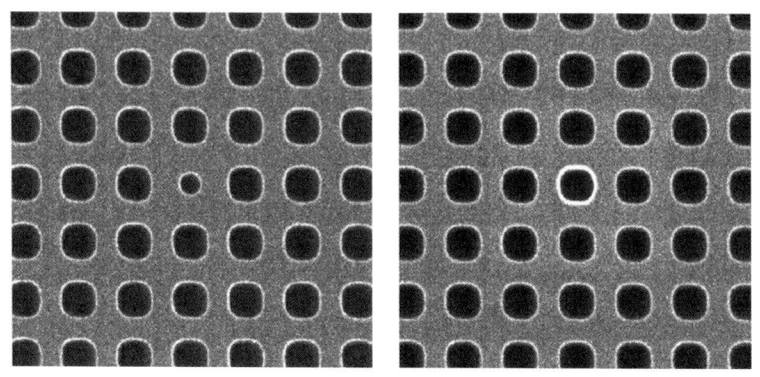

Figure 5.43 Opaque film defect: the repair of an undersized contact hole [71].

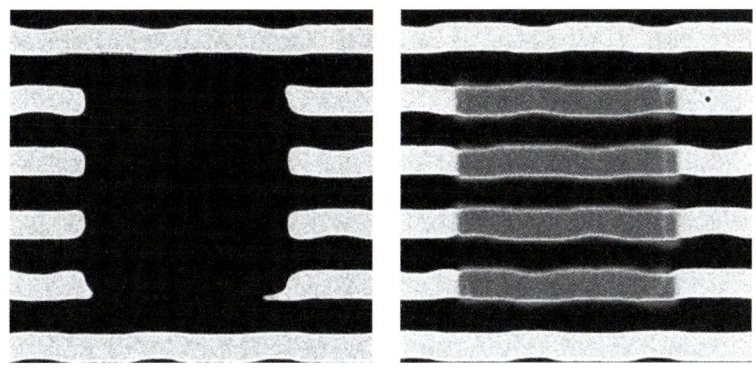

Figure 5.44 Pattern copy: the repair of missing lines [71].

while single stencil masks with fourfold exposure could enable complex patterns, but require half-sized features [54]. A comprehensive discussion of IPL, in addition to the details of a proposed IPL system, are presented in Refs. [54, 72].

IPDS can, for example, be applied to structure magnetic media for high-density data storage in a single exposure by inter-mixing the films of a multilayer structure [73]. For such an application a single stencil mask is sufficient [74].

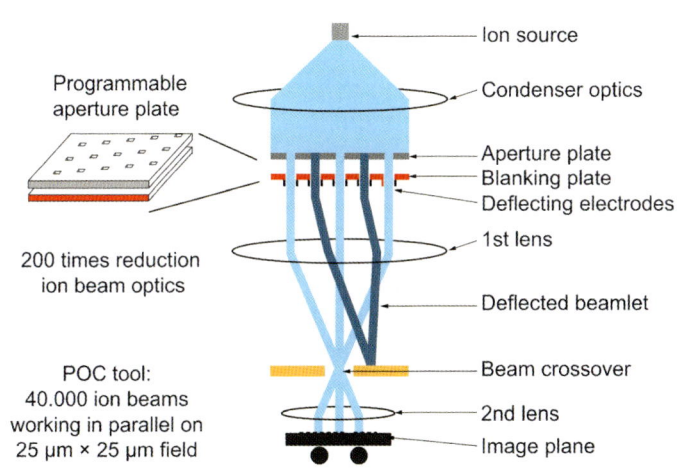

Figure 5.45 Schematic of multi-ion-beam column demonstrator [53, 75].

Considerable effort has also been made towards developing IPDS for volume production. The most-often investigated approach is to use the technique for multiple EBL (see Section 5.2.3.3), to split a broad ion beam into multiple beams, and to image with a programmable APS. This multiple ion beam projection maskless patterning (PMLP) technique is currently being developed within the project CHARPAN (CHARged PArticle Nanotech) [53, 75], and the column of the demonstration system used is shown schematically in Figure 5.45 [53, 75].

A possible multi-ion-beam tool resulting from CHARPAN would have a wide range of applications. In addition to IPDS, several other ion-beam-induced patterning processes, such as beam-assisted etching, deposition, polishing, nanometer-resolved ion implantation, and ion-beam-induced mixing, are possible. All of these processes are considered to be fundamental for the fabrication of emerging nanoscale devices.

5.3.2.2 Imprint Templates

As discussed in Section 5.2.3, NIL requires the use of templates which are currently fabricated by EBL. However, this method requires a full process sequence similar to photomask making, such as resist coating, exposure, development, and etch to be applied to a blank. The use of a FIB tool enables direct-structuring of the chromium film on the blank, by sputtering.

Whilst FIB tools, compared to state-of-the-art EBDWL tools, lack the throughput required for sensible template fabrication, a reliable multi-ion-beam tool would clearly take over from EBL, and consequently the resistless fabrication of templates has been included within the CHARPAN project.

5.4 Conclusions

The continuous improvement of EBL has always placed it one step ahead of the most advanced optical lithography in integrated circuit fabrication. Although mask fabrication for optical lithography is still its principal application, EBL has become a time- and cost-effective technique for early device and technology development. Further, with mask costs currently showing huge increases, EBL represents a viable option for small-volume production, despite its comparatively low throughput. Additionally, even in medium-volume production, EBL is employed for writing critical layers within mix-and-match and hybrid lithography. Because of ever-increasing device complexity, applying EBDWL, using shaped-beam writing tools in combination with advanced CA resists, is mandatory. In parallel, efforts are being continued in the investigation of parallel electron beam writing systems (ML2), which show the potential almost to match the throughput of optical lithography, and thus may in time complement or even replace the latter process for high-volume production.

In contrast, in its current form, IBL is not applicable for integrated circuit fabrication, although further improvements in FIB tools towards IPDS, as well as the development of parallel ion beam writing systems (PMLP), may lead to its feasible application.

Integrated circuit fabrication aside, both EBL and IBL techniques are currently being used and continuously improved for the fabrication of special devices in low volume, such as nano-electromechanical systems, nanophotonics, nanomagnetics, and molecular nanotechnology devices.

Acknowledgments

The authors thank L. Markwort of Carl Zeiss AG, H. Loeschner of IMS Nanofabrication GmbH, and H.J. Doering and T. Elster of Vistec Electron Beam GmbH, for their support. Special thanks are due to R. Waser of Forschungszentrum Jülich GmbH for carefully reviewing this chapter.

References

1. Levinson, H.J. (2001) *Principles of Lithography*, SPIE, The International Society for Optical Engineering, Bellingham, WA.
2. Rai-Choudhury, P. (1997) *Handbook of Microlithography, Micromachining and Microfabrication Vol. 1*, SPIE, The International Society for Optical Engineering, Bellingham, WA.
3. Attwood, D. (2000) *Soft X-rays and Extreme Ultraviolet Radiation, Principles and Applications*, Oxford University Press, Oxford.
4. International Technology Roadmap for Semiconductors, www.itrs.net.
5. Craighead, H.G. (1984) 10 nm resolution electron-beam lithography. *Journal of Applied Physics*, **55**, 4430.
6. Breton, B. (2004) *Fifty Years of Scanning Electron Microscopy*, Academic Press.
7. Hawkes, P.W. (1989) *Electron Optics*, Academic Press.
8. Ximen, J. (1990) Canonical aberration theory in electron optics. *Journal of Applied Physics*, **68**, 5963.
9. Hu, K. and Tang, T.T. (1998) Lie algebraic aberration theory and calculation method for combined electron beam focusing-deflection systems. *Journal of Vacuum Science & Technology B*, **16**, 3248.
10. Rose, H. and Wan, W. (2005) Aberration correction in electron microscopy. *IEEE Particle Accelerator Conference Proceedings*, p. 44.
11. Chu, H.C. and Munro, E. (1998) Computerized optimization of electron-beam lithography systems. *Journal of Vacuum Science & Technology B*, **19**, 1053.
12. Uno, S., Honda, K., Nakamura, N., Matsuya, M. and Zach, J. (2005) Aberration correction and its automatic control in scanning electron microscopes. *Journal for Light and Electron Optics*, **116**, 438.
13. Bas, E.B. and Cremosnik, G. (1965) Experimental Investigation of the Structure of High-Power-Density Electron Beams. First Electron and Ion Beam Science and Technical Conference Proceedings, p. 108.
14. Thomson, M.G.R., Collier, R.J. and Herriot, D.R. (1978) Double-aperture method of producing variably shaped writing spots for electron lithography. *Journal of Vacuum Science & Technology*, **15**, 891.
15. Pfeiffer, H. (1979) Recent advances in electron-beam lithography for the high-volume production of VLSI devices. *IEEE Transactions on Electron Devices*, **26**, 663.
16. Vistec Electron Beam, www.vistec-semi.com.
17. JEOL, www.jeol.com.
18. NuFlare, www.nuflare.co.jp.
19. Chang, T.H.P. (1975) Proximity effect in electron beam lithography. *Journal of Vacuum Science & Technology*, **12**, 1271.
20. Owen, G. and Rissman, P. (1983) Proximity effect correction for electron beam lithography by equalization of background dose. *Journal of Applied Physics*, **54**, 3573.
21. Murai, F., Yoda, H., Okazaki, S., Saitou, N. and Sakitani, Y. (1992) Fast proximity effect correction method using a pattern area density map. *Journal of Vacuum Science & Technology B*, **10**, 3072.
22. Hudek, P. and Beyer, D. (2006) Exposure optimization in high-resolution e-beam lithography. *Microelectronic Engineering Elsevier*, **83**, 780.
23. Haller, I., Hatzakis, M. and Srinivasan, R. (1968) High-resolution positive resists for electron-beam exposure. *IBM Journal of Research and Development*, **12**, 251.

24 Nippon Zeon Chemical Corp., www.zeon.co.jp.

25 Berger, L., Dieckmann, W., Krauss, C., Dress, P., Waldorf, J., Cheng, C.Y., Wei, S.L., Chen, W.S., Kao, M.J. and Tsai, M.J. (2005) E-beam direct-write lithography for the 45 nm node using a novel single substrate coat-bake-develop track. *SPIE Proceedings*, Volume 5751, p. 609.

26 Pacansky, J. and Waltman, R.J. (1988) Solid-state electron beam chemistry of mixtures of diazoketones in phenolic resins: AZ resists. *Journal of Physical Chemistry*, **92**, 4558.

27 Katoh, K., Kasuya, K., Sakamizu, T., Satoh, H., Saitoh, H. and Hoya, M. (1999) Chemically amplified positive resist for the next generation photomask fabrication. *SPIE Proceedings*, Volume 3873, p. 577.

28 Technical Bulletin, Fujifilm Arch Corp . (2001) EB Positive Resist for Mask Process FEP-171.

29 Technical Bulletin, Sumitomo Chemical Corp . (2001) Negative-type photoresist for electron beam lithography NEB22.

30 Irmscher, M., Beyer, D., Butschke, J., Constantine, Ch. Hoffmann, Th. Koepernik, C., Krauss, Ch. Leibold, B., Letzkus, F., Mueller, D., Springer, R. and Voehringer, P. (2002) Comparative evaluation of e-beam sensitive chemically amplified resists for mask making. *SPIE Proceedings*, Volume 4754, p. 175.

31 Irmscher, M., Butschke, J., Koepernik, C., Mueller, D., Springer, R., Voehringer, P., Beyer, D., Hudek, P., Tschinkel, M., Berger, L. and Dress, P. (2003) Investigation of e-beam sensitive negative-tone chemically amplified resists for binary mask making. *SPIE Proceedings*, Volume 5130, p. 168.

32 Schwersenz, A., Beyer, D., Boettcher, M., Choi, K.H., Denker, U., Hohle, C., Irmscher, M., Kamm, F.M., Kliem, K.H., Kretz, J., Sailer, H. and Thrum, F. (2006) Evaluation of most recently chemically amplified resists for high resolution direct write using a Leica SB350 variable shaped beam writer. *SPIE Proceedings*, 6153, 47.

33 Qimonda, www.qimonda.com.

34 Lutz, T., Kretz, J., Dreeskornfeld, L., Ilicali, G. and Weber, W. (2005) Comparative study of calixarene and HSQ resist systems for the fabrication of sub-20 nm MOSFET device demonstrators. *Microelectronic Engineering Elsevier*, **479**, 78–79.

35 Advanced Mask Technology Center, www.amtc-dresden.com.

36 ASML, www.asml.com.

37 Koepernik, C., Becker, H., Birkner, R., Buttgereit, U., Irmscher, M., Nedelmann, L. and Zibold, A. (2006) Extended process window using variable transmission PSM materials for 65 nm and 45 nm node. *SPIE Proceedings*, Vol. 6283, p. 1D.

38 Beyer, D., Löffelmacher, D., Goedel, G., Hudek, P., Schnabel, B. and Th. Elster, (2001) Tool and process optimization for 100 nm mask making using a 50 kV variable shaped e-beam system. *SPIE Proceedings*, Vol. 4562, p. 88.

39 Dameron, D.H., Fu, C.C. and Pease, R.F.W. (1988) A multiple exposure strategy for reducing butting errors in a raster-scanned electron beam exposure system. *Journal of Vacuum Science & Technology B*, **6**, 213.

40 Berger, L., Dress, P., Gairing, T., Chen, C.J., Hsieh, R.G., Lee, H.C. and Hsieh, H.C. (2004) Global CD uniformity improvement for mask fabrication with nCARs by zone-controlled post-exposure bake. *Journal of Microlithography, Microfabrication, and Microsystems*, **3**, 203.

41 Ehrlich, C., Edinger, K., Boegli, V. and Kuschnerus, P. (2005) Application data of the electron beam based photomask repair tool MeRiT MG. *SPIE Proceedings*, Vol. 5835, p. 145.

42 Weber, W., Ilicali, G., Kretz, J., Dreeskornfeld, L., Roesner, W., Haensch, W. and Risch, L. (2005) Electron beam lithography for nanometer-scale planar double-gate transistors. *Microelectronic Engineering Elsevier*, **206**, 78–79.

43 Kretz, J., Dreeskornfeld, L., Hartwich, J. and Roesner, W. (2003) 20 nm electron beam lithography and reactive ion etching for the fabrication of double gate FinFET

devices. *Microelectronic Engineering Elsevier*, **763**, 67–68.
44 Kretz, J., Dreeskornfeld, L., Schroeter, R., Landgraf, E., Hofmann, F. and Roesner, W. (2004) Realization and characterization of nano-scale FinFET devices. *Microelectronic Engineering Elsevier*, **803**, 73–74.
45 Seidel, R.V., Graham, A.P., Kretz, J., Rajasekharan, B., Duesberg, G.S., Liebau, M., Unger, E., Kreupl, F. and Hoenlein, W. (2005) Sub-20 nm short channel carbon nanotube transistors. *Nano Letters*, **5**, 147.
46 Parker, N.W., Brodie, A.D. and McCoy, J.H. (2000) High-throughput NGL electron-beam direct-write lithography system. *SPIE Proceedings*, Vol. 3997, p. 115.
47 Pickard, D.S. (2003) Distributed axis electron beam technology for maskless lithography and defect inspection. *Journal of Vacuum Science & Technology B*, **21**, 2834.
48 Berger, S. et al. (1994) The SCALPEL System. *SPIE Proceedings*, Vol. 2322, p. 434.
49 Okamoto, K. et al. (2000) High throughput e-beam stepper lithography. *Solid State Technology*, **5**, 118.
50 Harriot, L.R. et al. (1995) Space charge effects in projection charged particle lithography systems. *Journal of Vacuum Science & Technology B*, **13**, 2404.
51 Utsumi, T. (2006) Present status and future prospects of LEEPL. *Microelectronic Engineering*, **83**, 738.
52 Lin, Q. et al. (1998) Extension of 248 nm optical lithography: a thin film imaging approach. *SPIE Proceedings*, Vol. 333, p. 278.
53 IMS Nanofabrication, www.ims.co.at.
54 Loeschner, H., Platzgummer, E. and Stengl, G. (21–22 March 2002) *Projection-ML2 with programmable aperture plate, International Mask-Less Lithography Workshop, Erfurt, Germany*, (see also Ref. [55]).
55 Loeschner, H. et al. (2003) Large-field particle beam optics for projection and proximity printing and for maskless lithography. *Journal of Microlithography, Microfabrication, and Microsystems*, **2**, 34.
56 Doering, H.-J., Elster, T., Heinitz, J., Fortagne, O., Brandstaetter, C., Haugeneder, E., Eder-Kapl, S., Lammer, G., Loeschner, H., Reimer, K., Eichholz, J. and Saniter, J. (2005) Proof-of-concept tool development for projection mask-less lithography (PML2). *SPIE Proceedings*, Vol. 5751, p. 355.
57 RIMANA project, www.rimana.org.
58 Mapper Lithography, www.mapperlithography.com.
59 Lin, Burn J. (2006) The ending of optical lithography and the prospects of its successors. *Microelectronic Engineering*, **83**, 604.
60 Chou, S.Y. (1995) Imprint of sub-25 nm vias and trenches in polymers. *Applied Physics Letters*, **67**, 3114.
61 Sasaki, S., Itoh, K., Fujii, A., Toyama, N., Mohri, H. and Hayashi, N. (2005) Photomask process development for next generation lithography. *SPIE Proceedings*, Vol. 5853, p. 277.
62 Hudek, P., Beyer, D., Groves, T., Fortagne, O., Dauksher, W.J., Mancini, D., Nordquist, K. and Resnick, D.J. (2004) Shaped beam technology for nano-imprint mask lithography. *SPIE Proceedings*, Vol. 5504, p. 204.
63 Dauksher, J., Mancini, D., Nordquist, K., Resnick, D.J., Hudek, P., Beyer, D. and Fortagne, O. (2004) Fabrication of step and flash imprint lithography templates using a variable shaped-beam exposure tool. *Microelectronic Engineering Elsevier*, **75**, 345.
64 Institut für Mikroelektronik Stuttgart, www.ims-chips.de.
65 Pain, L. et al. (2006) Transitioning of direct e-beam write technology from research. and development into production flow. *Microelectronic Engineering Elsevier*, **83**, 749.
66 Narihiro, M., Wakabayashi, H., Ueki, M., Arai, K., Ogura, T., Ochiai, Y. and Mogami, T. (2000) Intra-level mix-and-match lithography process for fabricating sub-100-nm complementary metal-oxide-semiconductor devices using the JBX-9300FS point-electron-beam system. *Journal of Applied Physics*, **39**, 6843.
67 Steen, S.E. et al. (2006) Hybrid lithography: The marriage between optical and e-beam

lithography method to study process integration and device performance for advanced device nodes. *Microelectronic Engineering Elsevier*, **83**, 754.
68 Steen, S.E. *et al.* (2005) Looking into the crystal ball: future device learning using hybrid e-beam and optical lithography. *SPIE Proceedings*, Vol. 5751, p. 26.
69 Pain, L. *et al.* (2004) Manufacturing concerns for advanced CMOS circuit realization: EBDW alternative solution for cost and cycle time reductions. *SPIE Proceedings*, Vol. 5374, p. 590.
70 Melngailis, J. (1998) A review of ion projection lithography. *Journal of Vacuum Science & Technology B*, **16**, 927.
71 FEI Company, www.fei.com.
72 Kaesmaier, R., Wolter, A., Loeschner, H. and Schunk, S. (2000) Ion-projection Lithography Status and sub-70 nm Prospects. *SPIE Proceedings*, Vol. 4226, p. 52.
73 Loeschner, H. *et al.* (2002) Ion projection direct-structuring for nanotechnology applications. *MRS Proceedings*, Vol. 739.
74 Dietzel, A., Berger, R., Loeschner, H., Platzgummer, E., Stengl, G., Bruenger, W.H. and Letzkus, F. (2003) Nanopatterning of magnetic discs by single-step Ar^+ ion projection. *Advanced Materials*, **15**, 1152.
75 CHARPAN project, www.charpan.com.

6
Extreme Ultraviolet Lithography

Klaus Bergmann, Larissa Juschkin, and Reinhart Poprawe

6.1
Introduction

6.1.1
General Aspects

The ongoing reduction of structure sizes in semiconductor devices such as memory chips or microprocessors means that conventional optical lithography is reaching its physical and technological limits. This technology makes use of the demagnified imaging of structures on a mask onto a photo resist. Currently, optical lithography utilizes deep ultraviolet (DUV) light at 193 nm and a high numerical aperture (NA) optical system consisting of transmitting lenses. Generally, the achievable resolution at wafer level (RES) and the depth of focus (DOF) can be expressed by the Rayleigh formulas:

$$\text{RES} = k_1 \frac{\lambda}{\text{NA}}, \quad \text{DOF} = k_2 \frac{\lambda}{(\text{NA})^2} \qquad (6.1)$$

Here, λ is the wavelength, NA is the numerical aperture ($= n \sin\alpha$, where n is the index of refraction of the medium between the wafer and the last optical element, and α is the half-opening angle of the beam), and k_1 and k_2 are process-dependent constants with values typically of approximately 0.5. Today, the best resolution is in the region of 60–65 nm, operating at a wavelength of 193 nm, a NA of 0.93, and a k_1-value of 0.31 [1]. Currently, the semiconductor industry is investigating all the possibilities for further reducing the structure size offered by Equation 6.1. That is to say, by reducing the wavelength, increasing the NA above unity, and operating with lower k_1-values. In this way, ASML – one of the leading stepper manufacturers – has successfully demonstrated a water-based immersion system with NA = 1.20 and $k_1 = 0.28$ for printing 45-nm structures. Another strategy is that of *double patterning*, where two masks are imaged successively onto the wafer, which permits a smaller k_1-value of 0.2. However, structure sizes below 32 nm are considered only to be achievable with a large reduction of the wavelength into the extreme ultraviolet

Nanotechnology. Volume 3: Information Technology I. Edited by Rainer Waser
Copyright © 2008 WILEY-VCH Verlag GmbH & Co. KGaA, Weinheim
ISBN: 978-3-527-31738-7

(EUV) range. A reduction from 193 to 13.5 nm in EUV lithography relieves the situation with the process constants and the numerical aperture. Thus, 16 nm is expected to be printable with a NA of 0.35 and k_1 equal to 0.41 [1].

According to the current roadmap for semiconductors [2], EUV lithography will be introduced into the production process at the 45 nm node during the year 2011, together with improved 193 nm technologies. In moving towards ever smaller features below 22 nm and approaching the physical limits of silicon-based chips, EUV seems to be the only photon-based solution from today's point of view [1].

The step from DUV to EUV, however, implies a variety of technological changes compared to the conventional technology. Operating in the EUV range requires that all components are held in a vacuum in order to avoid absorption in an ambient gas. EUV radiation has the strongest interaction with matter – that is, the highest cross-section for absorption. Typical penetration depths of EUV radiation into solids are in the range of a few hundreds of nanometers. The optical system and the mask to be imaged onto the wafer consist of reflecting multilayer mirrors, and the source is no longer a UV laser but rather an incoherent plasma source emitting isotropically with a wavelength around 13.5 nm. The technology still requires further development before reaching industrial maturity, which is expected to be achieved by about 2010.

A simple discussion of Equation 6.1 allows an estimation to be made of the required wavelength region aiming at a resolution of, for example, better than 70 nm and a depth of focus of more than several hundreds of nanometers. This consideration leads to a wavelength below 20 nm. Multilayer-based mirrors with a high normal incidence reflectivity at a wavelength of about 13 nm are currently available, and the semiconductor industry has fixed the wavelength to 13.5 nm as a standard to maintain lithium-based plasmas as an option, which have a strong line emission at this wavelength. With our current knowledge of all components, such as mirror reflectivity, sensitivity of the resist, parameters of the optical system and the desired wafer throughput, it is possible to specify the requirements for the source, which can be considered as the least known component of the system. Although a solution for the final concept has not yet been found, some early examples of source and system specifications have been identified, for example in Refs. [3, 4]. In addition, the actual requirements are updated continuously, taking into account new aspects and increasing knowledge [5].

The present chapter provides an overview of the system architecture of an EUV scanner, together with the demands made on each component, namely the light source, optical components for beam propagation and imaging, masks and resists. The current status of development and future challenges are also addressed.

6.1.2
System Architecture

A variety of EUV lithography systems have been installed during the past few years in order to test the whole chain, beginning at the source and ending at the wafer level or using simpler, high-NA systems with small imaging fields for printing fine structures [4, 6, 7]. The principle of an EUV scanner will be explained with the example of

Figure 6.1 Schematic view of the ASML alpha demo tool, the first full-field scanner for extreme ultraviolet (EUV) lithography. The collector and source are not shown clearly; the beam propagation at the respective location of the source collector module is shown on the left.

the ASML alpha demo tool demonstrator, which can be regarded as the latest development and which is close to the future lithography tool with respect to design. A schematic of the alpha demo tool is shown in Figure 6.1 (taken from Ref. [8]). Using a collector – preferably a Wolter-type nested shell collector – the light of a plasma source is focused into the so-called *intermediate focus* as the second focal point of the collector (Figure 6.1 shows only the beam propagation from the source and the collector to the second focus, but not the hardware itself). The light is fed into the illuminator, which consists of a set of spherical multilayer mirrors and is used to produce a banana-shaped illumination of the mask (the top optical element in Figure 6.1). Another set of mirrors is used to image this field onto the wafer (the bottom optical element in Figure 6.1), with a typical magnification of 0.25. Wafer and mask are moved continuously to scan the whole mask and to transfer the structures onto a wafer of typically 300 mm diameter. In contrast to conventional DUV scanners, all of the components are contained inside a vacuum in order to achieve a high optical transmission for the EUV light. The etendue of the current optical system is around 3.3 mm^2 sr, which leads directly to a specification for the source size [9]. The optical system is able to use all the light from a spatially extended plasma source of around 1.6 mm in length along the optical axis and 1 mm in diameter in the radial direction, which is emitted into the solid angle of the collector. The plasma source is operated in a pulsed mode, ultimately requiring repetition rates of 7–10 kHz to guarantee a sufficiently homogeneous illumination of the resist. There are requirements on all components of the scanner – that is, the source, collector, optical system and components, masks, resist and on the system itself, concerning, for example, vacuum conditions and contamination issues. Today, the source is regarded as the most critical component, although this might also be due to the fact that other components were developed prior to of the plasma source, and consequently less experience is available with this component.

The wafer throughput is furthermore dependent on the mechanical properties of the mask and the wafer handling system. According to Ref. [10], a simplified wafer throughput model can be formulated:

$$\begin{aligned} T &= T_{scan} N + T_{oh} \\ &= N(t_{acc} + t_{settle} + t_{exp} + t_{settle} + t_{dec}) + T_{oh} \\ &= N\left[\frac{2P}{a_w W R} + 2t_{settle} + \frac{(L+H)WR}{P}\right] + T_{oh} \end{aligned} \quad (6.2)$$

where T_{scan} is the scanning time per field, N is the number of fields per wafer, T_{oh} is the overhead time (wafer exchange, wafer alignment, ...), t_{acc} (t_{dec}) is the acceleration (deceleration) time, t_{exp} is the field exposure time, t_{settle} is the stage settling time after acceleration and before deceleration, P is the EUV intensity on wafer, a_w is the acceleration of the wafer stage, $W(H)$ is the field width (arc height + slit width) of the banana-shaped field, L is the field height, and R is the sensitivity of the resist. With this model the wafer throughput as a function of the reticle stage acceleration has been estimated for different illumination power levels ranging from 160 to 640 mW on the wafer [5]. The higher dose leads to a higher throughput only if the acceleration is increased; for example, an 80 wafers per hour throughput can only be achieved for the 640 mW, if the acceleration is more than 1.5 G. In the 160 mW case, a lower acceleration is required. With higher acceleration values and higher power levels at the wafer throughput is, of course, higher. These numbers are based on a $R = 5$ mJ cm^{-2} resist, a stage settling time of 25 ms, an overhead time of 11.5 s, a field size of 25×25 mm, a number of 89 fields, and an exposure slit of 2×25 mm^2.

The first results have been obtained with the alpha demo tool, and the possibility of full-field imaging has been successfully proven. An example of different printed lines of 50 nm down to 35 nm and the corresponding line edge roughness (LER) using a resist of 18 mJ cm^{-2} sensitivity, is shown in Figure 6.2. Full-field imaging

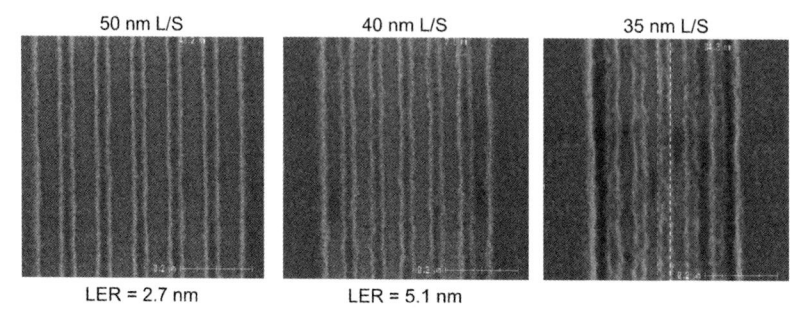

Figure 6.2 First results obtained with the ASML alpha demo tool of printed lines and spaces with resolution down to 35 nm and the respective LERs. The sensitivity of the resist was $\sim$18 mJ cm^{-2}.

over more than 20 mm slit height at the wafer level with a depth of focus of more than 240 nm has been demonstrated experimentally [6, 8].

6.2
The Components of EUV Lithography

6.2.1
Light Sources

According to Ref. [5], some of the requirements for the source and lifetime of the system for a production tool are as follows:

• Central wavelength (nm)	13.5
• Usable bandwidth (nm)	0.27
• Throughput (wafers h^{-1})	100
• EUV power at IF (W)	>115
• Repetition rate (kHz)	7–10
• Collector lifetime (months)	12
• Source electrode lifetime (months)	12
• Projection optics lifetime (h)	30 000
• Etendue of source output (mm^2 sr)	<3.3
• Spectral purity (% of EUV)	to be determined

The use of a multilayer mirror system (see Section 6.2.3) restricts the usable bandwidth to 2% around 13.5 or 0.27 nm, which is termed *inband radiation*. The throughput model is based on a 5 mJ cm^{-2} sensitivity of the resist, which has not yet been achieved (as discussed below). A less-sensitive resist would lead to higher source power specifications. The incoherent plasma source emits not only light but also debris in the form of particles, at least from the EUV-emitting plasma, irrespective of the source concept. Thus, some type of debris mitigation element is required between the source and collector appropriate to the actual source design. Typical collector half-opening angles range up to 70–80°. The total overall efficiency of the collector and the debris mitigation system can be estimated as around 20% of all the inband light emitted in the hemisphere of 2πsr [12], assuming a transmission of the debris mitigation system of 50%. This requires an inband emission of the source of at least 600 W/(2% b.w. 2πsr). Reasonable conversion efficiencies in the range of, at maximum, a few percent of the input energy for usable EUV radiation require a power input in the range of several tens of kilowatts. This imposes strict thermal demands, especially for the cooling of the debris mitigation system and the collector, which is the closest optical element to the source.

The multilayer mirrors have a finite transmission in the DUV range of 130 to 400 nm, which means that the emission from the source should not be too great within this wavelength region, as the resists are also sensitive in the DUV. The final specification is still under consideration and is, for example, dependent on progress in spectral purity filters with minimum losses for EUV radiation.

6.2.1.1 Plasmas as EUV Radiators

Extreme ultraviolet radiation sources can be divided into thermal and non-thermal emitters. Non-thermal emitters are X-ray tubes or synchrotron radiation sources, where the radiation is generated by deflecting charged particles. Thermal emitters, based on the generation of hot plasmas, are a cost-effective and compact solution for EUV lithography. Generally, for thermal emitters matter is heated up to a high temperature, T, where the limit of the emission of light can be described by Planck's law of radiation:

$$B_\lambda(T) = \frac{2hc^2}{\lambda^2} \left(e^{\frac{hc}{k_B T \lambda}} - 1 \right)^{-1} \tag{6.3}$$

with speed of light, c, Planck's constant, h, and Boltzmann constant, k_B. For a blackbody radiator, the temperature and the wavelength of maximum emission are related by Wien's law, which is derived from Equation 6.3:

$$\lambda_{max} T = 250 \text{ nm eV} \tag{6.4}$$

By aiming at a wavelength of $\lambda = 13.5$ nm, we obtain a temperature of around $T = 20$ eV (1 eV = 11 605 K). Such a high temperature is associated with matter in the plasma state. Usually, the emission of a plasma does not reach the Planck limit over the whole wavelength range, but only in individual strong emission lines of highly charged ions. Furthermore, for real plasmas the optimum emission is achieved at somewhat higher temperatures, depending on a variety of conditions, as discussed elsewhere [13]. The emission spectrum is characteristic of the respective element. Typical candidates discussed for EUV lithography are hydrogen-like lithium ions; that is, twofold ionized lithium with a strong single emission line at 13.5 nm, or tin and xenon as broadband emitters around 13.5 nm. In the case of xenon, the radiation around 13.5 nm arises from a transition of 10-fold ionized ions. For tin, the spectral efficiency is better compared to xenon (see Figure 6.3). With tin, more ionization levels exhibit transitions around 13.5 nm, leading to a more pronounced emission in the spectral range of interest.

Two concepts are pursued for generating such plasmas: laser-induced plasmas and discharge-produced plasmas. With laser-induced plasmas a pulsed laser beam is focused onto the target to be heated up. In the other case, the energy is taken from a pulsed electrical discharge. Many reports have been made concerning the different concepts, and discussing their special advantages and drawbacks [11]. In the next subsection, attention will be focused on the physical fundamentals of laser-induced and discharge-produced plasmas.

Irrespective of the individual concept, such plasmas are not only a source of light but also of debris consisting of fast ions and neutrals, clusters, droplets, and also heat. Sophisticated strategies are required to protect the optical system against this debris in order to avoid the deposition of matter onto the optics surface, or sputtering.

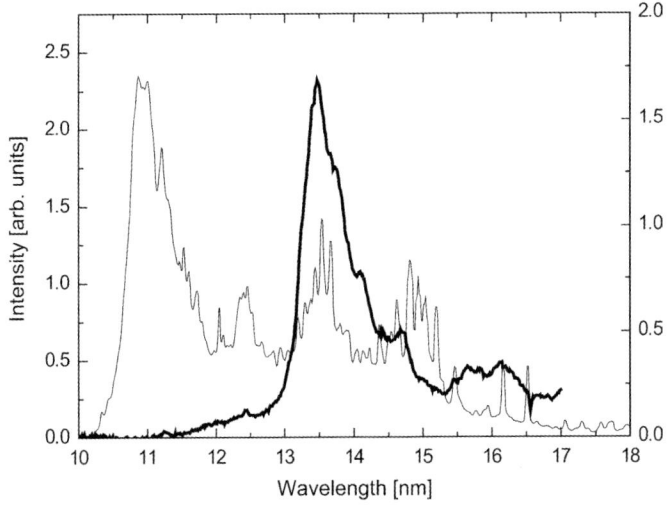

Figure 6.3 Typical emission spectra in the EUV for tin- and xenon-based gas discharge plasma sources. In the case of tin (bold line), the laser-induced plasma appears similar, whereas for xenon the laser-induced emission spectrum is smoother due to overlapping emission lines. The transitions of tin around 13.5 nm are iso-electronic to those of xenon around 11.0 nm.

6.2.1.2 Laser-Induced Plasmas

Hitting a target that is either solid, liquid or gaseous with a high-intensity laser beam leads to a plasma, where the laser energy is converted to thermal energy by inverse bremsstrahlung as the dominant process. Electrons are accelerated in the electrical field of the laser and transfer their energy to the ions. The laser energy is coupled to the plasma in a region where the plasma has the critical density, n_{crit}, which is dependent on the laser wavelength, λ_{laser} (ε_0 is the permittivity of free space, m_e the electron mass, e the electron charge, and ω_{laser} the laser frequency, that equals the plasma frequency at the critical density):

$$n_{crit} = \frac{\varepsilon_0 m_e \omega_{laser}^2}{e^2} = 1.11 \times 10^{21} \text{ cm}^{-3} \left(\frac{\mu m}{\lambda_{laser}}\right) \tag{6.5}$$

The temperature of the resulting plasma roughly scales with the laser intensity, I_{Laser}, according to $T_e \sim I_{Laser}^{4/9}$ [14, 15]. For a Nd:YAG laser with $\lambda_{laser} = 1.064\,\mu m$, the electron temperature, T_e, can be estimated as described in Ref. [16]:

$$T_e = 2.85 \times 10^{-4} \text{ eV}(I_{Laser}/(W/cm^2))^{\frac{4}{9}} \tag{6.6}$$

Thus, in order to achieve an electron temperature of around 30 eV a laser intensity of 2×10^{11} W cm^{-2} is required, which is also observed experimentally as an optimum laser intensity [17, 18]. Typical pulse durations of a laser-induced plasma are in the range of nanoseconds or even less. The spatial extension of the EUV-emitting region is below 100 μm; thus, the etendue requirement of <3.3 mm² sr is easily fulfilled with this type of plasma. Maximum conversion efficiencies of 5%/(2πsr 2% b.w.) for solid tin targets have been reported in the literature [18]. The conversion efficiency is defined as the ratio of usable inband EUV radiation into 2πsr to the incident laser light energy. In order to meet the source power requirement of 115 W in the intermediate focus, an average laser power of more than 5 kW is required, assuming an optimistic efficiency of 50% for the collector and the debris mitigation system. Obtaining high-power pulsed lasers at this level is an issue in current research and development activities. Different laser concepts are under discussion, such as pulsed CO_2 lasers or solid-state diode pumped laser, as reported elsewhere [19, 20, 22, 24]. However, it has not yet been shown that laser-induced plasmas can operate continuously on this power level. Further details on laser-induced plasma are available elsewhere [21, 23].

Besides the availability of the laser itself, the target is still an issue. Currently, different target concepts are under discussion, such as mass-limited targets to reduce debris production to a minimum level, and gaseous or droplets targets from frozen liquids or gases [18]. Most of the effort is currently being expended on tin-based targets, which have the highest expected conversion efficiencies.

6.2.1.3 Gas Discharge Plasmas

Producing the hot plasma by an electrical discharge is another well known method for the generation of light. For EUV-emitting plasmas, a pulsed electrical current is fed into an electrode system, which is filled with the working gas to be heated up at a neutral gas pressure of several tens of Pa. In a simplified concept, the current can be assumed to flow through a plasma cylinder, which is compressed by the self-magnetic field of the current to a high density of up to typically $n_e \sim 10^{19}\,\mathrm{cm}^{-3}$. The plasmas also experience ohmic heating, finally resulting in a dense and hot plasma column of several tens of eV electron temperature, and with a typical diameter of several hundreds of micrometers and a length in the range of few millimeters. The necessary current, I_o, can be approximated by assuming a Bennet equilibrium of the magnetic force and the plasma pressure [25]:

$$\frac{\mu_0\, I_0^2}{8\pi^2\, r_p^2} = (n_i + n_e) k_B T_e \tag{6.7}$$

where μ_0 is the magnetic field constant, r_p is the radius of the compressed plasma column, and n_i and n_e are the electron and ion density, respectively. The term $r_p^2 \cdot (n_i + n_e)$ can be expressed by the starting radius, a, of the neutral gas column and the neutral gas pressure, p, by pa^2 [26]. As an example, for a xenon plasma with 10-fold ionized ions ($\langle Z \rangle = 10$, $n_e = \langle Z \rangle n_i$) and a desired electron temperature of 35 eV, a current of 8 kA results, which is also characteristic of the devices under investigation. This pulsed current is usually produced in a fast discharge of a charged capacity, C, which is connected in a low-inductive manner to the electrode system. Typical values for the inductance of the system are around 10 nH and few 100–1000 nF for the capacity. Stored pulse energies are in the range from 1 to 10 J.

A variety of different concepts exist for discharge-based plasmas, which differ mainly in the special geometry of the electrode system and the ignition of the plasma. For further information the reader is referred to numerous other reports [11, 27, 28].

6.2.1.4 Source Concepts and Current Status

During recent years, many industrial laboratories have increased their efforts into the development of sources for EUV lithography, and consequently a variety of concepts of either laser-induced plasmas or gas-discharge plasmas have been investigated, or are currently under investigation. An excellent overview of the current "players" and of the technological progress made can be found in Refs. [11, 27, 29]. An overview of the current status and progress, compared to the year 2000, for different concepts such as the hollow-cathode-triggered pinch plasma [30, 31], the plasma focus [32], different Z-pinch-like concepts [33–36] and laser-induced plasma [37], is shown in Figure 6.4 (from Ref. [28]). Of note, it is clear that rapid progress is being made, and that more powerful sources will be available in the near future. The overview in Figure 6.4 refers only to source power and repetition rate, which is of course not sufficient to assess a certain concept. For example, the source powers achieved do not generally refer to continuous operation but rather to short-term operations ranging down to a few seconds. Furthermore, other specifications such as the plasma source size or the lifetime must also be considered more closely. Often, only the best values are presented and the current status for the simultaneous achievement of specifications is difficult to identify.

One promising concept, which is also used in the ASML alpha demo tool described above, is the Philips laser-triggered vacuum arc [39] (Figure 6.5). This concept makes use of two electrodes, which rotate in a liquid tin bath to continuously re-cover the electrode surface. The system set-up is illustrated schematically in Figure 6.6. Both

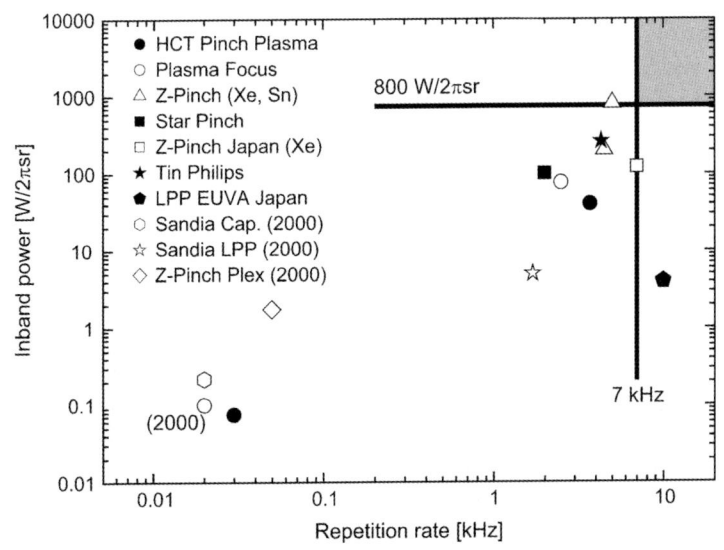

Figure 6.4 Currently achievable radiation power at 13.5 nm into 2% spectral bandwidth for different source concepts, and the corresponding repetition rates. Some data from 2000, taken from Ref. [38], are also shown to illustrate the rapid progress in source power.

190 | *6 Extreme Ultraviolet Lithography*

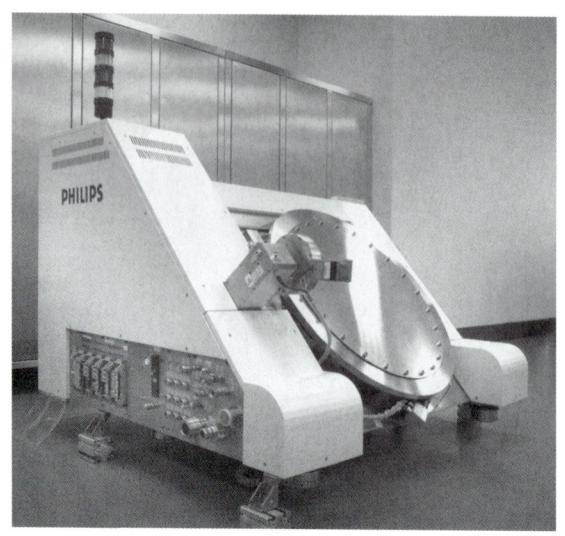

Figure 6.5 The Philips NovaTin EUV source based on the vacuum arc concept, which is used in ASMLs alpha demo tool.

electrodes are connected to a charged capacity, whereupon a laser pulse is used to evaporate a certain amount of tin, which also closes the electrical circuit in the gap between the two electrodes. The rapid discharge of the capacity heats up the tin plasma, which is used as an emitter of EUV radiation. Conversion efficiencies of up to 2.5%/(2πsr 2% b.w.) have been reported for this concept, with an average power of up to 300 W/(2πsr 2% b.w.) [40]. This is not too far away from the final specification for the source power. This concept has advantages with respect to cooling due to the rotating electrodes and electrode lifetime arising from covering the electrode surface with liquid tin and liquid metal cooling. However, a large amount of tin is also produced and emitted towards the collector. Thus, sophisticated means for debris

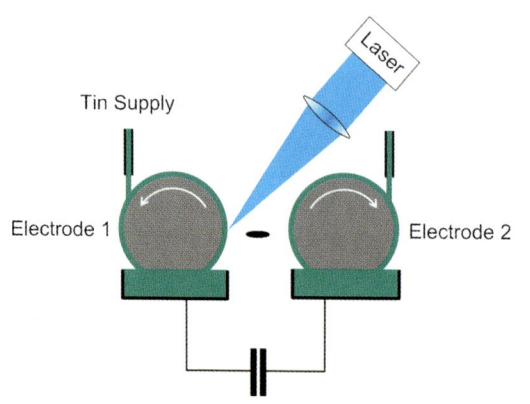

Figure 6.6 Scheme of the working principle of Philips NovaTin EUV source, based on the vacuum arc concept.

mitigation and cleaning strategies are required to guarantee a sufficient lifetime of the optical system. Recently, collector integration with a lifetime of more than 500 Gshots was successfully demonstrated using such a tin source [39]. Details of the collector design and the debris mitigation system are provided in the following section.

6.2.2
Collectors and Debris Mitigation

The currently preferred technical solution for collecting the light of the isotropically emitting plasma is the nested Wolter-type multi-shell collector [41]. A single collector shell consists of a hyperboloid and an ellipsoidal shell with identical focal points. The source is located within this common focal point, and is focused into the other focal point (intermediate focus), where the collector, although not an imaging element, leads to a "magnification" of the source by a factor of about 10 in the intermediate focus. The first optical element is about 50–100 cm behind this second focus. This type of collector allows light to be collected over a relatively large opening angle with a moderate gracing incidence angle at the reflecting surface, which is of particular importance for high reflectivity in the EUV. Such gracing incidence optics usually have a ruthenium coating with large reflectivity up to angles of $\sim 20°$, which is typical of applications in EUV lithography. An example of a multi-shell collector which is used for modeling light distribution after the intermediate focus at the first optical element of the illuminator is shown in Figure 6.7. A collection angle of more than 80° half-opening angle with a total efficiency, including the finite reflectivity of the ruthenium coating, of more than $40\%/(2\pi sr)$ is reported from the collector supplier [12].

Figure 6.8 shows the theoretical angular-dependent reflectivity of a ruthenium coating for different surface roughnesses, σ, and the typical range of operation for the collector. It should be noted that two reflections occur for the hyperboloid and the

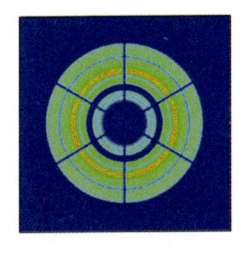

Figure 6.7 Schematic diagram of a Wolter-type nested shell collector with eight shells, as used in a ray-tracing calculation. This collector has opening angles between 11° and 45° corresponding to a collected solid angle of 1.7 sr or 27% of 2πsr. The right-hand diagram shows a simulated distribution in the far field for a spherical source of 50 μm FWHM.

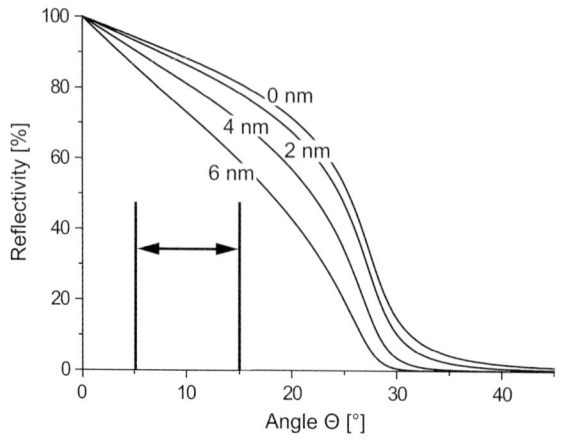

Figure 6.8 Calculated grazing incidence reflectivity of ruthenium for different roughness values, according to the CXRO database. The typical angle range for a multi-shell grazing incidence collector is also indicated. For collectors to be used in EUV lithography, the roughness is below 1 nm, leading to a reflectivity close to the theoretical limit.

ellipsoid. The data points are based on the atomic data for the refractive index published by CXRO [42]. The state of the art at the collector manufacturers – for example, Zeiss or Media Lario – involves surface roughness below 1 nm and chemically clean surfaces based on a physical vapor deposition (PVD) coating technology. Thus, the transmissions of the collectors are close to the theoretical limit. In addition, there is no longer any difficulty in fabricating substrates for the shells with diameters of several tens of centimeters.

Figure 6.7 also shows a typical light distribution for an eight-shell collector and a point source in a plane after the intermediate focus; this indicates the contributions of the different shells and the shadow of the mechanical support structure for the shells. Usually, the collector is located at a distance of only a few tens of centimeters from the plasma, which is also a thermal source in the 10 kW range. Cooling of the collector is achieved by using water-cooled lines around each collector shell. Results relating to the cooling capabilities are reported in Ref. [43], with a temperature increase of less than 1° when operated in the vicinity of a high-power source. Currently, research is going on in order to clarify whether this approach is sufficient, or whether more sophisticated cooling strategies must be applied, including for example a homogeneous temperature increase of the shell surfaces.

Another option is to have a normal-incidence collector based on a Schwarzschild design, with two spherical, multilayer coated mirrors. Some possible solutions to this problem are presented in Refs. [44–46].

As the closest optical element to the source, the collector experiences most of the heat load and debris emitted from the source. Consequently, overcoming the problems of a limited collector lifetime represents some of the major issues in current EUV lithography development activities. Today, such investigations are under way not only in industry but also at various academic institutes, all of which have encountered this problem [47, 48]. Within the present chapter, the details of various debris mitigation schemes, and the results obtained, are restricted to the activities at Philips, whose clear aim is to integrate the above-mentioned tin-based gas discharge source.

The minimum number of particles necessary for the effective generation of an EUV-emitting pinch plasma is about 10^{15} atoms per pulse. These particles can be assumed to be emitted into 4πsr, partly redeposited onto the electrodes, and also emitted towards the optical system. For a 1-hour operation at 5 kHz this will require approximately 3.5 g of tin. Such an amount is clearly excessive, assuming that this will be deposited on the collector surface, where even a few nanometers' thickness is unacceptable due to the reduced reflectivity of a tin-coated surface. Hence, both the deposition of material and sputtering of the optical coating by the emitted particles must be avoided. The particles are emitted in the form of fast ions with energies exceeding at least 10 keV, as neutrals, and also as droplets from the wet electrode surfaces. One highly effective method of stopping and removing particles is the so-called "foil trap concept", which is described in more detail in Ref. [48]. The process, which is shown schematically in Figure 6.9, includes a system of lamellas located between the source and the collector. The foil trap is operated in combination with a buffer gas, usually argon with high transmission in the EUV. The emitted particles are deflected and finally stopped by the ambient argon atoms, and then stick to the walls of the foil trap. In the case of tin, the foil trap is heated above the tin melting point in order to avoid an accumulation of tin in the system of the lamellas. Using only this technique, a collector lifetime of more than 10^9 shots has been reported for an operation with a tin-based discharge source [39]. However, the foil trap concept does not permit complete suppression of the emission of particles towards the collector, and consequently for longer operating times a deposition of tin on the

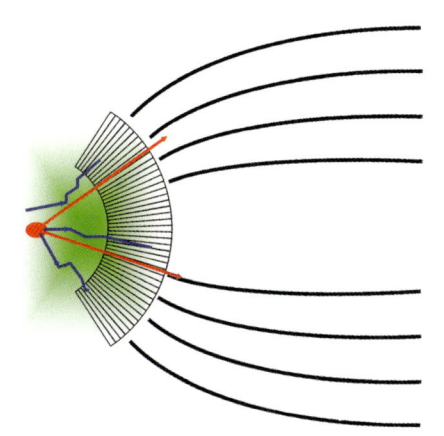

Figure 6.9 Schematic diagram of a foil trap system including buffer gas to protect the multi-shell collector against debris from the source. The foil trap has a high optical transmission, while the particles are efficiently stopped.

nanometer scale would be expected. In an attempt to overcome this problem several cleaning strategies for the collector have been proposed. One possibility would be to flood the collector chamber with a halogen gas, such as chlorine or iodine; the gas reacts with the tin to form volatile tin halides, which can be pumped away, while the ruthenium coating is unaffected. Using this technique permitted the complete recovery of a tin-coated ruthenium surface [49].

6.2.3
Multilayer Optics

For EUV radiation, the index of refraction is close to unity, and the absorption in matter is relatively high. A high reflectivity at surfaces is only achieved for incidence angles of the light of typically below 20°. This feature is, for example, exploited with gracing incidence optics as presented above. A high reflectivity for normal incidence is only achieved with multilayer systems, as shown schematically in Figure 6.10. Such multilayer systems consist of alternating layers of so-called "spacer material" and "absorber material", which have different indices of refraction and are thus reflective at their boundaries. Part of the incident light is reflected at each layer boundary, and the superimposed beam exhibits a high intensity. A well-known example, which is also used in EUV lithography, is a system consisting of silicon and molybdenum with high peak reflectivity around a central wavelength of 13–14 nm. A transmission electron microscopy (TEM) image of a real mirror is also shown in Figure 6.10. The

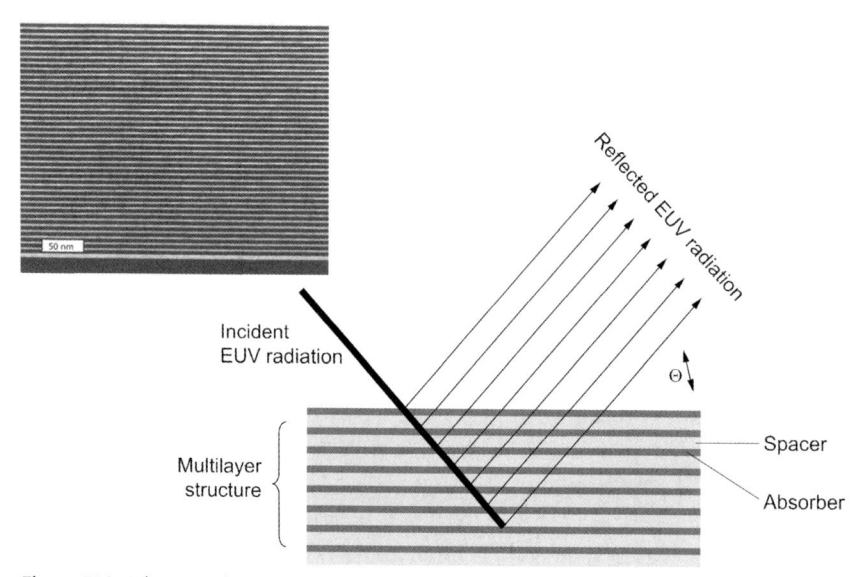

Figure 6.10 Schematic diagram of a multilayer mirror consisting of spacer (e.g., silicon) and absorber (e.g., molybdenum), where a high reflectivity is achieved by superimposing all rays reflected at the boundaries. A transmission electron microscopy image of a real Mo/Si system is shown in the top left section of the figure.

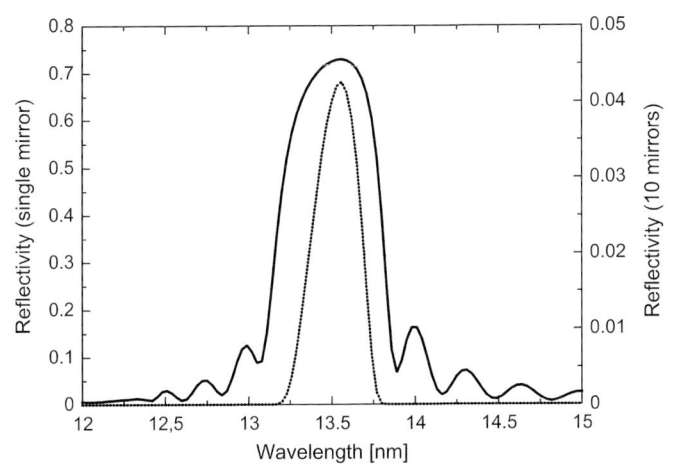

Figure 6.11 Wavelength-dependent reflectivity (solid line) of an ideal Mo/Si multilayer mirror according to the CXRO database, and the resulting reflectivity of a 10-mirror system (dotted line).

center wavelength of the maximum reflectivity of such a system can be expressed in terms of the total thickness, d, of a bi-layer, the incident angle, θ, ($\theta = 90°$ corresponds to normal incidence) and a material constant δ' by the Bragg equation:

$$m\lambda = 2d \sin\theta \sqrt{1 - \delta'/\sin^2\theta} \tag{6.8}$$

where δ' is the weighted material constant for both elements with a complex index of refraction, $n = 1 - \delta + i\beta$. By using the atomic data of the CXRO database [43], Figure 6.11 shows the reflectivity of an ideal Mo/Si multilayer system with zero roughness and no intermixing of the layers with a peak reflectivity of more than 70%. Values of approximately 70% are also achieved for real mirrors. A multilayer reflectivity close to this maximum is achieved at different locations. In order for systems to be used in an EUVL scanner, some losses occur due to the capping layers necessary for avoiding contamination, or additional layers for improving the thermal stability of the multilayer system. Figure 6.11 also shows, graphically, the transmission of a corresponding system of ten multilayer mirrors, which are typically used in EUV lithography. The overall transmission is only 4%, as the theoretical limit and the bandwidth decrease to below 0.3 nm FWHM. This is also the main reason for the restriction to only 0.27 nm or 2% of the 13.5-nm bandwidth, as discussed for the source specifications.

A variety of activities have been developed to improve the multilayer coatings to be used in EUV lithography systems with respect, for example, to increasing the reflectivity or achieving better thermal stability. In order to achieve a higher reflectivity and better thermal stability, additional layers of boron carbide (B_4C) are introduced to reduce the interdiffusion of silicon and molybdenum at their boundaries [50]. This diffusion leads effectively to a higher surface roughness, and thus to a

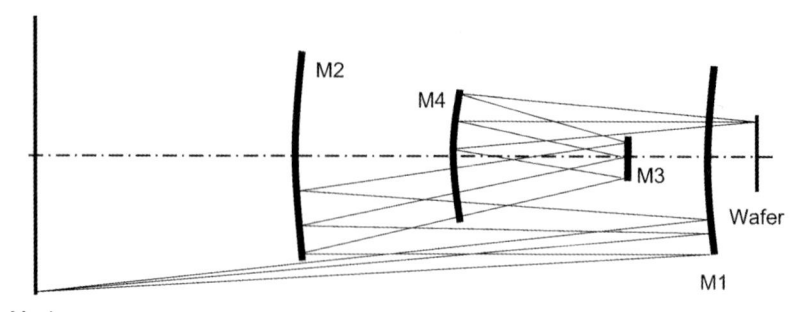

Figure 6.12 Schematic drawing of the projection optics of the ETS, consisting of four multilayer-coated reflective mirrors.

reduction in reflectivity. Furthermore, pure silicon molybdenum interfaces tend to be unstable and show even higher diffusion for temperatures above 100 °C. Here, ruthenium inter-layers are discussed as an alternative to boron carbide for increasing thermal stability [51]. Such protective layers are of special interest if multilayer coated components are to be used as collectors, as these must be heated for debris mitigation purposes [52]. Other activities are aimed at an improved coverage of the multilayer coating to protect against contamination or oxidation or at the suppression of the deep UV (100–200 nm) reflectivity in comparison to the EUV reflectivity. The latter strategy is one of several such approaches, including special thin filters for the deep UV, which are intended to fulfill the specification of spectral purity at the resist [53].

An imaging system for an EUV scanner consists of a number of multilayer-coated reflective mirrors. The design and specifications are discussed here as examples of the optical system of the Engineering Test Stand (ETS), which was the first full-field imaging system based on a plasma source. Details of the optical system can be found in Ref. [54]. A schematic of the projection optics consisting of four mirrors with $NA = 0.1$, a magnification of 0.25 and a resolution of 100 nm, is shown in Figures 6.12 and 6.13. Mirrors M1 and M3 are convex, while M2 and M4 are concave. The beam propagates off-axis, as indicated in Figure 6.13. In this special case, the mirror

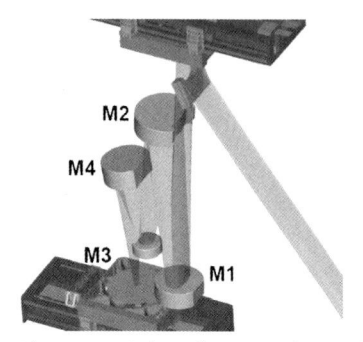

Figure 6.13 A three-dimensional view of the ETS projection optics system.

diameters are 165 mm for M1, 209 mm for M2, 104 mm for M3, and 170 mm for M4. The corresponding radii are −3055 mm for M1, +1088 mm for M2, −389 mm for M3, and +504 mm for M4 [55], where "+" indicates a concave and "−" a convex surface. Usually, a system of several mirrors is chosen in order to have sufficient degrees of freedom for the correction of aberrations and other imaging errors.

Typical diameters of the mirrors reach 200 mm for the ETS system, and even more for other optical systems. In order to meet the imaging specifications, the root mean square (RMS) figure error and the roughness must be below a certain level. The surface specifications will be discussed in more detail. Usually, the surface topology is described by a function $z(x, y)$. For simplicity, the following discussion and definitions are for the one-dimensional case $z(x)$. The extension to two dimensions is described elsewhere [56].

The average of the surface height is defined:

$$\bar{z} = \lim_{L \to \infty} \frac{1}{L} \int_{-L/2}^{L/2} z(x) dx \qquad (6.9)$$

with L being the spatial extension under consideration of the surface. The surface roughness, σ, is given by:

$$\sigma^2 = \lim_{L \to \infty} \frac{1}{L} \int_{-L/2}^{L/2} (z(x) - \bar{z})^2 dx \qquad (6.10)$$

It is useful to discuss the Fourier transform of the surface in terms of the spatial frequency, f_x:

$$Z(f_x, L) = \int_{-L/2}^{L/2} z(x) e^{-2\pi i f_x x} dx \qquad (6.11)$$

The power spectral density (PSD) function is often used for characterization of a surface, which can also be directly measured in scatterometry [56] and can be related to the Fourier transform $Z(f_x, L)$:

$$\text{PSD}(f_x) = \lim_{L \to \infty} \frac{1}{L} |Z(f_x, L)|^2 \qquad (6.12)$$

As defined in Equation 6.10, the roughness is the integral over all spatial frequencies of the PSD function. Often, different regions are defined in the specifications depending on the respective frequency interval f_{min} to f_{max}

$$\sigma^2_{\Delta f} = 2 \int_{f_{min}}^{f_{max}} \text{PSD}(f_x) df_x \qquad (6.13)$$

The surface figure error corresponds to frequencies typically ranging from the inverse aperture to $1\,\text{mm}^{-1}$. This type of error is responsible for aberrations. The roughness in the mid-spatial frequency range (MSFR), from $1\,\mathrm{mm}^{-1}$ to $1\,\mathrm{\mu m}^{-1}$, determines flare and contrast. The high spatial frequency range (HSFR) includes all frequencies above $1\,\mathrm{\mu m}^{-1}$. The HSFR roughness influences the EUV reflectivity; for the ETS optical system, a surface figure roughness of <0.25 nm RMS, a MSFR roughness of <0.2 nm RMS, and a HSFR roughness of <0.2 nm RMS are specified. The specifications for the Zeiss projection optics system for the ASML alpha demo tool are similar [57]. Here, the figure error should be <0.2 nm RMS, and the MSFR and the HSFR roughnesses should be between 0.1 and 0.2 nm RMS. Different analysis methods for determining the PSD function show that these specifications are fulfilled [57], which is also confirmed by the successful printing of small structures with diffraction-limited resolution.

6.2.4
Masks

In contrast to conventional DUV lithography, masks are also based on multilayer-coated reflective mirrors. A cross-section of a mask is shown schematically in Figure 6.14. The mask blank is defined as that part including the substrate and a protective layer of, for example, SiO_2 necessary for the patterning process. The structures to be imaged onto the wafer are written onto the surface using an absorber layer of typically 100 nm thickness. The preferred absorber materials are Cr, TaN, Al or W. However, as the mask is imaged, there are additional specifications in comparison to the multilayer mirrors for the optical system [58, 59], and these will be addressed in the following.

The substrate must have a low thermal expansion coefficient (CTE) of typically less than 5 ppb K^{-1}, as approximately 40% of the incident EUV light is absorbed and heats up the mask. A low thermal expansion is required in order to avoid any magnification correction between the changing of a wafer, and also to minimize image placement distortion due to thermal expansion of the mask. With respect to multilayer optics, the roughness specification is divided into high spatial frequency roughness (HSFR) and mid-spatial frequency roughness (MSFR). HSFR ($\lambda_{spatial} < 1\,\mathrm{\mu m}$) should be

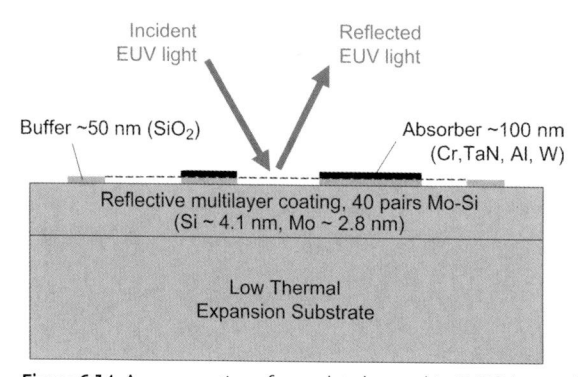

Figure 6.14 A cross-section of a mask to be used in EUV lithography.

below 0.10–0.15 nm (RMS) in order to reduce the losses due to scattering of light out of the entrance pupil of the optical system. In order to reduce the small angle scattering and image speckles, MSFR (1 µm < $\lambda_{spatial}$ < 10 µm) should also be below 0.1–0.2 nm (RMS). A peak reflectivity of more than 67% with a centroid wavelength uniformity across the mask of below 0.03 nm is required.

Another specification refers to the defect density of below 0.003 defects cm^{-2} for defects larger than 30 nm. This is the most challenging demand for the masks, and is one of the most critical issues in EUV lithography. As EUV light has a strong interaction with matter, and thus a short penetration depth of typically <100 nm, defects on the masks have a much higher probability of being printed, in contrast to other wavelengths as in the UV region. Thus, special care must be taken to reduce the defects on the masks to a level of 0.003 per cm^2 or, in other words, to less than a few defects per mask.

Many different categories of defect have been defined, and many activities are required simply to reduce the number on masks, mask blanks and the substrate by cleaning and repair techniques [60], detecting printable defects [61] and simulation of their influence on the picture at the wafer level [62]. Although defects on the substrate will be buried after the multilayer coating is applied, the various types of defect may lead to phase errors of the reflected light. Once such defect has been localized the absorber structure can be appropriately aligned to cover these defect and thus reduce its influence. The influence of defects (particles) on the mask depends on their size. If they are sufficiently small, they are not seen in the de-magnified image on the wafer; hence, only defects larger than 20–30 nm are of interest. The influence of defect size on image is discussed in Ref. [62].

In order to obtain an impression of the current status, the defect densities achieved on mask blanks are taken from Ref. [1]. For defects larger than 120 nm the density is 0.03 defects cm^{-2}, while for >60 nm a density of 0.3 defects cm^{-2} is achieved, this being more than two orders away from the final specification. As yet, no appropriate metrology is available for smaller defects.

6.2.5
Resist

The use of higher photon energies and the printing of ever-smaller features requires the development of a new generation of photo resists compared to those currently used in DUV lithography. According to the International Roadmap for Semiconductors (ITRS), a resist thickness of between 40 and 80 nm is required for EUV, while the line edge roughness (LER) and the critical dimension control (resolution) should be below 1 nm (3σ) [63, 64]. It should be noted that these specifications have near-atomic-scale resolution, which is not achievable with sufficiently high sensitivity when using the resists and concepts currently available. To ensure a certain wafer throughput, the resist sensitivity – that is, the number of photons or energy per unit area required to convert the resist molecules into solvable components – should be in the range of a few mJ cm^{-2}. The required thickness of less than 80 nm is lower than for DUV resists because of the short absorption length for EUV radiation in polymers. To ensure an approximately homogeneous illumination as a function of the penetration depth, the permissible resist thickness is limited to these values below 100 nm. This also implies a loss of usable photons by having a rather large portion of transmitted light.

There is a trade-off between high sensitivity, low LER and high resolution, which means that improving one feature will lead to a worsening of the other features. This fact, and the special challenges involved in applications for EUV lithography, are discussed in more detail below.

Conventional lithography makes use of a chemically amplified resist (CAR). An incident photon releases a H^+ ion (acid), which serves as a catalyst to react with other molecules to form solvable components, volatile products and another acid to trigger this reaction again. This mechanism is used to increase the sensitivity of the resist, but has an impact on the achievable resolution and LER. The processes are shown schematically in Figure 6.15. The amplification of soluble production by the acids is accompanied by a diffusion process into the unexposed regions. This diffusion process takes place during the post-exposure baking process, as indicated in Figure 6.15, and is determined by the diffusion constant, D, and the duration of the process, t_f. The higher the diffusion, the more sensitive is the resist, but the achievable resolution decreases. In order to quantify this dependence, a one-dimensional exposure with a sinusoidal modulation of photo acids with pitch, p, is considered for simplicity. A measure of the achievable resolution is the modulation transfer function (MTF_{diff}) of this initial distribution altered by the diffusion process [65, 66]:

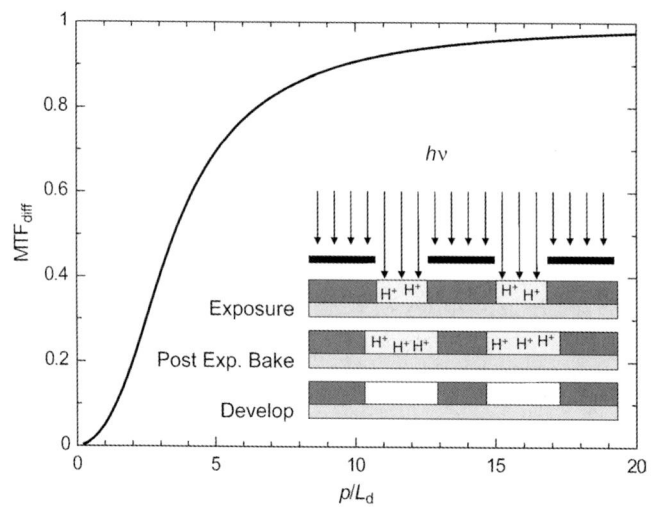

Figure 6.15 Modulation transfer function (MTF) for a sinusoidal exposure of lines and spaces with pitch, p, as a function of diffusion length and a schematic drawing of the exposure, post-exposure baking and developing process of a chemically amplified resist.

$$\text{MTF}_{\text{diff}} = \frac{p^2}{4\pi^2 Dt_f}\left\{1 - e^{-\frac{4\pi^2 Dt_f}{p^2}}\right\} \tag{6.14}$$

The respective diffusion length, L_d, is defined by $L_d^2 = 2Dt_f$. The modulation transfer function is shown as a function of the ratio p/L_d in Figure 6.15, where $\text{MTF}_{\text{diff}} = 1$ implies no change of the initial distribution. For example, if a deterioration to 70% is accepted, the diffusion length should not exceed a value of $0.2p$. This limitation of the diffusion length also implies a limitation in resist sensitivity. It should be noted that, with decreasing pitch, the absolute value of the diffusion length must also decrease, which therefore means less sensitivity in the transition from DUV to EUV.

Another parameter describing the quality of a resist is the LER, which is distinct from the achievable resolution in terms of the above discussion. Generally, the LER is dependent on the spatial distribution of solvable components after exposure with a number density, A. The LER is given by the ratio of the standard deviation of this density and its gradient, leading to $\text{LER} \propto \sigma_A/\nabla A$. For a sufficiently low number of photons, both parameters are determined by the incident number of photons. Here, a variation due to the Poisson statistic (shot noise) of the absorbed photons and, in the case of CA resist, the number of produced acids also comes into play. In general, the standard deviation, σ_N, of the number of photons, N, in a certain volume is proportional to $\sqrt{N}$, while $A \propto N$. Consequently, the LER scales as $\text{LER} \propto 1/\sqrt{N}$ or $1/\sqrt{E}$, where E is the incident dose. For a chemically amplified resist, the volume – which is relevant to the estimation of the number of photons – is the diffusion sphere. Thus, for a low diffusion length the LER is proportional to $L_d^{-3/2}$ when the dose is kept constant. With increasing diffusion length and lower variation due to photon statistics, the diffusion process and the MTF become dominant. The scaling of LER with the diffusion length can be expressed [66]:

$$\text{LER} \propto \left(\frac{1}{L_d}\right)^{3/2} / \text{MTF}_{\text{diff}}\left(\frac{L_d}{p}\right) \tag{6.15}$$

The LER scaling factor according to Equation 6.15 is shown in Figure 6.16 as a function of the diffusion length relative to the pitch. Two regions can be distinguished: for $L_d/p < 0.33$, the scaling is dominated by the photon statistics, whereas for $L_d/p > 0.33$ region the acid diffusion process is relevant for the LER. These two scaling regions are also observed experimentally [66].

The absolute value of LER is still dependent on the resist sensitivity or the necessary dose, which leads to the scaling with $1/\sqrt{E}$. This is illustrated in Figure 6.17, which shows the LER achieved for resists of different sensitivity [67]. The estimated shot noise limit is also indicated. The theoretical limit has clearly not yet been achieved, as the experimental data are slightly higher compared to this limit.

A number of other parameters, such as resist thickness, molecular size or outgassing, are also relevant to use in EUV lithography (see discussion in Ref. [65]). In summary, it is somewhat challenging to meet the specifications for a chemically amplified resist for use in EUV lithography, and in fact such a resist does not yet exist. In terms of LER and resolution, the specifications may be achieved with a

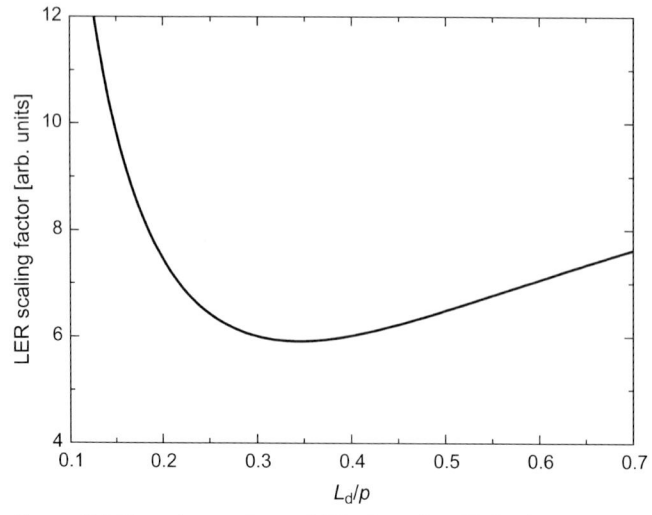

Figure 6.16 Line edge roughness (LER) of a resist with fixed sensitivity as a function of the diffusion length relative to the pitch.

non-chemically amplifying resist. One example is polymethyl methacrylate (PMMA), although this has a rather low sensitivity (in the range of 50–100 mJ cm^{-2}), and is therefore not acceptable for EUV lithography. Some printed lines down to 17.5 nm half pitch [68] are illustrated in Figure 6.18; the smallest currently achieved structure size is a half pitch of 12.5 nm [69].

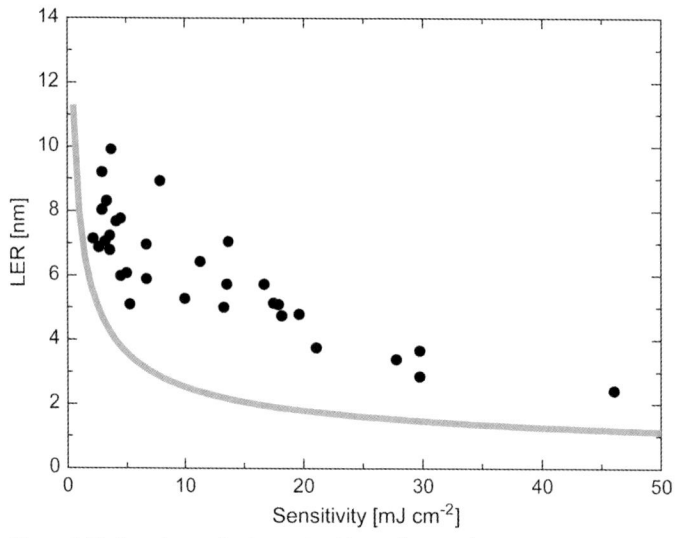

Figure 6.17 Experimentally determined line edge roughness (LER) as a function of sensitivity for different resists. The line gives an estimation of the shot noise limit, which has not yet been achieved.

Figure 6.18 Printed lines and spaces with a non-chemically amplified resist (polymethyl methacrylate), with a half-pitch down of 17.5 nm.

Finally, it is illustrative to discuss the role of shot noise simply by estimating the number of photons involved in the exposure process. The incident number of photons per unit area, I, can be rewritten in terms of the necessary dose, E, and the wavelength, λ, of the photons:

$$I = 5.0 \times 10^{-2} \frac{N_{Ph}}{nm^2} \frac{\lambda}{nm} E \frac{cm^2}{mJ} \tag{6.16}$$

In the transition from DUV with 193 nm to EUV at 13.5 nm, the reduced wavelength alone leads to a more serious influence of the photon statistics. For EUV radiation and an envisioned resist sensitivity of $E = 5$ mJ cm^{-2}, we obtain 3.4 Ph nm^{-2}. With regards to the specifications of LER and CD control below 1 nm, it is clear that shot noise becomes a limiting factor in EUV lithography. This not only requires the development of new resist materials, but also implies that many research investigations will be necessary over the next few years.

6.3
Outlook

Intensive research and development activities conducted during the past decade have shown that EUV lithography has the potential to provide a solution for the high-volume manufacture of semiconductor devices. Moreover, the technique has the potential to decrease structure sizes to 11 nm, into the range of the physical limits of silicon-based semiconductor technology. Several machines have been installed to demonstrate the capability of printing small structures using EUV radiation; most notably, the ASML alpha demo tool exhibits the full architecture with respect to the optical system, wafer and mask handling for a scanning operation and full-field imaging. The diffraction-limited printing of small structures down to 29 nm was also successfully demonstrated, though major efforts are still needed to meet the requirements of the components and to drive the technology to its theoretical limits in different areas. These challenges extend not only to the source power but also to the simultaneous high reliability and long lifetime of the source, and this is valid for both laser-induced and discharge-based plasmas. Further issues here include debris mitigation in order to increase the collector lifetime, collector thermal issues, and increase the opening angle. Additional studies are also required on the lifetime and contamination of the optical system by oxygen and hydrocarbons under EUV radiation, on defect-free masks, and on resists with a sufficiently high sensitivity at high resolution and low LER.

Despite the final specifications not having yet been met for several components, progress is nonetheless being made in all fields. For example, activities during the past few years have led to plasma sources which emit inband radiation into the hemisphere on a power level of several hundred watts – close to final specification that in the past was believed to be the most critical issue in EUVL. Major progress is also being achieved in improving the lifetime of both the source and the collector, using sophisticated debris mitigation techniques. In fact, a collector lifetime of more than 1 Gshot has recently been demonstrated, operating with a tin-emitting plasma source, and an ultimate lifetime of 100 Gshot seems feasible.

References

1 van den Brink, M. (2006) The only cost effective extendable lithography option: EUV, Third International Symposium on EUV Lithography, Barcelona.

2 *International Technology Roadmap for semiconductors (ITRS)*. The current version is available from www.sematech.org or www.itrs.net.

3 Ceglio, N.M., Hawryluk, A.M. and Sommargren, G.E. (1993) Front-end design issues in soft-X-ray projection lithography. *Applied Optics*, **32** (34), 7050–7056.

4 Gwyn, C.W., Stulen, R., Sweeney, D. and Attwood, D. (1998) Extreme ultraviolet lithography. *Journal of Vacuum Science & Technology B*, **16** (6), 3142–3149.

5 Ota, K., Watanabe, Y., Banine, V. and Franken, H. (2006) EUV source requirements for EUV lithography, in *EUV Sources for Lithography* (ed. V. Bakshi), SPIE Press, Bellingham, Washington, pp. 27–43.

6 Meiling, H. Meijer, H., Banine, V., Moors, R., Groeneveld, R., Voorma, H.-J., Mickan, U., Wolschrijn, B., Mertens, R., van Baars, G., Kürz, P., Harned, N., (2006) First performance results of the ASML alpha demo tool, in Emerging Lithographic Technologies X, Proceedings of SPIE, Vol. 6151, San Jose, USA (ed. Lercel, M.J.), pp. 615108.

7 Booth, M., Brioso, O., Brunton, A., Cashmore, J., Elbourn, P., Ellner, G., Gower, M., Greuters, J., Grünewald, P., Gutierrez, R., Hill, T., Hirsch, J., Kling, L., McEntee, N., Mundair, S., Richards, P., Truffert, V., Wallhead, I., Whitfield, M. and Hudyma, R. (2005) High-resolution EUV imaging tools for resist exposure and aerial image monitoring. Proc. SPIE 5751, 78–89.

8 Groeneveld, R., Harned, N., Zimmermann, J., Meijer, H., Meiling, H., Mickan, U., Voorma, H.J. and Kuerz, P. (2006) Full Field Imaging by the ASML

Alpha Demo Tool, International EUVL Symposium, Barcelona, Spain, Proceedings available at www.sematech.org.

9 Derra, G., and Singer, W. (2003) Collection efficiency of EUV sources. Proc. SPIE 5037, 728–741.

10 Ota, K., Tanaka, K. and Kondo, H. (2003) Throughput model considerations and impact of throughput improvement request on exposure tool, Second International EUVL Symposium, Antwerp, Belgium. Proceedings available from www.sematech.org.

11 Bakshi, V. (2006) EUV Sources for Lithography, SPIE Press, Bellingham, Washington.

12 Rigato, V. (2006) Evolution from current demonstrated α-hardware collector to full HVM, EUV Source Workshop, Barcelona, Spain. Proceedings available from www.sematech.org.

13 For example: Krücken, T., Bergmann, K., Juschkin, L. and Lebert, R. (2004) Fundamentals and limits for the EUV emission of pinch plasma sources for EUV lithography. *Journal of Physics D-Applied Physics*, **37** (23), 3213–3224.

14 Puell, H. (1970) Heating of laser produced plasmas generated at plane solid targets. *Zeitschrift für Naturforschung*, **A25**, 1807–1815.

15 Wood, O.R., II, Silvfast, W., Macklin, J. and Maloney, P. (1986) Comparison of extreme-ultraviolet flux from 1.06- and 10.6-μm laser-produced plasma sources for pumping photoionization lasers. *Optics Letters*, **11**, 198–200.

16 Schriever, G., Mager, S., Naweed, A., Engel, A., Bergmann, K. and Lebert, R. (1998) Laser-produced lithium plasma as a narrow-band extended ultraviolet radiation source for photoelectron spectroscopy. *Applied Optics*, **37** (7), 1243–1248.

17 Spitzer, R., Orzechowski, T., Phillion, D., Kauffman, R. and Cerjan, C. (1996) Conversion efficiencies from laser produced plasmas in the extreme ultraviolet regime. *Journal of Applied Physiology*, **79**, 2251–2258.

18 Richardson, M., Koay, C.-S., Takenoshita, K., Keyser, Ch., George, S., Al-Rabban, M. and Bakshi, V. (2006) Laser plasma EUV sources based on droplet target technology, in *EUV Sources for Lithography* (ed. Bakshi, V.), SPIE Press, Bellingham, Washington, pp. 687–718.

19 Takahashi, A., Tanaka, H., Akinaga, K., Matsumoto, A., Uchino, K. and Okada, T. (2005) Laser-wavelength dependence of laser produced plasma EUV emission, Third International Symposium on EUV Lithography, November 2004, Miyazaki, Japan. Proceedings available at: www.sematech.org.

20 Stamm, U., and Gäbel, K. (2006) Technology for LPP sources, in *EUV Sources for Lithography* (ed. V. Bakshi), SPIE Press, Bellingham, Washington, pp. 537–561.

21 (a) Eidmann, K., and Schwanda, W. (1991) *Laser Particle Beams*, **9**, 551.(b) Sigel, R. (1989) *Proceedings of SPIE*, **1140**, 6. (c) Sigel, R., Eidmann, K., Lavarenne, F. and Schmalz, R.F. (1990) *Physics of Fluids B*, **2**, 199.(d) Eidmann, K., Kühne, M., Müller, P. and Tsakiris, G.D. (1990) *Physics of Fluids B*, **2**, 208.

22 Hertz, H.M., Rymell, L., Berglund, M. and Malmqvist, L. (1996) Debris-free liquid-target laser-plasma soft X-ray source for microscopy and lithography, in *X-ray Microscopy and Spectromicroscopy* (eds J. Thieme, G. Schmahl, E. Umbach and D. Rudolph), Springer, Heidelberg.

23 Bakshi, V.(ed.) (2006). Section IV Laser-Produced Plasma (LPP) Sources, in *EUV Sources for Lithography*, SPIE Press, Bellingham, Washington, pp. 535–718.

24 Hansson, B.A.M., and Hertz, H.M. (2006) Liquid-xenon-jet LPP source, in *EUV Sources for Lithography* (ed. V. Bakshi), SPIE Press, Bellingham, Washington, pp. 619–648.

25 For example: Krall, N.A., and Trivelpiece, A.W. (1986) *Principles of Plasma Physics*, San Francisco Press, New York.

26 Bergmann, K., Lebert, R. and Neff, W. (1997) Scaling of the K-shell line

emission in transient pinch plasmas, *Journal of Physics D-Applied Physics*, **30** (6), 990.
27 Lercel, M.J. (ed.), Emerging Lithographic Technologies X, in Proceedings of SPIE, Vol. 6151, San Jose, USA, February 2006.
28 Juschkin, L., Derra, G. and Bergmann, K. (2007) EUV light sources, in *Low-Temperature Plasma Physics* (ed. Hippler, R.), Springer Verlag, pp. 619–654.
29 (2004) Special cluster on extreme ultraviolet light sources for semiconductor manufacturing. *Journal of Physics D-Applied Physics*, **37** (23), 3207–3284.
30 Bergmann, K., Schriever, G., Rosier, O., Müller, M., Neff, W. and Lebert, R. (1999) Highly repetitive, extreme-ultraviolet radiation source based on a gas-discharge plasma. *Applied Optics*, **38**, 5413–5417.
31 Pankert, J. Bergmann, K.,Klein, J., Neff, W., Rosier, O., Seiwert, S., Smith, C., Apetz, R., Jonkers, J., Loeken, M., Derra, G., (2002) Physical properties of the HCT EUV source, SPIE 27th International Symposium on Microlithography, Santa Clara, USA, 3–8 March.
32 Fomenkov, I.V., Partlo, W.N., Böwering, N.R., Khodykin, O.V., Rettig, C.L., Ness, R.N., Hoffman, J.R., Oliver, I.R. and Melnychuk, S.T. (2006) Dense plasma focus source, in *EUV Sources for Lithography* (ed. V. Bakshi) SPIE Press, Bellingham, Washington, pp. 373–394.
33 Stamm, U., Kleinschmidt, J., Bolshukin, D., Brudermann, J., Hergenhan, G., Korobotchko, V., Nikolaus, B., Schürmann, M.C., Schriever, G., Ziener, C. and Borisov, V.M. (2006) Development status of EUV sources for use in beta-tools and high volume chip manufacturing, in Proceedings of SPIE, Vol. 6151, San Jose, USA. (ed. M.J. Lercel) p. 61510O.
34 Teramoto, Y., Sato, H. and Yoshioka, M. (2006) Capillary Z-pinch source, in *EUV Sources for Lithography* (ed. Bakshi, V.) SPIE Press, Bellingham, Washington, pp. 505–522.
35 McGeoch, M. (1998) Radio-frequency preionized xenon z-pinch source for extreme ultraviolet lithography. *Applied Optics*, **37**, 1651.
36 McGeoch, W. (2006) Star pinch EUV source, in *EUV Sources for Lithography* (ed. V. Bakshi) SPIE Press, Bellingham, Washington, pp. 453–476.
37 Endo, A. (2006) Driver Laser, Xenon Target, and System Development for LPP Sources, in *EUV Sources for Lithography* (ed. V. Bakshi), SPIE Press, Bellingham, Washington, pp. 607–618.
38 Stuik, R., Fledderus, H., Bijkerk, F., Hegeman, P., Jonkers, J., Visser, M., Banine, V., Flying Circus EUV Source Comparison, Second International EUVL Workshop, San Francisco, 19–20 October 2000, Presentation available from www.sematech.org.
39 Pankert, J., Apetz, R., Bergmann, K., Damen, M., Derra, G., Franken, O., Janssen, M., Jonkers, J., Klein, J., Kraus, H., Krücken, T., List, A., Loeken, M., Mader, A., Metzmacher, C., Neff, W., Probst, S., Prümmer, R., Rosier, O., Schwabe, S., Seiwert, S., Siemons, G., Vaudrevange, D., Wagemann, D., Weber, A., Zink, P. and Zitzen, O. (2006) EUV sources for the alpha-tools, in Proceedings of SPIE, Vol. 6151, San Jose, USA. (ed. M.J. Lercel) p. 61510Q.
40 Corthout, M., Bergmann, K., Derra, G., Jonkers, J., Pankert, J. and Zink, P. (2006) The Philips Extreme UV Sn Source: Recent progress in power, lifetime and collector lifetime, International EUVL Symposium, Barcelona, Spain. Proceedings available at www.sematech.org.
41 Thompson, P.L. and Harvey, J.E. (2000) Systems engineering analysis of aplanatic Wolter type X-ray telescopes. *Optical Engineering*, **39** (6), 1677–1691.
42 Center for X-Ray Optics, Berkeley, USA, www-cxro.lbl.gov.
43 Zocchi, F.E., Bianucci, G., Rigato, V., Pirovano, G., Cassol, G.L., Salmaso, G.,

Bind, P., Zink, P., Bergmann, K., Nikolaus, B. and Schürmann, M.C. (2006) Experimental validation of collector's thermo-optical design, EUV Source Workshop, Barcelona, Spain. Proceedings available from www.sematech.org.

44 Kortright, B. and Underwood, J.H. (1991) Design considerations for multilayer coated Schwarzschild objectives for the XUV, Proceedings of SPIE, Vol. 1343, X-ray/EUV optics for astronomy, microscopy, polarimetry, and projection lithography, pp. 95–103.

45 Artioukov, I.A., and Krymski, K.M. (2000) Schwarzschild objective for soft X-rays. *Optical Engineering*, **39** (8), 2163–2170.

46 Geyl, R. (2006) Near normal incidence collectors for easier debris mitigation, EUV Source Workshop, Barcelona, Spain. Proceedings available from www.sematech.org.

47 Klebanoff, L.E., Anderson, R.A., Buchenauer, D.A., Fornaciari, N.R. and Kimori, H. (2006) Erosion of condensor optics exposed to EUV sources, in *EUV Sources for Lithography* (ed. V. Bakshi) SPIE Press, Bellingham, Washington, pp. 995–1031.

48 Ruzic, D.N. (2006) Origin of debris in EUV sources and its mitigation, in *EUV Sources for Lithography* (ed. V. Bakshi), SPIE Press, Bellingham, Washington, pp. 957–993.

49 Pankert, J., Apetz, R., Bergmann, K., Derra, G., Janssen, M., Jonkers, J., Klein, J., Krücken, T., List, A., Loeken, M., Metzmacher, C., Neff, W., Probst, S., Prümmer, R., Rosier, O., Seiwert, S., Siemons, G., Vaudrevange, D., Wagemann, D., Weber, A., Zink, P. and Zitzen, O. (2005) Integrating Philips' extreme UV source in the alpha-tools. Proceedings of SPIE 5751, 260–271.

50 Böttger, T., Meyer, D.C., Paufler, P., Braun, S., Moss, M., Mai, H. and Beyer, E. (2003) Thermal stability of Mo/Si multilayers with boron carbide interlayers. *Thin Solid Films*, **444**, 165–173.

51 Rigato, V., Mattarello, V., Nannarone, S. and Borgatti, F. (2005) Thermal stability of Mo/Si multilayers with ruthenium interlayers, International EUVL Symposium, San Diego, CA USA. Proceedings available from www.sematech.org.

52 Bajt, S., Dai, Z.R., Nelson, E.J., Wall, M.A., Alamenda, J.B., Nguyen, N.Q., Baker, S.L., Robinson, J.C. and Taylor, J.S. (2006) Oxidation resistance and microstructure of ruthenium-capped extreme ultraviolet multilayers. *Journal of Microlithography, Microfabrication, and Microsystems*, **5** (2), 023004.

53 Van de Kruijs, R.W.E., Yakshin, A.E., van Herpen, M.M.J.W., Klunder, D.J.W., Louis, E., Alonso van der Westen, S., Enkisch, H., Müllender, S., Bakker, L., Banine, V. and Bijkerk, F. (2006) Multilayer optics with spectral purity layers for the EUV wavelength range, Conference on Physics of X-Ray Multilayer Structures, Sapporo, Japan.

54 Sweeney, D.W., Hudyma, R., Chapman, H.N. and Shafer, D. (1998) EZV optical design for a 100 nm CD imaging system, in Proceedings of the SPIE 23rd Annual International Symposium on Microlithography, Santa Clara, CA, USA, February 22–27.

55 Montcalm, C., Grabner, R.F., Hudyma, R.M., Schmidt, M.A., Spiller, E., Walton, C.C., Wedowski, M. and Folta, J.A. (1999) Multilayer coated optics for an alpha-class extreme ultraviolet lithography system, in Proceedings of the 44th Annual Meeting of the International Symposium on Optical Science, Engineering and Instrumentation, Denver, CO, USA, July 18–23.

56 Stover, J.C. (1995) *Optical Scattering: Measurement and Analysis*, SPIE.

57 Kuerz, P., Böhm, T., Müllender, S., Bollinger, W., Dahl, M., Lowisch, M., Münster, C., Rohmund, F., Stein, T., Louis, E. and Bijkerk, F. (2005) Optics for EUV lithography, International EUVL Symposium, San Diego, CA, USA. Proceedings available from www.sematech.org.

58 Vernon, S.P., Kerney, P.A., Tong, W., Prisbrey, S., Larson, C., Moore, C.E.,

Weber, F., Cardinale, G., Yan, P.Y. and Hector, S. (1998) Masks for extreme ultraviolet lithography, Proceedings, 18th Annual BACUS Symposium on Photomask Technology and Management, Redwood City, CA, USA, September 16–18.

59 Tong, W. (1999) EUVL Mask Substrate Specifications (wafer type), UCRL-ID-135579, Report of U.S. Department of Energy, Lawrence Livermore National Laboratory.

60 Yu, Y.S., Kim, T.G., Lee, S.H., Park, J.G., Kim, T.H., Busnaina, A. and Lee, J.M. (2006) Removal of nano particles on EUV mask buffer and absorber layers by laser shockwave cleaning, International EUVL Symposium, Barcelona, Spain. Proceedings available from www.sematech.org.

61 Barty, A., Liu, Y., Gullikson, E., Taylor, J.S. and Wood, O. (2005) Actinic inspection of multilayer defects on EUV masks, Proceedings of the SPIE Microlithography, San Jose, CA, USA, April 2–3.

62 Lin, Y., and Bokor, J. (1997) Minimum critical defects in extreme-ultraviolet lithography masks. *Journal of Vacuum Science & Technology B*, **15** (6), 2467–2470.

63 Leeson, M., Cao, H., Yueh, W., Meagley, R., Sharma, G. and Sharma, S. (2006) EUV Resist Materials, Properties and Performance, International EUVL Symposium, Barcelona, Spain. Proceedings available from www.sematech.org.

64 Oizumi, H., Tanak, Y., Kumise, T., Nishiyama, I., Shiono, D., Hirayama, T., Hada, H., Onodera, J. and Yamaguchi, A. (2006) Performance of new molecular resist in EUV lithography, International EUVL Symposium, Barcelona, Spain. Proceedings available from www.sematech.org.

65 Okoroanyanwu, U. and Lammers, J.H. (2004) Resist Road to the 22 nm Technology Node. *Future Fab International*, **17**, Available at www.future-fab.com.

66 Zandbergen, P., Domke, W.D., Cantu, P., Thony, P., Postnikov, S. and Robic, J.Y. (2005) EXCITE, the MEDEA+ Extreme UV Consortium for Imaging Technology, International EUVL Symposium, San Diego, CA, USA. Proceedings available from www.sematech.org.

67 Nalleau, P., Rammeloo, C., Cain, J.P., Dean, K., Denham, P., Goldberg, K.A., Hoef, B., La Fontaine, B., Pawloski, A., Larson, C. and Wallraf, G. (2005) Investigation of the current resolution limits of advanced EUV resists, International EUVL Symposium, San Diego, CA, USA. Proceedings available from www.sematech.org.

68 Solak, H.H., He, D., Li, W. and Cerrina, F. (1999) Nanolithography using extreme ultraviolet lithography interferometry: 19 nm lines and spaces. *Journal of Vacuum Science & Technology B*, **17** (6), 3052–3057.

69 Solak, H.H., Ekinci, Y., Käser, P. and Park, S. (2007) Photon-beam lithography reaches 12.5 nm half pitch resolution. *Journal of Vacuum Science & Technology B*, **25** (1), 91–95.

7
Non-Optical Lithography

Clivia M. Sotomayor Torres and Jouni Ahopelto

7.1
Introduction

In the quest to use nanofabrication methods to exploit the know-how and potentials of nanotechnology, one major roadblock is the high cost factor which characterizes high-resolution fabrication technologies such as electron beam lithography (EBL) and extreme ultraviolet (EUV) lithography. The need to circumvent these problems of cost has inspired research and development in alternative nanofabrication, also referred to as "emerging" or "bottom-up" approaches. Hence, it is within this context that the status and prospects of nanoimprint lithography (NIL) are presented in this chapter.

Nanofabrication needs are highly diverse, not only in the materials used but also in the range of applications. Within the physical sciences, the drive is to realize nanostructures in order to produce artificial electronic, photonic, plasmonic or phononic crystals. This, in turn, depends on an ability to realize periodic or quasi-periodic arrays of nanostructures, on the one hand to meet the stringent demands of periodicity, order and critical dimensions to obtain the desired dispersion relation and, on the other hand, to identify a reproducible, cost-effective and reliable way in which such materials may be fabricated, using a suitable form of nanopatterning.

Nanopatterning covers a wide range of methods from top-down approaches, as well as bottom-up approaches (for discussions, see Chapters 5, 6, 8 and 9 of this volume). In fact, the needs for lithography are found in several fields:

- In nano-CMOS (complementary metal-oxide semiconductor) for example, to produce pattern gates of lengths down to a few nanometers in order to reach the technology nodes of the semiconductor industry roadmap [1], whilst at the same time complying with the most strict lithography demands.

- In (nano)photonics, a field in which – in addition to packaging – the cost of fabrication of III-V semiconductor optoelectronic devices containing nanostructures in the form of photonic crystals, is prohibitive.

Nanotechnology. Volume 3: Information Technology I. Edited by Rainer Waser
Copyright © 2008 WILEY-VCH Verlag GmbH & Co. KGaA, Weinheim
ISBN: 978-3-527-31738-7

- In nanobiotechnology, to fabricate a variety of sensors and lab-on-a-chip platforms based on micro- and nano-fluidics.

- In organic opto- and nano-electronics, where the lifetime issue of the organic materials is compounded with that of a cost-effective volume production with lateral resolution down to a few hundreds of nanometers for electrodes and pixels.

- In micro electro-mechanical systems (MEMS) and nano electro-mechanical systems (NEMS), where the fabrication of resonators, cantilevers and many other structures with and without direct interface to Si-based electronics, requires the control of 3-D nanofabrication with minimum damage to the underlying electronic platform.

Moreover, nanopatterning methods act as enabling technologies to facilitate the progress of research in chemistry, such as the realization of nanoelectrodes to monitor electric activity; in biology, to connect electrically to cells; in physics, to realize nanostructures commensurate with the De Broglie wavelength of a given excitation; in material sciences, through research on novel nanostructured artificial materials; and also in several other engineering disciplines.

In 2003, the state of the art covering most bottom-up emerging nanopatterning methods was collected in Ref. [2] under the title *Alternative Lithography: unleashing the potentials of Nanotechnology*. Of these methods, probably the most advanced is polymer molding or nanoimprint lithography [3]. Other emerging bottom-up methods are those based on scanning probes [4, 5], self assembly (see Chapters 9 and 10 in this volume), micro-contact printing or soft-lithography [6, 7] and stenciling [8], as well as atom lithography [9] and bio-inspired lithography [10].

In this chapter, attention is focused primarily on NIL as an example of non-optical lithographies, as it covers the 1 µm to few nanometer lateral resolution range. Here, the basic principles of this method are described, the state of the art is reviewed, and the main scientific and engineering issues are addressed.

7.2
Nanoimprint Lithography

7.2.1
The Nanoimprint Process

Historically, NIL has been preceded by some remarkable events. In the twelfth century, metal type printing techniques were developed in Korea; for example, in 1234 the "Kogumsangjong-yemun" (Prescribed Ritual Text of Past and Present) appeared, while in 1450 Gutenberg introduced his press and printed 300 issues of the two-volume Bible. Somewhat strangely, an extensive time lapse then occurred until the early twentieth century, when the first vinyl records were produced by using hot embossing [11]. The next major step occurred during the 1970s, when compact discs were fabricated by injection molding.

The term "nanoimprint lithography" was most likely used for the first time by Stephen Y. Chou, when referring to patterning of the surface of a polymer film or resist with lateral feature sizes below 10 nm [3]. Previously, within a larger lateral size range, the method was referred to as "hot embossing". At about the same time, Jan Haisma reported the molding of a monomer in a vacuum contact printer and subsequent curing by UV radiation, which was known as "mold-assisted lithography" [12]. The first comprehensive review of these two approaches appeared in 2000 [13], but since then several excellent reviews of NIL have been produced [14, 15]. During recent years these methods have developed further and have become to be known as "thermal nanoimprint lithography" and "UV-nanoimprint lithography" (UV-NIL), respectively.

The question must be asked, however, what is NIL? Nanoimprint lithography is basically a polymer surface-structuring method which functions by making a polymer flow into the recesses of a hard stamp in a cycle involving temperature and pressure. In order to nanoimprint a surface, three basic components are required: (i) a stamp with suitable feature sizes; (ii) a material to be printed; and (iii) the equipment for printing with adequate control of temperature, pressure and control of parallelism of the stamp and substrate. The NIL process is illustrated schematically in Figure 7.1. In essence, the process consists of pressing the solid stamp using a pressure in the range of about 50 to 100 bar, against a thin polymer film. This takes place when the polymer is held some 90–100 °C above its glass transition temperature (T_g), in a time scale of few minutes, during which time the polymer can flow to fill in the volume delimited by the surface topology of the stamp. The stamp is detached from the printed substrate after cooling both it and the substrate. The cycle, which is illustrated graphically in Figure 7.2, involves time, temperature, and pressure. Here, we have the main issues of NIL: polymer flow and rheology. Although these points have been addressed from the materials point of view in Refs. [16, 17], they remain a serious challenge for feature sizes below 20 nm. These aspects will be discussed in the following sections.

In UV-NIL the thermal cycle is replaced by curing the molded polymer by UV light through a transparent stamp. This requires different polymer properties, as will be discussed in the next section.

7.2.2
Polymers for Nanoimprint Lithography

The polymers used in NIL play a critical role. A comparison of the 10 most-often used polymers (resists) used in thermal NIL is provided in Ref. [15]. The resists determine both the quality of printing and the throughput. Quality is achieved via the thickness uniformity of the spin-coated film, the strong adhesion to the substrate, and the weak adhesion to the stamp. Throughput is achieved via the duration of the printing cycle, which in turn is determined by several time scales including:

- the time needed to reach the printing temperature (the higher the T_g, the longer the cycle, unless there is a pre-heating stage)

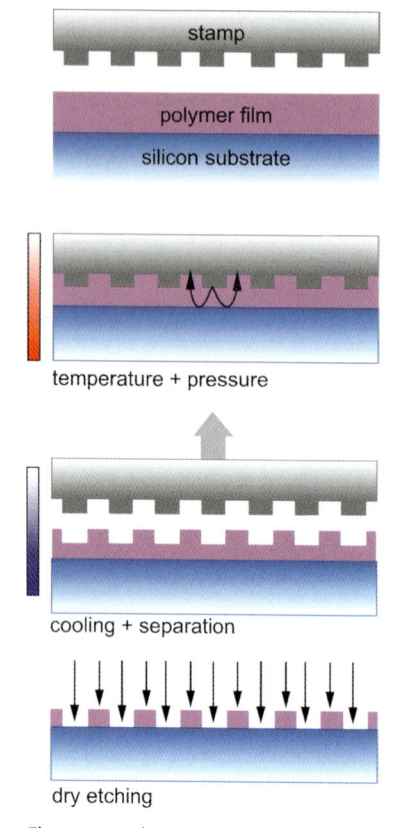

Figure 7.1 Schematics of the thermal nanoimprint concept. Top to bottom: The polymer layer on a solid substrate is heated to a temperature above the glass transition temperature (T_g). The stamp and polymer layer are brought into contact. Pressure is applied to start the polymer flow into the cavities of the stamp. The sample and stamp are cooled down for demolding or separation at a temperature below T_g. The residual polymer layer is removed, typically by dry etching. The end result is a patterned polymer layer on a substrate.

- the hold time for optimum flow at the printing temperature (the more viscous the polymer, the longer the hold time)
- the time needed for cooling and demolding.

Different criteria must be met for thermal NIL and for UV-NIL [18]. For thermal NIL, the polymer is used as a thin film of a few hundred nanometers thickness which is spin-coated onto the support substrate. The key properties are of a thermodynamic nature, and therefore these polymers are of the thermoplastic and thermosetting varieties with varying molecular weights, chemical structures, and rheological and mechanical behaviors. The dependence of the viscosity of a thermoplastic polymer as a function of temperature is shown graphically in Figure 7.3, and illustrates the region where thermal NIL takes place. The mechanical behaviors of polymers in different temperature regimes, in relation to the molecular mobility, are listed in Table 7.1.

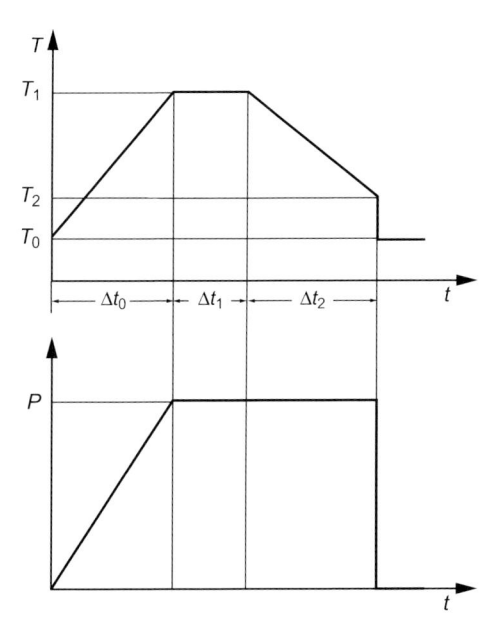

Figure 7.2 Temperature and pressure cycles as a function of time in the thermal NIL process. Typical parameters used for thermoplastic polymers are: Printing temperature T_1 (°C) = 185; demolding temperature T_2 (°C) = 95; printing pressure P (bar) = 30; time to reach printing temperature allowing polymer to go from solid to viscous regime Δt_0(s) = 60; molding time Δt_1 (s) = 60; cooling time Δt_2 (s) = 160.

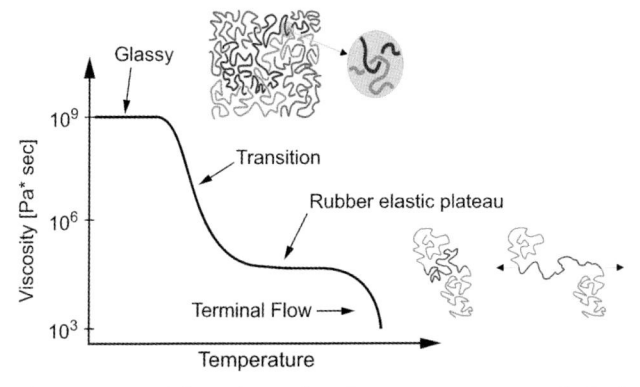

Figure 7.3 Typical dependence of a polymer viscosity on temperature. At room temperature, the polymer is in its solid (glassy, brittle) state. As the temperature increases the short-chain segments become disentangled and the polymer rapidly undergoes a transition from its solid to a rubbery state, changing its viscosity by several orders of magnitude around the glass transition temperature, T_g. Further temperature increases lead to disentanglement of the long polymer chains and resulting in a terminal flow of the viscous melt. Printing takes place in the region where the flow is optimum for filling of the stamp cavities, depending on the molecular weight and stamp design.

Table. 7.1 Relationship between molecular mobility and mechanical behavior of polymers in different temperature regimes.

Temperature regime		$T_{subtransition}$		T_g		T_{flow}	
State	Glassy		Rubber elastic		Plastic		
Mechanical appearance	Brittle		Hard elastic, rigid		Rubber elastic		Viscoelastic
Young's modulus (N mm^{-2})	about 3000		about 1000		about 1		Too small to measure
Molecular mobility	Molecular conformation completely fixed.		Molecular conformation largely fixed. Occasional change in molecular positions of side groups and chain segments.		Entanglement and physical junction zones prevent movement of entire macromolecules. Entropy-elastic change of molecular position of chain segments. Micro-Brownian motion. Creep, no plastic flow.		No restricted rotation around single bonds. Whole macromolecules change their positions gliding past each other. Plastic flow. Macro-Brownian motion.
Effect of stress	Energy-driven elastic distortions.		Energy-driven elastic distortions.		Entropy-driven elastic distortion. Besides temperature, the deformation rate affects the mechanical behavior.		Pseudoplasticity, shear thinning.
Suitability for printing					Imprinting is possible, but will have memory effects.		Best printing temperature range.

[a] Adapted from Ref. [19]; reproduced with permission.

One of the polymer strategies used to reduce the printing temperature and to improve thermal stability has been to cure prepolymers (special precursors of crosslinked polymers). Here, the term "curing" refers to the photochemical (UV)- or thermal-induced crosslinking of macromolecules to generate a spatial macromolecular network. The prepolymers are low-molecular-weight products, with a low T_g, which are soluble and contain functional groups for further polymerization. Thus, lower printing temperatures of about 100 °C can be used. Curing can take place during the printing time, or thereafter, with the thermal stability enhancement arising from the crosslinking process of the macromolecules.

Polymers for UV-NIL must be suitable for liquid resist processing; that is, they are characterized by a lower viscosity than the polymers for thermal NIL. Naturally, they must also be UV-curable over short time scales [19]. The characteristics of these polymers after printing for their direct use, as in polymer optics or microfluidics, or as a mask for subsequent pattern transfer, by means of dry etching, demand high mechanical, thermal and temporal stability. In photonic applications, stability in terms of optical properties, such as refractive index, is also essential. Recently, micro resist technology GmbH [20] has developed a whole range of polymers for thermal and UV-NIL (for a discussion, see Ref. [21]). Moreover, tailoring the polymer properties to increase the control of critical dimensions remains an area where, although rapid progress has recently been made [18, 19, 21], further research investigations are still required. The importance of this research may be appreciated especially in a one-to-one filling of the stamp cavities, thereby making the printed polymer features resilient to residual layer removal. Moreover, polymer engineering is also a determinant in larger throughputs, in terms of shorter times for the curing and pressure cycles.

In recent years several reports detailing mechanical studies of thermal NIL have been made, and the interested reader is referred to the data of Hirai [22], the review of Schift and Kristensen [15], and to a recent review of the research on the simple viscous squeeze flow theory [23].

7.2.3
Variations of NIL Methods

To date, four main variations of the NIL process have been developed, and these are briefly described below.

7.2.3.1 Single-Step NIL
This is the most commonly used method to print a polymer in one temperature–pressure cycle, and has been extended to the printing of 150-mm [24] and 200-mm wafers [25]. A scanning electron microscopy (SEM) image of an array of lines of 200 nm width printed over a 200-mm silicon wafer is shown in Figure 7.4. Although one-step thermal NIL can be performed using regular laboratory-scale equipment, commercially available tools include, among others, those of OBDUCAT [26] and EVG [28], which are available in Asia and the Americas. Thermal expansion may cause distortions in the imprinted pattern and to avoid this, strategies for room-temperature NIL have been investigated [28, 29].

Figure 7.4 A scanning electron microscopy image of 200-nm lines printed in polymer on a 200-mm wafer [25].

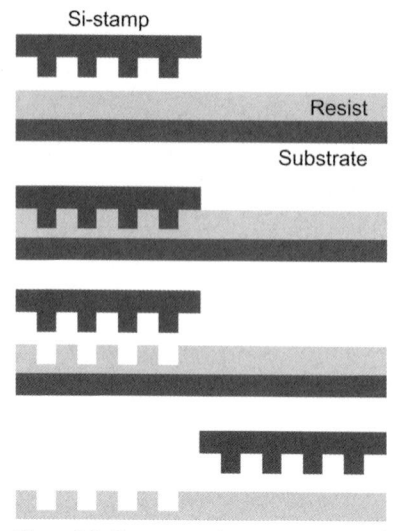

Figure 7.5 The step-and-stamp imprinting lithography process shown schematically. The substrate is patterned in a sequential process by stepping and imprinting across the surface. During the process, the substrate temperature is kept below the glass transition temperature of the resist polymer, while the temperature of the stamp is cycled up and down, above and below the glass transition temperature [30].

7.2.3.2 Step-and-Stamp Imprint Lithography

Step-and-stamp imprint lithography (SSIL) is a sequential process, pioneered by Tomi Haatainen and Jouni Ahopelto [30], and is depicted schematically in Figure 7.5. Basically, the system employs thermal NIL and uses a small stamp to print, step and print again, in order to nanostructure the desired area. Initially, SSIL was developed using a commercially available flip-chip bonder, but a dedicated wafer-scale tool from SUSS MicroTec is now available for SSIL [31]. One advantage of SSIL is its capability to achieve a high overlay accuracy, which makes it possible to pattern several consecutive layers or to mix and match with other lithography techniques [32]. An example of a full-patterned wafer is shown in Figure 7.6.

7.2.3.3 Step-and-Flash Imprint Lithography

Step-and-flash imprint lithography (SFIL) is also a sequential process, and uses UV radiation instead of temperature to generate relief patterns with line widths below 100 nm. Like NIL, SFIL does not use projection optics but, unlike NIL, it functions at room temperature. SFIL, which is depicted schematically in Figure 7.7, was pioneered by the team of Grant Wilson in the USA [33, 34], with an initial target of meeting the needs of front-end CMOS process fabrication. One of its attractive features is the ability to print over already patterned surfaces. Molecular Imprints Inc. has developed a range of tools for SFIL [35]. As with UV-NIL, SFIL requires transparent stamps

7.2 Nanoimprint Lithography | 217

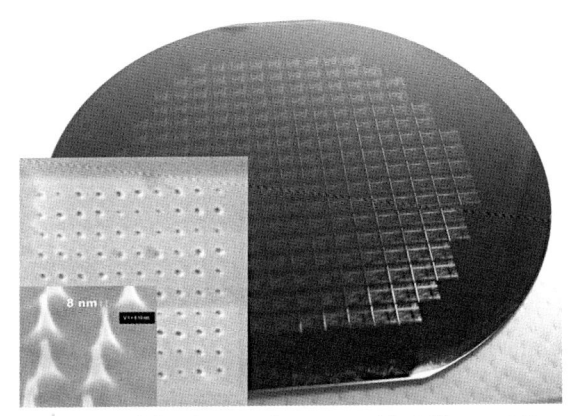

Figure 7.6 A full 100-mm wafer patterned by SSIL, consisting of a matrix of more than 200 imprints into mr-I 7030 resist. The inset shows scanning electron microscopy images of a silicon stamp with sub-10-nm pillars, together with the corresponding imprint.

(usually quartz), the fabrication of which is at present less straightforward than for thermal NIL.

7.2.3.4 Roll-to-Roll Printing

This is an advanced sequential method stemming from production method used in, for example, the newspaper industry. Roll-to-roll nanoimprinting is a versatile method that can be combined with other continuous printing techniques, as shown schematically in Figure 7.8. Its extension to 100-nm lateral resolution has been reported [36]. Recent developments suggest that roll-to-roll nanoimprinting is the closest to an industrial technology for organic opto- and nano-electronics, as well as for lab-on-chip device fabrication. The challenge is to fabricate the round stamps; that is, the printing rolls with nanometer-scale features. Moreover, due to the nature of the continuous process, some restrictions may arise in applications requiring multilevel patterning with high alignment accuracy between the layers. Examples of the feature size that can be obtained with a laboratory-scale roll-to-roll printer are shown in Figure 7.9.

7.2.4
Stamps

Stamps for NIL have been extensively discussed in Ref. [15]. The main considerations from the materials aspect include:

- Hardness (e.g., typically from 500 to thousands of kg mm^{-2}), which determines the stamp lifetime and the way in which it wears out.

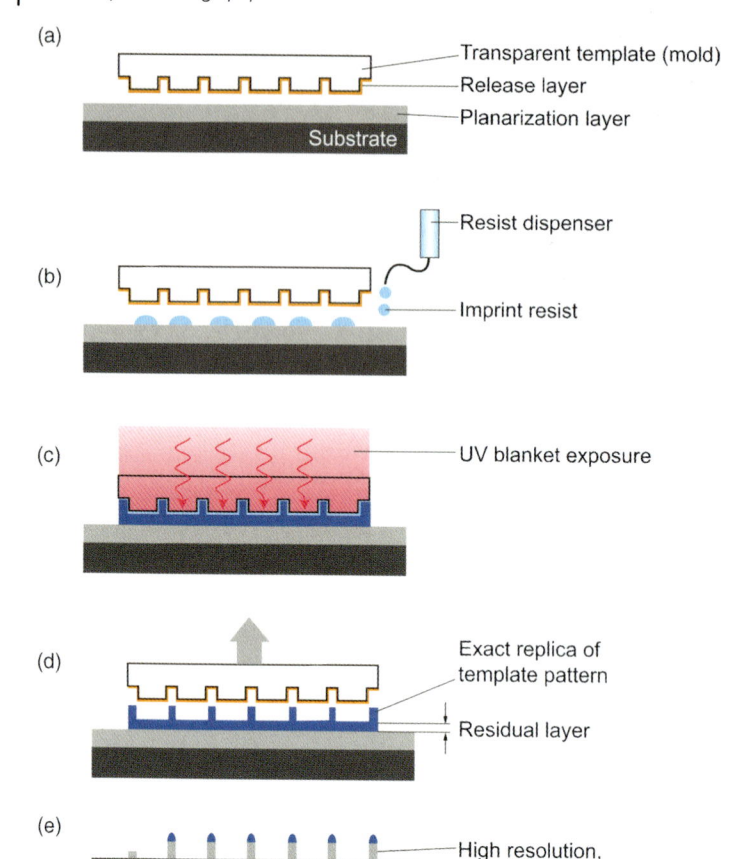

Figure 7.7 Schematics of the step-and-flash imprint lithography (SFIL) process. (a) The pre-planarized substrate and treated stamp are oriented parallel to each other. (b) Drops of UV-curable, low-viscosity imprint resist are dispensed on specified places. (c) The stamp is lowered to fill the patterns and the imprint fluid is polymerized (cured) with UV light at room temperature and low pressure. (d) The stamp is separated from the imprinted substrate. (e) A halogen breakthrough etch to remove the residual layer is performed, followed by an oxygen reactive ion etch [35].

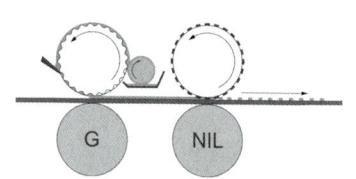

Figure 7.8 The advanced roll-to-roll nanoimprinting process, shown schematically. The gravure unit on the left (G) spreads a film of the conducting polymer on the web. This is followed by patterning of the film, using the nanoimprinting unit (NIL) on the right. The combining of different techniques allows the fabrication of complex layered structures in a single pass [36].

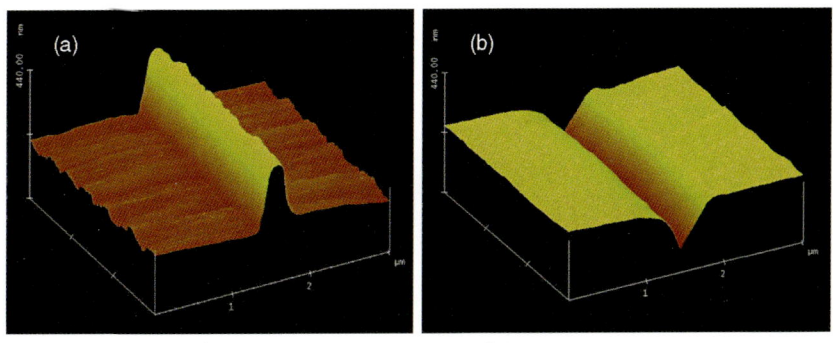

Figure 7.9 (a) Atomic force microscopy (AFM) image of a 100 nm-wide and 170 nm-high ridge on an electroplated roll-to-roll nanoimprinting stamp. (b) AFM image of a trench imprinted into cellulose acetate using the stamp shown in (a). The process temperature is 110 °C and the printing speed 1 m min^{-1}. There was no significant difference between the results obtained at speeds ranging from 0.1 to 5 m min^{-1}.

- Thermal expansion coefficient (e.g., typically from 0.6 to 3×10^{-6} K^{-1}), as well as Poisson's ratio (e.g., typically from 0.1 to 3.0), which will have a strong impact on distortion while demolding.
- Surface smoothness (e.g., better than 0.2 nm), as a rough surface will require large demolding forces and may lead to stronger than needed adhesion.
- Young's modulus (e.g., typically from 70 to hundreds of GPa), which in turn will control possible stamp bending. The latter effect may lead to uneven residual layer thickness, and thus compromise critical dimensions.
- Thermal conductivity (e.g., typically from 6 to hundreds of Wm^{-1} K^{-1}), which determines the duration of the heating and cooling cycles.

With regards to fabrication, the parameters to be considered include the minimum lateral feature size or resolution, the aspect ratio (feature lateral size: feature height), the homogeneity of the feature height across the stamp, as well as depth homogeneity, sidewall roughness, and inclination. For thermal NIL, stamps are usually fabricated in silicon by using EBL and reactive ion etching for the highest resolution and versatility. Unfortunately, these procedures are rather expensive, so that strategies for lower-cost replication have been developed, such as SSIL, the use of a master stamp and a (negative) first-generation replica using NIL or a (positive) second-generation replica, again using NIL. Replication while maintaining a resolution of 100 nm or better requires electroplating to replicate the original in metal. Current developments in the replication of a master stamp in thermosetting polymers show great promise, as they are expected greatly to reduce the cost of stamp replication [37]. As nanoimprint is a 1-to-1 replication technology, it is essential that the stamp has the correct feature sizes required on the wafer, thus emphasizing the need for quality stamps.

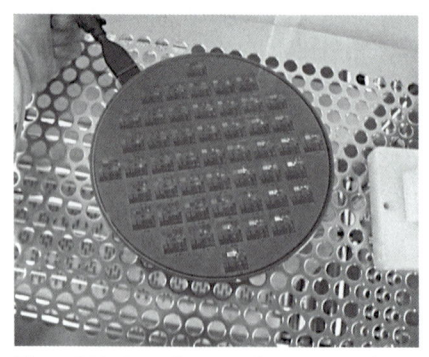

Figure 7.10 Optical image of a 200-mm silicon stamp fabricated by electron beam lithography [25].

In the past, stamps have been realized for wafer-scale thermal NIL (see Figure 7.10). In addition, electron-beam-written silicon stamps for thermal NIL are commercially available from NILTechnology [38], and an example is shown in Figure 7.11.

In recent years, the adhesion between the stamp and the printed polymer film has been the subject of significant research effort in thermal NIL. Here, the main issue is to ensure that the interfacial energy between the stamp and the polymer film to be printed is smaller than the respective interfacial energy between the substrate and the polymer film [39]. However, based on the materials commonly used, this matching is not sufficient for easy detachment, in which the frozen strain also plays a role. The normal practice here in order to facilitate demolding and to prolong the stamp

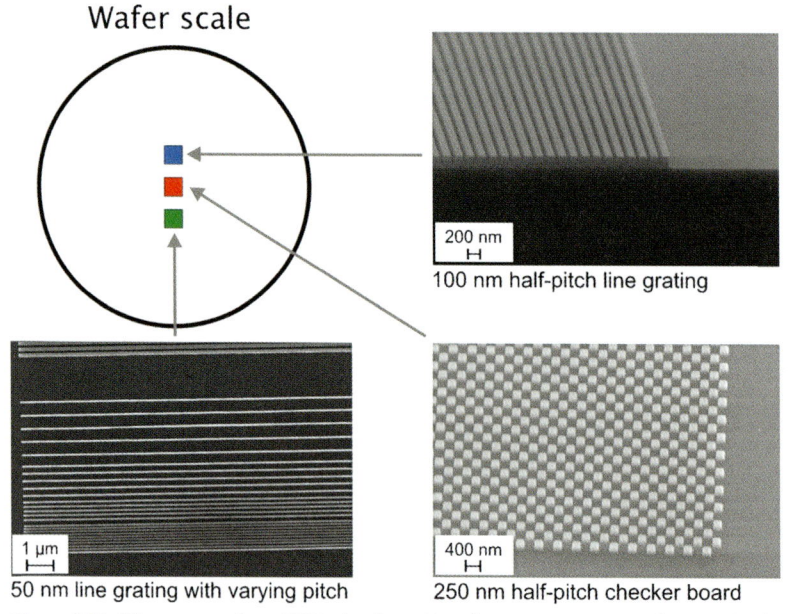

Figure 7.11 Silicon stamp from NILTechnology [38]. (Illustration courtesy of NIL Technology.)

Table. 7.2 Surface energies of common materials used in nanoimprint lithography.

Material	Surface energy (mN m^{-1})
PMMA	41.1
PS	40.7
PTFE	15.6
–CF3 and –CF2	15–17
Silicon surface	20–26

Values are taken from Reference [39].

lifetime, is to coat the stamp with an anti-adhesive layer to minimize the interfacial energy and, therefore, the adhesion. Values of the surface energies of materials commonly used in the NIL process are listed in Table 7.2. These data show that a fluorinated compound can dramatically reduce the surface energy and minimize adhesion while demolding a stamp from the printed polymer.

For both UV-NIL and SFIL, UV-transparent stamps are required, and these are typically constructed from quartz. Although, the fabrication of quartz stamps for high resolution has not yet been standardized, various efforts have been made to use photomask fabrication methods to prepare stamps or templates for UV-NIL [40]. A schematic overview of the stamp fabrication process is shown in Figure 7.12. In a first lithography step, the stamp is structured using EBL (first level writing), while in a second lithography step the pedestal requirement for imprint is made (second level

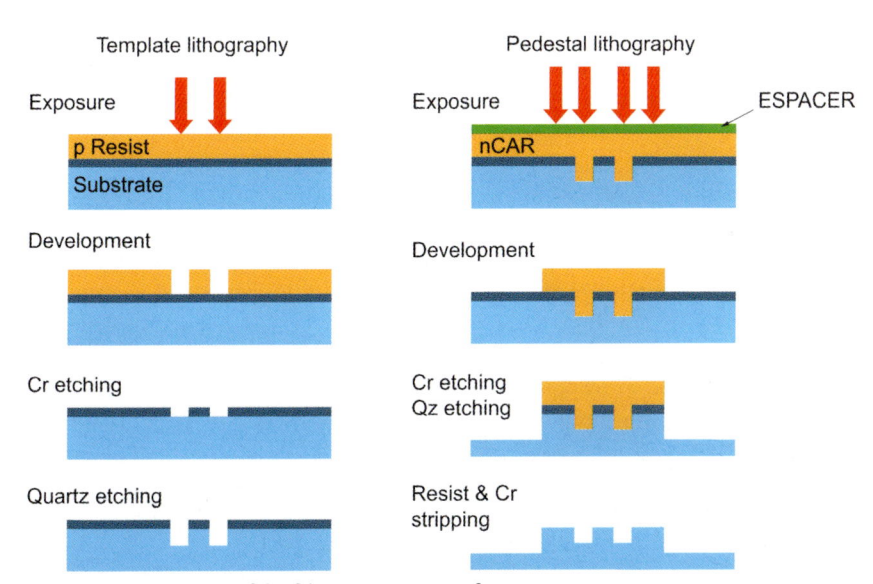

Figure 7.12 Schematics of the fabrication process of two-dimensional stamps for step-and- stamp and or step-and-flash UV-NIL [41, 42].

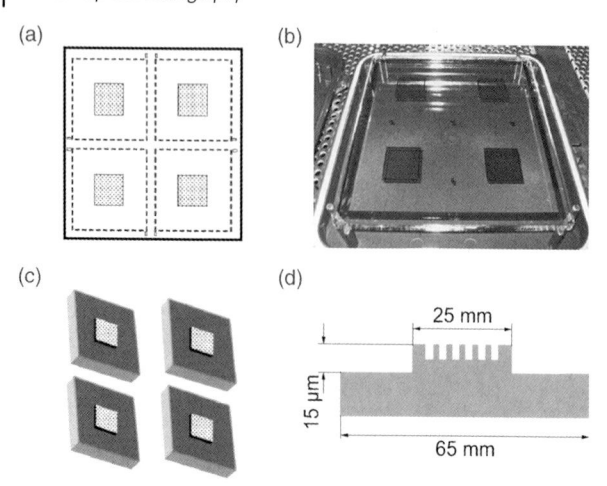

Figure 7.13 Photomask fabrication methods for UV-NIL stamp fabrication [43].

writing). Further details on the process flow for UV-NIL stamps can be found in Refs. [41, 42]. By using photomask fabrication methods, four imprint stamps or templates may be structured on a single photomask blank (see Figure 7.13a and b). The photomask blank is then diced into separate stamps (Figure 7.13c). As dicing introduces some contamination and mechanical strains, a modified fabrication process must be introduced before step-and-repeat- and step-and-flash- UV-NIL can be employed in volume production. The size standard for stamps resembles the exposure field of current optical lithography steppers (see Figure 7.13d).

Recently, stamps for 3-D structuring tests of several layers of functional films using UV-NIL targeting the back-end CMOS processes have been developed, using stamps similar to that shown schematically in Figure 7.13d.

The fabrication of stamps with high-resolution features for roll-to-roll nanoimprinting is more complicated because patterning of the curved surfaces is not straightforward. One possibility way to overcome this is to make a bendable shim that is wrapped around the printing roll. Such bendable large area stamps can be fabricated by electroplating, and exploiting SSIL in large-area pattering has been shown to reduce the fabrication time remarkably [44]. A 100 mm-diameter bendable Ni stamp is shown in Figure 7.14; this figure also shows that sub-100-nm features can be easily reproduced by using an electroplating process.

The details of stamps used for 3-D printing are discussed later in the chapter.

7.2.5
Residual Layer and Critical Dimensions

Most of the processes described above yield a nanostructured polymer layer (as shown schematically in Figure 7.1), with a residual layer under the features of the stamp. If the desired nanostructured surface is the polymer itself, with no material

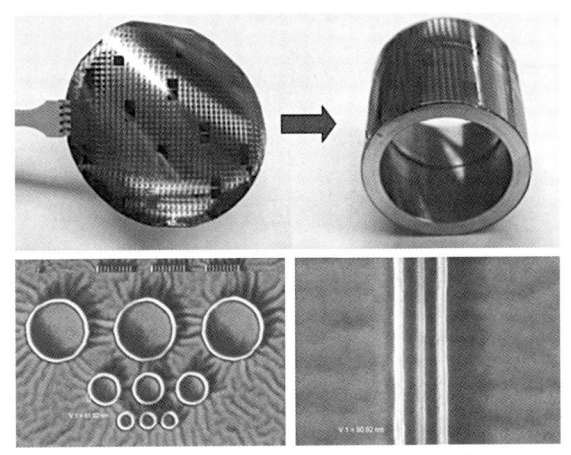

Figure 7.14 (a) An electroplated, bendable 100 mm-diameter Ni shim. The thickness of the shim is 70 µm. The roll-to-roll stamp is made by wrapping the shim around a stainless steel roll.
(b) Scanning electron microscopy images of various 80 nm-wide features on an electroplated Ni shim patterned by SSIL. The surface metal layer on the stamp is TiW.

between the features, then the residual layer must be removed. The same applies [44] if the patterned polymer or resist is to be used directly as a mask for pattern transfer into the substrate by, for example, reactive ion etching, or if a metal lift-off step will be needed that will result in a metal mask for further pattern transfer [45]. An etching step of the printed polymer, whether to remove the residual layer or to be used as a mask, necessarily results in the feature sizes experiencing change. This was shown in the variation in the width of printed Aharonov–Bohm ring leads following removal of the residual layer by etching, and after metal lift-off. The leads increased by 15 nm in width from the targeted width of 500 nm, taken over an average of 20 samples [32]. This means that, in order to control the critical dimensions of the printed features, the residual layer thickness uniformity must also be controlled, as its removal leads to a size fluctuation of the resulting nanostructures. Significant efforts have been made to develop non-destructive metrology for nanoimprinted polymers. One of the most salient approaches is to use scatterometry as applied to NIL [46], in order to determine both the feature height and residual layer thickness. Being based on the principles of ellipsometry, a laser spot is used, which is scanned over the region of interest. An example of this is shown in Figure 7.15 (right panel), with a cross-sectional SEM image of the printed ridges and the corresponding scatterometry data and curve fitting. The left panel of Figure 7.15 depicts an optical reflection image of a printed wafer, showing the thickness variation of the residual layer across the wafer [47]. An *in-situ* and non-destructive method was demonstrated by adding chromophores to the printed polymer and using their emission as an indicator of stamp deterioration (such as missing features), mirrored in the printed fields. Although the resolution of this method was poor, it did at least demonstrate the feasibility of the in-line monitoring of printed arrays of nanostructures [48].

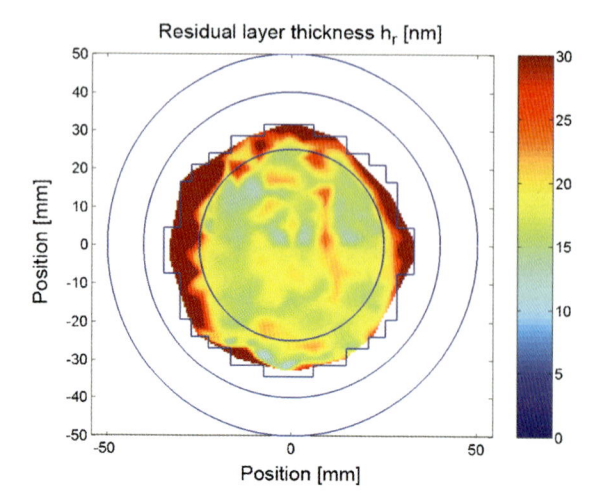

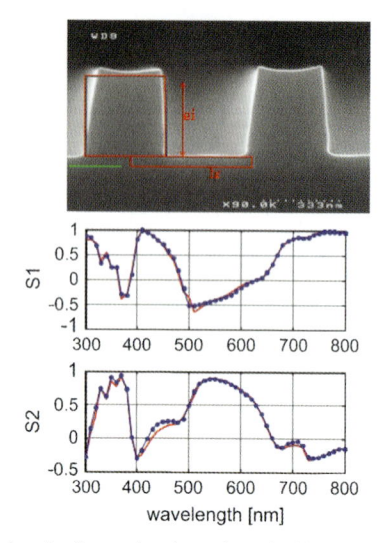

Figure 7.15 Two approaches to metrology. Left: Optical imaging of the contrast resulting from variations of the residual layer thickness over a 100-mm wafer, showing a good uniformity over most of the wafer, except at the edges [47]. Right: The upper image is a cross-section micrograph showing the feature height and residual layer thickness; the lower images show experiments and fit of a scatterometry spectrum recorded on a printed sample, from which the residual layer as well as the feature size can be obtained [46].

Control of the residual layer is necessary due to a need to fill in completely the stamp cavities, whilst achieving as thin and as uniform a residual layer as possible. This is a non-trivial issue which depends not only on nanometer-scale polymer rheology but also on the stamp and substrate deformation.

The polymer challenge in thermal NIL is basically four-fold: (i) to obtain complete filling of the cavities or to ensure a one-to-one transfer; (ii) to obtain as thin a residual layer as possible to control critical dimensions; (iii) to ensure that the printed features do not relax mechanically; and (iv) to achieve a reasonable throughput.

A typical curve of the dependence of viscosity on temperature (e.g., Figure 7.3) shows that T_g occurs in a regime where the viscosity is changing by several orders of magnitude. This poses a non-negligible challenge to understanding polymer flow in the context of NIL. Initially, the polymer flow has been approximated to that of a Newtonian fluid in the gap between two parallel disks of radius R (as discussed in Ref. [49]). The discussion of Ref. [49], which is summarized below, is probably the most complete account to date covering the simple case and providing an insight into the scope of the problem. By using the Stefan equation for the quasi steady-state solution (this is a simplified version of the non-stationary Navier–Stokes equation), the force is found to be proportional to the viscosity, the fourth power of the disk radius, the speed of the disks coming together, and inversely proportional to the cube of the initial layer thickness. In other words, a huge force is needed for a fast fluid motion in thin films over large distances. There are two basic considerations to this point:

- The force has only a linear dependency on viscosity, which changes by orders of magnitude when the temperature is in the vicinity of T_g.
- Although a Newtonian fluid, or a fluid in the limit of small shear rates, the viscosity does not depend on the shear rate. However, at moderate or high shear rates, a non-linear flow can lead to a decrease in viscosity by several orders of magnitude. The effect of this on the NIL process would be seen as a reduction either in the pressure needed or in the processing time.

The question is, therefore, what are the contributions to the force from pressure and shear stress of the fluid motion? Clearly, pressure is related to the contact area of the stamp and the fluid (polymer above T_g), whereas the shear stress is related to the flow velocity, which in turn depends on the distances over which the fluid must be transported, and therefore on the particular stamp design. At any given time, the condition of continuity and the conservation of momentum of an incompressible liquid requires that the velocity must increase with radial distance, which would result in a parabolic velocity profile in the z-direction. The velocity would be least at the interface with the disk walls, and greatest in the middle of the gap. To this, the inversely proportional cubic dependence on liquid layer thickness must be added. The calculated values of viscosity and transport thickness tend to agree with the observed experimental values for polymethyl methacrylate (PMMA) and, rather simplistically, some basic trends can be obtained:

- Thermal NIL works best for smallest features (sub-100 nm) which are close together and in which a local flow takes place, allowing easy and reliable filling of the stamp cavities.
- Conversely, large features (>10 μm) separated by large distances require a large displacement of material, and larger forces.

Here, the force–displacement curve results reviewed in Ref. [23] are highly illuminating, as a more complete model requires the consideration of several flow fields arising from the different shapes, depths, and separation of cavities in the stamp. Schift and Heyderman carried out a thorough analysis in this respect in the linear micrometer regime [50].

In the linear regime, the temperature dependence of the viscosity is viewed as a thermally activated process [49, 50], following a formalism of amorphous polymers and remaining within the limit of small shear rates. Such a non-linear regime is substantially more complex, and is basically exemplified by shear thinning and extrudate swelling. Hoffmann suggested that shear thinning, with its inherent shear rate-dependent viscosity, may influence the thermal NIL process, especially for small features [49].

A key remaining issue is the understanding of how the stored deformation energy depends on the rate at which the temperature and pressure are applied and released, and to what degree these influence the mechanical stability of the printed polymer features in time scales of weeks, months, and years.

In practice, on order to gain an understanding of the filling dynamics of a stamp cavity under the combined effects of squeezing flow, polymer rheology, surface tension and contact angle in typical NIL experiments, full fluid–solid interaction models based on the continuum approach have been devised [51, 52]. In these, both the fluid bed and the solid stamp are represented and a continuity of displacement and pressure is applied at the interface. As these are based on finite elements, there are almost no limits to the choice of the materials' constitutive behaviour, and these clearly reflect the effects of stamp anisotropy and the shear thinning behavior of the polymer. In particular, they are especially efficient at predicting the shape of the polymer in partially filled cavities.

Coarse grain methods have proven powerful in computing the residual layer thickness of the embossing process [53]. Based on the Stokes equation, they solve the simple squeeze flow equation for Newtonian fluids and embossed areas of up to

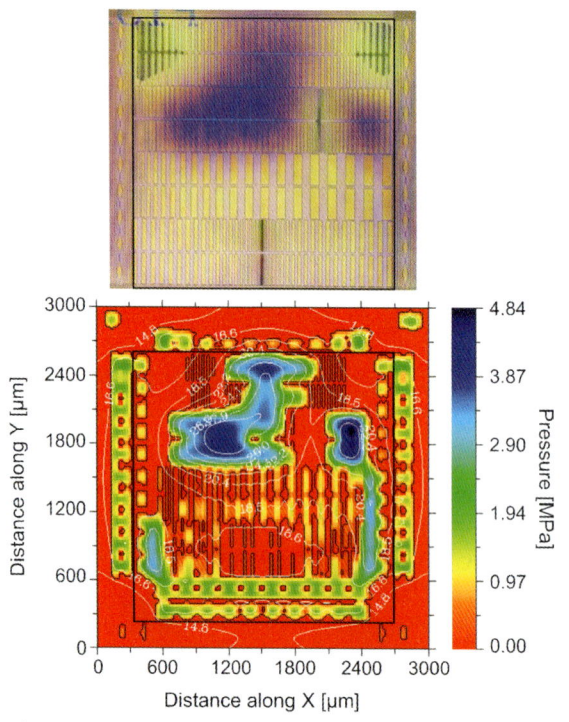

Figure 7.16 Experimental (top) and simulated (bottom) residual layer thickness. The colors correspond to different thicknesses, as observed in an optical microscope. The self-consistent coarse-grain model considers that the stamp is flexible; thus, the resulting contour lines are the variation with respect to the imposed average residual layer thickness. (Illustration courtesy of D.-A. Mendels and S. Zaitsev.).

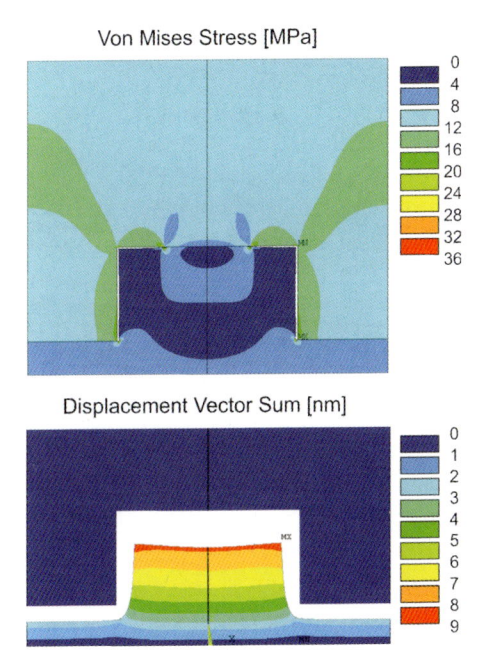

Figure 7.17 Upper: Von Mises stress and stamp/polymer interfacial separation during cool-down of a 200 × 100 nm² single polymer cavity obtained by embossing, with a poor interfacial adhesion. Lower: Residual displacement and shape of the same structure after stamp removal in the case of high interfacial adhesion. (Illustration courtesy of D.-A. Mendels.)

several square millimeters within a matter of minutes. Here, the calculation is based on the determination of a homogenized depth which is representative of the average pressure applied to the area. The quantitative agreement has proven excellent, and is generally acceptable when the polymer layer is embossed well above T_g [54]. Freezing of the embossed structures through the T_g has also been modeled [55, 56], and has provided precious insight into the build-up of internal stresses prior to stamp release and of the polymer–stamp interface. It has also been possible to simulate the demolding process, and thus the final shape of the embossed structures, both after stamp release and after relaxation for a given period of time [57]. Two examples of the models described in this section are shown in Figures 7.16 and 7.17.

7.2.6
Towards 3-D Nanoimprinting

One special aspect of NIL and SFIL is their ability to pattern in three dimensions compared to other lithographies. Several applications require this ability, from MEMS to photonic crystals, including a myriad of sensors. One of the first demonstrations of 3-D patterning by NIL was the realization of a T-gate for microwave transistors with a footprint of 40 nm by a single-step NIL and metal lift-off [58]. SFIL also showed its 3-D patterning ability in the fabrication of multitiered structures, maintaining a high aspect ratio [59].

If metal lift-off is to be avoided, then 3-D NIL requires 3-D stamps. These are produced by gray-scale lithography with sub-100 nm resolution, but are limited in depth and volume production due to the sequential nature of EBL [60]. One recent variation of this approach consisted of using inorganic resists and low-acceleration electron-beam writing, thus allowing the control of the depth to tens of nanometers [61].

A combination method which was based on focused ion beam and isotropic wet etching has been demonstrated by Tormen et al. [62], and resulted in tightly controlled 3-D profiles in the range from 10 nm to 100 µm.

Within the microelectronics industry, one of the main expectations from NIL was its application as a lithography method in the dual damascene process for back-end CMOS fabrication [1], and this process is still undergoing testing today.

Bao et al. showed that it is possible to print over non-flat surfaces using polymers with different mechanical properties using thermal NIL and polymers with progressively lower T_g-values for each subsequent layer [63]. This meant that a different polymer must be used for each layer. In order to overcome this situation, several other variations and combinations of methods based on NIL have been developed. One such development is that of reversed contact ultraviolet nanoimprint lithography (RUVNIL) [64], which combines the advantages of both reverse nanoimprint lithography (RNIL) and contact ultraviolet (UV) lithography. In this process, a UV crosslinkable polymer and a thermoplastic polymer are spin-coated onto a patterned

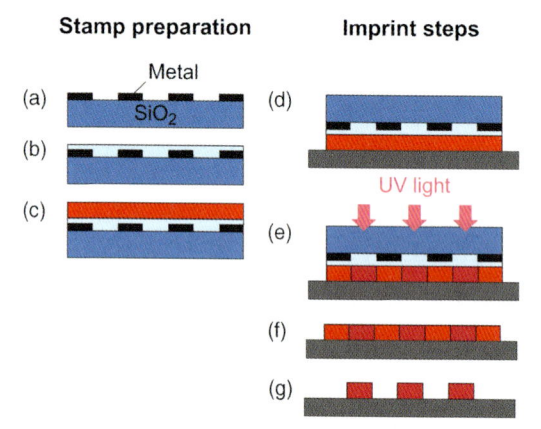

Figure 7.18 Schematics of the reverse contact UV NIL (RUVNIL) process. Left panel: Steps to prepare a stamp. (a) A hybrid mask of SiO$_2$ with metal feature; (b) a thermoplastic polymer, for example, mr-I 7030, is spin coated; (c) a UV crosslinkable polymer, for example, mr-I 6000 is spin-coated. Right panel: Steps to obtain nanostructures by this method. (d) Reverse imprinting on a Si substrate is carried out; (e) the silicon substrate is heated to heat the polymer above T_g, and pressure is applied; (f) the polymer is cooled down and exposed to UV light; (g) the stamp is separated from the substrate; (h) the exposed polymer layer is developed in acetone, resulting in a polymer pattern with no residual layer. The fabrication time per printed layer is just under 2 min [64].

hybrid metal–quartz stamp. The thin polymer films are then transferred from the stamp to the substrate by contact at a suitable temperature and pressure, after which the whole assembly is exposed to UV light. Following separation of the stamp and substrate, the unexposed polymer areas are rinsed away with a suitable developer, leaving behind the negative features of the original stamp. The process is shown schematically in Figure 7.18.

By using the same UV-curable polymer for each layer, 3-D nanostructures have been obtained (Figure 7.19). This technique offers a unique advantage over reverse-contact NIL and thermal NIL, as no residual layer is obtained by controlling the UV light exposure. This avoids the normal post-imprinting etching step, and therefore results in a much better control of the critical dimensions. Another interesting feature here is that it is not necessary to treat the stamp with an anti-adhesive coating.

Three-dimensional UV-NIL can be potentially used in the fabrication of modern integrated circuits, which employ several layers of copper interconnects, separated by an interlayer dielectric (ILD) and connected by copper vias (Figure 7.20). An imprintable and curable UV-ILD material is deposited on an existing interconnect layer (Figure 7.20a), this material having been structured by a 3-dimensional

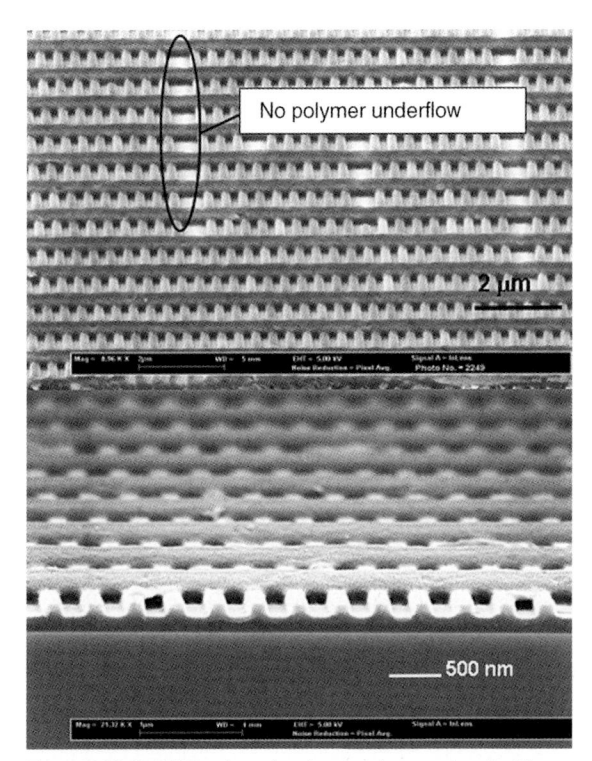

Figure 7.19 RUVNIL prints showing two layers printed without leaving a residual layer, and avoiding polymer overflow of the second layer [64].

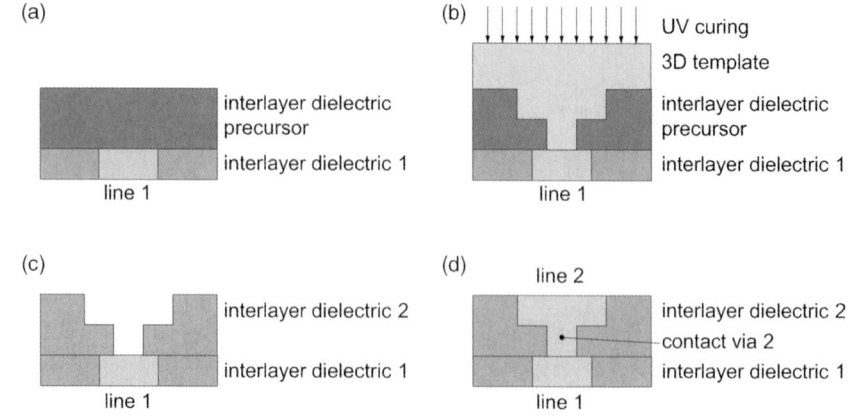

Figure 7.20 Schematics of direct-printing of interconnect layers using a three-dimensional stamp. (Illustration courtesy of L. Berger.)

stamp (Figure 7.20b). After UV-curing, the materials resembles a structured ILD (Figure 7.20c), which is then filled with copper to form two layers of vias and interconnects (Figure 7.20d).

7.2.7
The State of the Art

A comparison of the methods discussed to date, in addition to some relevant data, are displayed in Table 7.3. This information forms part of the studies of the European integrated project "Emerging Nanopatterning Methods (NaPa)" [65], which is exploring several non-optical lithographic methods with the purpose of gathering a library of processes that employ some of these newly emerging patterning technologies.

7.3
Discussion

Nanoimprint lithography, as an example of non-optical lithographies, has proven to be a versatile patterning method in several fields of application where a rather rapid development has been demonstrated, in addition to sole pattern transfer, notably in the areas of optics [66] (some of them at 200 mm wafer scale [67]) and microfluidics [68]. The versatility of NIL opens new possibilities for the nanostructure of various types of functional material, such as conducting polymers [69], light-emitting polymers [70], polymers loaded with nanocrystals [71], and biocompatible polymers [72]. An example of photonic applications is shown in Figure 7.21, which depicts a printed two-dimensional photonic crystal in polymer. The patterning of functionalized materials may be difficult when using traditional methods such as optical or electron beam

Table 7.3 Comparison of the different printing techniques.

Technique	Smallest/largest features in same print	Min pitch (nm)	Largest wafer printed (mm)	Overlay Accuracy (nm)[*]	T align, T print, T release, T cycle	No. of times stamp used	Materials
NIL	5 nm[a]/N/A	14	200[b]	500	Minutes, 10 s, Min, 10–15 min	>50[c]	Various
SSIL[d]	8 nm min features. 50 nm/5 μm on same stamp	50	200	<250	Full cycle 2.5 min with, 20 s without full auto-collimation.	1000	mr-I 8000, mr-I 7000
SFIL[e]	25 nm[f]/μm	50[f]	300[g,h,m] stamp size: ~26×26 mm²	50[i] (about 20[j])	20 wafers/h[j]	800[g]	Various NILTM105, AMONIL, PAK 01
UV-NIL[k]	9 nm/100 μm[l]	12[m]	200[n]	about 20[o]	20 s/step[p] three wafers/h[q]	>1000[r]	MRT07xp, PAK01, AMONIL1, AMONIL2, NXR, Laromer
Soft UV-NIL	25 nm/20 μm[s]	150[t]	200	1–50 μm[u]	4–5 min; about 12 wafers/hr[v]	>50[w]	AMONIL1, NXR-Mod, Laromer

Data from several sources.
[a]Depends more on the equipment than on the imprinting method.
[b]M.D. Austin, et al., Appl. Phys. Lett. 2004, 84, 5299.
[c]From Ref. [25].
[d]This value is from manual tests. A cassette-loading tool will have better values.
[e]Step-and-stamp imprint lithography is based on thermal NIL using the step-and-stamp imprinting tool, NPS300 by SUSS MicroTec.
[f]Step-and-flash imprint lithography.
[g]D.J. Resnick, G. Schmid, E. Thompson, N. Stacey, D.L. Olynick and E. Anderson, Step and Flash Imprint Lithography Templates for the 32 nm Node and Beyond, NNT 06, San Francisco, US, November 15–17, 2006.
[h]M. Miller, G. Schmid, G. Doyle, E. Thompson and D. J. Resnick, S-FIL Template Fabrication for Full Wafer Imprint Lithography, NNT 06, San Francisco, US, November 15–17, 2006.

(Continued)

[i] T.-Wei Wu, M. Best, D. Kercher, E. Dobisz, Z. Bandic, H. Yang and T. R. Albrecht, *Nanoimprint Applications on Patterned Media*, NNT 06, San Francisco, US, November 15–17, **2006**.
[j] S. V. Sreenivasan, P. Schumaker, I. McMackin and J. Choi, *Nano-Scale Mechanics of Drop-On-Demand UV Imprinting*, NNT 06, San Francisco, US, November 15–17, **2006**.
[k] R. Hershey, M. Miller, C. Jones, M. G. Subramanian, X. Lu, G. Doyle, D. Lentz and D. LaBrake, *SPIE* **2006**, 6337, 20.
[l] UV-NIL includes Single-Step & Step&Repeat on a EVG770 tool.
[m] B. Vratzov, et al., *J. Vac. Sci. Technol. B* **2003**, 21, 2760; and http://www.amo/de.
[n] S. Y. Chou, et al., *Nanotechnology* **2005**, 16, 10051.
[o] 4-inch (10-cm) Single-Step. 8-inch (20-cm) Step&Repeat on EVG770.
[p] A. Fuchs, et al., *J. Vac. Sci. Technol.* **2004**, 22, 3242–3245, and to be published.
[q] Without fine alignment nor automation yet.
[r] http://www.molecularimprints.com.
[s] M. Otto, et al., *Microelectronic Eng.* **2004**, 73–74, 152.
[t] U. Plachetka, et al., *Microelectronic Eng.* **2006**, 83, 944.
[u] Pitch for Soft UV-NIL tested so far and to be published.
[v] Only coarse alignment available; depending on stamp material used.
[w] 70–80% of the given time is to cure the resist.
[x] Based on laboratory tests to date.

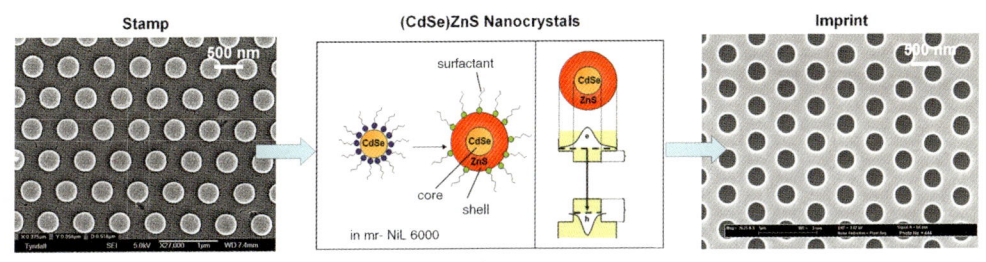

Figure 7.21 Two-dimensional photonic crystals printed in a polymer containing semiconductor quantum dots to control the spontaneous emission. This single-step imprint resulted in a 200% increase of the emission efficiency [71].

lithography, because it may not be possible to incorporate photosensitive components without degrading the functionality; alternatively, the materials may not tolerate the chemical processes associated with these pattern-transfer technologies. As described above, NIL requires only a fairly moderate temperature cycle in order to mediate the patterning process.

One exciting extension of NIL is the potential for patterning curved, 3-D surfaces, and this is yet to be exploited both in research and commercial applications. Somewhat surprisingly, it has been the lack of straightforward ways to provide curved surfaces that has hindered progress in this area. Nonetheless, NIL provides a simple means of realizing various types of curved 3-D surface which, of course, require the fabrication of a master stamp. This ability can be used, for example, in optics [72], cell cultivation [73], and plasmonics [74, 75]. The possibility of aligning to already existing patterns in SSIL and SFIL allows the use of mix-and-match approaches and the combination of more than one technique to build up multifunctional structures. The promise of low-cost and high-throughput uses of NIL in nanofabrication may be fulfilled by the roll-to-roll type of continuous approaches. During the late 1990s, the tools used for nanoimprinting were mainly commercial presses with heating units, but some time later tools based on modified optical mask aligners emerged, both for thermal and UV NIL. Today, several commercially available machines are dedicated to the nanoimprinting processes. The development of materials intended for NIL has also witnessed similar progress since the late 1990s, with not only methods but also instruments and software for non-destructive characterization and metrology having been introduced [46]. Clearly, whilst NIL is becoming a mature and capable technology for nanofabrication, its prospective roles are reaching even further, and have been included among the top ten technologies considered capable of "changing the world" [76]. This situation is reflected in the dissemination of information pertaining to NIL, with the numbers of published reports increasing at breath-taking pace, along with numerous conferences and discussion sessions of microfabrication and nanofabrication systems dedicated to nanoimprinting.

Despite these many advances, much remains to be understood and achieved. One such example is the need to pin down design rules based on an understanding of non-linear processes in viscous flow. If NIL is to improve its throughput, then by necessity faster processes will have to be partly non-linear and undergo concomitant modeling challenges. Here, the impact will be upon stamp layout, and on the design rules. With regards to critical dimensions, those applications with a strict control of periodicity and smoothness of features will serve as the "acid test" for NIL. For example, if NIL is to be used as a mask-making method for the transfer of 2-D photonic crystal patterns into a high-refractive index material then, in addition to alignment, the "disorder" must be controlled to better than a few nanometers after pattern transfer – that is, after reactive ion etching. Although these critical dimensions are more relaxed in lower-refractive index photonic crystals, such as those printed directly in polymers [71], the verticality of the side walls is still of paramount importance. While the current resolution of NIL is already of a few nanometers, such demands will need to be even more stringent in order to fabricate hypersonic phononic crystals and nanoplasmonic structures, where the relevant wavelengths are of the order of only a few nanometers.

The versatility of the printable polymers, the resolution of NIL and the ability to realize 3-D structures open further possibilities. One such advance is the use of 3-D templates or scaffolds to provide not only the support but also input and output contacts for supramolecular structures. To be successful, this has two requirements: first, the feature sizes must be commensurate, and the structured polymer surface site-selectively functionalized. Whilst the complete proof of concept is still missing, some degree of progress has been made towards spatially selective functionalization by means of NIL, chemical functionalization and lift-off, and this has resulted in electrical contacts to 150 nm-wide arrays of polypyrrole nanowires [76]. The second requirement is in the use of 3-D nanostructures, beyond the face-centered cubic (fcc) and cubic symmetries, the properties of which can be modified by subsequent surface treatment, while preserving the symmetry. Modifications may include coating with oxides, and also removal of the polymer template, followed by subsequent infilling with another material. In this way, an artificial 3-D superlattice may be realized, thereby providing a periodic or quasi-periodic arrangement for electronic and or optical excitations.

7.4
Conclusions

In this chapter we have reviewed some of the key developments of NIL, as a non-optical lithography method, paying particular attention to the schematics of the process, and discussing: (i) materials issues and their impact on the process; (ii) the variations of NIL (among which roll-to-roll appears particularly promising for volume production); (iii) stamp design, both in terms of robustness and adhesion; (iv) the issue of residual layer thickness and its impact on critical dimensions; and (v) finally briefly reviewing the latest progress in 3-D NIL. In addition, we have discussed the importance of understanding polymer flow as an enabling knowledge to optimize stamp design, and thereby throughput. When discussing 3-D NIL, the scope for further progress was outlined as to date this is largely unexplored.

NIL has been said to have short-term prospects in back-end CMOS fabrication processes, but more so in the fabrication of photonic structures and circuits, with a proviso in case of photonic crystals where stringent tolerances are still to be met. On the other hand, applications in less-demanding areas, such as gene chips for diagnostic screening, appear to have a very bright future.

Acknowledgments

he authors gratefully acknowledge the support of the EC-funded project NaPa (Contract no. NMP4-CT-2003-500120) and of Science Foundation Ireland (Grant No. 02/IN.1/172). They are also grateful to A. Kristensen, H. Schift, M. Tormen, T. Haatainen, P. Majander, T. Mäkelä, N. Kehagias, V. Reboud, M. Zelsmann, F. Reuther, D.-A. Mendels and many other colleagues of the NaPa project for fruitful discussions and joint investigations over the years. Note: The content of this work is the sole responsibility of the authors. Part of the sections on stamps (Photomasks) and on 3-D stamps for integrated electronic circuits were kindly provided by L. Berger.

References

1. http://www.itrs.net/Links/2006/Update/Final/ToPost/08_Lithography2006.Update.pdf.
2. Sotomayor Torres, C.M. (Ed.) (2003) *Alternative Lithography: Unleashing the Potentials of Nanotechnology*, Kluwer Academic Plenum Publishers, New York.
3. (a) Chou, S.Y., Krauss, P.R. and Renstrom, P.J. (1995) *Applied Physics Letters*, **67**, 3114. (b) Chou, S.Y. (1998) United States Patent, No. 5,772,905, 57.
4. Garcia, R. (2003) *Alternative Lithography: Unleashing the Potentials of Nanotechnology* (ed. C.M. Sotomayor Torres), Kluwer Academic Plenum Publishers, New York, p. 213.
5. Piner, D., Zhu, J., Xu, F., Hong, S. and Mirkin, C.A. (1999) *Science*, **283**, 661.
6. Xia, Y., Zhao, X.-M. and Whitesides, G.M. (1996) *Microelectronic Engineering*, **32**, 255.
7. Michel, B., Bernard, A., Bietsch, A., Delamarche, E., Geissler, M., Juncker, D., Kind, H., Renault, J.-P., Rothuizen, H., Schmid, H., Schmidt-Winkel, P., Stutz, R. and Wolf, H. (2001) *IBM Journal of Research and Development*, **45**, 697.
8. Brugger, J., Berenschot, J.W., Kuiper, S., Nijdam, W., Otter, B. and Elwenspoek, M. (2000) *Microelectronic Engineering*, **53**, 403.
9. Muetzel, M., Tandelr, S., Haubrich, D., Meschede, D., Peithmann, K., Flaspoehler, M. and Buse, K. (2002) *Physical Review Letters*, **88**, 083601.
10. Zubarev, E.R., Xu, J., Sayyad, A. and Gibson, J.D. (2006) *Journal of the American Chemical Society*, **128**, 15098.
11. Ruda, J.C. (1977) *Journal of the Audio Engineering Society*, **11/12**, 702.
12. Haisma, J., Verheijen, M., van den Heuvel, K. and van den Berg, J. (1996) *Journal of Vacuum Science & Technology B*, **14**, 4124.
13. Scheer, H.-C., Schulz, H., Hoffmann, T. and Sotomayor Torres, C.M. (2002) *Handbook of Thin Film Materials* (ed. H.S. Nalwa,), Vol. 5 Academic Press, New York, p. 1.
14. Guo, L.J. (2004) *Journal of Physics D-Applied Physics*, **37**, R123.
15. Schift, H. and Kristensen, A. (2007) *Handbook of Nanotechnology*, (ed. B. Bhushan), 2nd edn., Springer, Berlin, Heidelberg, p. 239.
16. Hirai, Y., Fujiyama, M., Okuno, T., Tanaka, Y., Endo, M., Irie, S., Nakagawa, K. and

Sasago, M. (2001) *Journal of Vacuum Science & Technology B*, **19**, 2811.

17 (a) Heyderman, L.J., Schift, H., David, C., Gobrecht, J. and Schweizer, T. (2000) *Microelectronic Engineering*, **54**, 229. (b) Schift, H., Heyderman, L.J., Auf der Maur, M. and Gobrecht, J. (2001) *Nanotechnology*, **12**, 173.

18 (a) Reuther, F., Fink, M., Kubenz, M., Schuster, C., Vogler, M. and Gruetzner, G. (2005) *Microelectronic Engineering*, **78–79**, 496. (b) Pfeiffer, K., Reuther, F., Carlsberg, P., Fink, M., Gruetzner, G. and Montelius, L. (2003) *Proceedings of SPIE*, **5037**, 203.

19 Reuther, F. (2005) *Journal of Photopolymer Science and Technology*, **18**, 523.

20 http://microresist.de/, micro resist technology GmbH, Koepenicker Str. 325, 12555 Berlin, Germany.

21 Vogler, M., Wiedenberg, S., Mühlberger, M., Glinsner, T. and Grützner, G. (2006) poster P_NIL_31 3C-6, presented at the Micro- and Nano-Engineering International Conference MNE 2006, 17–20 September, Barcelona, Spain.

22 Hirai, Y., Konishi, T., Yoshikawa, T. and Yoshida, S. (2002) *Journal of Vacuum Science & Technology B*, **22**, 3288.

23 Cross, G.L.W. (2006) *Journal of Physics D-Applied Physics*, **39**, R363.

24 Heidari, B., Maximov, I., Sarwe, E.-L. and Montelius, L. (2000) *Journal of Vacuum Science & Technology B*, **18**, 3552.

25 Gourgon, C., Perret, C., Tallal, J., Lazzarino, F., Landis, S., Joubert, O. and Pelzer, R. (2005) *Journal of Physics D-Applied Physics*, **38**, 70.

26 http://www.obducat.com, Obducat AB, Box 580, 20125 Malmö, Sweden.

27 http://evgroup.com, EV Group, E. Thallner GmbH, DI Erich Thallner Strasse 1, A-4782 St. Florian/Inn, Austria.

28 Matsui, S., Igaku, Y., Ishigaki, H., Fujita, J., Ishida, M., Ochiai, Y., Komuro, M. and Hiroshima, H. (2001) *Journal of Vacuum Science & Technology B*, **19**, 2801.

29 Matsui, S., Igaku, Y., Ishigaki, H., Fujita, J., Ishida, M., Ochiai, Y., Namatsu, H. and Komuro, M. (2003) *Journal of Vacuum Science & Technology B*, **21**, 688.

30 Haatainen, T., Ahopelto, J., Gruetzner, G., Fink, M. and Pfeiffer, K. (2001) *Proceedings of SPIE*, **3997**, 874.

31 http://www.S.E.T.-SAS.fr, S.E.T.S.A.S., 131, impasse Bartheudet, BP24, F-74490, Saint-Jeoire, France.

32 Ahopelto, J. and Haatainen, T. (2003) *Alternative Lithography: Unleashing the Potentials of Nanotechnology* (ed. C.M. Sotomayor Torres), Kluwer Academic Plenum Publishers, New York, p. 103.

33 Colburn, M., Johnson, S.C., Stewart, M.D., Damle, S., Bailey, T.C., Choi, B., Wedlake, M., Michaelson, T.B., Sreenivasan, S.V., Ekerdt, J.G. and Grant Willson, C. (1999) *Proceedings of SPIE*, **3676** (I), 379.

34 Bailey, T.C., Colburn, M., Choi, B.J., Grot, A., Ekerdt, J.K., Sreenivasan, S.V. and Wilson, C.G. (2003) *Alternative Lithography: Unleashing the Potentials of Nanotechnology* (ed. C.M. Sotomayor Torres), Kluwer Academic Plenum Publishers, New York, p. 117.

35 http://www.sfil.org and http://www.molecularimprints.com.

36 Mäkelä, T., Haatainen, T., Majander, P. and Ahopelto, J. (2007) *Microelectronic Engineering*, **84**, 877–879.

37 Schultz, H., Lyebyedyev, D., Scheer, H.-C., Pfeiffer, K., Bleidiessel, G. and Gruetzner, G. (2000) *Journal of Vacuum Science & Technology B*, **18**, 3582.

38 http://www.nilt.com/NILTechnology, Oersteds Plads, DTU-Building 347, DK-2800 Kongens Lyngby, Denmark.

39 Brandrup, J. and Immergut, E.H. (1975) *Polymer Handbook*, 2nd edn., John Wiley & Sons, New York.

40 Sasaki, S., Itoh, K., Fujii, A., Toyama, N., Mohri, H. and Hayashi, N. (2005) *Proceedings of SPIE*, **5853**, 277.

41 Hudek, P., Beyers, D., Groves, T., Fortagne, O., Dauksher, W.J., Mancini, D., Nordquist, K. and Resnick, D.J. (2004) *Proceedings of SPIE*, **5504**, 204.

42 Dauksher, J., Mancini, D., Nordquist, K., Resnick, D.J., Hudek, P., Beyer, D. and

Fortagne, O. (2004) *Microelectronic Engineering*, **75**, 345.
43 Institut fuer Mikroelektronik Stuttgart, www.ims-chips.de.
44 Haatainen, T., Majander, P., Riekkinen, T. and Ahopelto, J. (2006) *Microelectronic Engineering*, **83**, 948.
45 Arakcheeva, E.M., Tanklevskaya, E.M., Nesterov, S.I., Maksimov, M.V., Gurevich, S.A., Seekamp, J. and Sotomayor Torres, C.M. (2005) *Technical Physics*, **50**, 1043. (Translated from Zhurnal Tekhnichesko Fiziki **2005**, 75, 80–84).
46 Fuard, D., Perret, C., Farys, V., Gourgon, C. and Schiavone, P. (2005) *Journal of Vacuum Science & Technology B*, **23**, 3069.
47 Nielsen, T., Pedersen, R.H., Hansen, O., Haatainen, T., Tollki, A., Ahopelto, J. and Kristensen, A. (2005) Technical Digest 18th IEEE Conference Micro Electro Mechanical Systems, MEMS 2005, Miami, FL, USA, January 30–February 3, 2005, pp. 508.
48 Finder, Ch., Beck, M., Seekamp, J., Pfeiffer, K., Carlberg, P., Maximov, I., Reuther, F., Sarwe, E.-L., Zankovych, S., Ahopelto, J., Montelius, L., Mayer, C. and Sotomayor Torres, C.M. (2003) *Microelectronic Engineering*, **67–68**, 623.
49 Hoffmann, T. (2003) *Alternative Lithography: Unleashing the Potentials of Nanotechnology* (ed. C.M. Sotomayor Torres), Kluwer Academic Plenum Publishers, New York, p. 103.
50 Schift, H. and Heyderman, L. (2003) *Alternative Lithography: Unleashing the Potentials of Nanotechnology* (ed. C.M. Sotomayor Torres), Kluwer Academic Plenum Publishers, New York, p. 47.
51 Rowland, H.D., Sun, A.C., Schunk, P.R. and King, W.P. (2005) *Journal of Micromechanics and Microengineering*, **15**, 2414.
52 Mendels, D.-A. (2006) *Proceedings of SPIE*, **6151**, 615113.
53 Sirotkin, V., Svintsov, A., Zaitsev, S. and Schift, H. (2006) *Microelectronic Engineering*, **83**, 880.
54 Sirotkin, V., Svintsov, A. and Zaitsev, S. Paper presented at the Micro and NanoEngineering 2006, MNE'06, 17–20 September 2006, Barcelona Spain.
55 Mendels, D.-A. (2006) in: Proceedings Nanoimprint and Nanoprint Technology (NNT'06), San Francisco, USA, November 17–20.
56 Worgull, M., Heckele, M., Hétu, J.F. and Kabanemi, K.K. (2006) *Journal of Microlithography, Microfabrication, and Microsystems*, **5**, 011005.
57 Ishii, Y. and Taniguchi, J. (2007) *Microelectronic Engineering*, **84**, 912–915.
58 Li, M., Chen, L. and Chou, S.Y. (2001) *Applied Physics Letters*, **78**, 3322.
59 (a) Colburn, M., Grot, A., Amistoso, M.N., Choi, B.J., Bailey, T.C., Ekerdt, J.G., Sreenivasan, S.V., Hollenhorst, J. and Willson, C.G. (2000) *Proceedings of SPIE*, **3997**, 453. (b) Johnson, S., Resnick, D.J., Mancini, D., Nordquist, K., Dauksher, W.J., Gehoski, K., Baker, J.H., Dues, L., Hooper, A., Bailey, T.C., Sreenivasan, S.V., Ekerdt, J.G. and Willson, C.G. (2003) *Microelectronic Engineering*, **67–68**, 221.
60 Yamazaki, K. *et al.* (2004) *Microelectronic Engineering*, **73–74**, 85.
61 (a) Jun Taniguchi, *et al.* (2004) *Applied Surface Science*, **238**, 324. (b) Ishii, Y. and Taniguchi, J. (2006) Paper P-NIL09, presented at the Micro and NanoEngineering 2006, MNE,(06, 17–20 September, Barcelona Spain.
62 Tormen, M. *et al.* (2005) *Journal of Vacuum Science & Technology B*, **23**, 2920.
63 Bao, L.-R., Cheng, X., Huang, X.D., Guo, L.J., Pang, S.W. and Lee, A.F. (2002) *Journal of Vacuum Science & Technology B*, **20**, 2881.
64 (a) Kehagias, N., Zelsmann, M., Pfeiffer, K., Ahrens, G., Gruetzner, G. and Sotomayor Torres, C.M. (2005) *Journal of Vacuum Science & Technology B*, **23**, 2954. (b) Kehagias, N., Reboud, V., Chansin, G., Zelsmann, M., Jeppesen, C., Schuster, C., Kubenz, M., Reuther, F., Gruetzner, G. and Sotomayor Torres, C.M. (2007) *Nanotechnology*, **18**, 175303.
65 http://www.phantomsnet.net/NAPA/index.php.

66 Merino, S., Retolaza, A. and Lizuain, I. (2006) *Microelectronic Engineering*, **83/4-9**, 897.
67 Chaix, N., Landis, S., Gourgon, C., Merino, S., Lambertini, V.G., Repetto, P.M., Durand, G. and Perret, C. (2006) Paper P_NIL01, presented at the Micro- and Nano-Engineering International Conference MNE 2006, 17–20 September, Barcelona, Spain.
68 Kristensen, A., Balsev, S., Gersbrog-Hansen, M., Bilenberg, B., Rasmussen, T. and Nilsson, D. (2006) *Proceedings of SPIE*, **6329**, 632901.
69 Mäkelä, T., Haatainen, T., Ahopelto, J. and Isotalo, H. (2001) *Journal of Vacuum Science & Technology B*, **19**, 487.
70 Kim, C., Stein, M. and Forrest, S.R. (2002) *Applied Physics Letters*, **80**, 4051.
71 Reboud, V., Kehagias, N., Zelsmann, M., Striccoli, M., Tamborra, M., Curri, M.L., Agostiano, A., Fink, M., Reuther, F., Gruetzner, G. and Sotomayor Torres, C.M. (2007) *Applied Physics Letters*, **90**, 011115.
72 Bilenberg, B., Hansen, M., Johansen, D., Ozkapici, V., Jeppesen, C., Szabo, P., Obieta, I.M., Arroyo, O., Tegenfeldt, J.O. and Kristensen, A. (2005) *Journal of Vacuum Science & Technology B*, **23**, 2944.
73 Martines, E., Seunarine, K., Morgan, H., Gadegaard, N., Wilkinson, C.D.W. and Riehle, M.O. (2005) *Nano Letters*, **5**, 2097.
74 Tormen, M., Carpentiero, A., Ferrari, E., Cabrini, S., Cojoc, D. and Di Fabrizio, E. (2006) *Proceedings of SPIE*, **6110**, 611055.
75 Pedersen, R.H., Boltasseva, A., Johansen, D.M., Nielsen, T., Jørgensen, K.B., Leosson, K., Østergaard, J.E. and Kristensen, A. Paper P_NIL04, presented at the Micro- and Nano-Engineering International Conference MNE 2006, 17–20 September, 2006, Barcelona Spain.
76 MIT Technology Review (2003) February, 33.
77 Dong, B., Lu, N., Zelsmann, M., Kehagias, N., Fuchs, H., Sotomayor Torres, C.M. and Chi, L. (2006) *Advanced Functional Materials*, **16**, 1937.

8
Nanomanipulation with the Atomic Force Microscope
Ari Requicha

8.1
Introduction

The Scanning Probe Microscope (SPM) provides a direct window into the nanoscale world, and is one of the primary tools that are making possible the current development of nanoscience and nanoengineering. The first type of SPM was the Scanning Tunneling Microscope (STM), invented at the IBM Zürich laboratory by Binnig and Rohrer [1], who received the Nobel Prize for it only a few years later (1986). The STM provided for the first time the ability to image individual atoms and small molecules, and it is still widely used, especially to study the physics of metals and semiconductors. Much of the STM work is conducted in ultra high vacuum (UHV) and often at low temperatures. The STM main drawback is the need for conductive samples, which rules out many of its potential applications in biology and other important areas.

The next instrument to be developed in the SPM family was the Atomic Force Microscope (AFM), sometimes also called Scanning Force Microscope [2]. The AFM has become the most popular type of SPM because, unlike the STM, it can be used with non-conductive samples, and therefore has broad applicability. Today, there are many other types of SPMs. All of these instruments scan a surface with a sharp tip (with apex radius on the order of a few nm), placed very close to the surface (sometimes at distances ~ 1 nm), and measure the interaction between tip and surface. For example, STMs measure the tunneling current between tip and sample, and AFMs measure interatomic forces – see Section 8.2 below for a lengthier discussion of SPM principles.

It was noticed from the beginnings of SPM work that scanning a sample with the tip often modified the sample. This was initially considered undesirable, but researchers soon recognized that the ability to modify a surface could be exploited for nanolithography and nanomanipulation. In SPM nanolithography one writes lines and other structures directly on a surface by using the SPM tip. A well-known technique is called local oxidation, first demonstrated by passing an STM tip in air over a surface of hydrogen-passivated silicon [3]. Other materials can be used, as well as a conductive AFM tip in lieu of an STM [4]. Another STM method involves removing atoms from a silicon surface by applying voltage pulses to the tip [5]. Lines as narrow as a silicon dimer have been produced by this method. Lithography by material deposition, as opposed to material removal, has also been demonstrated in early work. For example, in [6] atomic-level structures of germanium were deposited on a Ge surface in UHV by pulsing the voltage on an STM tip, whereas in [7] gold clusters were deposited on a gold surface also by applying voltage pulses to the STM tip, but in air and at room temperature. Many other SPM nanolithography approaches have been demonstrated – see [8] for a survey of early work.

More recently, several other SPM nanolithographic techniques have been developed. Some examples follow. Dip Pen Nanolithography [9] involves depositing material on a surface much like one writes with a pen on paper. A pen (the AFM tip) is inked by dipping it into a reservoir containing the material to be deposited, and then it is moved to the desired locations on the sample. As the tip approaches the sample, a capillary meniscus is formed, which drives the material onto the sample. Other approaches are discussed for example in [10–15]. For a recent review see [16].

Nanomanipulation is defined in this chapter as the motion of nanoscale objects from one position to another on a sample under external control. Precise, high-resolution nanolithography shares with nanomanipulation the need for accurately positioning the tip on the sample. This is a challenging issue, which we will discuss later in this article.

Given the atomic resolution achieved by SPM imaging, one would expect also that atoms might be moved individually. This is indeed the case, and it was demonstrated in the early 1990s [17]. At the IBM Almadén laboratory, Eigler's group has been able to precisely position xenon atoms on a nickel surface, platinum atoms on platinum, carbon monoxide molecules on platinum [18], iron on copper [19], and so on, by using a sliding, or dragging process. The tip is brought sufficiently close to an adsorbed atom for the attractive forces to prevail over the resistance to lateral motion. The tip then moves over the surface, and the atom moves along with it. Tip withdrawal leaves the atom in its new position.

Eigler also has succeded in transferring to and from an STM tip xenon atoms on platinum and nickel, platinum on platinum, and benzene molecules. This was done by approaching the atoms or molecules with the tip until contact (or near contact) was established. In addition, xenon atoms on nickel were transferred to the tip by applying a voltage pulse to the tip. All of Eigler's work cited above has been done in ultra high vacuum (UHV) at 4 K.

Avouris group, at the IBM Yorktown laboratory, and Aono's group in Japan have transferred silicon atoms between an STM tip and a surface in UHV at room temperature, by aplying voltage bias pulses to the tip [20, 21].

Atomic manipulation with SPMs continues to be studied today and is providing new insights into nanoscience. For example, in the late 1990s, Rieder's group in Berlin conducted a series of experiments in which they showed that xenon atoms can be pushed and pulled across a copper surface, in UHV and at low temperature, and that the tunnel current during the motion has distinct signatures that correspond to the pushing and pulling modes [22]. As an example of very recent work, a NIST group has shown that a cobalt atom can be moved on a copper surface by exciting it electronically with and STM tip in UHV and at low temperature [23].

Molecules have been arranged into prescribed patterns at room temperature by Gimzewski's group at IBMs Zürich laboratory (now at UCLA). They push molecules at room temperature in UHV by using an STM. They have succeeded in pushing porphyrin molecules on copper [24], and they have arranged bucky balls (i.e., C_{60}) in a linear pattern, using an atomic step in the copper substrate as a guide [25]. They approach the molecules, change the voltage and current values of the STM so as to bring the tip closer to the sample than in imaging mode, and push with the feedback on. C_{60} molecules on silicon also have been pushed with an STM in UHV at room temperature by Maruno and co-workers in Japan [26], and Beton and co-workers in the UK [27]. In Maruno's approach the STM tip is brought closer to the surface than in normal imaging mode, and then scan across a rectangular region with the feedback (essentially) turned off. This may cause probe crashes. In Beton's approach the tip also is brought close to the surface, but the scan is done with the feedback on and a high value for the tunneling current; the success rate is low.

It should be clear from this brief review, which is not meant to be exhaustive, that much work has been done in nanomanipulation and related topics. In this article our focus is on manipulation by using AFMs, in air or a liquid, of objects such as nanoparticles or nanowires, which are larger than atoms or small molecules. The remainder of the chapter is organized as follows. First we address in some detail the principles of operation of the AFM, and the spatial uncertainties associated with the instrument. Next we present various protocols for moving nanoobjects with the AFM tip and discuss research aimed at building nanoassemblies. Section 8.4 addresses systems for nanomanipulation, both interactive and automated. We draw conclusions in a final section.

Before we embark on the main discussion of this chapter, we point out that SPM manipulation is not the only way of positioning nanoobjects on a surface. A variety of other approaches has been reported in the literature, using principles from optics, magnetics, electrophoresis and dielectrophoresis, which are beyond the scope of this chapter. The pros and cons of these other approaches are not fully understood. For example, optical, laser traps are normally used to move micrometer-sized particles, but they can also manipulate smaller objects. They can achieve 3-D (three-dimensional) positioning, unlike AFMs, but cannot resolve objects which are at a distance significantly below the wavelength radiation used, which is typically in the hundreds of nanometers.

8.2
Principles of Operation of the AFM

8.2.1
The Instrument and its Modes of Operation

The AFM links the macroworld inhabited by users, computers and displays to the nanoworld of the sample by using a microscopic *cantilever* that reacts to the interatomic forces between its sharp *tip* and the sample. Cantilevers are typically built from silicon or silicon nitride by using MEMS (microelectromechanical systems) mass fabrication techniques. They have typical dimensions on the order of $100 \times 20 \times 5$ µm. Tips are built at one of the ends of the cantilevers and are usually pyramidal or conical with apex diameters on the order of 10–50 nm. The tip apex has dimensions comparable to those of the sample's features and can interact with them effectively.

Figure 8.1 shows diagramatically the interaction between atoms in the tip and in the sample. The main forces involved are the long-range electrostatic force (if the tip or sample are charged), the relatively short-range van der Waals force, the capillary force (when working in ambient air, which always contains some humidity), and the repulsive force that arises when contact is established [28]. The cantilever bends under the action of these forces, and its deflection is usually measured by an optical system, as shown in Figure 8.2. Laser light bounces off the back of the cantilever, opposite to the tip, and is collected in the two halves of a photodetector. In the AFM jargon, the electrical signals output from the two photodetector halves are called A and B, and the diferential signal is called A–B. For zero deflection and a calibrated instrument, the differential signal from the detector is also zero, and in general A–B is approximately proportional to the cantilever deflection. The force between tip and sample is simply the product of the measured deflection and the cantilever's spring constant k, which can be determined in several ways [29].

Relative motion between the tip and the sample is accomplished by means of a *scanner*, which consists of piezoelectric actuators capable of imparting x, y, z displacements to the tip or, more commonly, to the sample. (We assume that the sample is on the x, y horizontal plane, and the tip is approximately aligned with the z axis.) Piezo drives react quickly and are very precise, but require high voltages on the order of 100–200 V, and have a small range of motion. They also are highly nonlinear, as we will discuss later.

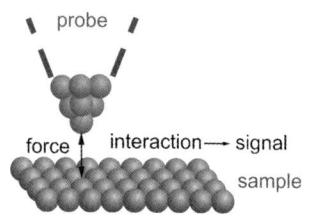

Figure 8.1 Tip sample atomic interactions.

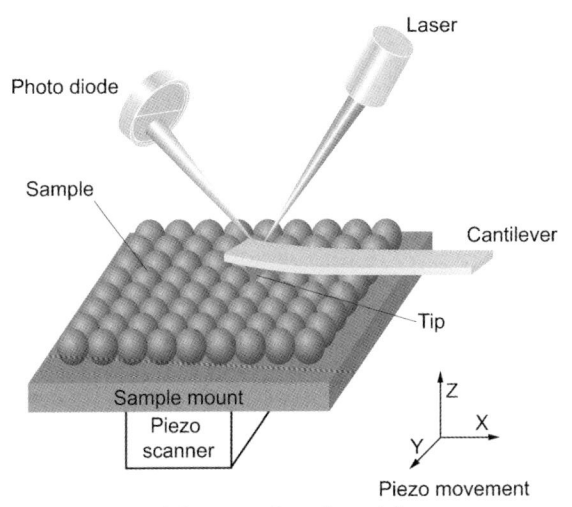

Figure 8.2 Optical detection of cantilever deflection.

In *contact mode* operation, the tip first moves in z until it contacts the sample and a desired value of the tip-sample force, called the *force setpoint*, is achieved. Then it scans the sample by moving in x, y in a raster fashion, while moving also in z under feedback control to maintain a constant force. The feedback circuitry is driven by the deviation (or error) between the (scaled) photodetector signal A–B and the force setpoint. Suppose that the tip moves in straight line along the x direction, at a constant y. When it encounters a change of height Δz of the sample, the scanner must move the sample by the same Δz in the z direction to maintain the contact force (i.e., cantilever deflection). Therefore, the amount of motion of the scanner in z at point x gives us exactly the height of the surface $z(x)$, usually called the *topography* of the surface. Now we move the tip back to the beginning of the line, increment y by Δy and scan again in the x direction. If we do this for a large number of y values we obtain a series of *line scans* that closely approximate the surface of the sample $z(x, y)$. In this example, x is called the *fast scan* direction, and y the *slow scan* direction. In practice, line scan signals are sampled (perhaps after some time-averaging) and discretized, and the output signal becomes a series of values $z(x_i, y_j)$, often called *pixel* values, since the output is a digital image. These images are normally displayed by encoding the height values z as intensities. Of course, other display options are also available, such as perspective images that provide a better feel for the three-dimensional structure of the sample, and so on. A typical AFM image contains 256×256 pixels. For this resolution and a square scan of size 1×1 μm, pixels are $\sim 4 \times 4$ nm, which is a rather large size. Therefore, very small scan sizes are necessary for precise operations.

In contact mode, essentially, the tip is pushed against the sample and dragged across it. This may damage the sample and the tip, and may dislodge nanoobjects from the surface on which they are deposited, thereby making it impossible to image them. An alternative mode of AFM operation that avoids the drawbacks of contact mode and is kinder to the tip and sample is the *dynamic* mode, also known by other designations such as non-contact, tapping, intermitent-contact, AC, and oscillatory. Here the cantilever is vibrated at a frequency near its resonant frequency, which is typically on the order of 100–300KHz in air and 1–30KHz in water. The vibration is usually generated by a dedicated piezo drive installed at the base of the cantilever. The piezo moves the cantilever endpoint opposite to the tip up and down at a frequency near resonance, and the vibration is mechanically amplified by the cantilever, resulting in a much larger amplitude of oscillation at the tip. Alternatively, the cantilever can be coated with a magnetic material and oscillated by means of an external electromagnet. This is usually called *MAC* mode (for magnetic AC), and does not require operation near the resonance frequency since it does not rely on mechanical amplification.

The amplitude of the vibration at the output of the photodetector is computed, typically by a lock-in amplifier or an analog or digital demodulation technique, and compared with an amplitude setpoint A_{set}. Feedback circuitry drive the z piezo and adjust the vertical displacement to keep the amplitude constant – see Figure 8.3.

The principles of dynamic mode operation may be explained by approximating the vibrating tip with a harmonic oscillator in a nonlinear force field, as follows. Suppose initially that the tip is at some distance z_0 to the sample, with a force $F(z_0)$ between tip and sample. Then, the following equation of motion must be satisfied:

$$m\ddot{z}_0 + c\dot{z}_0 + kz_0 = F(z_0),$$

where m is the effective mass, c is the damping coefficient, k is the spring constant, and the dots denote derivatives with respect to time. Now, consider deviations from this point that are small compared to the tip-sample distance. Denote by z the

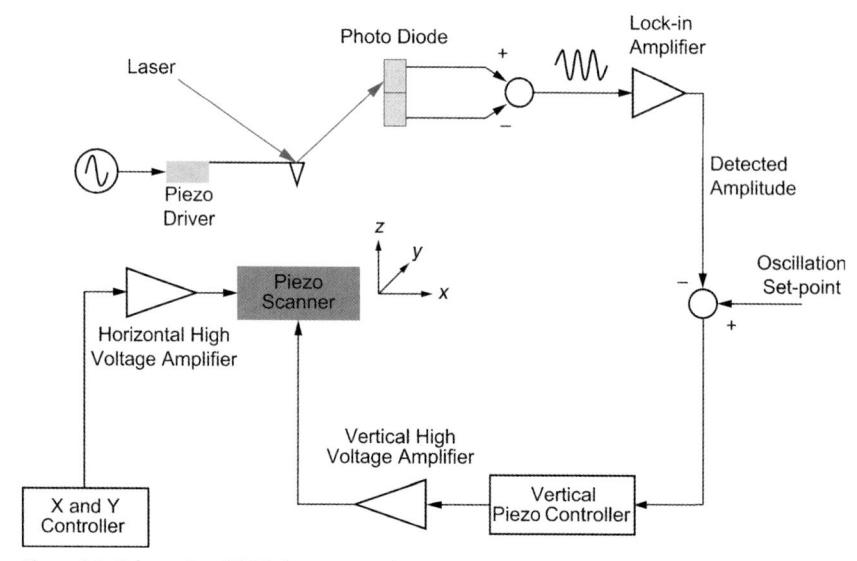

Figure 8.3 Schematic of AFM dynamic mode system.

deviation from the value z_0. The equation of motion may now be written as

$$m(\ddot{z}_0 + \ddot{z}) + c(\dot{z}_0 + \dot{z}) + k(z_0 + z) = F(z_0 + z).$$

Subtracting the two previous equations, expanding F in Taylor series and keeping only the first term yields

$$m\ddot{z} + c\dot{z} + kz = zF'(z_0),$$

where F' denotes the derivative of F with respect to z. This equation may be written as

$$m\ddot{z} + c\dot{z} + k'z = 0,$$

where

$$k' = k - F'(z_0).$$

This means that small deviations of the cantilever satisfy the equations of motion of a simple harmonic oscillator with a spring constant k', which depends on the actual spring constant of the cantilever and the gradient of the tip-sample force at the equilibrium point. Therefore, the resonance frequency changes from its initial value to

$$\omega'_0 = \sqrt{k'/m}.$$

When the cantilever is at a large distance from the sample, the interaction forces between the two are negligible, the cantilever has a resonance frequency f_{res} corresponding to its spring constant k, and has a resonance curve (amplitude vs. frequency) as shown by the red curve in Figure 8.4.

Suppose now that we drive the cantilever at a frequency f_{drive}, which generally is near f_{res}. If the tip is sufficiently far from the sample for the interation force to be negligible, $F = 0$, the cantilever oscillates with the frequency f_{drive} and an amplitude A_{free} that we can read directly from the resonance curve. This is called the *free amplitude*. Now we specify an amplitude setpoint, smaller than A_{free}. For the cantilever to oscillate at this amplitude, the resonance curve must shift as shown by the blue curve in the figure. Therefore, the feedback system must move the

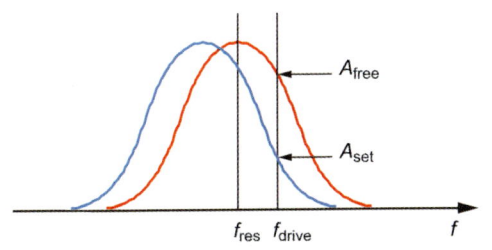

Figure 8.4 Amplitude vs. frequency curves when the cantilever is at a large distance from the sample (thick) and when the distance is smaller so that there is significant interaction and a concomitant shift of resonance frequency (thin).

cantilever closer to the sample until the force gradient is such that the spring constant and corresponding resonant frequency shift appropriately. We see that the DC, or average, position of the cantilever is controlled in a rather indirect manner, via the f_{drive}, A_{free}, and A_{set} parameters. (The free amplitude essentially scales the resonance curve.) A_{set} is usually specified as a percentage of A_{free}. Typical values of A_{set} are on the order of 80%. Lower setpoints imply large damping, which means that a considerable amount of the cantilever's oscillation energy is being transferred to the sample. In essence, we are tapping hard on the sample, and this is usually undesirable.

The theory just presented is simple and provides an intuitive understanding of the dynamic mode operation. Unfortunately, it is predicated on a linearization about an operating point, and is valid only for small oscillation amplitudes. Usually, however, the AFM is operated with a setpoint that implies a relatively large amplitude and causes the tip to hit ("tap" on) the surface of the sample at the lower part of each oscillation cycle. The actual behavior of the cantilever when tapping is involved is rather complicated – see for example [30, 31]. But in normal imaging conditions the oscillation amplitude varies approximately linearly with the DC tip position, as shown in Figure 8.5. Note that such A-d (amplitude-distance) curves vary from cantilever to cantilever and depend on several parameters.

The feedback circuitry in dynamic mode maintain a constant amplitude, and therefore a constant distance to the sample. (Here we are assuming that the sample is of a homogeneous material, or at least that the tip-sample forces do not vary over the sample's extent.) Scanning the tip over the sample in dynamic mode produces a topographic image of the sample. There are other modes of AFM operation, but the two discussed above are the most common and important.

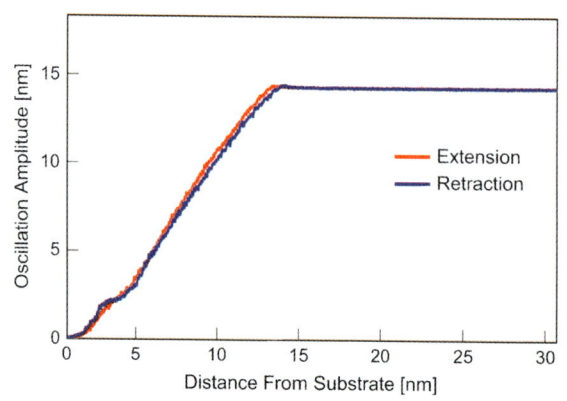

Figure 8.5 Experimental amplitude-distance curve obtained with the MFP-3D AFM (Asylum Research). Cantilever resonance frequency 240.654KHz, drive frequency 240.556KHz, spring constant ~25 N/m, free amplitude 14.2 nm. The zero point for the distance from the surface was inferred by extrapolating the A-d curve to the zero amplitude point.

More information on AFM theory and practice are available for example in [32–34].

8.2.2
Spatial Uncertainties

Let us now turn our attention to the sources of positional errors in AFM operation. These give rise to spatial uncertainty and are important for accurate imaging and especially for nanomanipulation. User intervention is normally used to compensate for spatial uncertainties in nanomanipulation, but extensive user interaction is slow and labor intensive, and therefore is severely limited in the complexity of structures it can construct. Automatic operation is highly desirable but cannot be accomplished without compensating for spatial uncertainties, as we will show below. Compensation techniques are described later in this chapter, in Section 8.4.2.

We noted earlier that the output of a line scan along the x direction is the topography $z(x)$. In the actual implementation this is not quite true. If no error compensation is used, the output is $V_z(V_x)$, where V_z and V_x are the voltages applied to the z and x piezos. In an ideal situation these voltages would be linearly related to the piezo extensions, and the signals $V_z(V_x)$ and $z(x)$ would coincide modulo scale factors. But in practice they don't.

There are many nonlinearities involved. Some of these are normally taken into account by AFM vendors' hardware and software, for example, non-linearities in the voltage-extension relationship for the piezos, coupling between the different axes of motion, and so on. The most pernicious are drift, creep and hysteresis. As far as we know, at the time of this writing (2006), drift is not adequately compensated for in any commercial instrument, and creep and hysteresis are negligible only in top of the line AFMs that have feedback control for the x, y directions, and whose controller noise r.m.s. is under 1 nm. The vast majority of AFMs in use today either have no x, y feedback or have noise levels on their feedback circuitry that are too large for precise lithography and manipulation. Open-loop operation with a small scan size (e.g., 1×1 µm) is preferable for nanomanipulation operations with such instruments.

Drift is caused by changes of temperature in an instrument made from several materials with different coefficients of expansion. At the very low temperatures often used for atomic manipulation the effects are negligible, but at room temperature they can be quite large. Figure 8.6 shows four AFM images of gold nanoparticles with 15 nm diameters, taken at successive times, 8 min apart. The particles appear to be moving, but in reality they are fixed on the substrate surface. The piezos are driven by the same voltage signals in all the panels of the figure, and have the same extensions, but the position of the tip relative to the sample is the sum of the piezo extension and the drift, and therefore changes with time. Experimental observations in our lab indicate that drift is a translation with speeds on the order of 0.01–0.1 nm/s. The drift velocity remains approximately constant for several minutes, and then appears to change randomly to another value.

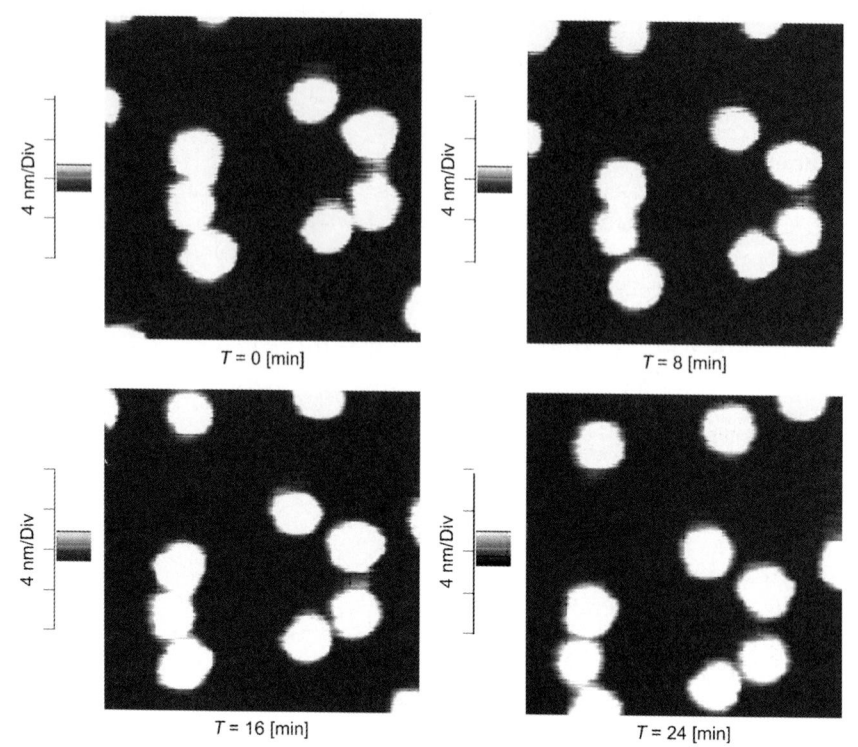

Figure 8.6 Successive images taken at 8 min intervals showing the effects of thermal drift. Gold nanoparticles with nominal diameter 15 nm on mica coated with poly-L-lysine, 1 Hz scan rate, Autoprobe CP-R AFM (Veeco). Reproduced with kind permission from [71].

Suppose now that we were trying to manipulate a large set of nanoparticles similar to those in Figure 8.6, and that the manipulation operations would take a total time of 1 hour. Assuming an average drift velocity of 0.05 nm/s, the total drift after 1 hour would be 180 nm. If we relied on the original images and did not compensate for drift, it is clear that as time went by we would completely miss most of the 15 nm particles.

After drift, creep is another major source of spatial uncertainty in AFMs. A piezo actuator commanded to move by a certain distance first responds very quickly and moves by ∼70–90% of the commanded distance, and then slowly "creeps" to the final position – see Figure 8.7. Creep is especially noticeable for large motions. Successive images of the (nominally) same area taken after a large tip displacement show an apparent motion of the features in the area. The effects of creep can last several minutes and are sufficiently large to foil manipulation attempts, especially for small particles with dimensions on the order of 10 nm. Creep can be avoided by waiting for several minutes after any large tip motion, but this is obviously a very inefficient approach.

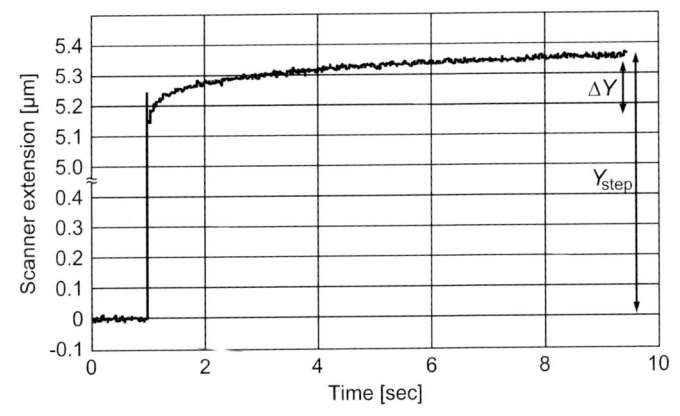

Figure 8.7 Experimental curve showing a scanner response to a 5.4 μm step function. Reproduced with kind permission from [95].

Hysteresis is also present in piezo actuators and has non-negligible effects – see Figure 8.8. Hysteresis is a nonlinear process with memory. The extension of a piezo depends not only on the currently applied voltage, but also on past extremal values.

Finally, the images of the features that appear in a topography scan differ from the actual physical features in the sample because of tip effects. The tip functions as a low-pass filter, and broadens the images. To a first approximation, the image's lateral dimensions of a feature equal the true dimensions plus the tip diameter. Algorithms are known for combating this effect [35]. Note that the vertical dimensions of a feature's image are not affected by the tip's dimensions.

Compensation of spatial uncertainties due to drift, creep and hysteresis in AFMs will be discussed later, in Section 8.4.2, in the context of automated nano-manipulation.

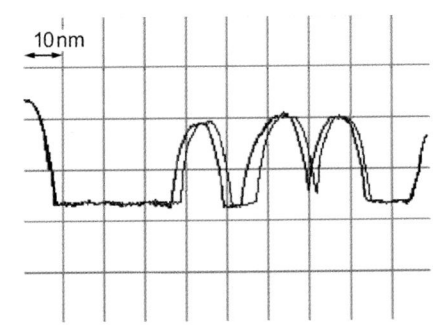

Figure 8.8 Hysteresis effects. Left-to-right vs. right-to-left single-line scans of 15 nm Au particles on mica. Scan size 100 nm.

8.3
Nanomanipulation: Principles and Approaches

8.3.1
LMR Nanomanipulation by Pushing

Here we discuss the approach to nanomanipulation that has been under development at USCs Laboratory for Molecular Robotics (LMR) over the last decade. It was first presented at the fourth International Conference on Nanometer-Scale Science and Technology, Beijing, P.R. China, September 8–12, 1996, and later reported in a string of papers [36–41]. Other approaches are considered in the next subsection.

We begin by preparing a sample with nanoparticles or other structures to be manipulated. A typical sample consists of a mica surface coated with poly-L-lysine, on which we deposit Au nanoparticles. The coating is needed because freshly cleaved mica is negatively charged, and so are the nanoparticles; the poly-L-lysine attaches to the mica and offers a positively-charged surface to the nanoparticles. We have also used other surfaces such as (oxidized) silicon, glass and ITO (indium tin oxide), other coatings such as silane layers [42], other particles such as latex, silver or CdSe, and rods or wires of various kinds. We typically manipulate particles with diameters between 5 and 30 nm, but have occasionally moved particles as small as 2 nm and as large as 100 nm. In all cases the structures to be moved are weakly attached to the underlying surfaces and cannot be imaged by contact mode AFM. We image them in dynamic mode, apply a flattening procedure to remove any potential surface tilt, and then proceed with the manipulation. The bulk of our experiments have been conducted in ambient air at room temperature and without humidity control, but we also have demonstrated manipulation in a liquid environment [43]. We use stiff cantilevers, with spring constants >10 N/m.

The nanomanipulation process is very simple. We move in a straight line with an oscillating tip towards the center of a particle and, before reaching the particle, turn off the z feedback. We turn the feedback on when we reach the desired end of the particle trajectory. With the feedback off, the tip does not move up to keep constant distance to the sample when it encounters a nanoparticle. Rather, it hits the particle and pushes it. We use the same dynamic AFM parameters for pushing as for imaging, but sometimes we force the tip to approach the surface by applying directly a command to move by Δz immediately after turning off the feedback.

Figure 8.9 shows experimental data acquired during a pushing operation for a 15 nm Au particle on mica. The two vertical dashed lines indicate the points where the feedback is turned off and on. The top trace (A) is simply the topography signal acquired by a single line scan in dynamic mode. The next trace (B) is the topography signal during the push. The topography signal is flat while the feedback is off because the tip does not move up and down to follow the sample topography. Observe that as soon as the feedback is turned back on we immediately get a non-null topography signal that indicates that the tip was somewhat below the top of the particle at the end

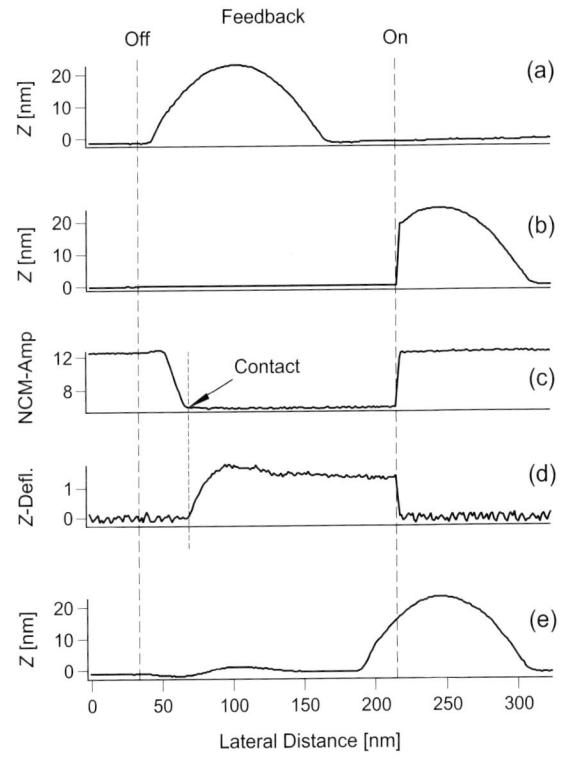

Figure 8.9 Data acquired during a manipulation operation. (a) Dynamic-mode single line scan image of the particle before manipulation. (b) Topography signal during the push. (c) Vibration amplitude during the push. (d) Average position of the antilever during the push. (e) Image of the particle after the manipulation. AutoProbe AFM, 15 nm Au particles on mica with poly-L-lysine, cantilevers with (nominal) spring constant 13 N/m. Reproduced with kind permission from [37].

of the manipulation. We conclude from these data that the tip is pushing the particle forward rather than dragging it behind itself.

Trace C shows the amplitude of the vibration during the manipulation. The amplitude is constant at the setpoint value when the feedback is on, but it decreases as the tip approaches the particle with the feedback off, and eventually reaches zero and stays at zero for the reminder of the push. At the same time that the amplitude decreases, the average (DC) value of the cantilever deflection increases and then reaches an approximately constant level – trace D in the figure. Finally, trace E is a single line scan after the push.

We interpret the data in Figure 8.9 as follows. When the tip approaches the particle with the feedback off, it starts to exchange energy with the particle and the vibration amplitude decreases, much like in standard A-d curves (see Section 8.2.1). When the vibration goes to zero, the tip touches the particle, and remains in contact with it until the feedback is turned on. While the tip contacts the particle the cantilever starts to "climb" the particle, and the DC deflection increases. When enough force is exerted on the particle for it to overcome the surface adhesion forces, the particle moves, and the deflection (and hence the force) remains approximately constant. Our experiments reveal that there is a deflection (or force) threshold above which the particle moves.

For successful pushing, the trajectory has to pass close to the center of the particle. Here the spatial uncertainties discussed in Section 8.2.2 are a major source of problems. We address these problems in the interactive version of our Probe Control Software (PCS) as follows. The user draws a line over the image and instructs the AFM to scan along that line and output the corresponding topography signal. The user moves the line over (or perhaps near) the particle to be pushed until he or she detects the maximum height of the particle – see Figure 8.10. This indicates that the line is going through the center of the particle. Usually this is several nm away from the apparent center because of drift and other spatial uncertainties. After the center is found, the user sets two points along the trajectory for turning the feedback off and on, and instructs the AFM to proceed with the push. The result can be assessed immediately by looking at the single line scan after the push – see trace E in Figure 8.9.

The amplitude and deflection signals – see traces C and D in Figure 8.9 – are useful to assess whether a pushing operation is proceeding normally. For example, we have observed experimentally that when we "loose" a particle (i.e., when it does not move as far as specified) the deflection drops to zero prematurely. In principle, one could monitor the amplitude and deflection signals while pushing and, for example, stop and locate the particle when the signals are not as expected. In practice, however, this may require substantial modifications to the controller, if decisions are to be made automatically based on this information while the manipulation operation is taking place.

How reliably can particles be moved by the LMR aproach? Figure 8.11 attempts to answer this question. Observe that operations in which the commanded motion is below 50 nm are very successful, whereas for distances ~80 nm the actual and

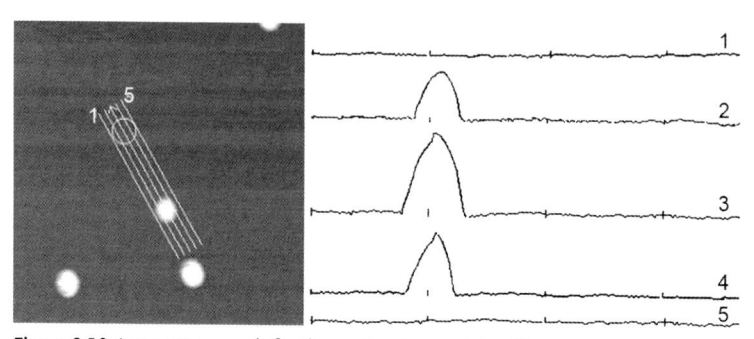

Figure 8.10 Interactive search for the center of a particle with single line scans. Line 3 has the largest peak and is chosen as trajectory for the pushing operation. Reproduced with kind permission from [36].

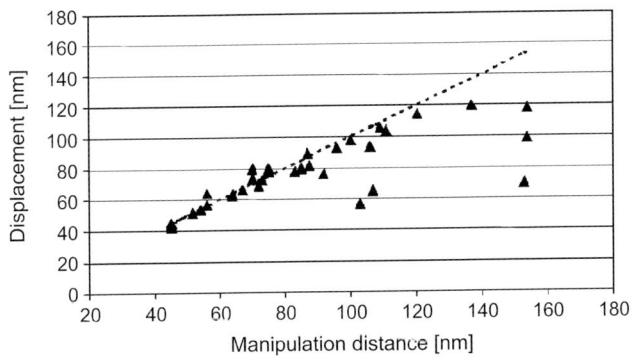

Figure 8.11 Reliability plot, showing the actual distance a particle moved as a function of the commanded manipulation distance. AutoProbe CP-R, 15 nm Au particles on mica, interactive pushing. The dashed line corresponds to perfect pushing, with equal actual and commanded displacements. Reproduced with kind permission from [90].

desired displacements of the particles begin to differ considerably. The reasons why pushing over large distances is unsuccessful are not fully understood.

8.3.2
Other Approaches

At the LMR we have experimented with several other approaches to pushing nanoparticles. We had occasional success with all of these approaches, but did not achieve reproducible, controlled manipulation. The reliability of these protocols has been until now much lower than that of the standard method discussed in the previous subsection. Here is a short description of these various protocols. They all begin with imaging in dynamic mode and finding a particle's center interactively, as discussed earlier.

- As the tip approaches the particle, instead of turning the feedback off and on, change the amplitude setpoint so that the tip gets closer to the surface.
- Move towards the particle while tapping hard on the substrate and then turn the feedback off an on. This appears to induce a "lateral push", in which the cantilever deflection does not increase as in the standard pushing protocol of the previous subsection.
- Approach the particle while moving in a zig–zag pattern, in a direction normal to the desired trajectory and with the feedback off. This appears to simulate pushing with a linear edge rather than a round tip, and has been reported in [44, 45].

The first published reports that demonstrated manipulation of nanoparticles with the AFM came from Purdue University in the US [46] and the University of Lund in

Sweden [47]. The Purdue group pushed 10–20 nm gold clusters on graphite or WSe_2 substrates with an AFM, in a nitrogen environment at room temperature [46]. They image with non-contact AFM, but then stop the cantilever oscillation, approach the substrate until contact, disable the feedback, and push. Samuelson's group at the University of Lund succeeded in pushing galium arsenide (GaAs) nanoparticles of sizes in the order of 30 nm on a GaAs substrate at room temperature in air [47]. They use an AFM in non-contact mode, approach the particles, disable the z feedback and push. This is the protocol investigated in detail later on by the LMR, and discussed in the previous subsection. Essentially the same protocol is used in [48] to push Ag nanoparticles. They observe that the vibration amplitude decreases as the particle is approached, and then essentially vanishes during pushing, which agrees with the findings in our own laboratory.

Lieber's group at Harvard has moved nanocrystals of molybdenium oxide (MoO_3) on a molybdenium disulfite (MoS_2) surface in a nitrogen environment by using a series of contact AFM scans with large force setpoints [49]. The nanoManipulator group at the University of North Carolina at Chapel Hill moves particles by increasing the contact force, under user control through a haptic device [50]. Sitti's group reports manipulation of Au-coated latex particles with nominal diameters 242 and 484 nm on a Si substrate with accuracies on the order of 20–30 nm [51]. First, they move the tip until it contacts the surface, and then move it horizontally to a point near the particle, and up by a predetermined amount. Next, they move against the particle using feedback to maintain either constant height or constant force on the particle. In constant-height pushing, the force signal exhibits several characteristic signatures that may be interpreted as signifying that the particle is sliding, rolling or rotating. Constant-force pushing is equivalent to contact-mode manipulation. Xi's group at Michigan State University has demonstrated pushing of latex nanoparticles with 110 nm diameters on a polycarbonate surface by two methods [52]. The first consists of scanning in contact mode with a high force. The second disables the feedback and moves the tip open loop along a computed trajectory based on a model of the surface acquired by a previous scan. This requires an accurate model of the surface.

Theil Hansen and coworkers moved Fe particles on a GaAs substrate by approaching a particle in dynamic mode, switching to contact mode and pushing with the feedback on [53]. This has the advantage that the pushing force can be controlled by the AFM. However, switching modes is a non-trivial operation that can cause damaging transients. (In the AutoProbe CP-R AFM that we use routinely for nanomanipulation at LMR, such a switch is not allowed by the vendor's software.) Furthermore, switching from tapping to contact mode implies that the tip in contact mode does not touch the same point of the surface it was tapping on, because the cantilever normally is not horizontal – see Figure 8.12.

The LMR and related protocols are essentially "open-loop", because it is virtually impossible to incorporate the force sensed by the cantilever into a feedback loop during actuation for commercial AFMs. For example, in the AutoProbe CP-R we are completely "blind" during a pushing operation. We can

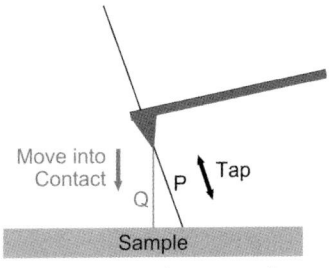

Figure 8.12 A cantilever is initially tapping on a sample at point P. When the AFM is switched into contact mode and the vibration stops, the tip must approach the sample to maintain contact. However, the contact will be at point Q, not P.

record the force (deflection) signal while pushing, but cannot do anything with it until the motion stops. To do otherwise would require a major change to the controller. On the other hand, it is not difficult to make the force signal available for visualization in interactive pushing, and several research groups have reported such capabilities. By developing their own controllers, Sitti and co-workers have been able to use the force signal during pushing, primarily to determine when an operation should be stopped because it is not going to succeed. When the tip-particle force drops to zero the particle is no longer being pushed. One should stop the motion, locate the particle and schedule a new operation to deliver it to its target position.

The AFM is both an imaging device and a manipulator but not both simultaneously. For example, it would be useful to see a particle while it is being pushed, but this cannot be done solely with an AFM. An interesting approach that provides real-time visualization consists of operating the AFM within the chamber of a Scanning Electron Microscope (SEM), or sometimes a Transmission Electron Microscope (TEM). The motion of the tip can then be monitored by the electron microscope and known techniques for visual feedback developed at larger spatial scales can be deployed [54].

The AFM-SEM approach was pioneered by Sato's group for microscopic objects [55, 56], and has been used successfully by several groups [50, 57–59]. In some of this work an AFM cantilever is used as an end effector for a specially-built micromanipulator. Working inside an SEM has its drawbacks: electron microscopes are expensive instruments, they are less precise than AFMs, require more elaborate sample preparation, and normally operate in a vacuum environment, which precludes their use for certain applications, for example, in biology.

All of the work on AFM nanomanipulation discussed above involves essentially pushing objects on a flat surface. Pushing nanoparticles over steps [37] and onto other particles [60] has been demonstrated by our group, but this is a very rudimentary 3-D capability. More sophisticated 3-D tasks would be feasible if there was the equivalent of a macroscopic "pick-and-place" operation for nanoparticles. (Pick-and-place is possible with atoms and small molecules, as noted in the

Introduction.) We know of only one report in which nanoparticles are controllably picked up by the AFM tip and then deposited elsewhere [61]. They succeed in picking up Si nanocrystals deposited by silane CVD (chemical vapor deposition) on a Si surface. They place the tip in contact with the particle and apply successive voltage pulses of opposite polarity to the AFM tip. The tip is then moved to a target location, lowered until there is contact with the surface, and again a series of pulses is applied. This work requires a dry atmosphere and the experiments were performed in a nitrogen environment. The process appears to have limited reproducibility. Diaz and co-workers report a pick-and-place process that uses redox reactions on the tip to pick up and deposit particles, but they only demonstrate it for large, micrometer scale particles [10]. Note that it is not difficult to pick up a nanoobject with a tip, even by just using van der Waals forces. What is very hard to do is to controllably release the object at a target location.

8.3.3
Manipulation and Assembly of Nanostructures

The majority of the experimental work on nanomanipulation has been conducted with nanoparticles. These are simple but interesting nanobjects because many types of them are available (metalic, semiconducting, magnetic, etc.), they often are monodisperse (i.e., have the same size), they can be smaller and more uniform than structures made by other means such as electron-beam lithography, and can be arranged into arbitrary patterns by nanomanipulation. Patterns of nanoparticles may be useful in themselves, or they may serve as templates for constructing other structures [62]. We present examples of both below.

Figure 8.13 illustrates the use of particle patterns to store digital information. The particles are located at the nodes of a uniform grid. We interpret the presence of a particle at a node as a digital "1" and its absence as a "0". Each row of particles represents an ASCII character. From top to bottom, this structure encodes the characters "LMR". These are 15 nm Au nanoparticles, and the internode distance along a row or column is ~100 nm, which corresponds to a density of 10 Gbit/cm^2. This is considerably higher than the current compact disk (CD) density. By using smaller particles and closer spacing, higher densities are achievable. In addition, this structure is editable, simply by moving the particles.

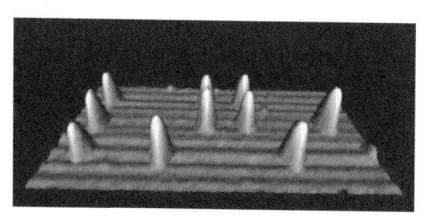

Figure 8.13 The ASCII characters "LMR" encoded in the positions of 15 nm Au nanoparticles on mica.

Nanoparticle patterns can also be used as a resist, as shown by research at the University of Konstanz, Germany [63]. They started with a self-assembled regular pattern and deposited material so as to fill the space between the particles. By etching the particles away, they obtained the complement of the original pattern. In a similar vein, a Japanese/English group used regular nanoparticle patterns as templates in a process that involves both etching and growth [64]. They deposited nanoparticles on a Si substrate and then etched the Si. As the substrate was etched away, reaction products condensated around the nanoparticles and formed regular patterns of pillars with diameters on the order of a few nm. Similar results were obtained in [65]. Another use of Au nanoparticles as a mask is reported in [66], where they demonstrate that the particles can serve as an anti-oxidation mask to prevent the AFM-tip induced oxidation of a modified Si surface. Subsequent etching produces Si nanopillars. In all of these cases, it should be possible to perform similar operations for arbitrary patterns constructed by nanomanipulation. Nanoparticle manipulation may also be an effective way to build templates or molds, for example for imprinting techniques that construct a large number of structures in a parallel fashion by pressing a template against a substrate [67]. Nanoparticle-based templates can be more uniform and have smaller features than those built by other means such as electron-beam lithography. Applications to template or mold making, however, have not yet been demonstrated.

Nanoparticle manipulation with the AFM has been used to build prototype structures for several nanodevices such as single-electron transistors [44, 68], plasmonic waveguides [69, 70], and quantum-dot cellular automata gates [71].

Most nanoparticles are approximately spherical, although some of the nanoparticles used in the research cited above (e.g., [47]) are more like "islands", and have vertical dimensions smaller than their horizontal counterparts. Structures with other shapes have also been nanomanipulated. Rods, wires and tubes have been investigated extensively. They normally require a series of pushes to reach a target position because they tend to rotate. And sometimes they deform or even break, rather than simply move on the surface. At the LMR we have manipulated Au rods with diameters ~ 10 nm and lengths ~ 70 nm deposited on a SiO_2/Si (1 1 1) substrate modified with MPMDMS (3-mercaptopropylmethyldimethoxysilane) [72]. Xi's group has demonstrated manipulation of Ag rods with diameters ~ 110 nm on a polycarbonate surface [73].

Several groups have manipulated carbon nanotubes (CNTs). Avouris and co-workers at the IBM Yorktown laboratories moved multiwall CNTs on a hydrogen-passivated Si surface by using contact mode AFM [74]. They also showed that the nanotubes can be bent and cut by the AFM tip. The nanoManipulator group in North Carolina reported rolling and sliding of CNTs on a graphite surface [75]. Other work in CNT manipulation include [57, 58, 76, 77], Roschier an co-workers [78] who built a single-electron transistor by manipulating a multiwall CNT so as to connect it to electrodes made by electron-beam lithography. Several of these groups also used the AFM tip to probe the mechanical properties of the CNTs – see for example [50, 57]. Nanowires can also be

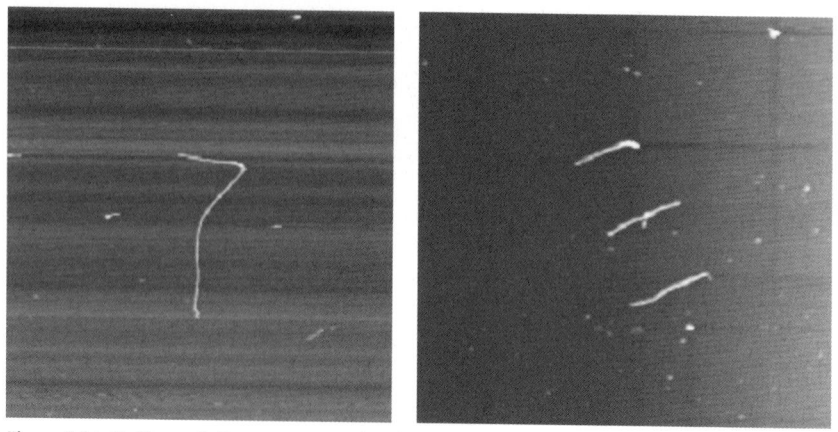

Figure 8.14 Cutting a SnO$_2$ nanowire and making an array with the resulting three pieces by using the AFM tip.

manipulated by the AFM. Figure 8.14 shows on the left a SnO$_2$ wire with a diameter of ~10 nm and a length of ~9 μm, and on the right the result of cutting the wire in two spots and then moving the three smaller wires. The manipulation was done at the LMR by using our standard protocols.

For many applications, particles and other nanoscale objects must be linked together to form a single, larger object. Linking may be accomplished by various methods: chemically, by using material deposition, sintering, and "welding". We showed that Au nanoparticles can be connected chemically by using linkers with thiol functional ends [79]. This can be done in two ways: (1) the particles are first functionalized with the di-thiols, then deposited and manipulated against one another to form the target structure, or (2) the manipulation is done first and then the thiol treatment is applied. In either case the result appears to be the same. The resulting assemblies can then be manipulated by using the same protocols, and joined to make larger assemblies. Therefore, we have demonstrated that hierarchical assembly is possible at the nanoscale [80]. Figure 8.15 shows on the left clusters of 2 and 3 particles, which were constructed by manipulating individual particles (initial configuration not shown). On the right is a ring-like structure obtained by moving the clusters on the left.

A different approach is reported in [81]: nanoparticles are manipulated to form a target structure, which is then grown by deposition of additional material. Growth is accomplished essentially by electroless deposition, by immersing the sample in a hydroxilamine seeding solution. Figure 8.16 illustrates the results. On the top left is a "wire" made by manipulating Au particles, and on the top right is a single line scan through the centerline of the structure, showing that the height of the particles is ~8 nm. On the bottom left is the structure after deposition by hydroxilamine seeding, and on the bottom right is a single line scan, which now shows a particle size of ~20 nm. The initial structure looks like a continuous wire

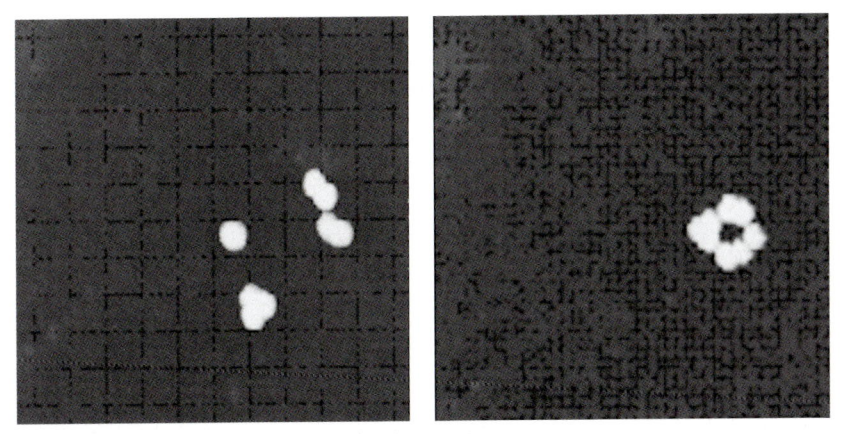

Figure 8.15 Hierarchical assembly. Thiolated 27 nm Au nanoparticles on Si coated with poly-L-lysine, 800×800 nm scan size. Reproduced with kind permission from [79].

in the figure but is not mechanically stable; touching it with the tip causes the structure to fall apart. In contrast, the final structure is a solid wire. One disadvantage of this method is that after the seeding the structures can no longer be moved on the substrate surface.

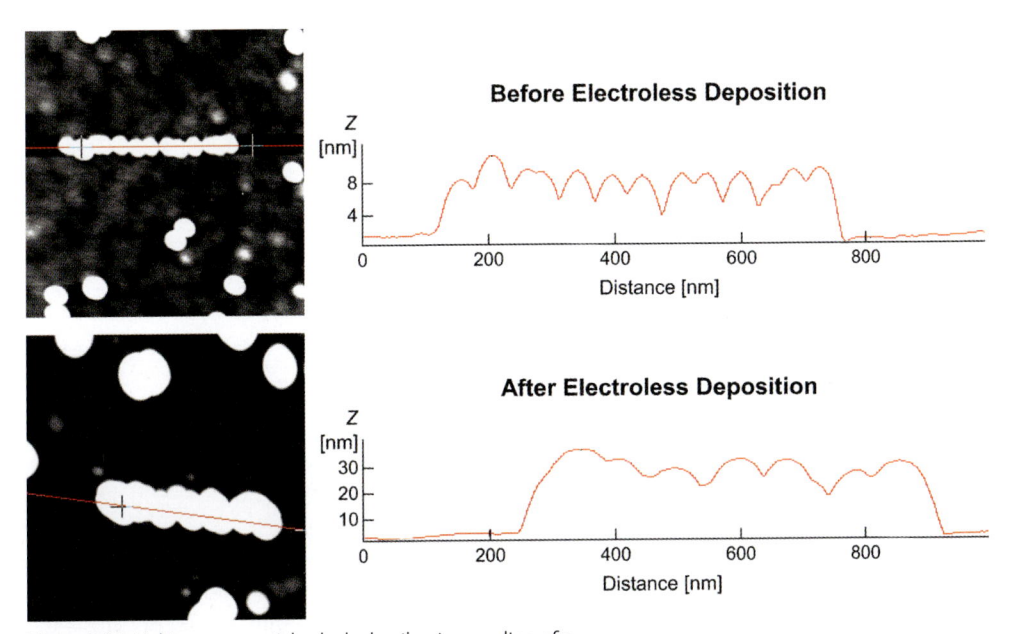

Figure 8.16 Linking nanoparticles by hydroxilamine seeding of a template built by manipulation. 8 nm Au particles on Si modified with aminopropyltrimethoxysilane (APTS). Reproduced with kind permission from [81].

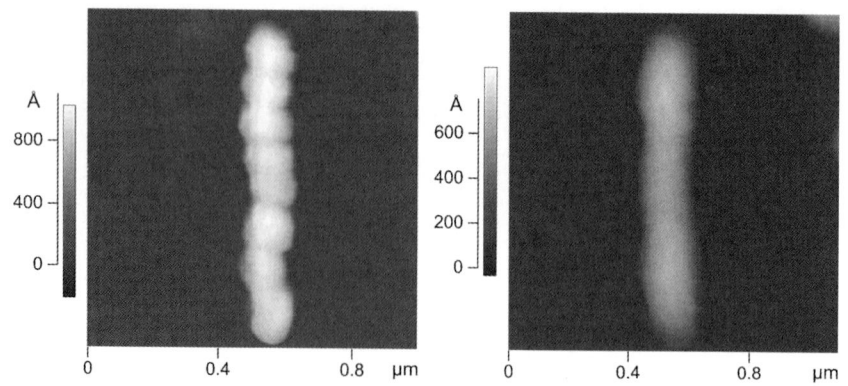

Figure 8.17 Linking nanoparticles by sintering a template built by nanomanipulation. 100 nm latex particles on Si. Reproduced with kind permission from [82].

Finally, we have demonstrated an even simpler approach to particle linking, which is based on sintering a structure after its template is built by nanomanipulation [82]. Figure 8.17 shows on the left a "wire" built by manipulation of latex particles with 100 nm diameter. On the right is the result of heating the sample at ∼160° for ∼10 min. The particles form a contiguous, solid structure. This process works very well for latex particles, but, unfortunately does not appear to be applicable to Au (and perhaps other metalic) particles.

Another interesting approach to joining nanoobjects is "welding", usually done within an SEM by using the electron beam. Contamination within the SEM chamber usually suffices to generate carbonaceous residues at the beam's target, and this can be used to link objects such as nanotubes [57, 76]. Similar approaches but using an environmental SEM and a field emission SEM are reported, respectively, in [77, 83]. Here they "solder" by inducing deposition of conductive materials with the electron beam in the presence of a source of a precursor.

8.4
Manipulation Systems

8.4.1
Interactive Systems

We discussed briefly in Section 8.3.1 the interactive manipulation capabilities of the LMR software. Ours is an unsophisticated system which provides a set of minimal capabilities a user needs to manipulate nanoobjects with the AFM. Since the beginnings of the LMR we have focused on automation – see the next subsection – and therefore did not invest resources in user interfacing.

Much more sophisticated interfaces have been developed by others. Hollis' group at the IBM Yorktown laboratory (now at Carnegie Mellon University) built an interface to an STM in which the user could drive the tip over the sample by moving a mechanical wrist [84]. The z servo signal is fed back to the wrist so that the user feels the topography of the surface as wrist vertical motions. This force, however, is not (a scaled version of) the actual force between tip and sample.

The nanoManipulator group in North Carolina has developed virtual reality user interfaces, first for STMs and then for AFMs [45, 50, 85, 86]. In the AFM interface a user can either be in imaging or manipulation mode. During imaging, the topographical data collected by the AFM is presented to the user in virtual reality, as a 3-D display. In addition, the user can feel the surface by using a haptic device, as if moving a stylus over a hard surface. Note, however, that the forces felt through the haptic device are not the cantilever-sensed forces, but rather are forces computed by standard virtual reality techniques so as to simulate the feel of a surface that approximates the measured topography of the sample. In the imaging mode, the user haptic input does not control the actual motion of the instrument, but rather the position of a virtual hand over the image of the surface. In contrast, the hand can be used to move the tip over the sample in manipulation mode. As the hand moves in virtual space and the tip moves correspondingly over the sample, the topography data generated by the AFM is used on the fly to compute a planar approximation to the surface. The user feels this approximated surface through the haptic stylus. Although the user does not feel the actual forces sensed by the cantilever, he or she can control the force applied to the sample during manipulation by using a set of knobs.

Sitti and co-workers also implement a virtual reality graphics interface, and add a one degree-of-freedom haptic device [51, 87]. Through a bilateral feedback system based on theoretical models of the forces between tip, sample and particle, the user can drive the tip over the sample by using a mouse, while at the same time feeling with the haptic device the forces experienced by the cantilever.

Xi's group has developed an augmented reality system in which cantilever forces are reflected in a haptic device [88, 89]. They develop a theoretical model for the interaction forces between tip, object and surface, and use it to compute the position of the tip based on the real-time force being measured. The visual display in a small window around the point of manipulation is updated in real-time to reflect the computed particle position. Thus, a user can follow the (computed) motion of the particle in real time during the manipulation.

8.4.2
Automated Systems

The automatic assembly of nanoobject patterns with the AFM consists of planning and executing the motions required for moving a set of objects from a given initial configuration into a goal configuration. The initial state usually corresponds to nanoobjects randomly dispersed on a surface.

As far as we know, there are only two systems today that are capable of building nanoobject patterns *automatically* by AFM nanomanipulation. One is being developed by Xi's group at Michigan State University, and the other at the LMR [73, 90]. The two systems use different planning algorithms and pushing protocols. Xi's system addresses at length the issues that arise in nanorod manipulation, and therefore is more general than ours, which focuses on nanoparticles. On the other hand, the Michigan State system has a more rudimentary drift compensator than ours, and does not compensate for creep or hysteresis, which are important for the manipulation of small objects, with dimensions $\sim$10 nm or less. The manipulation tasks demonstrated in [73] involve objects which are roughly one order of magnitude larger than those we normally manipulate, and the positional errors in the final structures shown in the figures of [73] also appear to be similarly larger than ours.

In the remainder of this section we describe the LMR automatic manipulation systems, from the top down, starting with high-level planning and ending with the system software architecture.

The input to the planner consists of a specification for a goal assembly of nanoparticles, and an initial arrangement that is obtained by imaging a physical sample with a compensated AFM (compensation is discussed below). In an initial step the planner assigns particles to target locations by using the Hungarian algorithm for bipartite matching, which is optimal [91]. It uses direct, straight-line paths if they are collision free, or indirect paths around obstacles computed by the optimal visibility algorithm [92]. Next, the planner computes a sequence of positioning paths, to connect the locations of the tip at the end of a push (determined in the previous step) and at the beginning of another push. This is done by a greedy algorithm, which sequentially selects the shortest paths. It is sub-optimal but performs well in practice. In a general case collisions between particles may arise. The planner handles collisions by exploiting the fact that all particles are assumed identical. It simulates the sequence of operations previously computed, at each step updating the state of the particle arrangement. If a collision is detected, it swaps operations, and does this recursively because solving one collision problem may generate new ones.

The planner just outlined is the second one we write. Our first planner, developed several years ago [93], was more complicated and slower, and did not perform better than the current one. However, we abandoned work on planning at that time not because of planner problems, but rather because we could not implement reliably the primitive operation assumed by the planner, which is simply to move from an initial point P to another goal point Q. The spatial uncertainties associated with the AFM – see Section 8.2.2 – were such that after a few operations we could no longer find the particles and push them without user interaction, and the task could not be completed automatically. We embarked on a research program aimed at compensation of uncertainties, and developed the compensators described briefly below. Details are available in [71, 94, 95].

The drift compensator is based on Kalman filtering, a standard technique in robotics and dynamic systems. We assume a simple (but incorrect) model for the time evolution of the drift. The model can be used to predict future values of the drift, but these values will become increasingly wrong as time goes by, because the model is not perfect. The Kalman equations provide us with means to estimate the prediction error. When this value exceeds a threshold, drift measurements are scheduled, and the measured and predicted values are combined to produce better estimates, again by using the Kalman equations. A decade-long series of experiments indicates that the drift is accurately approximated as a translation, with a direction and speed that vary slowly. The estimated drift values obtained from the Kalman filter are added as offsets to all the motion commands of the AFM, thus compensating for this translation. Drift measurement techniques require that the AFM tip move on the sample to acquire data. Therefore, manipulation operations must be suspended when a measurement is needed. In contrast, when the filter is in prediction mode the offsets can be calculated very quickly and used to update the coordinates without interrupting the manipulation task.

Creep and hysteresis compensation is achieved through a feedforward scheme. A model for the two phenomena together is constructed as explained below, using a Prandtl-Ishlinskii operator [95]. This operator has the important property of invertibility. The inverse operator is computed and the desired trajectory is fed to the inverse system. The result is the signal required to drive the AFM piezos so as to follow the goal trajectory, assuming that the model is perfect. The model, of course, is not perfect, but experimental results show that it is sufficiently accurate for obtaining very good results – an order of magnitude decrease on the effects of creep and hysteresis has been verified experimentally. Creep is modeled by a linear term plus a superposition of exponentially decaying terms, with different time constants. Hysteresis is modeled by a superposition of operators which are essentially simple hysteresis loops. The piezo extension is the sum of the values of creep and hysteresis obtained from their models, and can be expressed in terms of a Prandtl-Ishlinskii operator. The combined model depends on several parameters, which can be estimated by analyzing the AFM topography signal for a line scan over a few particles. The line should span the entire region in which the manipulations will take place. The parameters are valid as long as the scan size of the AFM is not changed, and can be computed rapidly by running the tip back and forth a few times with the compensator on. Details may be found in [95].

Running both compensators together results in a software-compensated AFM with sufficiently low spatial uncertainties to provide a reliable implementation of the most basic robotic primitive "Move from point P to point Q" on the sample. However, this is not sufficient to reliably push particles between arbitrary points because long pushes tend to be unreliable – see Figure 8.11. Therefore, we break down any long pushing trajectory into smaller segments, currently $\sim$30 nm long. Having a reliable pushing routine, the output of the high-level planner can now be executed also with high reliability.

Now that we have discussed the high-level planner and its primitive commands, let us turn our attention to the software needed to implement the system. We found in the beginnings of the LMR, in 1994, that commercial AFM software was designed for imaging and not suitable for manipulation. Therefore we designed and implemented a manipulation system, called Probe Control Software (PCS), running on top of the vendor-supplied Application Programming Interface (API) [36]. It was implemented on AutoProbe AFMs (Park Scientific Instruments, which later on became Thermomicroscopes and now Veeco), which to our knowledge were the only instruments sold with an available API. PCS evolved as time went by, and was the workhorse for the interactive manipulation research in our laboratory until recently. Research often moves in unpredictable ways, and we found that the ability to easily modify the software was fundamental to our experimental work. Unfortunately, the API was written for a 16-bit Windows system, which is far from being convenient to program. We concluded that we were spending too much time fighting an inhospitable programming environment and launched a re-write of the whole system, which has been "completed" recently. (We find that research software is in a permanent state of flux.)

The new system is called PyPCS, for Python PCS. It is written in C++ and Python, which is a scripting language that greatly facilitates program development. For example, new modules may be added to the system without the need to recompile and link the whole system. PyPCS has a client-server architecture. The server is written in C++ for 16-bit Windows and runs in the PC that controls the instrument. The client is written in Python, communicates with the server via standard interprocess communication primitives, and may run in the AFM PC or in any computer that is connected to it by Ethernet [96].

We conclude this section with a complete example, including planning and execution in the AFM. Figure 8.18 shows on the left panel the initial random dispersion of nanoparticles on the sample. On the right is the goal configuration (yellow crosses) plus the result of planning, with the pushing paths in red, and the positioning paths between pushes in green. The particles marked with a black

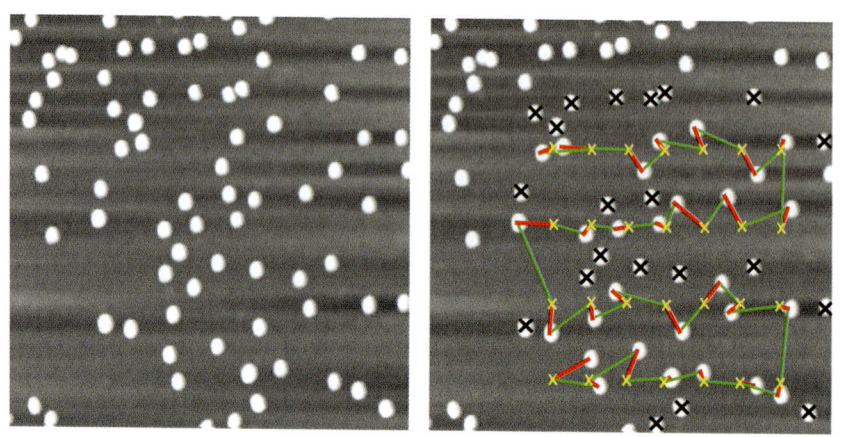

Figure 8.18 Left: Initial state. Right: Goal state (yellow) and planner output superposed on the intial image, showing pushing paths (red) and positioning paths (green). 15 nm Au particles on mica. Reproduced with kind permission from [90].

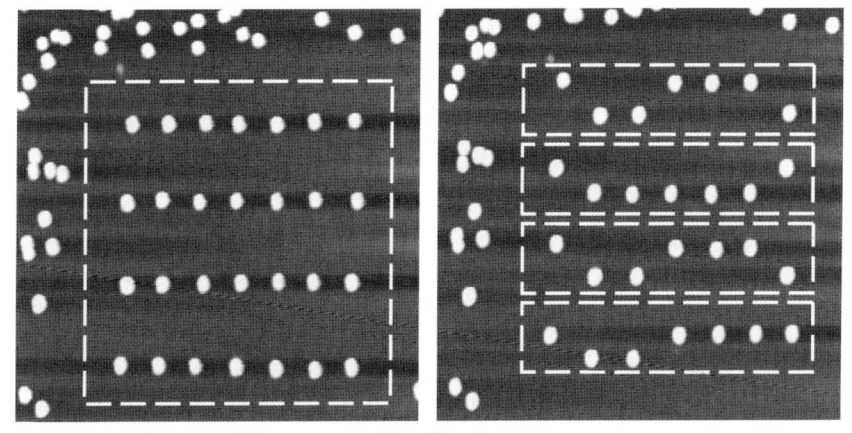

Figure 8.19 Left: Result of the execution of the plan of Figure 8.4.2.8. Right: An additional task, also planned and executed automatically. Reproduced with kind permission from [90].

cross are extraneous and should be removed from the area where the pattern is being built. (This is also done automatically.) Figure 8.19 shows on the left the result of executing the plan. On the right is the result of another similar operation also performed automatically. The pattern on the right of Figure 8.19 represents a different encoding of ASCI characters into nanoparticle positions – compare with Figure 8.13. Here a particle on the top row of each 2-row group signifies a "1" and a particle on the bottom row signifies a "0". The 4 groups of 2 rows read "NANO" in ASCI. This encoding uses twice as many particles as that of Figure 8.13 but has an interesting advantage: editing the stored data amounts simply to pushing particles up or down by a fixed amount, and could be achieved very efficiently by an AFM with a multi-tip array with spacing equal to that of the particle grid – see [97] and the discussion in the next section. The patterns shown in Figure 8.19 were built in a few minutes with the automated system. They would take at least one day of work by a skilled user if they had been built with our interactive system.

8.5
Conclusion and Outlook

Manipulation with the AFM of nanoscale objects with dimensions $\sim$5–100 nm has been under study for over a decade, and is now routinely performed in several laboratories. Nevertheless, some basic questions remain unanswered. For example, how far off center can we hit a particle for it to be pushed reliably? How high above the surface can we strike a particle for it to move? Does the size of a particle matter? Do the shape and size of a tip matter? Why do particles fall off the desired trajectories? What is the force threshold needed to move a particle? Are there preferred directions of motion? For a given surface, coating, nanoparticle, cantilever and environmental conditions, can we predict which operational parameters, if any, will result in reliable manipulation? And we could go on. In short, we lack a predictive understanding of the manipulation process, especially an understanding grounded on measurable parameters. However, there is enough experimental evidence on certain materials and systems for us to successfully complete non-trivial manipulation tasks, as shown in this chapter.

Nanomanipulation has been used until now for demonstrations or to prototype new devices. It is a useful prototyping tool because it can be used to build devices that cannot be made otherwise, and it greatly facilitates parametric studies. For example, the effect of spatial errors on a given device can be investigated by moving one of its constituent particles and recording the associated functional changes. The complexity of the structures built by nanomanipulation has been severely restricted by the sheer amount of labor and time needed to construct them interactively. Automated systems such as those described in this chapter are beginning to appear, and may significantly impact what can be done by nanomanipulation. It is fair to say, though, that a "killer app" has not yet been found for nanomanipulation. A high value, low volume application seems most appropriate, because, even with automated operation, nanomanipulation with the AFM is a serial and relatively slow process, not very suitable for mass production. Perhaps repair of fabrication masks and other high cost devices will be an important application in the future.

Another technical advance that may have a strong impact on nanomanipulation is the development of instruments with multiple tips. There are currently several research efforts aimed at producing multi-tip arrays, but they tend to be unusable for manipulation because they do not provide individual control of height (z) for each tip – see for example [98]. Furthermore, whereas efficient algorithms for nanolithography with multi-tip arrays are known [97, 99], manipulation tasks are inherently more difficult because of registration problems between tips and particles. The tips are normally arranged in a regular array, while the particles are initially randomly dispersed, and when a tip is positioned near a particle for pushing it, the other tips are unlikely to be in positions where they can push other particles. If this happens, a multi-tip array will be no faster than a single tip.

In summary, much has been learned about nanomanipulation with AFMs, and new automated systems are a breakthrough improvement over their traditional, interactive counterparts, but we still lack a deep, predictive understanding of the manipulation phenomena, as well as a convincing demonstration of economic viability for practical applications. Massively parallel operation by using multi-tip arrays may be he next breakthrough.

Acknowledgment and Disclaimer

The LMR work on nanotechnology was supported in part by the NSF Grants EIA-98-71775, IIS-99-87977, EIA-01-21141, DMI-02-09678 and Cooperative Agreement CCR-01-20778; and the Okawa Foundation.

I would like to thank my LMR faculty colleagues, postdocs and students, too many to mention here, who did much of the LMR work reported in this chapter and from whom I learned much over the last decade. I wish to single out the students who built the probe control software who made possible all of our work on nanomanipulation. They were Cenk Gazen, who started it all, Nick Montoya who extended and maintained PCS for a couple of years, Jon Kelly, who wrote the first version of PyPCS, Babak Mokaberi, who built the drift, creep and hysteresis compensators, Dan Arbuckle, who is the architect of the current version of PyPCS, which integrates the old PCS with Mokaberi's work, and Jaehong Yun, who did the high level planning software and, with Arbuckle, has integrated it into PyPCS.

An exhaustive bibliography on nanomanipulation and related topics is beyond the scope of this Chapter. In many cases I have attempted to cite the pioneering works on specific subjects but I may have failed to acknowledge some of them. I offer my apologies to the colleagues whom I may not have cited, or whose work I may have misinterpreted. There is simply too much research in this area for me to be able to keep up with all of it.

References

1 Binnig, G., Rohrer, H., Gerber, Ch. and Weibel, E. (1982) Surface studies by scanning tunneling microscopy. *Physical Review Letters*, **49**, 57–61.

2 Binnig, G., Quate, C.F. and Gerber, Ch. (1986) Atomic force microscope. *Physical Review Letters*, **56**, 931–933.

3 Dagata, J.A., Schneir, J., Harary, H.H., Evans, C.J., Postek, M.T. and Bennett, J. (1990) Modification of hydrogen-passivated silicon by a scanning tunneling microscope operating in air. *Applied Physics Letters*, **56**, 2001–2003.

4 Snow, E.S. and Campbell, P.M. (1995) AFM fabrication of sub- 10-nanometer metal-oxide devices with in situ control of electrical properties. *Science*, **270**, 1639–1641.

5 Salling, C.T. and Lagally, M.G. (1994) Fabrication of atomic-scale structures on Si(001) surfaces. *Science*, **265**, 502–506.

6 Becker, R.S., Golovchenko, J.A. and Swartzentruber, B.S. (1987) Atomic-scale surface modifications using a tunneling microscope. *Nature*, **325**, 419–421.

7 Mamin, H.J., Guethner, P.H. and Rugar, D. (1990) Atomic emission from a gold scanning-tunneling-microscope tip. *Physical Review Letters*, **65**, 2418–2421.

8 Wiesendanger, R. (1994) *Scanning Probe Microscopy, and Spectroscopy*. Cambridge University Press, Cambridge, UK. Chapter 8.

9 Piner, R.D., Zhu, J., Xu, F., Hong, S. and Mirkin, C.A. (1999) Dip-Pen, Nanolithography. *Science*, **283**, 661–663.

10 Diaz, D.J., Hudson, J.E., Storrier, G.D., Abruna, H.D., Sundararajan, N. and Ober, C.K. (2001) Lithographic applications of redox probe microscopy. *Langmuir*, **17**, 5932–5938.

11 Mesquida, P. and Stemmer, A. (2001) Attaching silica nanoparticles from suspension onto surface charge patterns generated by a conductive atomic force microscope tip. *Advanced Materials*, **13**, 1395–1398.

12 Sun, S. and Legget, G.J. (2002) Generation of nanostructures by scanning near-field photolithography of self-assembled monolayers and wet chemical etching. *Nanoletters*, **2**, 1223–1227.

13 Davis, Z.J., Abadal, G., Hansen, O., Borisé, X., Barniol, N., Pérez-Murano, F. and Boisen, A. (2003) AFM lithography of aluminum for fabrication of nanomechanical systems. *Ultramicroscopy*, **97** (1–4), 467–472.

14 Garno, J.C., Yang, Y., Amro, N.A., Cruchon-Dupeyrat, S., Chen, S. and Liu, G.-Y. (2003) Precise positioning of nanoparticles on surfaces using scanning probe lithography. *Nanoletters*, **3**, 389–395.

15 Takeda, S., Nakamura, C., Miyamoto, C., Nakamura, N., Kageshima, M., Tokumoto, H. and Miyake, J. (2003) Lithographing of biomolecules on a substrate surface using an enzyme-immobilized AFM tip. *Nanoletters*, **3**, 1471–1474.

16 Wouters, D. and Schubert, U.S. (2004) Nanolithography and nanochemistry: probe-related patterning techniques and chemical modification for nanometer-sized devices. *Angewandte Chemie-International Edition*, **43**, 2480–2495.

17 Eigler, D.M. and Schweizer, E.K. (1990) Positioning single atoms with a scanning tunneling microscope. *Nature*, **344**, 524–526.

18 Stroscio, J.A. and Eigler, D.M. (1991) Atomic and molecular manipulation with the scanning tunneling microscope. *Science*, **254**, 1319–1326.

19 Crommie, M.F., Lutz, C.P. and Eigler, D.M. (1993) Confinement of electrons to quantum corrals on a metal surface. *Science*, **262**, 218–220.

20 Lyo, I.-W. and Avouris, Ph. (1991) Field-induced nanometer- to atomic-scale manipulation of silicon surfaces with the STM. *Science*, **253**, 173–176.

21 Uchida, H., Huang, D.H., Yoshinobu, J. and Aono, M. (1993) Single atom manipulation on the Si(111)7×7 surface by the scanning tunneling microscope (STM). *Surface Science*, **287/288** (Part 2), 1056–1061.

22 Bartels, L., Meyer, G. and Rieder, K.-H. (1997) Basic steps of lateral manipulation of single atoms and diatomic clusters with a scanning tunneling microscope tip. *Physical Review Letters*, **79**, 697–700.

23 Stroscio, J.A., Tavazza, F., Crain, J.N., Celotta, R.J. and Chaka, A.M. (2006) Electronically-induced atom motion in engineered CoCu$_n$ nanostructures. *Science*, **313**, 948–951.

24 Jung, T.A., Schlitter, R.R., Gimzewski, J.K., Tang, H. and Joachim, C. (1996) Controlled room-temperature positioning of individual molecules: molecular flexure and motion. *Science*, **271**, 181–184.

25 Cuberes, M.T., Schlittler, R.R. and Gimzewski, J.K. (1996) Room-temperature repositioning of individual C$_{60}$ molecules at Cu steps: operation of a molecular counting device. *Applied Physics Letters*, **69**, 3016–3018.

26 Maruno, S., Inanaga, K. and Isu, T. (1993) Threshold height for movement of molecules on Si(111)-7×7 with a scanning tunneling microscope. *Applied Physics Letters*, **63**, 1339–1341.

27 Beton, P.H., Dunn, A.W. and Moriarty, P. (1995) Manipulation of C$_{60}$ molecules on a Si surface. *Applied Physics Letters*, **67**, 1075–1077.

28 Israelachvili, J.N. (1992) *Intermolecular, and Surface Forces*. 2nd edn., Academic Press, San Diego, CA.

29 Burnham, N.A., Chen, X., Hodges, C.S., Matei, G.A., Thoreson, E.J., Roberts, C.J., Davies, M.C. and Tendler, S.J.B. (2003) Comparison of calibration methods for atomic-force microscopy cantilevers. *Nanotechnology*, **14**, 1–6.

30 Garcia, R. and San Paulo, A. (1999) Attractive and repulsive tip-sample interaction regimes in tapping-mode atomic force microscopy. *Physical Review B-Condensed Matter*, **60**, 4961–4967.

31 Garcia, R. and San Paulo, A. (2000) Amplitude curves and operating regimes in dynamic atomic force microscopy. *Ultramicroscopy*, **82**, 79–83.

32 Sarid, S. (1994) *Scanning Force Microscopy*, Oxford University Press, Oxford, UK.

33 Waser, R. (ed.) (2003) *Nanoelectronics, and Information Technology*, Wiley-VCH, Weinheim, Germany.
34 Meyer, E., Hug, H.J. and Bennewitz, R. (2004) *Scanning Probe Microscopy*, Springer Verlag, Heidelberg, Germany.
35 Villarrubia, J.S. (1994) Morphological estimation of tip geometry for scanned probe microscopy. *Surface Science*, **321**, 287–300.
36 Baur, C., Gazen, B.C., Koel, B., Ramachandran, T.R., Requicha, A.A.G. and Zini, L. (1997) Robotic nanomanipulation with a scanning probe microscope in a networked computing environment. *Journal of Vacuum Science & Technology B*, **15**, 1577–1580.
37 Baur, C., Bugacov, A., Koel, B.E., Madhukar, A., Montoya, N., Ramachandran, T.R., Requicha, A.A.G., Resch, R. and Will, P. (1998) Nano-particle manipulation by mechanical pushing: underlying phenomena and real-time monitoring. *Nanotechnology*, **9**, 360–364.
38 Bugacov, A., Resch, R., Baur, C., Montoya, N., Woronowicz, K., Papson, A., Koel, B.E., Requicha, A.A.G. and Will, P. (1999) Measuring the tip-sample separation in dynamic force microscopy. *Probe Microscopy*, **1**, 345–354.
39 Requicha, A.A.G., Baur, C., Bugacov, A., Gazen, B.C., Koel, B., Madhukar, A., Ramachandran, T.R., Resch, R. and Will, P. (1998) Nanorobotic assembly of two-dimensional structures. Proceedings IEEE International Conference on Robotics and Automation (ICRA '98), Leuven, Belgium, May 16–21, pp. 3368–3374.
40 Requicha, A.A.G., Meltzer, S., Terán Arce, P.F., Makaliwe, J.H., Sikén, H., Hsieh, S., Lewis, D., Koel, B.E. and Thompson, M.E. (2001) Manipulation of nanoscale components with the AFM: principles and applications. Proceedings 1st IEEE International Conference on Nano-technology, Maui, HI, October 28–30, pp. 81–86.
41 Resch, R., Bugacov, A., Baur, C., Koel, B.E., Madhukar, A., Requicha, A.A.G. and Will, P. (1998) Manipulation of nanoparticles using dynamic force microscopy: simulation and experiments. *Applied Physics A*, **67**, 265–271.
42 Resch, R., Meltzer, S., Vallant, T., Hoffmann, H., Koel, B.E., Madhukar, A., Requicha, A.A.G. and Will, P. (2001) Immobilizing Au nanoparticles on SiO_2 surfaces using octadecylsiloxane monolayers. *Langmuir*, **17**, 5666–5670.
43 Resch, R., Lewis, D., Meltzer, S., Montoya, N., Koel, B.E., Madhukar, A., Requicha, A.A.G. and Will, P. (2000) Manipulation of gold nanoparticles in liquid environments using scanning force microscopy. *Ultramicroscopy*, **82** (1–4), 135–139.
44 Requicha, A.A.G. (2003) Nanorobots, NEMS and nanoassembly. Proceedings IEEE, Special issue on nanoelectronics and nanoscale processing, Vol. 91, No. 11, November, pp. 1922–1933.
45 Taylor, R.M. II, Chen, J., Okimoto, S., Llopis-Artime, N., Chi, V.L., Brooks, F.P. Jr., Falvo, M., Paulson, S., Thiansathaporn, P., Glick, D., Washburn, S. and Superfine, R. (1997) Pearls found on the way to the ideal interface for scanned-probe microscopes. Proceedings IEEE Visualization '97, Phoenix, AZ, October 19–24, pp. 467–470.
46 Schaefer, D.M., Reifenberger, R., Patil, A. and Andres, R.P. (1995) Fabrication of two-dimensional arrays of nanometer-size clusters with the atomic force microscope. *Applied Physics Letters*, **66**, 1012–1014.
47 Junno, T., Deppert, K., Montelius, L. and Samuelson, L. (1995) Controlled manipulation of nanoparticles with an atomic force microscope. *Applied Physics Letters*, **66**, 3627–3629.
48 Martin, M., Roschier, L., Hakonen, P., Parts, U., Paalanen, M., Schleicher, B. and Kauppinen, E.I. (1998) Manipulation of Ag nanoparticles utilizing noncontact atomic force microscopy. *Applied Physics Letters*, **73**, 1505–1507.

49 Sheehan, P.E. and Lieber, C.M. (1996) Nanotribology and nanofabrication of MoO$_3$ structures by atomic force microscopy. *Science*, **272**, 1158–1161.

50 Guthold, M., Falvo, M.R., Matthews, W.G., Paulson, S., Washburn, S., Erie, D.A., Superfine, R., Brooks, F.P. Jr. and Taylor, R.M. II. (June 2000) Controlled manipulation of molecular samples with the nanoManipulator. *IEEE/ASME Transactions on Mechatronics*, **5** (2), 189–198.

51 Sitti, M. and Hashimoto, H. (June 2000) Controlled pushing of nanoparticles: modeling and experiments. *IEEE/ASME Transactions on Mechatronics*, **5** (2), 199–211.

52 Li, G., Xi, N., Yu, M. and Fung, W.K. (2003) 3-D nanomanipulation using atomic force microscopy. Proceedings IEEE International Conference on Robotics and Automation (ICRA '03), Taipei, Taiwan, September 14–19, pp. 3642–3647.

53 Theil Hansen, L., Kühle, A., Sørensen, A.H., Bohr, J. and Lindelof, P.E. (1998) A technique for positioning nanoparticles using an atomic force microscope. *Nanotechnology*, **9**, 337–342.

54 Vikramaditya, B. and Nelson, B.J. (1997) Visually guided microassembly using optical microscopes and active vision. Proceedings IEEE International Conference on Robotics and Automation, Albuquerque, NM, April 21–27, pp. 3172–3177.

55 Sato, T., Kameya, T., Miyazaki, H. and Hatamura, Y. (1995) Hand-eye system in the nano manipulation world. Proceedings IEEE International Conference on Robotics and Automation, Nagoya, Japan, May 21–27, pp. 59–66.

56 Miyazaki, H. and Sato, T. (1997) Mechanical assembly of three-dimensional microstructures from fine particles. *Advanced Robotics*, **11**, 169–185.

57 Yu, M.-F., Dyer, M.J., Skidmore, G.D., Rohrs, H.W., Lu, X.-K., Hausman, K.D., von Her, J.R. and Ruoff, R.S. (1999) Three dimensional manipulation of carbon nanotubes under a scanning electron microscope. *Nanotechnology*, **10**, 244–252.

58 Dong, L., Arai, F. and Fukuda, T. (2001) 3D nanorobotic manipulation of multi-walled carbon nanotubes. Proceedings IEEE International Conference on Robotics & Automation, Seoul, S. Korea, May 21–26, pp. 632–637.

59 Fatikow, S., Wich, T., Hülsen, H., Sievers, T. and Jähnisch, M. (2006) Microrobot system for automatic nanohandling inside a scanning electron microscope. Proceedings IEEE International Conference on Robotics & Automation (ICRA '06), Orlando, FL, May 15–19, pp. 1401–1407.

60 Resch, R., Baur, C., Bugacov, A., Koel, B.E., Madhukar, A., Requicha, A.A.G. and Will, P. (1998) Building and manipulating 3-D and linked 2-D structures of nanoparticles using scanning force microscopy. *Langmuir*, **14**, 6613–6616.

61 Decossas, S., Mazen, F., Baron, T., Brémond, G. and Souifi, A. (2003) Atomic force microscopy nanomanipulation of silicon nanocrystals for nanodevice fabrication. *Nanotechnology*, **14**, 1272–1278.

62 Requicha, A.A.G. (1999) Nanoparticle patterns. *J Nanoparticle Res*, **1**, 321–323.

63 Burmeister, F., Schäfle, C., Keilhofer, B., Bechinger, C., Boneberg, J. and Leiderer, P. (1988) From mesoscopic to nanoscopic structures: lithography with colloid monolayers. *Advanced Materials*, **10**, 495–497.

64 Tada, T., Kanayama, T., Koga, K., Seeger, K., Carroll, S.J., Weibel, P. and Palmer, R.E. (1998) Fabrication of size-controlled 10-nm scale Si pillars using metal clusters as formation nuclei. *Microelectronic Eng*, **41/42**, 539–542.

65 Lewis, P.A. and Ahmed, H. (1999) Patterning of silicon nanopillars formed with a colloidal gold etch mask. *Journal of*

Vacuum Science & Technology B, **17**, 3239–3243.

66 Zheng, J., Chen, Z. and Liu, Z. (2000) Atomic force microscopy-based nanolithography on silicon using colloidal Au nanoparticles as a nanooxidation mask. *Langmuir*, **16**, 9673–9676.

67 Chou, S.Y., Krauss, P.R. and Renstrom, P.J. (1996) Imprint lithography with 25-nanometer resolution. *Science*, **272**, 85–87.

68 Junno, T., Carlsson, S.-B., Xu, H., Montelius, L. and Samuelson, L. (1998) Fabrication of quantum devices by Ångström-level manipulation of nanoparticles with an atomic force microscope. *Applied Physics Letters*, **72**, 548–550.

69 Maier, S.A., Brongersma, M.L., Kik, P.G., Meltzer, S., Requicha, A.A.G., Koel, B.E. and Atwater, H.A. (2001) Plasmonics – a route to nanoscale optical devices. *Advanced Materials*, **13**, 1501–1505.

70 Maier, S.A., Kik, P.G., Atwater, H.A., Meltzer, S., Harel, E., Koel, B.E. and Requicha, A.A.G. (2003) Local detection of electromagnetic energy transport below the diffraction limit in metal nanoparticle plasmon waveguides. *Nature Materials*, **2**, 229–232.

71 Mokaberi, B. and Requicha, A.A.G. (2006) Drift compensation for automatic nanomanipulation with scanning probe microscopes. *IEEE Transactions on Automation Science and Engineering*, **3**, 199–207.

72 Hsieh, S., Meltzer, S., Wang, C.R.C., Requicha, A.A.G., Thompson, M.E. and Koel, B.E. (2002) Imaging and manipulation of gold nanorods with an Atomic Force Microscope. *The Journal of Physical Chemistry B*, **106**, 231–234.

73 Chen, H., Xi, N. and Li, G. (2006) CAD-guided automated nanoassembly using atomic force microscopy-based nanorobotics. *IEEE Transactions on Automation Science and Engineering*, **3**, 208–217.

74 Hertel, T., Martel, R. and Avouris, Ph. (1998) Manipulation of individual carbon nanotubes and their interaction with surfaces. *The Journal of Physical Chemistry B*, **102**, 910–915.

75 Falvo, M.R., Taylor, R.H. II, Helser, A., Chi, V., Brooks, F.P. Jr., Washburn, S. and Superfine, R. (1999) Nanometre-scale rolling and sliding of carbon nanotubes. *Nature*, **397**, 236–238.

76 Dong, L.X., Arai, F. and Fukuda, T. (2001) Three-dimensional nanoassembly of multi-walled carbon nanotubes through nanorobotic manipulations by using electron-beam induced deposition. Proceedings 1st IEEE International Conference on Nanotechnology, Maui, HI, October 28–30, pp. 93–98.

77 Dong, L., Arai, F., Nakajima, M., Liu, P. and Fukuda, T. (2003) Nanotube devices fabricated in a nano laboratory. Proceedings IEEE International Conference on Robotics & Automation, Taipei, Taiwan, September 24–29, pp. 3624–3629.

78 Roschier, L., Penttilä, J., Martin, M., Hakonen, P., Paalanen, M., Tapper, U., Kauppinen, E., Journet, C. and Bernier, P. (1999) Single-electron transistor made of multi-walled carbon nanotube using scanning probe manipulation. *Applied Physics Letters*, **75**, 728–730.

79 Resch, R., Baur, C., Bugacov, A., Koel, B.E., Echternach, P.M., Madhukar, A., Montoya, N., Requicha, A.A.G. and Will, P. (1999) Linking and manipulation of gold multi-nanoparticle structures using dithiols and scanning force microscopy. *The Journal of Physical Chemistry. B*, **103**, 3647–3650.

80 Requicha, A.A.G., Resch, R., Montoya, N., Koel, B.E., Madhukar, A. and Will, P. (1999) Towards hierarchical nanoassembly. Proceedings International Conference on Intelligent Robots and Systems (IROS '99), Kyongju, S. Korea, October 17–21, pp. 889–893.

81 Meltzer, S., Resch, R., Koel, B.E., Thompson, M.E., Madhukar, A., Requicha,

A.A.G. and Will, P. (2001) Fabrication of nanostructures by hydroxylamine-seeding of gold nanoparticle templates. *Langmuir*, **17**, 1713–1718.

82 Harel, E., Meltzer, S.E., Requicha, A.A.G., Thompson, M.E. and Koel, B.E. (2005) Fabrication of latex nanostructures by nanomanipulation and thermal processing. *Nanoletters*, **5**, 2624–2629.

83 Madsen, D.N., Mølhave, K., Mateiu, R., Bøggild, P., Rasmussen, A.M., Appel, C.C., Brorson, M. and Jacobsen, C.J.H. (2003) Nanoscale soldering of positioned carbon nanotubes using highly conductive electron beam induced gold deposition. Proceedings IEEE International Conference on Nanotechnology, S. Francisco, CA, August 12–14, pp. 335–338.

84 Hollis, R.L., Salcudean, S. and Abraham, D.W. (1990) Toward a tele-nanorobotic manipulation system with atomic scale force feedback and motion resolution. Proceedings IEEE International Conference on Microelectromechanical Systems, Napa Valley, CA, February 11–14, pp. 115–119.

85 Taylor, R.M. II, Robinett, W., Chi, V.L., Brooks, F.P. Jr., Wright, W.V., Williams, R.S. and Snyder, E.J. (1993) The nanomanipulator: a virtual reality interface for a scanning tunneling microscope. Proceedings ACM SIGGRAPH '93, Anaheim, CA, August 1–6, pp. 127–134.

86 Finch, M., Chi, V.L., Taylor, R.M. II, Falvo, M., Washburn, S. and Superfine, R. (1995) Surface modification tools in a virtual environment interface to a scanning probe microscope. Proceedings ACM Symposium on Interactive 3D Graphics, Monterey, CA, April 9–12, pp. 13–18.

87 Sitti, M. and Hashimoto, H. (1998) Tele-nanorobotics using atomic force microscope. Proceedings IEEE/RSJ International Conference on Intelligent Robots and Systems (IROS '98), Victoria, Canada, October 13–17, pp. 1739–1746.

88 Li, G., Xi, N., Yu, M. and Fung, W.K. (2003) Augmented reality system for real-time nanomanipulation. Proceedings IEEE International Conference on Nanotechnology, S. Francisco, CA, August 12–14, pp. 64–67.

89 Li, G., Xi, N., Yu, M. and Fung, W.K. (June 2004) Development of augmented reality system for AFM-based nanomanipulation. *IEEE/ASME Transactions on Mechatronics*, **9** (2), 358–365.

90 Mokaberi, B., Yun, J., Wang, M. and Requicha, A.A.G. (2007) Automated nanomanipulation with atomic force microscopes. Proceedings IEEE International Conference on Robotics and Automation (ICRA '07), Rome, Italy, April 10–14, pp. 1406–1412.

91 Knuth, D.E. (1993) *The Stanford GraphBase*, The ACM Press, New York, NY.

92 Latombe, J.-C. (1991) *Robot Motion Planning*, Kluwer, Boston, MA.

93 Makaliwe, J.H. and Requicha, A.A.G. (2001) Automatic planning of nanoparticle assembly tasks. Proceedings IEEE International Symposium on Assembly & Task Planning (ISATP '01), Fukuoka, Japan, May 28–30, pp. 288–293.

94 Mokaberi, B. and Requicha, A.A.G. (2004) Towards automatic nanomanipulation: drift compensation in scanning probe microscopy. Proceedings IEEE International Conference on Robotics and Automation (ICRA '04), New Orleans, LA, April 25–30, pp. 416–421.

95 Mokaberi, B. and Requicha, A.A.G. (in press) Compensation of scanner creep and hysteresis for AFM nanomanipulation. *IEEE Transactions on Automation Science & Engineering*. doi:10.1109/TASE.2007.895008.

96 Arbuckle, D.J., Kelly, J. and Requicha, A.A.G. (2006) A high-level nanomanipulation control framework. Proceedings International Advanced Robotics Programme (IARP) Workshop on Micro and Nano Robotics, Paris, France, October 23–24,

97 Requicha, A.A.G. (1999) Massively parallel nanorobotics for lithography and data storage. *International Journal of Robotics Research*, **18**, 344–350.

98 Vettiger, P., Cross, G., Despont, M., Drechsler, U., Dürig, U., Gotsmann, B., Häberle, W., Lantz, M.A., Rothuizen, H.E., Stutz, R. and Binnig, G.K. (March 2002) The millipede – nanotechnology entering data storage. *IEEE Transactions on Nanotechnology*, **1** (1), 39–55.

99 Arbuckle, D.J. and Requicha, A.A.G. (2003) Massively parallel scanning probe nanolithography. Proceedings, 3rd IEEE International Conference on Nanotechnology, San Francisco, CA, August 12–14, pp. 72–74.

9
Harnessing Molecular Biology to the Self-Assembly of Molecular-Scale Electronics

Uri Sivan

9.1
Introduction

Microelectronics and biology provide two distinct paradigms for complex systems. In microelectronics, the information guiding the fabrication process is encoded into computer programs or glass masks and, based on that information, a complex circuit is imprinted in silicon in a series of chemical and physical processes. This top-to-bottom approach is guided by a supervisor whose "wisdom" is external to the circuit being built. Biology adopts an opposite strategy, whereby complex constructs are assembled from molecular-scale building blocks, based on the information encoded into the ingredients. For example, proteins are synthesized from amino acids based on the instructions coded in the genome and other proteins. The assembled objects process further molecules to form larger structures capable of executing elaborate functions, and so on. This autonomous bottom-up strategy allows, in critical bottlenecks, for an exquisite control over the molecular structure in a way which is unmatched by man-made engineering. In other cases it allows for the errors that are so critical for evolution.

The fact that man-made engineering evolved so differently from "nature engineering" deserves a separate discussion that is beyond the scope of this chapter. Here, we will only comment that the perception of nature as a type of engineering is somewhat oversimplifying. While engineering aims at meeting a predefined challenge – namely, to execute a desired function – nature evolved with no aim. Yet, the hope behind biomimetics is that concepts and tools which evolved during several billions years of evolution may find applications in engineering.

Electronics is particularly alien to biology. With the exception of short-range electron hopping in certain proteins, biology relies on ion transport rather than electrons. The electronic conductivity of biomolecules is orders of magnitude too

---

Nanotechnology. Volume 3: Information Technology I. Edited by Rainer Waser
Copyright © 2008 WILEY-VCH Verlag GmbH & Co. KGaA, Weinheim
ISBN: 978-3-527-31738-7

small for implementing them as useful electronic components. For instance, albeit in earlier reports, DNA has been found to be an excellent insulator [1–3]. The foreseen potential of biology in the context of electronics is, therefore, in the assembly process rather than in electronic functionality *per se*. This observation is reflected in the scientific research described below; it concerns the bioassembly of electronic materials to form devices, rather than attempts to use biomolecules as electronic components.

The term "self-assembly" is widely used to describe a variety of processes which include the self-assembly of organic molecules to form uniform monolayers on substrates. This is not the type of self-assembly under consideration in this chapter, whereby the term refers to the construction of an elaborate object, namely, the embedment of a significant amount of information into the object being built. The subject of the intimate relationship between self-assembly, information, and complexity will be revisited in Section 9.4.

The term "complex self-assembly" deserves some introductory remarks. When looking back at nature, one realizes that complex objects are typically assembled in a modular way. Most protein machines, for instance, comprise several subunits, each made of a separate protein. Each such protein is synthesized in the cell from amino acids which are in turn synthesized from atoms. This example is identified in four levels of hierarchy, namely atoms, amino acids, proteins, and machines made of several protein subunits. This hierarchal or modular assembly is an essential ingredient of complex self-assembly, the reason being that none of the modules reflects a global minimal free energy of its elementary constituents. The protein machine, for instance, does not pertain to a minimal free energy of the collection of amino acids making it, and so on.

In many instances the system is guided to a certain configuration by auxiliary molecules (enzymes, chaperones, etc.) which at times consume energy. However, in the cases of interest here, where self-assembly is governed by non-covalent interactions and relatively simple configurations, each step can be driven by a down-hill drift in free energy towards a long-lived metastable state, thus rendering the module amenable for the next assembly step. Clearly, complex electronics cannot be assembled from its elementary building blocks in a single step, and so requires modular assembly.

The next comment concerns the unavoidable errors characterizing self-assembly. In order for molecular recognition to take place, the molecules should effectively explore multiple docking configurations with other parts of the target molecule or with other molecules. The free energy landscape corresponding to the collection of all such configurations should, therefore, facilitate thermally assisted hops between local minima, corresponding to "wrong" configurations, in addition to the desired configuration. Special measures must be devised in order to produce overwhelming discrimination in favor of the desired configuration at finite time experiments. In the absence of such measures, the yield of self-assembly is intrinsically limited by the same fluctuations that facilitate molecular recognition. Over time, biology has evolved sophisticated error suppression and correction tools, and equivalent methods will have to be developed if the self-assembly of molecular-scale electronics is to be taken seriously.

The effect of errors on modular assembly is of particular importance. As the yield in each assembly step is less than perfect, faulty modules are produced. An uncontrolled modular assembly therefore inevitably produces an exponentially larger fraction of faulty modules as the levels of hierarchy accumulate. One strategy for making useful circuits may thus rely on circuit architectures that are tolerant to faults. One such outstanding example is embodied in the Teramac machine developed at HP laboratories [4]. In the present chapter we adhere to conventional architectures requiring near-perfect circuits. The faulty modules in each step therefore need to be identified and either repaired or eliminated. Within the context of electronic circuits built by biology, the identification of faulty devices and their removal presents a remarkable challenge; that of devising a biomolecular machine that tests non-biological devices for electronic functionality, filters out non-functional devices, and then signals the system to proceed to the next assembly step. Although significant progress has been made towards the isolation of antibodies that sense the electric output presented to them by an electronic device, the discussion of electro-bio feedback loops is deferred to future publications, and focus here is on free-running assembly.

The conjecture behind the experiments described in this chapter may be summarized as follows. Simple functional devices can be assembled efficiently from electronic materials, taking advantage of the remarkable assembly tools provided by molecular biology. The realization of elaborate constructs necessitates hierarchical modular assembly, while the inevitable accumulation of errors with increasing levels of hierarchy requires error suppression and correction mechanisms, as well as biomolecular feedback switches that judge for electronic functionality and feedback to the bioassembly process.

In Section 9.2 the concept is of DNA-templated electronics [1, 5] is introduced, and expanded to include sequence-specific molecular assembly [6, 7] based on the recombinant protein, RecA. This section culminates in the bioassembly of a fully functional field effect transistor made from a carbon nanotube (CNT) [8]. While the topics of Section 9.2 rely on existing biotechnological tools, in Section 9.3 the toolbox is expanded to demonstrate how to isolate antibodies that recognize electronic materials directly [9]. The fact such antibodies can be isolated is encouraging with respect to the prospect of realizing a functional interface between molecular biology and nanoelectronics [10].

As DNA-templated electronics requires long DNA molecules with unique addresses, advantage is then taken of DNA computing to demonstrate the autonomous synthesis of DNA templates having these properties [11]. The synthesis algorithm, as outlined in Section 9.4, relies on the chemical realization of shift registers (SRs), and incorporates an error suppression scheme inspired by the redundancy codes employed in data communication. To the best of the present author's knowledge, these chemical SRs constitute the first embodiment of error suppression codes in chemical synthesis.

9.2
DNA-Templated Electronics

9.2.1
Scaffolds and Metallization

Double-stranded DNA (dsDNA) is chosen in most cases to template the assembly of molecular-scale electronics as well as other constructs of non-biological functionality. dsDNA is mechanically and chemically stable, easy to obtain at any desired sequence, and readily amenable to diverse enzymatic manipulations including restriction, digestion, replication, ligation, and recombination. In the schemes described below, dsDNA doubles as the information-carrying molecule and the physical support for the assembled electronic materials.

The assembly of DNA-templated electronics comprises two steps. First, the biological machinery is employed to construct a DNA scaffold with well-defined molecular addresses. Then, electronic functionality is instilled by the localization of electronic devices at specific addresses along the scaffold and conversion of the DNA template into a conductive network interconnecting devices to each other and to the external world.

An heuristic solution to some of the major challenges faced by molecular electronics, namely, the precise localization of a large number of molecular devices, inter-device wiring, and electrical interface between the molecular and macroscopic worlds, is depicted in Figure 9.1(a–d). The first step involves the definition of macroscopic electrodes on an inert substrate. As the electrodes are macroscopic, this process can be performed using standard photolithographic techniques (Figure 9.1a). The electrodes are provided with an identity by covering each of them

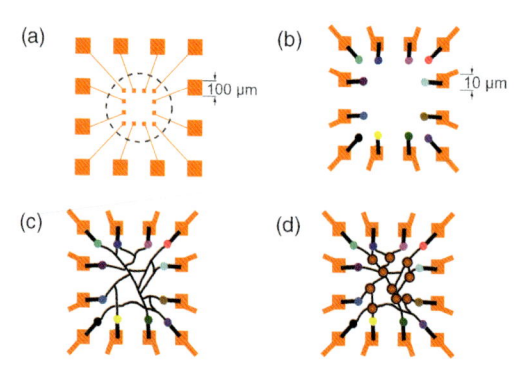

Figure 9.1 Heuristic scheme of a DNA-templated electronic circuit. (a) Gold pads are defined on an inert substrate. Panels (b–d) correspond to the circle of (a) at different stages of circuit construction. (b) Oligonucleotides of different sequences are attached to the different pads. (c) DNA network is constructed and bound to the oligonucleotides on the gold electrodes. (d) Metal clusters or molecular electronic devices are localized on the DNA network. The DNA molecules are finally converted into metallic wires, rendering the construct into a functional electronic circuit. Note that the figures are not to scale; the metallic clusters are nanometer-sized, while the electrode pads are micrometer-sized.

with a monolayer of a different short, single-stranded oligonucleotide using, for example, an ink-jet printer (Figure 9.1b). This step may still involve physical manipulations as the electrodes to be covered are macroscopic. After this step, each electrode is labeled with a monolayer of a unique oligonucleotide sequence and, hence, is able to recognize a specific complementary sequence in solution. In the third step, a network of well-defined connectivity is assembled using DNA hybridization and recombinant processes (see below). The network is then localized on the substrate using, for example hybridization of DNA molecules with the electrode-bound oligonucleotides (Figure 9.1c). The previous steps instill the formerly uniform substrate with well-defined molecular addresses based on distinct sequences of DNA molecules. This allows the subsequent positioning of functional electronic elements at molecularly accurate addresses (Figure 9.1d). At the end of this step, the network should bear functional elements at predesigned sites. However, as DNA molecules have insulating properties, the network should be functionalized (e.g., metallized) in order to render it conductive.

Now, the questions to be asked are how accurate is the topology of the assembled DNA network? Can it be inspected for structural integrity? Was the template assembled properly on the electrodes? Were the devices localized at their planned destinations? Did the devices connect electrically? Are they functional? These are just a few of the questions that must be addressed in any specific attempt to self-assemble molecular-scale electronics.

The experimental procedure used to demonstrate DNA-templated assembly and electrode attachment of a conductive silver wire [1, 5] are depicted in Figure 9.2. First, 12-base oligonucleotides, derivatized with a disulfide group at their 3′ end, were attached to the electrodes through a thiol–gold interaction. Each of the two electrodes was marked with a different oligonucleotide sequence. The electrodes were then bridged by hybridization of a 16 µm-long λ-DNA molecule containing two 12-base-long sticky ends, each of which was complementary to one of the two sequences attached to the gold electrodes.

The inset to Figure 9.3 presents a single DNA molecule bridge as observed by fluorescence microscopy. The measurements on the stretched DNA molecules indicated a resistance higher than the internal resistance of the measurement apparatus ($>10^{13}\,\Omega$). It was therefore concluded that, in order to instill electrical functionality, the DNA bridge must be coated with metal. Albeit contradicting results reported previously in the literature, it is now widely accepted that the intrinsic conductivity of DNA is indeed too small for direct application as a conducting element in a circuit [2].

The three-step silver-coating process (Figure 9.2c–e) was based on the selective localization of silver ions along the DNA molecule through Ag^+/Na^+ ion-exchange [1, 5], and the formation of complexes between the silver and the DNA bases. The silver ion-exchanged DNA was then reduced to form nanometer-sized silver aggregates bound to the DNA skeleton. These aggregates were further "developed" (much as in a standard photographic procedure) by using an acidic solution of hydroquinone and silver ions under low-light conditions [12, 13]. This solution was metastable, and spontaneous metal deposition was normally very slow,

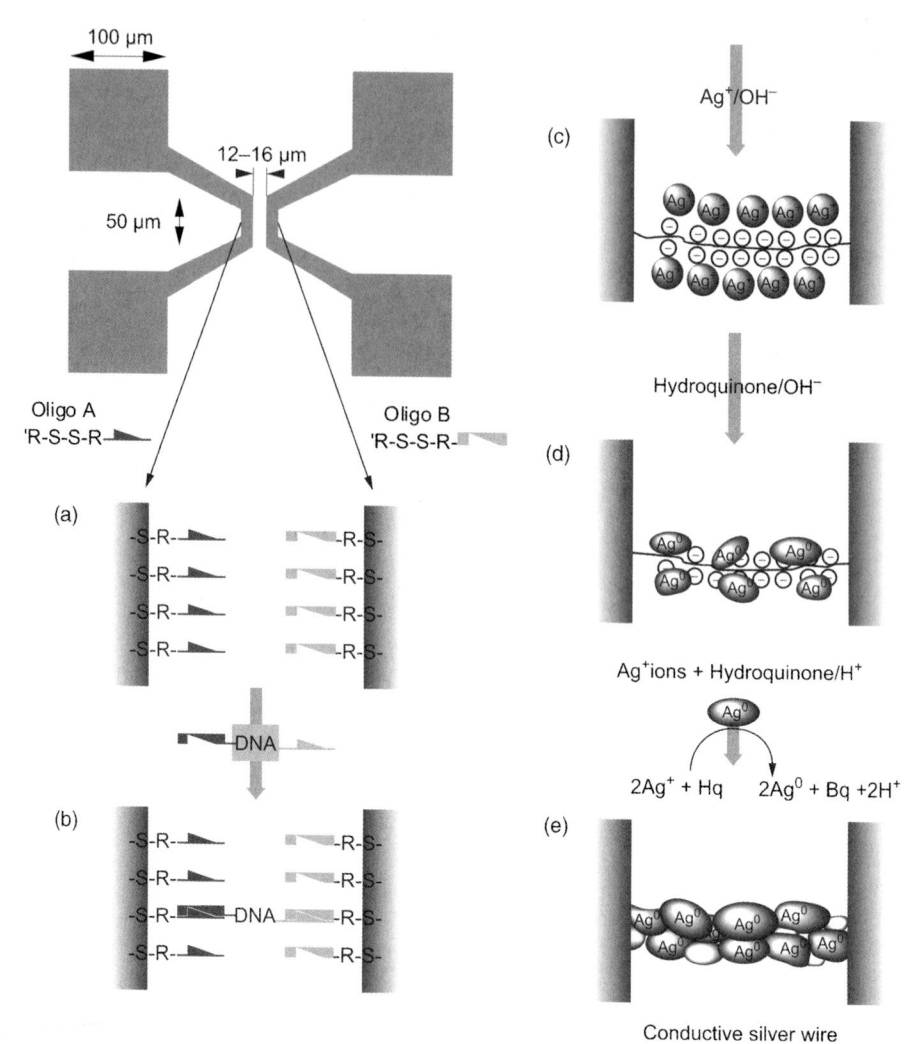

Figure 9.2 A gold pattern, 0.5 × 0.5 mm in size, was defined on a passivated glass using microelectronics techniques. The pattern comprised four bonding pads, each 100 μm in size, connected to two 50 μm-long parallel gold electrodes, 12–16 μm apart. (a) The electrodes were each wetted with a 10^{-4} μL droplet of disulfide-derivatized oligonucleotide solution of a given sequence (Oligos A and B). (b) After rinsing, the structure was covered with 100 μL of a solution of λ-DNA having two sticky ends that are complementary to Oligos A and B. A flow was applied to stretch the λ-DNA molecule between the two electrodes, allowing its hybridization. (c) The DNA bridge was loaded with silver ions by Na^+/Ag^+ ion exchange. (d) The silver ion–DNA complex was reduced using a basic hydroquinone solution to form metallic silver aggregates bound to the DNA skeleton. (e) The DNA templated wire was "developed" using an acidic solution of hydroquinone and silver ions. (Reprinted from Ref. [1]; © Nature, 1998.).

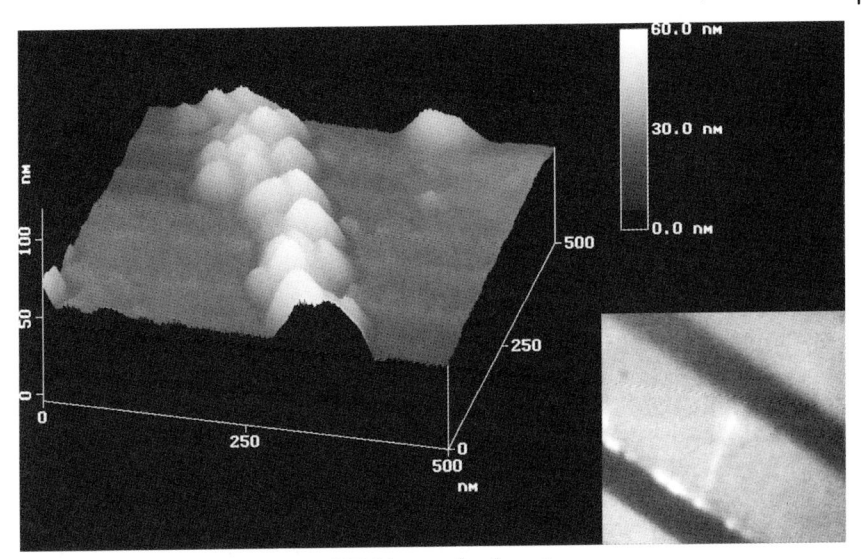

Figure 9.3 Atomic force microscopy (AFM) image of a silver wire connecting two gold electrodes 12 μm apart. Field size = 0.5 μm. Inset: Fluorescently labeled λ-DNA molecule stretched between two gold electrodes (dark strips), 16 μm apart. (Reprinted from Ref. [1]; © Nature, 1998.).

except on the silver aggregates attached to the DNA catalyzed the process. Under the experimental conditions, metal deposition therefore occurred only along the DNA skeleton, leaving the passivated substrate practically clean of silver. An atomic force microscopy (AFM) image of a segment of a 100 nm-wide, 12 μm-long silver wire prepared in this way is shown in Figure 9.3.

Since the publication of Ref. [1], the metallization scheme has been improved in two essential ways. First, silver has been replaced with gold [6] in the enhancing step and, after a few hours sintering at 300 °C, excellent wires were obtained. Second, the hydroquinone has been substituted for glutaraldehyde [6, 14] localized on the DNA itself. The confinement of the reducing agent to the DNA molecule suppressed nonspecific metal deposition on other objects in the system, leading to much cleaner circuits. A DNA-templated gold wire is depicted in the inset of Figure 9.4, together with its current–voltage (I–V) characteristics.

Other research groups have since extended the scope of the metallization of biomolecules to proteins, amyloid fibrils, protein S-layers, microtubules, actin fibers, and even complete viral particles. Today, the choice of metals includes Pd, Pt, Au, Cu, and Co. An account of biomolecules metallization can be found in Refs. [15–20].

9.2.2
Sequence-Specific Molecular Lithography

In analogy with photolithography in conventional microelectronics, the realization of DNA-templated devices and circuits requires tools for defining circuit architectures.

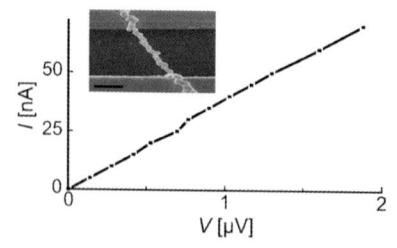

Figure 9.4 Two-terminal current–voltage (I–V) curve of a DNA-templated gold wire. The resistivity of the wire ($1.5 \times 10^{-7}\ \Omega \cdot m$) was only seven-fold higher than that of polycrystalline gold ($2.2 \times 10^{-8}\ \Omega \cdot m$). Inset: Scanning electron microscopy (SEM) image of a typical DNA-templated gold wire stretched between two electrodes deposited by electron-beam lithography. Scale bar = 1 µm. (Reprinted from Ref [6]; © 2002, American Association for the Advancement of Science.).

These include the formation of rich geometries, wire patterning at molecular resolutions, and molecularly accurate device localization. To that end, "sequence-specific molecular lithography" has been developed which enables the elaborate manipulation of dsDNA molecules, including patterning of the metal coating of DNA, the localization of labeled molecular objects at arbitrary addresses on dsDNA, and the generation of molecularly accurate stable DNA junctions [6, 8, 14].

The molecular lithography system developed at Technion utilizes homologous genetic recombination processes carried out by the RecA protein from *Escherichia coli*. The patterning information encoded in the DNA molecules replaces the masks used in conventional photolithography, while the RecA protein serves as the resist. The molecular lithography functions at high resolution over a broad range of length scales, from nanometers to many micrometers.

Homologous genetic recombination is one of several mechanisms that cells use to manipulate their DNA [21]. In this process, two parental DNA molecules which possess some sequence homology cross-over at equivalent sites. The reaction is based on protein-mediated, sequence-specific DNA–DNA interaction. Although RecA is the major protein responsible for this process in *E. coli*, it is also able to carry out the essential steps of the recombination process *in vitro*.

In the present author's procedure, RecA monomers are polymerized on a probe single-stranded DNA (ssDNA) molecule to form a nucleoprotein filament (Figure 9.5, step i). The nucleoprotein filament binds to a substrate molecule at an homologous probe–substrate location (Figure 9.5, step ii). RecA allows the addressing of an arbitrary sequence, from as few as 15 bases [22] to many thousands of bases, by the same standard reaction. This versatility presents an advantage over DNA-binding proteins which are restricted to particular DNA sequences. Moreover, unlike DNA hybridization, sequence-specific recognition can be performed on dsDNA, rather than ssDNA. Being chemically more inert and mechanically more rigid, the former provides a better substrate than the latter. The high efficiency and specificity of the recombination reaction, which evidently is essential for its biological roles, are beneficial in its utilization for molecular lithography.

(i) Polymerization

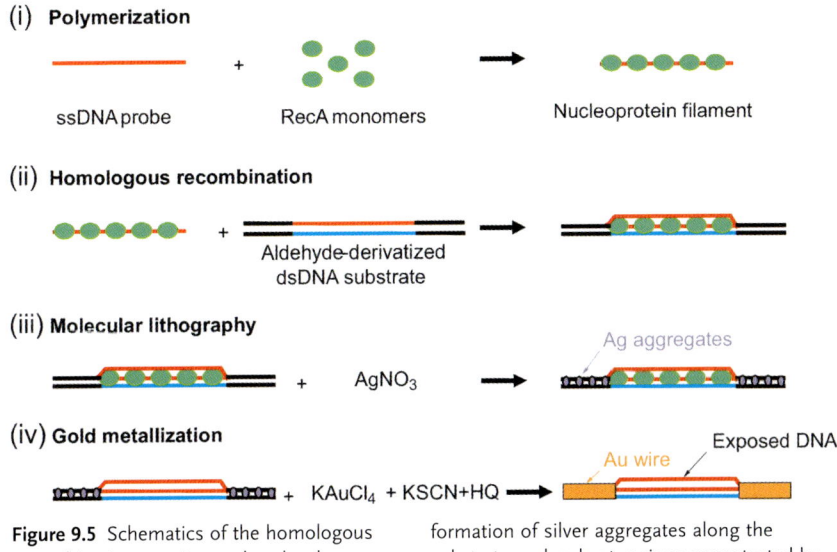

ssDNA probe + RecA monomers → Nucleoprotein filament

(ii) Homologous recombination

+ Aldehyde-derivatized dsDNA substrate →

(iii) Molecular lithography

+ AgNO$_3$ → Ag aggregates

(iv) Gold metallization

+ KAuCl$_4$ + KSCN+HQ → Au wire / Exposed DNA

Figure 9.5 Schematics of the homologous recombination reaction and molecular lithography. (i) RecA monomers polymerize on a ssDNA probe molecule to form a nucleoprotein filament. (ii) The nucleoprotein filament binds to an aldehyde-derivatized dsDNA substrate molecule at an homologous sequence. (iii) Incubation in AgNO$_3$ solution results in the formation of silver aggregates along the substrate molecule at regions unprotected by RecA. (iv) The silver aggregates catalyze specific gold deposition on the unprotected regions. A highly conductive gold wire is formed with a gap in the protected segment. (Reprinted from Ref [6]; © 2002, American Association for the Advancement of Science.).

The application of sequence-specific molecular lithography to the definition of a patterned gold wire is outlined in Figures 9.5 and 9.6. Here, the previously described DNA metallization scheme is employed in which DNA-bound glutaraldehyde is used as a localized reducing agent [6], and the RecA is used as a sequence-specific resist. RecA monomers polymerize on a single-stranded probe DNA to form a nucleoprotein filament (Figure 9.5, step i) which locates and binds to a homologous sequence on a dsDNA molecule (Figure 9.5, step ii). Once bound, the RecA in the nucleoprotein filament acts as a sequence-specific resist, physically protecting the aldehyde-derivatized substrate DNA against silver cluster formation in the bound region (Figure 9.5, step iii). Subsequent gold metallization leads to the growth of two extended DNA-templated wires separated by the predesigned gap (Figure 9.5, step iv).

Figure 9.6 depicts images of the products of the various steps leading to a patterned gold-coated λ-DNA. Extensive AFM and scanning electron microscopy (SEM) imaging confirmed that the metallization gap was located where expected. The position and size of the insulating gap could be tailored by choosing the probe's sequence and length. The ability to pattern DNA metallization facilitates modular circuit design, and is therefore valuable for the realization of DNA-templated electronics. Insulating and conducting regions can be defined on the DNA scaffold according to the underlying sequence, thus determining the electrical connectivity in the circuit. In addition, patterning DNA metallization is useful for the integration of molecular objects into a circuit. Such objects can be localized and electrically

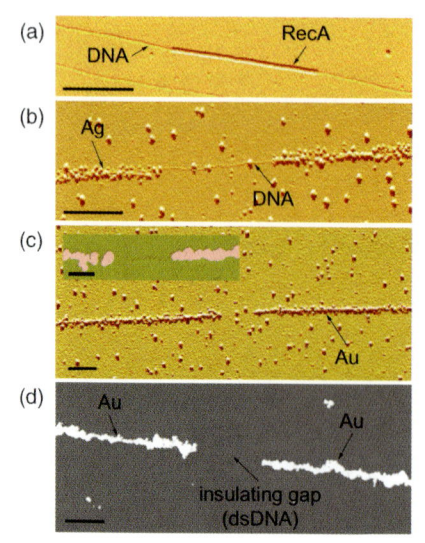

Figure 9.6 Sequence-specific molecular lithography on a single DNA molecule. (a) AFM image of a 2027-base RecA nucleoprotein filament bound to an aldehyde-derivatized λ-DNA substrate molecule. (b) AFM image of the sample after silver deposition. Note the exposed DNA at the gap between the silver-loaded sections. (c) AFM image of the sample after gold metallization. Inset: zoom on the gap. The height of the metallized sections is ∼50 nm. (d) SEM image of the wire after gold metallization. All scale bars = 0.5 μm; inset to (c) = 0.25 μm. The variation in the gap length is due mainly to variability in DNA stretching on the solid support. The very low background metallization in the SEM image compared with the AFM images indicates that most of the background is insulating. (Reprinted from Ref [6]; © 2002, American Association for the Advancement of Science.).

contacted within the exposed DNA sequences present in the unmetallized gaps. Further manipulations of DNA templates including the localization of man-made objects at specific addresses along the DNA molecule, the generation of three- and four-armed junctions, and elaborate metallization patterning can be found in Refs. [6, 7, 14, 16].

9.2.3
Self-Assembly of a DNA-Templated Carbon Nanotube Field-Effect Transistor

The superb electronic properties of CNTs [23], their large aspect ratio, and their inertness with respect to the DNA metallization process, make them an ideal choice for the active elements in DNA-templated electronics. The ability to localize molecular objects at any desired address along a dsDNA molecule and to pattern sequence-specifically the DNA metallization (as described above) facilitate the incorporation of CNTs into DNA-templated functional devices, and their wiring. In the assembly of the field-effect transistor (FET), a DNA scaffold molecule provided the address for the precise localization of a semiconducting single-wall carbon nanotube (SWNT), and templated the extended wires contacting it. The localization of the SWNT relied on homologous recombination by the RecA protein. The assembly of the SWNT–FET, which is shown schematically in Figure 9.7, employed a three-strand homologous recombination reaction between a long dsDNA molecule serving as a scaffold and a short auxiliary ssDNA. The short ssDNA molecule was synthesized so that its sequence was identical to the dsDNA at the designated location of the FET. RecA

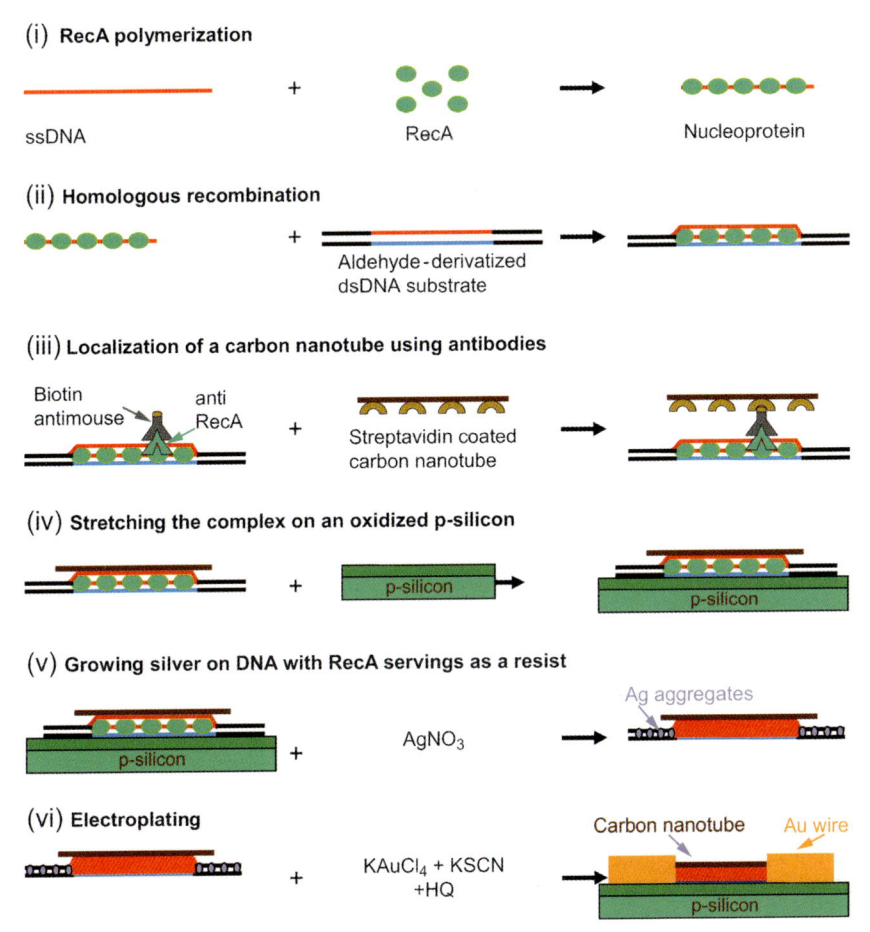

Figure 9.7 Assembly of a DNA-templated FET and wires contacting it. Steps are as follows. (i) RecA monomers polymerize on a ssDNA molecule to form a nucleoprotein filament. (ii) Homologous recombination reaction leads to binding of the nucleoprotein filament at the desired address on an aldehyde-derivatized scaffold dsDNA molecule. (iii) The DNA-bound RecA is used to localize a streptavidin-functionalized single-wall carbon nanotube (SWNT), utilizing a primary antibody to RecA and a biotin-conjugated secondary antibody. (iv) The complex is stretched on an oxidized p-type silicon wafer by dipping the substrate in a solution containing the complexes and pulling it out. (v) Incubation in an $AgNO_3$ solution leads to the formation of silver clusters on the segments that are unprotected by RecA. (vi) Electroless gold deposition, using the silver clusters as nucleation centers, results in the formation of two DNA-templated gold wires contacting the SWNT bound at the gap.

proteins were first polymerized on the auxiliary ssDNA molecules to form nucleoprotein filaments (Figure 9.7, step i), which were then mixed with the scaffold dsDNA molecules. The nucleoprotein filament bound to the dsDNA molecule according to the sequence homology between the ssDNA and the designated address on the dsDNA (Figure 9.7, step ii). The RecA later served to localize a SWNT at that address and to protect the covered DNA segment against metallization. A streptavidin-functionalized SWNT was guided to the desired location on the scaffold dsDNA molecule using antibodies to the bound RecA and biotin–streptavidin-specific binding (Figure 9.7, step iii). The SWNTs were solubilized in water by micellization in sodium dodecyl sulfate (SDS) [24] and functionalized with streptavidin by non-specific adsorption [25, 26].

Primary anti-RecA antibodies were reacted with the product of the homologous recombination reaction, and this resulted in specific binding of the antibodies to the RecA nucleoprotein filament. Next, biotin-conjugated secondary antibodies, having high affinity to their primary counterparts, were localized on the primary anti-RecA antibodies. Finally, the streptavidin-coated SWNTs were added, leading to their localization on the RecA via biotin–streptavidin-specific binding (Figure 9.7, step iii). The DNA/SWNT assembly was then stretched on a passivated oxidized silicon wafer. An AFM image of a SWNT bound to a RecA-coated 500-base-long ssDNA localized at the homologous site in the middle of a scaffold λ-DNA molecule is shown in Figure 9.8a. The conducting CNT can be clearly distinguished from the insulating DNA by the use of scanning conductance microscopy [27, 28]. The topographic and conductance images of the same area are depicted in Figure 9.8b and c, respectively. The evident difference between the two images identifies the SWNT on the DNA molecule. It should be noted that the CNT is aligned with the DNA, which is almost always the case due to the stiffness of the SWNT and the stretching process.

Following stretching on the substrate, the scaffold DNA molecule was metallized. The RecA, doubling as a sequence-specific resist, protected the active area of the transistor against metallization. The metallization scheme described above was employed, in which aldehyde residues, acting as reducing agents, were bound to the scaffold DNA molecules by reacting the latter with glutaraldehyde. Highly conductive metallic wires were formed by silver reduction along the exposed parts of the aldehyde-derivatized DNA (Figure 9.7, step v) and subsequent electroless gold plating using the silver clusters as nucleation centers (Figure 9.7, step vi). As the SWNT was longer than the gap dictated by the RecA, the deposited metal covered the ends of the nanotube and contacted it. A SEM image of an individual SWNT contacted by two DNA-templated gold wires is depicted in Figure 9.8d.

The extended DNA-templated gold wires were contacted by electron-beam lithography, and the device was characterized by direct electrical measurements under ambient conditions. The p-type substrate was used to gate the transistor. The electronic characteristics of the device are shown in Figure 9.9a and b. The gating polarity indicated p-type conduction of the SWNT, as is usually the case with semiconducting CNTs in air [29]. The saturation of the drain-source current for negative gate voltages indicated resistance in series with the SWNT; this

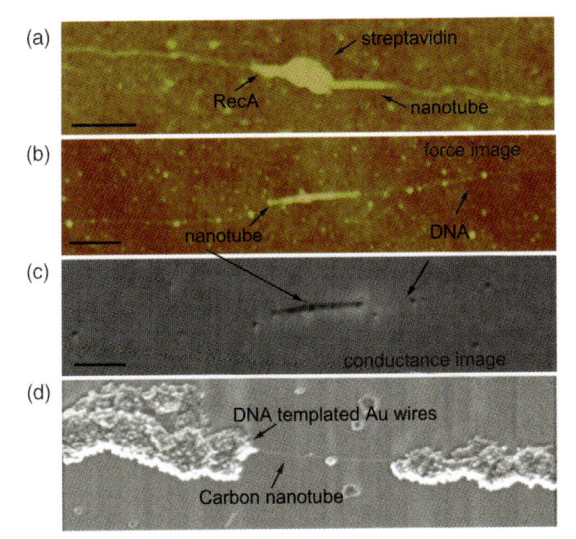

Figure 9.8 Localization of a single-wall carbon nanotube (SWNT) at a specific address on the scaffold dsDNA molecule using RecA. (a) An AFM image of a 500-base-long (~250 nm) RecA nucleoprotein filament localized at a homologous sequence on a λ-DNA scaffold molecule. Scale bar = 200 nm. (b) An AFM image of a streptavidin-coated SWNT bound to a 500-base-long nucleoprotein filament localized on a λ-DNA scaffold molecule. Scale bar = 300 nm. (c) A scanning conductance image of the same region as in (b). The conductive SWNT yields a considerable signal, whereas the insulating DNA is hardly resolved. Scale bar = 300 nm. (d) SEM image of the resulting device. The DNA-templated gold wires and the assembled nanotube are indicated by arrows. The DNA molecule itself is not resolved in this image.

resistance was attributed to the contacts between the gold wires and the SWNT as the four-terminal resistance of the DNA-templated gold wires was typically smaller than 100 Ω. Each of the different devices had somewhat different turn-off voltages.

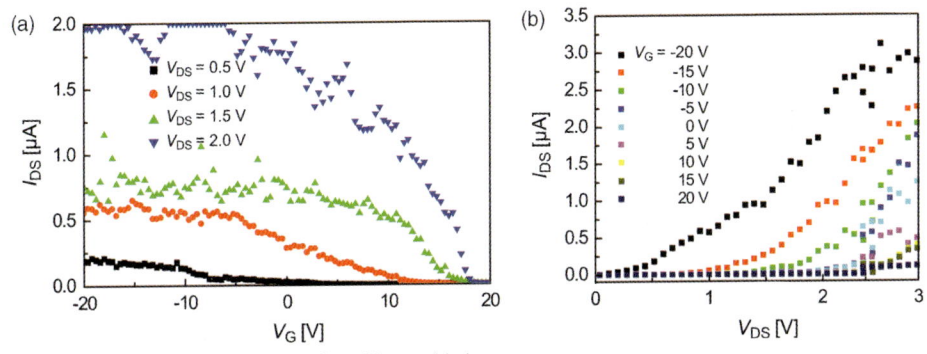

Figure 9.9 Electrical characteristics of a self-assembled p-type field effect transistor based on a semiconducting single wall carbon nanotube. (a) Drain-source current versus gate bias applied between the p-type substrate and the source electrode. (b) Same versus drain-source voltage for different gate voltages.

9.3
Recognition of Electronic Surfaces by Antibodies

The self-assembly described in Section 9.2 relies on existing biotechnological tools, but in this section the toolbox is expanded to show how to isolate antibody molecules that recognize electronic surfaces directly. As a specific example, the isolation of antibody molecules capable of discriminating between different crystalline facets of a GaAs crystal is reviewed. Beyond the potential application of such antibodies for the direct localization of molecular-scale objects at desired sites on an electronic substrate, the success in isolating these antibodies is encouraging with regards to the prospects of isolating antibodies that can "read" electrical signals presented to them by electronic devices. The latter constitute a critical milestone on the way to *functional* integration between molecular biology and nanoelectronics.

The mammalian immune system offers a vast repertoire of antibody molecules capable of binding, in selective manner, an immense number of molecules presented to the body by invading pathogens such as bacteria, viruses, and parasites. Although this repertoire has evolved to target mostly biomolecules, it may potentially contain selective binders to other targets, or it may be expanded to include such binders. Indeed, the injection into mice of cholesterol and 1,4-dinitrobenzene [30, 31] microscopic crystals, as well as C_{60} conjugated to bovine thyroglobulin [32] have resulted in the generation of antibodies against these materials by the immune system of the injected animal. Here, the scope of the system is expanded, and it is shown that human antibody libraries – specifically, single-chain Fv (scFv; [33]), which are the antibody variable binding domains – contain specific binders, capable of discriminating between different crystalline facets of a GaAs semiconductor crystal, which is an almost flat target and unfamiliar to the immune system. This selectivity is remarkable given the very simple structure of semiconductors compared with biomolecules.

By using phage display technology, the *in-vitro* isolation of scFv that bind GaAs (1 1 1A) facets almost 100-fold better than GaAs (1 0 0). is demonstrated. More generally, this finding implies that antibody molecules may find application in the assembly of nanoelectronics [1, 6, 8], in the production of templates for localizing nanoparticles [34], or for biosensors [35].

The isolation of short peptides that bind inorganic materials has been demonstrated for gold [36, 37], silver [38], silica [39], metal oxides [40, 41], minerals [42], CNTs [43], and various semiconductors [44–46]. Of these reports, Ref. [46] is particularly relevant to the present section, as the authors report the isolation of peptides (by phage display) that bind GaAs (1 0 0) preferentially to GaAs (1 1 1A) and (1 1 1B). However, all assays in Ref. [46] probed the peptides displayed on the phages rather than the free peptides. Indeed, when one of these peptides was later synthesized and applied to GaAs [47], no selectivity was found between the (1 0 0) and (1 1 0) facets (see Figure 5 in Ref. [47] and the following discussion).

As this discrepancy was difficult to comprehend, attention was towards studying the non-specific binding of M13 phages (which carried no peptides or antibodies) to GaAs (1 0 0), GaAs (1 1 1A) and GaAs (1 1 1B). As a consequence, M13 was found to bind preferentially to the (1 0 0) facet through its coat protein (Figure 1S, supplementary material to Ref. [9]). As those phages were identical to the library phages used in Ref. [46], and given the lack of selectivity displayed by the only free peptide tested thus far [47], it seemed that further experiments with free peptides would be needed in order to either confirm or disprove semiconductor facet recognition by short peptides. In contrast to Ref. [46], the present antibodies were also tested and found to be selective towards crystal orientation when detached from the phage.

The 7- and 12-mer peptides used in most *in-vitro* selections of binders to inorganic crystals are typically too short to assume a stable structure. Antibodies on the other hand, display a rigid three-dimensional (3-D) structure which is potentially essential for high-affinity selective binding [30, 31]. Moreover, the recognition site in the latter case involves six amino acid sequences grouped into three complementarity-determining regions (CDR). All together, these CDRs form a large, structured binding site spanning up to 3×3 nm. The critical role of the antibody 3-D structure for the recognition of organic crystal facets is well established [30, 31].

Another hint to the importance of rigidity for facet recognition is provided by the rigid structure characterizing antifreeze peptides that target specific ice facets [48]. It has also been shown that the stable helical structure of a 31-mer peptide catalyzing calcite crystallization is essential for inducing directed crystal growth along a preferred axis [49], possibly due to its differential binding to the various facets. Hence, structure rigidity may turn central to facet recognition by biomolecules, thereby underscoring the importance of antibody libraries as a promising source for selective binders.

Selective binding to specific crystalline facets can be directly utilized for numerous micro- and nanotechnological applications, including the positioning of nanocrystals at a well-defined orientation, governing crystal growth and forcing it to certain directions [49], and positioning nanometer-scale objects at specific sites on a substrate marked by certain crystalline facets. An application of one of these soluble antibodies to the latter task is demonstrated in Figure 9.10. By using conventional photolithography and $H_3PO_4 : H_2O_2 : H_2O$ etching, a long trench has been defined on a GaAs (1 0 0) substrate in the (1 1 0) direction (Figure 9.10a). Due to the slow etching rate of phosphoric acid in the (1 1 1A) direction, the process leads to slanted (1 1 1A) side walls and a flat (1 0 0) trench floor (Figure 9.10a). A SEM image of a cut across the trench, and proving that the slanted walls are indeed tilted in the (1 1 1A) direction (54.7° relative to the (1 0 0) direction), is depicted in Figure 9.10b. When the isolated scFv antibodies are applied to the GaAs substrate they attach themselves selectively to the (1 1 1A) slopes.

In order to image the bound antibody molecules, they were targeted with anti-human secondary antibodies conjugated to a fluorescent dye, Alexa Fluor. As shown in Figure 9.10c, fluorescence is limited solely to the (1 1 1A) slopes with practically no background signal coming from the (1 0 0) surfaces. Control experiments depleted of the scFv fragments exclude possible artifacts such as the natural fluorescence of the (1 1 1A) facets, selective binding of the fluorescent dye, or secondary antibodies to that facet.

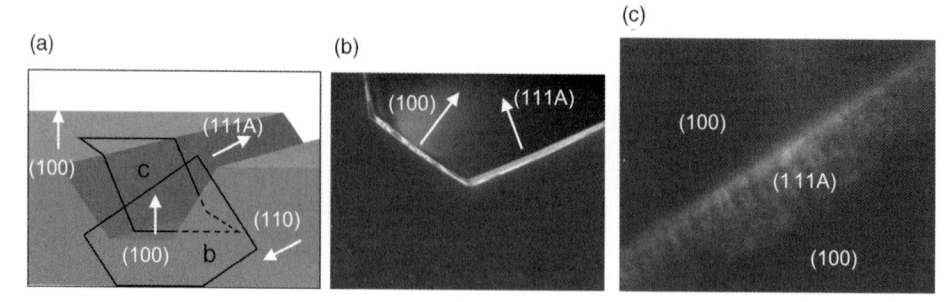

Figure 9.10 (a) Diagrams of the etched trench labeled with the various crystalline facets. The black frames correspond to the views depicted in panels (b) and (c). (b) SEM image of a cut across the trench. (c) Fluorescence image of the trench viewed from the top. Fluorescence is confined to the (1 1 1A) slopes, proving selective binding of the scFv fragments to that facet. Note the negligible binding of antibody molecules to the (1 0 0) facets. (Reprinted from Ref. [9]; © 2006, American Chemical Society.).

The images in Figure 9.10 prove that the selected scFv antibody molecules recognize and bind selectively GaAs (1 1 1A) as opposed to GaAs (1 0 0). As such, they can be used to localize practically any microscopic object on (1 1 1A) surfaces, with negligible attachment to other crystalline facets. The isolation of such binders using phage display technology, and the quantification of their selectivity, is described in the following section.

The Ronit1 scFv antibody phage library [50] used in the present study, is a phagemid library [51] comprising 2×10^9 different human semi-synthetic single-chain Fv fragments, where in-vivo-formed CDR loops were shuffled combinatorially onto germline-derived human variable region framework regions of the heavy (V_H) and light (V_L) domains.

To select scFv binders to GaAs (1 1 1A), approximately 10^{11} phages (≈100 copies of each library clone) were applied to the semiconductor crystal (panning step). After washing the unbound phages, the bound units were recovered by rinsing the sample in an alkaline solution. The recovered viruses were then quantified by infecting bacteria and plating dilution series on Petri dishes. The amplified sublibrary was applied again to the target crystal facet, and so on. Typically, three to four panning rounds were required to isolate excellent binders to the target. As is evident from Figure 9.11, the number of bound phages retrieved from the semiconductor grew 300-fold when panning was repeated three times. For comparison, the non-specific binding of identical phages (M13) carrying no scFv fragments remained low throughout the selection process. It was found experimentally that blocking with milk was essential to prevent the non-specific binding of phages to the GaAs targets. Interestingly, as shown in the supplementary material to Ref. [9], in the absence of blocking against non-specific binding (a step missing in Ref. [46]), the non-specific binding of phages through their coat protein to GaAs (1 0 0) was larger than to GaAs (1 1 1A). The data in Figure 9.11 prove the selection of increasingly better binders to GaAs (1 1 1A), but provide no indication of selectivity with respect to GaAs (1 0 0).

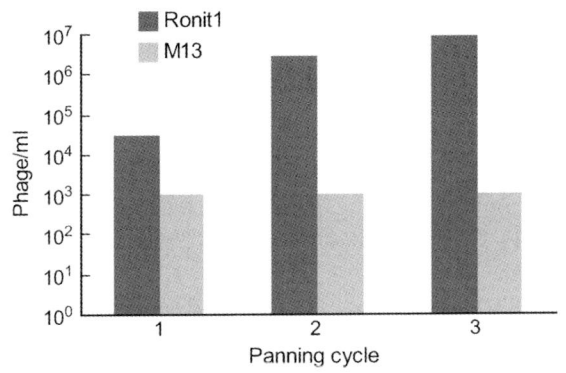

Figure 9.11 Enrichment of anti-GaAs (1 1 1A) phages carrying scFv fragments versus panning cycle. Phage concentration has been deduced by counting colonies of E. coli bacteria infected with different dilutions of the phages recovered after each cycle. The monotonic increase in binding of phages carrying scFv (Ronit1) is contrasted with the much weaker, non-specific binding of similar phages lacking the scFv antibody. The value of the latter (1000 phages mL^{-1}) sets an experimental upper limit on their binding; the actual values are likely to be smaller. (Reprinted from Ref. [9]; © 2006, American Chemical Society.).

Indeed, as indicated by the two left-hand columns of Figure 9.12, application of the polyclonal population of binders selected on GaAs (1 1 1A) to GaAs (1 0 0) shows similar binding to the latter crystalline facet. Hence, the process described above produced good, but non-selective, binders.

Preferential binding to a given crystalline facet was achieved by a slight modification of the process. The phages recovered from the first panning on GaAs (1 1 1A)

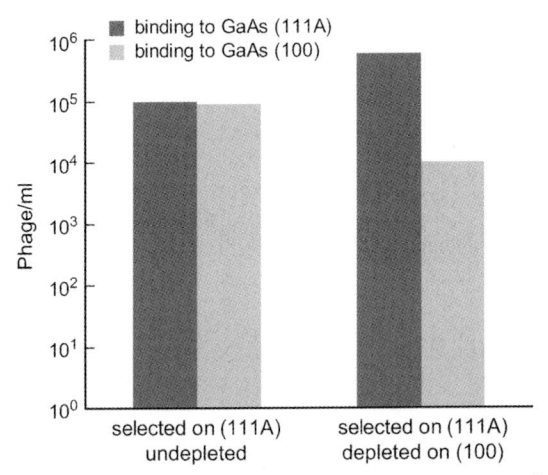

Figure 9.12 Density of recovered binders to GaAs (1 1 1A) after three panning cycles. The two right-hand (or left-hand) columns correspond to selection on GaAs (1 1 1A) with (or without) depletion on GaAs (1 0 0). (Reprinted from Ref. [9]; © 2006, American Chemical Society.).

were amplified in *E. coli* and then applied to GaAs (1 0 0). However, this time the *unbound* phages were collected and applied in a second panning step to GaAs (1 1 1A). As evident from the two right-hand columns of Figure 9.12, the "depletion" step on GaAs (1 0 0) is enriched for specific phage clones that both bind GaAs (1 1 1A) and lack binding to GaAs (1 0 0). On this occasion, binding of the selected phages to the (1 1 1A) facet was almost 100-fold higher than to the (1 0 0) facet. This depletion step, which was crucial to the present case, was missing in Ref. [46].

The polyclonal population of selected phages contains different scFv fragments, each characterized by different affinity and selectivity to the two crystalline facets. In order to correlate specificity with sequence, the binding selectivity of the individual clones was next analyzed. Monoclonal binders were isolated by infecting *E. coli* bacteria with the sublibrary and plating them on solid agar. As each bacterium can be infected by a single phage, all bacteria within a given colony carry DNA coding for the same scFv fragment. Infection of the colony with helper phages resulted in the release of phages displaying the same scFv on their PIII coat proteins. The isolated monoclonal phages were then analyzed with ELISA against GaAs (1 1 1A) and (1 0 0). The sequences of the light (V_L) and heavy (V_H) CDRs of ten monoclonal binders that were identified by the ELISA assay can be found in Ref. [9] and its supplementary material, together with a discussion of their main features.

Figures 9.11 and 9.12 correspond to the scFv fragments displayed on phage particles. For practical applications (such as that demonstrated in Figure 9.10) it is preferable to have soluble monoclonal scFv fragments detached from the phage coat proteins. The results of the ELISA assays of the scFv fragment of Figure 9.10, in its soluble form, are presented in Figure 9.13.

In Figure 9.13, bars 1–6 correspond to the six ELISA assays on GaAs (1 1 1A) and GaAs (1 0 0) pieces, each of 4×4 mm. After washing the substrates, the bound antibodies were reacted with anti-human horseradish peroxidase (HRP), and the binding was quantified by adding tetramethylbenzidine (TMB) as a colorimetric substrate, and reading the resulting optical density (OD) at 450 nm. Bars 7–9

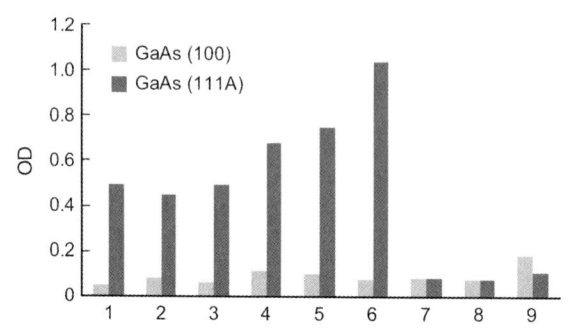

Figure 9.13 Bars 1–6 display the results of six comparative ELISA assays of the scFv molecule (detached from the phage) on GaAs (1 1 1A) and GaAs (1 0 0) substrates. The optical density (OD) reflects the number of bound molecules in arbitrary units. Bars 7–9 display the results of three control experiments (see text), and can be used to estimate the background signal (ca. 0.1 OD) coming from sources, other than selective binding of the scFv to the semiconductor substrates. (Reprinted from Ref. [9]; © 2006, American Chemical Society.).

provided the following controls. Bars 7 quantified the non-specific binding of the secondary anti-human HRP to the ELISA plate in the absence of the EB scFv and semiconductor substrates. Bars 8 corresponded to the non-specific binding of the scFv to the plate, and bars 9 to non-specific binding of the secondary antibodies to the semiconductor substrates.

The background ELISA signal, depicted by bars 7–9, accounts for most of the GaAs (1 0 0) signal in columns 1–6. When subtracting this background from columns 1 to 6, a remarkable preference is found to GaAs (1 1 1A) compared with (1 0 0). Interestingly, the binding of the secondary antibody to GaAs (1 0 0) was almost twice as large compared to its binding to GaAs (1 1 1A), in opposition to the selectivity of the isolated scFv fragments. Overall, the data in Figure 9.13 prove that the isolated scFv preserves its selectivity also when detached from the phage.

Little is known of the interaction between biomolecules and inorganic surfaces, let alone the recognition of such surfaces by antibody molecules. The GaAs surface is modified by surface reconstruction, oxidation, and possibly other chemical reactions. Moreover, it displays atomic steps and possibly surface defects. It is therefore difficult to estimate how much of the underlying crystalline order manifests itself in the recognition process. Unfortunately, as no experimental tools capable of determining these parameters with atomic resolution exist at present, the recognition mechanism is unclear, except for the accumulating indications of the importance of structural rigidity (as discussed in the introduction to this section). The discrimination between the two crystalline facets may reflect the different underlying crystalline structures, they may stem from the different surface chemistries of the two facets, or they may result from global properties such as atom density and different electronegativity. The latter factor has been found to be important for the differential binding of short peptides to different semiconductors [47]. The abundance of positively charged amino acids in the heavy chain of CDR1 and CDR3 and the light chain of CDR1 may indicate an affinity to the exposed gallium atoms. The negatively charged amino acid in CDR3 V_L (missing in anti-gold scFv isolated from the same library) combined with the positively charged CDR3 V_H may match the polar nature of GaAs.

The recognition of man-made materials by antibodies opens new opportunities for a *functional* interface between biology and nanotechnology, far beyond what was has been exercised to date.

9.4
Molecular Shift-Registers and their Use as Autonomous DNA Synthesizers [11]

9.4.1
Molecular Shift-Registers

The DNA-templated assembly of elaborate circuits requires distinct dsDNA molecules with non-recurring sequences. For the assembly of periodic structures, such as memories, a segment of non-recurring sequences should be replicated to form a periodic molecule, and the synthesis of such molecules presents a remarkable challenge to biotechnology. The two existing strategies for generating long molecules, namely PCR [52] and ligase assembly [53], utilize synthetic oligonucleotides which together span (with overlap) the full length of the desired molecule. Hence, when following any of these strategies, the assembly of an N-base long molecule with distinct p-long segments requires $O(N/p)$ oligonucleotides. These approaches therefore quickly become impractical when a rich variety of distinct molecules or addresses along a given molecule are needed for the construction of an elaborate template for molecular electronics [1, 5]. Motivated by the concept of DNA-templated electronics, the present author and colleagues were therefore forced to invent an exponentially more economic synthesis strategy based on the chemical realization of molecular SRs. The dramatic reduction in synthesis effort by SRs is facilitated by exploiting a novel concept in DNA synthesis; a sliding overlapping reading frame. Rather than the fixed frame that directs segment ligation or polymerization in the two schemes listed above or in hairpin-based DNA logic [54, 55] and programmed mutagenesis [56], the SRs utilize a previously synthesized sequence to dictate synthesis of the next bases. The automaton is an example of DNA computing where the result of the computation (tape) is a useful molecule.

An autonomous binary p-shift register (p-SR) is a computing machine with 2^p internal states represented by an array of p cells (Figure 9.14a), each occupying one bit, $x_i\{i=1\ldots p\}$. In each step a binary function, $f(x_1, x_2, \ldots x_p)$, is computed and its value is inserted into cell p. Simultaneously, x_j is shifted to cell $j-1$; $\{j=2\ldots p\}$. On printing x_1 to a tape, a long periodic binary sequence is generated. Electronic SRs are utilized in many applications including secure communication, small signal recovery, and sequence generation [57]. Here, it is shown that molecular SRs can be realized and utilized for the autonomous synthesis of DNA molecules the sequence of which is uniquely determined by a chemical embodiment of the function $f(x_1, x_2, \ldots x_p)$.

Consider a 3-SR with $x_{n+1}=f(x_{n-2}, x_{n-1}, x_n)=x_{n-2}\oplus x_n$ ($\oplus\equiv$ XOR) and an initial setting (seed) $x_1, x_2, x_3=001$. Repetitive application of f generates the sequence 001110100111010.... The sequence is periodic with a period seven and any of the seven $L\geq 3$ bit long consecutive strings in a period is different from the rest. In general, it is well known [57] that for any p, a SR can be found a with a linear feedback function [58], $f = \sum_{i=1}^{p} \alpha_i x_i; \alpha_i \in \{0, 1\}$ (the sum is mod 2), that generates a sequence

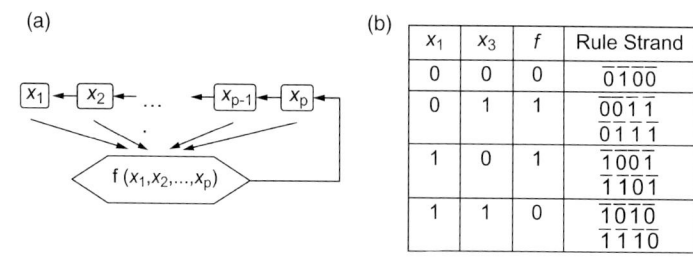

(a)

(b)

x_1	x_3	f	Rule Strand
0	0	0	$\overline{0100}$
0	1	1	$\overline{0011}$
			$\overline{0111}$
1	0	1	$\overline{1001}$
			$\overline{1101}$
1	1	0	$\overline{1010}$
			$\overline{1110}$

Figure 9.14 (a) An autonomous binary p-shift register. (b) Truth table and rule strands corresponding to the first example. (Reprinted from Ref. [11]; © 2006, American Physics Society.).

of maximal period $2^p - 1$ bits with no repetition of strings of lengths $L \geq p$ within a period. Such SRs are termed "maximal linear SRs" as they generate all possible permutations of a p-long sequence except the zero string [59]. Functions, f, can always be found such that the number of non-vanishing α_i is smaller than p (in the example above, $\alpha_2 = 0$). Consequently, $2^p - 1$ different addresses can be generated by an exponentially smaller truth table and, hence, as shown below, by an exponentially smaller synthesis effort compared with direct synthesis of all addresses.

We now show how to implement an autonomous molecular SR using DNA. Imagine a DNA molecule for which the Watson–Crick rules are that 1 binds exclusively to its complementary bit, $\bar{1}$, but not to 1, 0 or $\bar{0}$. Similarly, 0 binds to $\bar{0}$ but not to 0, $\bar{1}$, or 1. We translate the function $f(x_1, x_2, x_3) = x_1 \oplus x_3$ to an equivalent truth table (left three columns in Figure 9.14b) and embody it by the mixture of the seven [59] possible four-bit rule strands, $(\bar{x}_1, \bar{x}_2, \bar{x}_3, \overline{(x_1 \oplus x_3)})$, listed in the right-hand column of Figure 9.14b.

The SR sequence is generated by thermally cycling a mixture containing the seven rule strands, a "seed" strand (e.g., the strand 001), and a polymerase. For simplicity, it is assumed that the rule strands are synthesized with ddDNA at their 3' end and are therefore not elongated in the process. Each cycle comprises annealing, extension, and melting steps. In the first annealing step, some of the first, $\bar{0}\bar{0}\bar{1}\bar{1}$, rule strands bind to seed molecules, leaving an $\bar{1}$ overhang (Figure 9.15, steps i and ii) which is readily copied by the polymerase in the extension step (Figure 9.15, step iii). Next (Figure 9.15, step iv), the temperature is raised to 95 °C and the rule strand dissociates from the elongated seed (tape). In the second annealing step, a $\bar{0}\bar{1}\bar{1}\bar{1}$ rule strand binds to the tape (Figure 9.15, step v), leaving again an $\bar{1}$ overhang which is readily copied by the polymerase (Figure 9.15, step vi). At each additional cycle (Figure 9.15, steps vii–x) some

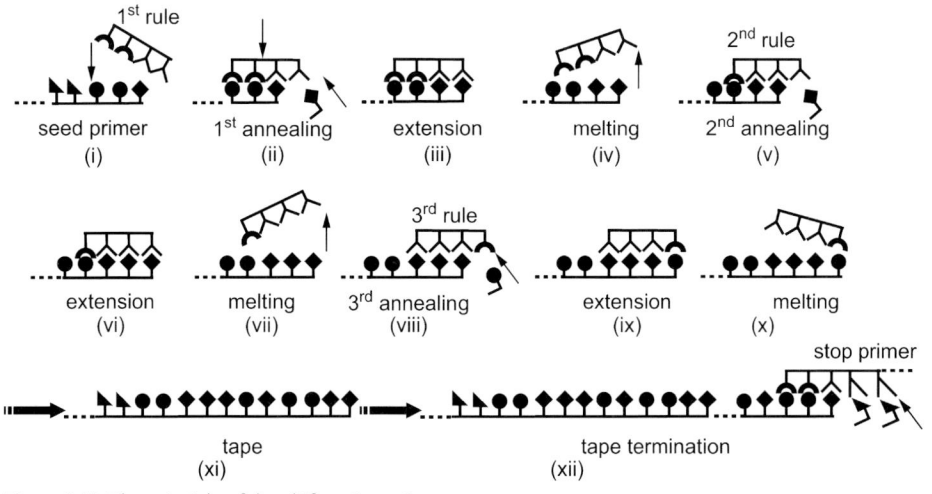

Figure 9.15 The principle of the shift register. ♦ ∧ ● ⌒ represent 1, $\bar{1}$, 0, $\bar{0}$, respectively. ▲ ⋏ represent sequences other than 0 or 1 and their complementary sequences, respectively. (Reprinted from Ref. [11]; © 2006, American Physics Society.).

of the tape molecules are elongated by one bit according to the rule $x_{n+1} = x_n \oplus x_{n-2}$. Elongation is terminated by addition of excess stop primers that intercept the tape molecules as soon as the latter display a desired tail (001 in the example of Figure 9.15 step xii). The polymerase then copies the stop primer and adds its alien sequence to the tape, which is unrecognizable by any rule strand. As a result, elongation terminates. The 5′ seed and 3′ stop primers tails are later used for PCR amplification of the tape. Elongation is guided by a sliding reading frame where all, except the first, shifted bits from the previous reading frame plus a single new bit provide the current reading frame. The sliding frame is the crux of our concept, as it facilitates exponentially smaller synthesis effort compared with any of the previous, fixed-frame approaches.

It should be noted that rule strands are not consumed during synthesis; rather, they only serve as enzymes to direct the reaction. Thus, synthesis in flow may be envisioned, where the rule strands are attached in synthesis order to subsequent segments of a tube or a column. While the reactants flow through the tube the correct sequence is generated, and this strategy is advantageous to straightforward synthesis in a DNA synthesizer as faulty strands are not recognized (and hence not elongated) by rule strands. Clearly, errors are doomed to be short.

Now, an actual demonstration of the concept may be described. In the first implementation each bit is realized by a sequence of three nucleotides, 5′TGC for "0" and 5′GCT for "1". These sequences were chosen as they minimize errors due to one and two base shifts in the annealing step. The demonstration starts with the three-bit maximal SR as discussed above. Such SR requires seven 4-bit (3-bit rules plus one function bit) strands (Figure 9.14b), but in order to suppress synthesis errors longer, redundant 6-bit rules (5-bit rules plus one function bit) are employed. Error suppression by redundancy is discussed in Section 9.4.2. The seven 6-bit rule strands [60] used in the synthesis comprise, 3′$00\bar{1}1\bar{1}0$, 3′$0\bar{1}1\bar{1}01$, 3′$\bar{1}1\bar{1}0\bar{1}0$, 3′$\bar{1}10\bar{1}00$, 3′$\bar{1}0\bar{1}001$, 3′$0\bar{1}00\bar{1}\bar{1}$, 3′$\bar{1}00\bar{1}\bar{1}\bar{1}$. The complementary bits, $\bar{0}$ and $\bar{1}$, correspond to 3′ACG and 3′CGA, respectively. The rule strands are synthesized with three nucleotides only (G, C, A) in order to prevent their extension by polymerase ("poor man's ddDNA"). The 2/3 GC content gives [61] $\Delta G \approx 8.5 \div 10.5\, k_B T$ free energy per bit (stacking included) which in a 5-bit realization of a 3-bit SR translates ideally to suppression of the error rate by a factor proportional to $\exp(-3\Delta G/k_B T) \leq \exp(-25.5)$ (see Section 9.4.2). The seed strand comprises a 5′ tail followed by a 5-bit sequence [62], 5′GCATGCGCCCGTCAGGCG00111. The tail is later used to amplify the SR sequence by PCR. The seed, the rule strands, and three nucleotides (dGTP, dCTP, dTTP) are mixed together and subjected to 45 thermal cycles [63], after which a stop primer, 3′$0\bar{1}00\bar{1}$GACGTC, is added in 10-fold excess compared with each rule strand. During an additional five to ten cycles the tape molecules are further elongated until in some cycle their last five bits read 01001. At that point a stop primer binds to the tape and its complementary sequence is added to the tape by the polymerase. The elongation now terminates as the sequence added by the stop primer is alien to all rule strands. The absence of dATP guarantees single strand synthesis. The expected synthesized sequences read

5′GCATGCGCCCGTCAGGCG00111 $(0100111)_n$ 01001CTGCAG   with $n = 0, 1, \ldots$
  seed primer   complementary to stop primer

$$(9.1)$$

Finally, the elongation products are PCR amplified with two primers, identical to the first 19 nucleotides of the seed (5'GCATGCGCCCGTCAGGCGT) and to the last 19 nucleotides of the stop primer (5'CTGCAGAGCGCAGCAAGCG).

The resulting PCR products, when run against a standard ruler in a polyacrylamide gel, are depicted in Figure 9.16a. Four bands corresponding to Equation 9.1 with $n = 0, 1, 2, 3$ are clearly resolved. Sequencing of the four bands with a primer identical to the first 19 nucleotides on the 5' end of the seed primer proves the bands identification with the respective n values in Equation 9.1. The high fidelity of the automaton is reflected in the perfect matching of the sequencing with Equation 9.1, and the absence of any unexpected bands.

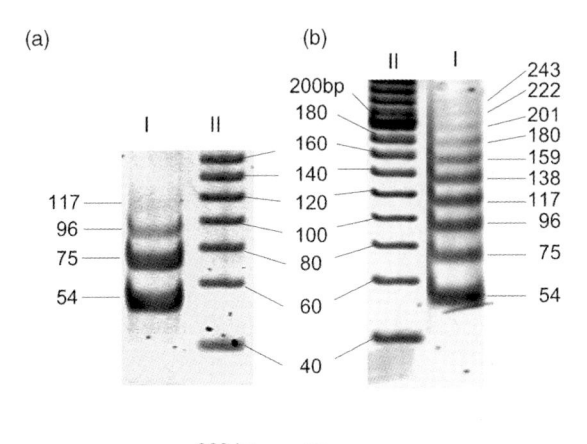

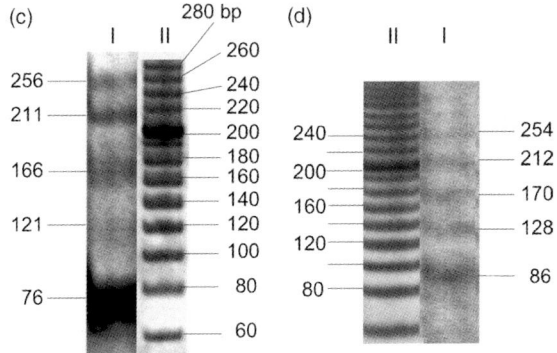

Figure 9.16 (a) Lane I, product after 45 elongation cycles, five cycles with stop primer, and PCR amplification. The four bands correspond to Eq. 9.1 with $n = 0, 1, 2, 3$, namely 54, 75, 96, and 117 base-long sequences. Lane II, ruler. (b) Lane I, same as (a), but with 100 elongation cycles followed by filtering out short sequences (Microcon YM-10; Millipore Corporation, Bedford, MA, USA). Ten bands corresponding to Eq. 9.1 with $n = 0, 1, 2, 3, 4, 5, 6, 7, 8, 9$ are resolved. Lane II, ruler. (c) Four-shift register with 45 bp periodicity realized with 7-bit rule strands; 2 h reaction time at a constant temperature of 72 °C. Lane I, shift register product. The five resolved bands are indicated. Lane II, ruler. (d) Same as (c) for partial 3-shift register with four-letter alphabet. The period comprises 14 bits (42 bp). (Reprinted from Ref. [11]; © 2006, American Physics Society.).

As shown in Figure 9.16b, after 100 elongation cycles it was possible to resolve 10 bands, $n = 0 - 9$, corresponding to 54, 75, 96, 117, 138, 159, 180, 201, 222, and 243 base-long sequences. The automaton thus synthesizes at least 204 bases at a remarkable fidelity. The $n = 9$ sequence comprises 10 periods, each of 21 bases, with exactly one repetition of each 3-bit (or longer) address per period. Direct sequencing of the bands confirmed the results up to $n = 6$. The small material quantities in the higher bands were insufficient for reliable sequencing. As PCR amplification favors shorter sequences, the relative band brightness cannot be taken as a measure for synthesis efficiency of molecules with different n-values.

The synthesis of longer period molecules, as well as of non-binary sequences, is demonstrated in Figure 9.16c and d, and details can be found in Ref. [11]. The synthesis of the last two examples was held in a thermal ratchet mode at a fixed temperature [11].

9.4.2
Error Suppression and Analogy Between Synthesis and Communication Theory

As emphasized in the introduction, errors are intrinsic to molecular assembly. Thus, the invention of error correction and suppression codes is critical for the realization of complex structures. Since the introduction of DNA computing by Adleman [53], the intimate relationship between self-assembly and computation has been slowly revealed. At this point, it may be beneficial to highlight another intriguing link between self-assembly and an engineering concept, this time "communication theory". This link draws an analogy between the synthesis of the long DNA molecule by the SR apparatus and the decoding of a long message transmitted over noisy lines. The addition of a wrong DNA base in synthesis is equivalent in that analogy to the assignment of a wrong value to a bit read in a message. This analogy has far-reaching consequences, as it suggests that some of the powerful strategies developed for suppressing and correcting errors in communication may be adapted to chemical synthesis. One such principle, the addition of degenerate bits to a message, is implemented in the synthesis of DNA molecules by the SRs.

The first stage is to classify any possible synthesis errors. At each annealing step, rule strands other than the correct ones may bind to the tape and affect the SR operation. These events may be divided into two groups: benign, and error. The benign events include all cases where rule strands bind to the tape either with no overhang or with a 2-bit overhang with the correct sequence. In the first case, the particular tape molecule remains idle throughout the cycle, whereas in the latter case it grows by two correct bits. Errors, on the other hand, are generated mostly by annealing of the wrong rule strand, shifted one bit to the right, to form a 2-bit overhang with the wrong sequence. It is easy to verify that, since the maximal register sequence contains all p-bit permutations, an error is manifested in a partial deletion of a period. The same fact guarantees that the tape is always "legal", namely it is available for elongation in the next cycle.

9.4 Molecular Shift-Registers and their Use as Autonomous DNA Synthesizers | 299

A SR is conveniently represented by a path on a corresponding de Bruijn graph [57], where the nodes depict all distinct internal states and the directed edges connecting them are labeled by the rules, notably by a string comprising the predecessor state plus a function bit. When a p-SR is realized with $p+1$ long rule strands, a maximal linear SR sequence passes exactly once through all nodes, except the zero node. A de Bruijn graph for a 3-SR is depicted in Figure 9.17, with arrows indicating the walk guided by Equation 9.1.

Although an elongation error corresponds to skipping some nodes, synthesis can always proceed as the rule strands recognize all nodes. When a p-SR is realized with rule strands of length $p'+1$; $p' > p$, as is the case here, the sequence passes exactly once through a subset of nodes in the much larger graph corresponding to p'-SRs. In the SR of Figure 9.16a and b, for instance, the 6-bit long rules correspond to a partial walk on de Bruijn graph of order 5 rather than 3. Two types of errors may then occur – a skip to a node in the sequence, or a skip to an alien node. In the first case, synthesis proceeds with partial deletion of the sequence. In the second case, the new node is not recognized by any rule strand and synthesis halts until that node is connected again to the SR sequence by an additional error. In both cases, each additional bit in the rule strands increases the Hamming distance for an error by at least 1 and, hence, suppresses the synthesis error rate by $\sim\exp(-\Delta G/k_B T)$. Optimization of the alphabet minimizes one- and two-base shift errors. Errors other than shifts, including hairpins, require further analysis.

The formation of an unwarranted 2-bit overhang can be minimized with respect to the desired 1-bit overhang by optimizing the temperature. Optimally, the error

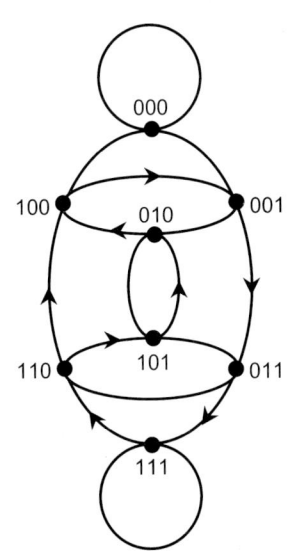

Figure 9.17 de Bruijn graph for a 3-shift register. The maximal path defined by Equation 9.1 corresponds to a walk on the graph (start from node 001 and follow the arrowheads).

rate (the ratio between incorrect and correct annealing) can be reduced in this way to $\approx\exp(-\Delta G/k_B T)$, where ΔG is the corresponding free energy per bit. The error rate may be systematically suppressed by using longer rule strands to generate the same sequence. By using de Bruijn graphs it can optimally be shown that each extra bit can reduce the error rate by an additional factor of $\approx\exp(-\Delta G/k_B T)$. This is the reason for the 6-bit long rule strands used in the realization of the 3-SR. The two extra bits are meant to suppress synthesis errors.

To the best of the present author's knowledge, this is the first incorporation of a redundancy code in chemical synthesis. The analogy drawn between chemical synthesis and transmission of messages over noisy lines suggests further applications of communication theory to chemical synthesis.

9.5
Future Perspectives

In Sections 9.2 to 9.4, a novel concept was outlined, namely the harnessing of the remarkable assembly strategies and tools of molecular biology to the self-assembly of molecular-scale electronics. Central issues such as instilling biomolecules with electrical conductance, molecular lithography for patterning metallization and localizing devices on DNA templates, the direct recognition of electronically relevant man-made objects by biomolecules, and the economic synthesis of DNA molecules characterized by non-recurring sequences have now been resolved to a point where the formidable challenge of complex self-assembly can be faced with confidence. However, harnessing the power of bioassembly presented here to the realization even of simple circuits requires more than mere optimization of the tools developed to date. As argued above, complex self-assembly will require a hierarchical, modular approach and, hence, the development of molecular switches that test for electronic functionality and feed back on the bioassembly process. Such switches will involve a functional interface between molecular biology and electronics, namely the ability of biomolecules to read electronic signals presented to them by the assembled devices and circuits, and then to effect the assembly process based on those findings. Only then can a full merging of biology and electronics be achieved.

Acknowledgments

The concepts and tools described in this chapter have been developed over the past decade by a significant group of researchers at Technion – Israel Institute of Technology. The author is especially grateful to Erez Braun, Kinneret Keren, Yoram Reiter, Arbel Artzi, Stav Zeitzev, Ilya Baskin, and Doron Lipson, whose contributions were immeasurable. Different areas of the research were funded by the Israeli Science Foundation, Bikura, the fifth EU program, the German Israeli DIP, the Rosenbloom family, and the Russell Berrie Nanotechnology Institute.

References

1. Braun, E., Eichen, Y., Sivan, U. and Ben Yoseph, G. (1998) DNA templated assembly and electrode attachment of conducting silver wire. *Nature*, **391**, 775–778.
2. Endres, R.G., Cox, D.L. and Singh, R.R.P. (2004) The quest for high-conductance DNA. *Reviews of Modern Physics*, **76** 195. and references therein.
3. Legrand, O., Côte, D. and Bockelmann, U. (2006) Single molecule study of DNA conductivity in aqueous environment. *Physical Review*, **E73**, 031925. and references therein.
4. Heath, J.R., Kuekes, P.J., Snider, G.S. and Stanley Williams, R. (1998) A defect-tolerant computer architecture: opportunities for nanotechnology. *Science*, **280**, 1716.
5. Eichen, Y., Braun, E., Sivan, U. and Ben Yoseph, G. (1998) Self assembly of nanoelectronics components and circuits using biological templates. *Acta Polymerica*, **49**, 663–670.
6. Keren, K., Krueger, M., Gilad, R., Ben-Yoseph, G., Sivan, U. and Braun, E. (2002) Sequence-specific molecular lithography on single DNA molecules. *Science*, **297**, 72.
7. Keren, K., Berman, R.S. and Braun, E. (2004) Patterned DNA Metallization by sequence-specific localization of a reducing agent. *Nano Letters*, **4** (2), 323–326.
8. Keren, K., Berman, R., Sivan, U. and Braun, E. (2003) DNA-templated carbon-nanotube field-effect transistor. *Science*, **302**, 1380–1382.
9. Artzy-Schnirman, A., Zahavi, E., Yeger, H., Rosenfeld, R., Benhar, I., Reiter, Y. and Sivan, U. (2006) Antibody molecules discriminate between crystalline facets of gallium arsenide semiconductor. *Nano Letters*, **6**, 1870.
10. Brod, E., Nimri, S., Turner, B. and Uri, Sivan (2008) Electrical control over antibody-antigen binding, *Sensors and Actuators B: Chemical*, **128**, 560.
11. Baskin, I., Zaitsev, S., Lipson, D., Gilad, R., Keren, K., Ben-Yoseph, G. and Sivan, U. (2006) A molecular shift register and its utilization for an autonomous DNA synthesis. *Physical Review Letters*, **97**, 208103.
12. Holgate, C.S. *et al.* (1983) Immunogold-silver staining: new method of immunostaining with enhanced sensitivity. *Journal of Histochemistry & Cytochemistry*, **31**, 938–944.
13. Birrell, G.B. *et al.* (1986) Silver-enhanced colloidal gold as a cell surface marker for photoelectron microscopy. *Journal of Histochemistry & Cytochemistry*, **34**, 339–345.
14. Keren, K. (2004) PhD thesis, *Self-assembly of molecular-scale electronics by genetic recombination*, Technion, Haifa Israel.
15. Braun, E. and Sivan, U. (2004) DNA templated electronics, in *Nano-Biotechnology, Concepts, Applications and Perspectives* (eds C.M. Nimeyer and C.A. Mirkin), Wiley-VCH, Weinheim, pp. 244–253.
16. Keren, K., Sivan, U. and Braun, E. (2004) DNA Templated electronics, in *Bio-electronics: From Theory to Applications* (eds I. Willner and E. Katz), Wiley-VCH, Weinheim, pp. 265–284.
17. Braun, E. and Keren, K. (2004) From, DNA to transistors. *Advances in Physics*, **53**, 441–496.
18. Gazit, E. (2007) Use of biomolecular templates for the fabrication of metal nanowires. *FEBS*, **274**, 317–322.
19. Gu, Q. *et al.* (2006) DNA nanowire fabrication. *Nanotechnology*, **17**, R14–R25.
20. Richter, J. (2003) Metallization of DNA. *Physica*, **E16**, 157–173.
21. Cox, M.M. (2000) *Progress in Nucleic Acids Research and Molecular Biology*, **63**, 311–366.
22. Hseih, P., Camerini-Otero, C.S. and Comerini-Otero, D. (1992) The synapsis

event in the homologous pairing of DNAs: RecA recognizes and pairs less than one helical repeat of DNA. *Proceedings of the National Academy of Sciences of the United States of America*, **89**, 6492–6496.
23 Dekker, C. (1999) Carbon nanotubes as molecular quantum wires. *Physics Today*, May, **52**, 22–28.
24 Liu, J. *et al.* (1998) Fullerene pipes. *Science*, **280**, 1253–1256.
25 Balavoine, F. *et al.* (1999) Helical crystallization of proteins on carbon nanotubes: a first step towards the development of new biosensors. *Angewandte Chemie-International Edition*, **38**, 1912–1915.
26 Shim, M., Kam, N.W.S., Chen, R.J., Li, Y. and Dai, H. (2002) Functionalization of carbon nanotubes for biocompatibility and biomolecular recognition. *Nano Letters*, **2** (4), 285–288.
27 Gomez-Navarro, C. *et al.* (2002) Contactless experiments on individual DNA molecules show no evidence for molecular wire behavior. *Proceedings of the National Academy of Sciences of the United States of America*, **99**, 8484–8487.
28 Bockrath, M. *et al.* (2002) Scanned conductance microscopy of carbon nanotubes and λ-DNA. *Nano Letters*, **2**, 187–190.
29 Avouris, P. (2002) Molecular electronics with carbon nanotubes. *Accounts of Chemical Research*, **35**, 1026.
30 Perl-Treves, D., Kessler, N., Izhaky, D. and Addadi, L. (1996) Monoclonal antibody recognition of cholesterol monohydrate crystal faces. *Chemistry & Biology*, **3**, 567–577.
31 Bromberg, R., Kessler, N. and Addadi, L. (1998) Antibody recognition of specific crystal faces; 1,4-dinitrobenzene. *Journal of Crystal Growth*, **193**, 656–664.
32 Braden, B.C. *et al.* (2000) X-ray crystal structure of an anti-Buckminsterfullerene antibody Fab fragment: Biomolecular recognition of C60. *Proceedings of the National Academy of Sciences of the United States of America*, **97**, 12193–12197.

33 Skerra, A. and Pluckthun, A. (1988) Assembly of a functional immunoglobulin Fv fragment in *Escherichia coli*. *Science*, **240**, 1038–1041.
34 Seeman, N.C. (2003) DNA in a material world. *Nature*, **421**, 427–431.
35 Mirkin, C.A., Letsinger, R.L., Mucic, R.C. and Storhoff, J.J. (1996) A DNA-based method for rationally assembling nanoparticles into macroscopic materials. *Nature*, **382**, 607–609.
36 Brown, S. (1997) Metal recognition by repeating polypeptides. *Nature Biotechnology*, **15**, 269–272.
37 Brown, S., Sarikaya, M. and Johnson, E. (2000) Genetic analysis of crystal growth. *Journal of Molecular Biology*, **299**, 725–732.
38 Naik, R.R., Stringer, S.J., Agarwal, G., Jones, S.E. and Stone, M.O. (2002) Biomimetic synthesis and patterning of silver nanoparticles. *Nature Mater*, **1**, 169–172.
39 Naik, R.R., Brott, L.L., Clarson, S.J. and Stone, M.O. (2002) Silica precipitating peptides isolated from a combinatorial phage display libraries. *Journal of Nanoscience and Nanotechnology*, **2**, 1–6.
40 Kjaergaard, K., Sorensen, J.K., Schembri, M.A. and Klemm, P. (2000) Sequestration of zinc oxide by fimbrial designer chelators. *Applied and Environmental Microbiology*, **66**, 10–14.
41 Brown, S. (1992) Engineering iron oxide adhesion mutants of *Escherichia coli* phage receptor. *Proceedings of the National Academy of Sciences of the United States of America*, **89**, 8651–8655.
42 Gaskin, D.J.H., Starck, K. and Wulfson, E.N. (2000) Identification of inorganic crystal-specific sequences using phage display combinatorial library of short peptides: a feasibility study. *Biotechnology Letters*, **22**, 1211–1216.
43 Wang, S. *et al.* (2003) Peptides with selective affinity for carbon nanotubes. *Nature Mater*, **2**, 196–200.
44 Willett, R.L., Baldwin, K.W., West, K.W. and Pfeiffer, L.N. (2005) Differential

adhesion of amino acids to inorganic surfaces. *Proceedings of the National Academy of Sciences of the United States of America*, **102**, 7817–7822.
45 Lee, S.W., Mao, C., Flynn, C.E. and Belcher, A.M. (2002) Ordering quantum dots using genetically engineered viruses. *Science*, **296**, 892–895.
46 Whaley, S.R., English, D.S., Hu, E.L., Barbara, P.F. and Belcher, A.M. (2000) Selection of peptides with semiconducting binding specificity for directed nanocrystal assembly. *Nature*, **405**, 665–668.
47 Goede, K., Busch, P. and Grundmann, M. (2004) Binding specificity of a peptide on semiconductor surfaces. *Nano Letters*, **4**, 2115–2120.
48 Knight, C.A., Cheng, C.C. and DeVries, A.L. (1991) Adsorption of α-helical antifreeze peptides on specific ice crystal surface planes. *Biophysical Journal*, **50**, 409.
49 DeOliviera, D.B. and Laursen, R.A. (1997) Control of calcite crystal morphology by a peptide designed to bind a specific surface. *Journal of the American Chemical Society*, **119**, 10627.
50 Azriel-Rosenfeld, R., Valensi, M. and Benhar, I. (2003) A human synthetic combinatorial library of arrayable single-chain antibodies based on shuffling in vivo formed CDRs into general framework regions. *Journal of Molecular Biology*, **335**, 177–192.
51 Smith, G.P. (1985) Filamentous fusion phage: novel expression vectors that display cloned antigens on the virion surface. *Science*, **228**, 1315–1317.
52 Stemmer, W.P. *et al.* (1995) Single-step assembly of a gene and entire plasmid from large numbers of oligodeoxyribonucleotides. *Gene*, **164**, 49–53.
53 Adleman, L.M. (1994) Molecular computation of solutions to combinatorial problems. *Science*, **266**, 1021–1024.
54 Hagiya, M. *et al.* (1997) Towards parallel evaluation and learning of Boolean µ-formulas with molecules, in *DIMACS Series in Discrete Mathematics and Theoretical Computer Science*, DNA Based Computers III, Volume **48**, pp. 57–72.
55 Sakamoto, K. *et al.* (2000) Molecular computation by DNA hairpin formation. *Science*, **288**, 1223–1226.
56 Khodor, J. and Gifford, D.K. (2002) Programmed mutagenesis is universal. *Theory of Computing Systems*, **35**, 483–499.
57 Golomb, Solomon W. (1982) *Shift Register Sequences*, Aegean Park Press.
58 Although linear shift registers are discussed here, the automaton should work equally well with non-linear feedback functions.
59 The zero string should be avoided as it maps onto itself by any linear feedback function.
60 A 5-bit degenerate rule is constructed by adding the two preceding bits of the sequence to the 3′ end of the corresponding 3-bit rule.
61 Sugimoto, N., Nakano, S., Yoneyama, M. and Honda, K. (1996) Improved thermodynamic parameters and helix initiation factor to predict stability of DNA duplexes. *Nucleic Acids Research*, **24**, 4501–4505.
62 For brevity, we use notation that mixes bases with bits. Each bit represents three bases.
63 Annealing at 54 °C for 30 s, extension at 72 °C for 1 min, melting at 95 °C for 30 s.

10
Formation of Nanostructures by Self-Assembly

Melanie Homberger, Silvia Karthäuser, Ulrich Simon, and Bert Voigtländer

10.1
Introduction

The increasing demand for high-density electronic devices has triggered – and continues to trigger – the development of new nanofabrication methods. Two conceptually different strategies are applied for the fabrication of nanostructures, namely: (i) the *top-down* strategy; and (ii) the *bottom-up* strategy.

The *top-down* approaches utilize lithographical methods to fabricate nanostructures starting from the bulk materials (see Chapters 5, 6, and 7), whereas in the *bottom-up* approaches nanostructures are built up from atoms, molecules, or nanoscale sub-units. The *top-down* methods enable the generation of a large variety of defined structures, but these are limited by the resolution of current lithography techniques. The *bottom-up* methods offer the opportunity to fabricate structures even in the single-digit nanometer range, but they suffer from the fact that it is still a great challenge to direct the functional sub-units into desired structures. One extreme approach in this context is the utilization of a scanning probe microscope for building up nanostructures atom by atom at low temperatures (see Chapter 9). However, although this approach is ultimate in terms of the size of the nanostructures, it is a very slow and sophisticated method. Compared to this method, processes based on self-organization or self-assembly have the key advantage that they enable the formation of billions of nanostructures with control over size, shape, and composition in a fast and parallel fashion. Due to entropic effects during the formation of nanostructures by self-assembly, defects are expected always to be present, and fault-tolerant architectures are required to cope with this problem. The combination of the self-assembly of atoms, molecules and nanoscale subunits could lead to well-ordered functional nanostructures. For example, inorganic nanostructures, generated by the self-assembly of atoms via epitaxial growth, may serve as templates for the selective adsorption of functional molecules, which themselves display "anchor-points" at which size-selected clusters could be attached, altogether leading to highly ordered functional nanostructures with applications in molecular electronics. One critical factor determining the benefits of this approach for electronic systems will be the surface-selective SAM formation – that is, the selective assembly of functional molecules on special device patterns forming an ordered array. In this context, in the following chapter attention is focused on the formation of nanostructures by self-assembly via epitaxial growth, the self-assembly of molecules, and the formation and self-assembly of nanoscale subunits. Basic physical principles and selected examples will be presented.

10.2
Self-Assembly by Epitaxial Growth

One approach for the fabrication of nanostructures is *epitaxial growth*. Such growth usually occurs under kinetic conditions, so that the sizes can be tuned down to the single-digit nanometer range by choosing appropriate growth conditions. However, size uniformity is the greatest challenge here. If the growth is taking place under (near) equilibrium conditions, then the size distribution of the nanostructures may be narrow, but is provided by the material system and cannot be varied easily. The formation of islands, wires and rods will be presented as examples of nanostructures grown by epitaxy. Subsequently, the growth of nanostructures on template substrates structured by step arrays or underlying dislocation networks will be considered. The combination of self-organized growth with lithography ("hybrid methods") allows the self-assembled nanostructures to be aligned relative to predefined patterns. It is possible that such inorganic nanostructured templates may be used in the future for the selective formation of molecular layers.

10.2.1
Physical Principles of Self-Organized Epitaxial Growth

10.2.1.1 Epitaxial Growth Techniques
The main methods used for semiconductor epitaxial growth are chemical vapor deposition (CVD) [1] and molecular beam epitaxy (MBE) [2, 3]. In CVD, growth gases containing compounds of the elements to be deposited are introduced into the growth chamber. When the gas molecules hit the substrate surface, they decompose (partially) and the gaseous products desorb from the surface. Different chemical reactions taking place at the surface, or even in the gas phase, lead to a quite complex nature of the fundamental processes of epitaxial growth in CVD. Molecular beam epitaxy is conceptually simpler; here, the elements to be deposited are heated in evaporators until they evaporate, whereupon the beam of the atoms hits the surface and the atoms diffuse over the surface and finally bind at surface lattice sites (Figure 10.1).

In spite of the fact that the MBE growth is, in principle, much easier than the CVD growth, there are still many different fundamental processes occurring during epitaxial growth by MBE [4]. Part of these are illustrated schematically in Figure 10.1. Atoms from the molecular beam arrive at the surface of the crystalline substrate (a)

(a) deposition on the terrace

(b) diffusion on the terrace

(c) island nucleation

(d) attachment of atoms at islands

(e) diffusion to step edges

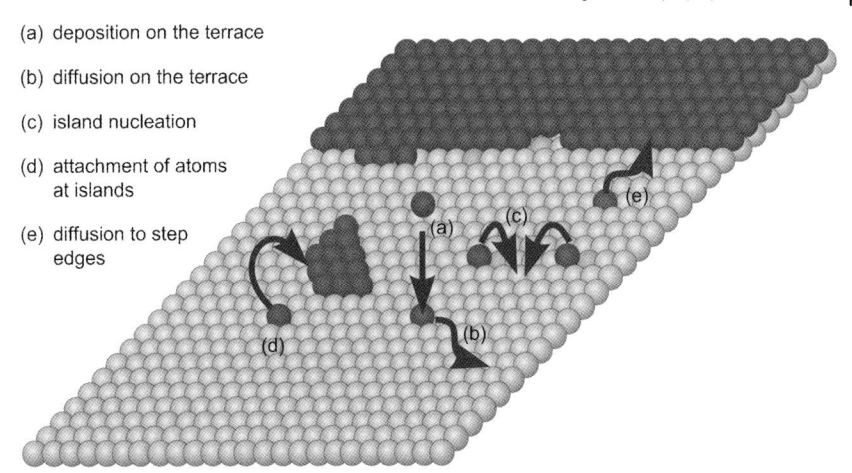

Figure 10.1 Scheme of different fundamental processes occurring during epitaxial growth, leading to a self-organization of two-dimensional islands.

and may diffuse over the surface when the activation energy for diffusion is overcome (b). When two atoms (or sometimes also more than two atoms) meet, they form a nucleus for a stable island (c). Such a nucleus may grow to a stable two-dimensional (2-D) island by attachment of further diffusing adatoms (d). The nucleus for which the probabilities to grow or decay are equal is called the *critical nucleus* [5]. Nuclei which are larger than the critical nucleus are termed stable 2-D islands, while nuclei smaller than the critical nucleus are called subcritical nuclei or *embryos*. Another process is the diffusion and attachment at pre-existing steps if the diffusion length is sufficient (e).

10.2.1.2 Kinetically Limited Growth in Homoepitaxy

In kinetically limited growth the system is governed by energetic barriers such as barriers for the diffusion of adatoms and barriers for incorporation of atoms into the crystal, and additionally by outer conditions such as the growth rate. The 2-D islands (which are one atomic layer high) represent the simplest example of the self-assembled growth of nanostructures. In the following section it will be shown how the density and size of these islands can be controlled by the kinetic parameters temperature and growth rate. First, the deposition temperature influences the island density strongly, as shown by the comparison of Figure 10.2a and b. The island density as function of temperature follows an Arrhenius law: $n \sim \exp(E_{act}/kT)$, where E_{act} is an effective activation energy consisting of a diffusion energy and binding energy component, having values around 1 eV in the case of semiconductors [5]. The temperature is one important parameter of growth kinetics, and the deposition rate is another. It has been found that the island density (n) scales with the deposition rate (F) in the form of a power law $n \sim F^\alpha$, with a scaling exponent α. Combining the temperature and the rate dependence results in the following scaling law: $n \sim F^\alpha$

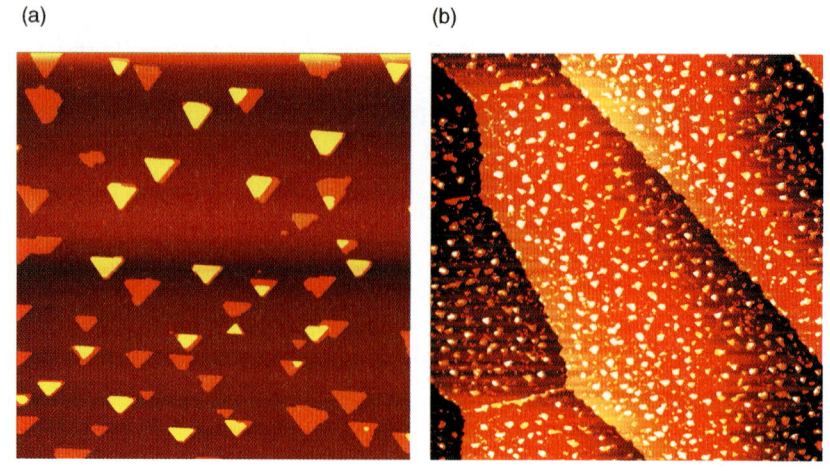

Figure 10.2 Scanning tunneling microscope images after the growth of 0.2 atomic layers of silicon on a Si(1 1 1) surface. The islands have triangular shape due to the symmetry of the substrate, and have a height of one atomic layer (orange) or two atomic layers (yellow). The island density depends on the temperature, as can be seen by comparison of growth at high temperatures of 770 K (a) to growth at a lower temperature 610 K (b). Both images have a size of 350 nm.

$\exp(E_{act}/kT)$ [5], which shows that the island density can be controlled over a wide range by adjusting the kinetic growth parameters of temperature and growth rate. The average island distance is simply the square root of the inverse of the island density $L = 1/\sqrt{n}$.

Although the nucleation of the islands is a random process, the distribution of the island sizes is centered around a mean value (Figure 10.3). This arises due to a saturation of the island nucleation, as will be explained in the following. During the early stage of growth (nucleation regime), the islands nucleate randomly on the surface and the distance between them decreases. If the distance between the islands is equal to the mean distance that an adatom travels before a nucleation event happens, then the incorporation of adatoms in existing islands becomes a more probable event than the nucleation of new islands; hence, a "capture zone" forms around each island. Adatoms deposited in this capture zone attach to the corresponding island. Without this effect the distribution of island sizes would be even broader. The nucleation of further islands is suppressed beyond a certain coverage (growth regime), and the average island size can be controlled by the deposited amount. The island size distributions for two different temperatures are shown in Figure 10.3, where it can be seen that the peak in the island size distribution scales towards larger sizes with higher temperatures. For very small islands, the surface reconstruction can also modify the island size distribution [4]. In the kinetic growth regime the island density of 2-D islands can be controlled by the kinetic parameters temperature and deposition rate, while the size distribution is quite broad due to the stochastic nature of the nucleation of the islands.

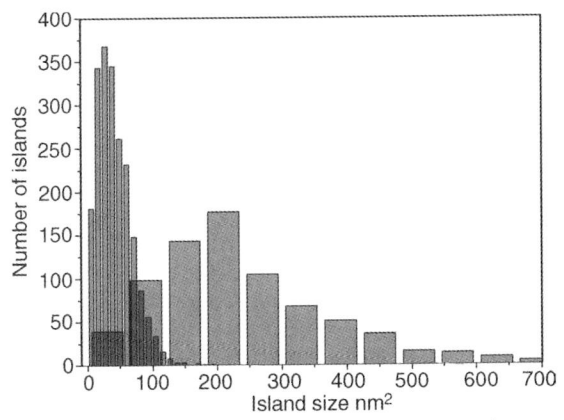

Figure 10.3 Island size distribution for two-dimensional Si islands on Si(1 1 1). The width of the distribution is of the order of the average size of the islands. Two distributions for two different temperatures are displayed. The narrow bins (peak at small island sizes) correspond to deposition at 610 K. The distribution with the wide bins peak at larger island sizes) corresponds to deposition at 710 K.

10.2.1.3 Thermodynamically Stable Nanostructures

If nanosized islands were to be thermodynamically stable, their size distribution could be narrow. A thermodynamically stable island size means that the energy (per atom) has a minimum for this stable size. For configurations with larger or smaller islands, the energy (per atom) would be higher, and therefore it only necessary to approach thermodynamic equilibrium in order to obtain a very narrow island size distribution. One way to achieve thermodynamic equilibrium is to heat a sample with different island sizes present and then to wait until equilibrium has established. The equilibrium configuration will be established by material transport between the islands, as the atoms will detach from islands with higher energy and attach to islands with a lower energy (per atom). However, as will be shown below, in the simplest case (considering only a surface or edge energy term) the thermodynamically stable island size is infinitely large. This behavior is not of any use for the formation of nanostructures with a narrow size distribution, and corresponds to the well-known Ostwald ripening. Only if additional terms in the energy are important (e.g., strain energy) will the energy per particle show a minimum for a finite particle size, while a narrow size distribution can be expected under equilibrium conditions.

In order to describe material transport in a system with a variable number of atoms, the chemical potential is used; this is the change of the energy (of an island) when the number of particles changes $\mu = dE/dN$. During the equilibration process atoms detach from islands where the chemical potential is highest, and attach to islands with a lower chemical potential. This lowers the total energy of the system, and consequently the material transport between different islands is governed by the chemical potential. A simple example is the chemical potential of quadratic 2-D islands of dimension L (Figure 10.4a). The energy difference between different-sized islands

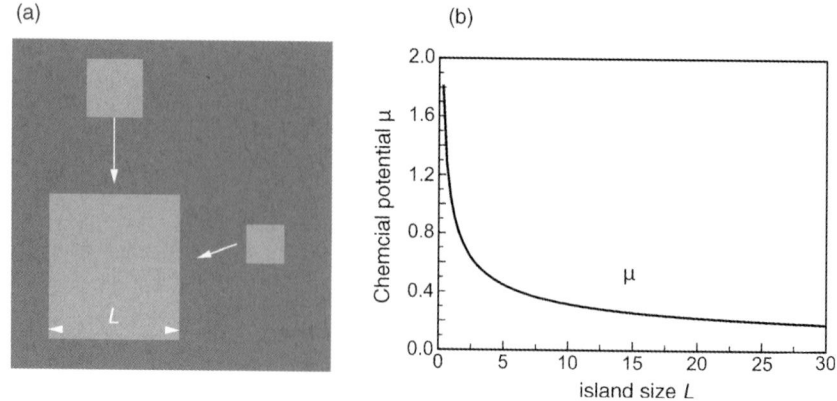

Figure 10.4 (a) Coarsening of a large island at the expense of small islands. (b) The chemical potential of an island.

comes from the edge energy (β is the edge energy per length). The energy of an island is $E = E_{\text{edge}} = 4\,L\beta$. The number of atoms in an island (N) depends on the dimension L, as $N = L^2/\omega$, with ω being the area per atom. The chemical potential is then

$$\mu = \frac{dE}{dN} = \frac{2\omega\beta}{L} \sim \frac{1}{L} \tag{10.1}$$

Since μ is decreasing for larger islands, infinite size islands have the lowest chemical potential (Figure 10.4b), which means that the stable island is infinitely large. In this case, the equilibration does not result in a stable finite island size; equilibration in this model by material transport between islands is also referred to as *coarsening* because it results in the shrinkage of small islands and a growth (coarsening) of large islands (Ostwald ripening).

An infinitely large stable island size is the result for homoepitaxial growth, taking into account only the edge energy. However, the situation becomes different when elastic stress is also taken into account, as it occurs in heteroepitaxy where two different materials grow onto each other. Here, stress is induced by the different lattice constants of the substrate material and the material of the islands. The elastic effect of strained 2-D islands can be approximated by that of a surface-stress domain – that is, the surface stress at the area of the island is different from that at the rest of the surface (Figure 10.5). The strain energy of a quadratic surface-stress domain can be calculated using the elastic theory as $E_{\text{strain}} = 2LC \ln L$ [6]. Adding the step edge energy results in a total energy of a strained island:

$$E = E_{\text{edge}} + E_{\text{strain}} = 2L[2\beta - C' \ln L] \tag{10.2}$$

This results in the following chemical potential:

$$\mu = \omega\left[\frac{2\beta - C'}{L} - \frac{C'}{L}\ln L\right] \tag{10.3}$$

which is illustrated in Figure 10.5. In this case, the chemical potential has a minimum at the size $L_{\min} = \exp(2\beta/C')$, which would mean that during coarsening the islands

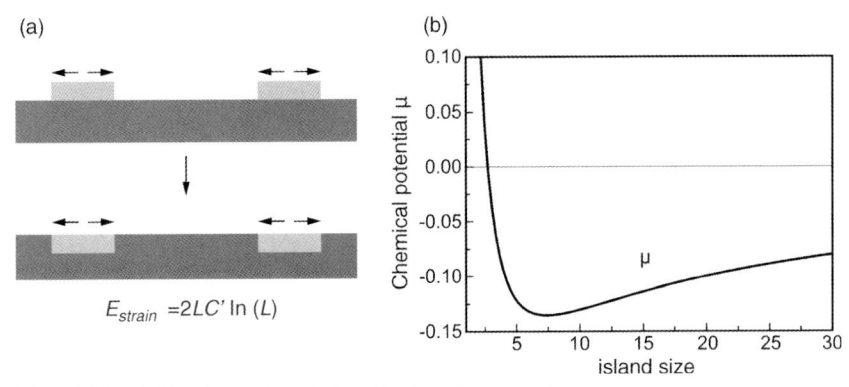

Figure 10.5 (a) The elastic stress induced by two-dimensional islands with a different lattice constant than the substrate can be approximated by surface stress domains. (b) Chemical potential of an island with an energy component due to elastic strain included.

$E_{strain} = 2LC' \ln(L)$

would approach this size. Larger islands would dissolve and smaller islands grow until all islands have the size L_{min}, that is, the lowest chemical potential, and this would result in a very narrow size distribution. Unfortunately, step energies are only very poorly known, so that it is not possible to predict a reliable number for the equilibrium island size. An experimental realization of thermodynamically stable islands has not yet been confirmed, apart from surface reconstructions with a relatively large unit cell.

If the formation of nanostructures in equilibrium is compared to the formation of nanostructures by growth kinetics, the following advantages and disadvantages occur. Nanostructures grown under equilibrium conditions have (under specific conditions) the advantage of a narrow size distribution around the optimum size. However, a disadvantage is that the size is determined by the material parameters (strain energy and step edge energy for instance), and cannot be tuned freely. The size and density of nanostructures formed under kinetic conditions can be tuned easily by variations of the growth parameters such as growth rate and temperature. On the other hand, the size uniformity of the islands grown under kinetic conditions is relatively poor.

10.2.1.4 Nanostructure Formation in Heteroepitaxial Growth

Semiconductor nanostructures can be fabricated by self-organization using heteroepitaxial growth, which is the growth of a material B on a substrate of different material A. In heteroepitaxial growth, the lattice constants of the two materials are often different. The lattice mismatch for the two most commonly used material systems, Si/Ge and GaAs/InAs, is 4.2% and 7%, respectively (shown schematically in Figure 10.6a). This lattice mismatch leads to a build-up of elastic stress in the initial 2-D growth in heteroepitaxy. In the case of Ge heteroepitaxy on Si, the Ge is confined to the smaller lattice constant of the Si substrate – that is, the Ge is strained to the Si lattice constant (Figure 10.6b). One way to relax this stress is via the formation of three-dimensional (3-D) Ge islands, in which only the bottom of the islands is

10 Formation of Nanostructures by Self-Assembly

(a) Ge is 4% larger then Si

(b) Build-up of strain energy

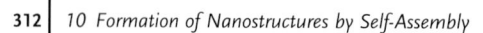

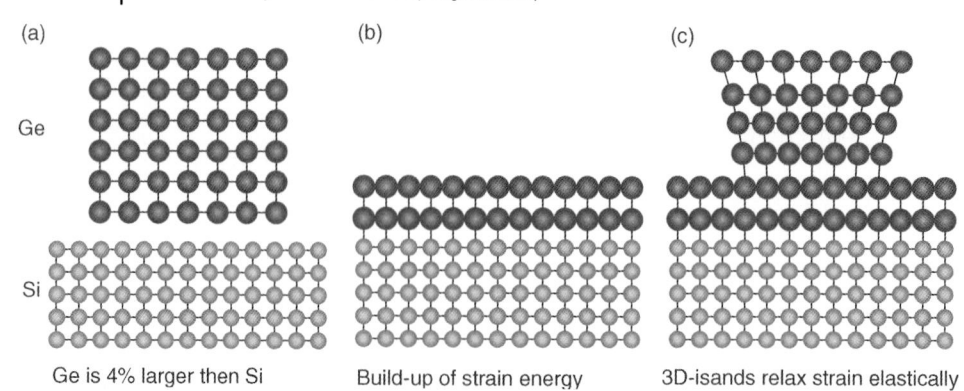

(c) 3D-isands relax strain elastically

Figure 10.6 (a) Schematic representation of Si and Ge crystals with different lattice constants. (b) Build-up of elastic strain energy during 2-D growth with Ge confined to the Si lattice constant, and (c) elastic relaxation by the formation of 3-D islands (Stranski–Krastanov growth). In the upper part of the 3-D island the lattice constant relaxes towards the Ge bulk constant. The usual form of the 3-D islands is a pyramid, and not like that shown in this schematic.

confined to the substrate lattice constant. In the upper part of the 3-D island the lattice constant can relax to the Ge bulk lattice constant and reduce the stress energy in this way (Figure 10.6c). This growth mode, which is characterized by the formation of a 2-D wetting layer and the subsequent growth of (partially relaxed) 3-D islands, is referred to as the Stranski–Krastanov growth mode, some examples are which are described shown in Section 10.2.2.

The driving force for the formation of self-organized nanoislands in heteroepitaxial growth is the build-up of elastic strain energy in the stressed 2-D layer. As a reaction to this, a partial stress relaxation by the formation of 3-D islands can lower the free energy of the system. The process of island formation close to equilibrium is a trade-off between elastic relaxation by the formation of 3-D islands, which lowers the energy of the system, and an increase of the surface area, which increases the energy.

In a simple model, where the islands are cubes with the length x, the additional surface energy for a film in an island morphology (compared to a strained film) is proportional to the island length squared (x^2). The gained elastic relaxation energy compared to that of a flat film is, in the simplest assumption, proportional to the volume of the island (x^3). For the same total volume in the film, the energy difference between the 3-D island morphology and the flat morphology is

$$\Delta E = E_{\text{surf}} - E_{\text{relax}} = C\gamma x^2 - C'\varepsilon^2 x^3 \tag{10.4}$$

where γ is the surface energy, ε is the lattice mismatch, and C and C' are constants. The contributions of E_{surf}, E_{relax} and the total energy difference between the 3-D island morphology and a flat film are shown in Figure 10.7, as a function of the island size x. For small sizes of the 3-D islands, the 3-D island morphology is unfavorable up until the point where the absolute value of the gained elastic relaxation energy ($\sim x^3$) becomes larger than the cost of the surface energy ($\sim x^2$). For islands larger than a

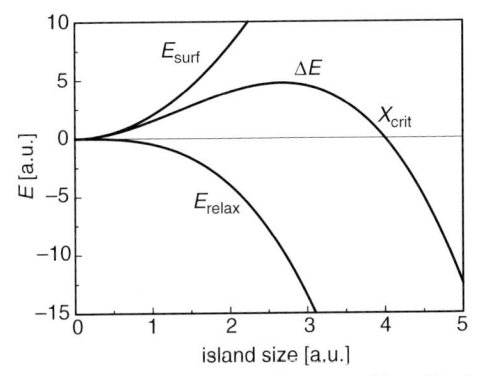

Figure 10.7 Energy difference between a film of flat 2-D morphology and a film morphology consisting of 3-D islands. The total energy difference and the contributions surface energy difference and relaxation energy are plotted.

critical island size, x_{crit}, the formation of 3-D islands is energetically preferred over the 2-D film morphology. While this simple model shows the basic driving forces for the 2-D to 3-D transition, it contains several simplifications. For example, in this simple model the island morphology is assumed as being cuboid, which does not correspond to the experimentally observed island shapes. Further, the simple model contains only energetic considerations of two final states. Kinetic effects, such as the required material transport necessary during the 2-D and 3-D transition are not considered.

Apart from the formation of 3-D islands, there is another process which can partially relax the stress of a strained 2-D layer, namely the introduction of misfit dislocations. This corresponds to the removal of one lattice plane of a compressively strained 2-D layer. If a lattice plane is removed in regular distances in the 2-D layer, then a misfit dislocation network forms. Depending on the growth parameters of temperature and growth rate, the self-organized growth can either be close to equilibrium or in the kinetically limited regime. At close to equilibrium (i.e., at high growth temperatures or low deposition rates), the occurring morphology (strained layer, 3-D islands, or a film with dislocations) is determined only by the energies of the particular configurations, and the morphology with the lowest energy will be formed. If the growth is kinetically limited, then the activation barriers are important. For instance, an initially flat strained layer can transform to a morphology with 3-D islands or to a film with dislocations. Yet, what actually happens depends on the kinetics of the growth process – that is, on the activation energy for the formation of 3-D islands compared to the activation energy for the introduction of misfit dislocations.

10.2.2
Semiconductor Nanoislands and Nanowires

10.2.2.1 Stranski–Krastanov Growth of Nanoislands
Stranski–Krastanov growth occurs, for example, in InGaAs/GaAs growth [7]. An example of InAs nanoislands grown on a GaAs substrate is shown in the transmission

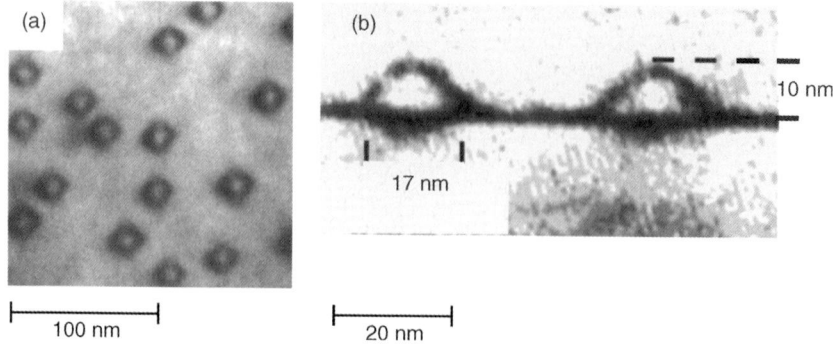

Figure 10.8 InAs nanoislands grown on a GaAs surface. (a) As imaged by plan-view transmission electron microscopy (TEM); (b) cross-sectional view with TEM [7].

electron microscopy (TEM) image in Figure 10.8. The GaAs islands were grown by MBE at a growth temperature of 775 K, and the density of the islands was 4.5×10^{10} cm^{-2}, with an average lateral size of 17.5 ± 0.5 nm. The challenges in the growth of these semiconductor islands are to grow islands of desired size and density, and with a high size uniformity. As in the case of the 2-D islands, a higher growth temperature generally leads to the formation of larger islands, while a higher growth rate leads to the formation of smaller islands. The size of the islands increases with coverage; often, the density of the islands saturates during an early stage of the growth. These are general trends which may depend on the material system and the particular deposition technique. In some cases (self-limiting growth), the size of the islands saturates while the density increases with coverage, and this type of growth mode leads to a high size uniformity of the islands. The size uniformity achieved in self-assembled growth of semiconductor islands may be as small as a small percent. The confinement of charge carriers in all three directions gives rise to atomic-like energy levels. Quantum dot lasers operating at room temperature have now been realized [8]. The islands grown on a flat substrate are usually not ordered laterally due to the random nature of the nucleation process. In the following section, it will be shown how nucleation at specific sites can be achieved.

10.2.2.2 Lateral Positioning of Nanoislands by Growth on Templates

An example of ordered nucleation at a prestructured substrate is shown in Figure 10.9a [9], where Ge islands nucleate above dislocation lines. However, when a SiGe film is grown on a Si(0 0 1) substrate, dislocations form at the interface between the SiGe film and the substrate. The driving force for the formation of dislocations is the relief of elastic strain, which arises due to the different lattice constants between the Si substrate and a Ge/Si film on this substrate. During annealing, the dislocations form a relatively regular network, due to a repulsive elastic interaction between the dislocations. The preferred nucleation of Ge islands above the dislocation lines (Figure 10.9a) can be explained by local stress relaxation above the dislocation lines providing a lattice constant closer to the Ge one. The nucleation does not occur

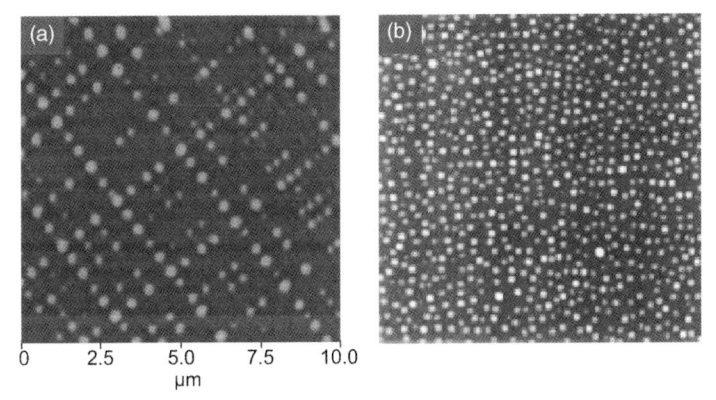

Figure 10.9 (a) Ordered nucleation of Ge islands on a template which is pre-structured by an underlying network of dislocations. (b) Germanium islands grown on a substrate without dislocations [9]. Image sizes 7 μm.

randomly at the surface, but rather occurs simultaneously at sites which have the same structure. This leads to a more narrow size distribution than that for the growth on unstructured Si(0 0 1) substrates (Figure 10.9b).

10.2.2.3 Silicide Nanowires

If the crystal structure of the deposited material is different from that of the substrate, then effects related to the anisotropic match of both crystal structures may appear. If the overlayer material has a crystal structure which is closely lattice-matched to the substrate along one major crystallographic axis, but has a significant lattice-mismatch along the perpendicular axis, this should allow unrestricted growth of the epitaxial crystal in the first direction but limit the width in the other direction. Such a strategy has been applied to grow silicide nanowires [10]. Here, the substrate is a Si(1 0 0) surface (Si has diamond crystal structure), and by deposition of Er and subsequent annealing, $ErSi_2$-oriented crystallites with a hexagonal AlB_2-type crystal structure were formed on the Si substrate. The [0 0 0 1] axis of the $ErSi_2$ was oriented along a [$\bar{1}$ 1 0] axis of the Si(0 0 1) substrate, and the [1 1 $\bar{2}$ 0] of the $ErSi_2$ was oriented along the perpendicular [1 1 0] axis, with lattice mismatches of +6.5% and −1.3%, respectively; this almost satisfies the proposed growth conditions for nanowires. $ErSi_2$ nanowires grown on the Si(1 0 0) surface are shown in Figure 10.10. The $ErSi_2$ nanowires align along one of the two perpendicular <1 1 0> Si directions, which are the small mismatch directions. In these directions the crystal can grow without much build-up of stress, while the width of the $ErSi_2$ nanowire is ∼4 nm, the height ∼0.8 nm, and the length is several hundred nanometers. Such self-assembled arrays of nanowires may also be used as conductors for defect-tolerant nanocircuits, or as a template for further nanofabrication.

10.2.2.4 Monolayer-Thick Wires at Step Edges

Monolayer-high surface steps can be used to fabricate Ge nanowires using step-flow growth. Here, pre-existing step edges on the Si(1 1 1) surface are used as templates for

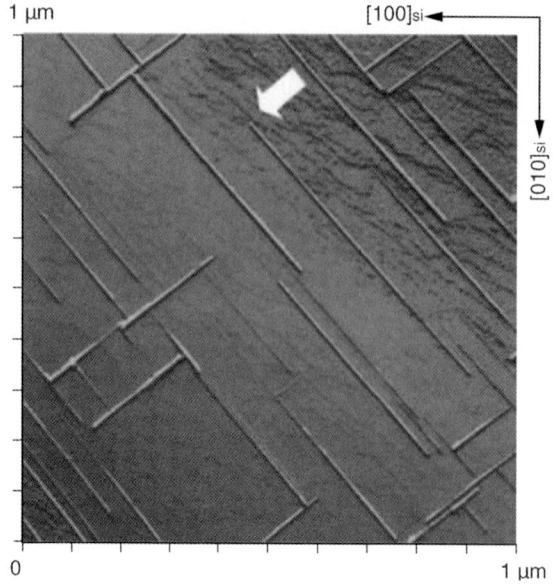

Figure 10.10 Scanning tunneling microscopy (STM) topograph showing ErSi₂ nanowires grown on a flat Si(0 0 1) substrate. The long direction of the nanowires is the one with the low lattice mismatch (1.3%), while the lattice mismatch in the perpendicular direction is 6.5%.

the growth of 2-D Ge wires at the step edges. When the diffusion of the deposited atoms is sufficient to reach the step edges, the deposited atoms are incorporated exclusively at the step edges, and the growth proceeds by a homogeneous advancement of the steps (step flow growth mode [4]. If small amounts of Ge are deposited, then the steps will advance only a few nanometers and narrow Ge wires can be grown.

One key issue for the controlled fabrication of nanostructures consisting of different materials is a method of characterization which can distinguish between the different materials on the nanoscale. If the surface is terminated with a monolayer of Bi, it is possible to distinguish between Si and Ge [11]. Figure 10.11a shows a scanning tunneling microscopy (STM) image after repeated alternating deposition of 0.15 atomic layers of Ge and Si, respectively. Due to the step-flow growth, the Ge and Si wires are formed at the advancing step edge. Whilst both elements can be easily distinguished by the apparent heights in the STM images, it transpired that the height measured by the STM was higher in areas consisting of Ge (red stripes) than in areas consisting of Si (yellow stripes). The apparent height of the Ge areas was ∼0.1 nm higher than that of the Si wires (Figure 10.11b), and the cross-section of a 3.3 nm-wide Ge nanowire was seen to contain only approximately 20 atoms (Figure 10.11c). The apparent height difference arises due to an atomic layer of Bi which is deposited initially and always floats on top of the growing layer. The different widths of the wires can easily be achieved by depositing different amounts of Ge and Si.

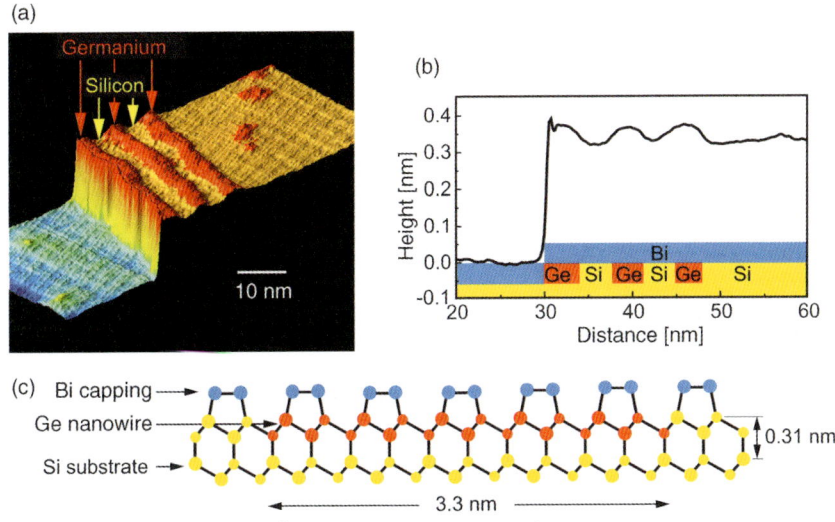

Figure 10.11 (a) STM image of 2-D Ge/Si nanowires grown by step-flow at a pre-existing step edge on a Si(1 1 1) substrate. The Si wires (yellow) and Ge wires (red) can be distinguished by different apparent heights. (b) A cross-section across the nanowires. (c) The atomic structure of a Ge wire on the Si substrate capped by Bi. The cross-section of the Ge wire contains only approximately 20 Ge atoms [11].

10.2.3
Hybrid Methods: The Combination of Lithography and Self-Organized Growth

In hybrid methods, self-organization is combined with lithographic patterning to form nanostructures on a smaller scale than are accessible by lithography. Most importantly, the hybrid methods provide a direct contact of nanostructures formed by self-organization to mesoscopic lithographically patterned structures. The self-organized growth of Ge islands in oxide holes is shown in Figure 10.12a–d. The starting surface is a silicon substrate with a thin oxide layer at the surface, and electron lithography is used to remove the oxide and form holes of a diameter of 0.5 μm where the bare Si surface is exposed [12]. The self-organized growth of Ge leads to the formation of Ge islands which may be smaller than the size scale of the electron beam lithography (EBL). The gas-phase growth of Ge is selective; that is, Ge will only grow on Si areas (inside the holes in the oxide), and not on the oxide itself. Figure 10.12 illustrates the nucleation of Ge islands in the holes in the oxide for different growth temperatures. At lower temperatures, the island density is so large that several islands nucleate in one oxide hole. However, if the temperature is increased, ultimately only one Ge island is able to nucleate in each oxide hole, and the size of the Ge island is smaller than the lithographically defined oxide hole. However, as seen in Figure 10.12d, the position of the Ge island inside the oxide hole is not defined but is rather randomly distributed within the oxide hole. Due to the fact that

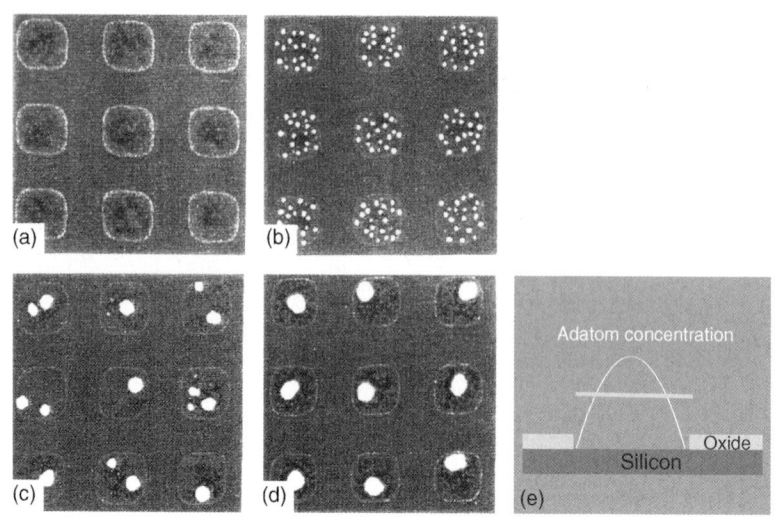

Figure 10.12 (a–d) Growth of Ge islands inside holes on an oxidized Si substrate [12]. (e) Adatom density in an oxide hole for those cases where the hole edges are sinks of adatoms (parabolic line), or for the case when the edges are not sinks for adatoms (horizontal line).

the Ge does not grow on the oxide, the edges of the oxide hole cannot serve as sinks for deposited Ge atoms, and therefore the Ge adatom concentration is homogeneous across the hole and nucleation of the Ge island is random within the oxide hole. If the edges of the hole were to serve as sinks for Ge atoms (e.g., if the edges of the hole were to consist of Si), then the adatom density would have a maximum at the center of the hole and the nucleation of Ge islands would occur preferentially at the center of the oxide holes (Figure 10.12e).

10.2.4
Inorganic Nanostructures as Templates for Molecular Layers

Several of the nanostructures discussed here can potentially be used as templates for the selective formation of molecular structures onto specific areas of the inorganic nanostructures generated by the self-assembly of atoms via epitaxial growth. The importance of the inorganic substrate for the formation of molecular layers, which is discussed in detail in the following section, is manifold. The role of the inorganic nanostructured template for the molecular self-assembly may be to steer the adsorption process kinetically, and to direct the molecules towards predefined adsorption sites. The first steps in this direction have been taken recently. Initially, special substrate surfaces were selected in accordance with their ability to adsorb molecules. For example, substrates with only weak adsorption properties are useful for molecular assemblies with weak intermolecular interactions, because such substrates allow for the necessary reorganization of molecules. In addition, it is necessary that the interatomic distances at the substrate surface correspond to the dimensions of molecular structure elements. One example of such an substrate, which allows for a weak adsorption of polycyclic aromatic compounds is the Ag/Si(1 1 1)-$\sqrt{3}\times\sqrt{3}$R30° surface (Figure 10.13a). This surface is described by the honeycomb-chain-trimer model, in which each surface Si atom is bound to one Ag atom. This structure is derived from a Si(1 1 1) bulk termination by removing the top half of the first bilayer of Si atoms, forming trimers from the remaining Si atoms, and then adding a full monolayer of Ag atoms in positions slightly distorted from the regular triangular lattice (Figure 10.13a). An STM image of this structure is shown in Figure 10.13b. In this empty states image (+1.6 V sample bias), the bright protrusions correspond to the center of three Ag atoms (Ag trimers indicated by A in Figure 10.13b), and the minima in this image, indicated by B, correspond to the Si trimers). In the following, for simplicity, this surface is represented by a hexagonal network also indicated in Figure 10.13a. A supramolecular 2-D honeycomb network,

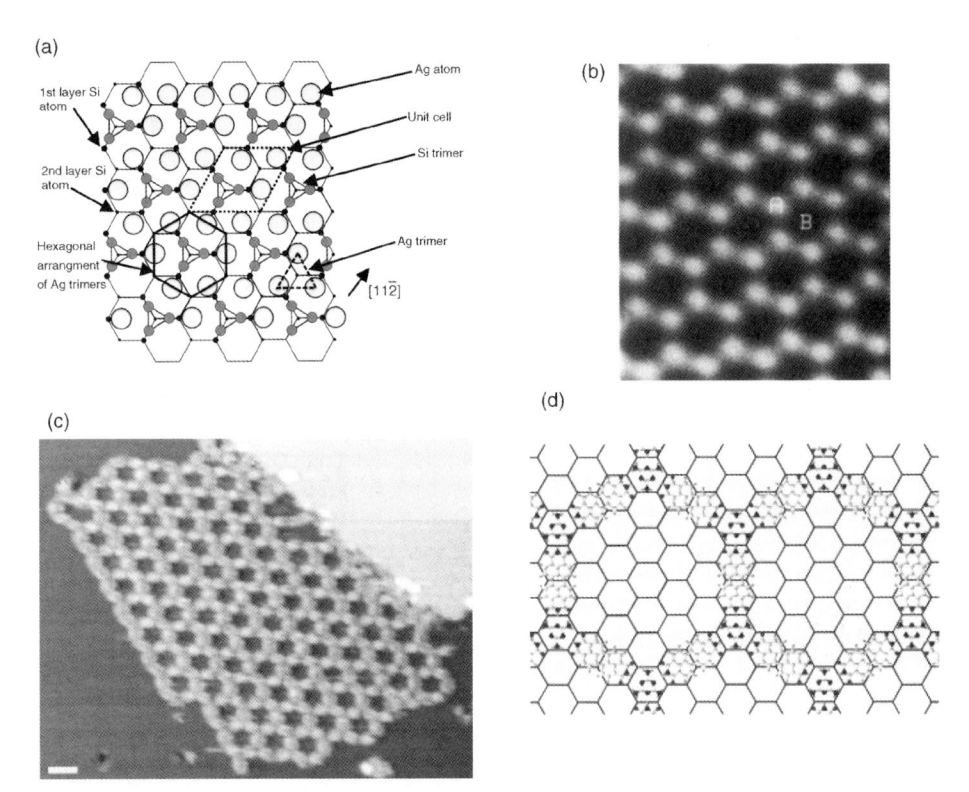

Figure 10.13 (a) Schematic showing the honeycomb-chain-trimer model for the Ag/Si (1 1 1)-$\sqrt{3}\times\sqrt{3}$R30° reconstruction [14]. STM image (empty states, +1.6 V sample bias) of the Ag/Si(1 1 1)-$\sqrt{3}\times\sqrt{3}$R30° substrate surface. The bright protrusions correspond to the center of three Ag atoms (image size 3 nm) [15]. (c) STM image of the hexagonal molecular network (see also Section 10.3.3) [13]. Scale bar = 3 nm. (d) Schematic diagram showing the registry of the molecular network with the underlying Ag/Si (1 1 1)-$\sqrt{3}\times\sqrt{3}$R30° surface reconstruction shown as hexagons.

with a larger periodicity (five times that of the Ag/Si(1 1 1)-$\sqrt{3}\times\sqrt{3}$R30° lattice constant; see Figure 10.13d) has been created by the assembly of two types of molecule on the Ag-terminated silicon surface [13]. This hexagonal molecular network is shown in Figure 10.13c, and is discussed in detail in Section 10.3.3. The registry of the molecules with respect to the underlying silver-terminated Si surface has been determined, and is shown schematically in Figure 10.13d. The calculated melamine–melamine separation has a near-commensurability with the surface lattice, showing the importance of the underlying inorganic template for the formation of the supramolecular structure.

In the future, the selective bonding of molecular species to inorganic template structures, which would enable site direction, will also represent a major challenge for the successful combination of inorganic templates and molecular structures.

10.3
Molecular Self-Assembly

Self-assembly is a bottom-up technique that uses the self-organization capabilities of molecular building blocks – that is, the ability to rearrange continuously until a complete ordered monolayer of molecules is formed – to assemble desired nanostructures. As a result of the self-assembly process, the molecular constituents form an ordered structure with a minimum global energy on well-defined, atomically flat surfaces. The term "molecular self-assembly" is reserved for the adsorption of molecular constituents onto surfaces and the spontaneous organization into regular arrangements. If only non-covalent interactions are used to direct the molecular constituents into the resulting surface pattern, these structures are termed "supramolecular" (supramolecular chemistry = the chemistry of the intermolecular non-covalent bond [16]). On the other hand, the term "self-assembled monolayer" (SAM) is reserved, according to Whitesides [17], for a 2-D film with the thickness of one molecule that is attached to a solid surface through covalent bonds.

The surface properties of metals, metal oxides or semiconductors can be changed in a desired way by the adsorption of SAMs onto these materials. Therefrom, a number of useful applications result, such as: (i) the modification of adhesion and wetting control [18]; (ii) an increase in corrosion resistance [19]; or (iii) the development of heterogeneous chiral catalysts [20]. By exploiting the chemical properties of the organic molecules used, additional functionalities can be created, and consequently the development of chemical sensors [21] and chemical force microscopy [22], the site-selective adsorption of nanoscale subunits (see also Section 10.4.2), or the fabrication of electronic devices [23, 24], is possible. Additionally, SAMs themselves are nanostructures with nanoscale dimensions useful in nanolithography [25]. Supramolecular surface patterns on the other hand can be used to create nanocavities, to provide well-defined reaction spaces, and they may also control host–guest chemistry or steer heterogeneous catalysis [26]. Further details of the present state of molecular self-assembly on planar substrates are provided in a series of reviews [20, 26–30].

10.3.1
Attaching Molecules to Surfaces

Bare surfaces of metals and metal oxides tend to adsorb organic materials because the adsorbates lower the free energy of the interface between the respective material and the ambient environment. The character of the chemical bond between the adsorbed molecules and the metal surface determines the interfacial electronic contact and the strength of the geometric fixation. Two main groups of links between molecules and solids can be distinguished: (i) covalent bonds, which result from the overlap of partially occupied orbitals of interacting atoms; and (ii) non-covalent bonds, which are based on the electrical properties of the interacting atoms or molecules.

Planar molecules with extended π-systems have been found to physisorb onto surfaces, such as highly oriented pyrolytic graphite (HOPG), Au(1 1 1), Cu(1 1 0), in a flat-lying geometry. This allows functional groups at the molecular periphery to approach each other easily and to build up intermolecular interactions, predominantly comprising hydrogen bonds and metal–ligand interactions. If the molecules are sufficiently mobile to diffuse on the surface, then the intermolecular interactions will guide the adsorbed molecules into 2-D supramolecular systems. Then, by adjusting the molecular backbone size and the position or number of the functional "recognition groups", complex supramolecular nanostructures can be designed [31].

Covalent bonds are established, if there is a significant overlap of the electron densities of the molecules and the metal, and this will result in a strong electronic and structural coupling. The spontaneous formation of SAMs on substrates through covalent bonds requires organic molecules with a chemical functionality or "headgroup" and a specific affinity for a selected substrate. There exists a number of headgroups, which bind to specific substrates forming directed covalent links. One frequently used covalent link is the bond between a thiol group on the molecular site and a noble metal substrate. Here, gold is favorable due to its proper non-oxidizing surface, although thiol or selenol bonds are also possible to Ag, Pt, Cu, Hg, Ge, Ni, and even semiconductor surfaces. The reason for the great success of the S–Au bond is its good stability at ambient temperature, and the ease of reorganization to form an ordered array. Both are elementary requirements for the building up of a self-assembled monolayer.

Besides the prominent thiolates, other functional molecules, such as alcohols (ROH) or acids, have been demonstrated to form organized monolayers on metals or metal oxide surfaces, such as Al_2O_3, TiO_2, ZrO_2, or HfO_2. SAMs of alkylchlorosilanes ($RSiCl_3^-$) and other silane derivatives require hydroxylated surfaces as substrates for their formation. The driving force for this self-assembly is the *in-situ* formation of polysiloxane, which is connected to surface silanol groups (−SiOH) via robust Si−O−Si bridges [32]. Substrates on which these monolayers have successfully been prepared include silicon oxide, aluminum oxide, quartz, glass, and mica.

During the past few years, significant advances have been made by coupling alkenes and alkines onto Si and Si−H surfaces. The covalent coupling of vinyl compounds on H-terminated silicon yields very stable Si−C covalent bonds [33], and recently a method for the direct assembly of aryl groups on silicon and gallium arsenide using aryl diazonium salts has also been developed. There is a spontaneous ejection of N_2 and direct carbon–silicon formation [34], but the C–Si bonds are so strong that a facile reconstruction in order to form a highly ordered SAM is implausible.

10.3.1.1 Preparation of Substrates

For the deposition of a SAM, a 2-D film with a thickness of one molecule, a high-quality surface with a very low surface roughness is required. Depending on any further use of the SAM, the quality of the surface must be adapted. For example, if the SAMs are applied as etch resists, protection layers, chemical sensors or model surfaces for biological studies, then polycrystalline films will mostly suffice as substrates. In contrast, if the properties of the SAMs themselves are to be studied in detail, such as their organization, structure or electronic properties, then oriented single crystalline surfaces are required as substrates.

Planar substrates for SAMs are either thin films or single crystals of metals, semiconductors, or metal oxides. Thin films can be grown on silicon wafers, glass, single crystals or mica by CVD, physical vapor deposition (PVD), electrodeposition, or electroless deposition. Metal films on glass or silicon are polycrystalline and composed of grains that can range in size from 10 to 1000 nm.

As pseudo "single crystals", thin films of metals on freshly cleaved mica are commonly used. Gold films grow epitaxially with a strongly oriented (1 1 1) texture on the (1 0 0) surface of mica. The films are usually prepared by thermal evaporation of gold at rates of 0.1–0.2 nm s^{-1} onto a heated (400–650 °C) sample [35]. By using an optimized two-step process, a surface roughness down to 0.4 nm over areas of 5×5 µm can be achieved [36]. Surfaces with almost comparable roughness can be created by a method known as "template stripping" [37]. Here, a glass slide or a silicon wafer is glued to the exposed surface of a gold film on mica, and subsequently the gold film is peeled from the mica to expose the surface that had been in direct contact with the mica. Typically, these methods result ultimately in surface roughnesses of 1 nm over areas of 200×200 nm^2. For fundamental studies of SAMs by ultra-high vacuum (UHV) methods, single-crystal metal substrates provide the highest quality with respect to surface roughness, orientation, and cleanliness. These substrates result in densely packed SAMs of the highest order.

10.3.1.2 Preparation of Self-Assembled Monolayers

In principle, there are two possibilities of preparing SAMs, namely deposition from solution, and deposition from the vapor phase. For deposition from solution, a clean, freshly prepared substrate is immersed into a highly diluted solution of the corresponding organic molecules. After only a few minutes of immersion, a dense molecular monolayer is built; however, to ensure that the film reaches equilibrium the substrates are kept in solution for several hours to allow reorganization (Figure 10.14). In particular, the structure of the adsorbate determines the highest achievable density of the SAM on a given surface, or whether a SAM can be formed at all. The other parameters, such as solvent, temperature, concentration and immersion time, should be chosen adequately to achieve the best possible result. The

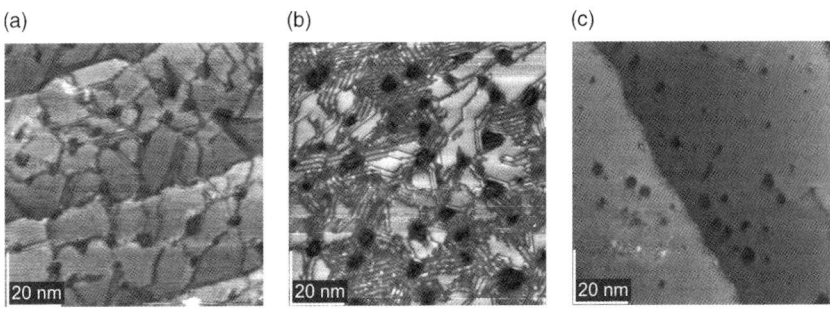

Figure 10.14 (a) Dodecanethiol SAM grown from solution. (b) SAM grown from solution. A 6.5-h annealing step at 78 °C in solution leads to a partial desorption of the dodecanethiol molecules, and results in the striped lying-down phase of alkanethiols [38]. (c) SAM grown from vapor phase. The domains extend over the whole gold terrace.

advantages of this method are the simplicity of the equipment and the ease of preparation.

In the case of alkanethiols on Au(1 1 1), the annealing procedure at elevated temperatures to increase the quality of the films has been studied extensively. Annealing the SAMs in a diluted solution of their molecules for short periods at 80 °C often results in a reduction in the number of vacancy islands, and an enlargement of the domain sizes due to *Ostwald ripening*. This behavior is explained by an intralayer diffusion of monovacancies towards larger holes, which grow at the expense of smaller holes. Furthermore, some vacancy islands diffuse towards the gold step edges and annihilate there, which explains the decrease in area occupied by the vacancy islands. In addition, the conformational defects in the SAMs decrease, and this will result in a higher order.

In the case of gas-phase deposition, UHV systems with base pressures in the range of 10^{-5} to 10^{-7} mbar are used. The amount of deposited molecules is controlled by the pressure, the temperature and the time. Vapor deposition has the advantage that absolutely clean surfaces can be used, a good control of the amount of deposited molecules is possible, and the SAM can be transferred to an analyzing tool without breaking the vacuum. By applying this method, submonolayers and highly ordered monolayers of extreme size can be created (Figure 10.14).

10.3.1.3 Preparation of Mixed Self-Assembled Monolayers

Mixed SAMs – that is, SAMs built up from different organic molecules and showing a well-defined structure – can be created in several ways; however, the two most widely used approaches will be described here.

The first method is coadsorption from solutions containing mixtures of selected organic molecules, and results in mixtures of molecular structures (Figure 10.15). This process allows the formation of SAMs with widely varying compositions and physical properties [39, 40].

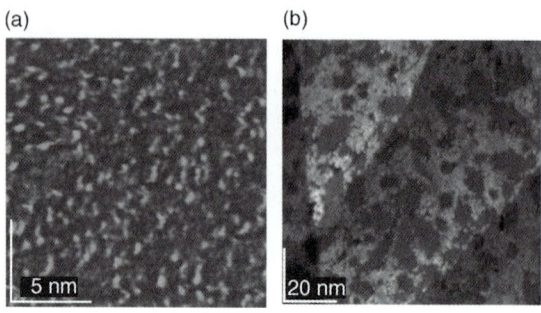

Figure 10.15 Scanning tunneling microscopy images of mixed monolayers. (a) By coadsorption of 11-mercaptoundecano-loctanethiol (=1:3) from solution [40]. (b) By insertion of a biphenylbutanethiol derivate into a closely packed SAM of dodecanethiol on Au(1 1 1). The film shows separate domains of the biphenylbutanethiol derivate (which appear higher) and dodecanethiol [43].

The second method is a two-step deposition process which begins with a full coverage monolayer of one organic species, which is used as the host matrix [41–43]. In a second step, the substrate covered with this host-matrix is immersed into the solution of an organic molecule of interest. Insertion of this "guest-molecule" takes place preferentially at defect sites such as pin holes or domain boundaries in the host matrix. The rate-determining step is the replacement of host molecules by guest-molecules. Depending on the immersion time, domains of inserted molecules, bundles or even single guest molecules can be identified in the resulting mixed monolayer (Figure 10.15). A well-ordered surrounding matrix can be used as the reference system for the analysis of the structural and electrical properties of the inserted molecules. This matrix-isolation method, in combination with scanning probe microscopy (SPM) techniques, is suitable for investigating series of organic molecules in order to determine new physical properties.

10.3.2
Structure of Self-Assembled Monolayers

The structure of SAMs is widely studied using spectroscopic methods including optical ellipsometry, reflectance absorption infrared spectroscopy (RAIRS), X-ray photoelectron spectroscopy (XPS), low-energy electron diffraction (LEED), and high-resolution electron energy loss spectroscopy (HREELS). In particular, the development of near edge X-ray absorption fine structure spectroscopy (NEXAFS) has led to new insights into the structure of SAMs. In addition, an increased understanding of the SAM structures has been achieved by the development and intense use of high-resolution topographic methods such as SPM.

During the self-assembly of organic molecules on planar substrates, complex hierarchical structures are formed involving multiple energy scales and multiple degrees of freedom (Figure 10.16). The geometric arrangement of organic molecules

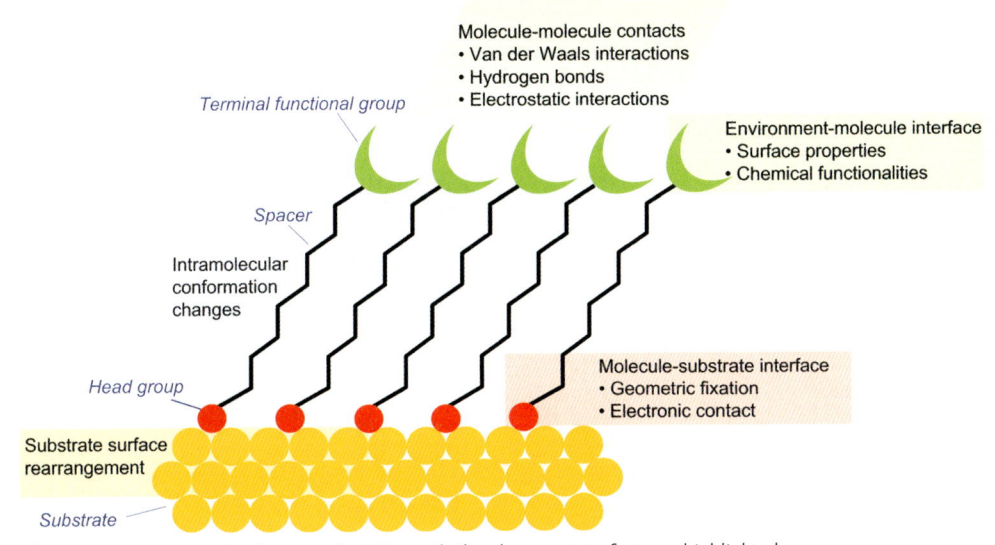

Figure 10.16 A schematic diagram of a SAM, with the characteristic features highlighted.

on a surface is determined in a first level of organization by the footprint of the molecule, the nearest-neighbor distances between the metal atoms at the surface, and the chemical bond formation of the molecules with the surface. The resulting 2-D density of the molecules on the surface may not correspond to the density that the same molecules can attain in crystalline form. In order to minimize the free energy of the organic layer, the molecules perform intramolecular conformation changes such as bond stretches, angle bends, or torsions, which in turn maximize the lateral interactions (e.g., van der Waals interactions, hydrogen bonds, or electrostatic interactions) in a second level of organization. The surface rearrangement of the substrate corresponds to a third level of organization. The balance of these forces determines the specific molecular arrangement, while the driving force is the minimization of the global energy.

10.3.2.1 Organothiols on Metals

The most studied – and probably best understood – SAM is the full-coverage phase of alkanethiols (R−SH) on Au(1 1 1) surfaces. The adsorbing species on the gold surface is the thiolate (RS$^-$), while the hydrogen atoms are desorbed in form of H_2 molecules with the gold surface acting as catalyst. The thus-formed Au−S bond that anchors the SAM is a strong homolytic bond with a strength on the order of approximately 200 kJ mol^{-1}. The alkanethiols are stabilized by van der Waals interactions between adjacent molecules. These dipole–dipole interactions are proportional to the alkyl chain length ($\sim$4.0 kJ mol^{-1} of stabilization to the SAM for each methylene group), and are responsible for the degree of order in the SAM.

A number of studies of alkanethiolate monolayers on gold have shown that the formed structure is commensurate with the sulfur atoms occupying every sixth hollow site on the Au(1 1 1) surface. The symmetry of the alkanethiolates is hexagonal with a ($\sqrt{3} \times \sqrt{3}$) R30° structure relative to the underlying Au(1 1 1) substrate, a S–S spacing of 0.4995 nm, and a calculated area per molecule of 0.216 nm^2. The alkanethiols are tilted $\sim$30° off the surface normal, and the hydrocarbon backbones are in all-*trans* configuration. Additionally, the alkanethiolates on Au(1 1 1) surfaces exhibit a c(4×2) superlattice which is characterized by a systematic arrangement of molecules showing a distinct height difference (Figure 10.17) [44]. The height differences in STM images are believed to be due to different conformations of the molecules.

Highly ordered SAMs can easily be built up from alkanethiols, although their structure is affected directly by the addition of any sterically demanding top-end group. The size and the chemical properties (e.g., high polarity) of additionally introduced surface functionalities may reduce the monolayer order.

10.3.2.2 Carboxylates on Copper

Compared to the extensive studies of organothiols on gold surfaces, very few investigations have been undertaken to study the self-assembly process of carboxylic acids on metal surfaces. The carboxyl group is known to be an anchoring group for the chemical bonding to metal surfaces. During the adsorption process of simple carboxylic acids the acid group is deprotonated into the carboxylate functionality, resulting in an upright adsorption configuration onto copper or nickel surfaces, as are observed for formic, acetic, and thiophene carboxylic acids [20]. The oxygen atoms in

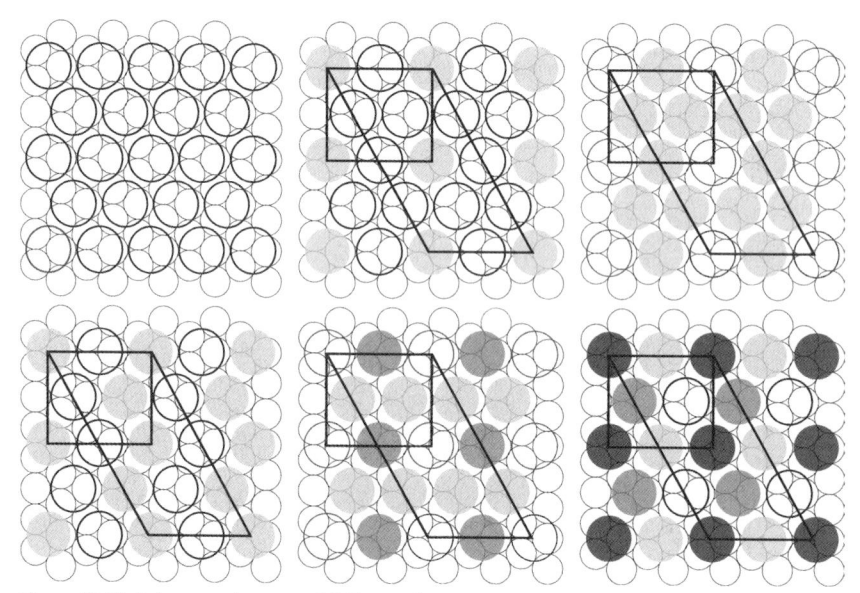

Figure 10.17 Schematic diagram of different phases for the superstructure of alkanethiols on Au(1 1 1) with the (3×2$\sqrt{3}$) and the c(4×2) superlattice unit cell outlined [44].

the carboxylate group are equidistant to the surface and form a rigid adsorption geometry.

More recently, tartaric acid adsorbed onto copper or nickel surfaces [20] has attracted interest because of a possible use in chiral technology. It is the aim of this technology to establish enantioselective catalytic methods to produce pure enantiomeric forms of materials such as pharmaceuticals and flavors. One way to create heterogeneous chiral catalysts is to adsorb chiral organic molecules at metal surfaces in order to introduce asymmetry. Tartaric acid has two chiral centers, and is therefore of potential interest as chiral modifier; indeed, recently it has been used successfully to stereodirect hydrogenation reactions with a yield of >90% of one enantiomer.

The self-assembly process of R,R-tartaric acid on Cu(1 1 0) under varying coverage and temperature conditions leads to a variety of different structures. Due to the two carboxylic acid functionalities, R,R-tartaric acid can adsorb in the monotartrate, the bitartrate, or the dimer form. It can be seen from Figure 10.18 that the bitartrate and the dimer–monomer assembly have no symmetry elements and create a chiral surface which is non-superimposable on its mirror image. This is a result of the inherent chirality of the R,R-tartaric acid molecules and their two-point bonding at the surface, which uniquely dictates the position of all its functional groups. Subsequently, the intermolecular interactions control the placement of the neighboring molecules. Due to the chirality of the adsorbates the lateral interactions are anisotropic and lead to organized chiral structures.

10.3.3
Supramolecular Nanostructures

Highly ordered 2-D supramolecular nanostructures can be created by using the MBE technique, or at the solid–liquid interface from solution [31] by tuning the molecular backbone size and controlling the supramolecular binding. This approach is based, in principle, on the concepts of supramolecular chemistry directing 3-D structures [16], but in the case of 2-D structures the influence of the substrate must also be taken into account [26].

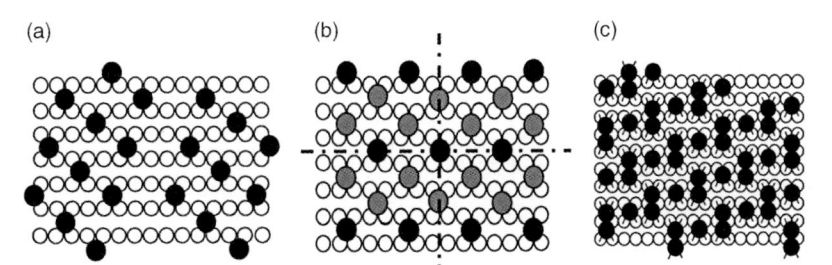

Figure 10.18 Adsorbate templates on Cu(1 1 0) surfaces created by (a) bitartrate (organizational chirality), (b) monotartrate (two symmetry planes), and (c) dimer–monomer assembly (organizational chirality) [20].

A supramolecular 2-D honeycomb network has been created by the assembly of two types of molecule on a silver-terminated silicon surface (this was discussed earlier, in Section 10.2.5) [13]. In the first step, a submonolayer of perylene tetra-carboxylic di-imide (PTCDI) was deposited by sublimation under UHV. Subsequently, melamine (1,3,5-triazine-2,4,6-triamine) was deposited while the sample was annealed at $\sim 100\,^\circ$C. The substrate allows a free diffusion of the molecules, and this makes formation of the supramolecular network possible. Furthermore, the compatibility of the molecular geometries results in three hydrogen bonds per melamine–PTCDI pair, an intentionally strong heteromolecular hydrogen bonding. In the 2-D honeycomb network the melamine molecules form three-fold connection sites, while the linear PTCDI molecules are used as one-dimensional linkers (Figure 10.19). This ordered array of pores can serve as traps for the co-location of several large molecules. By subliming C_{60} onto the hexagonal network, heptameric C_{60} clusters with a compact hexagonal arrangement are formed within the pores and are clearly stabilized by the PTCDI–melamine network.

The pronounced effect of the substrate in the self-assembly process of 2-D supramolecular structures becomes obvious, when prochiral molecules are used. Upon adsorption, these molecules lose their freedom of rotation and, in consequence, their symmetry and become chiral. Recently, the thermally induced switching of such prochiral molecules between different enantiomeric forms on the surface has been studied [45]. The molecule under investigation was a multiple-substituted phenylenethynylene oligomer (Figure 10.20a). The molecules align into rows and

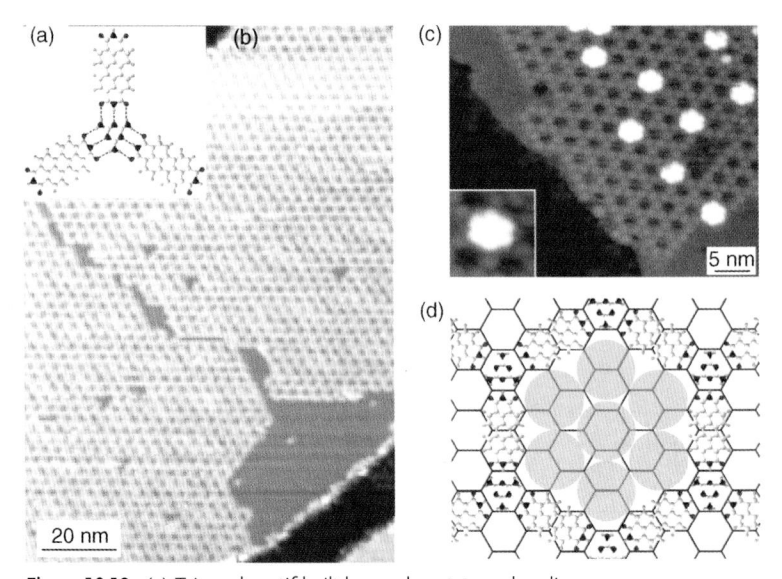

Figure 10.19 (a) Trigonal motif built by perylene tetracarboxylic di-imide and melamine. (b) STM image of a large-area PTCDI–melamine network. (c) STM image of C_{60} heptamers trapped in the pores of the PTCDI–melamine network. (d) Schematic diagram of C_{60} heptamer [13].

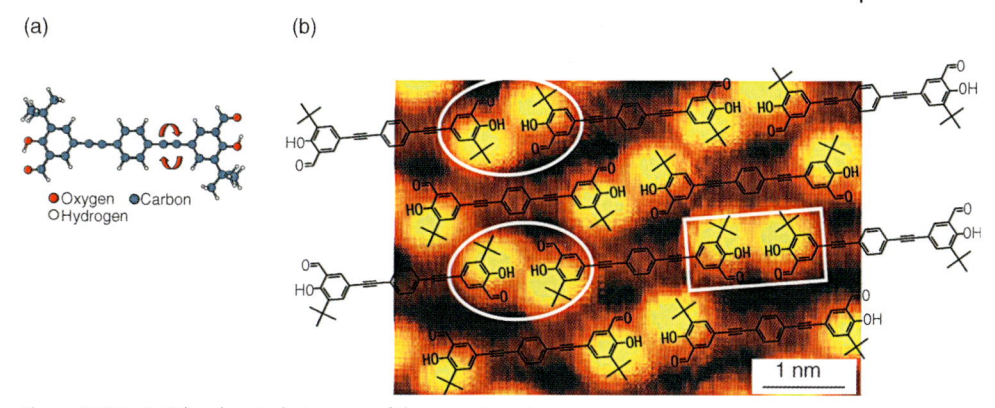

Figure 10.20 (a) The chemical structure of the investigated molecule. (b) A schematic model of the brick-wall adsorption structure superimposed on an STM image [45].

form a brick-wall structure on the Au(1 1 1) surface. The bright protrusions in the STM image (Figure 10.20b) may be attributed to *tert*-butyl groups at both ends of the molecules; the positions of these groups with respect to the molecular backbone identifies the enantiomeric form of the molecule. Two chiral enantiomers (LL and RR) and one achiral meso-form (LR/RL) of this molecule exist. The molecules are not completely stereochemically fixed by the substrate, but change between different surface conformers. An intermolecular *trans* configuration of the headgroups of two adjacent molecules has been found to exhibit the lowest potential energy ($\Delta \sim 4\,\mathrm{kJ\,mol^{-1}}$).

Supramolecular structures built by strong metal–ligand interactions are of high stability, and the incorporated metal centers offer additional functionalities. A system which forms a variety of 2-D surface-supported networks is based on iron (Fe) and aromatic dicarboxylic acids in different relative concentrations on copper surfaces. Mononuclear metal–carboxylate clusters are obtained from one Fe center per four tricarboxylic acid (TCA) molecules on Cu(1 0 0) surfaces [46]. The (Fe(TPA)$_4$) complexes form large, highly ordered arrays which are thought to be stabilized by substrate templating and weak hydrogen bonds between neighboring complexes (Figure 10.21a). From this a perfect arrangement of the Fe ions results, which cannot be achieved by using top-down methods.

A completely different network is obtained when two Fe atoms per three dicarboxylic acid (DCA) molecules are deposited onto the Cu(1 0 0) surface. The resulting array can be described as a ladder structure forming a regular array of nanocavities [26] (Figure 10.21b). The ladders are formed by metal–ligand interactions, while the connections between the ladders are formed by hydrogen bonds. If one Fe atom is deposited per linker molecule, a fully interconnected metal–ligand 2-D network results [47] (Figure 10.21c). By using DCAs of different length as linker molecules between the Fe centers, the size of the resulting nanocavities in the network can be tuned.

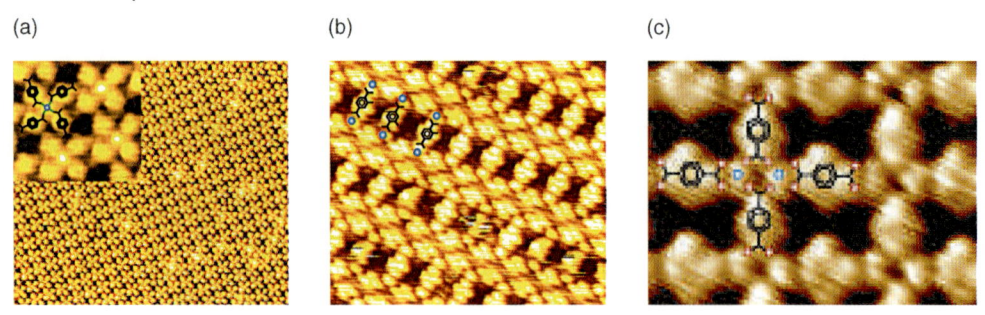

Figure 10.21 Supramolecular assembly of Fe-carboxylate coordination systems on Cu(100) substrates. STM images and schematic models: (a) mononuclear complexes [46]; (b) ladder structure [26]; (c) coordination network [47].

10.3.4
Applications of Self-Assembled Monolayers

10.3.4.1 Surface Modifications

One potential application for SAMs is the modification of surfaces in order to change the surface properties for special applications. For example, polar surfaces can be created by the adsorption of SAMs with terminal groups such as cyano (C≡N). These polar surfaces are useful for the investigation of dipole–dipole interactions in surface adhesion. On the other hand, SAMs with terminal OH groups can vary wetting behaviors, and are used in investigations to study the importance of H-bonding in surface phenomena. Additionally, surface OH and COOH groups – and especially acid chlorides – are very useful groups for chemical transformations. For example, reacting the acid chloride with a carboxylic acid-terminated thiol provides the corresponding thioester. The control of surface reactions opens up the way to chemical sensors [21], and is the basis of chemical force microscopy [22].

10.3.4.2 Adsorption of Nanocomponents

Mixed SAMs containing two or more constituent molecules can be used as test systems to study the interactions of surfaces with bioorganic nanocomponents (proteins, carbohydrates, antibodies). Usually, the SAM contains alkanethiols with a surface terminal group of interest (e.g., suitable for hydrophobic or hydrophilic interactions) and an alkanethiol with a reactive site for linking to a biological ligand. SAMs make it possible to generate surfaces with anchored biomolecules that remain biologically active and in their native conformations.

Additionally, it is possible to use the specific chemical binding properties of SAM surface groups to direct nanocomponents into desired structures. This approach has been used extensively to form selected assemblies of nanoparticles, and opens up a pathway to a variety of different structures (this subject is discussed in detail in Section 10.4.2). Another example of the fabrication of desired structures due to the binding properties of SAMs is the directed assembly of carbon nanotubes (CNTs). Recently, a method was developed which is based on the observation that CNTs are

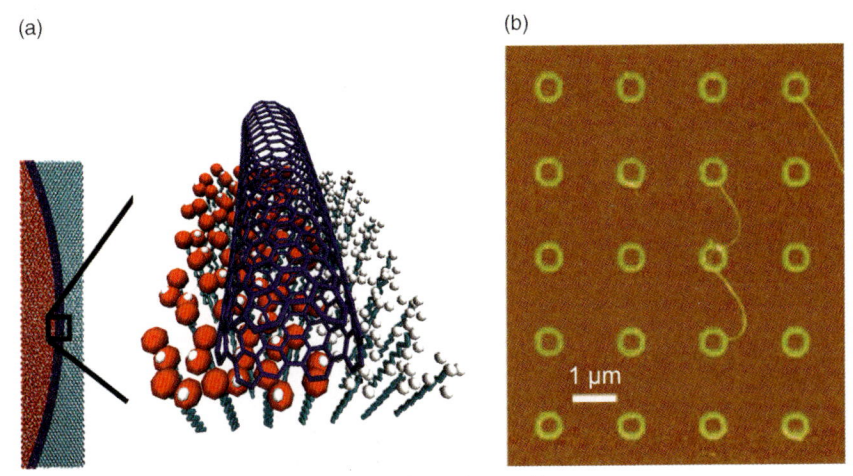

Figure 10.22 (a) Schematic diagram of the directed assembly process. (b) AFM tapping mode topographic image of carbon nanotubes assembled into rings [48].

strongly attracted to COOH-terminated SAMs or to the boundary between COOH- and CH$_3$-terminated SAMs. By using nanopatterned "affinity templates", desired structures of CNTs can be formed (Figure 10.22). Useful methods for the generation of appropriate templates include dip-pen nanolithography (see Chapter 8) and micro- or nano-contact printing [48].

10.3.4.3 Steps to Nanoelectronic Devices

Molecular electronics requires several structural elements such as wires, diodes, switches, and transistors in order to build up nanodevices. In the studies conducted by Weiss and colleagues [24], conjugated oligophenylene ethynylenes (OPE) have been investigated, which possess potentially interesting features, including negative differential resistance (NDR) (increased resistance with increasing driving voltage), bistable conductance states, and controlled switching under an applied electric field. Single OPEs have been studied in a 2-D isolation matrix of host SAMs of dodecanethiolate on a gold electrode. As a result, series of surface images (Figure 10.23) showed the conductance switching due to conformational changes of the OPE molecules with a low rate, if the surrounding matrix was well ordered. Conversely, when the surrounding matrix was poorly ordered, the inserted molecules switched more often [24]. The switching of OPE molecules can only be observed in arrays of small bundles of molecules. Therefore, it is assumed that the forming and breaking off of hydrogen bonds between adjacent molecules – and the consequent twisting of the molecule, which prevents conjugation of the π-orbitals of the molecular backbone – is responsible for the two conduction states. Such a device, constituted by a switching molecule attached to a bottom electrode and a conductive tip, represents a simple form of a memory.

According to Avriam and Ratner in 1974 [49], the working principle of a molecular diode should be based on two separated electron–donor and electron–acceptor

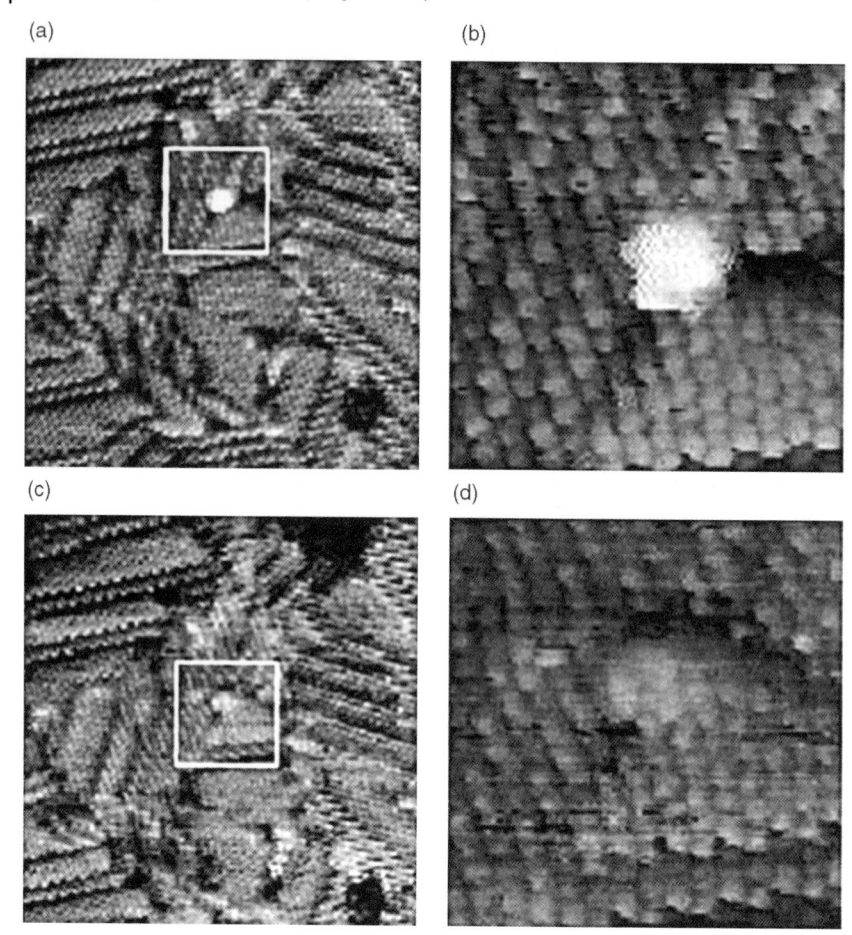

332 | *10 Formation of Nanostructures by Self-Assembly*

Figure 10.23 Topographic STM images of a molecular switch (OPE) inserted in a dodecanethiol SAM [24]. (a) A 20×20 nm image of the molecule in the ON state; (b) a 5×5 nm image of the same area; (c) 20×20 nm; and (d) 5×5 nm images of the same area, with the molecule in the OFF state.

π-systems (see Chapter 24). However, in recent years asymmetric electron transmission through symmetrical molecules has been also observed by SPM investigations on the molecular level [50]. Diode behavior is possible for symmetric molecules, if they are connected asymmetrically to the electrodes – that is, with two different molecule–electrode spacings, or if asymmetric electrodes are used. A further development of this idea leads to the assumption, that engineering the frontier orbitals of the molecule in the asymmetric junction should make it possible to control the orientation of the diode – that is, whether an alignment of the cathode to the LUMO of the molecule or of the anode to the HOMO is achieved at lower bias. If, additionally, the frontier orbitals of the molecule can be changed reversibly by an electrical pulse, then an optical pulse

10.3 Molecular Self-Assembly

or a chemical reaction a transistor would result. An example of such a molecular transistor device based on a supramolecular assembly is given by Rabe and coworkers [51], who used a hexa-peri-hexabenzocoronene (HBC) derivative with six electron-accepting anthraquinones (AQs) symmetrically attached to the HBC, and has the function of an electron donor. The resulting HBC–AQ$_6$ molecules (Figure 10.24a) were investigated at the HOPG/solution interface, where they form monolayers with an ordered structure. The identification of the conjugated HBC cores and the attached AQ molecules, as well as the recording of the current–voltage (I–V) curves through HBC cores, AQs and alkyl chains was possible by using STM/scanning tunneling spectroscopy (STS).

In a next step, the frontier orbitals of the HBC–AQ$_6$ molecules were intentionally changed in order to vary the electron transmission properties of the HBC cores. This was achieved by the addition of 9,10-dimethoxyanthracene (DMA) to the solution. DMA is an electron donor which is known to build a charge-transfer complex with AQ. It is remarkable that two different I–V curves through the HBC core are observed, depending on whether or not charge-transfer complexes are coadsorbed next to HBC.

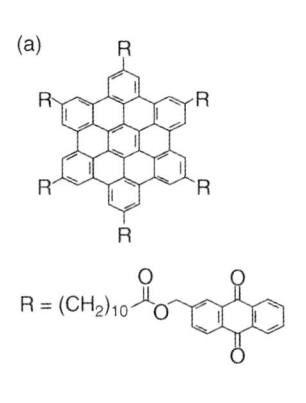

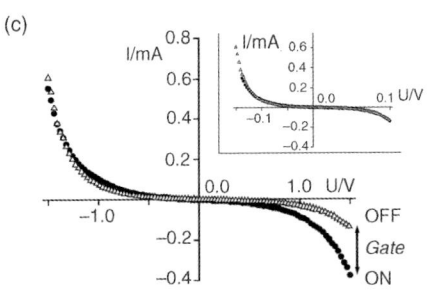

Figure 10.24 (a) Chemical formula of hexa-peri-hexabenzocoronene (HBC) decorated with six anthraquinone (AQ) functions. (b) STM current image of HBC–AQ$_6$ molecules with coadsorbed charge-transfer (CT) complexes. (c) Current–voltage (I–V) relationships through HBC cores in domains where the charge-transfer complexes are adsorbed, or where no charge-transfer complexes were present. (d) Schematic of a prototypical single-molecule chemical field effect transistor (CFET) [51].

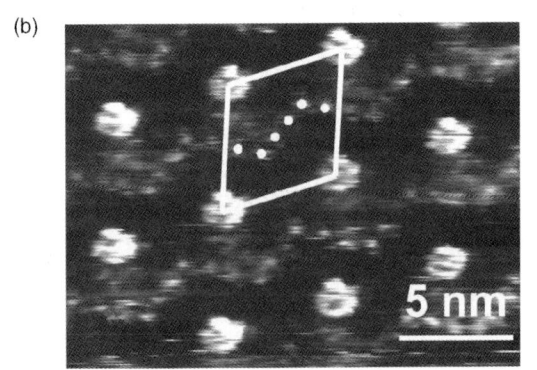

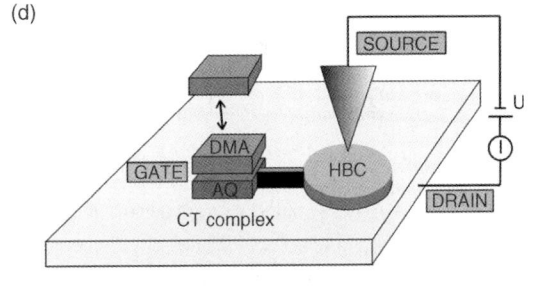

This set-up can be regarded as a "single-molecule chemical field-effect transistor", as the change in the I–V relationship results from the chemical formation/solution of a charge-transfer complex (= gate) which alters the electron transmittance through the covalently attached HBC (= channel) (Figure 10.24). Despite the fact, that the gates cannot be addressed selectively and the device structure changes simultaneously with the electron transmission properties, this approach is a major step towards monomolecular electronics with a complete transistor integrated into one molecule.

10.4
Preparation and Self-Assembly of Metal Nanoparticles

In addition to the previously discussed two nanofabrication methods of epitaxial growth and molecular self-assembly, metal nanoparticles also play an important role in the context of nanofabrication. The extraordinary size-dependent electronic, magnetic and optical properties [52] of nanoparticles have triggered many fascinating ideas for potential applications in breakthrough future technologies, including sensors, medical diagnostics, catalysis, and nanoelectronics. Therefore the preparation and the question of how to assemble metal nanoparticles remain objectives of great interest. The major challenges that are still to be overcome in this context are, on the one hand, the preparation of (ideally) monodisperse metal nanoparticles with simultaneous control over size, shape and composition; and on the other hand, the controlled assembly. Much effort has been expended in attempts to prepare metal nanoparticles of different sizes and shapes, to assemble them into three, two or even one dimensions, and to study and to understand their physical properties. The results of these investigations form the fundamental knowledge for potential applications in nanotechnology. Some of these synthetic routes and self-assembly patterns are outlined in the following paragraphs.

10.4.1
Preparation of Metal Nanoparticles

Generally, metal nanoparticles are prepared by the reduction of a soluble metal salt via suitable reducing agents, or via electrochemically [53–55] or physically assisted methods (e.g., thermolysis [56], sonochemistry [57], photochemistry [58]), or directly via the decomposition of labile zero-valent organometallic complexes. In all cases the synthesis must be performed in the presence of surfactants, which form SAMs on the nanoparticles surfaces (see also Section 10.3) and thus stabilize the formed nanoparticles. The stabilizing effects of the surfactants refer to: (i) steric effects, meaning stabilization due to the required space of the ligand shell; and (ii) electrostatic effects, implying stabilization due to coulombic repulsion between the particles. Furthermore, the surfactants influence the size, shape, and the physical properties and assembly patterns of the nanoparticles. In recent years, many excellent reviews have been produced providing detailed overviews on the preparation techniques, properties and surfactant influences [59–63]. In this context it should be mentioned that, because of the protecting SAM on the surface, ligand-stabilized nanoparticles are also sometimes denoted as monolayer-protected clusters (MPCs). However, in comparison to SAMs on planar surfaces, the structures of the SAMs on the nanoparticles surfaces differ greatly due to the surface curvature [64].

Important – and already more or less standardized – examples of the preparation of non-stoichiometrically composed metal nanoparticles via the reduction of a metal salt with a suitable reducing agent include the route of Turkevich et al. [65] and that of Brust et al. [66, 67]. Turkevich and colleagues were the first to introduce a standardized method for the preparation of gold nanoparticles with diameters ranging from 14.5 ± 1.4 nm to 24 ± 2.9 nm, via the reduction of $HAuCl_4$ with sodium citrate in water. Thereby, the nanoparticle size can be controlled by variation of the ratio $HAuCl_4$/sodium citrate. This route is often applied due to the fact that citrate-stabilized gold nanoparticles can simply be surface-modified because of a weak electrostatically bound, and thus easily exchangeable, citrate ligand. Brust et al. utilized sodium borohydride as a reducing agent, and took advantage of the high binding affinity of thiols to gold; this enabled the preparation of relatively stable nanoparticles that could be precipitated, redissolved, analyzed chromatographically, and further surface-modified without any apparent change in properties. This high stability represents an important property in terms of controlling nanoparticle assembly.

A recently published report described the surfactant-free synthesis of gold nanoparticles [68]. This approach is especially interesting in terms of the preparation of small gold nanoparticles with narrow dispersity, protected by ligands carrying functional groups that are typically not stable towards reducing agents. In this method, a solution of $HAuCl_4$ in diethyleneglycol dimethyl ether (diglyme) is reduced by a solution of sodium naphthalenide in diglyme to yield weakly solvent-molecule-protected gold nanoparticles. In the first step, these formed nanoparticles are further stabilized and functionalized simply by the addition of various ligands (1-dodecanethiol, dodecaneamine, oleylamine and triphenylphosphine sulfide). The size of the nanoparticles can be tuned within the range of 1.9 to 5.2 nm, with dispersities of 15–20% depending on the volume of the added reduction solution and the time between addition of the reduction solution and the ligand molecule solution.

Magnetic nanoparticles, such as cobalt or iron nanoparticles, are typically prepared via the decomposition of a zero-valent organometallic precursor, for example carbonyl metal complexes. One example is the synthesis of monodisperse ($\pm$ one atomic layer) Co and Fe nanoparticles with sizes of approximately 6 nm via thermal decomposition of the respective carbonyl compounds ($Fe(CO)_5$, $Co_2(CO)_9$) under an inert atmosphere [69]. In this way the nanoparticle size can be controlled by adjusting the temperature and the metal precursor:surfactant ratio. For example, higher temperatures and higher metal precursor:surfactant ratios produce larger nanoparticles. An additional control parameter is the ratio of the surfactants tributyl phosphine and oleic acid, both of which bind to the nanoparticle surface. Tributyl phosphine binds weakly, allowing rapid growth, while oleic acid binds tightly and favors slow growth to produce smaller particles. As might be expected, iron

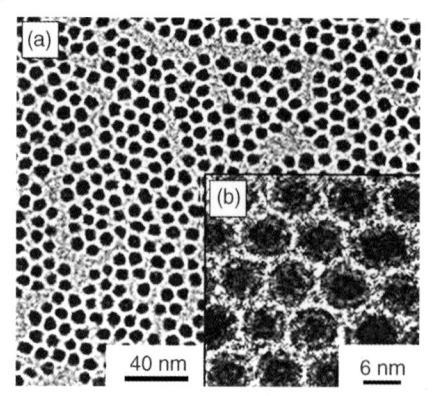

Figure 10.25 (a) TEM image of an ensemble of 6-nm iron nanoparticles. (b) At higher magnification, the surface oxide layer is clearly visible. (Illustration reprinted from Ref. [69], with kind permission.).

nanoparticles show great sensitivity towards oxidation in air; even a short contact of the nanoparticle surface with air resulted in the formation of an Fe_3O_4 layer with a thickness of approximately 2 nm (Figure 10.25).

The thermal decomposition of metal carbonyl complexes for the preparation of nanoparticles or nanostructured materials can also be achieved by treatment with ultrasound. Treatment of a liquid with ultrasound causes the formation, growth and implosive collapse of bubbles in the liquid, and this in turn generates a localized hot-spot [70]. As an example, amorphous Fe/Co nanoparticles are prepared by the sonolysis of $Fe(CO)_5$ and $Co(NO)(CO)_3$ in decanediphenylmethane at 293–300 K under an argon atmosphere, to produce pyrophoric amorphous Fe/Co alloy nanoparticles. Annealing of these particles in an argon atmosphere at 600 °C leads to growth of the Fe/Co particles, and this finally yields air-stable nanocrystalline Fe/Co particles due to carbon coating on the surface [70].

While the above-mentioned examples were all non-stoichiometrically composed gold nanoparticles, one famous example of a stoichiometrically composed gold nanoparticle – and thus great control over size-dependent properties – is the so-called *Schmid cluster* $[Au_{55}(PPh_3)_{12}Cl_6]$, which was introduced in 1981 [71]. The cluster is prepared by the reduction of $Au(PPh_3)Cl$ with *in-situ*-formed B_2H_6 in warm benzene. The relevance of this cluster refers to its quantum size behavior and to the fact that it can be regarded as a prototype of a metallic quantum dot [72, 73]. The defined stoichiometric composition of $Au_{55}(PPh_3)_{12}Cl_6$ is based on the so-called "full-shell cluster principle", whereby the cluster is seen as a cut-out of the metal lattice of the bulk metal. This implies that the cluster consists of a metal nucleus surrounded by shells of close-packed metal atoms, so that each shell has $10n^2 + 2$ atoms (n = number of shells) [59, 74]. Further examples in this context are $[Pt_{309}phen^*_{36}O_{30}]$ (four-shell cluster) and $[Pd_{561}phen_{36}O_{200}]$ (five-shell cluster) (phen* = bathophenanthroline; phen = 1,10-phenanthroline) [75–77].

10.4.2
Assembly of Metal Nanoparticles

As mentioned above, the unique physical properties of metal nanoparticles with diameters of between one and several tens of nanometers make them promising building blocks for the construction of functional nanostructures. Furthermore, it was found that assemblies of nanoparticles show physical properties that are situated between those of an isolated cluster and the bulk material; this in turn would lead to a new class of materials, the properties of which are affected by the nanostructure itself. Arrays of nanoparticles exhibit delocalized electron states that depend on the strength of the electronic coupling between the neighboring nanoparticles, whereby the electronic coupling depends on the particle size, including the particle size distribution, the particle spacing, the packing symmetry and the nature and the covering density of the stabilizing surfactant [78, 79]. Thus, major efforts are under way to organize nanoparticles into one to three dimensions in order to investigate electronic, magnetic and optical coupling phenomena within such assemblies, and even to utilize these coupling effects for the set-up of novel nanoelectronic, diagnostic, or nanomechanical devices [80].

The assembly principle to achieve large ordered arrangements of nanoparticles is the self-organization of ligand-protected nanoparticles due to weak ionic or van der Waals interactions, or due to strong covalent bond formation via respective functional ligand-protected nanoparticles. Some examples of building up 1-D to 3-D nanoparticle arrangements via self-organization are presented in the following sections.

10.4.2.1 Three-Dimensional Assemblies

The easiest achievable construction scheme is the self-assembly into three dimensions which, when occurring spontaneously, is in principle *crystallization*. For example, $Au_{55}(PPh_3)_{12}Cl_6$ crystallizes from dichloromethane solution if the solvent is evaporated, yielding small hexagonal microcrystals. Such a microcrystal obtained in this way is shown in Figure 10.26 [81].

The self-assembly of FePt nanocubes (see also Section 10.3.1) during controlled evaporation of the solvent from a hexane dispersion leads to (1 0 0) textured arrays. This assembly is energetically favored, as it gives the maximum van der Waals interaction energy arising from face-to-face interactions in a short distance of the cube assembly. The interparticle distance is approximately 4–5 nm, which is close to the simple thickness of the surfactant layer (2–2.5 nm, the length of oleate or oleylamine). Interestingly, thermal annealing induces an internal particles structure change and transforms the nanocube assembly from superparamagnetic to ferromagnetic. The study cited here is an example of the influence of particle shape on the assembly scheme.

Co, Ni, and Fe nanoparticles self-assemble from solution to form close-packed nanoparticle arrays on a variety of substrates if the dispersing solvent is evaporated. A TEM image of such a hexagonal superlattice with rows of 8-nm mt-fcc Co nanoparticles aligning to form facets is shown in Figure 10.27. Such 3-D magnetic

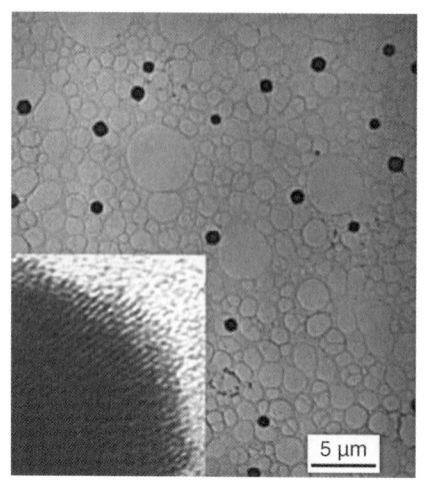

Figure 10.26 TEM image of typical microcrystals of $Au_{55}(PPh_3)_{12}Cl_6$ formed on the grid from dichloromethane solution. The inset shows a high-resolution TEM image of a Au_{55} $(PPh_3)_{12}Cl_6$ microcrystal. (Illustration reprinted from Ref. [81], with kind permission.).

nanoparticle arrays provide important models by which collective interparticle interactions may be explored [69].

Covalently linked gold nanoparticles can be obtained by using bifunctional ligands that are able to interconnect nanoparticles. Brust *et al.* were the first to describe the formation of gold nanoparticle networks by using dithiol molecules as ligands [82]. Their method involved the preparation of gold nanoparticles in a two-phase liquid–liquid system [83], whereby dithiols rather than monothiols were used. The use of dithiols leads directly to the formation of an insoluble precipitate of dithiol crosslinked clusters. The existence of self-assembled nanoparticles is deduced from TEM images (Figure 10.28).

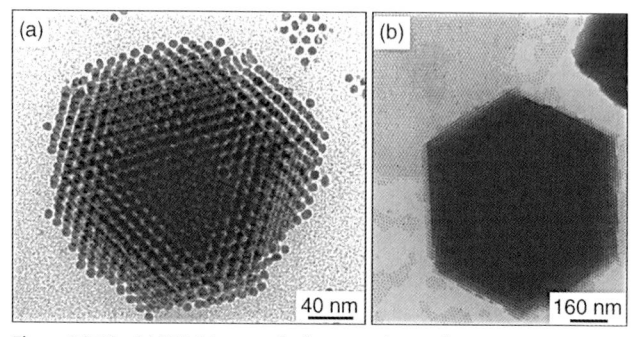

Figure 10.27 (a) TEM image of a hexagonal superlattice grown from 8-nm mt-fcc Co NPs. (b) Lower-magnification image of large hexagons formed in the sample. (Illustration reprinted from Ref. [69], with kind permission.).

10.4 Preparation and Self-Assembly of Metal Nanoparticles | 339

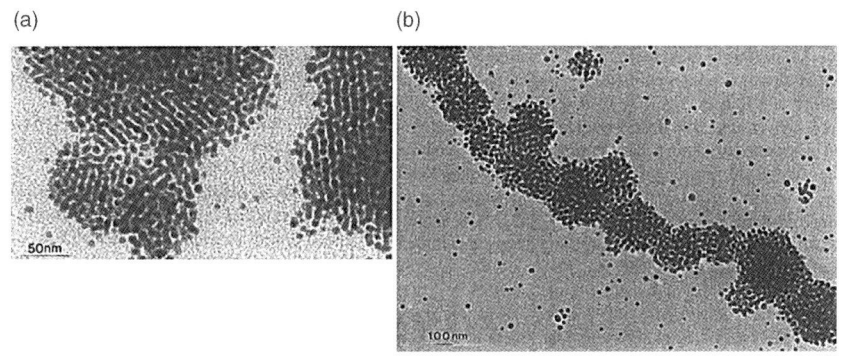

(a)　　　　　　　　　　　　　(b)

Figure 10.28 (a) TEM image of 8-nm gold nanoparticles crosslinked with 1,9-nonanedithiol, showing the parallel alignment of adjacent particles. (b) TEM image of 8-nm gold nanoparticles crosslinked with 1,9-nonanedithiol, showing a self-assembled string of "superclusters", which is a typical feature of these preparations. (Illustration reprinted from Ref. [82], with kind permission.).

The influence of the ligand, the type, and the linker length on charge-transport properties in metal nanoparticle assemblies has been studied for a variety of cases. For example, the insertion of bifunctional amines into the 3-D arrangement of $Pd_{561}phen_{36}O_{200}$ clusters yields an increased interparticle spacing compared to the closed sphere packing obtained from solution. The increased interparticle spacing is reflected in an increase of the activation energy of the electron transport through the material [84]. The influence of ligand type was recently discussed for the case of Au_{55}-cluster arrangements by Simon and Schmid [85]. When comparing the charge-transport properties of networks of Au_{55}-clusters interconnected by either weak ionic interaction or by covalent bond formation between bifunctional ligands, it transpired that both types showed characteristically different charge-transport properties. The non-covalently interconnected cluster systems showed a continuous increase of activation energy for the charge transport with increasing interparticle distance, whereas the covalently linked cluster systems showed a decrease in activation energy, to significantly lower values [85].

10.4.2.2 Two-Dimensional Assemblies: The Formation of Monolayers

In general, the assembly of nanoparticles into two dimensions is achieved by binding the metal nanoparticles onto a substrate surface. Thus, in order to direct the assembly into a defined pattern, the key point is the existence of functional groups on the substrate surface which enable a specific interaction between the nanoparticle and the surface. These interactions may be either weak electrostatic forces or weaker van der Waals forces; alternatively, covalent bonds may be formed between the monolayer-protected nanoparticle and the substrate surface. The weak interactions have the advantage that they allow a reasonable mobility on the surface, and so enable ordering due to self-organization. Covalent bond formation has the advantage of building more stable and durable arrays. However, in order to enable electrostatic or covalent interactions, the substrate surface must be adequately modified (see also Section 10.2). The formation of 2-D gold nanoparticle arrays is presented in the following examples.

One example of a 2-D assembly due to electrostatic interactions is the assembly of hydrophilic $Au_{55}(Ph_2PC_6H_4SO_3H)_{12}Cl_6$-clusters on a poly-(ethyleneimine)-coated TEM-grid, as reported by Schmid and coworkers [86]. The driving force for the deposition of colloids on the surface is the acid–base interaction between the -SO_3H group of the ligand and the NH-group of the imine. This procedure leads to a close packing which can be visualized by using TEM (Figure 10.29).

A recently published example for the controlled assembly of gold nanoparticles into two dimensions, and which utilizes electrostatic forces, is the assembly of citrate-protected gold nanoparticles on carbon surfaces [87]. The key point here is that the carbon surfaces are electrochemically modified with primary amines (n-hexylamine, tetraethylene glycol diamine). This electrochemical surface modification method allows the number of amine functionalities on the surface to be controlled, and in turn allows control to be exerted on the density of the nanoparticle assembly.

Another example, which is within the context with the electrical characterization of 2-D arrays of gold nanoparticles, is the electrostatic adsorption of 15-nm gold nanoparticles on 3-aminopropyltrimethoxysilane (APTS)-modified silicon substrates. Thereby, the particles are deposited from an aqueous solution (pH 5) to yield a densely packed monolayer of gold colloids with an average density of

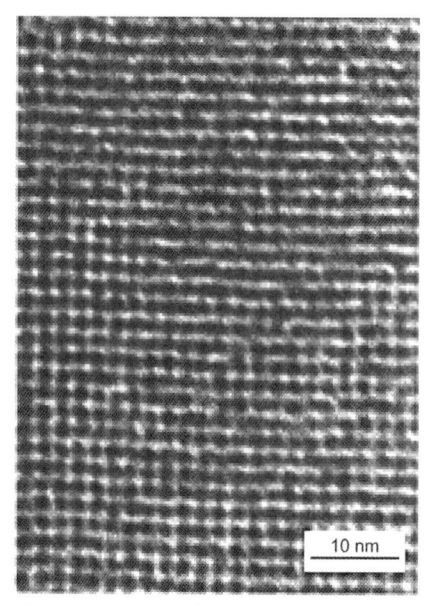

Figure 10.29 $Au_{55}(PPh_3)_{12}Cl_6$-clusters fixed on a PEI-coated grid, as imaged with TEM. (Illustration reprinted from Ref. [86], with kind permission.).

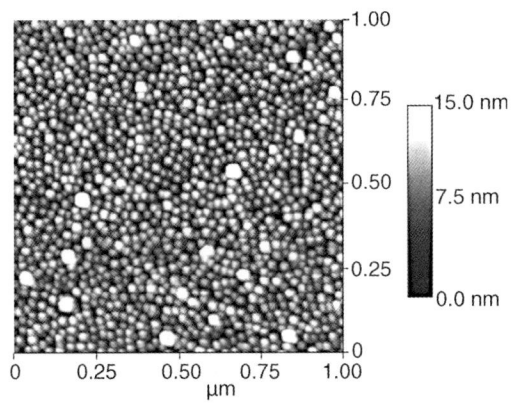

Figure 10.30 AFM image of densely packed citrate stabilized gold nanoparticles on amino-functionalized silicon surfaces.
(Illustration reprinted from Ref. [88], with kind permission.).

approximately 1500 particles per μm^2 (Figure 10.30) [88]. The average interparticle spacing is adjusted by the thickness of the citrate shell.

SAMs based on the covalent attachment of nanoparticles were presented by Schmid and coworkers, who described the formation of 2-D arrangements of ligand-stabilized gold clusters and gold colloids on various inorganic conducting and insulating surfaces [89]. For this purpose, oxidized silicon as well as quartz glass surfaces were treated with (3-mercaptopropyl) trimethoxysilane to generate monolayers of the SH-functionalized silane. When dipped into an aqueous solution of 13-nm gold colloids, stable covalent S–Au bonds were formed, thus fixing the colloids in a highly disordered arrangement. The coating of the surface was visualized using AFM.

Microcontact printing (μCP) is also used to create nanoparticle surface assemblies based on covalent forces, and with predefined positions of the nanoparticles within these arrays. A recent example of μCP use was the chemically directed assembly of monolayer-protected gold nanoparticles on lithographically generated patterns [90]. Here, gold surfaces patterned with mercaptohexadecanoic acid (MHA) were prepared using μCP and dip-pen nanolithography. A diamine molecule was used to link the mercaptoundecanoid acid-coated gold nanoparticles onto the MHA-defined patterns. Sonication was then used to remove the non-specifically absorbed nanoparticles, and the nanoparticle assembly was proven by using AFM, with height increases of 2–6 nm with respect to the preformed MHA SAM (Figure 10.31).

10.4.2.3 One-Dimensional Assemblies

The 1-D assembly of nanoparticles remains a major challenge, as 1-D (or at least quasi-1-D) assemblies require appropriate nanoparticle surface modification, appropriate templates, or special techniques (including special substrate modifications, e.g., AFM-based methods).

One recently published example is the spontaneous quasi-1-D arrangement of spherical Au nanoparticles protected by a liquid crystal ligand (the 4′-(12-mercapto-

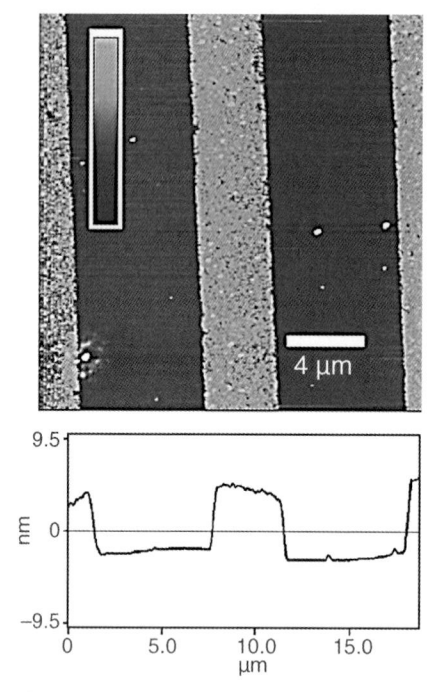

Figure 10.31 Contact mode AFM height image (top image) and average height profile (bottom image) of gold nanoparticles chemically directed onto mercaptohexadecanoic acid (MHA) features patterned by microcontact printing (μCP). The lines show an average height of 6.5 nm (the average height profile is shown at the bottom of the figure), and a 6 nm increase over the feature height after nanoparticle assembly. The inset shows the height scale bar of 20 nm). (Illustration reprinted from Ref. [90], with kind permission.).

dodecyloxy)biphenyl-4-carbonitrile). Thereby, gold nanoparticles protected by a ligand consisting of a liquid crystal mesogen unit and an alkane thiol unit spontaneously ordered themselves just by a simple thermal treatment, without the sue of any templates. The length of the arrangement was 1 to 60 nm, and the inter-array distance was approximately 7 nm (Figure 10.32) [91].

Another approach that utilizes covalent forces is the chemically directed assembly of gold nanoparticles on a thiol-patterned silicon surface [92]. The patterning of the silicon surface is predefined using AFM, and achieved via *chemical force microscopy* [22]. The process is as follows. A silicon surface is modified with a monolayer of 3,5-dimethoxy-α,α-dimethylbenzyloxycarbonyl (DZZ)-protected thiol (DZZ is a photocleavable group typically used in organic synthesis and lithography). It transpired that the application of a voltage between an AFM tip and a selected location on the monolayer yielded deprotection (Figure 10.33), and consequently it was possible to "write" a thiol-end-group pattern into the given monolayer. Following treatment with a 10-nm citrate-stabilized gold nanoparticle solution, the particles were chemisorbed via thiol–gold bond formation. In this way, single lines of gold nanoparticles can be produced, even with a defined particle separation (Figure 10.34) [92].

10.4 Preparation and Self-Assembly of Metal Nanoparticles

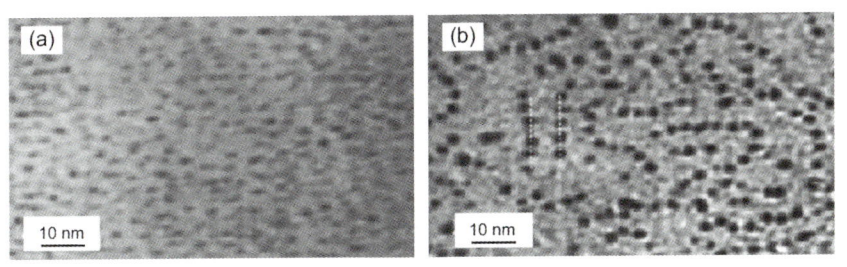

Figure 10.32 TEM images of Au nanoparticles with liquid crystalline ligands, (a) before and (b) after thermal treatment. (Illustration reprinted from Ref. [91], with kind permission.).

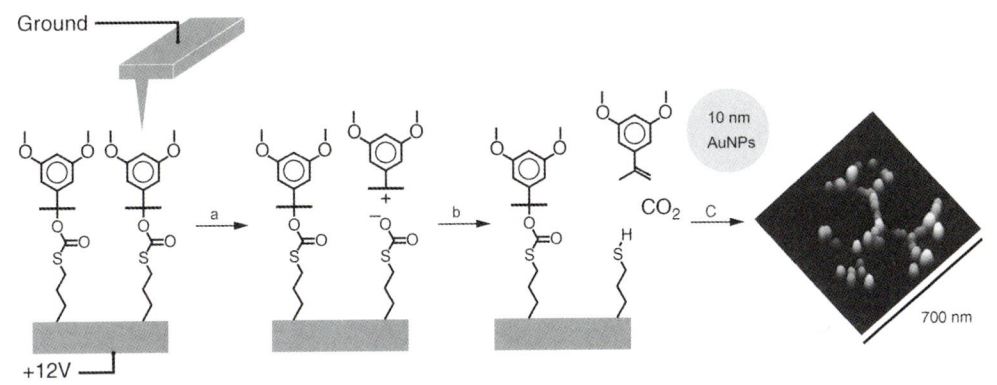

Figure 10.33 (Step a) Electrically stimulated bond cleavage. (Step b) Elimination of carbocation and release of carbon dioxide to produce surface-bound thiol. (Step c) Directed self-assembly of gold nanoparticles into a dendrimer pattern. (Illustration reprinted from Ref. [92], with kind permission.).

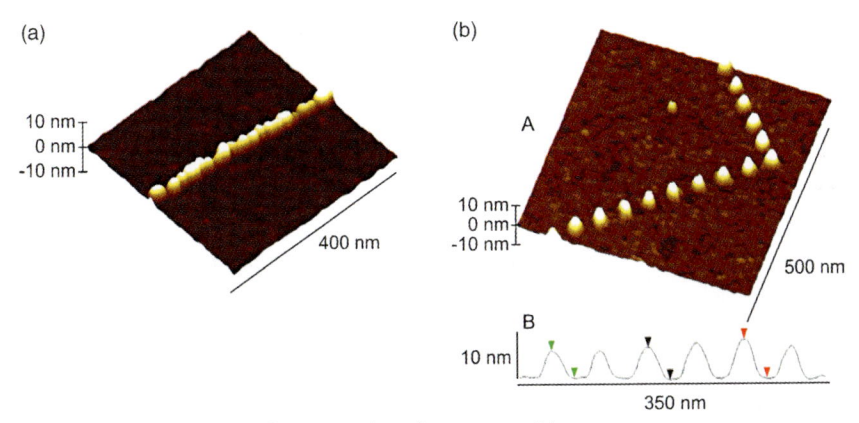

Figure 10.34 (a) A line of gold nanoparticles, of one nanoparticle width. (b) (A) Gold nanoparticles patterned with a 50-nm spacing between individual particles. (B) Cross-sectional analysis of the pattern. (Illustration reprinted from Ref. [92], with kind permission.).

10.5
Conclusions

In this chapter, we have discussed the basic principles and selected examples of the formation of nanostructures via the self-assembly of atoms by epitaxial growth, of molecules, and of metal clusters. The examples presented have provided an impressive demonstration that these methods represent powerful tools for the controlled preparation of highly ordered nanostructures. In future, a combination of the three methods should represent the next key step towards building up well-defined functional nanostructures with suitable properties for applications in molecular electronics. Moreover, besides the applications of self-assembled structures discussed here, such nanoscale structures are of growing interest in the development of new sensors and catalysts. Inorganic substrates with epitaxially grown nanostructures display suitable starting points for the site-selective attachment of molecules, and such molecules may serve as intelligent adhesives for the binding of nanoscale subunits. The future goal of fabrication of complex functional nanoscale structures will be achieved only by employing a hierarchical self-assembly approach, using different scales and materials.

References

1 Vescan, L. (1995) *Handbook of Thin Film Process Technology*, D.A. Glocker and S.I. Shah (Eds.), IOP, Bristol.
2 Kasper, E. (1988) *Silicon Molecular Beam Epitaxy*, Vol. **1–2**, CRC Press.
3 Herman, M.A., Richter, W. and Sitter, H. (2004) *Epitaxy – Physical Principles and Technical Implementation*, Springer.
4 Voigtländer, B. (2001) *Surface Science Reports*, **43**, 127.
5 Venables, J.A. (1994) *Surface Science*, **299/300**, 798.
6 Jesson, D.E., Voigtländer, B. and Kästner, M. (2000) *Physical Review Letters*, **84**, 330.
7 Shchukin, V.A., Lendentsov, A.A. and Bimberg, D. (2003) *Epitaxy of Nanostructures*, Springer, Heidelberg.
8 Heinrichsdorff, F., Ribbat, Ch., Grundmann, M. and Bimberg, D. (2000) *Applied Physics Letters*, **76**, 556–558.
9 Shiryaev, S.Yu., Jensen, F., Lundsgaard Hansen, J., Wulff Petersen, J. and Nylandsted Larsen, A. (1997) *Physical Review Letters*, **78**, 503.
10 Chen, Y., Ohlberg, D.A.A., Medeiros-Ribeiro, G., Chang, Y.A. and Williams, R.S. (2000) *Applied Physics Letters*, **76**, 4004.
11 Kawamura, M., Paul, N., Cherepanov, V. and Voigtländer, B. (2003) *Physical Review Letters*, **91**, 096102.
12 Kim, E.S., Usami, N. and Shiraki, Y. (1998) *Applied Physics Letters*, **72**, 1617.
13 Theobald, J.A., Oxtoby, N.S., Phillips, M.A., Champness, N.R. and Beton, P.H. (2003) *Nature*, **424**, 1029–1031.
14 Butcher, M.J., Nolan, J.W., Hunt, M.R.C., Beton, P.H., Dunsch, L., Kuran, P., Georgi, P. and Dennis, T.J.S. (2001) *Physical Review B-Condensed Matter*, **64**, 195401.
15 Wan, K.J., Lin, X.F. and Nogami, J. (1992) *Physical Review B-Condensed Matter*, **45**, 9509.
16 Lehn, J.M. (1995) *Supramolecular Chemistry*, Wiley-VCH Weinheim, Germany.
17 Prime, K.L. and Whiteside, G.M. (1991) *Science*, **252**, 1164–1167.
18 Arte, S.V., Liedberg, B. and Allara, D.L. (1995) *Langmuir*, **11**, 3882–3893.

19 Scherer, J., Vogt, M.R., Magnussen, O.M. and Behm, R.J. (1997) *Langmuir*, **13**, 7045–7051.
20 Barlow, S.M. and Raval, R. (2003) *Surface Science Reports*, **50**, 201–341.
21 Rickert, J., Weiss, T. and Göppel, W. (1996) *Sensors and Actuators B: Chemical*, **31**, 45–50.
22 Schönherr, H. and Vansso, G.J. (2006) Chemical force microscopy, in *Scanning Probe Microscopies Beyond Imaging* (ed. P. Samori), Wiley-VCH Weinheim, Germany.
23 Joachim, C., Gimzewski, J.K. and Aviram, A. (2000) *Nature*, **408**, 541–548.
24 Donhauser, Z., Mantooth, B., Kelly, K., Bumm, L., Monnell, J., Stapleton, J., Price, D., Rawlett, A., Allara, D., Tour, J. and Weiss, P. (2001) *Science*, **292**, 2303–2307.
25 Gölzhäuser, A., Geyer, W., Stadler, V., Eck, W., Grunze, M., Edinger, K., Weimann, Th. and Hinze, P. (2000) *Journal of Vacuum Science & Technology*, **B18**, 3414–3418.
26 Barth, J.V., Costantini, G. and Kern, K. (2005) *Nature*, **437**, 671–679.
27 Blinov, L.M. (1988) *Soviet Physics Uspekhi*, **31**, 623–644.
28 Metzger, R.M. (2003) *Chemical Reviews*, **103**, 3803–3834.
29 Schreiber, F. (2000) *Progress in Surface Science*, **65**, 151–256.
30 Yang, G. and Liu, G. (2003) *Journal of Physical Chemistry. B*, **107**, 8746–8759.
31 De Feyter, S. and De Schryver, F.C. (2005) *Journal of Physical Chemistry. B*, **109**, 4290–4302.
32 Sugimura, H., Hanji, T., Hayashi, K. and Takai, O. (2002) *Ultramicroscopy*, **91**, 221–226.
33 Buriak, J.M. (2002) *Chemical Reviews*, **102**, 1271–1308.
34 Kosynkin, D.V. and Tour, J.M. (2001) *Organic Letters*, **3**, 993–995.
35 DeRose, J., Thundat, T., Nagahara, L.A. and Lindsay, S.M. (1991) *Surface Science*, **256**, 102–108.
36 Lüssem, B., Karthäuser, S., Haselier, H. and Waser, R. (2005) *Applied Surface Science*, **249**, 197–202.
37 Wagner, P., Hegner, M., Güntherodt, H.J. and Semenza, G. (1995) *Langmuir*, **11**, 3867–3875.
38 Müller-Meskamp, L., Lüssem, B., Karthäuser, S. and Waser, R. (2005) *Journal of Physical Chemistry. B*, **109**, 11424–11426.
39 Bain, C.D. and Whitesides, G.M. (1989) *Journal of the American Chemical Society*, **111**, 7164–7175.
40 Li, L., Cheng, S. and Jiang, S. (2003) *Langmuir*, **19**, 3266–3271.
41 Bumm, L.A., Arnold, J., Cygan, M.T., Dunbar, T.D., Burgin, T.P., Jones, L., Allara, D.L., Tour, J.M. and Weiss, P.S. (1996) *Science*, **271**, 1705–1707.
42 Moth-Poulsen, K., Patrone, L., Stuhr-Hansen, N., Christensen, J.B., Bourgoin, J.-P. and Bjornholm, T. (2005) *Nano Letters*, **5**, 783–785.
43 Lüssem, B., Müller-Meskamp, L., Karthäuser, S., Homberger, M., Simon, U. and Waser, R. (2006) *Langmuir*, **22**, 3021–3027.
44 Lüssem, B., Müller-Meskamp, L., Karthäuser, S. and Waser, R. (2005) *Langmuir*, **21**, 5256–5258.
45 Weigelt, S., Busse, C., Petersen, L., Rauls, E., Hammer, B., Gothelf, K., Besenbacher, F. and Linderoth, T.R. (2006) *Nature Mater*, **5**, 112–117.
46 Dmitriev, A., Spillmann, H., Lingenfelder, M., Lin, N., Barth, J.V. and Kern, K. (2004) *Langmuir*, **20**, 4799–4801.
47 Stephanow, S. et al. (2004) *Nature Mater*, **3**, 229–233.
48 Wang, Y., Maspoch, D., Zou, S., Schatz, G.C., Smalley, R.E. and Mirkin, C.A. (2006) *Proceedings of the National Academy of Sciences of the United States of America*, **103**, 2026–2031.
49 Avriam, A. and Ratner, M. (1974) *Chemical Physics Letters*, **29**, 277–283.
50 Stabel, A., Herwig, P., Müllen, K. and Rabe, J.P. (1995) *Angewandte Chemie-International Edition*, **34**, 1609–1611.
51 Jäckel, F., Watson, M.D., Müllen, K. and Rabe, J.P. (2004) *Physical Review Letters*, **92**, 188303-1–188303-4.

52 Halperin, W.P. (1986) *Reviews of Modern Physics*, **58**, 533–606.
53 Yu, Y., Chang, S., Lee, C. and Wang, C.R.C. (1997) *Journal of Physical Chemistry. B*, **101**, 6661–6664.
54 Mohamed, M.B., Ismail, K.Z., Link, S. and El-Sayed, M.A. (1998) *Journal of Physical Chemistry. B*, **102**, 9370–9374.
55 Ma, H., Yin, B., Wang, S., Jiao, Y., Pan, W., Huang, S., Chen, S. and Meng, F. (2004) *ChemPhysChem*, **5**, 68–75.
56 Nakamoto, M., Kahiwagi, Y. and Yamamoto, M. (2005) *Inorganica Chimica Acta*, **358**, 4229–4236.
57 Okitsu, K., Mizukoshi, Y., Bandow, H., Maeda, Y., Yamamote, T. and Nagata, Y. (1996) *Ultrasonics Sonochemistry*, **3**, S249–S251.
58 Mallick, M., J. Witcomb, M. and Scurrell, M. (2004) *Journal of Materials Science*, **39**, 4459–4463.
59 Schmid, G. (2004) *Nanoparticles – From Theory to Applications*, Wiley-VCH, Weinheim.
60 Richards, R. and Bönnemann, H. (2005) *Nanofabrication Towards Biomedical Applications: Techniques, Tools, Applications, and Impact* (eds C.S.S.R. Kumar, J. Hormes, C. Leuschner), Wiley-VCH, Weinheim.
61 Daniel, M. and Astruc, D. (2004) *Chemical Reviews*, **104**, 293–346.
62 Burda, C., Chen, X., Narayanan, R. and El-Sayed, M.A. (2005) *Chemical Reviews*, **105**, 1025–1102.
63 Pileni, M. (2003) *Nature Mater*, **2**, 145–150.
64 Love, J.C., Estroff, L.A., Kriebel, J.K., Nuzzo, R.G. and Whitesides, G.M. (2005) *Chemical Reviews*, **105**, 1103–1169.
65 Turkevich, J., Stevenson, P.C. and Hiller, J. (1951) *Discuss Faraday Soc*, **11**, 55–75.
66 Brust, M., Walker, M., Bethell, D., Schiffrin, D.J. and Whyman, R. (1994) *Journal of the Chemical Society: Chemical Communications*, **7**, 801–802.
67 Brust, M., Fink, J., Bethell, D., Schiffrin, D.J. and Kiely, C. (1995) *Journal of the Chemical Society: Chemical Communications*, **24**, 1655–1656.
68 Schulz-Dobrick, M., Srathy, K.V. and Jansen, M. (2005) *Journal of the American Chemical Society*, **127**, 12816–12817.
69 Murray, C.B., Sun, S., Doyle, H. and Betley, T. (2003) *MRS Bulletin*, **26**, 985–991.
70 Li, Q., Li, H., Pol, V.G., Bruckental, I., Koltypin, Y., Calderon-Moreno, J., Nowik, I. and Gedanken, A. (2003) *New Journal of Chemistry*, **27**, 1194–1199.
71 Schmid, G., Boese, R., Pfeil, R., Bandermann, F., Meyer, S., Calis, G.H.M. and van der Velden, J.W.A. (1989) *Chemische Berichte*, **114**, 3634–3642.
72 Schmid, G. (1998) *Journal of The Chemical Society-Dalton Transactions*, **7**, 1077–1082.
73 Simon, U., Schön, G. and Schmid, G. (1993) *Angewandte Chemie*, **105**, 264–267. (b) Simon, U., Schön, G. and Schmid, G. (1993) *Angewandte Chemie-International Edition*, **32**, 250–254.
74 Schmid, G., Lehnert, A., Kreibig, U., Damczyk, Z.A. and Belouschek, P. (1990) *Zeitschrift Fur Naturforschung Section B-A Journal of Chemical Sciences*, **45b**, 989–994.
75 Schmid, G., Morun, B. and Malm, J. (1989) *Angewandte Chemie*, **101**, 772–773. (b) Schmid, G., Morun, B. and Malm, J. (1989) *Angewandte Chemie-International Edition*, **28**, 778–780.
76 Vargaftik, M.N., Zagorodnikov, V.P., Stolyarov, I.P., Moiseev, I.I., Likholobov, V.A., Kochubey, D.I., Chuvilin, A.L., Zaikovsky, V.I., Zamaraev, K.I. and Timofeeva, G.I. (1985) *Journal of the Chemical Society: Chemical Communications*, **14**, 937–939.
77 Moiseev, I.I., Vargaftik, M.N., Chernysheva, T.V., Stromnova, T.A., Gekhman, A.E., Tsirkov, G.A. and Makhlina, A.M. (1996) *Journal of Molecular Catalysis A-Chemical*, **108**, 77–85.
78 Remacle, F. and Levine, R.D. (2001) *ChemPhysChem*, **2**, 20–36.
79 Mote, L., Courty, A., Ngo, A., Sisiecki, I. and Pileni, M.,(Eds.), (2005) *Self-Organization of Inorganic Nanocrystals*, in *Nanocrystals Forming Mesoscopic Structures*,

Wiley-VCH, Weinheim. (b) Willner, I. and Katz, E. (2004) *Angewandte Chemie*, **116**, 6166–6235.

80 Willner, I. and Katz, E. (2004) *Angewandte Chemie-International Edition*, **43**, 6042–6108.

81 Schmid, G., Pugin, R., Sawitowski, T., Simon, U. and Marler, B. (1999) *Chemical Communications*, **14**, 1303–1304.

82 Brust, M., Bethell, D., Schiffrin, D.J. and Kiely, C.J. (1995) *Advanced Materials*, **7**, 795–797.

83 Brust, M., Walker, M., Bethell, D., Schiffrin, D.J. and Whyman, R. (1994) *Journal of the Chemical Society: Chemical Communications*, **7**, 801–802.

84 Simon, U., Flesch, R., Wiggers, H., Schön, G. and Schmid, G. (1998) *Journal of Materials Chemistry*, **8**, 517.

85 Schmid, G. and Simon, U. (2005) *Chemical Communications*, 697–710. (b) Schmid, G., Bäumle, M. and Beyer, N. (2000) *Angewandte Chemie*, **112**, 187–189.

86 Schmid, G., Bäumle, M. and Beyer, N. (2000) *Angewandte Chemie-International Edition*, **39**, 181–183.

87 Downard, A.J., Tan, E.S.Q. and C. Yu, S.S. (2006) *New Journal of Chemistry*, **30**, 1283–1288.

88 Koplin, E., Niemeyer, C.M. and Simon, U. (2006) *Journal of Materials Chemistry*, **16**, 1338–1344.

89 Schmid, G., Peschel, S. and Sawitowski, T. (1997) *Zeitschrift für Anorganische und Allgemeine Chemie*, **623**, 719–723.

90 Barsotti, R.J. Jr, and Stellacci, F. (2006) *Journal of Materials Chemistry*, **16**, 962–965.

91 In, I., Jun, Y., Kim, Y.J. and Kim, S.Y. (2005) *Chemical Communications*, 800–801.

92 Fresco, Z.M. and Frechet, J.M.J. (2005) *Journal of the American Chemical Society*, **127**, 8302–8303.

III
High-Density Memories

11
Flash-Type Memories

Thomas Mikolajick

11.1
Introduction

The trend towards mobile electronic devices drives an increasing demand for non-volatile memories [1]. In 2007, the market for NAND Flash memories approached the size of the dynamic random access memory (DRAM) market with regards to bit volume, and continues to grow. The hierarchy of today's non-volatile memories is illustrated schematically in Figure 11.1. From this, with the technologies available today, a trade-off between cost and flexibility must be made.

At the low end of flexibility there is the read-only memory (ROM), which can only be programmed during production, but delivers the lowest cost per bit. In a practical application the most important feature today is the *electrical rewritability*. The electrical erasable and electrical programmable ROM allows for this reprogrammability on a Byte level. To achieve this, each memory cell must be constructed from two transistors – the storage transistor and a select device – but this leads to a large cell size. The electrical programmable memory (EPROM), on the other hand, consists of only one memory transistor, but does not allow an electrical erase. The Flash-type memory combines the small cell size of the EPROM with the electrical erasability of the EEPROM simply by allowing the erase operation not on a Byte level but only on large blocks of 16 kB to 1 MB (see Figure 11.2). At the high end of flexibility there is a memory that allows random access-like operation as in DRAMs or static random access memories (SRAMs), and is non-volatile. Today, this non-volatile RAM can only be realized by the combination of DRAM or SRAM [2] with EEPROM or Flash, or by ferroelectric RAMs [3].

Today's standalone Flash memories can be divided into memories for code applications and for data applications. In the code application, the memory must allow a fast random access to enable real-time code execution. In the data application, the focus is on highest density and fast program and erase throughput. The implications of this difference on the array architecture and cell construction will be explained in Sections 11.3.2 and 11.3.3. Additionally a number of applications

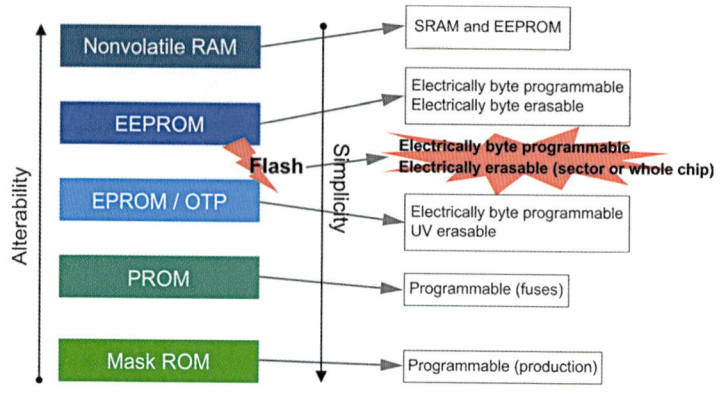

Figure 11.1 Hierarchy of today's non-volatile memory devices. With the available technologies a trade-off between simplicity (reflecting cost) and alterability exists. Flash memories that are programmable on a single byte or page level and erasable on large blocks have evolved as the best compromise for many mobile applications, such as cell phones.

such as "smart cards" call for Flash memory embedded into a high-performance logic circuit [4]. In the embedded Flash segment the density is typically much lower than in standalone memories. Therefore, the focus lies on easy integration into the standard complementary metal oxide–semiconductor (CMOS) flow and low design circuit overhead for the memory module. The requirements, however, are dependent on the actual application, which in turn leads to the development of a large number of different concepts for embedded Flash memories. In the standalone segment of the market, in contrast, one mainstream solution for code and one mainstream solution for data Flash memories have evolved.

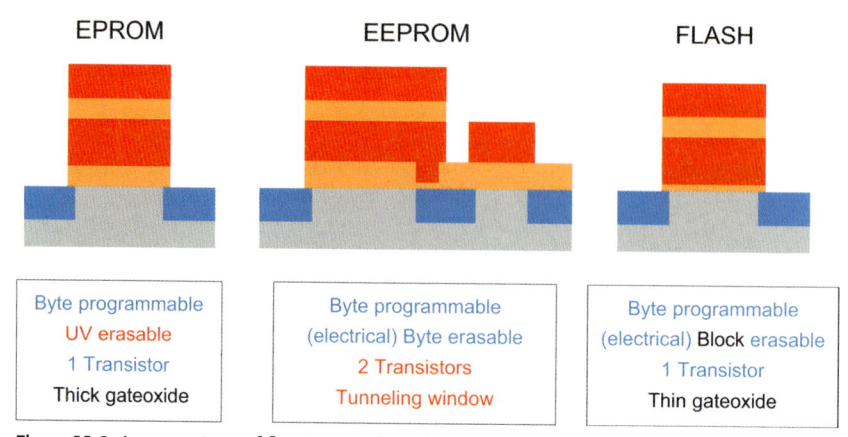

Figure 11.2 A comparison of floating-gate-based EPROM, EEROM and Flash memory cells. By sacrificing on the erase flexibility, a Flash memory cell can be realized with one transistor only.

At the heart of every non-volatile memory today there is either a floating-gate or a charge-trapping transistor, both of which were invented in 1967 [5, 6]. Due to the low oxide quality that was available during the late 1960s, a floating-gate transistor with good retention was difficult to achieve, and charge trapping was therefore very successful until the 1980s. However, with the improvement of oxides, the FAMOS cell [7], which is based on a floating gate that is programmed by avalanche injection and constructed similar to the cell shown at the left of Figure 11.2, was successfully introduced in 1971 for EPROM-type memories. Early EEPROM memories were based on charge-trapping devices using a trigate cell [8]. In 1980, the FLOTOX cell [9] (which is similar to the cell shown in the center of Figure 11.2) was demonstrated and became the mainstream for EEPROM memories. The first Flash memory was introduced at IEDM in 1985 [10], and in 1988 the ETOX cell [11] – which today is the mainstream for NOR-type memories – was proposed. Finally, NAND Flash [12] – the standard for data Flash memories – was first reported at the 1988 VLSI Technology Symposium. In 1987, charge trapping was revived by introducing a cell programmed by hot electrons and erased by hot holes [13]. This allowed the solution of the basic retention issue of charge-trapping devices, and in 1999 it was first demonstrated that this concept could be used to store two physically separated bits in one memory cell, thus reducing the effective cell size below the lithographic limit [14].

11.2
Basics of Flash Memories

In order to integrate a Flash memory into a product, two basic elements are required. First, a memory cell that can perform the program, erase and read operation with the required parameters is necessary. A generic charge storage memory cell, illustrating the program, erase and read operation, is shown in Figure 11.3. To build a large memory, these memory cells must be connected into memory arrays, with the final memory parameters being governed by the combination of memory cell and array architecture.

11.2.1
Programming and Erase Mechanisms

In programming and erase, the charge must be transferred to and from the charge storage layer, overcoming the large potential barrier of the bottom or tunneling dielectric. In principle, two different mechanisms are possible (see Figure 11.4). In the *hot carrier injection mode*, the energy of the carriers is heated up to a level which is sufficient to overcome the barrier. In the *tunneling mode* a large voltage is applied to the barrier in order to reduce its effective width. Variants of both effects are used in different type of Flash concepts. The ways in which several combinations of these effects are realized in Flash concepts are listed in Table 11.1, and the most important concepts will be explained in Section 11.3. Details of other concepts may be found in the references listed in Table 11.1. At this point it should be noted that in Table 11.1

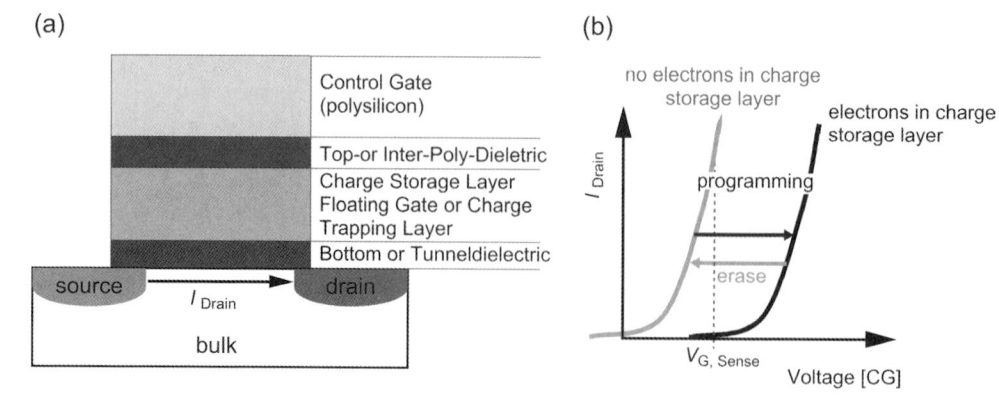

Figure 11.3 Generic charge storage memory cell (a) and basic cell operation (b). The charge storage layer, which can be either a floating gate (Section 11.3) or a charge-trapping layer (Section 11.4) is separated from both the control gate as well as the transistor channel by an insulator. By placing electrons or holes inside of the charge storage layer, the threshold voltage of the transistor can be controlled leading to a significant difference in drain current at a given gate voltage.

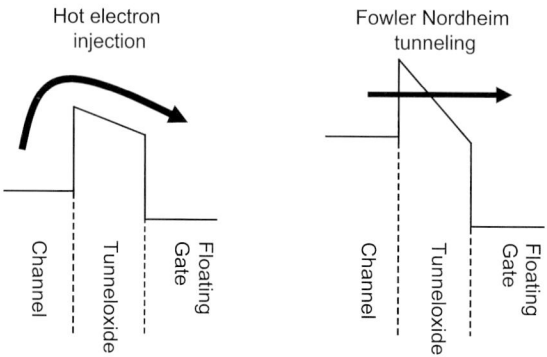

Figure 11.4 Two ways to overcome the silicon/silicon dioxide barrier. In the hot electron injection, carriers are accelerated until they have enough energy to surmount the barrier. In Fowler–Nordheim tunneling, a large electric field is applied to the barrier, leading to a reduction of the effective thickness of the barrier.

and the remainder of this chapter, the operation performed on a byte, word or page level is called "programming" and the operation performed on a block level is called "erase".

11.2.1.1 Hot Carrier Injection

In channel hot electron programming, the electrons are accelerated until they have enough energy to surmount the barrier between silicon and silicon dioxide. For electrons this barrier is about 3.1 eV [18], whilst for holes the silicon bandgap must be

11.2 Basics of Flash Memories

Table 11.1 Combination of program and erase mechanisms used in different Flash cell concepts.

Erase		Programming	Hot Electrons	Hot Electrons with secondary impact ionization	Source Side Injection	Hot Holes (BBT)	Fowler Nordheim Tunneling		
							from channel	from Drain	from Source and Drain
Fowler Nordheim Tunneling	to Source	ETOX till 0.18 μm [11]	CHISEL [19]						
	to Drain			HMOS [21]		AND [32] DINOR [16]	FLOTOX [9]	HiCR [17]	
	to Channel	ETOX below 0.18 μm [15]		AG-AND [44]	PHINES [88]	NAND [12] UCP [34] SONOS [70]			
	to Poly	Triple Poly Split gate [31]		Field Enhancing Tunneling Injector [47, 48]	PHINES [89]				
Hot Holes (BBT)		NROM [14], TwinFlash [78], MirrorBit [76]		Twin MONOS [85, 86]					

added, resulting in a barrier of 4.2 eV. To achieve this energy, a high field must be generated in the channel by applying a sufficiently high drain voltage. Additionally, a gate voltage that attracts the generated carriers must be applied. This method has the advantage of microsecond programming speed for a single bit, as well as the fact that it is a three-terminal operation making the disturb optimization easy in a NOR-type architecture. On the other hand, the mechanism has the problem of being very ineffective, as typically approximately 10^5 to 10^6 channel electrons are needed to inject one electron into the storage layer. This leads to a current consumption which is in the range of 100 µA per cell. The consequence is a limited parallelism of cells during programming, and therefore a limited programming throughput.

In order to reduce both the programming current as well as the drain voltage required during programming, two alternative approaches are possible (see Figure 11.5). First, it is possible to apply a bulk voltage during programming, and in this case the field between drain and bulk is increased. As hot electrons at the drain side will lead to electron hole pair creation by impact ionization, the generated holes can be accelerated to the bulk and thereby create a second impact ionization. The so-generated tertiary electrons again may be accelerated towards the charge storage layer. By using this approach the drain voltage can be reduced below 3 V and the current can also be significantly reduced [19]. However, care must be taken to maintain the benefit when scaling down the channel length [20].

Another approach is to use a so-called "split gate transistor", where the channel region is divided into two serial regions with individual gates. The storage layer is present only below one of the two gates. The gate voltage at the gate close to the source is chosen at a value slightly above the threshold voltage, which limits the current flowing trough the channel. The voltage of the second gate is set to a high enough voltage to accelerate the carriers into the charge storage layer. By doing this, a high

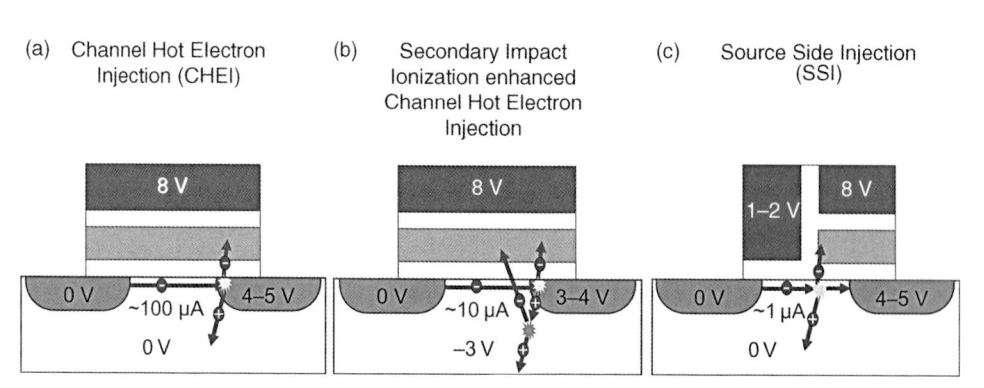

Figure 11.5 Channel hot carrier injection mechanisms. In the classical channel hot electron injection, the injection is mainly by primary channel electrons or secondary electrons (a). With applied back bias, the injection current is significantly increased by the carriers additionally generated during the secondary impact ionization event (b). In source side injection mode the channel current is limited by a second control gate and the field for hot carrier generation is decoupled from the field that attracts the carriers to the storage layer (c). This enhances the efficiency by about 2 orders of magnitude.

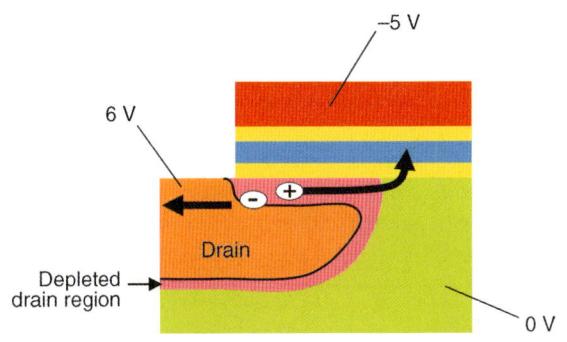

Figure 11.6 Generation of hot holes by band to band tunneling. The band bending in the highly doped region of the n⁺ junction leads to generation of carriers by band to band tunneling. The carriers are heated by the lateral electrical field and attracted to the storage layer via the vertical electrical field.

field is created at the region between the two gates, such that the electrons will become hot in that area of the device. As the carriers are injected at a location close to the source, this method is referred to as "source side injection" (SSI) [21]. The channel current can be reduced to the single µA range. Due to the necessity of a second gate electrode, however, there is a cell size drawback and a circuit overhead associated with this source side injection.

Besides channel hot electron generation, carriers generated by band to band tunneling [27] may also be used for programming and erase. In this case, band-to-band tunneling in the drain junction is induced by applying a high drain potential while the gate is turned off. The so-generated carriers are accelerated towards the channel by the electrical field, collecting enough energy to surmount the potential barrier. This is illustrated in Figure 11.6 for the case of hot hole generation in a n-channel device. A modified version uses a highly doped buried layer to generate the band-to-band tunneling [28].

11.2.1.2 Fowler–Nordheim Tunneling

In Fowler–Nordheim (FN) tunneling, a high electric filed applied to the barrier creates a trapezoidal barrier which significantly reduced the effective barrier for the carriers (see Figure 11.4). The current can be calculated according to the well-known equation [29]:

$$I_G = A_{FN} E_{ox}^2 \exp\left(-\frac{B_{FN}}{E_{ox}}\right) \tag{11.1}$$

where E_{ox} is the electrical field in the oxide and A_{FN} and B_{FN} are material-specific constants. If a dielectric charge-trapping layer is used for charge storage, the material stack relevant for the tunneling will also depend on the applied field [33]. For low fields, the carriers must tunnel through part of the trapping layer in addition to the tunneling dielectric. For higher fields and very thin layers, direct tunneling is dominant. Finally, for high fields the FN tunneling [as given by Equation 11.1] is the most important mechanism.

Until now, we have been considering charge transfer between the transistor channel and the charge storage layer. However, by using tunneling the charge can also be transferred to the gate electrode [30] or to a specially designed erase gate [31]. In this case the modified properties of the tunneling barrier created on a polysilicon electrode, as well as field enhancement by poly tips, must be taken into account in the FN tunneling current.

11.2.1.3 Array Architecture

When combining memory cells to a memory array, different versions are possible. The straightforward way is to connect the gate of every memory cell to a wordline, the drain of every memory cell to the bitline, and to connect all the sources of the memory cells to ground. By using this construction all n cells on one bitline are connected in parallel to each other. As this resembles the n-channel portion of a n-input NOR gate, this architecture is referred to as NOR, or more precisely common ground NOR. Other types of NOR architecture will be discussed later in this section. In contrast, it is also possible to connect the cells on the bitline in a serial connection, leading to a NAND-type array. In practical applications the number of cells in series must be limited to keep the read current on an acceptable level; therefore, typically 32 cells are placed between two select transistors and the so-constructed NAND strings are connected to the bitline in a similar manner as the individual cells in the standard NOR arrangement. When comparing NAND and NOR architectures, it is clear that the random access time is much faster in the NOR-type array, as every cell is directly accessed by a bitline. In the NAND architecture, in contrast, the serial connection of cells results in a high resistance through which the current must flow to the bitline. The result is a much lower read current – and therefore a much slower random access. On the other hand, the NAND arrangement has a distinct size advantage, as the contacts are shared between all cells of the string, whereas one contact for every two cells is necessary in common ground NOR.

The program and erase mechanism used in the cell also has an impact on the possible array architecture. For a cell programmed by channel hot electron injection, the common ground NOR arrangement is an ideal fit, as the necessary voltages can be precisely applied to the cell, thus minimizing any disturbance to other cells. For other injection mechanisms, or to minimize the cell size, the common ground NOR architecture may be modified. An overview of the possible array architectures is shown in Figure 11.7. In NOR, as well as the above-mentioned common ground NOR architecture, a separate source line may also be used for every bitline; such as array is commonly known as an AND [32] array. Although the most compact realization uses buried bitlines, some versions with metal bitlines are also used [34] if access time has to be minimized rather than cell size. Finally, in such an array each pair of neighboring bit line and source line can be combined to one line. Since here the ground is defined only by the operation of the array, this architecture is referred to as a "virtual ground NOR array" [35]. This has the advantage of a very small cell size (similar to NAND), and also enables a symmetrical operation of the cell, which is essential in multi bit operation (see Section 11.4.2).

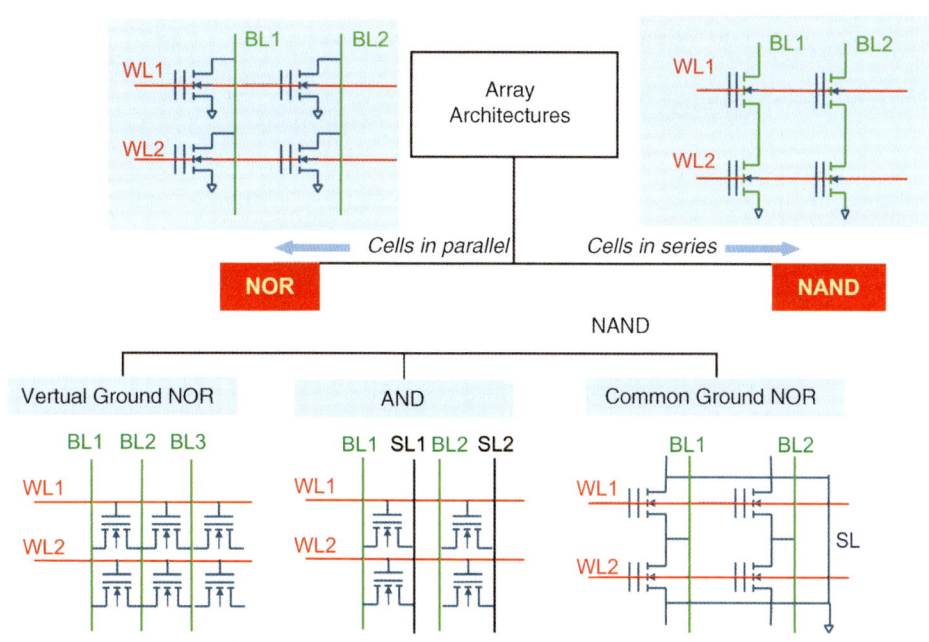

Figure 11.7 Architectures for Flash memory arrays. In principle, it is possible to connect the cells on one bitline in parallel leading to the NOR-type array and in series leading to the NAND-type array. For NOR-type arrays, many different variants have been proposed, whereas for NAND there is only one mainstream solution.

11.3
Floating-Gate Flash Concepts

11.3.1
The Floating-Gate Transistor

In essence, every flash cell is a metal oxide silicon (MOS) transistor with the charge storage layer placed in between the control gate and the channel. The drain current I_D of a MOS transistor can be expressed by the set of Equation 11.2 [36]

$$I_D = \beta \left((V_G - V_T)V_D - \frac{V_D^2}{2} \right) \quad \text{for} \quad V_G - V_T > V_D > 0$$

$$I_D = \frac{\beta}{2}(V_G - V_T)^2 \quad \text{for} \quad 0 < V_G - V_T \leq V_D \quad (11.2)$$

$$I_D = 0 \quad \text{for} \quad V_G - V_T < 0$$

where V_G is the gate voltage, V_D is the drain voltage, V_T is the threshold voltage, and β is the transconductance. β and V_T are given by Equation 11.3:

$$\beta = \frac{W}{L} C_{IS} \mu$$

$$V_T = \Phi_{MS} - 2\phi - \frac{Q_S}{C_{IS}} - \frac{Q_{IS}}{C_{IS}} \qquad (11.3)$$

where W and L are the channel width and channel length of the device, C_{IS} is the total capacitance of the gate insulator, μ is the channel mobility, Φ_{MS} is the workfunction difference between the gate electrode and the channel, ϕ_B is the Fermi potential, Q_S is the charge in the depletion layer, and Q_{IS} is the total charge in the insulator normalized to the silicon/insulator interface. In principle, the stored charge in a non-volatile memory cell can be modeled by the insulator charge Q_{IS}. For a charge-trapping device (as introduced in Section 11.4) this holds true, without further modifications.

In the floating-gate device, however, the gate controlling the device is the floating gate (FG) where, from the outside world, only the control gate (CG) can be accessed. Therefore, the gate potential V_G in Equation 11.2 can only be controlled according to the capacitive coupling of the floating gate to the external terminals. The capacitive coupling of the floating gate to the external accessible terminals is illustrated schematically in Figure 11.8; from this figure, under the assumption that the bulk and source of the device are grounded, Equation 11.4 can readily be deduced giving the floating-gate voltage V_{FG} as a function of the gate and drain voltage.

$$V_{FG} = \alpha_G (V_{CG} + f V_D)$$

$$f = \frac{\alpha_D}{\alpha_G} = \frac{C_D}{C_C} \qquad (11.4)$$

In Equation 11.4 α_G and α_D are the gate-coupling and the drain-coupling coefficients, which are a function of the respective capacitances according to Equation 11.5:

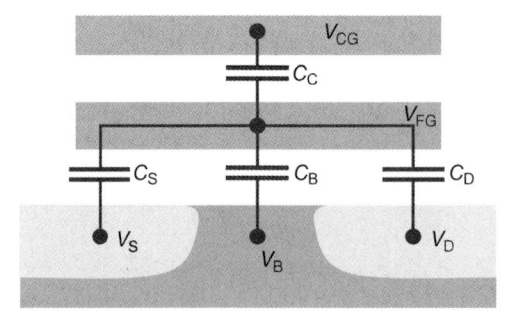

Figure 11.8 Schematic drawing of a floating-gate cell, illustrating the capacitive coupling of the floating gate to the external accessible terminals.

$$\text{gate coupling}: \quad \alpha_G = \frac{C_C}{C_T}$$
$$\text{gate coupling}: \quad \alpha_D = \frac{C_D}{C_T} \qquad (11.5)$$

$C_T = C_C + C_S + C_B + C_D$ is the total capacitance, where the individual capacitance are defined in Figure 11.8. By substituting Equation 11.4 into Equations 11.2 and 11.1 and rearranging, the threshold voltage with respect to the control gate is obtained

$$V_{T,CG} = \frac{V_{T,FG}}{\alpha_G} - \frac{\alpha_D}{\alpha_G} V_D - \frac{Q_{FG}}{C_C}. \qquad (11.6)$$

The CG threshold voltage is a function of the floating-gate threshold voltage, the charge in the floating gate, and also a function of the applied drain voltage. This means that the threshold voltage is lowered when the drain voltage is increase or, in other words, the device can be turned on by the drain terminal. This effect must be considered when designing floating-gate memory arrays.

11.3.2
NOR Flash

Whilst, in Section 11.2.2 it was shown that a significant number of variants of different NOR-type concepts exist, this section will focus on the mainstream NOR concept. This uses a common ground NOR architecture (see Figure 11.7) in combination with the so-called EPROM tunnel oxide (ETOX) cell. The ETOX cell is, in principle, a stacked gate cell (see Figure 11.2, right and Figure 11.3a) that is programmed using channel hot electrons, and is erased using FN tunneling. In older generations the erasing was carried out towards the source terminal of the device, but as this calls for a large underdiffusion of the source junction under the gate in newer generations it was replaced by tunneling towards the channel.

The cell layout and cross-sections through a cell fabricated in 90-nm technology is illustrated in Figure 11.9. It can be seen that the wordline (WL) pitch is larger than the bitline (BL) pitch, as the channel length cannot be scaled to the minimum feature size using channel hot electron programming and the contact that must be placed in between two cells. The source line, in contrast, is fabricated self- aligned to the word line. The BL pitch on the other side is close to twice the minimum feature size, which is accomplished by self-aligning the bottom part of the floating gate to the shallow trench isolation (STI) and using an unlanded contact. Note that the top part of the floating gate is overlapping the STI, leading to a floating gate space below the minimum feature size. This, in combination with the portion of the control gate that is between the floating gate, provides more area to increase the gate coupling coefficient (see Figure 11.9).

The voltages applied to the terminals for programming, erasing and reading the cell are listed in Table 11.2. Programming is carried out using channel hot electron injection; optionally, the efficiency may be improved by applying a low bulk voltage.

Another important aspect that must be considered in the operation of every flash memory array are *disturbs*. These can cause an unwanted change of the cell content.

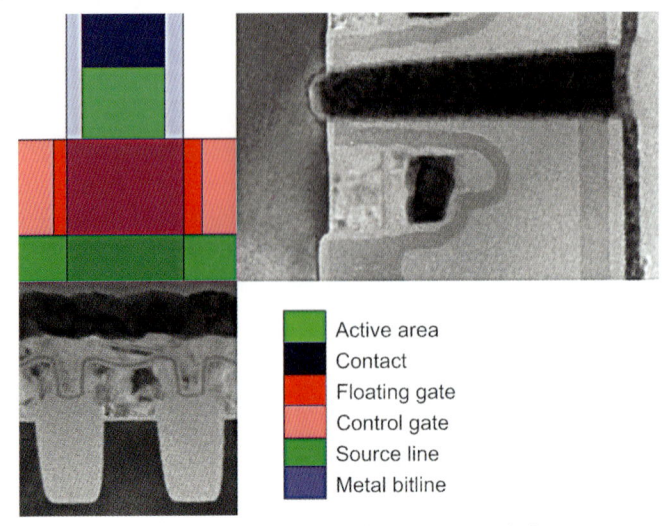

Figure 11.9 Layout and cross-sections of an ETOX-type Flash memory cell. The scanning electron microscopy cross-sections are taken from a 90-nm technology [37].

- Active area
- Contact
- Floating gate
- Control gate
- Source line
- Metal bitline

Table 11.2 Voltage conditions in read, program and erase for a typical ETOX-type cell.

	Gate	Drain	Bulk
Read	4 V	<1 V	0 V
Program	7 to 10 V	3 to 6 V	0 to −1.5 V (optional)
Erase	−6 to −8 V	float	6 to 8 V

A read disturb is caused by the fact that, for very low drain voltages, there is a probability of hot carriers being injected into the floating gate. Additionally, during read the applied gate voltage may cause FN tunneling into the floating gate. Both of these effects give rise to an unwanted programming of a cell during read. During programming, all cells connected to the same bitline are seeing a drain disturb, while all cells connected to the same wordline are seeing a gate disturb (see Figure 11.10). In

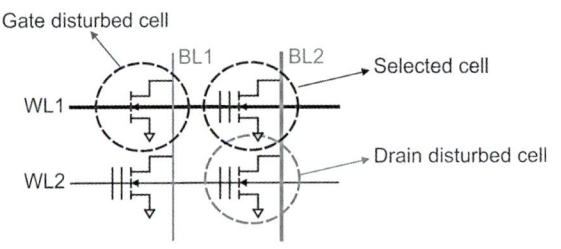

Figure 11.10 Gate and drain disturb in a common ground NOR-type Flash memory array.

11.3.3
NAND Flash

In NAND, only one concept exists, as illustrated in Figure 11.7. As with the NOR-type memory, the cell is a stacked gate but, due to the serial connection of cells, hot electron programming is not practical. The cells are therefore programmed and erased by FN tunneling between channel and floating gate. A schematic overview of a typical NAND string, including cell cross-sections taken from a 60-nm cell, is shown in Figure 11.11 [38]. Today, 32 cells are connected between two selects; this number has increased from eight cells via 16 cells, and may increase to 64 cells in the future [59]. The ground select connects the string to a sourceline, while the bitline select connects each string to a metal bitline.

In the wordline direction the cell resembles the ETOX cell described earlier. In the bitline direction, the cell is much denser due to a lack of contacts, as well as the channel length which can be scaled down to the minimum feature size due to the fact that the cell only has to isolate very low voltages. This becomes clear from Figure 11.12, where the main operations – read, write and erase – are explained in more detail. Erase is achieved simply by applying a sufficiently high voltage for FN tunneling to the well and applying 0 V to all the wordlines in the sector that must be erased, while leaving the wordlines of the non-erased sectors floating.

In the read operation a voltage higher than the highest V_T of a programmed cell must be applied to all the non-selected wordlines. In the erased state, the threshold voltage of the cells is chosen to be negative. Therefore, 0 V can be used on the wordlines that must be read. If a voltage of about 1 V is applied to the bitline, the selected cell will conduct if it is in the erased state and will be below threshold in the programmed state. For writing, a voltage high enough for FN tunneling (e.g., 15–20 V) is applied to the selected wordline. The channel of the selected cell is set to 0 V by applying 0 V to the selected bitline, turning the bitline selector on, and applying a high enough voltage to all the other wordlines; this will allow all other cells in the string to be turned on.

Care must be taken to avoid programming of the cells on the same wordlines. In early NAND Flash implementations a voltage of about 7 V was applied to the unselected bitlines and transferred to the channel of the cells on the selected wordline [39]. This, however, had two significant drawbacks. First, the junctions of the cells had to withstand the high voltage applied to the unselected bitlines, which restricts the scalability of the cell. Second, all the unselected bitlines had to be charged to a high voltage during programming. When the supply voltage was reduced from 5 to 3.3 V a new solution was implemented [40]. In that solution the bitline selects of the unselected wordlines are switched off by applying V_{DD} to the unselected bitlines. The channel of the inhibited cell will then raise its potential by capacitive coupling via the tunnel oxide capacitance C_{Tunnel} and the ONO capacitance C_{ONO} to the high voltage applied to the wordline. A high voltage applied to the passing wordlines will

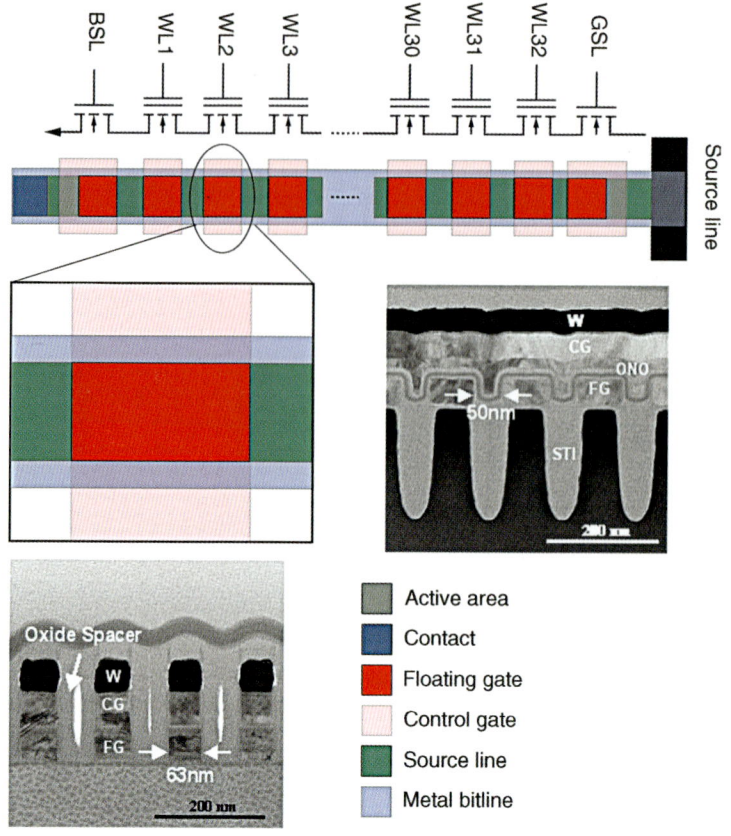

Figure 11.11 Layout and cross-sections of a string of NAND memory cells. In today's technologies, 32 cells are placed between a ground select and a bitline select. The ground select connects the string to a ground line, that in turn connects all strings to ground. The bitline select connects every string to a metal bitline via a bitline contact. The cross-sections are taken from a 60-nm technology [38].

further help to raise the channel potential in the disturbed cell. Now, the programming voltage and the voltage applied to the passing wordlines must be optimized to minimize the disturb effect. Examples for 120 and 90 nm Technology generations can be found in Ref. [41].

A large number of floating-gate concepts other than standard NOR or NAND have been proposed, and some of these are – or were – in production. Good overviews can be found in Refs. [42, 43]. Embedded flash memories have somewhat different requirements than standalone memories. Normally, much smaller memory densities are required than in standalone memories, and therefore the cell size is not as important but rather the size of the complete memory module including charge pumps, decoding, and so on must be minimized. This results in a high incentive to minimize the voltage requirements. Besides, every application focuses on different

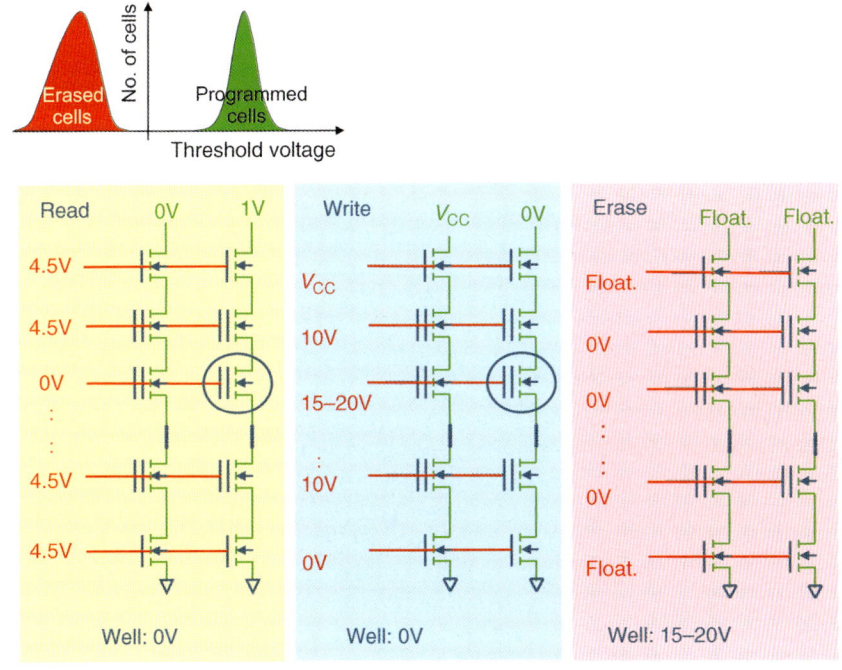

Figure 11.12 Read, write and erase operations of a NAND memory array.

requirements such as very fast random access, high endurance, high reliability, or low power, and therefore a large number of concepts coexist. In general, the ETOX is a good fit to many requirements, as long as the power consumption during programming can be tolerated. Among the large number of different concepts (some of these are referred to in Table 11.1) the field-enhancing tunneling injector cell [45, 47] is very popular and has been adopted by many foundries. Here, source side injection is used for the programming.

11.3.4
Reliability Aspects of Floating-Gate Flash

Every charge-based non-volatile memory inherently faces two reliability challenges. The first challenge is that the stored charge may be lost or charge may be gained during storage, leading to a loss of information; this phenomenon is referred to as *retention loss*. The second challenge is that the memory cell will degrade during repeated programming and erase cycles. Therefore, the endurance that a memory cell can achieve is a decisive quality criterion. Figure 11.13 illustrates how these basic properties manifest themselves in a 90 nm NAND Flash memory cell [46]. As program and erase cycling degrades the properties of the memory cell, the retention properties will also be affected by the precycling.

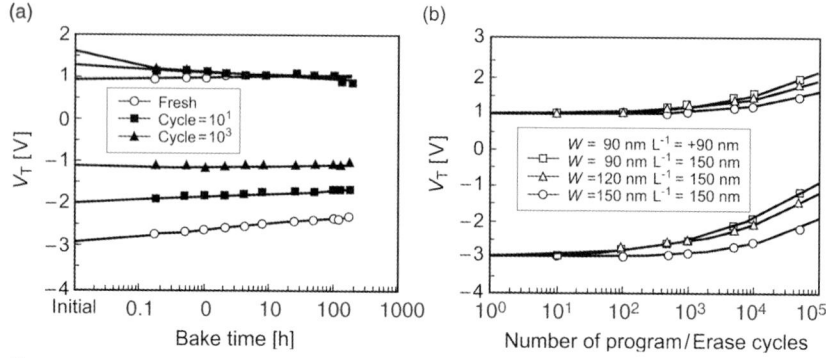

Figure 11.13 Retention (a) and endurance (b) characteristics of a 90-nm NAND Flash memory cell [49]. The dependence on precycling is shown for the retention characteristics, while the influence of device geometry is illustrated for the endurance curve.

In floating-gate memories, two specific effects must be mastered. The first is an effect that causes an abnormal erase behavior, leading occasionally to a much faster erase in some cells (see Figure 11.14a). As this effect is erratic in nature and affects single bits rather than the whole population, this phenomenon is called *erratic bit* [47]. The second phenomenon is also statistical in nature, in that during storage some cells may lose charge much faster than the main population. This effect occurs at lower temperatures, disappears at high storage temperatures, and becomes more pronounced after cycling (Figure 11.14b and c). This second effect is referred to as *anomalous SILC* or *moving bit effect*. Erratic bits are attributed to hole trapping in the bottom oxide. A small probability exists that clusters of three or more trapped holes exist. The overlap of the electrical field of the trapped holes leads to a strong increase in tunneling current [48], and thus to a much faster erase. The anomalous SILC or moving bit is explained by charge loss via neutral traps that are generated while cycling the cell. As a percolation path of such traps must exist for a charge loss to occur, the phenomenon is a strong function of oxide thickness, with thicker tunneling oxides showing much lower moving bit rates than their thinner counterparts [49].

11.3.5
Scaling of Floating-Gate Flash

The most severe issue in scaling down floating-gate devices is the non-scalability of the tunnel oxide and the inter poly dielectric. In order to achieve non-volatile retention, the tunneling dielectric must be at least 6 nm thick [50], although in practical memories thickness of 8–10 nm are used, based on the concrete reliability specification. This margin allows the covering of extrinsic effects such as moving bits. Scaling of the oxide-nitride-oxide (ONO) layer used as the interpoly dielectric is limited to an electrical effective thickness of about 13 nm due to retention and V_t stability constraints [51].

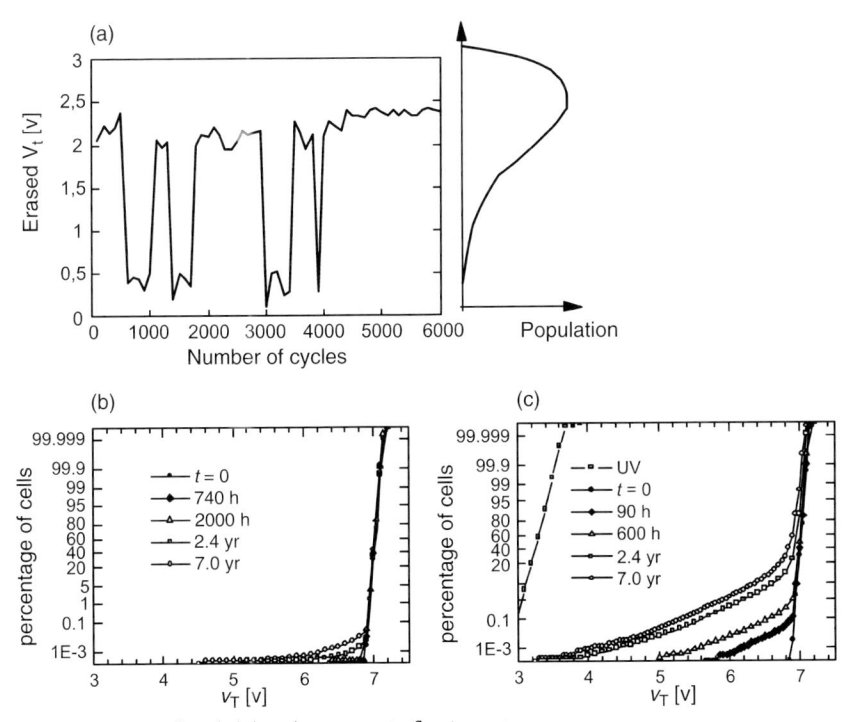

Figure 11.14 Specific reliability phenomena in floating gate memory cells [50]. Erratic bits (a) occasionally have a much faster erase than the majority of the population. Some bits show a pronounced low-temperature retention loss which is much higher after cycling (c: after 10 k cycles) than before cycling (b: after 10 cycles).

Further scaling of the tunneling oxide can only be obtained by radically re-engineering the tunnel barrier. Materials with a higher dielectric constant (e.g., HfO_2, ZrO_2) that are currently investigated for logic transistors can help. Crested barriers [52] could further improve the basic memory cell by increasing the ratio between the on and off current, leading to much faster write times as well as lower programming voltages. The principle of such an approach is shown schematically in Figure 11.15. The triangular shape of the barrier maintains the maximum barrier height if no voltage is applied (retention case), but drastically reduces the effective barrier in case of an applied voltage (programming or erase case). As a crested barrier is not achievable with those materials that have the required barrier heights, a staircase approximation using three layers with different band offsets, as well as different dielectric constants, is a reasonable approach. In the optimum structure the center layer would have a high band offset and a high dielectric constant, while the surrounding layer has lower band offset as well as a lower dielectric constant (Figure 11.15c). In most materials, however, a high band offset is correlated with a low dielectric constant and vice versa, making the optimum choice very difficult. A stack consisting of $Si_3N_4/Al_2O_3/Si_3N_4$ could be a reasonable and producible

(a) Conventional barrier

(b) Crested barrier

(c) Staircase approximation

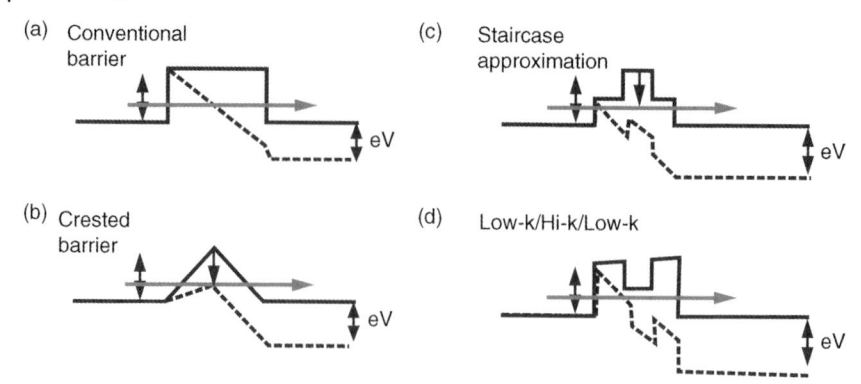

(d) Low-k/Hi-k/Low-k

Figure 11.15 Different approaches to re-engineer the tunneling barrier of a floating-gate memory cell. (a) Conventional barrier; (b) ideal crested barrier; (c) staircase approximation of a crested barrier; (d) hi-k barrier sandwiched in between two low-k layers.

compromise [53]. An alternative – and perhaps better – manufacturing approach would be to sandwich a hi-k layer in between two low-k layers [54], and encouraging data have recently been demonstrated by using such an approach [55].

The non-scalability of the inter-poly dielectric thickness will eventually lead to a situation where the control gate will not fit into the space between two floating gates (Figure 11.16a). With the conventional ONO dielectrics this would lead to a situation where the control gate to floating gate coupling is significantly degraded [56]. To compensate for this, a high-k coupling dielectric is required, which has a k-value high enough to achieve the coupling only via the top of the floating gate (Figure 11.16b). Typical materials are very similar to those used for gate dielectrics, including HfO_2 and Hf/Al micro laminates [57].

(a) Today's device

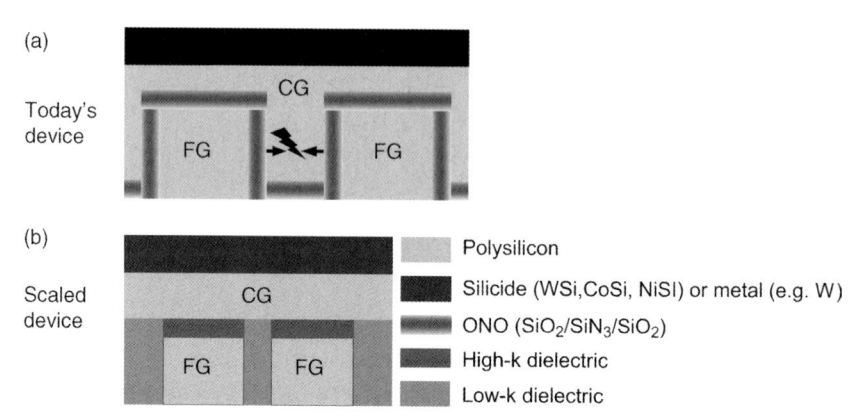

(b) Scaled device

- Polysilicon
- Silicide (WSi, CoSi, NiSi) or metal (e.g. W)
- ONO ($SiO_2/SiN_3/SiO_2$)
- High-k dielectric
- Low-k dielectric

Figure 11.16 Possible scaling path for floating-gate memory cells. (a) Today's device architecture. (b) A scaled device using high-k inter-poly dielectrics as well as low-k dielectrics between floating gates. CG, control gate; FG, floating gate.

The reduced spacing between floating gates will also lead to higher capacitive coupling between floating gates, and result in severe crosstalk between cells [55]. This calls for a material with a lower dielectric constant between the floating gates, as shown in Figure 11.16b. This must also be implemented in the area between the word lines. Replacement of the silicon nitride spacer of the cell transistor with a silicon dioxide spacer (as shown in Ref. [58]) may help to significantly reduce the effect, but in the long term real low-k materials will be necessary. The recently demonstrated air gaps between the floating gates may represent an ultimate solution [59].

Although the scaling challenges presented so far are valid for all types of floating-gate flash memory devices, in the NOR-type architecture two more limitations must be considered [60]. First, a contact is required for every two cells, and this results in a significant area overhead. The overhead can be minimized by using a contact that is self-aligned to the control gate [61], or by using a virtual ground NOR array rather than a common ground NOR array [62]. The second limitation is that the channel length scaling is limited by the high voltages required during channel hot electron programming. However, a vertical device may be required to overcome this issue [63].

Instead of reducing the feature size, the cost reduction and density increase of a flash memory can also be achieved by increasing the number of bits stored on the same surface area, rather than reducing the size of a physical cell. As the charge storage is analogue in nature, more than two levels can be stored on one floating gate. To code n bits, 2^n levels are required. Figure 11.17 illustrates the corresponding V_T distributions for both NOR and NAND devices. For NOR devices, the multi-level approach was introduced to the market back in 1997 [64], and today in NAND devices

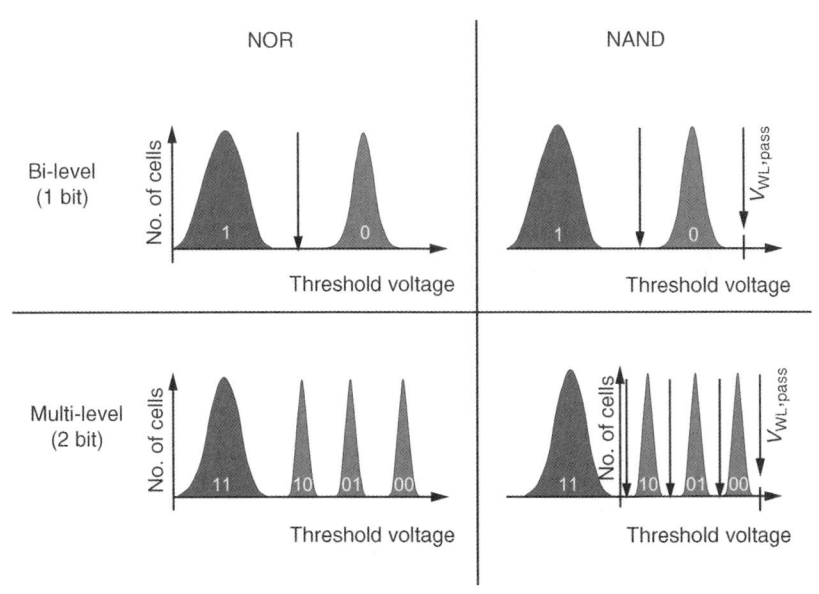

Figure 11.17 Multi level storage. The V_T distributions of a 1-bit and a 2-bit cell are compared for NOR as well as NAND Flash memories.

multi-level is also becoming increasingly popular as cost reduction is the most important aspect in data storage [65]. In the long term, the three-dimensional stacking of memory cells may also be an option; an example of this, namely the vertical integration of a NAND string, was demonstrated in Ref. [66].

11.4
Charge-Trapping Flash

Instead of using a floating gate to store the charge, an insulator with a high density of traps – a so-called *charge-trapping layer* – may also be used. Although this approach enjoyed some success during the 1970s and 1980s for EPROM and EEPROM memories [67, 68], in Flash memories the floating gate, in its ETOX and NAND versions, became dominant during the 1990s. The charge-trapping concept has some interesting advantages over floating-gate devices. As the charge is highly localized, no capacitive coupling effects (as described in Section 11.3.1) need to be considered. The cell can be described by Equations 11.2 and 11.3 with the storage charge included in Q_{IS}, which means that there is no drain turn on effect. The localization of charge also makes the cell less sensitive to any local defects responsible for erratic and moving bits in floating-gate cells. The coupling between cells is also much less pronounced. A few years ago the localization was utilized to store two physically separated bits in the same memory cell to create a multi-bit cell [69] (see Section 11.4.2), and this led to a revival of charge-trapping devices in the Flash world.

11.4.1
SONOS

Today, charge-trapping devices typically use a stack consisting of a polysilicon control gate electrode, a silicon dioxide topoxide, a silicon nitride storage layer, and a silicon dioxide bottom oxide placed on top of the active silicon. This stack (see Figure 11.18, left side) is referred to as a SONOS structure. Figure 11.18 also illustrates the programming and erase operation of such a structure using tunneling. During the programming and erasing operation, electrons or holes can be injected into or ejected out of the nitride storage layer. In programming, it is mainly electron injection from the silicon channel over the bottom oxide barrier into the nitride that takes place. Generally, for very long programming times and high V_T shifts, the hole injection from the control gate becomes significant and reduces any further programming. In the initial phase of the erase, the tunneling of electrons from traps via the nitride conduction band into the channel region is the dominant process, but this is increasingly being replaced by hole tunneling from the channel to the nitride as the erase progresses [70]. The erase process will result in less-negative charge in the trapping layer, leading to a reduction of the trapping layers potential. Therefore, the field over the bottom oxide is reduced and the field over the top oxide is increased, leading to an onset of electron tunneling from the gate to the trapping layer and a

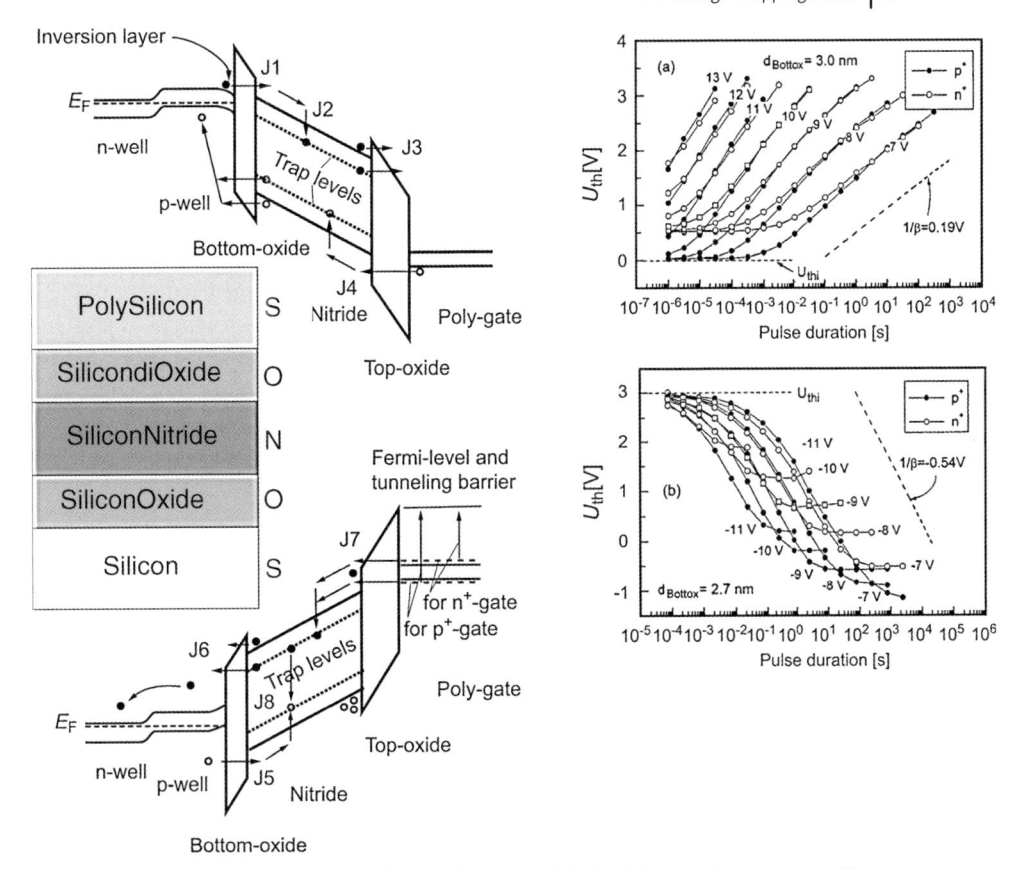

Figure 11.18 Programming (top row) and erase (bottom row) operation of a SONOS-type structure illustrated in the band diagram and a V_T over time characteristic [70]. During erase, the field over the bottom oxide will gradually decrease, while the field over the top oxide will increase; this will lead to a steady state and therefore to a saturation effect. The asymmetry of the two interfaces makes the saturation occur at higher V_T levels for higher gate voltages.

reduction of hole tunneling from the channel to the trapping layer. Finally, a steady state of the two processes – and therefore a saturation of the erase – will occur.

Due to the asymmetry of the two interfaces, the increase in erase voltage will lead to a strong increase in the injection from the gate, and hence to an even higher saturation level. This can be seen in the lower right-hand graph in Figure 11.18. In order to increase the erase speed, it is possible to increase the field over the bottom oxide, to reduce the field over the top oxide, or to increase the barrier for electrons at the control gate–top oxide interface. Although the simplest choice is to reduce the bottom oxide thickness, which will lead to a significantly increased field over the bottom oxide, it will also degrade the retention properties. For many years, therefore, the SONOS development was caught in the trade-off between bad retention and slow erase.

The increase in the barrier for electron injection from the gate electrode is very effective, as can be seen in Figure 11.18 (lower right), where n^+ doped gates and p^+ doped gates are compared. However, this alone is insufficient to achieve an erase performance that will be fast enough for a data Flash memory, yet still be secure after 10 years of retention. The field over the top oxide can be reduced by replacing the silicon dioxide with a high-k material such as Al_2O_3 [71]. Indeed, recently the combination of an Al_2O_3 topoxide and a high-workfunction TaN gate electrode was implemented in order to achieve NAND Flash-compatible performance on a 63 nm demonstrator using a bottom oxide as thick as 4 nm [72]. The structure used was referred to as a TANOS (Tantalum-nitride/Silicon-nitride/Silicon-dioxide/Silicon). The cross-section of the used stack, as well as the erase curves demonstrating a memory window of 5 V, are shown in Figure 11.19. This device represents a very promising candidate for NAND Flash memories with sub-40 nm ground rules.

11.4.2
Multi-Bit Charge Trapping

If the localized charge storage properties of a charge-trapping layer are combined with the localized injection by channel hot electrons, then the stored charge can be

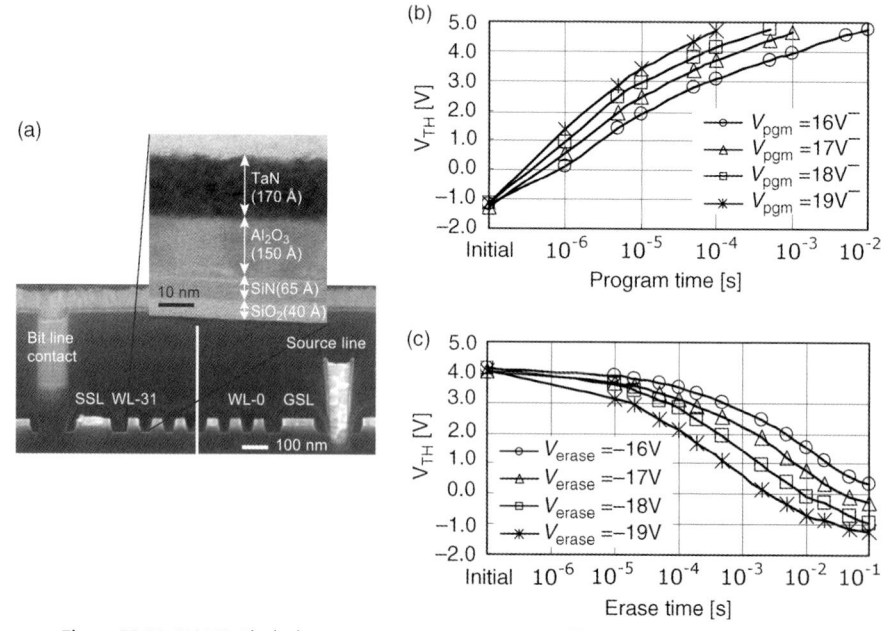

Figure 11.19 NAND Flash demonstrator using a TANOS cell on 63-nm ground rules [72]. The cross-section (a) shows the similarity to a floating-gate NAND. Program and erase performance is good enough to achieve 5 V memory window using a 4 nm-thick bottom oxide which ensures non-volatile retention.

11.4 Charge-Trapping Flash

(a) Programming: Hot Electron Injection

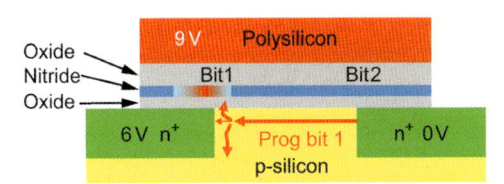

(c) Reverse read

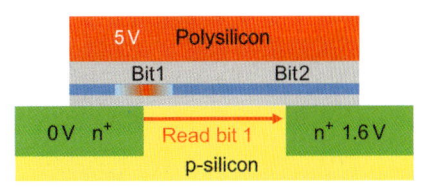

(b) Erase: Hot Hole Injection

(d) Potential: Vg = +2.0 V   Vd = +2.0 V

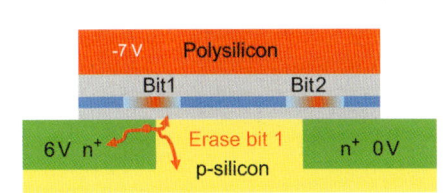

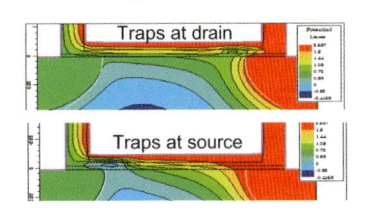

Figure 11.20 Multi-bit charge trapping memory cell. To inject charge self-aligned to the drain junction, programming is done by channel hot electron injection (a). Hot hole erase (b) enables the use of a thick bottom oxide. In reverse read (c), the charge of the second bit in the same cell is screened by applying a sufficiently high drain potential, while the role of drain and source are interchanged as compared to the programming conditions. The potential diagrams (d) show how the potential of the region below the charge is controlled by the drain potential rather than the stored charge if the drain potential is high. In (d), in red = high potential; blue = low potential.

placed in very narrow region which is self-aligned to the drain junction of the device. By interchanging the source and drain of such a device, two physically separated bits can be stored (see Figure 11.20a and Ref. [69]). In order to read the bits separately from one another, the source and drain must be interchanged compared to the programming conditions, and a drain voltage large enough to punch through the region below the charge above the drain region used in read must be applied (Figure 11.20c and d). As the charge is localized in a region close to the drain junction during programming, hot holes generated by band-to-band tunneling can be used for erase to compensate the stored charge (Figure 11.20b; see also Figure 11.7 and Ref. [13]). This allows for a sufficiently thick bottom oxide layer so as to avoid vertical charge loss and circumvent the erase saturation issue described in Section 11.4.1.

This device is implemented under different trade names such as NROM [13], MirrorBIT [73], NBit [74] and Twin Flash [75], both in code and data Flash products.

In order to operate the multi-bit charge-trapping cell described above, it is essential to have an architecture where no difference between bitlines and sourceline exists. The virtual ground NOR array (see Figure 11.7) therefore is the natural choice. This can be constructed by the structuring of a ONO layer, implanting the bitlines through the openings, growing an oxide for isolation, and finally forming the wordlines perpendicular to the bitlines. This method was used in the first-generation systems (see Figure 11.21a), but it has the drawback of a high thermal budget that must be applied to the bitline implants.

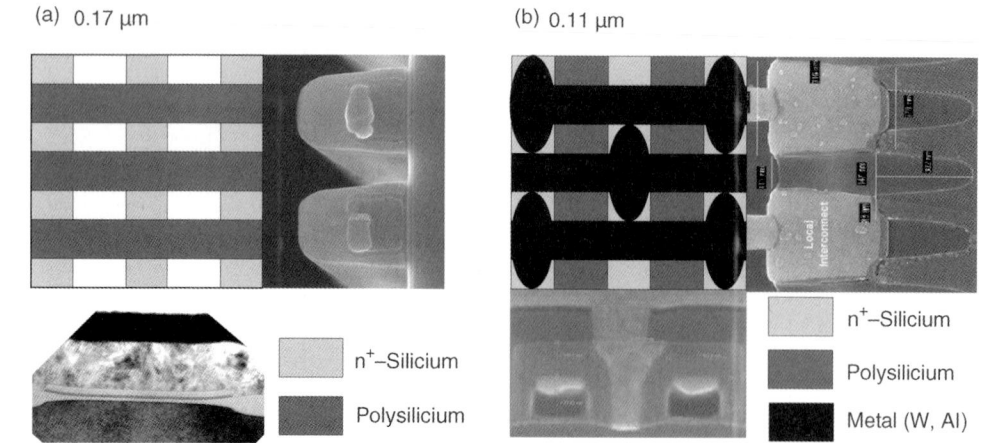

Figure 11.21 Multi-bit charge-trapping memory cells. In the 0.17 μm generation a buried bitline with crossing wordlines was used [79]. A more advanced version uses a device that resembles a standard MOS transistor [80, 81]. To end up with a virtual ground NOR architecture, a local interconnect must connect two neighboring devices over an isolation region. This local interconnect is then connected to the metal bitline.

Although the multi-bit charge-trapping type of memory cell has all the advantages described in Section 11.4.1, plus the inherent two bit per cell operation, there remain two challenges that must be considered. First, the unique mechanism of localized charge storage makes an understanding of the reliability-governing factors more complex. On an empirical basis, all effects that are necessary to create a reliable product are well understood and under control [76]. The physical basis for the observed results, however, remains the subject of debate among the scientific community. In principle, two effects may occur: First, the injection of hot holes during erase may damage the bottom oxide, leading to traps that can cause a vertical loss of the stored electrons [77]. Second, due to the fact that in programming and erase two localized mechanisms are used that will not be totally aligned to each other, a dipole will be created. This dipole may lead to lateral charge movement and therefore a change in V_T. In practice, however, both mechanisms may be involved, leading to a well-behaved and predictably reliable unit [78].

A number of modifications of the above-described multi-bit charge-trapping cell have been reported. For example, by adding an assist gate programming can be carried out using source side injection, which significantly reduces the cell current during programming. Examples of multi-bit charge-trapping cells using source side injection may be found in Refs. [82–84]. Another interesting variant is created by programming the cell with hot holes rather than hot electrons. In that case, the erase may be performed by FN tunneling either to the channel or to the gate [85, 86]. This concept, which is referred to as PHINES (programming by hot-hole injection nitride electron storage), has one main drawback in that programming of the second bit on

11.4.3
Scaling of Charge-Trapping Flash

When further scaling down the planar type of SONOS cell, the cell properties will suffer from low gate control, low read current, and a small number of electrons. In principle, the options of tailoring the tunneling barrier described for floating-gate device scaling can also be applied to the bottom oxide of a charge-trapping cell. Charge trapping, however, also enables another scaling path, by utilizing a FinFET device [88]. In principle, this could also be achieved with a floating-gate device, but two problems are encountered: First, the stack of tunnel oxide, floating gate and interpoly dielectric is too thick to fit the space between two neighboring cells; and second, the high coupling of the floating gate to the channel in a FinFET device will cause a deterioration in the gate coupling ratio [see Equation 11.2]. In a charge-trapping device the implementation of a FinFET device is straightforward; the general concept, as well as the excellent programming and erase curves that may be achieved with devices as short as 20 nm [89], are illustrated in Figure 11.22.

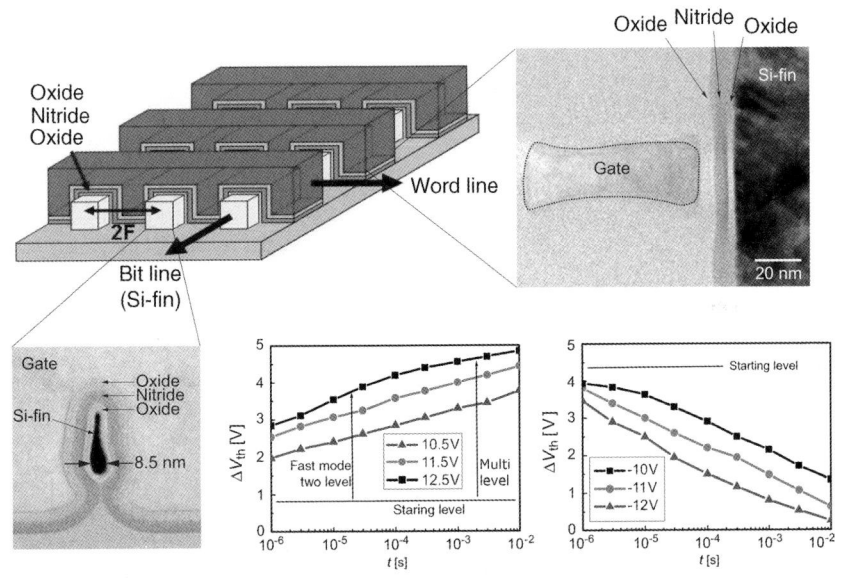

Figure 11.22 FinFET-based charge-trapping NAND [89]. The SEM images show cross-sections of a fabricated device with a channel length of 20 nm and a fin width below 10 nm. The programming and erase characteristics demonstrate a memory window of more than 4 V with fast program and erase.

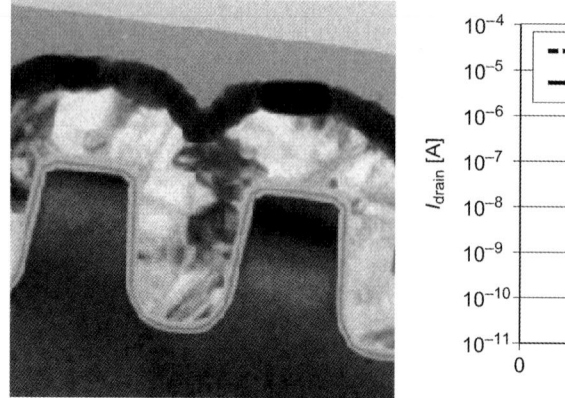

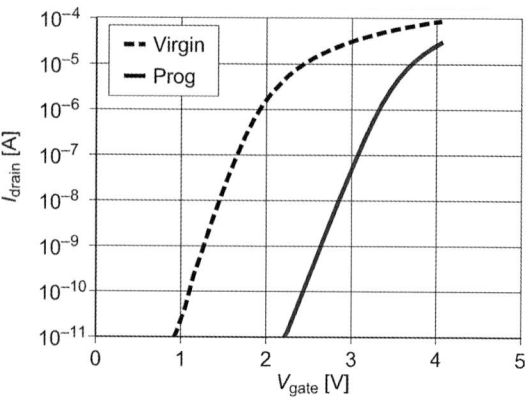

Figure 11.23 U-shaped multi-bit charge-trapping cell [93]. The U-shaped channel allows the surface area usage to be reduced, without any reduction of the effective channel length. The current–voltage (I–V) curves demonstrate the excellent separation of the two bits. Virgin = unprogrammed; Prog = programmed.

In order to further increase the memory density, a 3-D NAND-type memory would be very beneficial. Although the straightforward approach here would be to use thin-film transistors, charge trapping is again the favored solution, as it is much easier to integrate in a stacked manner compared to a floating-gate device. The first results of a thin-film transistor-based charge-trapping memory cell have recently been published by Walker and colleagues [90].

For the multi-bit charge-trapping memory cell it is essential to scale down the channel length, and indeed some excellent cell properties have been demonstrated down to 60 nm generation with the type of cell shown in Figure 11.21b [75, 81]. For further scaling down, a cell which uses structured nitride areas to store the charge was proposed. This has the advantage of controlling the cross-talk between bits, as well as being able to us thinner gate dielectrics between the ONO layers [91, 92]. A more radical approach that utilizes the third dimension by adopting a U-shaped device [93] is shown in Figure 11.23. Another approach to increase the storage density and reduce the area usage per bit is to combine the multi-bit concept with a multi-level approach. As a result, with four levels on each side of the cell (a total of eight levels), 4 bits can be stored [94], whereas by comparison a floating-gate cell requires 16 levels to store 4 bits (see Section 11.3.5).

11.5
Nanocrystal Flash Memories

In addition to floating gates and charge-trapping layers, nanocrystals are also currently under investigation for future use as flash devices [95]. A typical

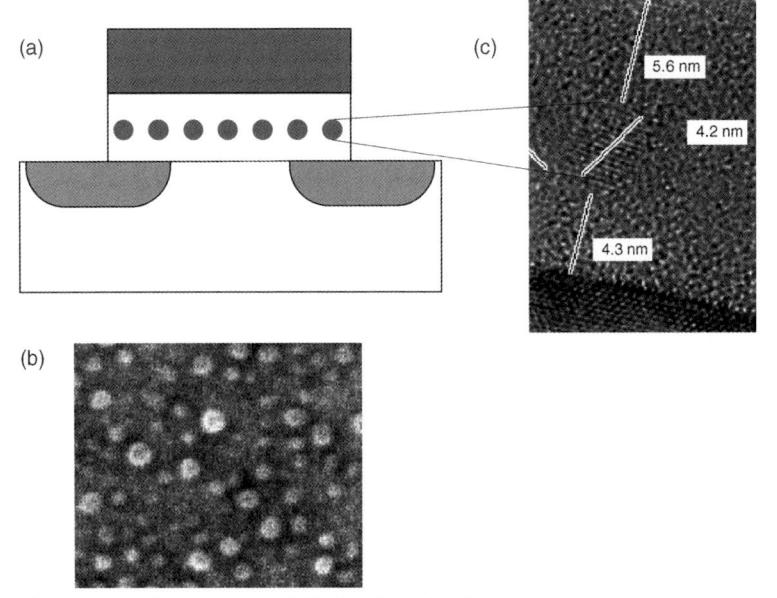

Figure 11.24 Silicon nanocrystal Flash cell. (a) Conducting nanocrystals are embedded into the gate dielectrics of a MOS device. The SEM images show (b) a top view and (c) a cross-section of a typical silicon nanocrystal layer [99].

nanocrystal device is illustrated schematically in Figure 11.24, whereby small conductive crystals with a diameter of about 2–5 nm are embedded into a silicon dioxide layer. Most of these investigations have been conducted on silicon nanocrystals, although germanium [96] or metal [97, 98] nanocrystals have also been studied. As with charge-trapping devices, nanocrystal devices are much more robust with respect to local defects in the bottom oxide layer. They also show the unwanted erase saturation that is observed in SONOS. However, in the case of silicon nanocrystals there are two identical interfaces/barriers between the control electrode and topoxide, as well as between the nanocrystal and bottom oxide. Hence, a higher voltage will not lead to an even higher erase saturation level, and the erase can be accelerated to the appropriate level by using a higher erase voltage [100].

Nanocrystal devices can be programmed and erased either by tunneling [101] or by hot electron/hot hole injection [102]. In the latter case, as with charge-trapping memories, a multi-bit device can be realized, whereas in the former version nanocrystals may be candidates for future NAND-type memories. Although the erase saturation is somewhat relaxed, the trade-off between fast program and erase and sufficient non-volatile retention is a key issue for nanocrystal devices which are programmed and erased using tunneling. One way to improve the retention is to use a self-aligned double stack of nanocrystals [103]; in this way, by utilizing the coulomb blockade effect and quantum confinement, the retention can be improved if a small nanocrystal embedded into a thin oxide layer is placed below the larger nanocrystal which actually carries the stored charge.

In most cases, either a low-pressure chemical vapor deposition (LPCVD) from SiH_4 [104] or ion-implantation with subsequent thermal treatment [105] are used to fabricate nanocrystals. Although other techniques have shown promise [106], LPCVD and ion implantation are the easiest procedures for integration into a standard CMOS process. In both cases, the nanocrystal formation is a statistical process leading to controllability issues in scaled down devices [107]. Methods for controlled fabrication of nanocrystal size and distance by using templates or self-organization would, therefore, significantly improve the outcome [108]. When further scaling down the nanocrystal device, this path may in time lead to a single-electron memory [109].

11.6
Summary and Outlook

The trend towards mobile electronic devices has created – and continues to create – a rapidly increasing demand for non-volatile memories. Today, Flash memories represent the best solution for most of these applications, where coded Flash applications are typically covered by NOR Flash devices and data Flash applications by NAND Flash devices. Currently, the floating-gate transistor is seen as the "workhorse" of those cell devices used in many of today's technologies. Indeed, floating-gate technology shows a scaling potential for further generations if innovations such as high-k coupling dielectrics or low-k isolation oxides can be mastered. By contrast, charge-trapping devices are possibly due to make a return, with multi-bit charge-trapping having recently emerged in a number of applications. In fact, a modified version of the classical SONOS device, programmed and erased by tunneling, may replace the floating-gate transistor in future generations of NAND Flash. Nanocrystals represent another option to replace the floating gate, although at present the challenges that they face seem much more severe than for the charge-trapping case. However, in the long term this development may lead to a single-electron device.

Unfortunately, flash-type memories based on charge storage in either floating gates or charge-trapping layers still suffer from important drawbacks, including limited endurance, slow write/erase, and no direct overwrite. Hence, for many years research groups have sought new storage mechanisms that could supply a non-volatile memory without such shortcomings. To achieve this goal, new materials with innovative switching effects must be integrated into the CMOS flow [110, 111]. Although these technologies are beyond the scope of this chapter, it is important to note that although they may have distinct advantages over Flash memories, and indeed some have now reached the production stage (see Chapters 13–16), the scaling of Flash memories has to date been much more successful. Such scaling possibilities provide Flash with a major competitive advantage in terms of system cost.

References

1. Niebel, A. (2004) *Proceedings of the 20th Nonvolatile Semiconductor Memory Workshop*, p. 14.
2. Harari, E., Schmitz, L., Troutman, B. and Wang, S. (1978) *ISSCC Digest of Technical Papers*, 108.
3. Mikolajick, T. *et al.* (2001) *Microelectronics Reliability*, **7**, 947.
4. Yoshikawa, K. (1999) *VLSI Symposium on Technology, Systems and Applications*, p. 183.
5. Kahng, D. and Sze, S.M. (1967) *BELL System Technical Journal*, **46**, 1288.
6. Wegener, H.A.R. *et al.* (1967) *IEDM Digest of Technical Papers*, 70.
7. Frohman-Bentchkowsky, D. (1971) *ISSCC Digest of Technical Papers*, 80.
8. Cricchi, J.R., Blaha, F.C. and Fitzpatrick, M.D. (1974) *IEDM Digest of Technical Papers*, 204.
9. Johnson, W. *et al.* (1980) *ISSCC Digest of Technical Papers*, 152.
10. Masuoka, F. *et al.* (1984) *IEDM Digest of Technical Papers*, 464.
11. Kynett, V.N. *et al.* (1988) *ISSCC Digest of Technical Papers*, 132.
12. Shirota, R. *et al.* (1988) *Symposium on VLSI Technology*, 33.
13. Chan, T.Y., Young, K.K. and Hu, C. (1987) *IEEE Electron Device Letters*, **8**, 93.
14. Eitan, B. *et al.* (1999) *Proceedings SSDM*, 522.
15. Keeney, S. (2001) *IEDM Digest of Technical Papers*, 41.
16. Onoda, H. *et al.* (1992) *IEDM Digest of Technical Papers*, 599.
17. Hisamune, Y.S. *et al.* (1993) *IEDM Digest of Technical Papers*, 19.
18. Sze, S.M. (1981) *Physics of Semiconductor Devices*, John Wiley & Sons, New York. p. 397.
19. Bude, J.D. *et al.* (1997) *IEDM Digest of Technical Papers*, 279.
20. Mahapatra, S., Shukuri, S. and Bude, J.D. (2002) *IEEE Transactions on Electron Devices*, **7**, 1296.
21. Van Houdt, J. *et al.* (1992) *IEEE Transactions on Electron Devices*, **39**, 1150.
22. Ingrosso, G. *et al.* (2002) *European Solid-State Device Research Conference*, 187.
23. Tam, K., Ko, P.K. and Hu, C. (1984) *IEEE Transactions on Electron Devices*, **31**, 1116.
24. Mikolajick, T. *et al.* (2004) *Proceedings of the Nonvolatile Semiconductor Memory Workshop*, p. 98.
25. Hagenbeck, R. *et al.* (2004) *Journal of Computational Electronics*, **3**, 239.
26. Meinerzhagen, B. (1988) *IEDM Digest of Technical Papers*, 504.
27. Wolf, S. (1995) *Silicon Processing for the VLSI Era. Volume 3: The Submicron MOSFET*, Latice Press, Sunset Beach, p. 198.
28. Sim, J.S. *et al.* (2005) *Symposium on VLSI Technology*, 122.
29. Lenzliner, M. and Snow, E.H. (1969) *Journal of Applied Physics*, **40**, 278.
30. Kianian, S. *et al.* (1994) *Symposium on VLSI Technology*, 71.
31. Mehrotra, *et al.* (1992) *Symposium on VLSI Circuits*, 24.
32. Kume, H. *et al.* (1992) *IEDM Digest of Technical Papers*, 991.
33. Libsch, F.R. and White, M.H. (1990) *Solid State Electronics*, **33**, 105.
34. Peters, C. *et al.* (2004) *Proceedings of the 20th Nonvolatile Semiconductor Memory Workshop*, p. 55.
35. Ohi, M. *et al.* (1993) *Symposium on VLSI Technology*, 57.
36. Wolf, S. (1995) *Silicon Processing for the VLSI Era. Volume 3: The Submicron MOSFET*, Latice Press, Sunset Beach, p. 134.
37. Song, Y. *et al.* (2003) *Symposium on VLSI Technology*, 91.
38. Park, J.H. (2004) *IEDM Digest of Technical Papers*, 873.
39. Yoshihisa, I. *et al.* (1995) *IEEE Journal of Solid State Circuits*, **30**, 1157.
40. Suh, K.-D. *et al.* (1995) *IEEE Journal of Solid State Circuits*, **30**, 1149.

41 Kim, D.-C. et al. (2002) *IEDM Digest of Technical Papers*, 919.
42 Paven, P. et al. (1997) *Proceedings of the IEEE*, **85**, 1248.
43 Eitan, B. and Roy, A. (1999) *Flash Memories* (eds P. Cappelletti, C. Golla, P. Olivo, and E. Zanoni), Kluwer, Boston, p. 91.
44 Kianian, S. et al. (1994) *Symposium on VLSI Technology*, 71.
45 Kotov, A. et al. (2002) *Proceedings of the NVMTS*, 110.
46 Lee, J.-D. et al. (2003) *IEEE International Reliability Physics Symposium*, 497.
47 Cappelletti, P., Bez, R., Modelli, A. and Visconti, A. (2004) *IEDM Digest of Technical Papers*, 489.
48 Cappelletti, P., Bez, R., Cantarelli, D. and Fratin, L. (1994) *IEDM Digest of Technical Papers*, 291.
49 Degraeve, R. et al. (2004) *IEEE Transactions on Electron Devices*, **51**, 1392.
50 Lai, S. (1998) *Proceedings of the Seventh Biennial IEEE International Nonvolatile Memory Technology Conference*, p. 6.
51 Mori, S. et al. (1991) *IEEE Transactions on Electron Devices*, **38**, 386.
52 Likharev, K. (1998) *Applied Physics Letters*, **73**, 2137.
53 Casperson, J. (2002) *Journal of Applied Physics*, **92**, 261.
54 Specht, M., Staedle, M. and Hofmann, F. (2002) *European Solid-State Device Research Conference*, 599.
55 Liu, R. et al. (2005) *IEDM Digest of Technical Papers*, 22.3.1.
56 Kim, K. (2006) *Proceedings of the 21st Nonvolatile Semiconductor Memory Workshop*, p. 9.
57 Lee, W.-H. (1997) *VLSI Technology Digest of Technical Papers*, 117.
58 Lee, J.-D. (2002) *IEEE Electron Device Letters*, **23**, 264.
59 Kang, D. et al. (2006) *Proceedings of the 21st Nonvolatile Semiconductor Memory Workshop*, p. 36.
60 Atwood, G. (2004) *IEEE Transactions on Device and Material Reliability*, **4**, 3001.
61 Watanabe, H. et al. (1998) *IEDM Digest of Technical Papers*, 975.
62 Koval, R. (2005) *Symposium on VLSI Technology Digest of Technical Papers*, 2004.
63 Pein, H. and Plummer, J.D. (1993) *IEEE Electron Device Letters*, **14**, 415.
64 Atwood, G. et al. (1997) *INTEL Technology Journal*, **1**, 1.
65 Byeon, D.S. (2005) *ISSCC Digest of Technical Papers*, 46.
66 Endoh, T. et al. (2003) *IEEE Transactions on Electron Devices*, **50**, 945.
67 Libsch, F.R. and White, M.H. (1998) *Nonvolatile Semiconductor Memory Technology* (ed. W.D. Brown and J.E. Brewer), IEEE Press, New York. p. 309.
68 Jones, F. (1983) *Wescon Conference Record*, **27**, 28.1.1.
69 Eitan, B. et al. (2000) *IEEE Electron Device Letters*, **21**, 543.
70 Bachhofer, H. et al. (2001) *Journal of Applied Physics*, **89**, 2791.
71 Specht, M. et al. (2003) *European Solid-State Device Research Conference*, 155.
72 Shin, Y. et al. (2005) *IEDM Digest of Technical Papers*, 13.6.1.
73 van Buskirk, M. (2006) *Proceedings of the 21st Nonvolatile Semiconductor Memory Workshop*, p. 8.
74 Zous, N.K. et al. (2004) *IEEE Electron Device Letters*, **25**, 649.
75 Stein, E. et al. (2005) *Proceedings of the NVMTS*, 5.
76 Janai, M. (2003) *Proceedings of the IRPS*, 502.
77 Tsai, W.J. et al. (2002) *Proceedings of the IRPS*, 34.
78 Janai, M. et al. (2004) *IEEE Transactions on Device Materials Reliability*, **4**, 404.
79 Maayan, E. et al. (2002) *ISSCC Digest of Technical Papers*, 100.
80 Willer, J. et al. (2004) *Symposium on VLSI Technology*, 76.
81 Nagel, N. et al. (2005) *Symposium on VLSI Technology*, 120.
82 Hayashi, Y. et al. (2000) *Symposium on VLSI Technology*, 122.

83 Ogura, T. et al. (2003) *Symposium on VLSI Technology*, 207.
84 Tomiye, H. et al. (2002) *Symposium on VLSI Technology*, 206.
85 Yeh, C.C. et al. (2002) *IEDM Digest of Technical Papers*, 931.
86 Yeh, C.C. et al. (2005) *IEEE Transactions on Electron Devices*, **52**, 541.
87 Yeh, C.C. et al. (2006) *Proceedings of the 21st Nonvolatile Semiconductor Memory Workshop*, p. 76.
88 Hisamoto, D. et al. (2000) *IEEE Transactions on Electron Devices*, **47**, 2320.
89 Specht, M. et al. (2004) *IEDM Digest of Technical Papers*, 1083.
90 Walker, A.J. et al. (2003) *Symposium on VLSI Technology*, 29.
91 Lee, Y.K. (2004) *Proceedings of the 20th Nonvolatile Semiconductor Memory Workshop*, p. 96.
92 Choi, B.Y. (2006) *Proceedings of the 21st Nonvolatile Semiconductor Memory Workshop*, p. 72.
93 Willer, J. et al. (2003) *Proceedings of the 19th Nonvolatile Semiconductor Memory Workshop*, p. 42.
94 Eitan, B. (2005) *IEDM Digest of Technical Papers*, 22.1.1.
95 Tiwari, S. (1996) *Applied Physics Letters*, **68**, 1377.
96 Bostedt, C. et al. (2004) *Applied Physics Letters*, **84**, 4056.
97 Tseng, J.Y. et al. (2004) *Applied Physics Letters*, **85**, 2595.
98 Samanta, S.K. et al. (2005) *Applied Physics Letters*, **87**, 113110.
99 Muralidhar, R. et al. (2003) *IEDM Digest of Technical Papers*, 26.1.1.
100 Sadd, M. et al. (2004) *Proceedings of the 20th Nonvolatile Semiconductor Memory Workshop*, p. 75.
101 Compagnioni, C.M. et al. (2005) *IEEE Transactions on Electron Devices*, **52**, 569.
102 Perniola, L. et al. (2005) *IEEE Transactions on Nanotechnology*, **4**, 360.
103 Ohba, R. et al. (2002) *IEEE Transactions on Electron Devices*, **59**, 1392.
104 Gerardi, C. et al. (2004) *IEEE International Conference on Integrated Circuit design and Technology*, p. 37.
105 Borani, J.v. et al. (2002) *Solid State Electronics*, **46**, 1729.
106 Zacharias, M. et al. (2002) *Applied Physics Letters*, **80**, 661.
107 Perniola, L. et al. (2003) *Solid State Electronics*, **47**, 1637.
108 Guarini, K.W. et al. (2003) *IEDM Digest of Technical Papers*, 22.2.1.
109 Kim, I. et al. (1999) *IEEE Electron Device Letters*, **20**, 630.
110 Mikolajick, T. and Pinnow, C.U. (2002) *Proceedings of the NVMTS*, 3.
111 Pinnow, C.-U. and Mikolajick, T. (2004) *Journal of the Electrochemical Society*, **151**, K12.

12
Dynamic Random Access Memory

Fumio Horiguchi

12.1
DRAM Basic Operation

Dynamic random access memories (DRAMs) use the charge stored in a capacitor to represent binary digital data values. They are called "dynamic" because the stored charge leaks away after several seconds, even with power continuously applied. Therefore, the cells must be read and refreshed at periodic intervals. Despite this complex operating principle, their advantages of small cell size and high density have made DRAMs the most widely used semiconductor memories in commercial applications. In 1970, the three-transistor cell used for the 1 kbit DRAM was first reported [1], and the one-transistor (1T-1C) cell became standard use in 4 kbit DRAMs [2]. During the following years, the density of DRAMs increased exponentially, with rapid improvement to the cell design, its supporting circuit technologies, and fine patterning techniques.

The equivalent circuit of the 1T-1C DRAM cell is shown in Figure 12.1. The array transistor acts as a switch and is addressed by the word line (WL), which controls the gate. The storage capacitor, C_S, represents the charge storage element containing the information and is connected to the bit line, BL, via the array transistor. When the array transistor switch is closed, the voltage level $+V_{DD}/2$ or $-V_{DD}/2$ is applied to C_S via the bit line. The corresponding charge on C_S represents the binary information, "1" or "0". After this write pulse, the capacitor is disconnected by opening the array transistor switch.

The memory state is read by turning on the array transistor and sensing the charge on the capacitor via the bit line, which is precharged to $V_{DD}/2$ (where V_{DD} is the power supply voltage). The cell charge is redistributed between the cell capacitance, C_S, and the bit line capacitance, C_B, leading to a voltage change in the bit line. This voltage change is detected by the sense amplifier in the bit line and amplified to drive the input/output lines. Because a read pulse destroys the charge state of the capacitor, it must be followed by a rewrite pulse to maintain the stored information. The plate, PL, is kept at $V_{DD}/2$ to reduce the electric voltage stress on the capacitor dielectric, which is charged to $+V_{DD}/2$ or $-V_{DD}/2$ instead of being discharged to 0 V and charged to the full power supply voltage, V_{DD}.

Nanotechnology. Volume 3: Information Technology. Edited by Rainer Waser
Copyright © 2008 WILEY-VCH Verlag GmbH & Co. KGaA, Weinheim
ISBN: 978-3-527-31738-7

12 Dynamic Random Access Memory

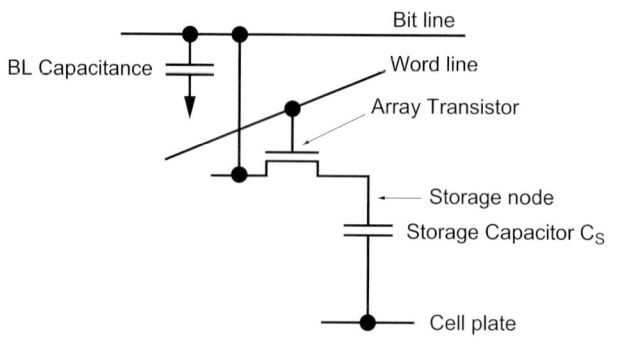

Figure 12.1 DRAM memory cell equivalent circuit. See text for details.

DRAM has required almost constant storage capacitance (more than 40 fF, i.e., 4×10^{-14} F) among the generations, despite the scaling to smaller cell sizes, and this is the reason for the requirement of three-dimensional (3-D) structures such as trench or stacked capacitor. A trench capacitor uses the inner surface of a Si hole to store charge, while a stacked capacitor uses a poly Si capacitor above the array transistor and bit line (see Figure 12.2).

12.2
Advanced DRAM Technology Requirements

Historically, the cost of DRAM has been forced to decrease to retain its share in the huge and competitive market for high-density memory. This has resulted in a

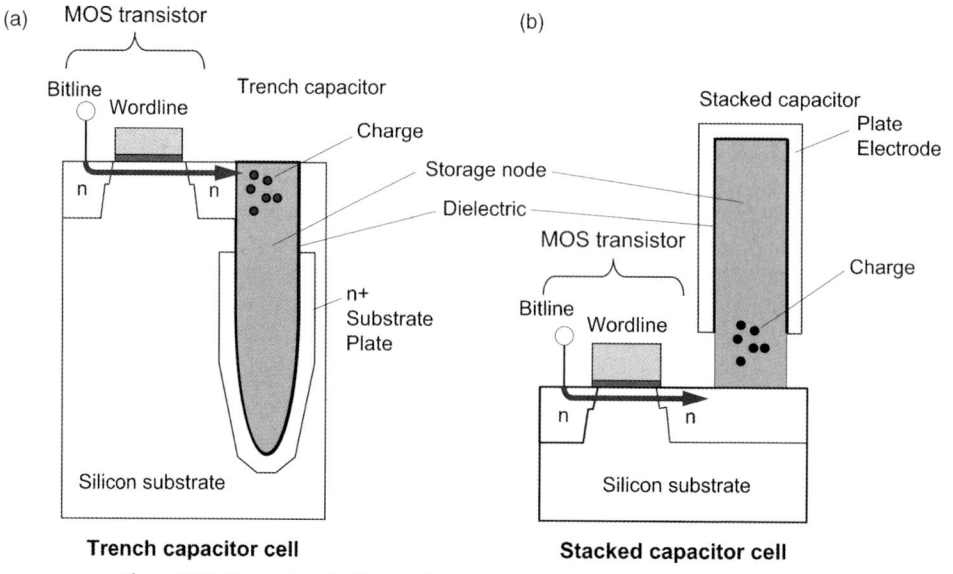

Figure 12.2 Conventional DRAM cells.

Table 12.1 Technology requirements for a DRAM memory cell.

Cell area	Smaller cell area $<8\,F^2$
Storage capacitor	$>40\,fF$, low leakage $<1\times 10^{-16}\,A$
Array transistor	High drivability, low leakage current ($<1\times 10^{-16}\,A$)
Bit line contact	Self-aligned to the word line
Storage node contact	Self-aligned to the word line (and trench capacitor or bit line)

decrease in DRAM cell size because the chip cost is directly related to the cell area. Thus, every part of the cell is required to be as small as possible, and the cell size must be less than or equal to $8\,F^2$ (where F is the feature size). A summary of the technological requirements for a DRAM memory cell is provided in Table 12.1. The most critical part in the cell shrinkage is the capacitor; thus, a 3-D structure such as a trench or stacked capacitor has been adopted to retain sufficient capacitor area within a limited space.

As the DRAM cell size shrinks to sub-100 nm, it becomes critically important to realize a sufficient on-off-current ratio in the array transistor. In general, the scaling approach implies that the transistor sizes L and W, the gate oxide thickness T_{ox}, the supply voltage, and the threshold voltage V_{th} should be reduced by a factor of $1/k$ (k: scale factor), and that channel doping should be increased by a factor of k in order to sustain or improve the transistor performance. In a DRAM cell, the charge must be stored in storage capacitors; therefore, an extremely low off-current in the array transistor is required for data retention. Thus, V_{th} should be made as small as possible to decrease the channel leakage current, and the supply voltage should be minimized so that sufficient charge can be written into the capacitor. Gate oxide thickness must also be reduced to maintain a sufficient breakdown voltage for the gate dielectrics. All this means that the array transistor cannot be scaled down in a conventional manner. On the other hand, a sufficient on-current in the array transistor is required for fast writing characteristics. This suggests a short L and large W, but scaling difficulties prevent the reduction of L and the cell size limits W. Thus, maintaining sufficient on-current in the array transistor is difficult and, moreover, increasing channel doping degrades the channel mobility, which further decreases the on-current in the array transistor.

In view of these considerations, a different structural approach for array transistors in DRAMs is necessary.

12.3 Capacitor Technologies

The first generation of DRAMs used a planar storage capacitor for memories of up to 1 Mbit. However, from the 4 Mbit generation onwards, trench or stacked capacitors were used for maintaining the same storage capacitance within a limited cell area. A comparison between the stacked capacitor cell and the trench capacitor cell is shown

12 Dynamic Random Access Memory

Table 12.2 Stacked versus trench cells.

	Stacked	Trench
Complexity of capacitor formation	×	×
Decrement of memory cell parasitics	○	△
Shrinkability of memory cell	△	○
Compatibility with logic process integration	△	○
Compatibility with logic device characteristics	○	◎
Compatibility with logic layout design rules	△	◎
Additional mask steps: @130 nm node		6–9

in Table 12.2. The differences arise mainly from the transistor formation process compared with the capacitor formation process. The array transistor in a stacked capacitor is formed before the capacitor; thus, the transistor source/drain junctions can easily be extended after the stacked capacitor is fabricated by a thermal process, which results in a degradation in transistor performance. Figures 12.2 to 12.4 show

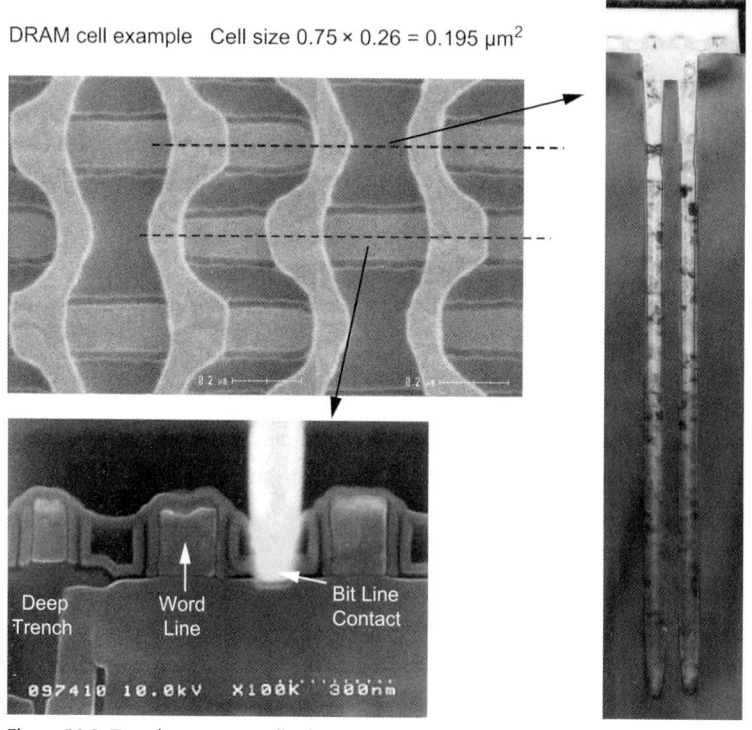

Figure 12.3 Trench capacitor cell. This has a 8 μm-deep trench capacitor and a sidewall storage-node contact for enough storage capacitance and small contact within a small area. (After Yamada, VLSI Tech. Short Course 2003 and Ref. [3]).

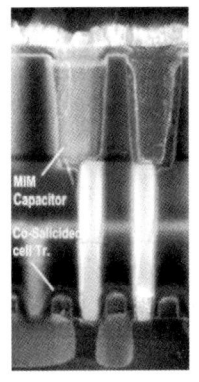

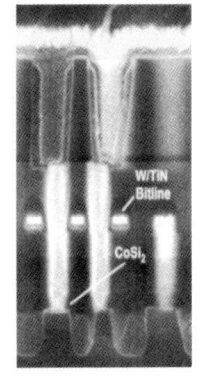

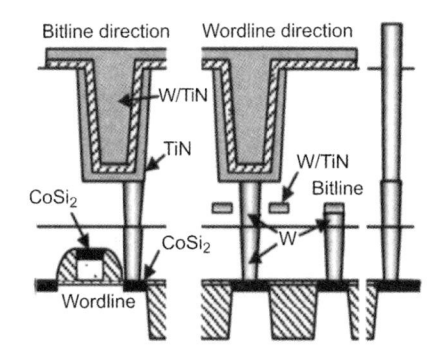

Figure 12.4 Stacked capacitor cell. This has a high-aspect-ratio storage node over the bit line and word line. The capacitor is composed of two electrodes and nitride-oxide or high-k dielectrics [4].

examples of trench and stacked capacitor cells, where each cell has high-aspect-ratio storage nodes under (trench) and over (stacked) the array transistor [3, 4].

Figure 12.5 shows the relationship between C_S and the trench diameter. C_S becomes smaller than 40 fF for a design rule of less than 90 nm because of the smaller capacitor area. To overcome the capacitor area reduction, hemispherical grains (HSG) [5, 6] can be used to enhance the surface area of the storage node in a deep trench cell (see Figure 12.6). HSG technology is an approach more typically used in the stacked capacitor cell. Another technique to enhance the capacitance is to use "high-k" dielectric materials such as Al_2O_3 or Ta_2O_5, where k denotes the dielectric permittivity. An example of a trench capacitor with a high-k Al_2O_3 dielectric is shown in Figure 12.7 [7]. The capacitance is increased by more than 30% compared with the standard nitride-oxide (NO) dielectric, which must be scaled to 1–2 nm thickness in a typical 65-nm process. As capacitor dielectrics are scaled, the leakage current increases (1.2 nm SiO_2 consists of only five atomic layers). Consequently, high-k dielectrics have been

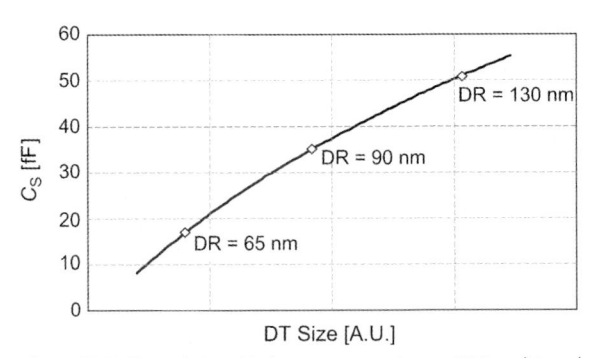

Figure 12.5 The relationship between capacitance (C_S) and trench diameter (DT).

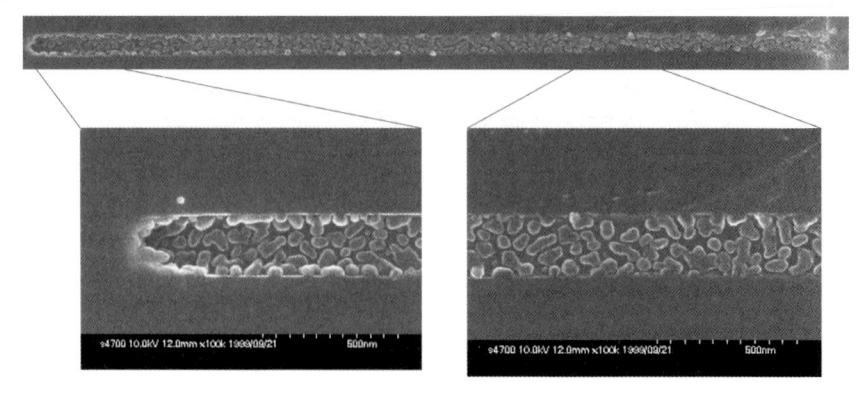

HSG is formed uniformly from top to bottom in trench.

Figure 12.6 Trench capacitor with hemispherical grains (HSG).

proposed as alternatives for capacitor dielectrics to address the leakage problem. The most common high-k capacitor dielectrics used for gate dielectrics are Al_2O_3, Ta_2O_5 and hafnium-based dielectrics (e.g., HfO_2, $HfSixOy$). However, high-k gate dielectrics are not compatible with the polysilicon gate electrodes commonly used in today's integrated circuit technology. Their combination leads to threshold voltage uncontrollability and on-current reduction. Thus, metal gate electrodes with appropriate work functions must be used. Despite this, the implementation of high-k

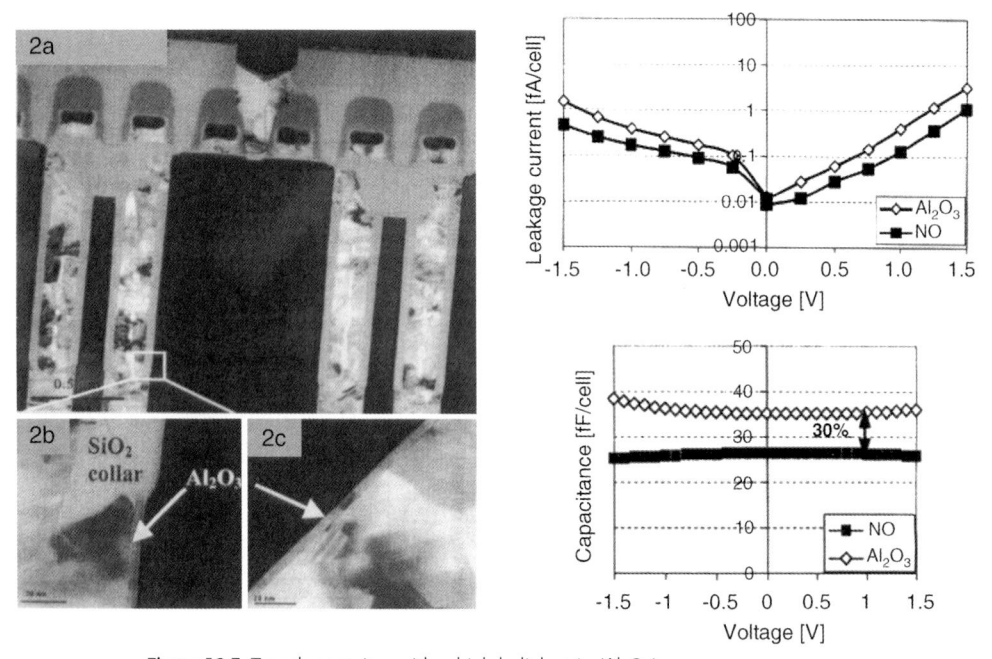

Figure 12.7 Trench capacitor with a high-k dielectric (Al_2O_3).

dielectrics and metal gate electrodes into complementary metal oxide–semiconductor (CMOS) technology is difficult, and involves many technical issues such as deposition methods, dielectrics reliability, charge trapping and interface quality. For capacitor dielectrics, it is much easier to implement high-k dielectrics because the threshold voltage shift induced by the charge trapping and the interface quality do not affect the capacitor characteristics compared with their effect on gate dielectrics. Thus, Al_2O_3 or Ta_2O_5 have been used as capacitor high-k dielectrics for a few DRAM products. For future DRAMs, high-k dielectrics will most likely be used not only for capacitor dielectrics but also for peripheral transistor gate dielectrics to overcome scaling problems.

12.4
Array Transistor Technologies

The recess-channel-array transistor (RCAT) [8] is used to reduce the electrical field near the drain to achieve a long data retention time in a stacked capacitor cell. Figure 12.8 shows the RCAT structure, which increases the effective gate length of the array transistor and mitigates the short-channel effect without increasing area.

Usually, channel doping enhances the electric field near the drain and degrades data retention characteristics because of the increased drain leakage current. To overcome this effect, the RCAT is used to reduce the electric field by separating the channel from the drain using an engraved channel region. This is effective in reducing the electric field and short-channel effects; however, the longer channel results in a small on-current in the array transistor. Thus, the RCAT is not suitable for high-speed writing.

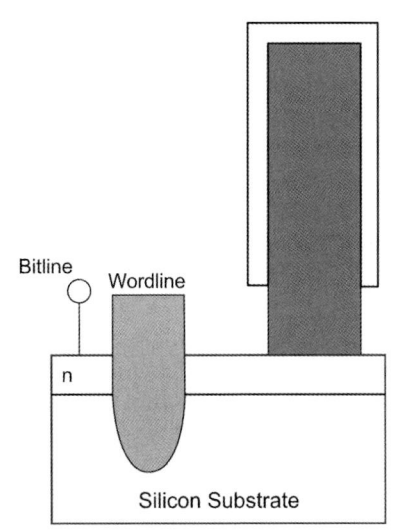

Figure 12.8 The recess-channel-array transistor (RCAT).

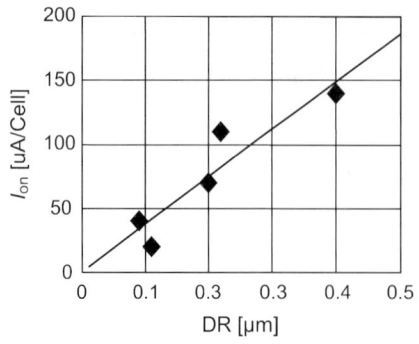

Figure 12.9 On-current (I_{on}) trend of array transistors.

The RCAT can be used down to around the 50 nm node, but below this another 3-D approach will be needed to satisfy the requirement of current drivability and to reduce the short-channel effect.

Figure 12.9 shows the on-current (I_{on}) trend of array transistors. The I_{on} decreases with transistor size in accordance with the design rule scaling; this in turn increases the signal delay in data sensing on the bit line (see Figure 12.10), in the case of reading a "1". When I_{on} is small, the signal appears on the bit line with a delay and approaches the "1" target level slowly.

To overcome these constraints in the array field-effect transistor (FET), a trench isolated transistor using sidewall gates (TIS) or a fin-array-FET can be adopted to improve the transistor performance, as in the case of silicon-on-insulator (SOI) transistors [9–12]. Figure 12.11 shows a "bird's-eye view" of a TIS-array FET where the TIS gate structure, which consists of a top gate and a sidewall gate enables a high on-current and a low off-current simultaneously because of the double-gate structure and high gate controllability. Figure 12.12 shows the T_{fin} (the width of the fin) dependence of the minimum gate length (L_g). A thinner T_{fin} would be expected to result in a marked reduction of off-current, which means that the TIS gate structure is very suitable for array transistors.

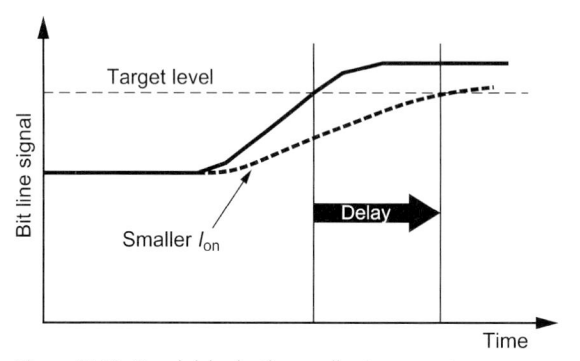

Figure 12.10 Signal delay by the smaller I_{on} current.

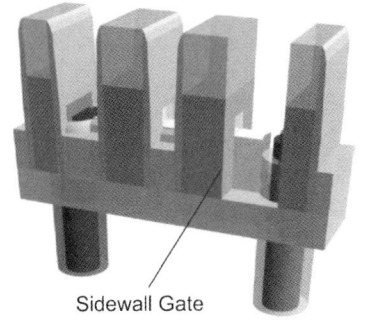

Figure 12.11 TIS/Fin array field-effect transistor (FET) DRAM.

In the TIS structure, the fin substrate is fully depleted and the double side gates contribute to the potential of each side channel. The subthreshold swing of the TIS transistor is smaller than that of the conventional planar transistor because of the strong effect of the sidewall gates. Thus, a small gate voltage difference can rapidly change the drain current from a small off-current to a large on-current. Moreover, the constant threshold voltage characteristics without a back-gate bias effect contribute to the large on-off-current ratio.

A more advanced array transistor is the *vertical transistor*, in which the source, gate and drain are arranged vertically. There are two types of vertical transistors. One uses the inner sidewall of the trench hole, while the other uses the outer sidewall of a silicon pillar for the channel. The former is suitable for trench capacitor cells [13], while the latter is known as a surrounding gate transistor (SGT) [14, 15]. The gate electrode of the SGT surrounds a pillar of silicon, and the gate length of the SGT is

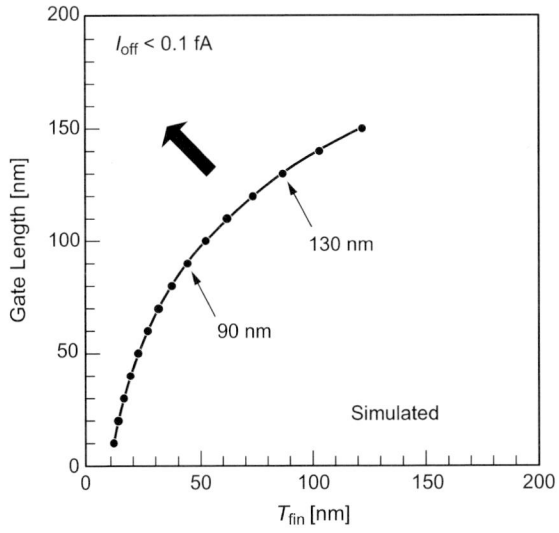

Figure 12.12 The dependence of fin width (T_{fin}) on minimum gate length (L_g).

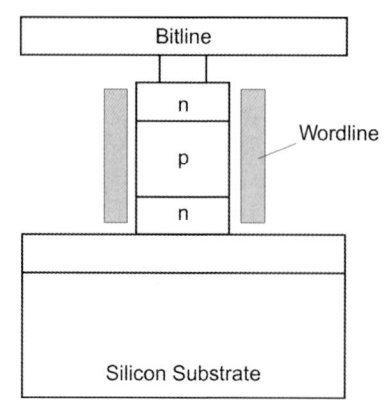

Figure 12.13 A surrounding gate transistor. The gate electrode of SGT surrounds a silicon pillar, with the gate length being adjusted by the pillar height.

adjusted by the pillar height, as shown in Figure 12.13. Therefore, the SGT has the merits of short-channel-effect immunity and superior current drivability resulting from the excellent gate controllability.

Planar array transistors cannot easily be scaled down (as noted above), and the TIS has good on-off-current ratio characteristics. The vertical transistor is different from the planar type in that the channel length is defined by the depth of the hole or the height of the pillar. Thus, the gate length is free from the minimum design rule and the cell area limitations, and can be selected to be sufficiently large so as to avoid the short-channel effect. Similar to the trench-type capacitor, the vertical transistor and the capacitor are formed in the same hole, and this contributes to the small cell size of less than $6\,F^2$. In the SGT cell, it is more difficult to form the capacitor and array transistor, although it has ideal array transistor characteristics. The SGT substrate is fully depleted and the surrounding gate contributes to the potential of the pillar surface channel. The subthreshold swing of the SGT cell is smaller than that of the conventional planar transistor and the TIS because of the stronger effect of the surrounding gate. Also, surrounding gate structures contribute to the large width of the transistor by using the entire perimeter of the pillar. Thus, a large on-off-current ratio can be attained without a back-gate bias effect, and the SGT can be used for $4\,F^2$ small-cell-size DRAMs.

DRAM scaling will continue to enable the integration of many advanced technologies in view of the huge size of the DRAM market. Thus, these advanced technologies will be used in future-generation DRAMs.

For the array transistor, the TIS/fin type structure is expected to be adopted using p^+ poly for obtaining a suitable threshold voltage with low channel doping. For the capacitor, a high-k dielectric (e.g., barium strontium titanate, BST) may be used in future DRAMs. For the peripheral transistor, mobility enhancement technologies such as the use of SiGe or a linear strain technique and high-k gate dielectrics will be adopted to achieve large driveability for a high-speed operation. A $4\,F^2$ cell layout is

12.5 Capacitorless DRAM (Floating Body Cell)

The difficulties of DRAM integration are mainly attributable to the necessity for constant capacitance, even when the cell size is reduced. For this reason, the integration of capacitors is very complicated for trench or stacked capacitors. The floating body cell (FBC) is a new concept of a DRAM without a capacitor. Because the cell is composed of one transistor, the FBC has a simple and compact structure.

Figure 12.14 shows the principle of the FBC, which involves the storage of the signal charge in the body of the cell transistor. To write "1", V_{WL} is biased to 1.5 V and V_{BL} to 2 V, so that the body potential (V_{body}) is increased by the holes that accumulate by impact ionization. To write "0", V_{WL} is biased to 1.5 V and V_{BL} to -1.5 V, so that V_{body} is decreased by ejecting holes from the body. The body potential difference (ΔV_{body}) is stored by setting V_{WL} to -1.5 V and V_{BL} to 0 V. In order to read the stored data, V_{BL} is biased to 0.2 V and V_{WL} is swept up to a certain level, while the bit line current (I_{read}) is measured. The I_{read}–V_{WL} characteristics are shown in Figure 12.15. The threshold voltage difference between a "0" cell and a "1" cell (ΔV_T), which is an index of the data reading margin, is about 0.32 V. In order to increase ΔV_T or ΔI_{read}, C_S (the body capacitance for data storage) plays an important role, because the ΔV_{body} of the hold state is reduced by WL-body and BL-body capacitance coupling. A back gate is used to enable charge accumulation in the body [16–18], and also to increase C_S, which stabilizes the body potential. The structure of the FBC has a modified double-gate configuration. A transmission electron microscopy cross-section of a fully depleted FBC, with thin SOI and BOX layers, is shown in Figure 12.16.

The body capacitance is small compared to the standard DRAM capacitor, typically by two orders of magnitude. However, the leakage current of the FBC storage node is small because of the small p–n junction area, which is located only at the channel-side

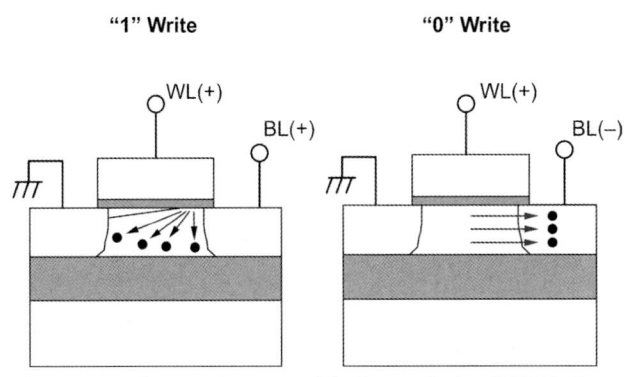

Figure 12.14 The write operation of the floating body cell (FBC). See text for details.

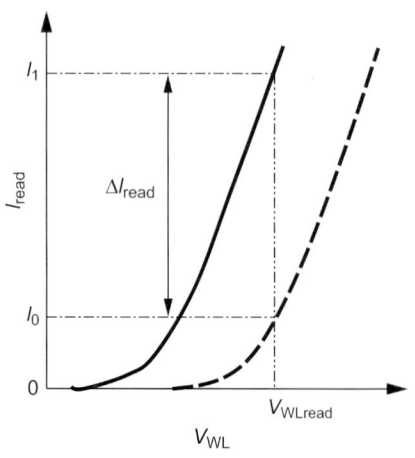

Figure 12.15 The read operation of the floating body cell (FBC). See text for details.

edges of the source and drain. Thus, the retention time of the FBC is reduced slightly compared to the standard DRAM. As a consequence, and because of the short retention time, the FBC is suitable for high-performance embedded DRAM applications rather than low-power applications.

The SOI structure has been widely used for high-performance applications, particularly game processors, and is expected to be used in the embedded DRAM for on-processor caches. An FBC using a SOI substrate can easily be used for these applications with the same compatibility as the SOI substrate. As the FBC is composed of one transistor and has no capacitor, it is scalable down to the 32 nm node. Details of this structure are provided in Ref. [17]. An image of an FBC with 128 Mb DRAM, along with the chip features, is shown in Figure 12.17. The FBC, which has dimensions of 7.6×8.5 mm, contains all of the necessary circuits (including internal voltage generators) and operates using a single 3.3 V power supply [18].

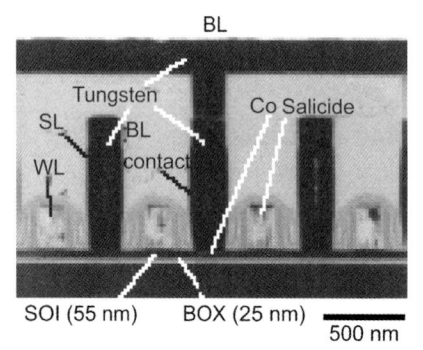

Figure 12.16 Transmission electron microscopy cross-section of a fully depleted floating body cell (FBC), showing the thin SOI and BOX layers.

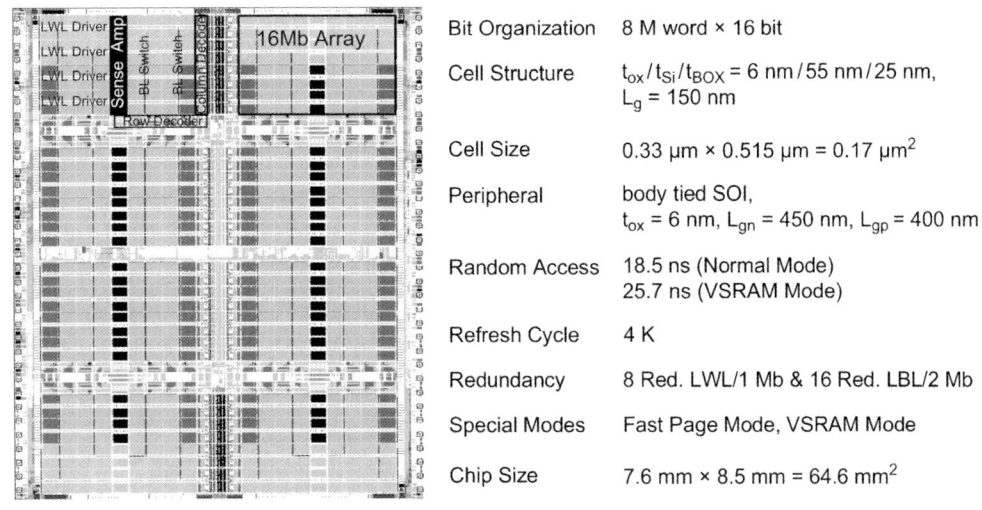

Figure 12.17 Floating body cell (FBC) 128 Mb DRAM and its features.

Bit Organization	8 M word × 16 bit
Cell Structure	$t_{ox}/t_{Si}/t_{BOX}$ = 6 nm / 55 nm / 25 nm, L_g = 150 nm
Cell Size	0.33 μm × 0.515 μm = 0.17 μm²
Peripheral	body tied SOI, t_{ox} = 6 nm, L_{gn} = 450 nm, L_{gp} = 400 nm
Random Access	18.5 ns (Normal Mode) 25.7 ns (VSRAM Mode)
Refresh Cycle	4 K
Redundancy	8 Red. LWL/1 Mb & 16 Red. LBL/2 Mb
Special Modes	Fast Page Mode, VSRAM Mode
Chip Size	7.6 mm × 8.5 mm = 64.6 mm²

12.6 Summary

Today, while the demand for DRAM remains greater than for any other type of memory, the capacity of DRAM is continually increasing such that variations are now becoming available for both low-power and high-speed applications. Because of the scaling limitations, the TIS/fin array-transistor is expected to be used in future-generation DRAMs, with more advanced DRAMs – such as vertical transistors such as the SGT – most likely being used for DRAMs with a smaller cell size. In addition, the capacitorless DRAM – the FBC – shows great promise as a candidate for next-generation embedded DRAMs offering both high density and high speed. Clearly, the use of 3-D structures should help to overcome the scaling problems likely to be encountered in future-generation memories.

References

1 Regitz, W.M. et al. (1970) *IEEE Journal of Solid-State Circuits*, **SC-5**, 181–186.
2 Boonstra, L. et al. (1973) *IEEE Journal of Solid-State Circuits*, **SC-8**, 305–310.
3 Yanagiya, N. et al. (2002) *IEDM Technical Digest*, 58–61.
4 Arai, S. et al. (2001) *IEDM Technical Digest*, 403–406.
5 Saida, S. et al. (2000) *Proceedings of ISSM*, 177.
6 Amon, J. et al. (2004) A highly manufacturable deep trench based DRAM cell layout with a planar array device in a 70 nm technology. *IEDM Technical Digest*, 73–76.
7 Seidl, H. et al. (2002) *IEDM Technical Digest*, 839–842.
8 Kim, J.Y. et al. (2003) The breakthrough in data retention time of DRAM using Recess-Channel-Array Transistor (RCAT)

for 88 nm feature size and beyond. *VLSI Technical Digest*, 11–12.
9 Hieda, K. *et al.* (1987) New effects of trench isolated transistor using side-wall gates. *IEDM Technical Digest*, 736–737.
10 Hisamoto, D. *et al.* (2000) FinFET – a self-aligned double-gate MOSFET scalable to 20 nm. *IEEE Transactions on Electron Devices*, **47**, 2320–2325.
11 Katsumata, R. *et al.* (2003) Fin-Array-FET on bulk silicon for sub-100 nm trench capacitor DRAM. *VLSI Technical Digest*, 61–64.
12 Weis, R. *et al.* (2001) A highly cost efficient $8F^2$ DRAM cell with a double gate vertical transistor device for 100 nm and beyond. *IEDM Technical Digest*, 415–418.
13 Lee, D.-H. *et al.* (2007) Improved cell performance for sub-50 nm DRAM with manufacturable bulk FinFET structure. *VLSI Technical Digest*, 164–165.
14 Sunouchi, K. *et al.* (1989) A surrounding gate transistor (SGT) cell for 64/256 Mbit DRAMs. *IEDM Technical Digest*, 23–26.
15 Goebel, B. *et al.* (2002) Fully depleted surrounding gate transistor (SGT) for 70 nm DRAM and beyond. *IEDM Technical Digest*, 275–278.
16 Shino, T. *et al.* (2004) Fully-depleted FBC (floating body cell) with enlarged signal window and excellent logic process compatibility. *IEDM Technical Digest*, 281–284.
17 Shino, T. *et al.* (2006) Floating body RAM technology and its scalability to 32 nm node and beyond. *IEDM Technical Digest*, 569–572.
18 Ohsawa, T. *et al.* (2005) An 18.5 ns 128 Mb SOI DRAM with a floating body cell. *ISSCC Technical Digest*, 458–459.
19 Ranica, R. *et al.* (2004) A capacitor-less DRAM cell on 75 nm gate length, 16 nm thin fully depleted SOI device for high density embedded memories. *IEDM Technical Digest*, 275–280.

13
Ferroelectric Random Access Memory

Soon Oh Park, Byoung Jae Bae, Dong Chul Yoo, and U-In Chung

13.1
An Introduction to FRAM

An ideal non-volatile memory should possess the required characteristics such as high density (high scalability and compact cell size), high reliability (excellent retention and endurance), low cost, and high performance (random access, high read/write speed, and low power consumption) [1, 2]. Although Si-based Flash memory with high density and low cost is the leading non-volatile memory, it cannot basically meet the needs of high endurance and performance characteristics. Therefore, new concepts for non-volatile memory such as FRAM (Ferroelectric RAM), PRAM (Phase change RAM), MRAM (Magnetoresistive RAM), and RRAM (Resistive RAM) have been demonstrated as strong candidates for an ideal non-volatile memory.

Among these emerging memories, PRAM uses phase-change material as a storage element, and shows high scalability and compact cell size owing to its simple cell structure [3]. However, it has disadvantages such as long crystallization time, high power consumption for phase-change switching, and low endurance performance. MRAM uses magnetic material for data storage and shows excellent high speed and good reliability performance, but it requires a large cell area to make a unit cell [4]. Recently emerged RRAM uses resistive switching material as a storage element, but its technology is not yet matured [5]. Finally, FRAM uses ferroelectric materials as a storage element and has been a strong candidate to a universal memory since the late 1980s [6–8]. Because of its similar structure and operation scheme to DRAM (Dynamic RAM) and additional non-volatility, FRAM has been developed as a universal memory for one-chip solution. The operation scheme, reliability of ferroelectric capacitor, and the technology of high-density FRAM for a universal memory will be introduced in the following sections.

Nanotechnology. Volume 3: Information Technology I. Edited by Rainer Waser
Copyright © 2008 WILEY-VCH Verlag GmbH & Co. KGaA, Weinheim
ISBN: 978-3-527-31738-7

13.1.1
1T1C and 2T2C-Type FRAM

The operation and architecture of capacitor-type FRAM is almost identical to that of dynamic random access memory (DRAM). It should be noted that every cell has its own separate plate line in capacitor-type FRAM, whereas DRAM uses common plate-line in the level of a half V_{dd}. Necessarily, a ferroelectric capacitor replaces the linear capacitor in a metal-insulator-metal (MIM) storage element.

A schematic view of the two-transistor–two-capacitor (2T2C) -type FRAM and the one-transistor–one-capacitor (1T1C) -type FRAM [9] are shown in Figure 13.1. In the 2T-2C type, the switching and non-switching charges of two adjacent ferroelectric capacitors are used as data "1" or "0" charges, which have a large sensing window and uniform cell operation. However, the cell area is too large to be used for high-density ferroelectric devices because two capacitors can store only a single bit. On the other hand, the 1T1C type provides an advantage of small cell area because single capacitor stores single bits. However, the sensing margin of the 1T1C type is reduced as a half of that of 2T2C type by setting a reference level in the middle of data "1" and "0". The sensing margin might be further reduced due to the variation of reference ferroelectric capacitors. Nevertheless, the 1T1C type is used as the cell structure due to its small cell size in most high-density ferroelectric devices.

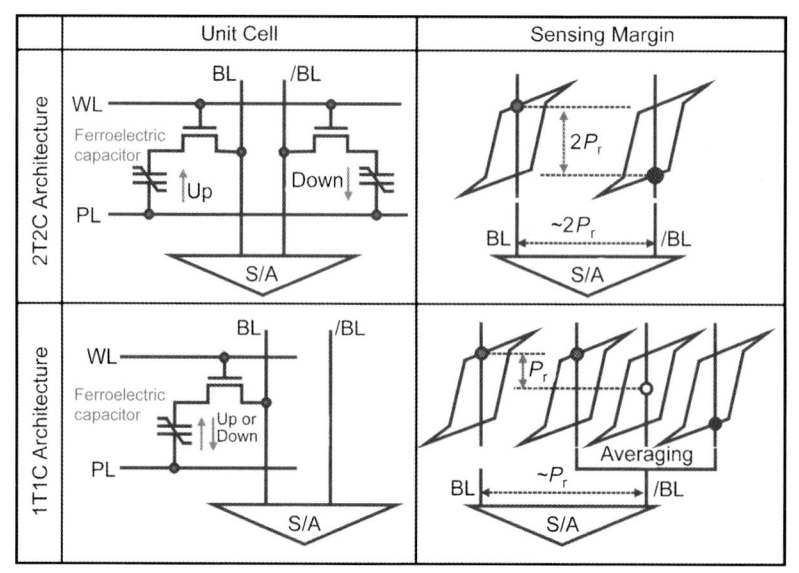

Figure 13.1 Comparison of 2T2C and 1T1C FRAM architectures in respect of cell size and sensing margin.

13.1.2
Cell Operation and Sensing Scheme of Capacitor-Type FRAM

Figure 13.2 illustrates the writing operation of 1T1C-type FRAM. Figure 13.2a is the schematic of 1T1C FRAM which is composed of the word line (WL), bit line (BL), and plate line (PL). Figure 13.2b shows the charges preserved in the hysteresis curve, while Figures 13.2c and d show the timing diagrams of writing data "1" and "0". To write "1" into the memory cell, the BL is raised to V_{pp} and PL is kept as ground (GND). The polarization directions are from PL to BL and the $-P_r$ value is preserved. To write "0", the BL is kept as GND and the PL is kept high as V_{pp}. Thus, the opposite direction of the polarization is generated and $+P_r$ value is preserved.

Figure 13.3 illustrates the reading operation of 1T1C-type FRAM [10]. A read access begins by precharging the BL to GND, after which the PL is raised to V_{pp}. This establishes serial two capacitors consisting of C_s and C_{BL} between the PL and the GND, where C_s is the capacitance of ferroelectric storage element and C_{BL} is the parasitic capacitance of BL. Therefore, the V_{pp} is divided into V_f and V_{BL} between C_s and C_{BL} according to their relative capacitance. Depending on the data stored, the voltage developed on the ferroelectric capacitor and BL can be approximated as follows:

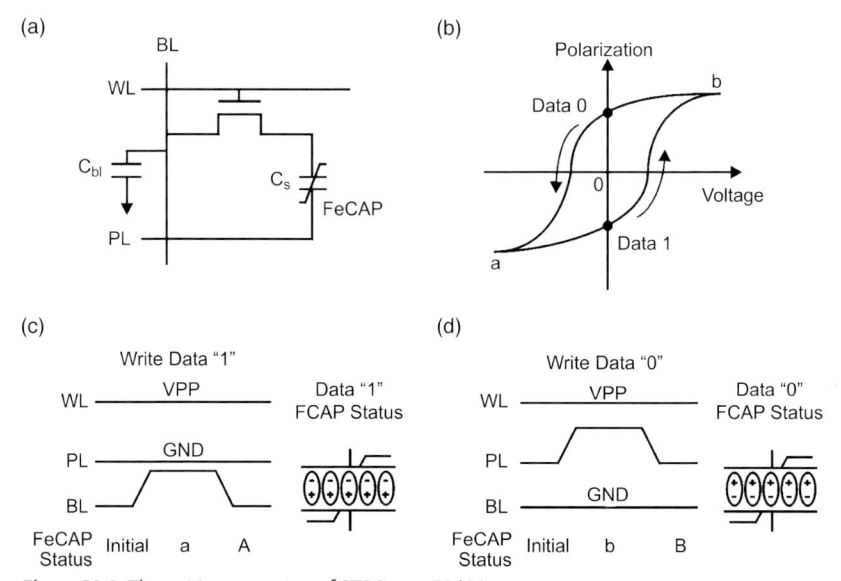

$$V_f = C_{BL} \times V_{PP}/(C_s + C_{BL}) \tag{13.1}$$

$$V_{BL}(\text{Data"1"}) = dQ_{sw}/(C_{sw} + C_{BL}) \tag{13.2}$$

Figure 13.2 The writing operation of 1T1C-type FRAM. (a) Schematic of 1T1C FRAM which is composed of word line (WL), bit line (BL), and plate line (PL). (b) Charges preserved in hysteresis curve. (c,d) Timing diagrams of (c) writing data "1" and (d) writing data "0".

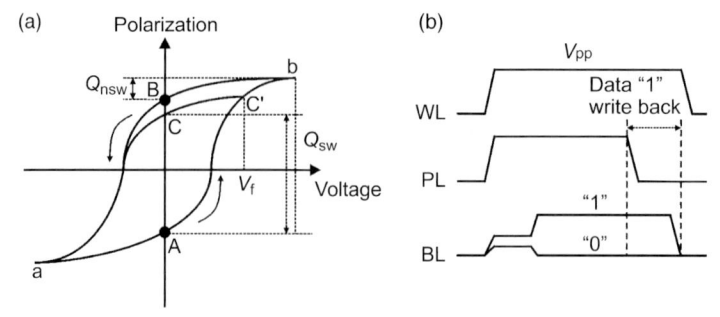

Figure 13.3 Schematic illustration of read operation procedures in 1T1C FRAM.

$$V_{BL} (\text{Data“0”}) = dQ_{nsw}/(C_{nsw} + C_{BL}) \tag{13.3}$$

In general, the voltage developed in the BL is too small to sense charge differences. Therefore, a sensing amplifier should be used in order to drive the BL to full V_{pp} if the data is "1", or to 0 V if the data is "0". The structure of the sensing amplifier of capacitor-type FRAM includes the cross-coupled latch sense amplifier of DRAM. It can be classified to a folded bit line and an open bit line according to the cell array, as shown in Table 13.1 and Figure 13.4 [11]. The open bit-line scheme is applicable to 1T1C structure, and the folded bit-line scheme can be applied to both 1T1C and 2T2C structures.

Table 13.1 Comparison of FRAM sense amplifier types.

	Sense amplifier	Realization	Noise immunity	Sensibility
Folded bit-line	1 ea/2 BL	Easy layout	Same noise environment	Good
Open bit-line	1 ea/1 BL	Difficult layout	Different noise environment	Not good

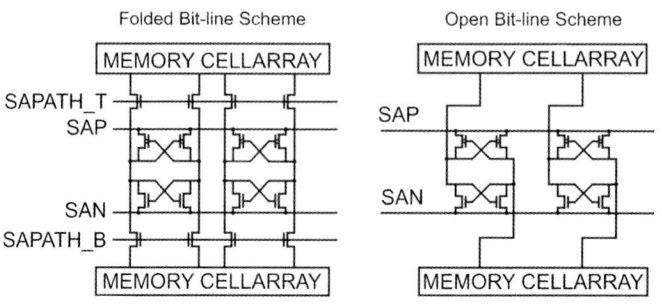

Figure 13.4 Schematic diagrams of FRAM sense amplification.

13.2
Ferroelectric Capacitors

Similar to the DRAM device, the capacitor technology concerning ferroelectrics serves as a guideline to the development of FRAM devices. In this regard, the material characteristics and reliability of typical ferroelectric capacitors will be considered in this section.

13.2.1
Ferroelectric Oxides

Representative ferroelectric materials for complementary metal oxide–semiconductor (CMOS) integration can be divided to two groups, including perovskite-structured (PZT and $BiFeO_3$) and Bi-layer structured (SBT, BLT, and BTO) materials. The characteristics of these are summarized in Table 13.2 [12, 13]. The crystal structure of PZT can be either tetragonal or rhombohedral according to the Zr/Ti composition ratio below Curie temperature. In SBT, two $SrTaO_3$ perovskite blocks and one $(Bi_2O_2)^{2+}$ layer constitute one unit cell, as shown Figure 13.5b. In BLT, La atoms are partially substituted to Bi atom in $Bi_4Ti_3O_{12}$ (BTO) crystal which is composed of three TiO_6 octahedra and one $(Bi_2O_2)^{2+}$ layer leading the wanted crystal structure. These differences of unit cell structure largely determine the characteristics of corresponding ferroelectrics. Therefore, the typical properties of PZT, SBT and BLT can be compared from this point of view.

First, the remanent polarization value (P_r) determines the sensing margin between data "0" and "1" (the larger $2P_r$, the better sensing-margin), while the coercive voltage (V_c) or coercive field (E_c) decides the operating voltage in an FRAM device (the smaller V_c, the better operation voltage). PZT shows large P_r and E_c values because of strong interactions between neighboring perovskite unit cells. In contrast, SBT and BLT request the anisotropic growth along the a-b axis to attain direct interactions between the neighboring perovskite unit cells toward electrical field direction. The smaller P_r and E_c values of SBT, compared to PZT, tend to increase by Nb doping.

Table 13.2 The features of typical ferroelectrics used for FRAM.

Ferroelectrics	$Pb(Zr,Ti)O_3$ (PZT)	$SrBi_2Ta_2O_9$ (SBT)	$(Bi,La)_4Ti_3O_{12}$ (BLT)
P_r [µC cm^{-2}]	10–40	5–10	10–15
E_c [Kv cm^{-1}]	50–70	30–50	30–50
Endurance	Poor on Pt electrode	Good on Pt electrode	Good on Pt electrode
	Good on oxide electrode		
Crystallization temperature [°C]	450–650	650–800	650–750
Curie temperature [°C]	~400	~400	~400

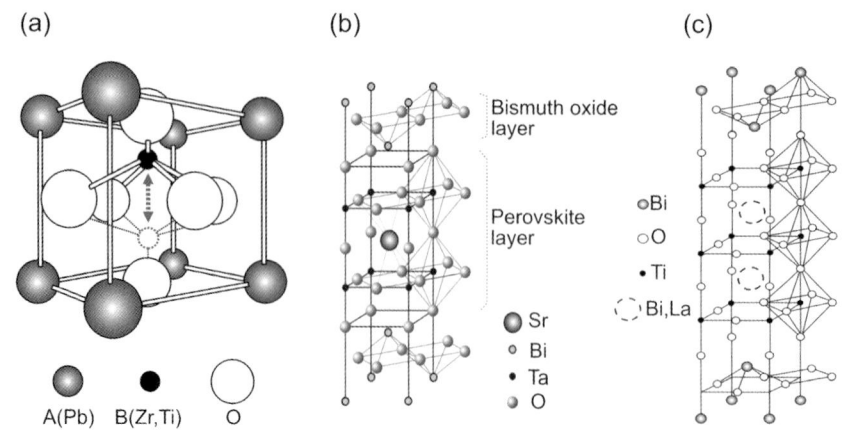

Figure 13.5 The crystal structures of ferroelectric (a) PZT, (b) SBT, and (c) BLT.

Ferroelectric films must attain a crystallized perovskite structure in order to show the polarization behavior. Therefore, the plentiful oxygen atmosphere and high substrate temperature in order to crystallize ferroelectric oxide lead to the demand for a noble metal electrode and oxidation barrier to achieve CMOS integration. In a stacked FRAM cell with COB (capacitor over bit-line) structure, the capacitor is located directly on the top of the MOSFET drain, which requires the low-temperature process to realize high-density CMOS integration. In this respect, the lower crystallization temperature of PZT than that of SBT and BLT is advantageous for the fabrication of future high-density FRAM.

To date, PZT has long been the leading material considered for ferroelectric memories, and has been superior to SBT and BLT. The higher P_r value and lower process temperature of PZT can act as strong merits for the fabrication of high-density COB cells in CMOS integration. On the other hand, lead-free SBT and BLT can be considered for environmentally friendly FRAMs. Recently renewed multiferroic $BiFeO_3$ exhibits both ferroelectric and magnetic properties, but it is unclear whether this is useful for memory application, or not [14]. In particular, the proper ferroelectrics for CMOS integration should be chosen by serious consideration for the degradation induced by hydrogen, plasma, stress, and heat in the succeeding integration processes [15].

13.2.2
Fatigue

"Fatigue" is a term describing the fact that the remanent polarization becomes small when a ferroelectric film experiences numerous polarization reversals. When PZT on Pt electrodes suffers from reading/writing cycles over 1E5 cycles, the P_r value shows a conspicuous reduction, which limits the repeated use of a memory. A few reports about non-fatigue phenomena of Pt/PZT/Pt capacitors should be regarded with great care because this type of behavior can be observed when the applied voltage is less

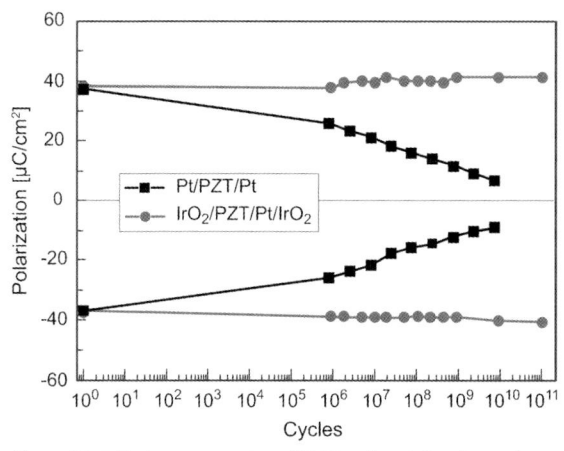

Figure 13.6 Fatigue properties of PZT and an IrO$_2$ electrode.

than $V_{(90\%)}$. Fatigue behavior is strongly related to the generation of oxygen vacancy by the repeated cycles, which induces dipole-pinning electrons for charge-neutrality; this is why oxygen vacancies are the only mobile ionic species in the lattice even at the room temperature on the basis of defect chemistry model. However, the fatigue problems of PZT can be almost solved at present by the use of conducting oxide electrodes (e.g., IrO$_2$, RuO$_2$, SrRuO$_3$, CaRuO$_3$, LaNiO$_3$, and LSCO), ensuring no degradation of the P_r value even up to 1E12 cycles. Figure 13.6 shows a comparison of fatigue properties in PZT capacitor with Pt and IrO$_2$ electrodes [16]. This improvement of fatigue property can be explained by the fact that oxygen in the IrO$_2$ electrodes reduces oxygen vacancies, which prevents fatigue degradation reducing the dipole-pinning effect. As Ir is stably converted into IrO$_2$ under oxygen at ambient temperature, the fatigue problem can be remarkably enhanced in the case of using an IrO$_2$ oxide electrode.

In contrast to the PZT film, an SBT film does not show the fatigue phenomenon up to 1E13 switching cycles, even if Pt electrodes are used. It was speculated that the $(Bi_2O_2)^{2+}$ interlayer can compensate the produced oxygen vacancy. However, similar Bi-layer structured BTO shows fatigue problems on a Pt electrode, which suggests that the simple charge-compensation role of the $(Bi_2O_2)^{2+}$ layers is not sufficient to make the fatigue-free films. This reduction in polarization could be much alleviated by using La-doped BTO (so-called BLT). Accordingly, the limited switching cycles of dipoles are no longer a serious problem for any ferroelectric materials, as shown in Figure 13.7 [13, 14].

13.2.3
Retention

Polarization retention is the ability of poled ferroelectric capacitors to preserve the poled state over time (generally 10 years into the future at 85 °C). The retention property represents an important reliability issue for non-volatile ferroelectrics

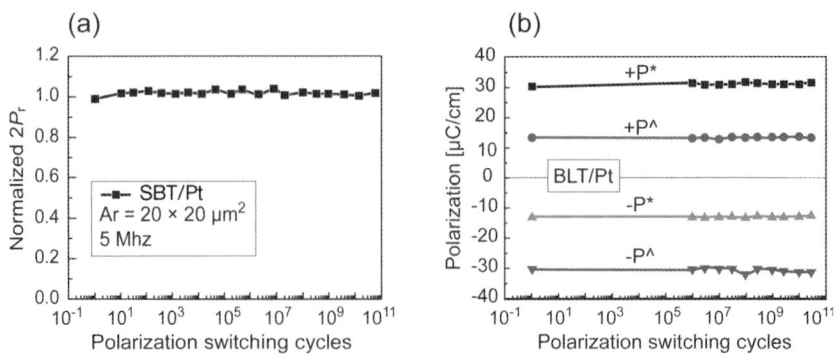

Figure 13.7 Fatigue properties of (a) SBT and (b) BLT film on Pt electrode. The symbols are indicated in Figure 13.9.

memories. Most commonly, the retention of ferroelectrics can be classified into the same-state and opposite-state retention.

The same-state retention, which is closely related to aging, represents the loss of polarizability when one first writes the datum of "0" or "1" in a capacitor with electrical pulses, and reads the datum again after long period without changing the initial status. Therefore, the same-state retention failure can occur when the relaxation component in the opposite-polarity state increases at the expense of the relaxation component in the stored polarity state. The stored polarity status can be stabilized by the use of ferroelectrics with "high" Curie temperature, after which the same-state retention loss can be improved from the viewpoint of thermodynamics. For instance, $BaTiO_3$ is not applicable for non-volatile FRAM because of its low Curie temperature (∼140 °C), although this can be raised to 500 °C by imposing biaxial compressive strain.

The opposite-state retention, which is closely related to imprint, represents the loss of polarizability when one first writes the datum of "0" or "1" in a capacitor with electrical pulses and reads the "changed" datum again. In words, the same-state retention is a longstanding problem of the read-only memory (ROM), while the opposite-state retention is that of the random-access memory (RAM), because information must be modifiable (as shown in Figure 13.8) [17].

The opposite-state retention failure occurs when a capacitor, which has aged considerably in one state, is switched to the opposite state. In this case, the capacitor behaves as if it would prefer to remain in the original state. The charge defects are activated by thermal energy and redistributed by the polarization field. Therefore, the resulting internal field causes a lower energy barrier and invokes polarization back-switching during the delay time, as shown in Figure 13.9 [18]. Accordingly, the opposite-state retention can be solved by minimizing the space charges, which results from defects inside the ferroelectrics, domain wall motion, or defects near the electrode-ferroelectric interfaces.

The thickness scaling of ferroelectric films is indispensable when pursuing a low switching voltage, making this suitable for integrated electronics applications. However, to date, thinner ferroelectric films have shown serious degradation of

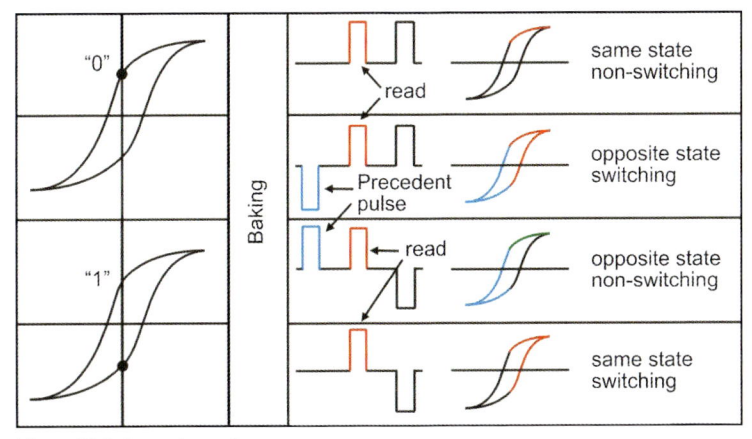

Figure 13.8 Retention pulse sequence.

opposite-state retention compared to same-state in cumulative studies. Consequently, during the past few years much attention has been focused on the failure mechanism of the opposite-state retention. In order to solve the opposite-state retention failure problem, it is necessary to utilize frequently used technologies such as seeding, metalorganic chemical vapor deposition (MOCVD), and perovskite oxide electrode in the case of PZT ferroelectrics. Thus, these technologies will be reviewed briefly in the following sections.

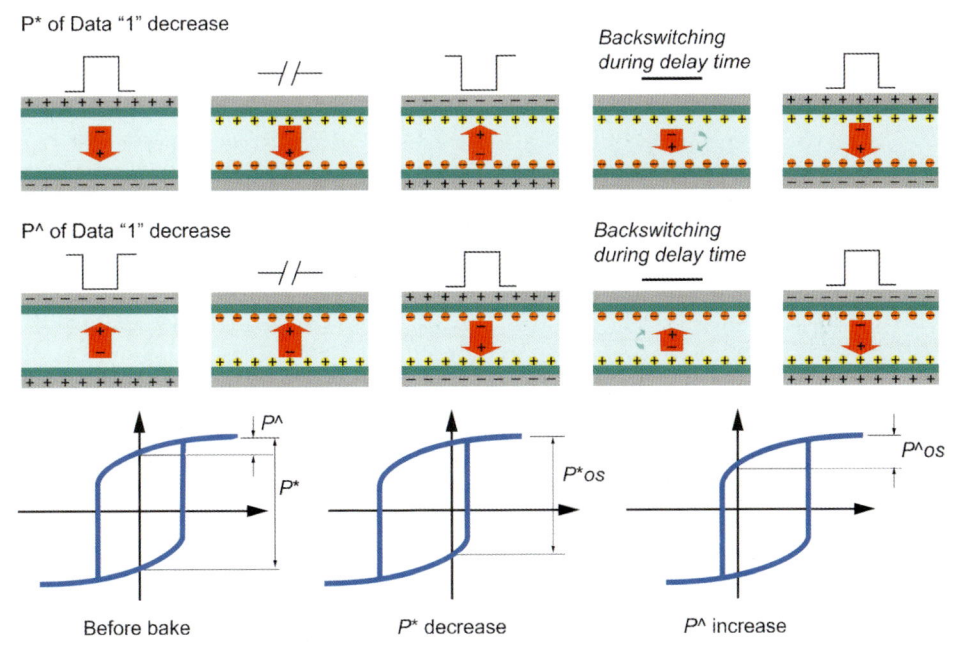

Figure 13.9 Retention failure mechanism.

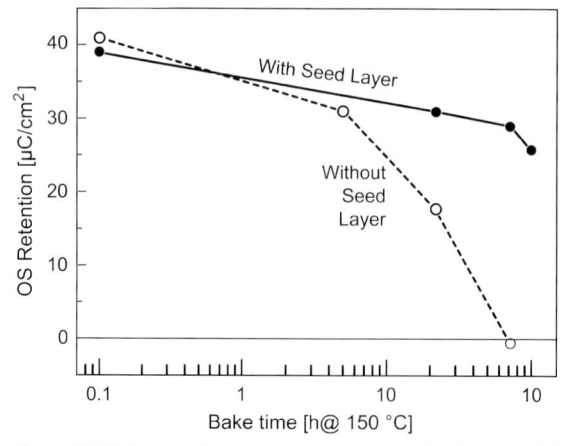

Figure 13.10 Improved opposite-state retention by the use of PbTiO$_3$ seed layer.

13.2.3.1 Crystallinity of PZT Film

Frequently, the perovskite-structured PZT phase can be generated by utilizing nucleation and the grain growth process from the pyrochlore phase. Because the nucleation process is strongly dependent on the substrates, an appropriate seed layer can supply nucleation sites in order to decrease activation energy for crystallization of the perovskite phase. A PbTiO$_3$ seed layer is very effective for supplying a high density of nuclei in the initial stage of deposition, because the crystallization temperature (350~680 °C) is lower than that of PZT (>650 °C). The succeeding PZT film shows a much enhanced crystallinity and preferred orientation, and thereby exhibits improved retention result (see Figure 13.10 [19, 20]. For this purpose, an optimum thickness of PbTiO$_3$ is essential because a thinner PbTiO$_3$ layer cannot play a sufficient role for the seed layer, while a thicker one may cause adverse effects on the electrical properties of the overall film. A PbTiO$_3$ seed layer is helpful in the initial stage of film growth, but still constitutes a portion of the ferroelectric films. Therefore, the use of a perovskite oxide electrode as a seed layer may provide a better means of preparing reliable ultrathin ferroelectric films, because the seed layer belongs to the electrode and not to the ferroelectrics.

13.2.3.2 The MOCVD Deposition Process

Most current FRAM cells below 64 Mb density are based on a planar capacitor stack. In CMOS integration, it is commonplace to use either chemical solution deposition (CSD) or a sputtering technique for the deposition of planar films, and chemical vapor deposition (CVD) for a conformal deposition, based largely on an economics viewpoint. Somewhat ironically, however, such common sense has caused a "dribbling" (slow movement, low amplitude) of technological developments in this field. For example, the current deposition method used for "planar" PZT films is mostly based on an "MOCVD process" in order to pursue the excellent opposite-state retention properties [21]. The comparative retention of a PZT film deposited by CSD

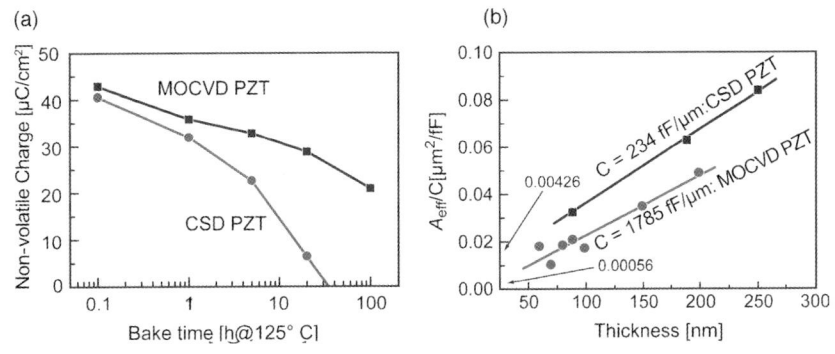

Figure 13.11 Improved opposite-state retention by using the metalorganic chemical vapor deposition (MOCVD) deposition process.

and MOCVD method is shown in Figure 13.11, where the MOCVD PZT film shows superior retention properties to those of CSD films due to the low defect density in ferroelectrics and/or interfaces. It may be speculated that an as-crystallized PZT film on an Ir electrode can be obtained by using the MOCVD process, such that the non-switching layer at the interface between the electrode and ferroelectrics is thinner, without the formation of Pt_3Pb alloys.

13.2.3.3 Perovskite Oxide Electrode

Most ferroelectric materials have a perovskite crystal structure, as outlined in the previous section. Therefore, if a conducting oxide electrode having a perovskite structure is use, then ferroelectric properties such as reliability can be greatly improved due to the reduction of any non-ferroelectric dead layer at the interface between the ferroelectrics and electrodes. Such remarkable improvement of retention properties by using an $SrRuO_3$ electrode with a perovskite structure is illustrated in Figure 13.12 [22].

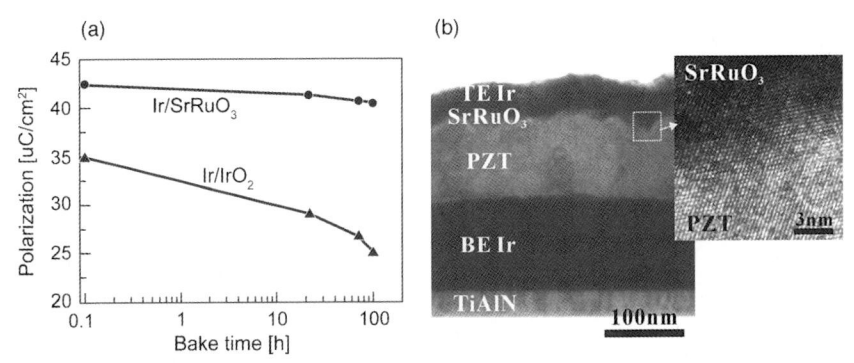

Figure 13.12 (a) Retention properties of a PZT capacitor with $Ir/SrRuO_3$ and Ir/IrO_2 electrodes. (b) Transmission electron microscopy image of the interface between $SrRuO_3$ and PZT films.

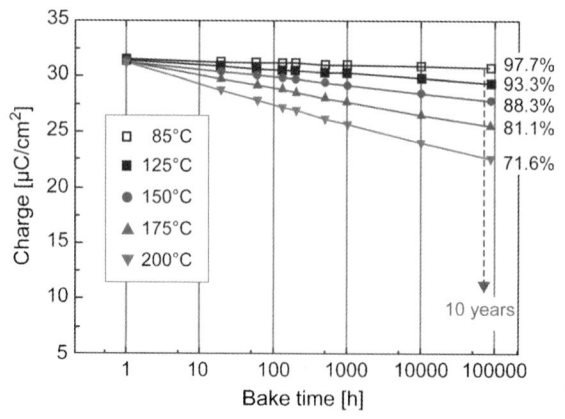

Figure 13.13 The ultrahighly reliable properties of the FRAM device.

During recent years, although perovskite oxide electrodes such as $SrRuO_3$, $LaNiO_3$ and $CaRuO_3$ have undergone intense investigation, the problems of high leakage currents – which inevitably are induced by high defect densities in the oxide electrode – remain to be overcome.

Recently, the successful development of an ultrahighly reliable FRAM device has been reported, and the retention properties of this fully integrated device, at different temperatures, are illustrated graphically in Figure 13.13 [23]. Based on these findings, the FRAM device could be expected to maintain >80% of any initial charge, even after 10 years at 175 °C.

13.3
Cell Structures

A vertical scanning electron microscopy image of the FRAM cell structure is shown in Figure 13.14 [24]. The cell is composed of a cell transistor, capacitor, buried contact, bit line, word line, and plate line. The cell structure can be divided into the CUB (capacitor under bit line) and COB (capacitor over bit line) structures, the merits and demerits of which are considered in the following section.

13.3.1
CUB Structure

In the CUB cell structure, the ferroelectric capacitor is formed beside the cell transistor, as shown in Figure 13.15 [25]. This requires a large cell area compared to the COB cell structure, in which the ferroelectric capacitor is formed over the cell transistor. The CUB scheme has no thermal budget limitations on the ferroelectric film deposition, and the subsequent anneal process for crystallization of the ferroelectric film, because the ferroelectric capacitor formation processes (including stack deposition and dry etching) are completed before the metallization process is

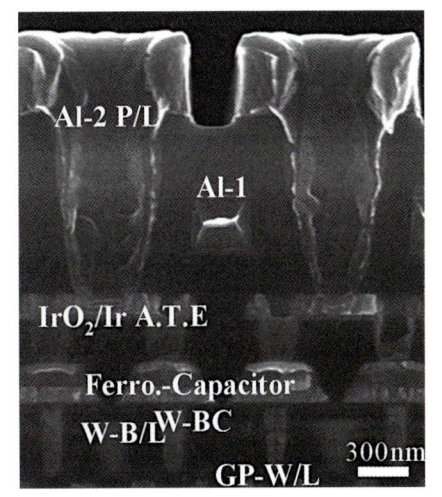

Figure 13.14 A vertical scanning electron microscopy image of the FRAM cell.

carried out. Due to technical difficulties in realizing a ferroelectric film with low thermal budget processes, and of identifying a suitable oxidation barrier metal which is stable above 600 °C, the early FRAMs were developed with a CUB cell structure, thereby sacrificing cell size efficiency.

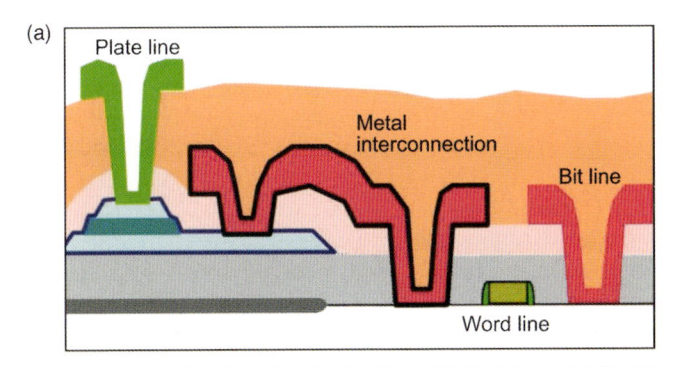

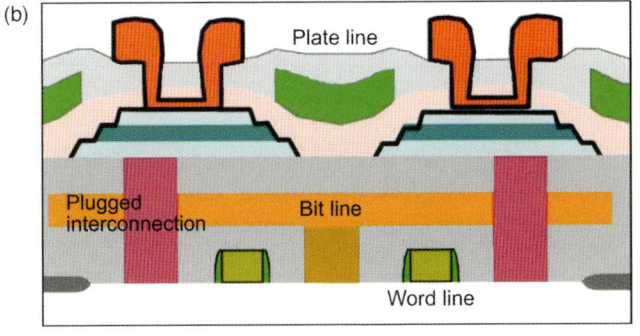

Figure 13.15 Schematic diagram of (a) capacitor under bit line (CUB) and (b) capacitor over bit line (COB) cell structures.

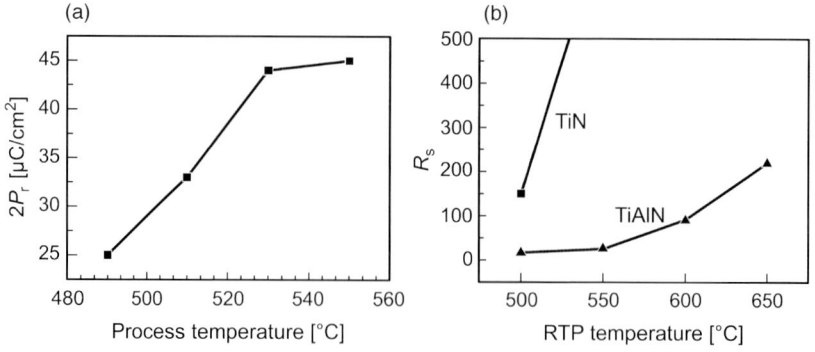

Figure 13.16 (a) $2P_r$ variation versus MOCVD process temperature and (b) comparison with oxidation resistance of TiAlN and TiN.

13.3.2
COB Structure

In the COB cell structure, the ferroelectric capacitor is formed over the bit line. Thus, the realization of a COB cell structure requires both a new buried contact (BC) plug and new metal technologies for the oxidation barrier. A stable contact between the BC plug and the bottom electrode must be provided when the ferroelectric capacitor has been processed at a high temperature of 600 °C or above [26]. As shown in Figure 13.16a, a high-temperature process is essential to obtain a sufficient polarization value in an MOCVD PZT process [27]. In order to prevent oxidation of the BC plug, various oxidation-barrier metals have been widely investigated; among these, a TiAlN film proved successful in preventing oxidation of the BC plug. The oxidation resistance properties of TiAlN and TiN thin films, as a function of temperature, are illustrated graphically in Figure 13.16b.

As mentioned above, as the COB structure is more beneficial with regards to high-density integration than are CUB structures, an increasing proportion of FRAM devices are today adopting the COB structure.

13.4
High-Density FRAM

In this section, the current status of planar capacitor technology, together with the technical issues involved in the development of 3-D capacitors for high-density FRAM device application, will be discussed.

13.4.1
Area Scaling

In order to achieve a high-density FRAM, the cell size must be scaled down as much as possible. Unfortunately, however, there exists a scaling limit because the

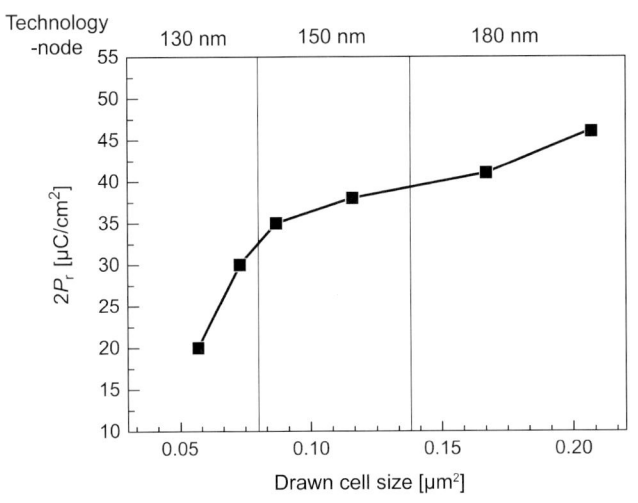

Figure 13.17 Polarization decay as a scale-down of the drawn cell size.

polarization value ($2P_r$) decreases in proportion to the cell size, such that etching damage on the capacitor becomes increasingly critical. The data in Figure 13.17 show that the polarization decays as the drawn cell size decreases. With a planar capacitor structure, although the polarization degradation is negligible down to the 150-nm technology node, the polarization value decreases rapidly below that level. This effect is mainly caused by the difference between the drawn area and the effective area, and indicates that the etched slope is no longer steep enough to provide both a designed top-electrode area and sufficient spacing between adjacent bottom electrodes at the 130-nm technology node. Therefore, in order to increase the effective capacitor area below the critical cell size, both thickness scaling and the high-etched slope of the capacitor stack should be guaranteed.

In order to maximize effective capacitor area, the most important technology is to achieve the high-etched slope of the capacitors, but this is difficult because both top and bottom electrodes are usually noble metals, and the noble metal etch process has remained an unanswered question since the initial stages of FRAM development. Even until quite recently, the limitation of the capacitor etched slope was about 60~65°, mainly owing to the loss of hard-mask from the sputtering condition of the noble metal etch. This lower capacitor slope can lead to a decrease in capacitor area of the top electrodes, or to a short circuit between the cap-to cap at the bottom electrode. However, based on some experimental findings (see Figure 13.18), new technology has been successfully developed in order to obtain a high-etched slope of about 80~85° [28]. This new etching scheme was tested at high temperature with chlorine and fluorine chemistry, and a dual hard mask (oxide and metal). As a result, the noble metal was successfully etched with a high slope by improving the reactivity between the noble metal and etch gases, and by increasing the process temperature and reinforcing the robustness of the hard-mask.

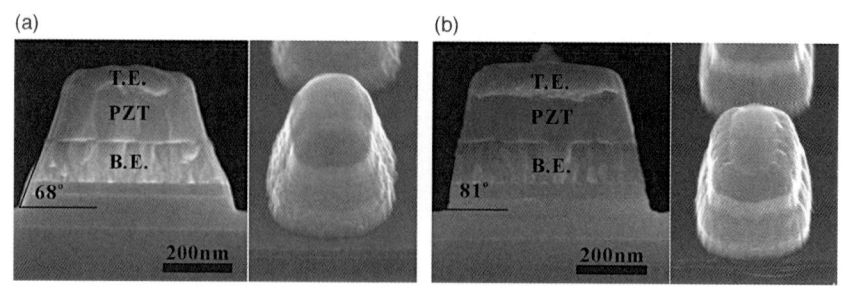

Figure 13.18 Scanning electron microscopy image showing the improvement of capacitor etch slope. (a) Normal cap etch condition; (b) enhanced cap etch condition.

13.4.2
Voltage Scaling

With the advent of the "mobile" era, low-voltage operation has become increasingly important in the reduction of power consumption. In the case of FRAM devices, the operation voltage is directly related to the thickness of the ferroelectric film; hence, the latter dimension should be minimized for low voltage application. As shown in Figure 13.19a, a PZT capacitor prepared by the CSD process shows a drastic degradation of ferroelectric properties below 100nm thickness. This is clearly a critical problem which must be solved in the case of high-density FRAM devices. As described above, both ferroelectric properties and reliability are greatly improved when the PZT films are prepared with MOCVD process; thus, even an 80 nm-thick PZT film prepared in this way demonstrates highly reliable ferroelectric properties [29] (Figure 13.19b).

It is difficult to prevent ferroelectric degradation at the interface between electrodes and ferroelectric material, even when the MOCVD process is employed. However, if perovskite oxide electrodes are used, the dead layer effect at the interface may be remarkably reduced. Recently, it has been reported that a high reliability can be achieved even with a 50 nm-thick PZT capacitor [30]. The charge-to-voltage (Q–V)

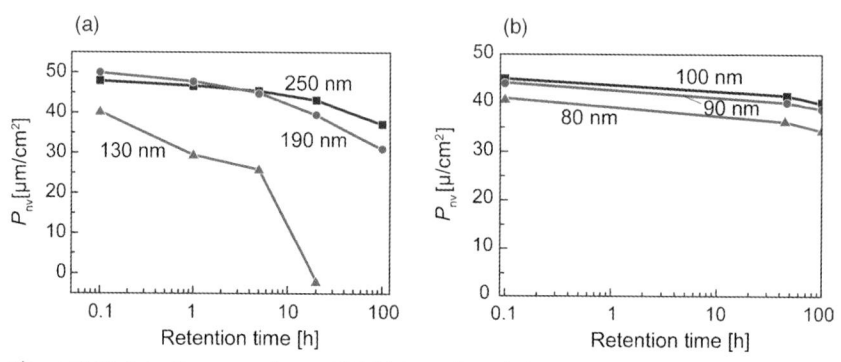

Figure 13.19 Retention properties as PZT thickness is scaled down. (a) CSD-processed PZT; (b) MOCVD-processed PZT.

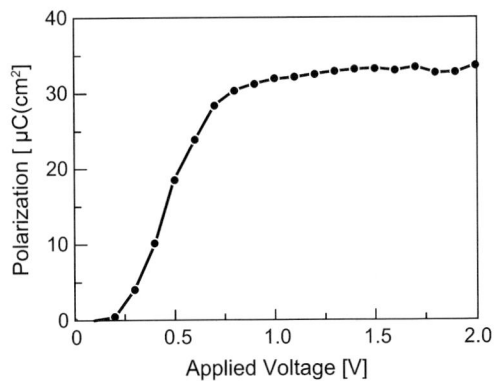

Figure 13.20 Charge–voltage (Q–V) diagram of 50 nm-thick PZT capacitor.

diagram of such as capacitor is illustrated graphically in Figure 13.20, with the capacitor being fully polarized well below an operation voltage of 1 V.

With thinner PZT films, however, several problems persist, including roughness and high leakage current. In order to overcome these difficulties, a chemical mechanical polishing (CMP) process has been introduced for PZT films. As the increase in leakage current for thin PZT film depends mainly on the surface roughness, a CMP process for PZT films can greatly reduce the leakage current [31]. The atomic force microscopy (AFM) findings and ferroelectric properties for PZT films, with or without the CMP process, are shown in Figure 13.21.

13.4.3
3-D Capacitor Structure

13.4.3.1 Limitation of Planar Capacitor
Today, many technical challenges remain to be solved for high-density planar capacitor structured FRAMs, including the limitation of capacitor stack thickness, the noble metal etch process, and thin PZT degradation. In addition, an optimum cell size is clearly required for a sufficient sensing window in FRAM devices (Figure 13.22) [32].

It can be seen from Figure 13.22 that it is difficult to achieve the 200 mV sensing margin which is required in 1T1C cell structure with sub-130 nm design rules. From this point of view, even if a thin capacitor stack and a high-etch slope were to be realized in the planar capacitor structure, it would appear difficult to embody a high-density FRAM device in excess of 256 Mb. Therefore, in order to overcome this limitation, FRAM development should ideally be pursued with a 3-D capacitor structure similar to the present-day DRAM.

13.4.3.2 Demonstration of a 3-D Capacitor
As mentioned above, the requirement for a 3-D capacitor structure is inevitable for high-density FRAM development, and the structure – together with the necessary technologies to develop a 3-D FRAM cell – are shown schematically in Figure 13.23 [33].

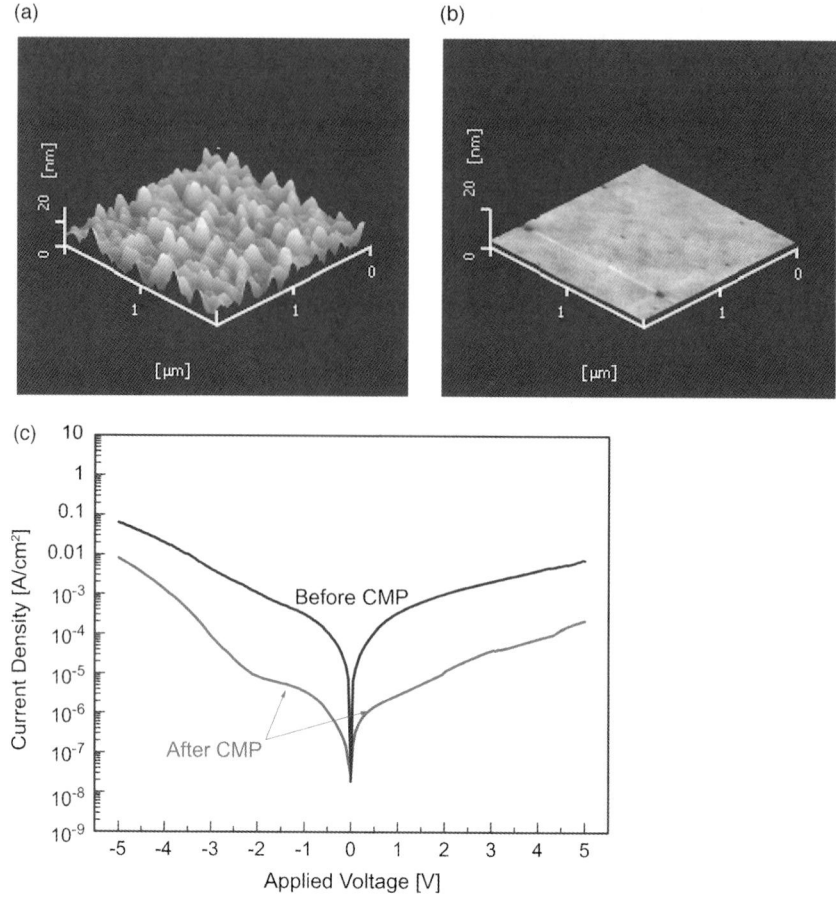

Figure 13.21 (a,b) Atomic force microscopy images and (c) leakage current characteristics of PZT films before and after CMP processing.

A prototype 3-D capacitor has recently been demonstrated, and a TEM image representing a 3-D PZT capacitor is shown in Figure 13.24. Although some pyrochlores remained in the trench capacitor, the columnar grains were well established at the side-wall of trench, with optimized deposition condition.

Figure 13.25 illustrates, graphically, the ferroelectric properties with different-sized trench structures. The polarization–voltage characteristics of a planar capacitor and trench capacitors are shown in Figure 13.25a. Under 2.1 V external bias, and an electric field of $350\,\text{kV}\,\text{cm}^{-1}$, these capacitors produced no current leakage and showed quite good hysteretic behavior compared to their planar counterpart. The remnant polarization ($2P_r$) plotted against the external maximum voltage is shown in Figure 13.25b; these data showed that $2P_r$ is very similar to that for the planar capacitor in the case of a 0.32 μm trench-diameter 3-D capacitor. However, a 0.25 μm trench-diameter capacitor showed a $2P_r$ value of $19\,\mu\text{C}\,\text{cm}^{-2}$ under an external

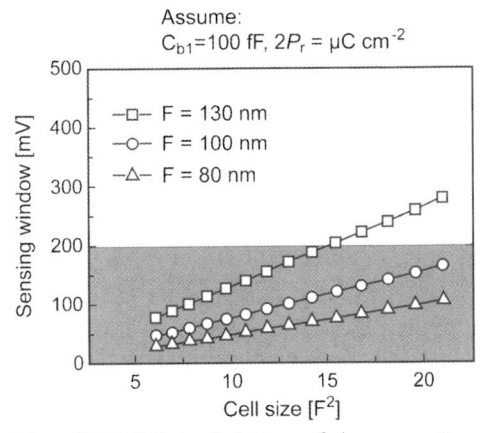

Figure 13.22 Cell size limitations of planar capacitors.

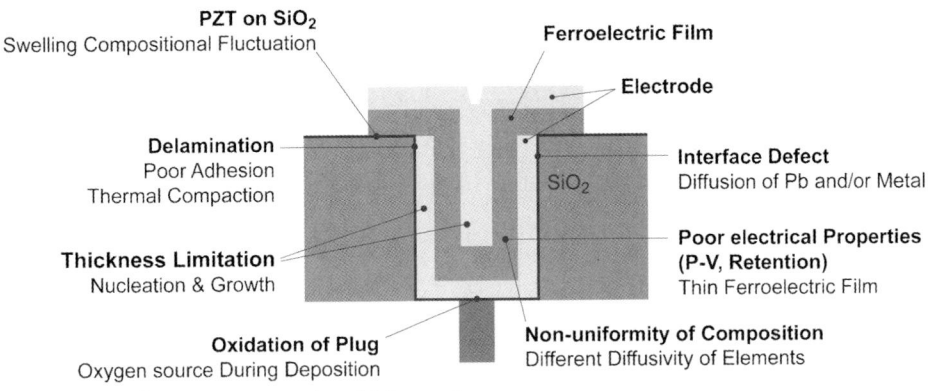

Figure 13.23 Schematic representation of 3-D capacitor structure, and the technical issues encountered.

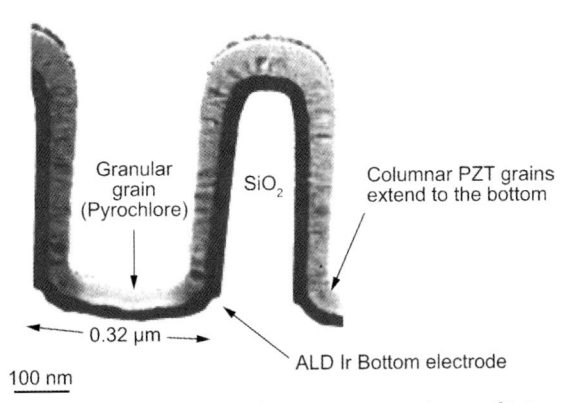

Figure 13.24 Transmission electron microscopy image of 3-D FRAM cell structure during the development of SAIT (Samsung Advanced Institute of Technology).

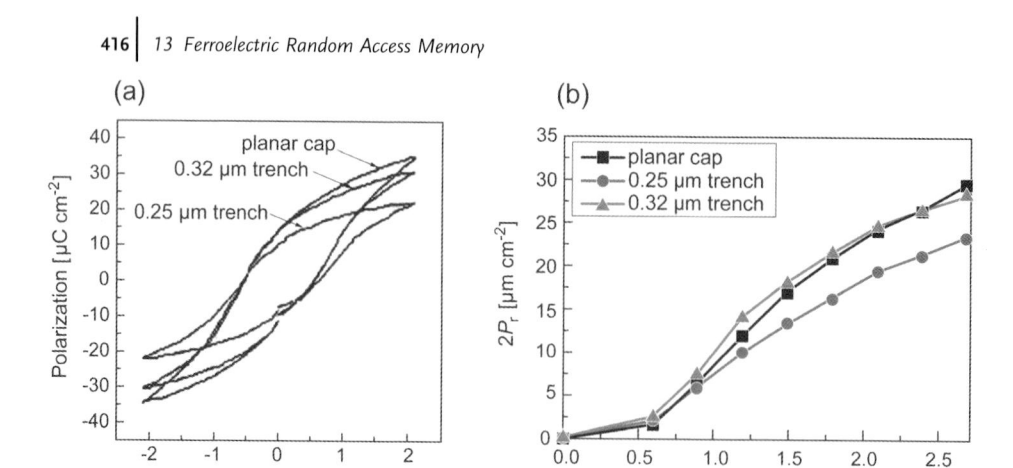

Figure 13.25 Ferroelectric properties of 3-D FRAM cell structure during the development of SAIT. (a) Polarization–voltage (P–V) loops and (b) charge–voltage (Q–V) results with different trench sizes.

maximum voltage of 2.1 V, which was 80% of the $2P_r$ values in either planar or 0.32 μm trench-diameter cases. This difference may be derived from an incomplete extension of the columnar grains on the 0.25 μm trench side-wall. Based on these findings, it is quite possible that the side-wall PZT film has the same ferroelectric properties as the planar PZT film.

In order to realize Giga-bit FRAMs with a 3-D capacitor, it has been necessary to develop the atomic layer deposition (ALD) process for the PZT and electrode material. As shown in Figure 13.23, the thickness of the ferroelectric material should be less than 50 nm because the bottom/top electrode and ferroelectric films may be formed inside a trench of 200 nm diameter. This means that the ferroelectric properties of sub-50 nm-thick PZT capacitors should be obtained for 3-D capacitor research. In addition, a step coverage of the PZT film becomes important as the aspect ratio of the capacitor increases. Because the PZT film should have a uniform composition at the bottom and side-wall, the ALD method is regarded as the best choice among other deposition methods, such as PVD and CVD. Although ALD for PZT has been investigated by many research groups, process optimization is still required. Moreover, both noble electrode metals and ferroelectric materials may be prepared using ALD. Recently, although iridium was successfully deposited using ALD, additional improvements of properties should also be investigated. In contrast, a CMP technology for noble metal electrodes may need to be introduced in order to separate each capacitor within this structure. Although noble metal CMP has not yet been achieved, it is currently undergoing extensive investigation.

Unfortunately, as of today several technical difficulties, including the reliability of the 3-D capacitor, have not been fully solved. Nonetheless, the activities of many research groups have provided much promise for 3-D FRAM development. It follows that, if some of the above-mentioned problems are solved in the near future, then the Giga-bit FRAM era will be well and truly opened.

13.5
Summary and Conclusions

FRAM technology, which has been undergoing continuous development since the early 1990s, has been used to target a universal memory in the semiconductor industry. Although reliability – notably endurance and retention – was initially a major challenge, recent findings have shown that this is no longer a key issue for FRAM devices. Rather, it is scalability which has become an important issue, following the development of 64 Mb FRAM through material and cell structural innovations. At this density, FRAM may be applied to low-density embedded memory (e.g., a smartcard), based on the demands of non-volatility, rapid access, high read/write endurance, low-power operation, and high security level. In order to produce high-density FRAM devices for use in major applications, a conventional planar-type capacitor technology is insufficient for further cell size scaling. Rather, breakthrough technologies such as the 3-D capacitor must be developed in order for the FRAM device to serve as an ideal, non-volatile memory in the future.

References

1 Kim, K.N. and Lee, S.Y. (2004) *Integrated Ferroelectrics*, **64**, 3–14.
2 Kim, K.N. (1999) *Integrated ferroelectrics*, **25**, 149–167.
3 Jeong, G.T., Hwang, Y.N., Lee, S.H., Lee, S.Y., Ryoo, K.C., Park, J.H., Song, Y.J., Ahn, S.J., Jeong, C.W., Kim, Y.T., Horii, H., Ha, Y.H., Koh, G.H., Jeong, H.S. and Kim, K.N. (2005) *IEEE International Conference on Integrated Circuit and Technology*, pp. 19–22.
4 Kim, H.J., Oh, S.C., Bae, J.S., Nam, K.T., Lee, J.E., Park, S.O., Kim, H.S., Lee, N.I., Chung, U.I., Moon, J.T. and Kang, H.K. (2005) *IEEE Transactions on Magnetics*, **41**, 2661–2663.
5 Baek, I.G., Lee, M.S., Seo, S., Lee, M.J., Seo, D.H., Suh, D.S., Park, J.C., Park, S.O., Kim, H.S., Yoo, I.K., Chung, U.I. and Moon, J.T. (2004) *IEDM Technical Digest*, pp. 587–590.
6 Ishiwara, H., Okuyama, M. and Arimoto, Y. (eds) (2004) *Ferroelectric Random Access Memories – Fundamentals and Applications*, Spinger-Verlag.
7 Rameash, R. (1997) *Thin Film Ferroelectric Materials and Devices*, Kluwer Academic Publishers.
8 Scott, J.F. and Paz De Araujo, C.A., (1989) *Science*, **246**, 1400.
9 Kim, K.N. (2001) *International Symposium on VLSI Technology*, pp. 81–84.
10 Ishiwara, H., Okuyama, M. and Arimoto, Y. (eds) (2004) *Ferroelectric Random Access Memories – Fundamentals and Applications*, Spinger-Verlag, pp. 149–163.
11 Choi, M.K. and Jeon, B.G. et al. (2002) *IEEE Journal of Solid-State Circuits*, **37**, 1472–1478.
12 Araujo, C.A., Cuchiaro, J.D., McMillan, L.D., Scott, M.C. and Scott, J.F. (1995) *Nature*, **374**, 627–629.
13 Park, B.H., Kang, B.S., Bu, S.D., Noh, T.W., Lee, J. and Jo, W. (1999) *Nature*, **401**, 682–684.
14 Wang, J., Neaton, J.B., Zheng, H., Nagarajan, V., Ogale, S.B., Liu, B., Viehland, D., Vaithyanathan, V., Schlom, D.G., Waghmare, U.V., Spaldin, N.A., Rabe, K.M., Wuttig, M. and Ramesh, R. (2003) *Science*, **299**, 1719–1722.

15 Joo, H.J., Song, Y.J., Kim, H.H., Kang, S.K., Park, J.H., Kang, Y.M., Kang, E.Y., Lee, S.Y., Jeong, H.S. and Kim, K.N. (2004) *International Symposium on VLSI Technology*, pp. 148–149.

16 Nakamura, T., Nakao, Y., Kamisawa, A. and Takasu, H. (1994) *Applied Physics Letters*, **65**, 1522–1524.

17 Kang, B.S., Yoon, J.G., Kim, D.J., Noh, T.W., Song, T.K., Lee, Y.K., Lee, J.K. and Park, Y.S. (2003) *Applied Physics Letters*, **82**, 2124–2126.

18 Shin, S., Hofmann, M., Lee, Y.K., Cho, C.R., Lee, J.K., Park, Y., Lee, K.M. and Song, Y.J. (2003) *Materials Research Society Symposium Proceedings*, **748**, U4.1.1–U4.1.10.

19 Bae, B.J., Lim, J.E., Yoo, D.C., Nam, S.D., Heo, J.E., Im, D.H., Cho, B.O., Park, S.O., Kim, H.S., Chung, U.I. and Moon, J.T. (2005) *Integrated Ferroelectrics*, **75**, 235–241.

20 Shimizu, M., Sugiyama, M., Fujisawa, H. and Shiosaki, T. (1994) *Japanese Journal of Applied Physics*, **33**, 5167.

21 Lee, M.S., Park, K.S., Nam, S.D., Lee, K.M., Seo, J.S., Joo, S.H., Lee, S.W., Lee, Y.T., An, H.G., Kim, H.J., Cho, S.L., Son, Y.H., Kim, Y.D., Jung, Y.J., Heo, J.E., Park, S.O., Chung, U.I. and Moon, J.T. (2002) *Japanese Journal of Applied Physics*, **41**, 6709–6713.

22 Heo, J.E., Bae, B.J., Yoo, D.C., Nam, S.D., Lim, J.E., Im, D.H., Joo, S.H., Jung, Y.J., Choi, S.H., Park, S.O., Kim, H.S., Chung, U.I. and Moon, J.T. (2006) *Japanese Journal of Applied Physics*, **45**, 3198–3201.

23 Lee, S.Y. and Kim, K.N. (2005) *International Symposium on Integrated Ferroelectrics 2005, Shanghai*.

24 Kang, S.K., Song, Y.J., Joo, H.J., Kim, H.H., Park, J.H., Kang, Y.M., Kang, E.Y., Lee, S.Y. and Kim, K.N. (2004) *Integrated Ferroelectrics*, **66**, 29–34.

25 Lee, S.Y., Kim, H.H., Jung, D.J., Song, Y.J., Jang, N.W., Choi, M.K., Jeon, B.K., Lee, Y.T., Lee, K.M., Joo, S.H., Park, S.O. and Kim, K.N. (2001) *International Symposium on VLSI Technology*, pp. 111–112.

26 Choi, D.Y., Park, J.H., Rhie, H.S., Joo, H.J., Kang, S.K., Kang, Y.M., Kim, J.H., Koo, B.J., Lee, S.Y., Jeong, H.S., and Kim, K.N. (2005) *International Symposium on Integrated Ferroelectrics 2005, Shanghai*.

27 Lee, J.K., Lee, M.S., Hong, S., Lee, W., Lee, Y.K., Shin, S. and Park, Y. (2002) *Japanese Journal of Applied Physics*, **41**, 6690–6694.

28 Ko, H.Y., Byun, K.R., Jung, Y.J., Im, D.H., Yoo, D.C., Joo, S.H., Ham, J.H., Park, S.O., Kim, H.S., Chi, K.K., Kang, C.J., Cho, H.K., Jung, U.I. and Moon, J.T. (2005) *AVS 52nd International Symposium & Exhibition*, Boston, USA.

29 Bae, B.J., Lee, K.M., Lim, J.E., Nam, S.D., Park, K.S., Yoo, D.C., Lee, C.M., Lee, M.S., Park, S.O., Kim, H.S., Chung, U.I. and Moon, J.T. (2004) *Integrated Ferroelectrics*, **68**, 123–128.

30 Yoo, D.C., Bae, B.J., Lim, J.-E., Im, D.H., Park, S.O., Kim, H.S., Chung, U.I., Moon, J.T. and Ryu, B.I. (2005) *Symposium on VLSI Technology Digest*, pp. 100–101.

31 Choi, S.H., Bae, B.J., Son, Y.H., *et al.* (2005) *Integrated Ferroelectrics*, **75**, 215–223.

32 Kang, Y.M., Kim, J.H., Joo, H.J., Kang, S.K., Rhie, H.S., Park, J.H., Choi, D.Y., Oh, S.G., Koo, B.J., Lee, S.Y., Jeong, H.S. and Kim, K.N. (2005) *Symposium on VLSI Technology Digest*, pp. 102–103.

33 Koo, J.M., Seo, B.S., Kim, S.P., Shin, S.M., Lee, J.H., Baik, H.S., Lee, J.H., Yang, M., Bae, B.J., Lim, J.E., Yoo, D.C., Park, S.O., Kim, H.S., Han, H., Baik, S., Choi, J.Y., Park, Y.J. and Park, Y. (2005) *Symposium on IEDM Technology Digest*, pp. 340–343.

14
Magnetoresistive Random Access Memory
Michael C. Gaidis

14.1
Magnetoresistive Random Access Memory (MRAM)

Through the merging of magnetics (spin) and electronics, the burgeoning field of "spintronics" has created MRAM memory with characteristics of non-volatility, high density, high endurance, radiation hardness, high-speed operation, and inexpensive complementary metal oxide–semiconductor (CMOS) integration. While MRAM is unique in combining all of the above qualities, it is not necessarily the best memory technology for any single characteristic. For example, SRAM is faster, flash is more dense, and DRAM is less expensive. Stand-alone memories are generally valued for one particular characteristic: speed, density, or economy. MRAM therefore faces difficult odds in competing against the aforementioned memories in a stand-alone application. However, embedded memory for application-specific integrated circuits or microprocessor caching often demands flexibility over narrow performance optimization. This is where MRAM excels: it can be called the "handyman of memories" for its ability to flexibly perform a variety of tasks at a relatively low cost [1]. Whilst one may hire a specialist to rewire the entire electrical circuitry of a house, or install entirely new plumbing, a handyman with a flexible toolbox is a much more reasonable option for repairing a single electrical outlet or a leaky sink. Moreover, the handyman may be able to repair a defective electrical circuit discovered while in the process of repairing leaky plumbing!

A semiconductor fabrication facility that has MRAM in its toolbox is more likely to tailor circuit designs to a customer's individual needs for optimal performance at reasonable cost. The ways in which the characteristics of MRAM compare to those of other embedded memory technologies at the relatively conservative 180 nm node are listed in Table 14.1. In the remainder of this chapter, the state of the art in MRAM technology will be reviewed: how it works; how its memory circuits are designed; how it is fabricated; the potential pitfalls; and an outlook for future use of MRAM as devices are scaled smaller.

Nanotechnology. Volume 3: Information Technology. Edited by Rainer Waser
Copyright © 2008 WILEY-VCH Verlag GmbH & Co. KGaA, Weinheim
ISBN: 978-3-527-31738-7

14 Magnetoresistive Random Access Memory

Table 14.1 Embedded memory comparison at the 180 nm node.

Parameter		eSRAM	eDRAM	eFlash	eMRAM
Size	Cell area (μm^2)	3.7	0.6	0.5	1.2
Size	Array efficiency	65%	40%	30%	40%
Cost	Additional process	0	20% (4 msk)	25% (8 msk)	20% (3 msk)
Speed	Read access	3.3 ns	13 ns	13 ns	15 ns
Speed	Write cycle	3.4 ns	20 ns	5000 ns	15 ns
Power	Data retention	400 μA	5000 μA	0	0
Power	Active read	15 pC b^{-1}	5.4 pC b^{-1}	28 pC b^{-1}	6.3 pC b^{-1}
Power	Active write	15 pC b^{-1}	5.4 pC b^{-1}	31 000 pC b^{-1}	44 pC b^{-1}
Endurance	Write	Unlimited	Unlimited	1e5 cycles	Unlimited
Rad Hard	—	Average	Poor	Average	Excellent

The shaded cells indicate where MRAM has a distinct advantage. Relative comparisons should hold through scaling to the 65 nm node [2].

14.2
Basic MRAM

MRAM (magnetoresistive RAM) differs from earlier incarnations of magnetic memory (magnetic RAM) in that MRAM tightly couples electronic readout with magnetic storage in a compact device structure. During the early second half of the twentieth century, the most widely used RAM was a type of magnetic RAM called *ferrite core memory*. These memories utilized tiny ferrite rings threaded by multiple wires used to generate fields to write or to sense the switching of the magnetic polarity in the rings [3]. Highly valued for its speed, reliability, and radiation hardness, approximately 400 kB of this core memory was used in early IBM model AP-101B computers on the space shuttle. However, with the advent of compact, reliable, and inexpensive semiconductor memory, the 1 mm^2 cell size of the core memory could no longer compete, and in 1990 the space shuttle converted to battery-backed semiconductor memory with around 1 MB capacity [4].

In order for magnetic memory to compete again in the RAM arena, miniaturization on the scale of semiconductor integrated circuitry had to be implemented. This was stimulated by the discovery in 1988 of giant magnetoresistance (GMR) structures which provided an elegant means of coupling a magnetic storage (spin) state with an electronic readout, thereby creating the field of spintronics [5]. Spintronics relies on the phenomenon wherein electrons in certain ferromagnetic materials will align their spins with the magnetization in the ferromagnet. In essence, this is a result of a greater electron density of states at the Fermi level for electrons with spin aligned parallel to the magnetization in the ferromagnet. The passing of a current along two ferromagnetic films in close proximity allows the transport of the electrons to be influenced by adjusting the relative orientation of the two films' magnetization. As shown in Figure 14.1, although for parallel orientation, electrons are less likely to suffer resistive spin-flip scattering events, for antiparallel orientation they will exhibit a stronger preference for scattering and thus an increase in resistance will be

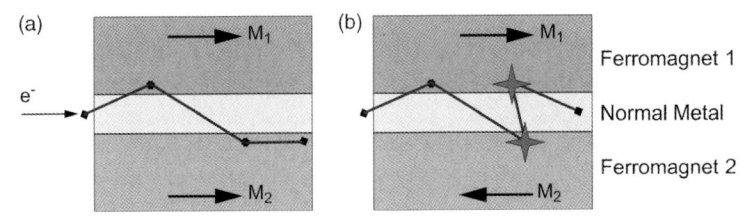

Figure 14.1 Illustration of the giant magnetoresistance (GMR) principle. For parallel alignment (a) of magnetizations M1 and M2, electron flow is subject to fewer resistive spin-flip scattering events than for antiparallel alignment (b).

apparent. The different resistance values for the high resistance state (R_{high}) and the low resistance state (R_{low}) can be used to define a magnetoresistance ratio (MR) as in Equation 14.1:

$$MR = \frac{(R_{high} - R_{low})}{R_{low}} \tag{14.1}$$

Typically, MR values for GMR devices are in the range of 5–10% for room-temperature operation.

By choosing different coercive fields for the two ferromagnets, it is possible to create a so-called *spin-valve* MRAM structure with a configuration similar to that shown in Figure 14.1. For example, ferromagnet 1 can be chosen to have a high coercivity, thus fixing its magnetization in a certain direction. Ferromagnet 2 can be chosen with a lower coercivity, allowing its magnetization direction to fluctuate. For a magnetic field sensor such as used in disk drive read heads, small changes in the magnetization angle of ferromagnet 2 induced by an external magnetic field can be sensed as changes in the resistance of the spin valve. Because the spin-valve sensitivity to external fields can be substantially better than inductive pickup, such devices have enabled dramatic shrinkage of the bit size in modern hard drives. An alternative use for the spin-valve structure is found if it is designed to utilize just two well-defined magnetization states of ferromagnet 2 (e.g., parallel or antiparallel to ferromagnet 1). Such spin-valve designs serve as a binary memory device, and have found application in rad-hard non-volatile memories as large as 1 Mb [6]. The drawbacks of this type of memory are:

- a relatively low magnetoresistance, providing only low signal amplitudes and thus longer read times
- a low device resistance, making for difficult integration with resistive CMOS transistor channels
- in-plane device formation which is more difficult to scale to small dimensions than devices formed perpendicular to the plane.

Solutions to these problems can all be found in the magnetic tunnel junction (MTJ) MRAM. The MTJ structure is similar to the GMR spin-valve in that it uses the property of electron spins aligning with the magnetic moment inside a ferromagnet. However, instead of passing current in-plane through a normal metal between

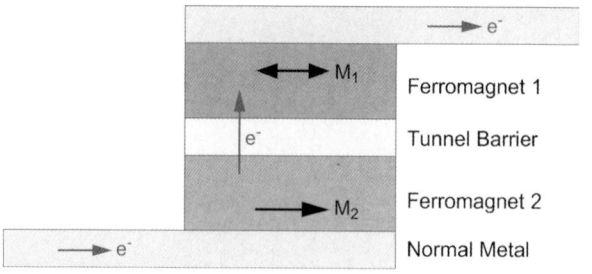

Figure 14.2 A simple magnetic tunnel junction structure. Ferromagnet 2 acts as an electron spin polarizer, and ferromagnet 1 as an electron spin filter, with magnetization either parallel or anti-parallel to the magnetization of ferromagnet 2. Parallel magnetizations generally result in a lower device resistance than anti-parallel magnetizations.

ferromagnets, the MTJ passes current perpendicular to the plane, through an insulating barrier separating two ferromagnets. An MTJ structure in its simplest form is shown in Figure 14.2. Here, one can envision the electric current impinging first on a ferromagnet which acts as a spin polarizer, then passing through the tunnel barrier and into a second ferromagnet which acts as a spin filter. The separation of polarizing and filtering functions is enabled by the physical thickness of the tunnel barrier, noting that the tunneling process preserves electron spin. The tunneling conductance will be proportional to the product of electron densities of states on each side of the barrier, and in general for ferromagnets there will be a larger density of states near the Fermi level for electrons polarized parallel to the magnetization of the ferromagnet as opposed to electrons polarized antiparallel. For polarizer and filter magnetizations aligned in the same direction, the density of states for spin-polarized electrons is large on both sides of the barrier, and the conductance of the structure is relatively high. For anti-parallel alignment of the polarizer and filter, the density of states available for spin-polarized electrons to tunnel into is somewhat reduced, and the conductance of the structure is relatively low. Proposed around 1974 [7], the first demonstrations of MTJs used Fe/Ge/Co multilayer stacks, but only showed appreciable MR (14%) at 4 K temperatures [8]. It was not until 1995 that improvements in materials processing techniques and the use of robust aluminum oxide tunnel barriers began to show reasonably large MR (18%) for MTJ devices at room temperature [9]. This breakthrough brought about huge investments from numerous companies, and ushered in a new era in the field of spintronics.

14.3
MTJ MRAM

The structure illustrated in Figure 14.2 can store binary information in the direction of magnetization within ferromagnet 1 (the "free layer"), provided that the magneti-

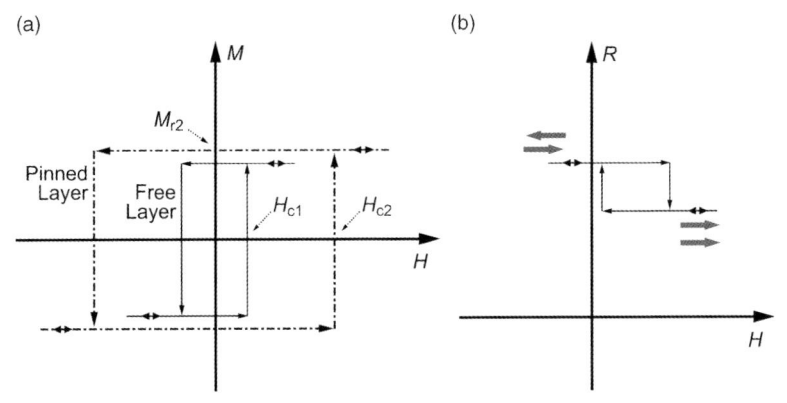

Figure 14.3 (a) Representative hysteresis curves of magnetization M versus applied field H, for a soft ferromagnet free layer and for a hard ferromagnet pinned layer in isolation. The coercive field H_{c2} is chosen large enough to keep the orientation of the pinned layer from switching while the free layer is being switched. M_{r2} represents the remanence from the pinned layer at zero applied field. (b) Resultant hysteresis of the MTJ resistance shown as a function of applied magnetic field. Due to the remanent magnetization from the pinned layer, the resistance loop is offset from the zero-applied field, and (as shown) can even result in but a single stable resistance at zero-applied field. The double arrows represent the magnetization state of the MTJ structure (anti-parallel or parallel).

zation within ferromagnet 2 (the "pinned layer") remains fixed in a predetermined direction. An asymmetry induced in the structure from device shape or intrinsic magnetic anisotropy can stabilize preferred orientations for the free layer to be one of either parallel to or anti-parallel to the pinned layer, thus maximizing the MR. A straightforward way to enable switching in the free layer without switching of the pinned layer is through the use of a material with a low coercive field H_c for the free layer, and a material with a high H_c for the pinned layer. This technique is illustrated in Figure 14.3a, with the hysteresis loops of a soft (low-H_c) free layer and a hard (high-H_c) pinned layer in isolation (i.e., not in the integrated MTJ stack structure). For operation at applied magnetic fields within the bounds set by H_c of the pinned layer, only the free layer will switch direction of magnetization. The hysteresis curve for the free layer demonstrates the necessary memory effect when the applied field is reduced to zero.

With the integrated multilayer structure of Figure 14.2, however, the hysteresis curves of the free and pinned layers in isolation are not straightforward predictors of the resistance states of the MTJ device. Because the pinned layer will maintain a remanence in a zero-applied field, there will be an offset imparted to the hysteresis loop of the free layer. (Note that there will be a similar offset of the pinned layer hysteresis loop imparted by the free layer's remanence, but for large enough H_{c2} there will be no effect on the device operation.) The effect of the pinned layer remanence on the magnetoresistive hysteresis loop R versus the applied field is illustrated graphically in Figure 14.3b. For a large remanence M_{r2}, the loop may shift so much that there is no longer a bistable memory for zero applied field. In principle, such an offset in memory product chips could be compensated by an external field

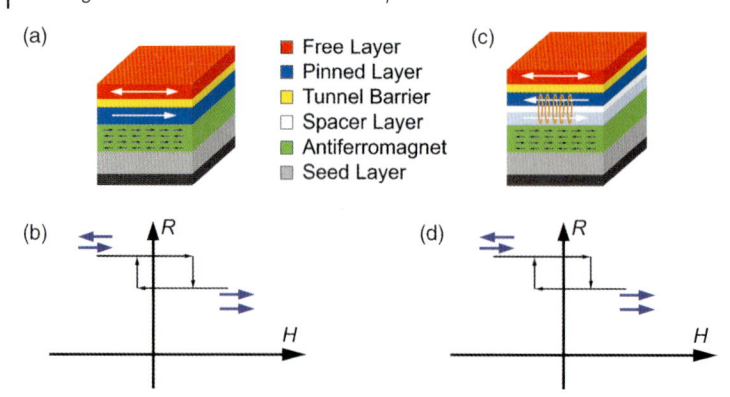

Figure 14.4 (a) Antiferromagnet-pinned reference layer structure with corresponding R versus H hysteresis loop (b). Also shown (c) is a flux-closed antiferromagnet-pinned reference layer structure with corresponding R versus H hysteresis loop (d) [2].

applied from a permanent magnet incorporated into the chip packaging. This is somewhat impractical, however, due both to packaging cost and to stringent requirements of across-chip uniformity.

Fortunately, clever manipulation of film properties has driven the evolution of several generations of MTJ structures, overcoming issues such as the offset field described above. Two such advances are illustrated in Figure 14.4. In Figure 14.4a, an antiferromagnet is exchange coupled to the pinned layer, thus providing a much larger effective coercive field for the pinned, or "reference" side of the tunnel junction [10]. With exceptional care to maintain a clean, smooth interface between the antiferromagnet and the pinned layer above it, one can obtain the strong exchange coupling between these films that is necessary to resist field switching. At least 1–1.5 nm of ferromagnetic pinned layer must still remain in the stack to act as an electron spin polarizer, but when coupled to the antiferromagnet it can be extremely well pinned even if the ferromagnet has a low H_c. By removing the need for a high-H_c ferromagnet in the pinned layer, this structure allows some additional flexibility in the choice of ferromagnet pinned layer material. One can optimize for maximum electron spin polarization for best magnetoresistance, and choose film qualities for low remanence and thus a lesser offset of the R versus H hysteresis curve. Correspondingly, Figure 14.4b illustrates a representative improvement in offset, for comparison with Figure 14.3b from the simpler stack structure.

Although there is much benefit in using the simple antiferromagnet (AF)-pinned structure of Figure 14.4a, best device operation often calls for reducing the R versus H hysteresis offset to an even smaller value. In this case, the flux-closed AF-pinned structure shown in Figure 14.4c can be tailored to give arbitrarily small offset fields. Here, a synthetic antiferromagnet (SAF) is formed from two ferromagnets separated by a thin spacer layer. For common spacer layers of 0.6–1.0 nm of Ru, one can obtain a strong antiparallel coupling between the two ferromagnets [11]. For reasonable external fields, this coupling forces them to be antiparallel, and thus the thicknesses of the two ferromagnets can be balanced such that the external magnetic flux is negligible. The pinning of one of these ferromagnet layers with an antiferromagnet gives a high effective H_c while at the same time causing negligible offset to the R versus H hysteresis loop (Figure 14.4d).

Flux-closing the reference layer ferromagnet works remarkably well in practice, particularly with recent advances in materials deposition tooling which enable tight control over film thicknesses for multilayer film structures covering entire 200- to 300-mm wafers [13]. A cross-section transmission electron microscopy (TEM) image of such a flux-closed reference layer MTJ stack is shown in Figure 14.5. Some interesting features of the magnetics-related elements can be discerned from the TEM image, and these are discussed below.

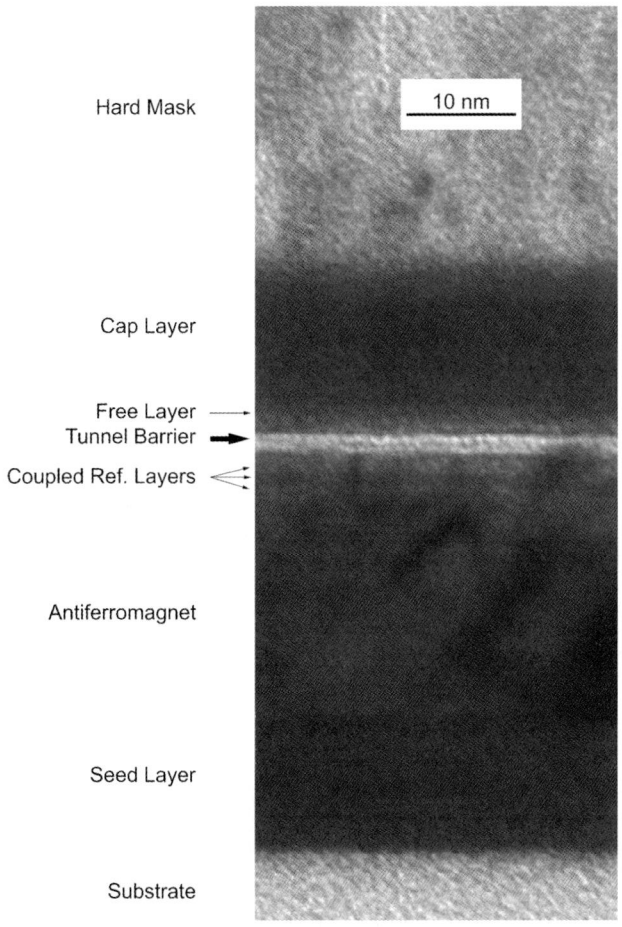

Figure 14.5 A transmission electron microscopy high-resolution, cross-sectional image of a MTJ stack with flux-closed, antiferromagnet-pinned reference layers.

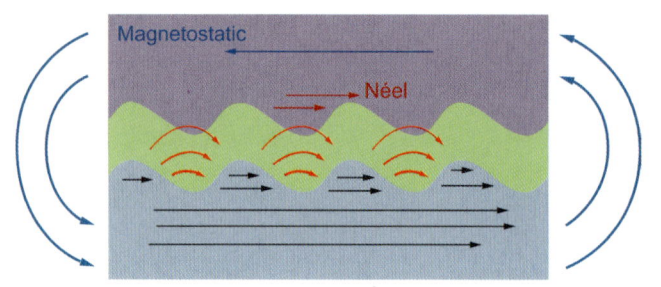

Figure 14.6 A schematic description of Néel coupling and how it relates to magnetostatic coupling. The rough-topped bottom film represents the pinned layer of Figure 14.4. Although exaggerated in the figure for clarity, an actual roughness greater than one atomic monolayer is cause for concern. The green intermediate layer represents the tunnel barrier, and the layer above is the free layer. Black arrows in the bottom film represent the internal magnetization of the pinned layer but, due to the rough surface, the magnetic poles are uncompensated in the region of the tunnel barrier. The resultant field from these poles creates a Néel field which favors parallel orientation of the free and pinned layers. The magnetostatic demagnetization field from the ends of the pinned layer favors antiparallel orientation of the free and pinned layers, but as this is non-local, it is less important in breaking the symmetry of devices with multiple layers [12].

14.3.1
Antiferromagnet

The antiferromagnet, which is generally a polycrystalline material such as FeMn, PtMn, or IrMn, is chosen and grown with several characteristics in mind:

- The interface roughness of the antiferromagnet must be sufficiently small that Néel coupling can be neglected (Figure 14.6), ensuring a smooth, pinhole-free tunnel barrier.

- The pinning strength must be large compared to the fields used to switch the free layer between its binary memory states.

- The blocking temperature of the antiferromagnet must be in a suitable range. In order to obtain an ideal pinning of the ferromagnet reference layer, the antiferromagnet/ferromagnet bilayer must be annealed above the blocking temperature T_B at which the exchange coupling between the films is zero. An applied magnetic field fixes the orientation of the ferromagnet, and then the bilayer is cooled. During cooling, the surface magnetization of the antiferromagnet aligns with the field-imposed ferromagnet magnetization. After cooling and removal of the field, exchange coupling across this interface keeps the ferromagnet pinned. Here, an antiferromagnet must be chosen with a blocking temperature T_B below approximately 300 °C in order to minimize material diffusion and tunnel barrier degradation. In addition, T_B must be sufficiently above the device operating temperatures, around 125 °C.

- The antiferromagnet must be able to withstand process temperatures of the ensuing circuit integration. Roughly, this translates into saying that the components of the antiferromagnet should not dissociate and diffuse out of the layer for process temperatures below about 250 °C.

14.3.2
Reference Layer

The reference layer closest to the tunnel barrier must act as an effective spin polarizer, and so it must be of thickness at least of order the electron spin-flip scattering length. This implies that 1–1.5 nm is the minimum thickness of the layer closest to the tunnel barrier. For best flux closure and minimal offset to the free layer, the reference layer adjacent to the antiferromagnet will be of a similar thickness, although a perfect zero free-layer offset may dictate small differences in the thicknesses. An upper limit to the thickness is set by the additional surface roughening and resultant Néel coupling that thicker films will generate. Reference layer materials are chosen for their best spin polarization properties and compatibility with device-processing techniques (e.g., minimal corrosion and thermal stability). Films of CoFe of the order of 2 nm thickness are typically used, separated by the 0.6- to 1.0-nm exchange-coupling Ru layer.

14.3.3
Tunnel Barrier

Aside from the requirement of reasonable magnetoresistive properties, the tunnel barrier is chosen primarily for robustness. It must be extremely thin to ensure that spin polarization is maintained during electron transit across the barrier, and the barrier must be able to survive under billions of cycles of electrical bias during its lifetime, without developing pinholes or any substantial shift in resistance. Aluminum oxide has proven an extremely suitable candidate for such tunnel barriers, and is known to offer reasonable magnetoresistance for suitable magnetic pinned and free layers. Recent developments in tunnel barrier engineering show that magnesium oxide tunnel barriers can offer MR near 500% at room temperature, although MgO-barrier devices have not yet proven to serve as robust, manufacturable layers in large arrays with good magnetic switching characteristics [14]. Aluminum oxide barrier devices can display MR near 100%, but trade-offs in the choice of magnetic materials for best switching characteristics, and in the choice of operating point for best CMOS integration, generally result in an MR less than 50%. Such MR is suitable for maintaining distinct resistance groupings of millions of devices in modern MRAM arrays, and increasing the MR is advantageous primarily in that it can reduce the necessary signal integration time to read the state of a device. Such a reduction is not a terribly strong driver at this time, as the array read time is set as much by the circuit overhead as by the device signal-to-noise ratio. Increasing the MR to 500% would likely result in only a 10–20% reduction in read duration. One area in which MgO barriers may soon establish a strong foothold is in the formation of highly transparent tunnel barriers. As device sizes shrink, the lower resistance–area product afforded by MgO will enable the best match to CMOS drive transistors, and thus the highest speed of operation. Today, even more highly transparent tunnel barriers are under development for a class of devices using electron spin current to switch the device state.

14.3.4
Free Layer

The free layer shown in the TEM is reasonably thin, rather like the underlying pinned layers. However, it does have a minimum thickness limit set by the spin filtering characteristics: for a thickness less than the approximate electron spin-flip scattering length, the magnetoresistance will begin to drop, and this again sets the thickness at around 1.5 nm or more. Thicker free layers require additional energy to switch, and so are undesirable for low-power operation. Of critical importance in the characteristics of the free layer is the need for well-defined magnetic states and well-behaved magnetic switching. As one cannot tailor the read or write circuitry to every individual device in megabit arrays of MRAM devices, it is critical that each device behave very much like all others in the array. Ill-defined magnetization states such as vortices, S-shapes, C-shapes, and multiple domains will add variability to the resistance measured by the circuitry, because electron spin polarization filtering may not be strictly parallel or antiparallel to the spin polarization imparted by the pinned layer. In addition, sensitivity of the film switching behavior to tunnel barrier and cap materials, or to device edge roughness or chemistry, can impart variability to the write operation of the individual bits in megabit arrays. NiFe alloys are preferred for good magnetic behavior with reasonable corrosion resistance. The addition of Co or Fe to the NiFe, or dusting with Co or Fe between the tunnel barrier and NiFe layer, can help to adjust the magnetic anisotropy and improve the MR. Layer thicknesses are typically in the 2- to 6-nm range for best low-power operation with good switching characteristics.

Several additional non-magnetic elements are visible in the TEM image, and these are discussed below.

14.3.5
Substrate

An ultra-smooth substrate is required as the starting point for smooth, uniform, and reliable tunnel barriers. Rough interfaces also result in increased Néel coupling, which is detrimental to device performance. Representative materials for the substrate are thermally oxidized silicon, or chemical-mechanical planarized (CMP) polished dielectrics such as silicon nitride, silicon oxide, or silicon carbide.

14.3.6
Seed Layer

An appropriate seed layer is required to obtain good growth conditions for the antiferromagnet, both to ensure a smooth top surface and to ensure good magnetic pinning strength. Given the high stress in some of the films in the MTJ stack, this seed layer is also critical for ensuring good adhesion to the substrate. It may be formed from tantalum nitride or permalloy (NiFe), for example.

14.3.7
Cap Layer

The proper choice of a cap layer is necessary to protect the free layer during further device fabrication processing. It is essential as a barrier or "getter" for contaminants, keeping the free layer clean and magnetically well behaved. Often-used materials for this layer include ruthenium, tantalum, and aluminum. The choice of this material may also depend on its effect on the magnetic behavior of the free layer: certain cap materials can discourage smooth switching between free layer states, and can result in substantial "dead layers" which must be compensated for by a thicker free layer.

14.3.8
Hard Mask

A hard mask (as opposed to a "soft" photoresist mask) is used to enable patterning of the MTJ with industry-standard etch techniques. It also eases integration with the surrounding circuitry by providing a contact layer to connect the MTJ to wiring levels above. The hard mask material is largely chosen for its compatibility with subsequent processing in the fabrication route, and can be chosen from any number of metallic or dielectric materials.

The processing of MTJ structures to integrate them with CMOS circuitry is discussed in greater detail later in the chapter.

14.4
MRAM Cell Structure and Circuit Design

14.4.1
Writing the Bits

The mechanism for switching the state of the free layer in MRAM lends itself well to an array layout with a conventional planar semiconductor design and fabrication. A typical rectangular MTJ array layout, with word lines (WLs) arrayed beneath the devices and bit lines (BLs) arrayed atop the devices, is illustrated in Figure 14.7. Current driven along the WLs or BLs generates a magnetic field which imparts a torque on the magnetization of the device. In normal operation, the superposition of properly-sized "write" fields from both WL and BL will enable a switching event to occur in the free layer of the device at the intersection of the two lines. The write fields are chosen small enough so as not to exceed the coercivity of the pinned layer. Potential pitfalls from this scheme include write errors from half-selected devices (i.e., those subjected to only a WL or a BL field, but not both) and, worse, write errors

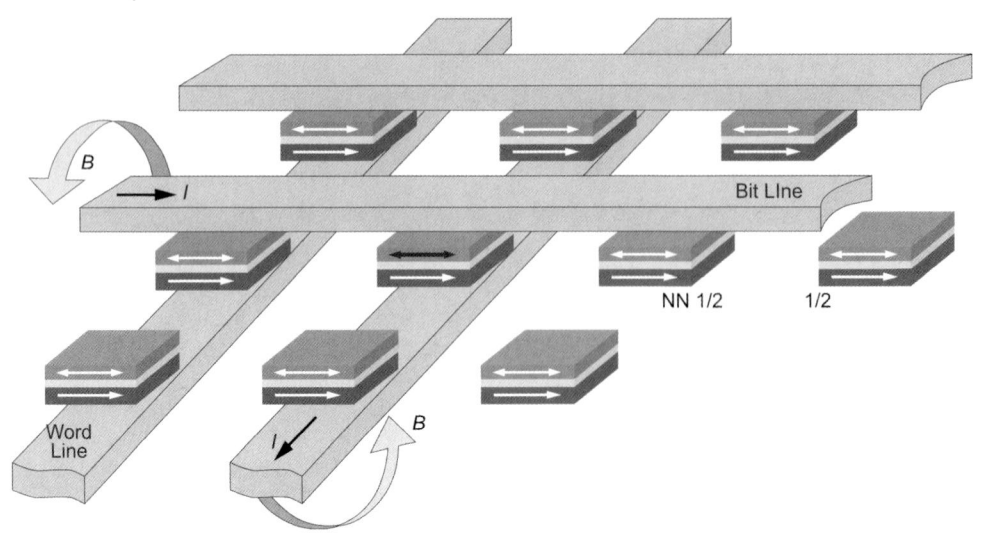

Figure 14.7 Schematic representation of a rectangular array of MTJ devices, with bit line and word line circuitry for writing the bits. Current-generated magnetic fields (B) from a given bit line and word line are sufficient only to switch the device at the intersection of the two wires. Write errors are typically worse for devices in the half-select state (MTJs labeled "1/2" in the figure), where a word line or a bit line is active, but not both. The situation is even worse for near-neighbor half-selected devices ("NN1/2" in the figure) where, for example, the device is in the column adjacent to the active word line, but is half-selected by the active bit line.

from near-neighbor half-selected devices (those subjected to a half-select field, but only one row or column away from another active line).

The diagrams in Figure 14.8 provide more details about the superposition of magnetic fields used to switch the active device. With the flux-closed antiferromagnet-pinned reference layer structure (see Figure 14.4c) forming the MTJ, the single-layer free layer is switched with characteristics first described by Stoner and Wohlfarth [15]. A simple case is that of an elliptical-shaped MTJ with shape anisotropy defining an easy axis (the major axis of the ellipse) and a hard axis (the minor axis of the ellipse). To switch the magnetization, a hard-axis field is applied to tilt the free layer magnetization away from the easy axis energy minimum, and an easy-axis field is applied to "set" the magnetization of the device in the desired easy-axis direction – parallel or antiparallel to the pinned layer. With this Stoner–Wohlfarth (S-W) switching, relatively small operating margins are illustrated by the closeness of the green and pink dots to the S–W boundary in Figure 14.8b. In addition to accounting for spreads in the switching characteristics between devices, one must also budget in extra operating window for thermal activation errors and the disturb effects of half-selects and near-neighbor field interaction. Circuit designers will try to tailor the operating window for at least 10 years of error-free operation. Without use of error-correction techniques, one generally aims for operating margins to keep the activation energy for a bit error to greater than 60 $k_B T$, where k_B is the Boltzmann constant and T is the temperature. This imposes extremely tight requirements on how uniform the array

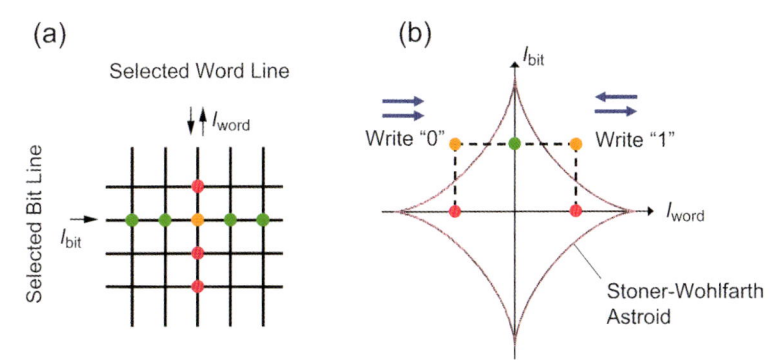

Figure 14.8 (a) Top-down schematic view of an MTJ array with rows and columns of bit lines and word lines with fields superposed to switch the device represented by a orange dot. Devices shown as green and pink dots are half-selected devices. Those green and pink devices adjacent to the orange device are near-neighbor half-selected devices. (b) A graph showing necessary bit line and word line current values needed to switch a desired device. The colored dots on the plot correspond to the devices represented by colored dots in Figure 14.8a. For suitable choice of word line and bit line currents, one can ensure switching of the desired device without switching half-selected devices.

must be in terms of switching, described in equation form by the array quality factor (AQF):

$$\text{AQF} = \frac{H_{sw}}{\sigma_{Hsw}} \qquad (14.2)$$

Here, H_{sw} is the average switching field of the devices and σ_{Hsw} is the standard deviation of the switching field distribution of all elements in the array. In rough terms, the AQF must be larger than about 30 in order to ensure a lifetime of 10 years, although some relief can be gained through the use of error correction techniques.

Toggle MRAM was invented to circumvent the difficulties faced by S–W MRAM in terms of the operating margin for half-selected bits [16]. As illustrated in Figure 14.9a, the structure has taken the flux-closed antiferromagnet-pinned reference layer structure (see Figure 14.4c) a step further by also flux-closing the ferromagnetic free layer. This is achieved by depositing a spacer layer atop the free layer ferromagnet, followed by a second ferromagnet. The spacer can be chosen (as in the pinned layer) to enhance antiparallel coupling, or the spacer can be chosen with zero or even with some parallel coupling characteristics to decrease the write field needed to switch the bit. The magnetizations of the two ferromagnets in the free layer will point in opposite directions, and their balance and proximity will flux-close the layers so there is little field seen emanating from the structure at a distance. The write operation of this toggle-mode structure is illustrated in Figure 14.9b. Noting the colors assigned to represent the magnetization of the free layers in Figure 14.9a (green for the top layer, red for the bottom layer), the plots at the top of Figure 14.9b show the relative orientation of the two magnetizations. Note that the initial state is such that the magnetization of the MTJ has easy (preferred) axis at 45° to the word and bit lines, rather than be aligned parallel to one of them as in S–W MRAM.

14 Magnetoresistive Random Access Memory

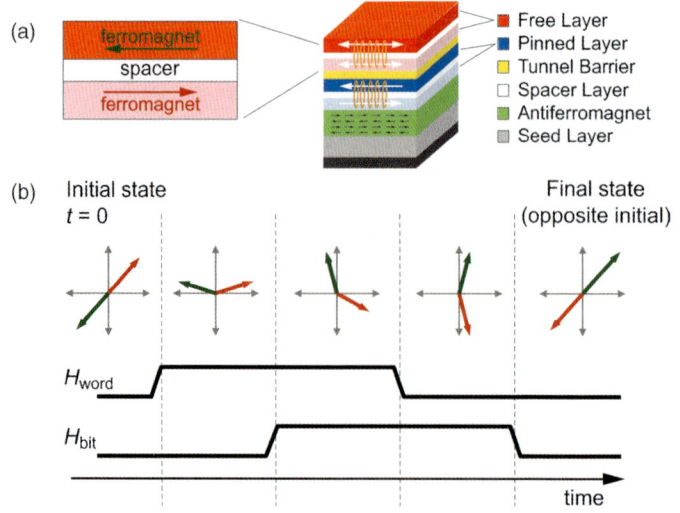

Figure 14.9 (a) Structure of the toggle-mode MTJ stack. (b) Time evolution of the free layer switching. See text for details.

Figure 14.9b illustrates the need for staggered timing of WL and BL write-field pulses. To switch the state of the free layers, a magnetic field is first applied from the WL along the positive y-direction. This magnetic field cants the magnetizations of both free layers as they try to align to the field. The antiparallel nature of the magnetic coupling between the free layers prevents the magnetizations from both fully lining up with the applied word field, as long as the field is not too large to overwhelm this antiparallel state. When the magnetizations are canted sufficiently, there is a net magnetic moment to the free layers, and this moment can be grabbed like a handle by the field now imparted by the BL. The BL applies a field in the positive x-direction, and the net moment of the two free layers follows this BL field. The WL field is then shut off, and the net moment continues to rotate around towards the applied BL field. As the BL field is shut off, the free layer magnetizations relax into their energetically favorable antiparallel configuration, but now with magnetizations exactly opposite to those at the start.

The name "toggle-mode device" is derived from the characteristic that cycling the WLs and BLs in this manner will always switch the state of the device. To set a bit in a particular state, a read operation must be performed to determine if a write "toggle" operation is required. Aside from this drawback, and the additional complexity of the magnetic stack, there are several advantages to the toggle-mode structure:

- As alluded to above, the write operating margins can be substantially larger than for devices with S–W switching. Rather than a S–W astroid boundary, the toggle-mode devices exhibit an L-shaped boundary that does not approach the WL or BL axes. The potential for half-select errors is dramatically reduced, and the requirement on AQF is approximately halved.

- In principle, shape anisotropy is not required to ensure that the bit has only two preferred states for binary memory. One can utilize the intrinsic anisotropy of the ferromagnetic free layers to define two such states. This allows the use of circular MTJ devices for the smallest memory cell size.
- The flux-closed nature of the free layers greatly reduces dipole fields emanating from the free layer. Such fields can affect the energetics of nearby devices, resulting in variability of switching characteristics, depending on the states of such devices. Thus, with flux-closed free layers, nearby devices can be packed in closer proximity for improved scaling.

14.4.2 Reading the Bits

The array structure illustrated in Figure 14.7 is often termed a "cross-point cell" (XPC) structure. More specifically, XPC refers to the case where the MTJ devices are located at the cross-points of the BLs and WLs, and are directly connected to the BLs and WLs above and below the MTJ stack. This structure offers an extremely high packing density for the lowest cost memory. The write mechanism is reasonably straightforward as described above, as long as the MTJ resistance is not so low that it shunts the write currents. More troublesome is that the read mechanism suffers from a reduced signal-to-noise ratio in this XPC structure. In order to read the resistance state of a XPC bit, a bias is applied between a desired BL and WL, and the resistance measured. However, due to the interconnected nature of the XPC structure, not only the resistance of the cross-point device is measured – there are parallel contributions of resistance from many other devices along "sneak paths" that include traversing additional sections of BL and WL. Due to the resulting loss of signal, the device must be read much more slowly to allow for integration to improve the signal-to-noise ratio. Device read times can be substantially longer for such XPC structures, making this type of memory far less desirable than one which can be read as fast as DRAM, for example.

The solution to the problem of sneak paths is to insert an isolation mechanism which ensures that read currents will only traverse a single MTJ device. For example, this can be achieved by placing a diode in series with each MTJ. Although this seems simple when drawn as a circuit schematic on paper, it is actually more straightforward to place a field effect transistor (FET) in series with each MTJ, and assign a second WL to control the read operation. The "FET cell" circuit structure is shown schematically in Figure 14.10, with separate WLs for the write and read operations. The BL is used for both read and write operations.

Figure 14.11 illustrates the implementation of the circuit structure shown in Figure 14.10, suitable for a densely packed array of MTJs. Structural additions to standard CMOS circuitry include:

- the via contact VJ between the bit line and the top of the MTJ stack
- the MTJ device

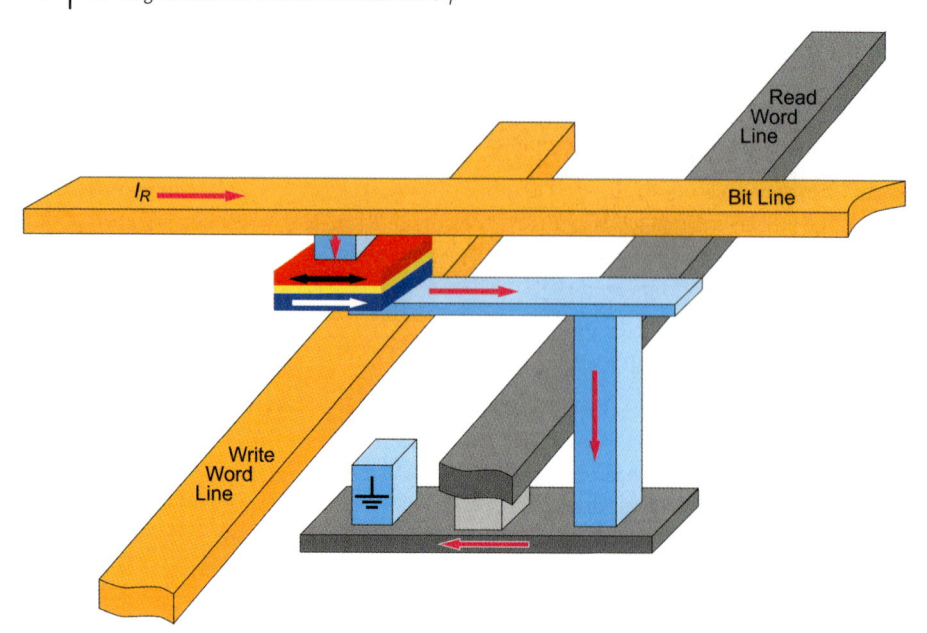

Figure 14.10 Field-effect transistor (FET) cell circuit topology, showing individual word lines for reading and for writing. A FET located in the silicon beneath the MTJ is used to switch on only the device being read, thus preventing leakage of read currents (purple arrows) through nearby MTJ devices. Additional conductor elements in this structure (compared to Figure 14.7) include a contact between the bit line and the top of the MTJ, a local metal strap (MA) connecting the base of the MTJ with a via chain that connects to the underlying FET.

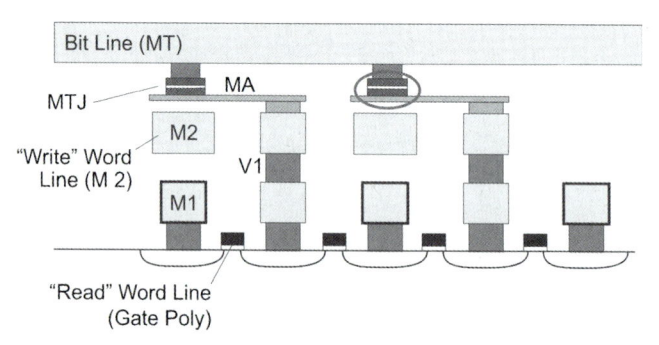

Figure 14.11 A cross-section of the FET cell topology, with two adjacent cells shown atop the silicon CMOS front-end of line (FEOL) structure. The oval encloses the critical components for MRAM implementation. As cell size is determined primarily by the MTJ and via chain above the via V1, two FETs can be used for each MTJ in order to achieve lower resistance and some redundancy. Thus, the FET gates on either side of a V1 via chain will be connected to the same "read" word line. Wires formed in the first level of metallization (M1) (outlined in bold) form a grid at a reference potential. M2 denotes the second level of metallization. The reader is referred to Ref. [17] for further details on such structures.

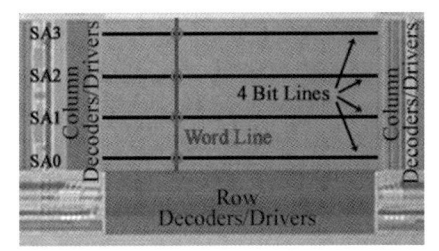

Figure 14.12 Photograph of a 128 Kb subarray, showing locations of the sense amplifiers (SA), the row and column decoders and drivers, and the concurrent activation of four bit lines with one word line for a ×4 organization of the block. A single MTJ cell is indicated by a circle at the intersection of a word line (WL) and bit line (BL).

- the local metal strap (MA) between the bottom of the MTJ stack and the via to M2
- the via VA between the MA strap and the M2 wiring, which serves to isolate the MTJ from the write WL while providing connection to the underlying FET structure for reading.

A slightly higher packing density may be achieved with a mirror-cell design, where adjacent bits mirror each other. The simple unmirrored design of Figure 14.11 is preferable to minimize any across-array non-uniformity due to inter-level misalignment and inter-cell magnetic interference. Megabit and larger MRAM memories are formed from multiple subarrays, with size determined largely by the resistance of the BLs and WLs. There is a desire always to keep applied voltage low, for CMOS compatibility and best array efficiency. The required current to generate the necessary MTJ switching fields then sets a maximum length on the BL or WL, depending on the resistive voltage drop. Bootstrapped write drivers can be used to allow smaller write drivers with improved write current control [18]. A 16 Mb MRAM under development at IBM utilizes 128 Kb subarrays (see Figure 14.12), with 512 WLs and 256 BLs of active memory elements.

The read operation is performed with sense amplifiers that compare the desired bit to a reference cell. The reference cell uses two adjacent MTJs fixed in opposite states in a configuration that acts like an ideal mid-point reference between the R_{high} and the R_{low} states [18]. Four BLs are activated in a given cycle, and are uniformly spaced along the height of the array to reduce magnetic interference between activated BLs during the write operation, and to minimize distance from the activated BLs to the sense amplifiers during a read operation. Additional reference BLs are located within the array, with one set shared by sense amplifiers 0 and 1, and one set shared by sense amplifiers 2 and 3.

The array driving circuitry for MRAM memories is commonly standardized to an asynchronous SRAM-like interface for easy interchangeability in battery-backed SRAM applications. The IBM 16 Mb chip uses a ×16 architecture that is prevalent in mobile and handheld applications with packaging intended for simple direct replacement of SRAM chips. As shown in Figure 14.13, the 16 Mb chip measures 79 mm² with individual memory cells of 1.42 µm², for an array efficiency of almost

436 | *14 Magnetoresistive Random Access Memory*

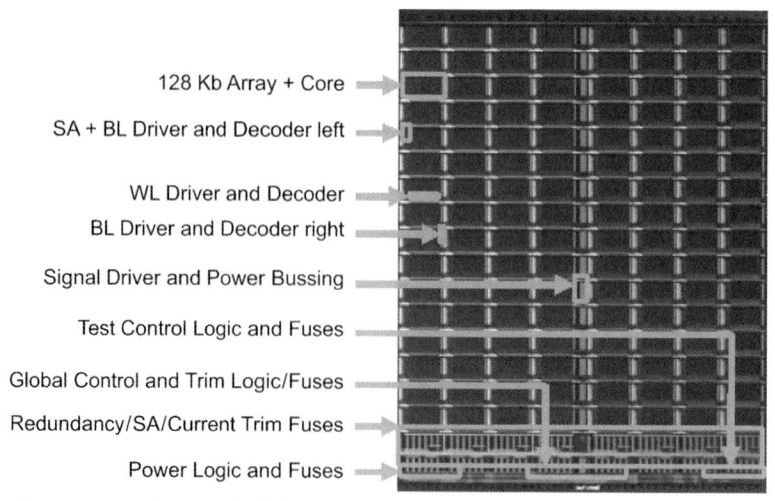

- 128 Kb Array + Core
- SA + BL Driver and Decoder left
- WL Driver and Decoder
- BL Driver and Decoder right
- Signal Driver and Power Bussing
- Test Control Logic and Fuses
- Global Control and Trim Logic/Fuses
- Redundancy/SA/Current Trim Fuses
- Power Logic and Fuses

Figure 14.13 A photograph of the 16 Mb MRAM chip, showing locations of 128 Kb array cores (eight columns of 16 rows) and support circuitry [18].

30%. The array efficiency may be improved by using more metal layers and by eliminating some of the developmental test mode structures used in this chip. A reduction of the standby current for power-critical applications is achieved through the extensive use of high threshold, long-channel FET devices and careful grounding of inactive terminals in the arrays and in the write driver devices [18].

Redundant elements are included in the chip to allow the correction of defective array elements. Such redundancy is implemented with fuse latches and address comparators in a manner consistent with industry-standard memory products. The CMOS base technology is quite mature, so the focus of the redundancy is on the MRAM features. Single-cell failures or partial WL failures (from MRAM reference cell defects) are considered the most likely defects. The redundancy architecture favors replacement of WLs to capture the partial WL fails from MRAM reference cell defects. Redundancy domains are implemented at a high level in the block hierarchy so as to span several blocks and be capable of effectively fixing any random defects [18].

14.4.3
MRAM Processing Technology and Integration

The implementation of MRAM hinges on complex magnetic film stacks and several critical steps in back-end-of-line (BEOL) processing. Cell size is presently limited by the size of the MTJ devices and driving wires, and older, mature CMOS front-end-of-line (FEOL) technology can be used without limiting performance. Fabrication of the FET-cell circuit, from the CMOS FEOL through the MRAM BEOL, can encompass several hundred process steps, resulting in the fully functional structure shown in Figure 14.14. The MRAM-critical portion of the circuit is a relatively small part of the entire configuration. After the last standard CMOS step (the M2 wire completion),

14.4 MRAM Cell Structure and Circuit Design

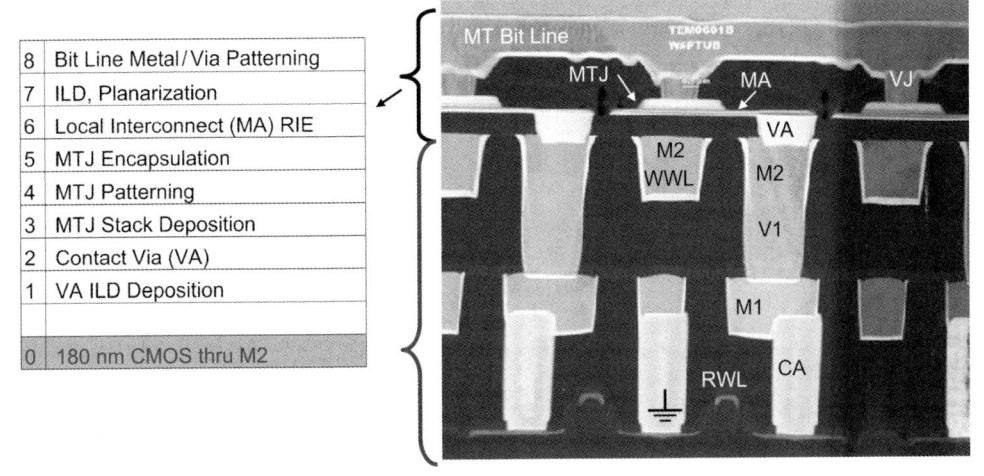

8	Bit Line Metal/Via Patterning
7	ILD, Planarization
6	Local Interconnect (MA) RIE
5	MTJ Encapsulation
4	MTJ Patterning
3	MTJ Stack Deposition
2	Contact Via (VA)
1	VA ILD Deposition
0	180 nm CMOS thru M2

Figure 14.14 Cross-section of a product cell, showing the integration of MRAM with CMOS, and the process steps used for the MRAM-specific layers. ILD is the interlayer dielectric.

there remains the need to pattern the shallow vias, the MTJs, the local interconnects, and at least one level of wiring with contact to MTJs and the functional circuitry below. Even for simple functional circuits, five or more photomask levels are required to complete the MRAM-centric portion of the structure.

14.4.3.1 Process Steps

In conjunction with the steps outlined in Figure 14.14, below is a discussion of the important considerations for the process steps in the fabrication of the MRAM-specific levels.

1. *VA contact via and ILD:* The VA via provides a path for read current to flow from the local (MA) metal strap down through a via chain to the underlying read transistor. The most critical aspect of this module is that it must form a substrate which is sufficiently smooth for good magnetic stack growth.

2. *Magnetic film stack deposition:* Arguably the most essential technological advance in enabling MTJ MRAM was the development of tooling for the large-area deposition of extremely uniform films with well-controlled thickness. Such tooling has proven suitable for the deposition of magnetic, spacer, and tunnel barrier films with sub-Ångström uniformity across 200 mm and even 300 mm wafers [13]. The critical aluminum oxide tunnel barrier is generally formed by depositing a thin aluminum layer, followed by exposure to an oxidizing plasma [19].

3. *Tunnel junction patterning:* A commonly used and straightforward approach to patterning the MTJs is with the use of a conducting hard mask. This is later utilized as a self-aligned stud bridging the conductive MT wiring to the active magnetic films in the device. A thick hard mask, however, introduces additional

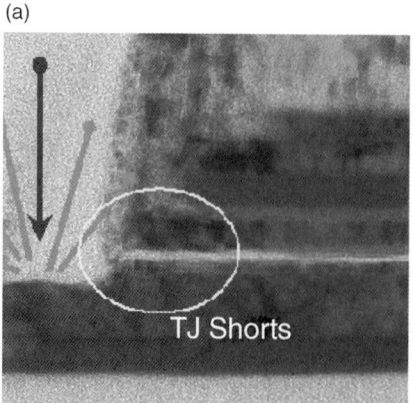

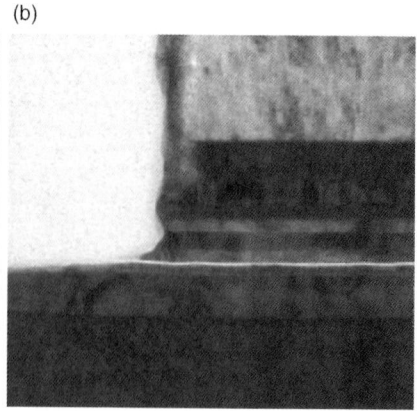

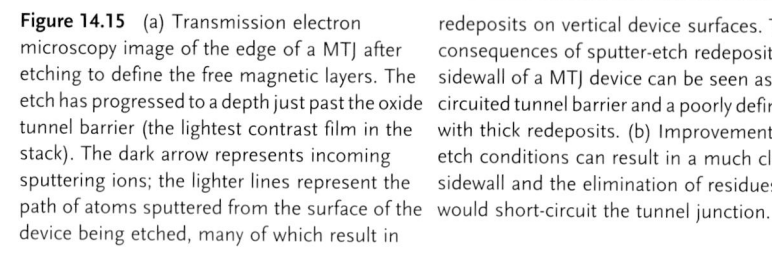

Figure 14.15 (a) Transmission electron microscopy image of the edge of a MTJ after etching to define the free magnetic layers. The etch has progressed to a depth just past the oxide tunnel barrier (the lightest contrast film in the stack). The dark arrow represents incoming sputtering ions; the lighter lines represent the path of atoms sputtered from the surface of the device being etched, many of which result in redeposits on vertical device surfaces. The consequences of sputter-etch redeposits on the sidewall of a MTJ device can be seen as a short-circuited tunnel barrier and a poorly defined edge with thick redeposits. (b) Improvements in the etch conditions can result in a much cleaner sidewall and the elimination of residues that would short-circuit the tunnel junction.

difficulties in that it can shadow the etch being used to pattern the magnetic devices. Such shadowing can add an element of variability into the size of the devices, and may also result in metal redeposits on the sidewall of the device structure. As illustrated in Figure 14.15, sidewall redeposits are particularly troublesome for commonly used MRAM stack materials because the materials do not readily form volatile RIE byproducts that provide some isotropic character to the etch. Directional physical sputtering is the main mechanism for etching of the stack materials [20]. Because the difficulties in etching the magnetic stack materials often outweigh the benefits of a simpler process integration scheme, it is often preferable to use a thinner hard mask for less etch shadowing, and an additional via level (VJ in Figure 14.14) to connect the top of the MTJ with the bit line wiring.

4. *MTJ encapsulation:* Silicon nitride and similar compounds are desirable for their adhesion to the MA and MTJ metal surfaces, and for strong interfacial bonds that inhibit migration of metal atoms along the dielectric/metal interfaces. Such metal migration is one well-documented cause of MTJ thermal degradation, and can limit processing temperatures in patterned MTJ devices to below 300 °C [21]. The use of tetra-ethyl-orthosilicate (TEOS) as a precursor in the deposition of silicon oxide films [22] is known to offer the benefits of a relatively inert depositing species which can readily diffuse into spaces adjacent to high-aspect ratio structures, even at temperatures below 250 °C.

5. *MA patterning:* For suitable thickness of seed and reference layers, the series resistance of layers remaining after MTJ etch is small enough to impart negligible

dilution of the MTJ MR signal. This simplifies the processing, as a dedicated film need not be created for the MA strap, and the reference or seed layers of the magnetic stack can perform double duty. As in the MTJ etch, the MA etch may be subject to the problem of non-volatile etch byproducts redepositing along the hard mask sidewalls.

6. *ILD deposition and planarization and wiring:* After the MA metal strap has been patterned, an interlayer dielectric is deposited in which to house the counterelectrode wiring layers VJ and MT. The counterelectrode wiring is formed with well-established semiconductor-industry Damascene techniques.

As alluded to in Figure 14.11, the MRAM-specific elements form but a small portion of the entire integrated circuit. For rapid characterization of these MRAM-specific elements, there is no need to perform a fully CMOS-integrated wafer build; rather, it is sufficient to perform a subset of the process integration steps to focus only on the critical magnetics issues [23].

14.5
MRAM Reliability

One of the strong selling points of MRAM is its reliability: write endurance is expected to be essentially infinite, the magnetics are intrinsically rad-hard, and its non-volatile memory storage can eliminate soft errors in many applications. As in any new technology struggling for successful commercialization, there are certain aspects of the new technology that are unproven and require demonstration of reliability. Areas of potential reliability risk include [24].

14.5.1
Electromigration

Electromigration in the write WLs and BLs, resulting from high write current density. Current pulses of 10 mA are typical for conservative wire cross-sections of $0.2\,\mu m^2$, corresponding to a current density of $5\,MA\,cm^{-2}$. This alone represents a serious challenge to the reliability in the array, and can potentially be worsened by local disruptions to the quality and thickness of wire material. The VJ vias of Figure 14.14, or direct connection between the BL and the MTJ hard mask in the thick hard mask integration scheme discussed above, can impact the BL wiring electromigration resistance.

Electromigration issues can potentially be improved through the use of bidirectional switching currents, which fit neatly into toggle-mode MRAM operation, but cost in terms of array efficiency. One promising method for reducing electromigration stress is through the use of ferromagnetic liners in a U-shape around the BLs and write WL. These liners serve to focus the magnetic field onto the MTJs in the desired row or column, and can increase the effective field by as much as a factor of 2 for a given current [25]. The use of ferromagnetic liners around the BL is illustrated in

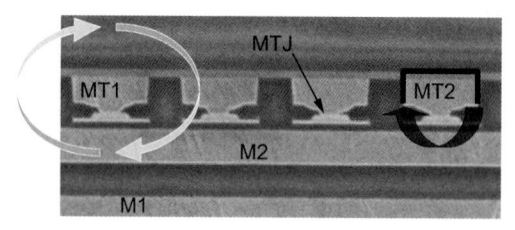

Figure 14.16 A cross-sectional image of a product array, from a viewpoint perpendicular to that of Figure 14.14. The arrows around bit line wire MT1 suggest the magnetic field configuration generated by a current through wire MT1; it is loosely contained, with only moderate magnitude at the MTJ free layers. Conversely, the wire MT2 exhibits an enhanced field magnitude due to its localization by the ferromagnetic film (lines) surrounding the copper MT2 wire.

Figure 14.16. Similar, but inverted, structures can be formed around the write WL (M2 in Figure 14.16) to enhance the field from that wire. The potential reduction in necessary current to obtain a required switching field can dramatically reduce electromigration issues. Not only do ferromagnetic liners offer potential reduction in current density, but they also improve electromigration performance relative to conventional copper processes. By reducing the interface diffusion of copper atoms, ferromagnetic cladding on the top surface of the MT wire enhances electromigration reliability to an extent similar to that seen in the industry by advanced Ta/TaN or CoWP capping processes [26].

One added benefit of the ferromagnetic liner field focusing is the reduction of any near-neighbor disturb effects. Because the field is better focused on devices along the desired WL and BL, adjacent devices are less likely to be switched by near-neighbor fields, or the combination of near-neighbor fields and thermal activation.

14.5.2
Tunnel Barrier Dielectrics

These are subject to reliability concerns because of the extremely thin nature of the barrier and related susceptibility to pinholes or dielectric breakdown. Aluminum oxide tunnel barriers have so far proven quite robust. Time-dependent dielectric breakdown (TDDB) and time-dependent resistance drift (TDRD) have been examined in 4 Mb arrays and found to exceed requirements for a 10-year lifetime [27]. The voltage stresses on the tunnel barrier are relatively modest, as the read operation takes place at 100–300 mV because the MR is higher for a lower voltage. The write operation is performed with one side of the MTJ floating, so there is no significant voltage stress on the MTJ during the higher-power write pulse.

14.5.3
BEOL Thermal Budget

The BEOL thermal budget for MRAM devices (<250–$300\,°C$) is significantly lower than for conventional semiconductor fabrication processes ($\sim 400\,°C$), in order to prevent degradation to the MTJs. This can affect the intrinsic quality of dielectrics being used in the BEOL, and can also worsen seam and void formation around the topographical features being encapsulated. A low thermal budget also prevents the use of certain post-processing passivation anneals, and packaging materials and processes. The move to lead-free solder with increased solder reflow temperatures represents a further challenge for MRAM.

14.5.4
Film Adhesion

This is a serious concern with the multiple new materials being introduced into the integrated process. The novel etch and passivation techniques being used also may leave behind poorly adherent layers which cannot be subjected to harsh wet cleans without MTJ exposure and degradation. Delamination risks must be mitigated through specially developed dry and wet cleans, the use of materials with tuned stress, and the choice of materials with compatible thermal expansion.

14.6
The Future of MRAM

As of July, 2006, MRAM products such as the 4 Mb memory shown in Figure 14.17 have been available from Freescale Semiconductor [28]. The market space targeted by Freescale includes networking, security, data storage, gaming, and printer data logging and configuration storage. From a customer viewpoint, this product means fewer part counts, a higher level of performance, higher reliability, greater environmental friendliness, and a lower cost solution than their current approaches, such as battery-backed SRAM.

Progressing downwards from the available 180 nm technology, future generations of MRAM are expected to utilize the same magnetic infrastructure with only evolutionary improvements, to below the 90 nm node. However, constraining the scaling are the following concerns:

Figure 14.17 Photograph of a MR2A16A 4 Mb MRAM chip atop a wafer filled with such chips, presently available from Freescale semiconductor. (Illustration courtesy of G. Grynkewich and Freescale Semiconductor.).

- *Near-neighbor interactions:* When packing devices closer together, magnetic fields emanating from a given device can affect the switching behavior of devices nearby, and this can also be dependent on the given device's free layer state. In addition, the write wires for switching a given device will perturb neighboring devices to a greater extent as the latter come closer. It remains unclear how well these effects can be suppressed with the use of flux-closed MTJ layers and ferromagnetic cladding of write wires. Additional techniques such as enhanced-permeability dielectric (EPD) encapsulating films may be required to overcome these problems [29].

- *Increased switching fields:* As devices are scaled to smaller volumes, the anisotropy field must be increased to compensate and maintain activation energy greater than $60\,k_B T$ [30]. Write fields will scale to be of similar magnitude to the anisotropy field, and will increase superlinearly with inverse device size. As with the MTJs, the write wires must scale to a smaller footprint, making it more difficult to accommodate the increasing switching fields. In addition, ferromagnetic cladding of the wires becomes less effective because of the bending energy of the flux inside the cladding as the wire corner radius sharpens. EPD device encapsulations will help in this regard.

- *Device-to-device variability:* Process-induced line-edge roughness will become a more substantial fraction of the total device width, so that edge irregularities may become more effective at pinning the domains so they do not switch smoothly. The total device area and aspect ratio will also exhibit larger spreads, both from line-edge roughness and from variability in lithography. A reduced aspect ratio for tighter packing density will also decrease AQF, as the anisotropy field is more sensitive to shape for devices with a smaller aspect ratio [30].

Each of these concerns is not a fundamental limitation, but rather a practical limitation that can most likely be overcome with sufficient – albeit perhaps prohibitively expensive – investment in materials development and processing techniques. Hard physical limits do not appear to set in until superparamagnetism becomes important – that is, for device sizes below 20 nm [30], a dimension substantially below the limits suggested by the aforementioned practical issues.

Even with the practical limits to scaling conventional MRAM, one can expect to see revolutionary modifications to standard MRAM cell such that MRAM will be available with far greater densities, lower cost, and faster operation. Beyond the scope of this chapter are the impressive developments and exciting new proposals in the areas of:

- thermally assisted MRAM for reduced power requirements [31]
- spin-momentum transfer (SMT) MRAM for scaling to advanced process nodes and extremely small active memory devices [32]
- domain-wall memory for very high density serial storage [33]
- embedded MRAM as a replacement for embedded flash and low-density on-chip SRAM, for high-performance microprocessor cache memory and other ASIC applications [34].

In summary, this chapter has provided an overview of the rapid developments in MRAM technology over the past decade. Many major hurdles for MRAM product development have been surmounted in the face of funding limits set by competition with the huge silicon industry. Now that MRAM devices have grasped a toe-hold in the marketplace, new applications will be identified and MRAM development will proceed at an even faster pace over the next decade. Perhaps soon we will again see magnetic RAM in spacecraft!

Acknowledgments

The author would like to thank: W.J. Gallagher for the figures and editorial assistance; IBMs Materials Research Laboratory (MRL) for process development and fabrication; P. Rice, T. Topuria, E. Delenia, and B. Herbst for the TEM imaging; J. DeBrosse, T. Maffitt, C. Jessen, R. Robertazzi, E. O'Sullivan, D.W. Abraham, E. Joseph, J. Nowak, Y. Lu, S. Kanakasabapathy, P. Trouilloud, D. Worledge, S. Assefa, G. Wright, B. Hughes, S.S.P. Parkin, C. Tyberg, S.L. Brown, J.J. Connolly, R. Allen, E. Galligan for various contributions; and M. Lofaro for the advances in CMP. It is also acknowledged that the studies summarized here were supported in part by the Defense Microelectronics Activity (DMEA) and built on prior investigations conducted with Infineon (now Qimonda) within the MRAM Development Alliance, and also on earlier MRAM studies at IBM that were supported in part by DARPA.

References

1 DeBrosse, J. personal communication.
2 Gallagher, W.J. and Parkin, S.S.P. (2006) Development of the magnetic tunnel junction MRAM at IBM: From first junctions to a 16-Mb MRAM demonstrator chip. *IBM Journal of Research and Development*, **50**, 5–23. Sincere thanks also to John DeBrosse, John Barth, Chung Lam, and Ron Piro.
3 Jones, J. (1976) Coincident current ferrite core memories. *Byte*, **1**, 6–22.
4 Hanaway, J.F. and Moorehead, R.W. (1989) Space Shuttle Avionics System, NASA SP-504 available at, http://klabs.org/DEI/Processor/shuttle/sp-504/sp-504.htm.
5 (a) Binasch, G. et al. (1989) *Physical Review B-Condensed Matter*, **39**, 4828–4830;(b) Baibich, M.N. et al. (1988) Giant magnetoresistance of (001)Fe/(001)Cr magnetic superlattices. *Physical Review Letters*, **61**, 2472;(c) Heiliger, C., Zahn, P. and Mertig, I. (2006) Microscopic origin of magnetoresistance. *Materials Today*, **9**, 46–54.
6 (a) Kaakani, H. (March 10–17 2001) Radiation Hardened Memory Development at Honeywell. IEEE Aerospace Conference, Big Sky, MT, vol. 5, pp. 2273–2279;(b) Katti, R.R. (2002) *Journal of Applied Physics*, **91**, 7245.
7 Slonczewski, J. IBM internal report.
8 (a) (1975) Juliere (CNR-France) – first MTJ demonstration Fe/Ge/Co, DR/R∼14% at 4.2K;(b) Julliere, M. (1975) Tunneling between ferromagnetic films. *Physics Letters A*, **54**, 225–226.
9 Miyazaki, T. and Tezuka, N. (1995) Giant magnetic tunneling effect in Fe/Al$_2$O$_3$/Fe junction. *Journal of Magnetism and Magnetic Materials*, **139**, L231–L234.

10 (a) Berkowitz, A.E. and Takano, K. (1999) Exchange anisotropy – a review. *Journal of Magnetism and Magnetic Materials*, **200**, 552–570;(b) Nogues, J. and Schuller, I.K. (1999) Exchange bias. *Journal of Magnetism and Magnetic Materials*, **192**, 203–232.

11 (a) Stiles, M.D. (2006) Exchange coupling in magnetic multilayers, in *Nanomagnetism: Ultrathin Films, Multilayers and Nanostructures. Contemporary Concepts of Condensed Matter Science*, Volume **1** (eds D. Mills and J.A.C. Bland), Elsevier, New York, pp. 51–77. (b) Parkin, S.S.P. and Samant, M.G. (2003) Magnetic random access memory with thermally stable magnetic tunnel junction cells, U.S. Patent 6,518,588;(c) Parkin, S.S.P., More, N. and Roche, K.P. (1990) Oscillations in exchange coupling and magnetoresistance in metallic superlattice structures: Co/Ru, Co/Cr, and Fe/Cr. *Physical Review Letters*, **64**, 2304–2306; (d) Slaughter, J.M., Dave, R.W., DeHerrera, M., Durlam, M., Engel, B.N., Janesky, J., Rizzo, N.D. and Tehrani, S. (2002) Fundamentals of MRAM Technology. *Journal of Superconductivity: Incorporating Novel Magnetism*, **15**, 19–25; (e) Parkin, S.S.P. et al. (1999) Exchange-biased magnetic tunnel junctions and application to nonvolatile magnetic random access memory. *Journal of Applied Physics*, **85**, 5828–5833.

12 Schrag, B.D. et al. (2000) Néel 'orange-peel' coupling in magnetic tunneling junction devices. *Applied Physics Letters*, **77**, 2373–2375.

13 (a) C-7100EX Sputter Deposition Tool from Canon Anelva Corporation, Tokyo, Japan (see http://www.canon-anelva.co.jp/english); (b) Timaris Sputter Deposition Tool from Singulus Technologies, Kahl, Germany (see http://www.singulus.de).

14 (a)Parkin, S.S.P. et al. (2004) Giant tunnelling magnetoresistance at room temperature with MgO (100) tunnel barriers. *Nature Materials*, **3**, 862–867;(b) Hayakawa, J., Ikeda, S., Lee, Y.M., Matsukura, F. and Ohno, H. (2006) Effect of high annealing temperature on giant tunnel magnetoresistance ratio of CoFeB/MgO/CoFeB tunnel junctions. *Applied Physics Letters*, **89**, 232510–232512.

15 Stoner, E.C. and Wohlfarth, E.P. (1948) A mechanism of magnetic hysteresis in heterogeneous alloys. *Philosophical Transactions of the Royal Society*, **A240**, 599–642.

16 (a) Savtchenko, L., Engel, B.N., Rizzo, N.D., DeHerrera, M.F. and Janesky, J.A. (2003) Method of writing to scalable magnetoresistance random access memory element, U.S. Patent 6,545,906; (b)Durlam, M. et al. (2003) A 0.18 μm 4 Mb toggling MRAM, *IEDM Technical Digest*, 995;(c) Worledge, D. (2004) Spin flop switching for magnetic random access memory. *Applied Physics Letters*, **84**, 4559–4561;(d) Worledge, D.C. (2006) Single-domain model for toggle MRAM. *IBM Journal of Research and Development*, **50**, 69–79.

17 Reohr, W., Hoenigschmid, H., Robertazzi, R., Gogl, D., Pesavenot, F., Lammers, S., Lewis, K., Arndt, C., Lu, Y., Viehmann, H., Scheuerlein, R., Wang, L.-K., Trouilloud, P., Parkin, S., Gallagher, W. and Mueller, G. (2002) Memories of Tomorrow. *IEEE Circuits & Devices*, **18**, 17–27.

18 (a) Maffitt, T.M., DeBrosse, J.K., Gabric, J.A., Gow, E.T., Lamorey, M.C., Parenteau, J.S., Willmott, D.R., Wood, M.A. and Gallagher, W.J. (2006) Design considerations for MRAM. *IBM Journal of Research and Development*, **50**, 25–39;(b) Gogl, D. et al. (2005) A 16 Mb MRAM featuring bootstrapped write drivers. *IEEE Journal of Solid State Circuits*, **40**, 902–908.

19 Zhu, J.G. and Park, C. (2006) Magnetic tunnel junctions. *Materials Today*, **9**, 36–45.

20 Pearton, S.J. et al. (2000) Dry etching of MRAM structures. *Materials Research Society Symposium Proceedings*, **614**, F10.2.1–F10.2.11.

21 Samant, M.G., Luning, J., Stohr, J. and Parkin, S.S.P. (2000) Thermal stability of

IrMn and MnFe exchange-biased magnetic tunnel junctions. *Applied Physics Letters*, **76**, 3097–3099.

22 Crowell, J., Tedder, L., Cho, H., Cascarano, F. and Logan, M. (1990) Model studies of dielectric thin film growth: chemical vapor deposition of SiO_2. *Journal of Vacuum Science & Technology*, **A8**, 1864–1870.

23 Gaidis, M.C., O'Sullivan, E.J., Nowak, J.J., Lu, Y., Kanakasabapathy, S., Trouilloud, P.L., Worledge, D.C., Assefa, S., Milkove, K.R., Wright, G.P. and Gallagher, W.J. (2006) Two-level BEOL processing for rapid iteration in MRAM development. *IBM Journal of Research and Development*, **50**, 41–54.

24 Hughes, B. (2004) Magnetoresistive random access memory (MRAM) and reliability. Proc. IEEE 42nd Annual International Reliability Physics Symposium, Phoenix, AZ, pp. 194–199.

25 Durlam, M. *et al.* (2003) A 1-Mbit MRAM based on 1T1MTJ bit cell integrated with copper interconnects. *IEEE Journal of Solid-State Circuits*, **38**, 769–773.

26 (a) Gajewski, D.A., Meixner, T., Feil, B., Lien, M. and Walls, J. (2004) Electromigration of MRAM-customized Cu interconnects with cladding barriers and top cap. IEEE Integrated Reliability Workshop Final Report, pp. 90–92; (b) Walls, J. *et al.* (2004) Improved electromigration resistance of copper interconnects using multiple cladding layers, unpublished; see www.IP.com Document ID IPCOM000009315D.

27 (a) Åkerman, J. *et al.* (2004) Demonstrated reliability of 4-Mb MRAM. *IEEE Transactions on Device Materials Reliability*, **4**, 428–435;(b) Åkerman, J. *et al.* (2005) Reliability of 4-Mbit toggle MRAM. *Materials Research Society Symposium Proceedings*, **830**, 191–200.

28 Engel, B.N., Åkerman, J., Butcher, B., Dave, R.W., DeHerrera, M., Durlam, M., Grynkewich, G., Janesky, J., Pietambaram, S.V., Rizzo, N.D., Slaughter, J.M., Smith, K., Sun, J.J. and Tehrani, S. (2005) A 4-Mbit toggle MRAM based on a novel bit and switching method. *IEEE Transactions on Magnetics*, **41**, 132; http://www.freescale.com/mram.

29 Pietambaram, S.V., Rizzo, N.D., Dave, R.W., Goggin, J., Smith, K., Slaughter, J.M. and Tehrani, S. (2007) Low-power switching in magnetoresistive random access memory bits using enhanced permeability dielectric films. *Applied Physics Letters*, **90**, 143510.

30 Cowburn, R.P. (2003) Superparamagnetism and the future of magnetic random access memory. *Journal of Applied Physics*, **93**, 9310.

31 (a) Abraham, D.W. and Trouilloud, P.L. (May 7, 2002) Thermally assisted magnetic random access memory, U.S. Patent 6,385,082;(b) Prejbeanu, I.L., Kula, W., Oundadjela, K., Sousa, R.C., Redon, O., Dieny, B. and Nozieres, J.-P. (2004) Thermally assisted switching in exchange-biased storage layer magnetic tunnel junctions. *IEEE Transactions on Magnetics*, **40**, 2625; (c) Redon, O., Kerekes, M., Sousa, R., Prejbeanu, L., Sibuet, H., Ponthennier, F., Persico, A. and Nozières, J.P. (May 21–24 2005) Thermo-assisted MRAM for low power applications. Proceedings 1st International Conference on Memory Technology and Design (ICMTD-2005), Giens, France, pp. 113–114.

32 (a) Stiles, M.D. and Miltat, J. (2006) Spin transfer torque and dynamics, in *Spin Dynamics in Confined Magnetic Structures III: Topics in Applied Physics 101* (eds B. Hillebrands and A. Thiaville), Springer, Berlin, pp. 225–308;(b) Huai, Y., Albert, F., Nguyen, P., Pakala, M. and Valet, T. (2004) Observation of spin-transfer switching in deep submicron-sized and low-resistance magnetic tunnel junctions. *Applied Physics Letters*, **84**, 3118; (c) Fuchs, G.D. *et al.* (2004) Spin transfer effects in nanoscale magnetic tunnel junctions. *Applied Physics Letters*, **85**, 1205.

33 (a) Parkin, S.S.P. (2004) Shiftable magnetic shift register and method of using the same, U.S. Patent 6,834,005; (b) Parkin, S.S.P. (2005) System and method for writing to a magnetic shift register, U.S. Patent 6,898,132.

34 Iyer, S.S., Barth, J.E. Jr., Parries, P.C., Norum, J.P., Rice, J.P., Logan, L.R. and Hoyniak, D. (2005) Embedded DRAM: Technology platform for the Blue Gene/L chip. *IBM Journal of Research and Development*, **49**, 333–350.

15
Phase-Change Memories
Andrea L. Lacaita and Dirk J. Wouters

15.1
Introduction

15.1.1
The Non-Volatile Memory Market, Flash Memory Scaling, and the Need for New Memories

During the past decade, the impressive growth of the market for portable systems has been sustained by the availability of successful semiconductor non-volatile memory (NVM) technologies, the key driver being the Flash memories. In the past 15 years, the scaling trend of these charge-based memories has been straightforward. The cell density of NOR Flash, which is adopted for code storage, has doubled every one to two years, following Moore's law; the memory cell size is 10–12 F^2, where F is the technology feature size. The NAND Flash, which is optimized for sequential data storage, has been aggressively scaled and, nowadays, has a cell size of about 4.5 F^2. However, further scaling of both NOR and NAND Flash is projected to slow down, due mainly to the tunnel oxide (NOR), which cannot be further thinned down without impairing data retention, and to electrostatic interactions between adjacent cells (NAND).

Moreover, as the scaling proceeds, the number of electrons stored on the floating gate and present in the device channel decreases. As few electrons are involved, effects such as the random telegraph noise arising from trapping processes are expected to cause threshold instabilities and reading errors [1], while the requirements on retention become even more challenging. At the 32 nm node, the maximum acceptable leakage over a 10-year period will be less than 10 electrons per cell [2]. All of these difficulties, arising from the fundamental limitation of the charge storage concept, are calling for novel approaches to non-volatile storage at the nanoscale.

In recent years a number of different alternative memory concepts have been explored. Most notably, memories based on switchable resistors are considered

Nanotechnology. Volume 3: Information Technology I. Edited by Rainer Waser
Copyright © 2008 WILEY-VCH Verlag GmbH & Co. KGaA, Weinheim
ISBN: 978-3-527-31738-7

promising; among these, the phase-change memory (PCM) technology is attracting growing interest.

15.1.2
PCM Memories

PCM-based memory devices were first proposed by J.F. Dewald and S.R. Ovshinsky who, during the 1960s, reported the observation of a reversible memory switching in chalcogenide materials [3, 4]. Chalcogenides are semiconducting glasses made from the elements of Group VI of the Periodic Table, such as sulfur, selenium and tellurium, and many of these demonstrate the desired material properties for possible use in PCM applications. Two different chalcogenide material systems may be discriminated, based on their switching properties [5]:

- *Threshold-switching* in so-called "stable" glasses that show negative differential resistance and a bistable behavior, requiring a minimum "holding voltage" to sustain the high-conductive state. The typical materials are three-dimensionally cross-linked chalcogenide alloy glasses.
- *Memory-switching* in "structure reversible films" that may form crystalline conductive paths. A typical composition is $Te_{81}Ge_{15}X_4$ close to the Ge-Te binary eutectic, with X being an element from Group V or VI (e.g., Sb). The latter materials also show threshold switching to initiate the high conduction in the glass state, followed by an amorphous to crystalline phase transition which stabilizes the high-conductive state.

A non-volatile and reprogrammable phase-change (256 bit) memory array based on chalcogenide materials originally was reported by R.G. Neale, D.L. Nelson and Gordon E. Moore as far back as 1970 [6]. In these memories the memory element is basically a resistor made from a chalcogenide material and, depending on whether the chalcogenide layer is amorphous or crystalline, the device resistance would be either high (RESET state) or low (SET state) [7]. Programming of the phase state is carried out by current-induced Joule heating: either the material is heated above the melting temperature, followed by fast quenching in the amorphous state; or the element is heated to a high temperature below the melting point, allowing crystallization of the amorphous material. However, the operation characteristics of these memories were still poor (e.g., 25 V, 250 mA, 5 µs for programming in the RESET state). Indeed, such a high programming power requirement led to the suggestion that these prototype memories should be called "Read-Mostly Memory".

Chalcogenide phase-change materials were instead successfully adopted in xerography, where the photoconductive properties of arsenic-selenide (As-Se) were exploited, and in optical recording, spurred on by the development of Ge-Te glasses capable of undergoing rapid crystalline-amorphous phase transformations [8, 9]. In particular, rewriteable optical media (e.g., CDs, DVDs) became a huge field of application. In the case of CDs, the selective crystallization/amorphization is induced by an external laser beam and not by Joule heating, while the binary information is read out by exploiting the change in optical reflectivity between the amorphous and the crystalline state, rather than the difference in electrical resistivity.

The advancements in the materials used for optical disks, coupled with significant technology scaling and a better understanding of the fundamental electrical device operation, eventually triggered the development of solid-state memory technology, which led initially to the Ovonic Unified Memory (OUM™) concept based on the use of the $Ge_2Sb_2Te_5$ chalcogenide compound [10, 11]. Since early 2000, the different semiconductor industries have considered the exploitation of the same concept for large-sized, solid-state memories [12–14]. Phase-change memories are known by different names. For example, the former OUM name was superseded by the terms PCM and phase-change RAM (PRAM). Today, PCM are considered promising candidates eventually to become the mainstream non-volatile technology, this being due to their large cycling endurance [15, 16], fast program and access times, and extended scalability [17, 18].

15.2
Basic Operation of the Phase-Change Memory Cell

15.2.1
Memory Element and Basic Switching Characteristics

The vertical OUM PCM memory element in the so-called Lance-like structure is shown schematically in Figure 15.1. The active phase-change material ($Ge_2Sb_2Te_5$; GST) is sandwiched between a top metal contact and a resistive bottom electrode (also called the heater). The programming current flows vertically from the bottom

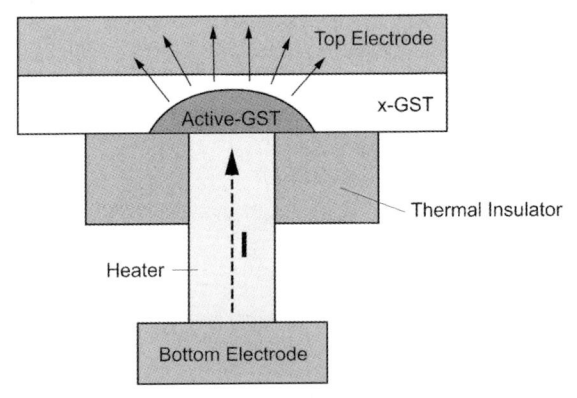

Figure 15.1 Schematic of the OUM™ vertical phase-change memory element. Due to the typical bias polarity, current flows vertically from the bottom electrode through the heater, through the $Ge_2Sb_2Te_5$ (GST) layer and to the top electrode. The current concentration near the (narrow) heater/GST contact results in local heating of the GST in a semispherical volume where the amorphous/crystalline phase change occurs. Amorphization of this region stops the low-resistive current path and results in an overall large resistance.

electrode through the heater, through the GST layer and to the top electrode. The current concentration near the (narrow) heater/GST contact results in a local heating of the GST in a semi-spherical volume where the amorphous/crystalline phase change occurs. Amorphization of this area stops the low-resistive current path and results in an overall large resistance.

The thermal and electrical switching characteristics of a vertical OUM PCM memory element are shown in Figure 15.2, with temperature evolution in the GST region above the heater contact in response to current pulses shown graphically in Figure 15.2a [12]. In order to form the amorphous phase, a 50- to 100-ns current pulse heats up the region until GST reaches its melting temperature (620 °C). The subsequent swift cooling, along the falling edge of the current pulse, freezes the undercooled molten material into a disordered, amorphous phase below the glass transition temperature. In order to recover the crystalline phase, Joule heating from another current pulse, with a lower amplitude (resulting in temperatures above the crystallization temperature but below the melting temperature), is used to speed-up the spontaneous amorphous-to-crystalline transition: the crystalline phase builds up in about 100 ns by a combination of nucleation and growth processes.

The typical current–voltage (I–V) curve of a cell for both states is shown in Figure 15.2b [19]. As the electrical resistivity of the two phases differs by orders of magnitude, at low bias, the resistance of the two memory states ranges from few kΩ (low resistance = ON or SET state) to some MΩ (high resistance = OFF or RESET state). Reading is accomplished by biasing the cell and sensing the current flowing through it; for example, a few hundreds of millivolts across the cell in the SET state generates 50–100 µA. This current is able to load the bit-line capacitances

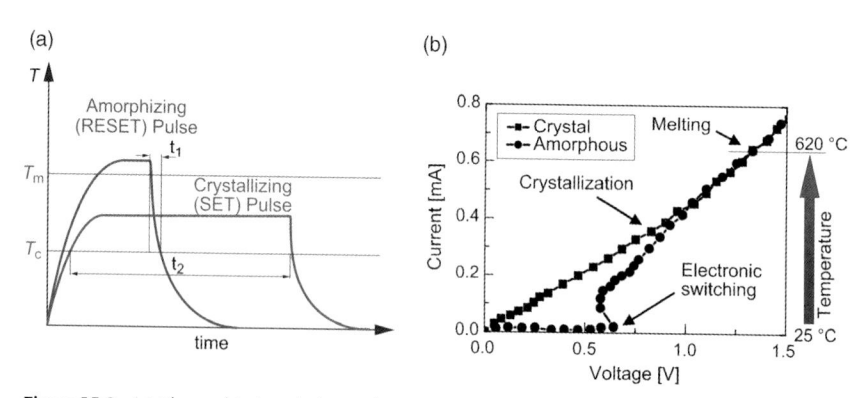

Figure 15.2 (a) Thermal-induced phase change of the material, either by melting and subsequent quenching in the amorphous phase, or by heating in the solid state inducing crystallization of the amorphous state. (Figure reproduced from Ref. [12]). (b) Current–voltage (I–V) curves for both the crystalline and amorphous states. (Figure reproduced from Ref. [19]). The high current levels required for the Joule heating can be obtained at low voltages, even for the amorphous state, on the basis of the electronic threshold switching phenomenon, which strongly increases the conductivity in the amorphous material above a certain threshold voltage.

of a memory array, making possible a reading operation in 50 ns. The same bias across the cell in the RESET state is not able to generate enough current to trigger the sensing amplifier, thus resulting in the evaluation of a "0".

It should be noted that the I–V curve in the high-resistance, amorphous state is quite peculiar. As the bias reaches a certain voltage (the threshold switching voltage) a "snap-back" takes place and the conductance abruptly "switches" to a high conductive state (see Figure 15.2b). The I–V curve of the crystalline GST does not feature threshold switching, and approaches the I–V of the amorphous state in the high current zone.

The occurrence of this "threshold switching" is a very important characteristic of PCM material. Indeed, without such a switching mechanism, which allows large currents to flow in the amorphous material at low voltages (∼few volts), very high voltages (∼100 V) would be required to switch the material to the "on" state, thus making electronic programming effectively non-practical.

The ratio of the threshold switching voltage and the thickness of the amorphous zone is usually referred to as the *critical threshold switching field*; for GST this quantity ranges between 30 and 40 V μm^{-1}. The critical threshold switching field can be taken as a guideline to compare different materials; for example, the lower the switching field the lower the switching voltage for the same thickness of the amorphous layer. However, as shown in Figure 15.3, even if the threshold voltage does scale with the memory resistance, which in turn depends on the amorphous layer thickness, the line does not cross the origin [20]. The concept of threshold switching field should, therefore, be handled with some care.

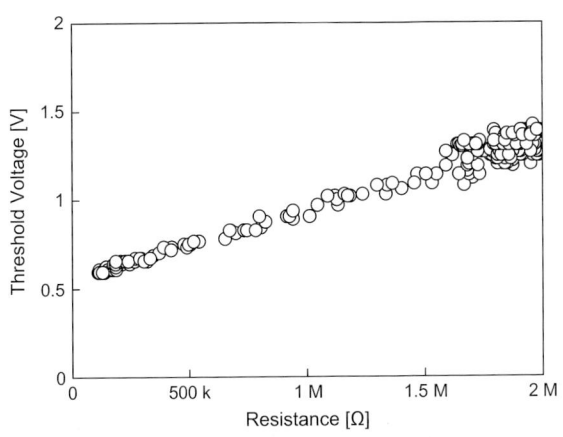

Figure 15.3 Experimental dependence of the threshold voltage on the low field resistance of the amorphous state. The threshold voltage scales with the device resistance, and therefore with the width of the amorphous zone. However, the line does not cross the origin, which highlights that a minimum voltage value of ∼0.50 V, close to the holding voltage value, is required for switching to occur. (Figure adapted from Ref. [20]).

15.2.2
SET and RESET Programming Characteristics

The programming characteristic of a PCM cell [20] – that is, the dependence of the cell resistance R as a function of the programming current – is shown in Figure 15.4. The open symbols in Figure 15.4 refer to the resistance obtained when driving a cell from the RESET state. During the measurement procedure, a 100-ns programming pulse is applied and the cell resistance after programming is read at 0.2 V. Before the subsequent measurement, the cell is brought again into the initial reference RESET state by using a proper current pulse. The measurement cycle is then restarted, driving the cell with a new 100-ns programming current pulse with a different amplitude.

During this procedure, three distinct regions can be recognized:

- For programming pulses below 100 µA, the ON-state conduction is not activated and the very small current does not provide any phase change.

- In the 100 to 450 µA range, the resistance decreases following the crystallization of the amorphous GST, reaching the minimum resistance in the SET state, as denoted by R_{set}.

- Above 450 µA, the programming pulse melts some GST close to the interface with the bottom electrode, leaving it in the amorphous phase.

The solid symbols in Figure 15.4 also show the R–I characteristics obtained for the same cell, but starting from the SET state. The resistance value changes only when the current exceeds 450 µA and the chalcogenide begins to melt. The current is therefore denoted as the *melting current*, I_{melt}. From thereon the curve overlaps to the R–I of the RESET state. For programming pulses above 700 µA, the resistance of the cell reaches an almost constant value. It transpires that the PCM cell can be switched between the two SET and RESET states using current pulses of 400 and 700 µA, respectively, these pulses

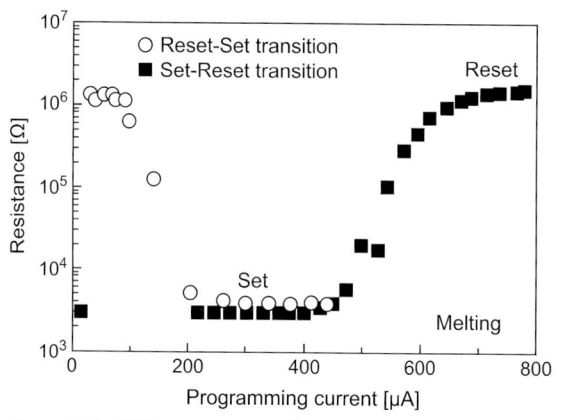

Figure 15.4 PCM programming characteristics, i.e., R as a function of the programming current I_P. Program pulses are applied to RESET cell (reset-set transition) or to a SET cell (set-reset transition). (Figure reproduced from Ref. [20]).

being independent of the initial cell state (resistance). Therefore, the cell can be rewritten with no need for any intermediate erase. The minimum current capable of bringing the cell into the full RESET state (700 μA in Figure 15.4) is denoted as reset current, I_{reset}.

The orders of magnitude difference between the cell resistance in the SET and RESET states makes the PCM memory ideally suitable for a multibit operation. In this scheme, the resistance of the cell may be set between the two extreme values, thus placing more than two levels per cell. This approach may become a viable option to further reduce the cost per bit of PCM devices.

15.3
Phase-Change Memory Materials

15.3.1
The Chalcogenide Phase-Change Materials: General Characteristics

The requirements for phase-change materials include easy glass formation during quenching from the melt, as well as congruent crystallizing compositions to avoid phase segregation during crystallization. Melting temperatures should be low to limit the switching power, whereas for non-volatility a good stability of the amorphous phase at application temperatures is required. It follows that the activation energy[1] for crystallization of the amorphous state should be high enough to enable long data retention times. On the other hand, crystallization rates, at least at elevated temperatures, should be high enough to allow for a rapid amorphous to crystal transition, preferably in the range of a few tens of nanoseconds.[2]

Such materials have now been under investigation for many years for their applications in DVD-RAM and DVD-R/W optical disk storage systems. Typically, metal alloys containing chalcogenide elements [by definition, elements of Group VI of the Periodic Table (O, S, Se, Te, Po)], and often referred to as "chalcogenide materials", are used. Chalcogenide elements are of interest as Se and Te compounds are easy glass-formers, because of their relatively high melt viscosities [22]. Compositions searched for are those that form a stable state in the solid phase ("polymorphic transformations"; i.e., where long-range diffusion is not required) [23].

The two typical chalcogenide material "families" used in PCM are both based on compositions of Ge, Sb and Te: (i) the pseudo-binary GeTe-Sb_2Te_3 compositions; and (ii) compositions based on the $Sb_{70}Te_{30}$ "eutectic" compound (see the Ge-Sb-Te ternary phase diagram in Figure 15.5) [24, 25].

---

1) The so-called "crystallization temperature" is not a uniquely defined material property, and varies depending on the time window of observation. "Activation energy" is therefore a better defined and more relevant physical parameter.

2) Recent discussions have indicated that the requirement of a rapid crystallization actually contradicts with easy glass formation, and rapid-crystallizing chalcogenides should be categorized rather as bad glass-formers based on their low glass transition to melt temperature (T_G/T_m) ratio compared to other easy glass formers such as SiO_2 [21].

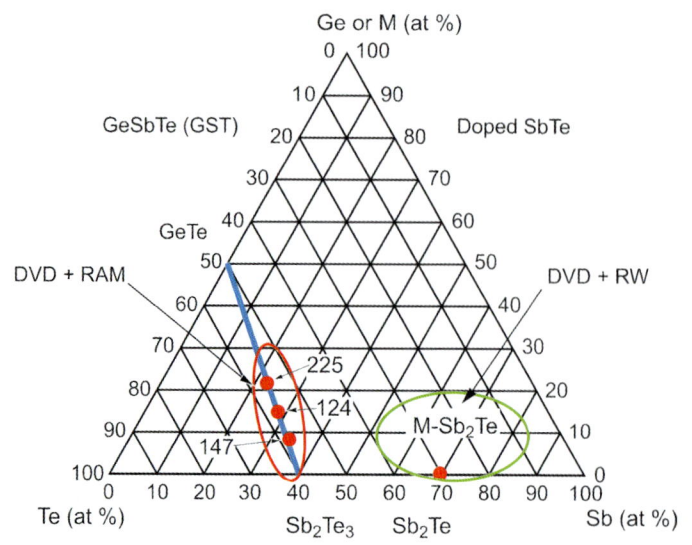

Figure 15.5 The Ge-Sb-Te (GST) ternary phase diagram, indicating the two classes of commonly used phase-change recording materials – that is, stoichiometric compositions along the GeTe-Sb$_2$Te$_3$ tie-line and compositions near the "eutectic" Sb$_2$Te. (Figure modified from Refs. [24, 25]).

15.3.1.1 The Pseudo-Binary GeTe-Sb$_2$Te$_3$ Compositions

The stoichiometric compositions around the GeTe and Sb$_2$Te$_3$ tie line are known as pseudo-binary compositions. These include the most widely used material Ge$_2$Sb$_2$Te$_5$, and they are used in the ovonic unified memory (OUM) [11], together with other compositions such as Ge$_1$Sb$_2$Te$_4$ [(1,2,4) material] and Ge$_1$Sb$_4$Te$_7$ [(1,4,7) material]. All of these materials are nucleation-controlled; that is, nucleation is dominant over growth [25], and are widely used in DVD-RAM applications. Along the tie line, the properties change from GeTe with high crystallization temperature (i.e., high stability) but slow crystallization speed, to Sb$_2$Te$_3$ that has a high crystallization speed but a low stability [26].

15.3.1.2 Compositions Based on the Sb$_{70}$Te$_{30}$ "Eutectic" Compound[3]

These compositions are more generally indicated as doped SbTe (M-SbTe) compounds. Variants include doping with In: In$_x$(Sb$_{70}$Te$_{30}$)$_{1-x}$, doping with Ag and In: Ag$_x$In$_y$(Sb$_{70}$Te$_{30}$)$_{1-x-y}$ (so-called "AIST"), and doping with Ge: Ge$_x$(Sb$_{70}$Te$_{30}$)$_{1-x}$ + Sb. These materials are so-called fast-growth materials [28]: the growth starting from the crystal regions surrounding the amorphous zone is the dominant crystallization mechanism rather than nucleation of new crystals inside the amorphous. The benefits of these materials are possibly faster switching (<20 ns), a better

---

3 It should be noted that the Sb$_{70}$Te$_{30}$ "eutectic" material composition (sometimes quoted as Sb$_2$Te, or more exactly as Sb$_{69}$Te$_{31}$) is actually not an eutectic but an azeotropic minimum [27]; that is, it fulfills the basic requirement of a congruent crystallizing material.

high-temperature retention, and a lower threshold field for conductivity switching in the amorphous phase (10–20 V μm^{-1} instead of 30–40 V μm^{-1}). On the other hand, cycle endurance and resistance ratio would be smaller [24].

15.3.1.3 Other Material Compositions

Some other material compositions, based on selenium rather than tellurium compounds, have more recently been investigated for possible application in phase-change memories, including antimony selenide (Sb_xSe_{1-x}; main attributes, lower T_m and faster crystallization speed) [29], and indium selenide (In_2Se_3; main advantage wider resistivity range) [30].

15.3.1.4 N- or O-Doped GST

Both, nitrogen and oxygen doping of GST have been used mainly to control the resistivity of the material.

N-doping has been used successfully used to increase the crystalline GST-resistivity (from 2 to 200 mΩ·cm for N concentration from 0 to 7 atom%), and furthermore results in a smaller grain size and an increased crystallization temperature (+50 °C) [31].

O-doping of GST is reported to increase the resistance ratio (from 100 to 1000), and to improve the high-temperature retention with an increase of the activation energy from 3.6 eV to 4.4 eV [32].

15.3.2
Material Structure

15.3.2.1 Long-Range Order: Crystalline State in GST and Doped Sb-Te

GST is characterized by two different crystal structures in the crystalline state – that is, a lower temperature (higher resistivity) fcc cubic state, and a higher temperature (lower resistivity) hexagonal state (see Figure 15.6) [24]. It should be noted at this point that the temperatures at which the crystallization and phase transitions occur are not fixed but rather depend on the heating rate. In a fast phase-change operation the amorphous GST probably crystallizes in the metastable fcc phase. The existence of the two different states may however influence the long-term, low-temperature stability and eventual resistance in the SET state. However, it has been reported that, for nano-sized structures, the temperature of the fcc to hexagonal phase moves up (from ∼360 °C to >450 °C for 65-nm patterns). The possible inhibition of the hexagonal state formation, together with the higher resistivity of the cubic state, may be a beneficial side effect of the scaling [34]. On the other hand, doped Sb-Te shows only one, low-resistive, hexagonal state.

15.3.2.2 Short-Range Order in Crystalline versus Amorphous State

Recent investigations have clarified the situation regarding the crystal-amorphous phase transition in these chalcogenide materials.

It has been noted [9, 35, 36] that, in contrast to the covalent semiconductors, in which amorphization does not changes the local ordering, in chalcogenide materials

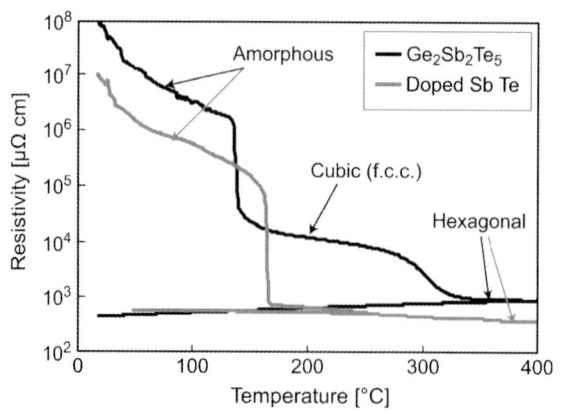

Figure 15.6 Resistive traces measured during temperature ramp of as-deposited amorphous GST and doped SbTe films, evidencing the occurrence of two different crystal states for GST but only one state for doped-SbTe. (Figure reproduced from Ref. [24]).

the amorphization induces a substantial increase in the local ordering and an important change in the resultant physical properties, such as an increased energy gap. This order–disordered transition is mainly due to a flip of the Ge atoms from an octahedral position into a tetrahedral position without rupture of strong covalent bonds (see Figure 15.7) [35]). Therefore, this class of materials is characterized by two competing structures with similar energy but different local order and different physical properties.

It is assumed that the nature of such a transformation ensures not only the large changes of physical properties (e.g., reflectivity and conductivity), but also the rapid performance and repeatable switching over millions of cycles.

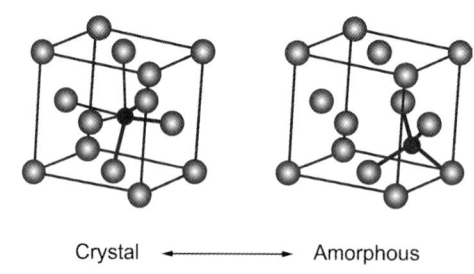

Figure 15.7 Fragments of the local structure of GST around Ge atoms in the crystallized (left) and amorphous (right) states. Upon heating the sample by a short intense pulse (above the melting point, T_m) and subsequent quenching, the Ge atoms flip from the octahedral to tetrahedral-symmetry position. Note that the stronger covalent bonds remain intact upon the umbrella-flip structural transition rendering the Ge lattice random. Exposure to light that heats the sample above the glass-transition temperature (T_g) – but below T_m – reverses the structure. (Figure reproduced from Ref. [35]).

15.3.3
Specific Properties Relevant to PCM

The majority of PCM materials investigated to date were developed for (re)writable optical discs, and existing knowledge of them, as well as experience of their reliability in products, should allow their rapid introduction into CMOS integration technology. However, their use in PC-RAM application may include various pitfalls:

- There exist certain *important operational differences* between DVD and PCM applications, that require the tuning/optimization of a number of specific material parameters for PCM that are not important for optical applications (e.g., resistivity in the on and off state, rather than reflectivity changes) (see Table 15.1). In addition, the amorphous phase should possess the particular property of threshold switching. The main material parameters for PCM applications are listed in Table 15.1.

- More importantly, however, the operating differences may have a major influence in the operation stability and repeatability. Indeed, in DVD applications, programming is achieved by laser pulse power coupling, and reading by reflectivity change. Data programming and storage relies on average material properties, such as reflectivity and absorption, that are not very sensitive to local variations. For example, they may be caused by small crystalline particles embedded in the amorphous region or vice versa, due to incomplete/inhomogeneous nucleation or amorphization. On the other hand, PCM relies on programming by Joule heating; that is, by current conduction through the device. Furthermore, the SET operation requires threshold current switching in the amorphous phase, which is a filamentary process. Also, the reading is based on a resistance (i.e., current) measurement. As the current conduction is greatly affected by the existence of local inhomogeneities, programming and reading may become highly sensitive to non-uniform/incomplete crystallization or amorphization. For example, low-resistive current paths in the incomplete amorphized state, or amorphous

Table 15.1 The major important material parameters for PCM, optimization direction, and value for GST material.

Symbol	Parameter	Optimization	GST 225 value	Reference(s)
T_m	Melting temperature	Minimal (low RESET power)	621 °C	[37]
T_c	Crystallization Temperature	Maximal (good retention)	155 °C	[37]
E_a	Activation energy	Maximal (good retention)	2.6–2.9 eV	[15, 38]
ρ_c	Resistivity crystalline state	"High" for low program current – "low" for fast read	$\sim$350 Ω μm	[18]
ρ_α	Resistivity amorphous state	High for good Resistance Ratio	$\sim$0.3 MΩ μm	[18]
E_c	Critical threshold switching field	Low for low SET program voltage	30–40 V–μm^{-1}	[24]

current-blocking regions in the incomplete crystallized state, may jeopardize proper programming or reading. In particular, the filamentary process of current switching during the SET state may induce strong inhomogeneous heating due to uncontrolled location of the filament that may results in only partial crystallization [20].

15.4
Physics and Modeling of PCM

From the basic device operation described in Section 15.2 it transpires that material-phase transitions (amorphization and crystallization dynamics) and conductance (threshold) switching mechanisms in the amorphous phase are the key processes involved in PCM. The physics of these mechanisms are outlined in greater detail below. The results of modeling studies implementing detailed microscopic descriptions of these effects have contributed greatly to an understanding of the subject, and have also supported the basis of design optimization of these devices.

15.4.1
Amorphization and Crystallization Processes

The different phase transitions of the PCM material during programming are illustrated graphically in Figure 15.8 [39]. Here, the crystalline material serves as an ideal starting point. During the programming transition from a SET to a RESET (high-resistance) state, the material is heated, begins to melt at solidus temperature,

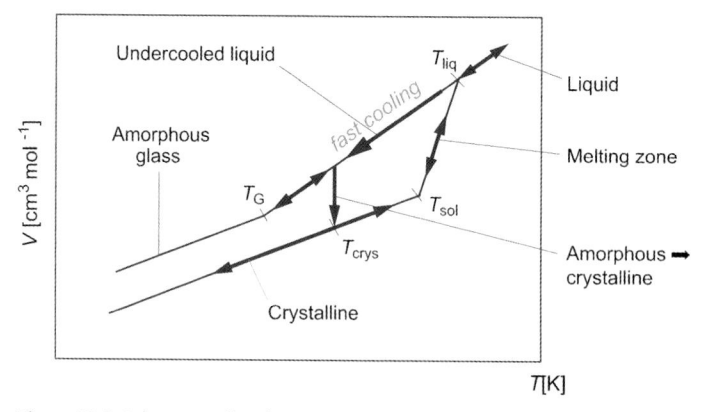

Figure 15.8 Schematic of molar volume (corresponding to 1/density) changes of phase-change material with temperature, showing the different possible phase transitions and critical temperatures. When heating a crystalline phase, melting will start at the solidus temperature T_{sol}, and will be complete at the liquidus temperature T_{liq}. Slow cooling will crystallize the material again following the same transition line; however, fast cooling results in an undercooling liquid that will be quenched in an amorphous glass below the glass transition temperature T_G. Heating the amorphous phase above T_G will result in crystallization at the crystallization temperature T_{crys}.

T_{sol}, and is completely molten at the melting temperature, T_m. When the cooling rate is higher than $10^9\,\mathrm{K\,s^{-1}}$ [40], the material does not begin to solidify at T_m, but remains an undercooled liquid. Below the glass transition temperature, T_G, the material freezes into the amorphous state.

The reverse situation is that, during the RESET to SET transition, the current pulse heats the material above T_G but below T_m, where it will begin to crystallize. There is no uniquely defined crystallization temperature, and even at relatively low temperatures (100–200 °C) crystallization may occur over long time scales (perhaps up to years). (These processes govern the basic long-term temperature retention of the RESET state, and will be discussed in Section 15.6.) For the SET programming, high-temperature rapid (<100 ns) crystallization processes are required, an understanding of which is based mainly on the general physical models for nucleation and growth. Different models have been proposed, for example by Peng et al. [41] and by Kelton [42], by which the temperature-dependent nucleation and growth rate can be calculated. Calculations based on these models using the different material properties of, for example GST and AIST, indeed confirm the nucleation, respectively growth-dominated crystallization mechanisms, that have been observed experimentally in these materials [43].

15.4.2
Band-Structure and Transport Model

The development of a comprehensive and quantitative framework to support the design and optimization of these devices is a challenging task. Today, PCM physics is extremely well developed, and a model should be capable of coupling a description of carrier transport in both crystalline and amorphous phases, together with the heat equation and phase-transition dynamics. The starting point here is to describe the electrical properties, thus aiming to reproduce correctly the electronic switching effect in the amorphous chalcogenide alloy.

Recently, it has been shown that the adoption of a semiconductor-like picture for both the amorphous and crystalline phases is quite effective, and may successfully account for the experimental I–V curves [33]. Moreover, this also allows the handling of simulations within the frame of codes already widely adopted by the semiconductor industries. Optical absorption data have shown that both crystalline and amorphous GST have gaps of 0.7 and 0.5 eV, respectively [33]. Moreover, there is no doubt that the carrier dynamics in the crystalline GST can be treated according to Bloch's theorem, as in a crystalline semiconductor. In contrast, investigations on amorphous compounds have demonstrated the existence of states with variable transport properties [44]. During the 1960s, it became common practice to describe these materials as shown in Figure 15.9 (left), where "mobility edges" separate fully conductive bands from the low mobility states. This picture can be easily translated in terms of a semiconductor-like framework. By assuming that low-mobility localized states behave like trapping centers, and that more conductive levels resemble delocalized states, a band structure can be defined and the amorphous GST modeled as a "very defective" crystalline semiconductor [33] (Figure 15.9, right). The material

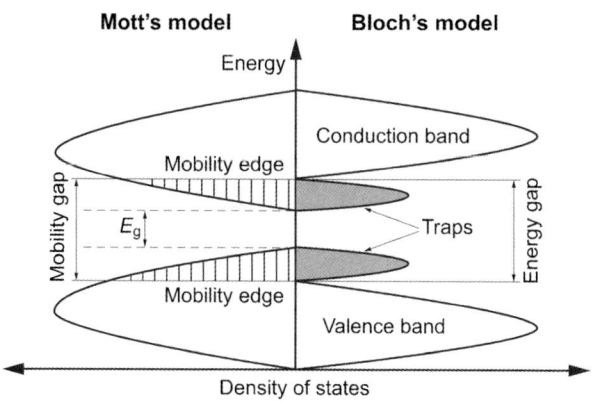

Figure 15.9 Comparison between the classical Mott and Davis's picture for the amorphous band diagram, and the scheme recently proposed by Pirovano et al. [33]. (Figure from Ref. [19]).

parameters, such as energy gap, trap densities and density of states for both phases, adopted in the numerical simulations, are listed in Table 15.2. It should be noted that the crystalline GST is p-type, due to a large density of vacancies (10% of the lattice sites), while the amorphous GST is characterized by a large density of donor/acceptor-like defects: the so-called Valence Alternation Pairs. This semiconductor-like picture is able to account for the peculiar conduction of the amorphous state and for the threshold switching [19].

The physics involved in the threshold switching remains a subject of debate. Since Ovshinsky first reported threshold switching [4], different models have been proposed, with many groups supporting the idea that switching is essentially a thermal effect and that the current in an amorphous layer rises above due to the creation of a hot filament [45, 46]. Later, Adler showed that the effect is not thermal (at least in thin chalcogenide films), in agreement with Ovshinsky's original picture. In their pioneering studies [47, 48], Adler and colleagues showed that a semiconductor resistor may feature switching, without any thermal effect. The condition for the threshold snap-back to occur is the presence of a carrier generation depending on

Table 15.2 Electronic parameters for both crystal and amorphous phases. (From Ref. [33]).

Property	GST crystalline	GST amorphous
E_{gap} [eV]	0.5	0.7
N_C [cm^{-3}]	2.5×10^{19}	2.5×10^{19}
N_V [cm^{-3}]	2.5×10^{19}	10^{20}
Vacancies [cm^{-3}]	5×10^{20}	—
C_3^+ [cm^{-3}]	—	10^{17}–10^{20}
C_1^- [cm^{-3}]	—	10^{17}–10^{20}
μ_n-μ_p [cm^2 V^{-1} s^{-1}]	0.1–23.5	5–200
F_C [V cm^{-1}]	3×10^5	3×10^5

field and carrier concentration (e.g., impact ionization) competing with a Shockley–Hall–Read (SHR) recombination via localized states.

The numerical model reported in Ref. [33] implements Adler's picture accounting for avalanche impact ionization in the amorphous and SHR recombination via the localized defects.

The schematic dependence of the band structure along a cross-section of a PCM device is shown in Figure 15.10, where the wide-gap region corresponds to the amorphous GST. At low bias, the quasi Fermi levels in the amorphous GST are close to their equilibrium position. As both the carrier density and their mobility is low (the average hole mobility is about $0.15\,\text{cm}^2\,\text{V}^{-1}\,\text{s}^{-1}$ [33]), the conduction regime is ohmic. By increasing the voltage, the applied field approaches the avalanche critical field of 3×10^5 [10], significantly increasing the carrier generation. The quasi Fermi levels thus split and move close to the band-edges (Figure 15.10, lower diagram). Carrier recombination mainly takes place in the region, close to the anode, where the electron Fermi level approaches the conduction band. At large bias, all defects available for recombination are full, and recombination may no longer be able to balance the exponentially rising generation rate. The system reacts by reducing the voltage drop in order to maintain the balance between recombination and generation, leading to the electronic switching. Hence, the snap-back takes place and, after switching, the GST is still amorphous but highly conductive. Generation is sustained by the large density of free carriers. According to this picture, the minimum voltage required for the switching to occur is of the order of the split between quasi Fermi levels (i.e., the

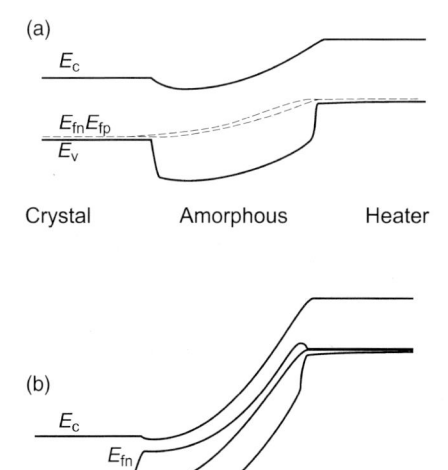

Figure 15.10 Band diagrams along the cross-section of a PCM cell in the RESET state (according to Ref. [33]). (a) At low bias, quasi-Fermi levels are close to the equilibrium value. (b) Close to threshold switching; generation by impact ionization is properly balanced by recombination though trap levels. When recombination saturates, generation finds a new stable working point reducing the voltage across the device. (Figure from Ref. [19]).

energy gap). This argument may justify why both, the holding voltage, V_H, and the asymptotic value at low R in Figure 15.2b, approaches approximately 0.5 V.

Although this picture so far has been successful in accounting for the experimental findings, it should be accepted with a degree of caution, and further investigations are needed to better assess the material properties. The quantitative description of impact ionization in these materials, as well as the role of the interfaces or of Poole–Frenkel mechanisms, deserve further investigation as many of the details still lack direct experimental verification. However, recent industrial interest in PCMs may lead to new experimental efforts and to the fabrication of devices purposely designed to test the validity of these key assumptions.

15.4.3
Modeling of the SET and RESET Switching Phenomena

The above conduction model was then coupled to heat equation and to phase-transition dynamics (nucleation and growth). When implemented in a three-dimensional (3-D) semiconductor device solver, it highlighted substantial differences in the two phase transitions [20]. The temperature maps during the SET–RESET transition, and the resulting phase distributions obtained with increasing current pulses with a plateau of 150 ns, are illustrated in Figure 15.11 [20]. In this figure, all of the pictures refer to a Lance PCM device in which a cylindrical metallic heater is in contact with the GST layer. The current flows almost uniformly across the polycrystalline GST, thus resulting in a roughly hemispherical shape of the final a-GST volume. As the programming current increases above melting, the volume left in the amorphous state increases.

On the other hand, Figure 15.12 [20] shows the calculated final phase distribution for a RESET to SET transition with programming current of 130 µA (a) and 160 µA (b), respectively. Figure 15.12a shows that the current first sparks by electronic threshold

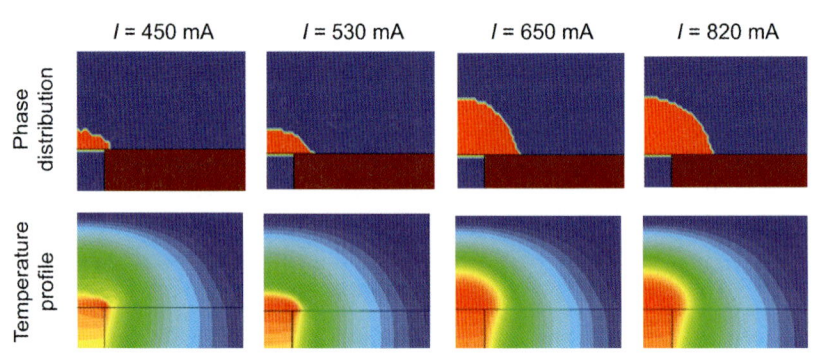

Figure 15.11 Homogeneous heating of the crystalline GTS region (blue) during SET to RESET transition (bottom row), resulting in homogeneous amorphous regions (red) at the end of a programming pulse (top row). Figures from left to right correspond to increasing the peak current value. A larger amorphous region will correspond to a higher resistance level. (Figure from Ref. [20]).

| Inital phase distribution | Current profile | Temperature profile | Final phase distribution |

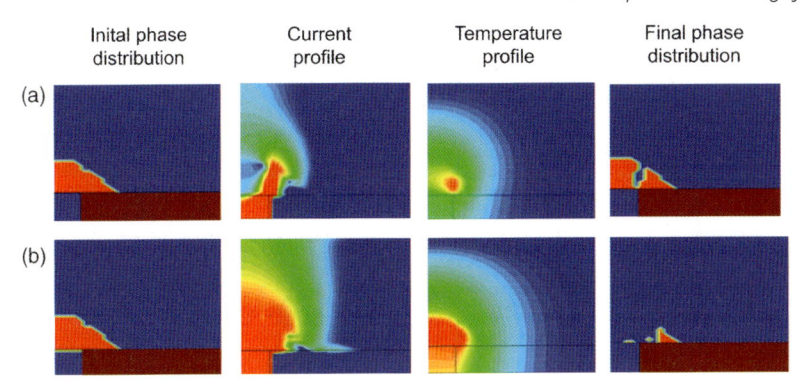

Figure 15.12 Programming operation from amorphous to crystal state (only half of the cell is shown because of cylinder-symmetry). (a) Top row: $I = 130\,\mu A$: electronic switching will occur first in the regions where the amorphous thickness is minimum. After the switching event, current spikes will increase the temperature only locally, and crystallization may only be induced in these hot zones resulting in a non-uniform final phase distribution with a major part of the amorphous zone remaining. (b) Bottom row: $I = 160\,\mu A$: only at higher voltages, eventually the hot filament extends in the whole the active area, leading to an homogenization of temperature and of the transformed volume. (Figure from Ref. [20]).

switching in the weakest a-GST region, locally triggering Joule heating and consequent crystallization processes. In case (a), the pulse amplitude/time is not sufficient to provide a complete crystalline path, and thus the active region features a residual a-GST layer causing the large measured R. In case (b), a further increase in the applied voltage eventually extends the hot filament in the whole of the active area. The initial amorphous volume has been almost completely crystallized by the programming pulse, resulting in a sufficiently low resistance. It should be noted that the localized phase transition is directly related to the details of the electronic switching mechanism, and thus can be reproduced only by a self-consistent model describing both electrothermal and phase-change dynamics.

Figures 15.11 and 15.12 also suggest that, by changing the programming pulse, the value of the cell resistance can be reliably placed in between the largest and the minimum SET value, thus opening the way to a multi-level operation. For example, four levels – each with different resistance values – might be programmed per cell, thus reducing the cost per bit.

15.4.4
Transient Behavior

V_{TH} and R are two key parameters of the memory cell. Figure 15.13 illustrates the time dependence of these parameters as measured soon after the current pulse programming the cell in the *RESET* state [19, 49]. The first fast component of the transient is referred to as *recovery*. On the longer time scale, in the so-called drift regime, the V_{TH} and R transients follow a slower power law. The recovery sets the minimum time needed after programming before reading (if the cell is read soon after being programmed in the *RESET* state, the read value might erroneously be "1").

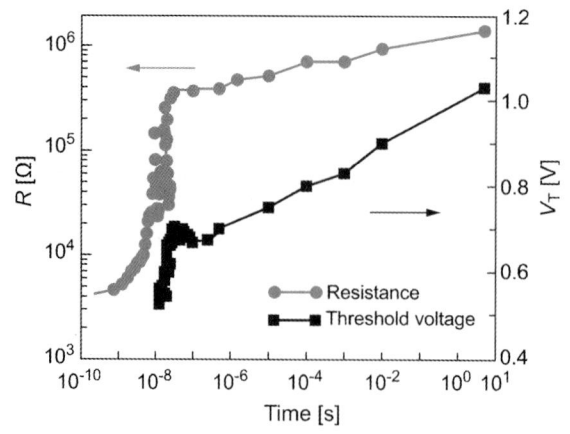

Figure 15.13 Low field resistance and threshold voltage for the amorphous phase as a function of time after reset programming operation. In about 50 ns, both low field resistance and threshold voltage are recovered, after which they continue to increase due to the drift phenomenon. (Figure from Ref. [19]).

Drift is instead a limit for multi-level operation, as the resistance of an intermediate level during ten years (3×10^8 s) might cause the bit to be erroneously decoded.

Different models have been proposed to justify these effects. Recovery is likely due to charge transients. After quenching, the newly formed amorphous region is full of trapped carriers, and some nanoseconds are required for these carriers to be released by trapping states and for the Fermi levels to recover the equilibrium value (see Figure 15.10a). During this transient stage, V_{TH} and R change from the *set* to the corresponding *RESET* values. Drift physics is more controversial, however, it having been suggested that drift might be due to mechanical stress release following the crystalline-to-amorphous phase transition. The resulting band-gap widening may reduce the mobile carrier density, thus contributing to the charge conduction. Another possible explanation links the effect to changes of the electronic states [50]. Variation of the density of states close to the band-edge already observed in other chalcogenide compounds [51] as the amorphous evolves towards to a more regular microscopic structure. In order for multi-level storage to be implemented in PCM memories, this effect should be fully understood and minimized.

15.5
PCM Integration and Cell Structures

15.5.1
PCM Cell Components

PCM elements may be organized in a memory matrix with or without adding a selection device to each phase-change element. Indeed, a raw cross-point matrix may be conceived [52], but this would suffer read errors due to leakage through neighboring "on" elements. Moreover, as important program disturbs may occur in half-select regimes, each PCM memory cell should contain both the phase-change element and a selection device. The choice and dimensions of the selection element is determined by cell size constraints and the RESET program current. A MOSFET transistor is the most evident choice for memory integration in CMOS technology (Figure 15.14) [53]). The need for Source and Drain contacts, such a "one transistor-one resistor" (1T1R) cell would require a minimum area of 8–10 F^2 (where F is the minimum feature size of the technology). However, MOSFETs have limited current drivability, so that minimum size transistors cannot be used, and cell sizes are much larger, up to $\sim$40 F^2 in 0.18 µm technology with 0.6 mA reset current [13].

An alternative is the use of a bipolar p-n-p select device. In that case, there are still two (Base and Emittor) contacts per cell, while the Collector is a collective substate contact. As the bipolar current drivability is much higher, the overall cell size is smaller (only $\sim$10 F^2 in 0.18 µm technology with 0.6 mA reset current [13]). The trade-off is of course a more complex integration scheme for fabricating these bipolar devices in a CMOS technology, restricting this solution for stand-alone memory applications only.

In order to obtain the smallest cell area, a diode selector may be used [17], as a diode would only need one contact and can handle large currents (a self-rectifying device would be the most ideal case, so that a raw cross-array could have the same functionality with minimum cell size). However, in the diode selection regime, both bitlines and wordlines have to carry relatively high currents, leading to a partitioning of the memory as well as to larger X-decoders. Moreover, isolation is also less perfect and parasitic resistances in the full signal path are important. Both, series resistances and leakage currents, contribute to read error, while diode integration may also result in the formation of a parasitic bipolar transistor.

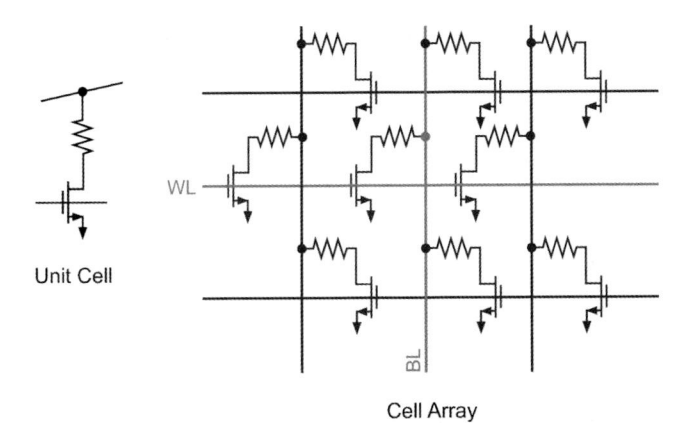

Figure 15.14 Schematic of 1T1R cell structure and memory array matrix. (Adapted from Ref. [53]).

15.5.2
Integration Aspects

Besides formation of the selection transistor, PCM fabrication requires the integration of a phase-change element in a CMOS technology. The memory element may be fabricated after the transistor processing (the so-called front-end-of-line processing; FEOL), and either before (e.g., in-between Si contacts and first metal interconnect layer) or after the first steps of the interconnect (e.g., on top of Metal 0 or Metal 1 interconnect levels). This latter scheme is the back-end-of-line or BEOL processing. A schematic state-of-the-art process flow is shown in Figure 15.15 [54, 55].

The PCM material is typically deposited by sputtering (physical vapor deposition; PVD) from a multi-element target with the desired composition. The as-deposited phase is either amorphous (for room- or low-temperature deposition), or crystalline if deposited above the crystallization temperature. In either case, due to the temperature budget of the following BEOL processing (up to 400–500 °C), the material will be fully crystallized after the integration. In a number of cell concepts, the conformal deposition of the PCM material and/or the ability to fill small pores is important [14], and these requirements would ideally call for a conformal deposition technique such as metalorganic chemical vapor deposition (MOCVD), rather than PVD. Critical elements for the integration are a good adhesion of the PCM material on the underlying substrate structure (typically a patterned metal electrode in a SiO_2 matrix), the material out-diffusion/inter-reaction/oxidation during high-temperature steps [56], and dry-etch patterning of the PCM [57, 58].

Furthermore, suitable electrode materials are needed for both the "heater" contact (where the metal will be in contact with the hot/molten PCM) and the (cold) top electrode. Material stability, low contact resistance, and good adhesion are important parameters. Poor electrode contact properties indeed have been identified as being responsible, for example, for "first fire effects" (see Section 15.6). Typical electrode materials are standard conductive barrier materials available in Si processing such as

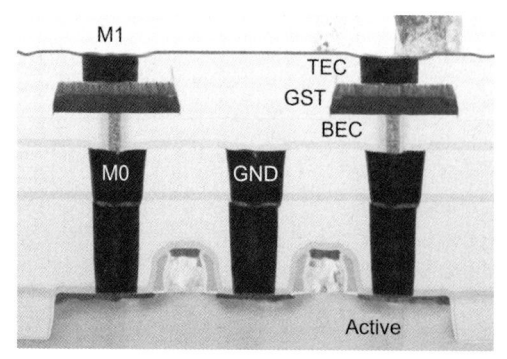

- CMOS transistor formation
- ILD formation
- Metal-0 formation
- BEC formation (< 70 nm)
- GeSbTe deposition
- Top electrode deposition
- Cell patterning
- IMD formation
- TEC/Metal-1/Metal-2 formation

Figure 15.15 Cross-sectional scanning electron microscopy image showing the cell integration scheme (from Ref. [54]), and schematic process flow for PCM cell formation (from Ref. [55]). In this case, the PCM cell is integrated between the M0 and M1 interconnect levels.

TiN (e.g., Ref. [31]), although W is also often used [59]. The top electrode typically defines the PCM area and can be used as (part of) a hard mask during the PCM patterning.

Stability and controllability of the different process steps of the integration technology are crucial for the preparation of large-density memory arrays. The most important array characterization technique is therefore the distribution of the ON and OFF program state resistances. Tight distributions are needed to maximize the sensing window and to avoid bit errors. By process optimization, excellent distributions have been obtained on 256 Mb PCM after full integration [60].

15.5.3
PCM Cell Optimization

In the basic OUM PCM cell, the PCM and top electrode are planar layers deposited on a plug-type, bottom heater contact. That part of PCM material effectively involved in the phase switching is basically a hemispherical volume on top of the heater. To reduce the heating power (or program current), it is important to try to confine the dissipated heat as much as possible. While many different cell structures have been proposed in literature (see Figure 15.16), the optimization of heat confinement is in fact based on two simple principles: (i) by concentrating the volume where effective Joule heating takes place; and/or (ii) by improving the thermal resistance to reduce the heat loss to the surroundings.

15.5.3.1 Concentrating the Volume of Joule Heating

The Joule heating volume can be confined by pushing the current through a small cross-section with high current density. One obvious way to do this is to reduce the contact area of the heater contact with the PCM material, for example by minimizing the heater plug diameter (as in the small "sub-litho" contact heater cell [31]), by using only a conductive liner as heater (as in the edge-contact cell [61]), or by filling the plug with isolating dielectric material (as in the μ-trench cell [13, 62] and ring-contact heater cell [63]). The main advantage of using only the conductive liner is that at least one dimension of the heater area is controlled by the liner thickness, and not by the lithography.

Another way of confining the Joule heating volume is by structuring the PCM material to a narrow cross-section (see bottle-neck cell [64], line heater cell [24], self-heating pillar cell [65]). Finally, further improvement can be made by increasing the resistivity of the PCM material, for example by N or O doping [31, 32, 53]. In a different approach, a highly resistive TiON layer is made between the TiN electrode and the PCM material, in which layer the Joule heating will be concentrated resulting in lower program power [66]. However, such an approach may negatively influence the contact resistance to the PCM.

15.5.3.2 Improving the Thermal Resistance

In the so-called confined cell structure [14], the PCM is deposed in a pore etched back in the heater. This not only concentrates the Joule heating region but also surrounds a

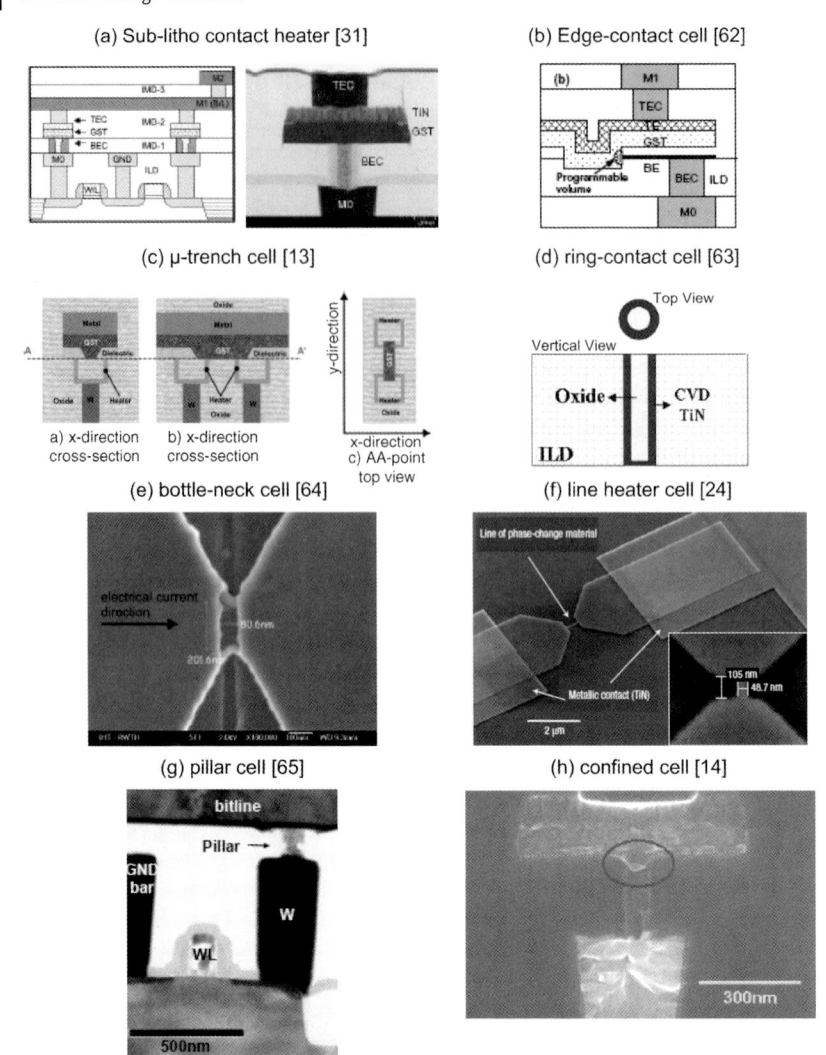

Figure 15.16 Different PCM cell structures for reducing the program power.

large part of this volume by a dielectric layer with reduced thermal conductivity. The drawback here is the topography which, ideally, would require a conformal deposition of the PCM. The alternative is to structure the PCM material, rather than the heater, leading to a plug or pillar. This approach is based on the same principle [18].

Another way to improve the thermal resistance is by increasing the PCM thickness (as this limits the heat flow to the top electrode heat sink) [14]. However, this option should be traded-off with the threshold voltage required for electronic switching during SET programming. One of the benefits of the horizontal line cell [24] is also the setting apart of the heated zone from a metal heater contact and a capping with thermal insulating dielectric.

Apart form changing only the cell structure, it is clear also that the correct material selection (use of different dielectric materials, and especially the use of porous dielectric materials) can improve the thermal heat confinement [53]. It should be noted, however, that the improved thermal isolation should not avoid the rapid quenching from the melt during RESET, otherwise the device cannot be programmed to the OFF-state.

15.6
Reliability

15.6.1
Introduction

As for any other non-volatile memory technology, reliability is one of the major concerns. The main specific reliability issues of PCM are: (i) data retention of the RESET, affected by the (limited) stability of the amorphous state; (ii) endurance, limited by the occurrence of stuck at RESET (open) or stuck at SET (short)-type defects; and (iii) program and read disturbs – that is, the stability of the amorphous phase due to repeated, though limited, thermal cycling caused by reading or programming neighboring cells.

While this section is based on reliability tests on the cell level (probing "intrinsic" reliability), for large memories the effect of reliability tests (e.g., of a temperature bake or of a large number of SET/RESET programming cycles) on the resistance distributions should also be evaluated to screen eventual "extrinsic" failures. However, until now only a few (preliminary) results on array reliability statistics have been reported [15, 67].

15.6.2
Retention for PCM: Thermal Stability

The most important requirement for a non-volatile memory is the ability to retain the stored information for a long time, the typical specification being 10 years (at a minimum of 85 °C). As the SET state is stable from a thermodynamic point of view, it has no problem of data retention. On the other hand, the RESET state, corresponding to the amorphous phase, is instead meta-stable and may crystallize following a dynamic which is heavily dependent on temperature. The retention of an amorphous state is, therefore, critical.

The retention performance of PCM technology is addressed by performing accelerated measurements at high temperatures. Figure 15.17 shows the typical failure time under isothermal conditions and with no applied bias, measured at several temperatures ranging from 150 to 200 °C [15]. The failure time is defined as the time required by a fully amorphized cell to lose the stored information. The resistance value for failure has been instead defined as the geometric average between set and reset resistances. The data clearly show an Arrhenius behavior

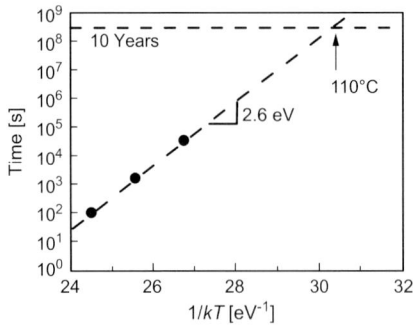

Figure 15.17 Experimental crystallization time of the RESET state as a function of temperature, shown in the Arrhenius plot [15]. A maximum temperature of 110 °C can be tolerated to guarantee 10 years' data retention.

with an activation energy of 2.6 eV, which extrapolates to a data retention capability of 10 years at 110 °C.

A wide range of activation energies have been reported. In general, quite high activation energies have been obtained from $E_a \sim 2.9$ eV for recent fully integrated cells [38] up to 3.5 eV for intrinsic material characterization of GST [17]. Even if such high activation energies are favorable for long retention, however, the physics underlying these experimental values is still not completely understood. The details of the material and integration processes apparently have a large effect, as activation energy was found to be dependent on the presence of capping layers (from a low 2.4 eV for uncapped GST to 2.7 eV for ZnS-SiO_2-capped GST [68]). Furthermore, material doping increases the activation energy (up to 4.4 eV has been reported for O-doped GST [32]).

In principle, the crystallization process during the accelerated retention tests should be described by the same theoretical models for crystal nucleation and growth as used to account for crystallization at much higher temperature (but at much shorter times) during the SET programming pulse (see Section 15.4). Although the conditions for the time window vary over many orders of magnitude (<100 ns during SET, but >10^2–10^3 s during retention tests), it has been shown [69] that – remarkably! – the models are indeed able accurately to describe the crystallization processes under both conditions. Moreover, the crystallization statistics also significantly impact the data retention measurements in high-temperature accelerated tests.

15.6.3
Cycling and Failure Modes

PCM cells have been shown to have an intrinsic long programming endurance – that is, up to 10^{12} SET–RESET program cycles [13, 17] (Figure 15.18), which is much superior with respect to Flash technology. In fact, cell endurance has been shown to depend heavily on the interface quality of the heater-GST system and on the possible interdiffusion between GST and adjacent materials. A non-optimized fabrication technology results in devices showing the so-called "first fire" effect [47], namely a

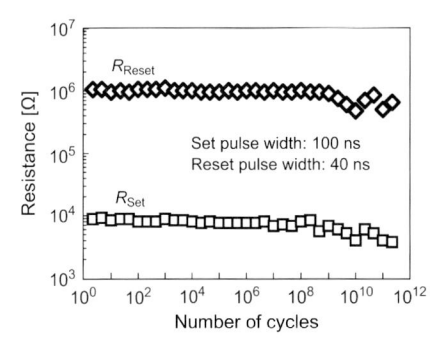

Figure 15.18 Program cycling of a PCM element. (From Ref. [13]).

higher initial programming pulse required for the first cycle of virgin devices. The same devices usually feature poor characteristics in terms of stability during cycling characterization, usually ending in a physical separation of the chalcogenide alloy from the heater (stuck at RESET).

A second failure mode has been also observed (usually called "short-mode failure" or "stuck at SET"), where the devices remain permanently in the highly conductive condition. This phenomenon requires an auxiliary physical mechanism that either forbids the phase-change transition of the GST or creates a conductive parallel path that shunts the cell electrodes. Both cases require a chemical modification of the chalcogenide alloy, suggesting that the interdiffusion of chemical species from adjacent materials plays a role. A careful definition of the materials belonging to the device active region is mandatory in order to achieve good reliability performance.

Finally, current density, as well as the high temperatures (>600 °C) reached in the active region during programming, must be considered as accelerating factors for the previously proposed failure mechanisms (i.e., poor quality interface and contamination). This explains the strong decrease in endurance as a function of the "overcurrent", or, equivalently with energy per pulse larger than required for switching [17]. The data in Figure 15.19 show that endurance at a constant programming current (~700 µA) scales inversely to the reset pulse [70]. The longer the pulse,

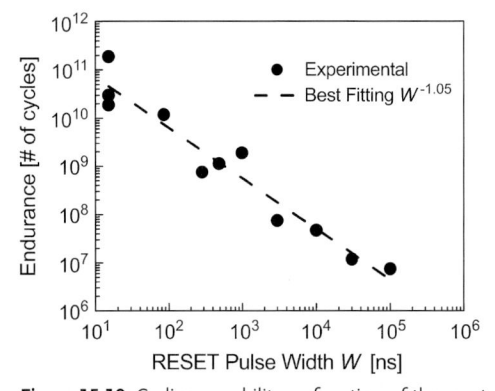

Figure 15.19 Cycling capability as function of the reset pulse window (programming energy per pulse; from Ref. [70]).

the larger the energy released per pulse, and the faster the interface degradation, while the overall energy released up to the bit failure remains almost constant.

15.6.4
Read and Program Disturbs

For most memory technologies, one important concern is the ability of the cell to retain data in the face of spurious voltage transients caused by reading and programming in the memory matrix. For PCM cells such disturbs are not directly induced by voltage pulses but rather by "thermal spikes" that can trigger the crystallization of the metastable amorphous state.

A first failure mechanism can be caused by multiple read accesses of a cell: the small current flowing through the device can induce a localized heating able to accelerate the spontaneous amorphous to crystalline transition. In a second failure mechanism, repeated programming operations on a cell can induce an unwanted heating of the adjacent bits (thermal cross-talk) that can lose the data stored.

Read disturb tests have been described in literature [15, 71], indicating that the repetitive reading of a cell in the high-resistance state with a current below 1 µA allows for a 10-year bit preservation. Such a current is one order of magnitude larger than the current flowing through the cell in standard reading conditions, thereby confirming the robustness to read-disturbs of PCM also in continuous reading, worst-case conditions. Results from program disturb tests on demonstrators have shown that cross-talk is not an issue down at the 90-nm technology node [15]. Thermal simulations, furthermore, confirm program disturb immunity up to at least the 45-nm technology node [18].

15.7
Scaling of Phase-Change Memories

A new memory technology, to be competitive with the existing Flash, must feature a small cell size combined with an excellent scalability beyond the 45-nm technology node. In this section, the scaling potential of the PCM memory is addressed. The main aspects are: (i) scaling of the thermal profiles in order to avoid thermal disturbs at shrinking cell separation; (ii) scaling of the program (RESET) current (and voltage), not only to reduce the program energy/cell, but also because it affects the cell size through the dimensions of the select transistor; and (iii) conservation of the basic material characteristics down to very small dimensions. These aspects will have a crucial impact on the scalability perspectives of PCM technology, and are still to be verified.

15.7.1
Temperature Profile Distributions

As programming of the PCM cell is based on strong heating up to high temperatures (>600 °C, above the PCM material melting point), yet on the other hand the amorphous RESET state can become unstable at much lower temperatures (<200 °C), it is crucial for the correct operation of a PCM memory that the heating remains very localized and does not affect neighboring cells. Whilst this has been proven for current technologies down to 90 nm (see Section 15.6), it may be less evident to maintain the high temperatures localized in much further scaled technologies, beyond the 45-nm node.

As far as all the linear dimensions are reduced isotropically, the temperature distribution profile indeed does scale. This property transpires from the heat equation:

$$\kappa \cdot \nabla^2 T = g = \rho J^2 \tag{15.1}$$

where κ is the thermal conductivity and g is the heat generation per unit volume. The latter is proportional, via the electrical resistivity, ρ, to the square of the current density, J.

Let us now assume that a new device is fabricated, by shrinking all the linear dimensions by a factor α, but keeping the same boundary conditions (e.g., $T(0) = T_0$) and material properties (e.g., κ, ρ). In the new device it is:

$$\kappa \cdot \nabla'^2 T' = g' = \rho J'^2$$

Since for all the spatial coordinates it is $x' = x/\alpha$, we obtain:

$$\kappa \cdot \nabla'^2 T' = \kappa \alpha^2 \cdot \nabla^2 T' = g' = \rho J'^2$$

That is:

$$\nabla^2 T' = \rho \frac{J'^2}{\alpha^2} \tag{15.2}$$

Provided that $J' = \alpha J$, Equation 15.1 and 15.2 coincide, leading to the same temperature profile, but on a spatial scale uniformly compressed by a factor α.

It follows that if two cells in the original device, at a distance d, do not suffer from cross-talk, then in the isotropically scaled device two cells at a distance d/α will be immune to thermal disturbs. The argument holds as far as $J' = \alpha J$, which will be demonstrated in the following section.

In a more aggressive scaling scheme, only the contact area of the cell is scaled down, without changing so much the thickness of the different layers. In such an anisotropically approach, aiming to more drastically reduce the programming current, cross-talk immunity is no longer granted. However, simulations results show that, without any specific care or materials, thermal disturbs are not expected to slow down cell scaling until the 45-nm node [18].

15.7.2
Scaling of the Dissipated Power and Reset Current

The highest power and programming current is required during the RESET operation, where locally the PCM material must be heated above the melting point.

By using simplified models, the RESET power can be calculated as follows. Even if the programming current pulses last only some tens of nanoseconds, the temperature rise ΔT of the hot spot, where the PCM material eventually melts down, may be computed at steady state. This assumption holds true since the thermal transients in such a small region are characterized by a nanosecond time constant, much faster than the typical current pulse width.

At steady state, the power dissipated by Joule heating (P_J) balances the heat loss ($P_{HL} = \Delta T/R_{TH}$), where R_{TH} is the thermal resistance to the thermal sinks at room temperature (i.e., top and bottom metal layers). On the other hand, the power P_J is proportional to the current, I^2, via the electrical resistance, $P_J = R \cdot I^2$. Using this simple model, the temperature rise ΔT_M, needed to reach the melting temperature, can be written as: $\Delta T_M = P_{J,M} R_{TH} = R_{TH} R \cdot I_M^2$, where I_M is the melting current. It follows:

$$I_M^2 = \Delta T/(R_{TH} \cdot R) \tag{15.3}$$

In the frame of the isotropic scaling rule[4], as the technology scales, the cell surface area decreases as F^2, but also the distances to the heat sinks decrease as F. The thermal resistance will therefore linearly increase with the scaling factor: $R_{TH} \sim \alpha$ or, equivalently, $R_{TH} \sim F^{-1}$. As for the thermal resistance, the electrical resistance of the PCM cell also increases linearly with geometry scaling ($R = \rho \cdot \text{length}/\text{Area} \sim F^{-1}$). It follows from Equation 15.3 that the melting current and the programming current, which is proportional to the latter, scales as F. The smaller the feature size, the smaller the programming current: $I_{reset} \sim F$ (or $I_{reset} \sim 1/\alpha$). Note that the current density $J = I/A$, will scale as α, as assumed above in deriving Equation 15.2 while discussing the immunity to thermal disturbs.

Although the scaling result is independent of the adopted cell architecture, this does not mean that cell architecture is not important. At a fixed technological node, the cell architecture should be optimized, by accurate design of the geometry and material engineering, to minimize the programming current and the dissipated power (examples of different cell optimizations were provided in Section 15.5).

A more aggressive reduction of the programming current may be obtained by scaling the contact area but not the other dimensions (e.g., the PCM thickness). This choice will mainly affect the thermal and electrical resistances (R and R_{TH} scale faster, i.e., $\sim F^{-2}$ instead of F^{-1}). It follows that $I_{reset} \sim F^2$ (or, $\sim 1/\alpha^2$). The scaling properties of the PCM cell are summarized in Table 15.3[18]. The more aggressive scaling will however pose some problems of manufacture, as the aspect ratio of some cell features (e.g., the thickness to cell size of the PCM material) will increase. Moreover, as discussed above, at this point thermal disturbs may begin to enter the game.

The scalability of the reset current has been addressed experimentally by measuring several test devices with different contact areas [17, 18]. An example of resulting values is given in Figure 15.20 [18]. From this figure it is clear that the reset current follows the reduction of the contact area, and values as low as 50 μA have been

---

4) Scaling can be indicated either by a dependence on technology feature size, F, or by a dependence on a scaling factor α, with $\alpha \sim 1/F$.

Table 15.3 Scaling rules of PCM cell. (Adapted from Ref. [18]).

	Isotropic	Aggressive
Parameters	Scaling factor	Scaling factor
Heater contact area A_{cell}	$1/\alpha^2$	$1/\alpha^2$
Vertical dimensions d	$1/\alpha$	1
Electrical/thermal resistances R	α	α^2
Power dissipation P_{cell}	$1/\alpha$	$1/\alpha^2$
Current I	$1/\alpha$	$1/\alpha^2$
Voltage V_{cell}	1	1
Current density J	α	1

α = scaling factor, $\alpha \sim 1/F$ with F the feature size.

achieved with a com1plete device functionality. The reset current reduction shown in Figure 15.20 can be mainly ascribed to the increase of the heater thermal resistance, R_{TH} caused by the reduction of the contact area.

Data published recently show that I_{reset} is indeed scaling in between $\sim F$ and $\sim F^2$ (Figure 15.21) [19]. A scaling behavior stronger than F^2 has been indeed reported by Cho et al. [72]. Such a steep dependence, which is well beyond the F^2 theoretical limit, is a strong indication that other cell parameters related to the cell structure and/or material characteristic have been changed.

15.7.3
Voltage Scaling

The required program voltage is determined by the threshold for the electronic switching of the amorphous phase (SET). This voltage value scales with PCM thickness, but is also strongly material dependent; for example, the threshold field is reported to be larger for GST (225) material (threshold field $\sim$30–40 V µm^{-1}) than for the fast-growth doped SbTe materials (threshold field $\sim$14 V µm^{-1}) [24].

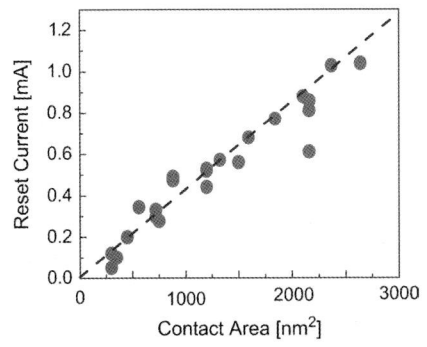

Figure 15.20 Reset current versus contact area (from Ref. [18]). Experimental values (dots), together with scaling trend line (dashed line).

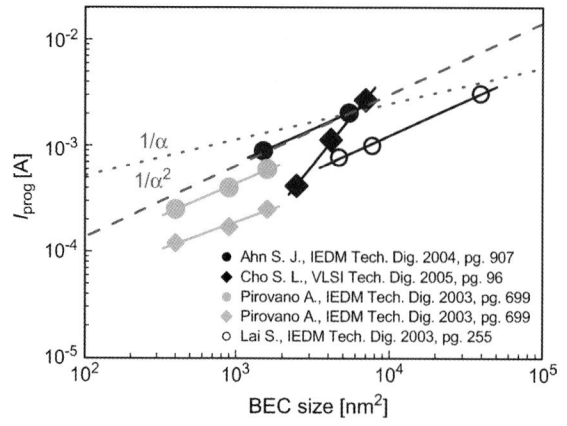

Figure 15.21 Plot of published experimental values of RESET program current versus bottom electrode contact (BEC) size, compared with the $1/\alpha$ and $1/\alpha^2$ scaling law (α = scaling factor, $\alpha \sim 1/F$ with F the feature size). The data are related to different cell structures. (Figure from Ref. [19]).

Furthermore, in Figure 15.3 there was shown to be a minimum value of the threshold switching voltage determined by material (bandgap) and/or contact characteristics. This is why, in Table 15.3, the voltage drop across the cell was considered not dependent on feature scaling. This limitation may pose a constraint on the maximum voltage which should be sustained by the bipolar selectors or by the gate oxide of the MOSFETs across the unselected cells of the array. Beyond the 45-nm technology node, an accurate design of the selecting device will therefore become mandatory.

15.7.4
Cell Size Scaling

In order to assess the scaling of PCM cell size, let us consider a typical MOS-select transistor PCM cell layout (Figure 15.22). The cell size can be calculated as $\sim (W + 1F) \times (4.5 F)$, where W is the width of the access transistor. For a minimal device, $W = 1 F$, the cell size would be $\sim 9 F^2$. However, this is the ideal case as in practice, $W/L \gg 1$ to obtain the required drive current to reset the cell through the access transistor. For $W/L > 1$, splitting the gate (the so-called "dual gate concept") is favorable to minimize the cell area [13], resulting in a cell area of $\sim 6W \times (1/2\, W + F)$.

Taking the I_{reset} values from the ITRS roadmap (predicting, in agreement with the scaling laws derived in Section 15.7.2 above, a scaling according $I_{reset} \sim \alpha^{1.5}$) [73], as well as the predicted scaling of the drive current with the technology node into account, the minimum access transistor size and the corresponding PCM cell size can be predicted as a function of the technology node (Figure 15.23). From these values it transpires that, at the 45-nm node, a cell size reduction to $<15 F^2$ would become possible (with $I_{reset} \sim 100\,\mu A$ and $W \sim 2.7 F$).

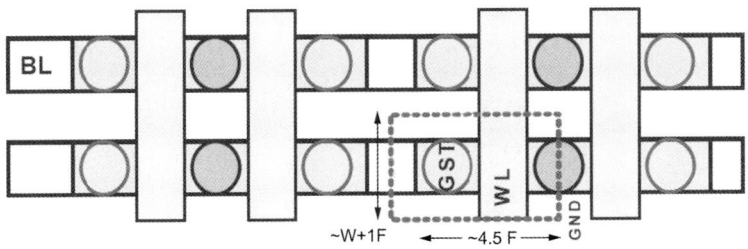

Figure 15.22 Layout schematic of PCM cell as shown in Figure 15.15. BL = bit line (M1); WL = word line (Poly-Si gate line); GND = ground connection (M0). The cell area (dashed rectangle) is approximately $(W + 1\,F) \cdot (4.5\,F)$, where W is the transistor width (in number of Fs) and F the technology feature size. (Adapted from Ref. [53]).

Further room may be obtained by using alternative solutions for ultra-scaled MOS transistors, such as ultra-thin body fully depleted on SOI or Double Gate (or FinFET). These devices are expected to have an improved driving current capability up to $2\,\mathrm{mA}\,\mu\mathrm{m}^{-1}$ (forecasted to be 2010) [73]. Long-term solutions may also involve the adoption of vertical MOS structures that take advantage of having only a single contact (and hence a reduced area on silicon) and an improved W (by about a factor of 3) with respect to planar solutions.

Figure 15.24 shows the projection of the PCM scaling trend. The PCM cell size will take full advantage of technology scaling, as no intrinsic limitations are expected to halt further scaling. On the other hand, both NOR and NAND scaling are expected to slow down due to scaling limitations (due to high program drain voltage for NOR and electrostatic field coupling for NAND). As a consequence, the PCM cell size will reach the NOR cell size at about the 45-nm node, and may even reach the NAND cell size at about the 32-nm node. Moreover, the memory is ideally suited to store more than two

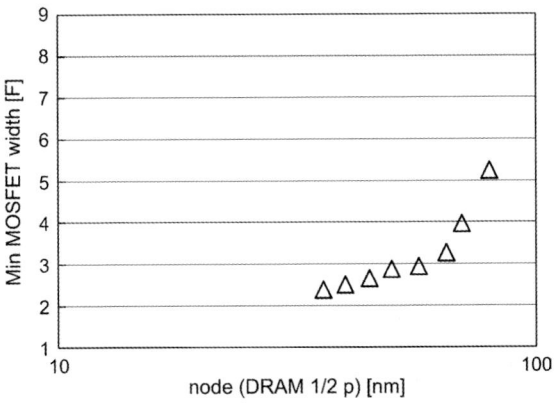

Figure 15.23 Predicted width W (in multiples of feature size F) of the PCM cell MOS select transistor required to drive the required reset current [73] as function of technology node (defined as DRAM half pitch).

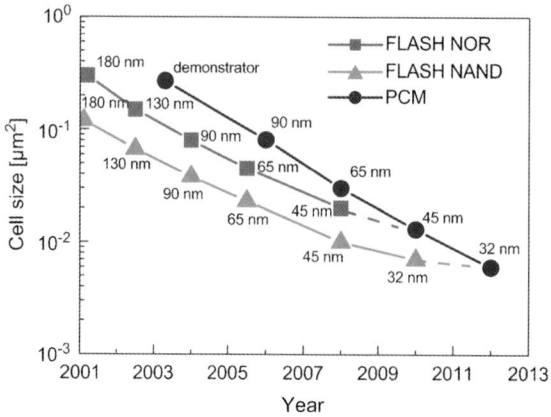

Figure 15.24 Scaling trends for NAND, NOR and PCM (with bipolar select transistor). The phase-change memory technology is expected to reach the same Flash-NAND size at the 32-nm technology feature size. (Figure from Ref. [19]).

levels per cell. This approach may become a viable option to further reduce the cost per bit of PCM devices, if reliable multi-level programming becomes feasible.

15.7.5
Scaling and Cell Performance: Figure of Merit for PCM

A low programming current can indeed be reached by tightly confining the current flow and the corresponding heat generation. However, this usually results in a high cell resistance, and there is an upper limit to the acceptable SET resistance value. The larger the resistance, the lower the cell current during the read operation, and the longer the time needed to charge the bit-line capacitance. In practice, for a read operation to occur within 50 ns the SET resistance should be kept at less than 50 kΩ (i.e., corresponding to a minimum read current in the SET state of a few µAs), but this constraint leads to a trade-off between programming current, and to the introduction of a new figure of merit, the product $R_{set} \cdot I_{melt}$, to compare different devices [19]. Figure 15.25 shows I_{melt} versus R_{set} as derived from published results, where the constant $R_{set} \cdot I_{melt}$ lines are highlighted by the dashed lines. Different cell architectures and material systems correspond to different $R_{set} \cdot I_{melt}$ values. The data clearly show the potential of PCM cells to reach programming currents of few tens of µAs.

15.7.6
Physical Limits of Scaling

To date, very few data are available regarding size effects on PCM behavior. However, from optical memory measurements, phase-change mechanisms seem scalable to at least 5 nm [74]. On the other hand, for both nanosized elements or films thinner than 20 nm, shifts in the crystallization behavior of GST (225) have been observed [75]. Yet, it is unclear if these are intrinsic effects or they are related to

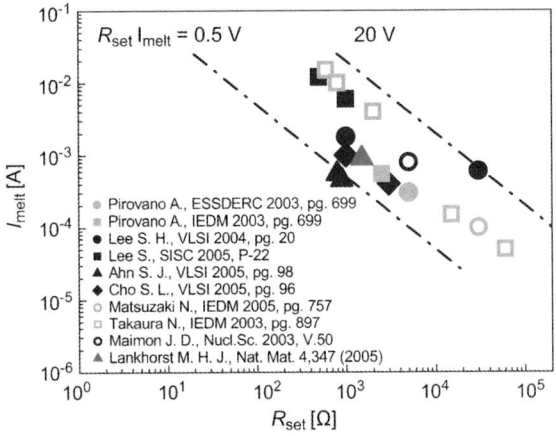

Figure 15.25 I_{melt} versus R_{set} as derived from published results. The constant $R_{set} \cdot I_{melt}$ lines are highlighted by the dashed lines. Different cell architectures and material systems correspond to different $R_{set} \cdot I_{melt}$ values, proposed as a figure of merit for PCM cells. The data clearly show the potential of PCM cells to reach programming currents of a few tens of μA. (Figure from Ref. [19]).

the preparation method (patterning damage and porosity of thinner films), or if they would inhibit phase switching. What is clear is that both the fabrication technology and behavior of nanosized PCM structures require further exploration in the near future.

15.8 Conclusions

During recent years, PCM have evolved from an interesting new concept to a viable memory technology, based on the use of improved and faster switching materials (<100 ns) and on an improved understanding of phase transformation and electronic switching processes in chalcogenide materials. PCM further shows excellent reliability properties, such as good data retention (10 years at 110 °C) and very high endurance (up to 10^{12} cycles, compared to $<10^6$ for FLASH), while the optimization of cell design has resulted in a drastic reduction of the program current (down to few hundred μA). In addition, integration technology has matured to a point where large demonstrator circuits (up to 256 Mb, in 100-nm technology) have already been built. Perhaps more importantly, the scaling potential of PCM has been assessed, and is expected to result in small cell sizes (in the range of 10 F^2 or smaller) for technologies of 45 nm and below, this being concomitant with a further reduction in program currents. For these reasons, PCM is expected eventually to replace the no-longer-scaling Flash technologies.

References

1 Kurata, H., Otsuga, K., Kotabe, A., Kajiyama, S., Osabe, T., Sasago, Y., Nerumi, S., Tosami, K., Kamohara, S. and Tsuchiya, O. (2006) The impact of random telegraph signals on the scaling of multilevel Flash memories. *IEEE Symp. VLSI Circuits, Tech. Digp*, pp. 140–141.

2 Shin, Y. (2005) Non-volatile memory technologies for beyond 2005. *IEEE Symp. VLSI Circuits, Tech. Digp*, pp. 156–159.

3 Dewald, J.F., Pearson, A.D., Northover, W.R. and Peck, W.F. (1962) *Journal of the Electrochemical Society*, **109**, 243.

4 Ovshinsky, S.R. (1968) Reversible electrical switching phenomena in disordered structures. *Physical Review Letters*, **21**, 1450–1453.

5 Ovshinsky, S.R. and Fritzsche, H. Amorphous semiconductors for switching, memory, and imaging applications. *IEEE Trans. Elec. Dev.*, Vol. ED-20, No. 2, February 1973, pp. 91–105.

6 Neale, R., Nelson, D. and Moore, G. (1970) Nonvolatile and reprogrammable, the read-mostly memory is here. *Electronics*, **43**, 56–60.

7 Maimon, J., Hunt, K., Rodgers, J., Burcin, L. and Knowles, K. Circuit demonstration of radiation hardened chalcogenide non-volatile memory. Proceedings of Aerospace Conference, Vol. 5, pp. 5_2373–5_2379.

8 Yamada, N., Ohno, E., Nishiuchi, K., Akahira, N. and Takao, M. (1991) Rapid-phase transitions of GeTe-Sb2Te3 pseudobinary amorphous thin films for an optical disk memory. *Journal of Applied Physiology*, **69**, 2849–2857.

9 Kolobov, A.V., Fons, P., Frenkel, A.I., Ankudinov, A.L., Tominaga, J. and Uruga, T. (2004) Understanding the phase-change mechanism of rewritable optical media. *Nature Materials*, **3**, 703–708.

10 Wicker, G. (1996) A comprehensive model of submicron chalcogenide switching devices, Ph. D. Dissertation, Wayne State University, Detroit, MI.

11 Wicker, G. (1999) Nonvolatile, high density, high performance phase change memory. in Proceedings SPIE Conference on Electronics and Structures for MEMS, Vol. 3891, SPIE, The International Society for Optical Engineering, Bellingham, WA, pp. 2–9.

12 Lai, S. and Lowrey, T. (2001) OUM-A 180 nm nonvolatile memory cell element technology for stand alone and embedded applications, *IEEE International Electron Devices Meeting Technical Digest*, pp. 803–806.

13 Pellizzer, F., Pirovano, A., Ottogalli, F., Magistretti, M., Scaravaggi, M., Zuliani, P., Tosi, M., Benvenuti, A., Besana, P., Cadeo, S., Marangon, T., Morandi, R., Piva, R., Spandre, A., Zonca, R., Modelli, A., Varesi, E., Lowrey, T., Lacaita, A., Casagrande, G., Cappelletti, P. and Bez, R., Novel μtrench phase-change memory cell for embedded and stand-alone non-volatile memory applications, 2004 Symposium on VLSI Technology Digest of Technical Papers, pp. 18–19.

14 Hwang, Y.N., Lee, S.H., Ahn, S.J., Lee, S.Y., Ryoo, K.C., Hong, H.S., Koo, H.C., Yeung, F., Oh, J.H., Kim, H.J., Jeong, W.C., Park, J.H., Horii, H., Ha, Y.H., Yi, J.H., Koh, G.H., Jeong, G.T., Jeong, H.S. and Kim, K. (2003) Writing current reduction for high-density phase-change RAM, *IEEE International Electron Devices Meeting Technical Digest*, pp. 893–896.

15 Pirovano, A., Redaelli, A., Pellizzer, F., Ottogalli, F., Ielmini, D., Lacaita, A.L. and Bez, R., Reliability study of phase-change nonvolatile memories. IEEE Transactions on Device and Materials Reliability, Vol. 4, No. 3, September 2004, pp. 422–427.

16 Kim, K. *et al.* (2005) Reliability investigations for manufacturable high density PRAM. *IRPS Tech. Dig.*, 157–162.

17 Lai, S. (2003) Current status of phase change memory and its future, *IEEE International Electron Devices Meeting Technical Digest*, pp. 255–258.

18 Pirovano, A., Lacaita, A.L., Benvenuti, A., Pellizzer, F., Hudgens, S. and Bez, R. (2003) Scaling analysis of phase-change memory technology. *IEDM Technical Digest*, 699–702.

19 Lacaita, A.L. Progress of phase-change non volatile memory devices, presented at European Phase Change Ovonic Science (Joint E*PCOS-IMST Workshop), Grenoble, France, May 29–31, 2006; http://www.epcos.org.

20 Lacaita, A.L., Redaelli, A., Ielmini, D., Pellizzer, F., Pirovano, A., Benvenuti, A. and Bez, R. (2004) Electrothermal and phase-change dynamics in chalcogenide-based memories, *IEEE International Electron Devices Meeting Technical Digest*, pp. 911–914.

21 Wuttig, M., Klein, M., Kalb, J., Lecner, D. and Spaepen, F., Ultrafast data storage with phase change media: from crystal structures to kinetics, Presented at the 5th European Phase Change Ovonic Science symposium (Joint E*PCOS-IMST Workshop), Grenoble, France, May 29–31, 2006; http://www.epcos.org.

22 Hudgens, S. and Johnson, B. (2004) Overview of phase-change chalcogenide nonvolatile memory technology. *MRS Bulletin*, **29**, 829–832.

23 Libera, M. and Chen, M. (1990) Multilayered thin-film materials for phase-change erasable storage. *MRS Bulletin*, **15**, 40–45.

24 Lankhorst, M.H.R., Ketelars, B.W.S.M.M. and Wolters, R.A.M. (2005) Low-cost and nanoscale non-volatile memory concept for future silicon chips. *Nature Materials*, **4**, 347–352.

25 Borg, H., Lankhorst, M., Meinders, E. and Leibbrandt, W. (2001) Phase-change media for high-density optical recording. Materials Research Society Symposium Proceedings, Materials Research Society, Vol. 674, V1.2.1–V1.2.10.

26 Miao, S.S., Shi, L.P., Zhao, R., Tan, P.K., Lim, K.G., Li, J.M. and Chong, T.C. Temperature dependence of phase change random access memory cell. Extended Abstracts, 2005 International Conference on Solid State Devices and Materials (SSDM), Kobe 2005, pp. 1052–1053.

27 Moffatt, W.G. *The Handbook of Binary Phase Diagrams*, Genum Publishing Companyp, p. 7/91.

28 Khulbe, P.K., Hurst, T., Horie, M. and Mansuripur, M. (2002) Crystallization behavior of Ge-doped eutectic Sb 70 Te 30 films in optical disks. *Applied Optics*, **41**, 6220–6229.

29 Yoon, S.-M., Lee, N.-Y., Ryu, S.-O., Choi, K.-J., Park, Y.-S., Lee, S.-Y., Yu, B.-C., Kang, M.-J., Choi, S.-Y. and Wuttig, M. Lower power and higher speed operation of phase-change memory device using antimony selenide (Sb_xSe_{1-x}), Extended Abstracts of the 2005 International Conference on Solid State Devices and Materials (SSDM), Kobe 2005, pp. 1050–1051.

30 Lee, H. and Kang, D.-H. Indium selenide based phase change memory, Extended Abstracts of the 2004 International Conference on Solid State Devices and Materials (SSDM), Tokyo, 2004, pp. 646–647.

31 Horii, H., Yi, J.H., Park, J.H., Ha, Y.H., Baek, I.G., Park, S.O., Hwang, Y.N., Lee, S.H., Kim, Y.T., Lee, K.H., Chung, U.-In. and Moon, J.T. A novel cell technology using N-dopes GeSbTe films for Phase change RAM, 2003 Symposium on VLSI Technology Digest of Technical Papers, pp. 177–178.

32 Matsuzaki, N., Kurotsuchi, K., Matsui, Y., Tonomura, O., Yamamoto, N., Fujisaki, Y., Kitai, N., Takemura, R., Osada, K., Hanzawa, S., Moriya, H., Iwasaki, T., Kawahara, T., Takaura, N., Terao, M., Matsuoka, M. andMoniwa, M. (2005) Oxygen-doped GeSbTe phase-change memory cells featuring 1.5-V/100-μA standard 0.13-μm CMOS operations, *IEEE International Electron Devices Meeting Technical Digest*, pp. 738–741.

33 Pirovano, A., Lacaita, A.L., Benvenuti, A., Pellizzer, F. and Bez, R. (2004) Electronic switching in phase-change memories.

IEEE Transactions on Electron Devices, **51**, 452–459.

34 Raoux, S., Rettner, C.T. and Jordon-Sweet, J.L., Crystallization behavior of phase change nanostructures. Presented at the European Phase Change Ovonic Science (EPCOS 2005) Symposium, 3–6 September, 2005, Cambridge, UK; http://www.epcos.org.

35 Kolobov, A.V., Fons, P., Tominaga, J., Frenkel, A.I., Ankudinov, A.L. and Uruga, T. (2005) Local structure of Ge-Sb-Te and its modification upon the phase transition. *Journal of Ovonic Research*, **1**, 21–24.

36 Welnic, W., Pamungkas, A., Detemple, R., Steimer, C., Bluegel, S. and Wuttig, M. (2006) Unraveling the interplay of local structure and physical properties in phase-change materials. *Nature Materials*, **5**, 56–62.

37 Kalb, J., Spaepen, F. and Wuttig, M. (2003) Calorimetric measurements of phase transformations in thin films of amorphous Te alloys used for optical data storage. *Journal of Applied Physiology*, **93**, 2389–2393.

38 Pellizzer, F., Benvenuti, A., Gleixner, B., Kim, Y., Johnson, B., Magistretti, M., Marangon, T., Pirovano, A., Bez, R. and Atwood, G., A 90 nm phase change memory technology for stand-alone non-volatile memory applications, 2006 Symposium on VLSI Technology Digest of Technical Papers, pp. 122–123.

39 Chen, H.S. (1980) Glassy metals. *Reports on Progress in Physics*, **43**, 353–432.

40 Wei, J. and Gan, F. (2003) *Thin Solid Films*, **441**, 292–297.

41 Peng, C., Cheng, L. and Mansuripur, M. (1997) Experimental and theoretical investigations of laser-induced crystallization and amorphization in phase-change optical recording media. *Journal of Applied Physiology*, **62**, 4183.

42 Kelton, K.F. (1991) Crystal nucleation in liquids and glasses. *Solid State Physics*, **45**, 75.

43 Gille, T., Goux, L., Lisoni, J., De Meyer, K. and Wouters, D.J. (2006) Impact of material crystallization characteristics on the switching behavior of the phase change memory cell. in Chalcogenide-Based Phase-Change Materials for Reconfigurable Electronics, Materials Research Society Symposium Proceedings 918E, (eds A.H. Edwards, P.C. Taylor, J. Maimon and A. Kolobov), Warrendale, PA, paper no. 0918-H06-02-G07-02.

44 Mott, N.F. and Davis, E.A. (1967) *Electronic processes in non-crystalline materials*, Clarendon Press, Oxford.

45 Popescu, C. (1975) The effect of local non-uniformities on thermal switching and high field behaviour of structures with chalcogenide glasses. *Solid-State Electron*, **18**, 671–681.

46 Owen, A.E., Robertson, J.M. and Main, C. (1979) The threshold characteristics of chalcogenide-glass memory switches. *Journal of Non-Crystalline Solids*, **32**, 29–52.

47 Adler, D., Henisch, H.K. and Mott, S.D. (1978) The mechanism of threshold switching in amorphous alloys. *Reviews of Modern Physics*, **50**, 209–220.

48 Adler, D., Shur, M.S., Silver, M. and Ovshinsky, S.R. (1980) Threshold switching in chalcogenide-glass thin films. *Journal of Applied Physiology*, **51**, 3289–3309.

49 Ielmini, D., Lacaita, A.L., Mantegazza, D., Pellizzer, F. and Priovano, A. (2005) Assessment of threshold switching dynamics in phase-change chalcogenide memories, *IEEE International Electron Devices Meeting Technical Digest*, pp. 877–880.

50 Pirovano, A., Lacaita, A.L., Pellizzer, F., Kostylev, S.A., Benvenuti, A. and Bez, R. (2004) Low-field amorphous state resistance and threshold voltage drift in chalcogenide materials. *IEEE Transactions on Electron Devices*, **51**, 714–719.

51 Kunghia, K., Shakoor, Z., Kasap, S.O. and Marshall, J.M. (2005) Density of localized electronic states in a-se from electron time-of-flight photocurrent measurements. *Journal of Applied Physiology*, **97**, 033706-1–033706-11.

52 Chen, Y., Chen, C.F., Chen, C.T., Yu, J.Y., Wu, S., Lung, S.L., Liu, R. and Lu, C. (2003) An access-transistor-free (0T/1R) non-volatile resistance random access memory (RRAM) using a novel threshold switching, self-rectifying chalcogenide device, *IEEE International Electron Devices Meeting Technical Digest*, pp. 905–908.

53 Kim, K., Jeong, G., Jeong, H. and Lee, S., Emerging memory technologies, Proceedings of the IEEE 2005 Custom Integrated Circuits Symposium, 18–21 September 2005, pp. 423–426.

54 Kim, Y.T. *et al.* (2004) (Samsung) Extended Abstracts of the 2004 International Conference on Solid State Devices and Materials, Tokyo D-3-2, pp. 244–245.

55 Yi, H., Ha, Y.H., Park, J.H., Kuh, B.J., Horii, H., Kim, Y.T., Park, S.O., Hwang, Y.N., Lee, S.H., Ahn, S.J., Lee, S.Y., Hong, J.S., Lee, K.H., Lee, N.I., Kang, H.K., Chung, U. and Moon, J.T. (2003) Novel cell structure of PRAM with thin metal layer inserted GeSbTe, *IEEE International Electron Devices Meeting Technical Digest*, pp. 901–904.

56 Alberici, S.G., Zonca, R. and Pashmakov, B. (2004) Ti diffusion in chalcogenides: a ToF-SIMS depth profile characterization approach. *Applied Surface Science*, **231/232**, 821–825.

57 Yoon, S.-M., Lee, N.-Y., Ryu, S.-O., Park, Y.-S., Lee, S.-Y., Choi, K.-J. and Yu, B.-G. (2005) Etching characteristics of $Ge_2Sb_2Te_5$ using high-density helicon plasma for the nonvolatile phase-change memory applications. *Japanese Journal of Applied Physics*, **44**, L869–L872.

58 Pellizzer, F., Spandre, A., Alba, S. and Pirovano, A. (2004) Analysis of plasma damage on phase change memory cells, 2004 IEEE International Conference on Integrated Circuit Design and Technology, p. 227.

59 Takaura, N., Terao, M., Kurotsuchi, K., Yamauchi, T., Tonomura, O., Hanaoka, Y., Takemura, R., Osada, K., Kawahara, T. and Matsuoka, H. (2003) A GeSbTe phase-change memory cell featuring a tungsten heater electrode for low-power, highly stable, and short-read-cycle operations, *IEEE International Electron Devices Meeting Technical Digest*, pp. 897–900.

60 Song, Y.J., Ryoo, K.C., Hwang, Y.N., Jeong, C.W., Lim, D.W., Park, S.S., Kim, J.I., Kim, J.H., Lee, S.Y., Kong, J.H., Ahn, S.J., Lee, S.H., Park, J.H., Oh, J.H., Oh, Y.T., Kim, J.S., Shin, J.M., Park, J.H., Fai, Y., Koh, G.H., Jeong, G.T., Kim, R.H., Lim, H.S., Park, I.S., Jeong, H.S. and Kim, Kinam, Highly reliable 256 Mb PRAM with advanced ring contact technology and novel encapsulating technology, 2006 Symposium on VLSI Technology Digest of Technical Papers, pp. 118–119.

61 Ha, Y.H., Yi, J.H., Horii, H., Park, J.H., Joo, S.H., Park, S.O., Chung, U.-In. and Moon, J.T., An edge contact type cell for phase change RAM featuring very low power consumption, 2003 Symposium on VLSI Technology Digest of Technical Papers, pp. 175–176.

62 Pirovano, A., Pellizzer, F., Redaelli, A., Tortorelli, I., Varesi, E., Ottogalli, F., Tosi, M., Besana, P., Cecchini, R., Piva, R., Magistretti, M., Scaravaggi, M., Mazzone, G., Petruzza, P., Bedeschi, F., Marangon, T., Modelli, A., Ielmini, D., Lacaita, A.L. and Bez, R. (2005) Trench phase-change memory cell engineering and optimization, Proceedings ESSDERC 2005, pp. 313–316.

63 Ahn, S.J., Hwang, Y.N., Song, Y.J., Lee, S.H., Lee, S.Y., Park, J.H., Jeong, C.W., Ryoo, K.C., Shin, J.M., Park, J.H., Fai, Y., Oh, J.H., Koh, G.H., Jeong, G.T., Joo, S.H., Choi, S.H., Son, Y.H., Shin, J.C., Kim, Y.T., Jeong, H.S. and Kim, K., Highly reliable 50 nm contact cell technology for 256 Mb PRAM, 2005 Symposium on VLSI Technology Digest of Technical Papers, pp. 98–99.

64 Haring-Bolívar, P., Merget, F., Kim, D.-H., Hadam, B. and Kurz, H. Lateral design for phase change random access memory cells with low-current consumption. Presented at the 3rd European Phase Change Ovonic Science Symposium (EPCOS 2004),

Balzers, Principality of Liechtenstein, September 4–7, 2004; http://www.epcos.org.

65 Happ, T.D., Breitwisch, M., Schrott, A., Philipp, J.B., Lee, M.H., Cheek, R., Nirschl, T., Lamorey, M., Ho, C.H., Chen, S.H., Chen, C.F., Joseph, E., Zaidi, S., Burr, G.W., Yee, B., Chen, Y.C., Raoux, S., Lung, H.L., Bergmann, R. and Lam, C., Novel one-mask self-heating pillar phase change memory, 2006 Symposium on VLSI Technology Digest of Technical Papers, pp. 120–121.

66 Kang, D.-H., Ahn, D.-H., Kwon, M.-H., Kwon, H.-S., Kim, K.-B., Seok Lee, K. and Cheong, B.-ki. (2003) Lower voltage operation of a phase change memory device with a highly resistive TiON layer. *Japanese Journal of Applied Physics*, **42**, 2382–2386.

67 Bedeschi, F., Resta, C., Khouri, O., Buda, E., Costa, L., Ferraro, M., Pellizzer, F., Ottogalli, F., Pirovano, A., Tosi, M., Bez, R., Gastaldi, R. and Casagrande, G., An 8 Mb demonstrator for high-density 1.8 V phase-change memories. 2004 Symposium on VLSI Circuits Digest of Technical Papers, pp. 442–445.

68 Friedrich, I., Weidenhof, V., Njoroge, W., Franz, P. and Wuttig, M. (2000) Structural transformations of $Ge_2Sb_2Te_5$ films studied by electrical resistance measurements. *Journal of Applied Physiology*, **87**, 4130.

69 Redaelli, A., Ielmini, D., Lacaita, A.L., Pellizzer, F., Pirovano, A. and Bez, R. (2005) Impact of crystallization statistics on data retention for phase change memories, *IEEE International Electron Devices Meeting Technical Digest*, pp. 742–745.

70 Ottogalli, F., Priovano, A., Pellizzer, F., Tosi, M., Zuliani, P., Bonetalli, P. and Bez, R., Phase-change memory technology for embedded applications. Proceedings, 34th European Solid-State Device Research Conference (ESSDERC 2004), Leuven, Belgium, 21–23 September 2004, pp. 293–296.

71 Osada, K., Kawahara, T., Takemura, R., Kitai, N., Takaura, N., Matsuzaki, N., Kurotsuchi, K., Moriya, H. and Moniwa, M., Phase change RAM operated with 1.5-V CMOS as low cost embedded memory, Proceedings, IEEE 2005 Custom Integrated Circuits Symposium, 18–21 September 2005, pp. 431–434.

72 Cho, S.L. *et al.* (2005) Symposium on VLSI Circuits Digest of Technical Papers, p. 96.

73 International Technology Roadmap for Semiconductors (ITRS), 2005 edition, Process Integration, Devices, and Structures, accessible through: http://www.itrs.net.

74 Wright, D., Aziz, M.M., Armand, M., Senkander, S. and Yu, W. Can we reach Tbit/sq.in. storage densities with phase-change media? Presented at the 3rd European Phase Change Ovonic Science symposium (EPCOS 2004), Balzers, Principality of Liechtenstein, September 4–7, 2004; http://www.epcos.org.

75 Raoux, S., Rettner, C.T., Jordan-Sweet, J.L., Deline, V.R., Philipp, J.B. and Lung, H.-L. Scaling properties of phase change nanostructures and thin films. Presented at the 5th European Phase Change Ovonic Science symposium (EPCOS 2006), May 29–31, 2006 Grenoble, France; http://www.epcos.org.

16
Memory Devices Based on Mass Transport in Solid Electrolytes
Michael N. Kozicki and Maria Mitkova

16.1
Introduction

As standard semiconductor nanoelectronic devices approach their scaling limits (see Chapters 00 on "FET", 11 on "Flash", and 12 on "DRAM"), concepts beyond traditional purely charge-based functionality offer additional opportunities. One such paradigm goes by the name "nanoionics". Whereas nanoelectronics involves the movement of electrons within their nanostructured settings, nanoionics concerns materials and devices that rely on ion transport and chemical change at the nanoscale. Rising interest in nanoionics has been fuelled by the wide range of demonstrated and potential applications so that the field has been equated in significance by some with nanoelectronics [1].

It is impossible to discuss nanoionics without introducing the basic principles of electrochemistry. As the name suggests, electrochemistry deals with the relationship between electricity and chemical change. In many respects, batteries are the prime example of the application of electrochemical principles; the movement of ions and the change in their oxidation state within the cell is used to release electrical energy over time. However, since ions not only carry charge but also have a significant mass, ion transport can be seen as a means to *move material* in a controlled manner. For example, a metal atom that becomes *oxidized* at one location can be moved as a cation through an electrolyte by an electric field. On receiving an electron at another location, the displaced ion is *reduced* and becomes a neutral metal atom again. In this situation, *the net change in the system is the redistribution of mass* – material is removed from one location and deposited at another using energy from an external power source. The world of electronics has benefited from such "deposition electrochemistry" for many decades. Electroplating, in which metal ions in a liquid solution are reduced to create a uniform metal film, is used in printed circuit boards and packages, and in the processes used to make copper interconnect within integrated circuits. In such cases, physical dimensions, such as electrode spacing, are typically quite large and the electric fields relatively small. The term nanoionics is applied when electrochemical effects occur in materials and devices with interfaces, for example electrodes or electrochemically different material phases, that are closely spaced – perhaps by a few tens of nanometers, or less. In this size regime, the functionality of ionic systems is quite different from the macro-scale versions, but in a highly useful manner. For example, internal electric fields and ion mobilities are relatively high in nanoionic structures and this, combined with the short length scales, results in fast response times. In addition, whereas deposition electrochemistry and many batteries use liquids as ion transport media, nanoionics can take advantage of the fact that a variety of solid materials are excellent electrolytes, largely due to effects which dominate at the nanoscale. This allows nanoionic devices based on *solid electrolytes* to be more readily fabricated using techniques common to the integrated circuit industry, and also facilitates the marriage of such devices with mainstream integrated electronics. Indeed, *in-situ* changes may be controlled by the integrated electronics, leading to electronic-ionic system-on-chip (SoC) hybrids.

In this chapter, we describe the basic electrochemistry, materials science and potential applications in information technology of mass-transport devices based on solid electrolytes and nanoionic principles. The electrodeposition of even nanoscale quantities of a noble metal such as silver can produce localized *persistent* – but *reversible* – changes to macroscopic physical or chemical characteristics; such changes can be used to control behavior in applications that go well beyond purely electronic systems. Of course, electrical resistance will change radically when a low resistivity electrodeposit (e.g., in the tens of $\mu\Omega \cdot cm$ or lower) is deposited on a solid electrolyte surface which has a resistivity many orders of magnitude higher. This resistance change effect has a variety of applications in memory and logic. Here, emphasis will be placed on low-energy, non-volatile memory devices which utilize such resistance changes to store information.

16.2
Solid Electrolytes

16.2.1
Transport in Solid Electrolytes

The origins of solid-state electrochemistry can be traced back to Michael Faraday, who performed the first electrochemical experiments with Ag_2S and discovered that this material was a good ion conductor [2]. Subsequently, greater emphasis was placed on liquid electrolytes and their use in plating systems and battery cells, until the 1960s and 1970s when a significant rise in interest was noted in solid-state electrochemistry. This renewed attention was spurred in part by the development of novel batteries which had a particularly high power-to-weight ratio due to the use of solid electrolyte, mainly beta-alumina, which is an excellent conductor of sodium ions [3]. Even though these solid materials were clearly different from their liquid counterparts, many of the well-known principles developed in the field of liquid electrochemistry were found to be applicable to the solid-state systems. One major difference between most solid and liquid ion conductors is that in solids, the moving ions are of only one polarity (cations or anions) and the opposite polarity species is fixed in the supporting medium. This has a profound effect on the types of structure that can be used for mass transport (this subject will be outlined later in the chapter). The solid electrolyte family currently includes crystalline and amorphous inorganic solids, as well as ionically conducting polymers. In general, the best solid electrolytes have high ionic but low electronic conductivity, chemical and physical compatibility with the electrodes used, thermodynamic stability over a wide temperature range, and the ability to be processed to form continuous mechanically stable thin film structures.

The mobile ions in a solid electrolyte sit in potential wells separated by low potential barriers, typically in the order of a few tenths of an electron-volt (eV), or less. The ions possess kinetic energy, governed by Boltzmann statistics, and so at finite temperature will constantly try (with around 10^{12} attempts per second) to leave their low-energy sites to occupy energetically similar sites within the structure. Thermal diffusion will result from this kinetic energy, driving ions down any existing concentration gradient until a uniform concentration is achieved. Subsequent movement of the ions produces no net flux in any particular direction. The application of an electric field to the electrolyte effectively reduces the height of the barriers along the direction of the field, and this increases the probability that an ion will hop from its current potential well to a lower energy site (see Figure 16.1) [4]. An ion current therefore results, driven by the field. It should be noted that, unlike electrons, ions in a solid are constrained to move through a confining network of narrow channels. These pathways may be a natural consequence of order in the material, as in the case of the interstitial channels present along certain directions in crystalline materials, or they may be a result of long-range disorder, as in amorphous (glassy) and/or nanoscopically porous materials. Glassy electrolytes, typically metal oxides, sulfides or selenides, are of particular interest as they can contain a wider variety of routes for cation transport than purely crystalline materials. This is a major reason for the interest in these materials – and for Group VI glasses in particular – and why they feature heavily in this chapter.

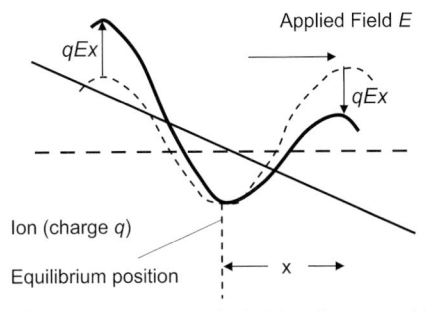

Figure 16.1 Change in the height of a potential barrier between shallow wells due to the application of an electric field. The effective barrier height is reduced by qEx, where E is the electric field and x is the barrier width.

Considering the above dependence of ion hopping on ion energy and barrier height, it should be no surprise that the expression for ion conductivity in the electrolyte is

$$\sigma = \sigma_0 \exp(-E_a/kT) \tag{16.1}$$

where k is Boltzmann's constant, T is absolute temperature, E_a is the activation energy for conduction, and the pre-exponential term σ_0 depends on several factors, including the mobile ion concentration (e.g., σ_0 is in the order of $10^4\,\Omega^{-1}\,\text{cm}^{-1}$ for >10 atom% Ag in Ge-S electrolytes [5]). As is evident from Equation 16.1, the activation energy is a major factor in determining ion conductivity. It is directly related to the structure of the host and to the existence of the conduction pathways, both of which govern the effective barrier height. Obviously, the smaller E_a is, the higher the conductivity and the better the electrolyte is. Since E_a is around 0.2–0.3 eV in Ag-saturated Ge-Se and Ge-S electrolytes, the above values lead to ion conductivities in the range 10^{-2}–$10^{-4}\,\text{S}\,\text{cm}^{-1}$. Just as with electron and hole conduction, we may also define ion conductivity as

$$\sigma = n_i q \mu \tag{16.2}$$

where n_i is the number of mobile ions per unit volume, q is the ionic charge (1.6×10^{-19} C for singly charged ions), and μ is the ion mobility. Interestingly, it is thought that n_i in high ion concentration solid electrolytes such as the Ag-Ge-S ternaries is fairly constant – around 10^{19} ions cm^{-3} [6]. This means that the ion mobilities in the solid electrolytes that are of interest to us are in the order of 10^{-2}–$10^{-4}\,\text{cm}^2\,\text{V}^{-1}\,\text{s}^{-1}$.

16.2.2
Major Inorganic Solid Electrolytes

As mentioned above, a variety of materials can act as solid electrolytes. Anion (oxide) conductors exist, such as ZrO_2, layered La_2NiO_4/La_2CuO_4 [7], or $Bi_{10}V_4MeO_{26}$, where Me is a divalent metal such as Co, Ni, Cu, or Zn [8]. However, for mass-transport devices there is a greater interest in electrolytes that conduct metallic cations as these can be used to form solid metal electrodeposits. In general, the smaller an ion is, the more mobile it should be as it will be able to slip more easily through the pathways in the solid electrolyte. This should be especially true for small-ionic radius elements such as the alkali metals (Li, Na, K). For example, Na^+ has been successfully used in beta-alumina and, to a lesser extent, in non-stoichiometric zirconophosphosilicate [9] to produce good solid ion conductors. The high conductivity in the beta-alumina compounds is a consequence of the structure which has open conduction pathways and a large number of partially occupied sites where cations can reside. Of course, Li^+ conductors in general are of great interest because of their use in high-voltage/high-power density lithium ion batteries, but highly stable Li^+ electrolytes are not easy to produce and there are not many examples of lithium solid electrolyte batteries (Li/LiI/I_2 is one of the few commercially available cells). Of course, the high chemical reactivity of these mobile elements makes them unsuitable for most mass transport/

solid electrolyte device applications, and so more stable alternatives must be considered.

The most widely studied solid ion conductors are those which contain silver. These tend to be less difficult to make than alkali metal ion electrolytes, and they have many desirable characteristics, including high ion mobility. Silver is especially appropriate for mass transport applications due to its nobility and ease of both reduction and oxidation. The crystalline Ag halides, principally AgI, and silver *chalcogenides* (e.g., Ag_2S, Ag_2Se, and Ag_2Te) are of particular interest as solid electrolytes. The phases of these materials that are stable at low temperature are semiconductors with moderate to low ion conductivity, but the high-temperature polymorphs (e.g., the cubic phase of Ag_2Se that is stable above 133 °C) are extremely good ion conductors [10]. The effect of this phase transition on the conductivity of AgI is shown in Figure 16.2. The Ag halides and chalcogenides possess a *bcc* structure formed by the covalently bonded halide or chalcogen atoms. An octahedral sublattice which can be occupied by Ag^+ in a multitude of ways is shown in Figure 16.3. The number of the octahedral states is typically much higher than the number of the available Ag ions, and this ensures that there are always non-occupied sites for ions to move into. This abundance of empty sites, in conjunction with the low potential barrier, results in the *superionic* nature of the materials. One other factor contributing to the high conductivity of these electrolytes is the low coordination that Ag^+ has to the immobile chalcogenide/halide sublattice (typically 2–3). This low coordination most likely plays a key role in reducing the activation energy for conduction.

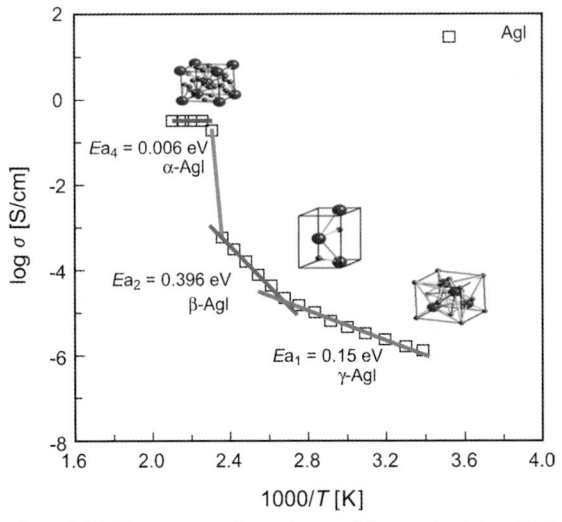

Figure 16.2 Temperature dependence of the conductivity of AgI related to its polymorph structure (Ref. [10]). The polymorph that is stable at high-temperature (denoted by α-AgI on the plot) has the highest conductivity and lowest activation energy compared with the lower temperature (β- and γ-AgI) polymorphs.

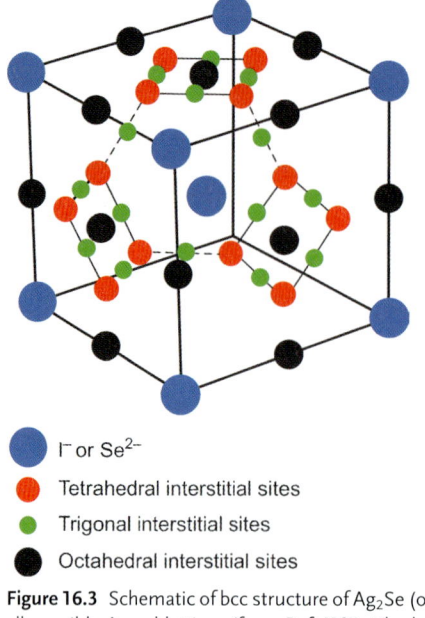

- I⁻ or Se²⁻ — (blue)
- Tetrahedral interstitial sites (red)
- Trigonal interstitial sites (green)
- Octahedral interstitial sites (black)

Figure 16.3 Schematic of bcc structure of Ag_2Se (or AgI) showing all possible Ag sublattices (from Ref. [10]). The large number of sites that can be occupied by cations such as Ag^+ lead to high ion mobility.

In practice, even deviations from stoichiometry, δ, as low as 1 part in 10 000 in some cases, lead to the existence of both mobile ions and conduction electrons. To illustrate this, in $\alpha Ag_{2+\delta}Se$, silver atoms are converted into silver ions and free electrons by

$$Ag_{2+\delta}Se \rightarrow (2+\delta)Ag^+ + \delta e^- + Se^{2-} \tag{16.3}$$

Both charge carriers (Ag^+ and e^-) contribute to the total conductivity, so that the material may be regarded as a *mixed conductor* [11]. Analogous effects are expected to occur in Cu-containing electrolytes (e.g., Cu_2S), although the situation will be more complex as participation of the electrons from the Cu d-orbital will result in a variety of bonding configurations. Extensive information on transport in superionic conductors is provided in a review [12]. The main issue with these binary materials is that, whereas they have been widely studied as superionic conductors, it is only their high-temperature phases that are of use in this respect, and this leads to severe practical limitations for electronic device applications.

16.2.3
Chalcogenide Glasses as Electrolytes

A chalcogenide glass is one that contains a large number of group VI or "chalcogen" atoms (S, Se, Te, and O, although oxide glasses are often treated separately from the others in the literature) [13]. These glasses have an astonishing range of physical characteristics and as such have found a multitude of uses. They lend themselves to a variety of processing techniques, including physical vapor deposition (evaporation and sputtering), chemical vapor deposition, spin casting, as well as melt-quenching. Stable binary glasses typically involve a Group IVB or Group VB atom, such as Ge–Se or As–S, with a wide range of atomic ratios possible. The bandgap of the Group VIB glasses rises from around 1–3 eV for the tellurides, selenides and sulfides, to 5–10 eV for the oxides. The tellurides exhibit the most metallic character in their bonding, and are the "weakest" glasses as they can crystallize very readily (hence their use in so-called phase change technologies as used in re-writable CDs and DVDs), and the others exhibit an increasing glass transition temperature on moving further up the Periodic Table column, with oxides having the highest thermal stability. The non-oxide glasses usually are more rigid than organic polymers but more flexible than a typical oxide glass, and other physical properties follow the same trend. This structural flexibility of these materials offers the possibility of the formation of voids through which the ions can readily move from one equilibrium position to another and, as will be seen later, allows the formation of electrodeposits *within* the electrolyte.

The addition of Group IB elements such as Ag or Cu transforms the chalcogenide glass into an electrolyte as these metals form mobile cations within the material. The ions are associated with the non-bridging chalcogen atoms, but the bonds formed are relatively long −0.27 nm in Ag-Ge-Se and 0.25 nm in Ag-Ge-S ternaries [14]. As with any coulombic attraction, the coulombic energy is proportional to the inverse of the cation–anion distance, so long bonds lead to reduced attractive forces between the charged species. The Ge-chalcogenide glasses are therefore among the electrolytes with the lowest coulombic energies [14]. The slightly shorter Ag^+–S^- bond length leads to a higher coulombic attraction, which is a factor contributing to the observed lower mobility of Ag in germanium sulfides versus selenides of the same stoichiometry. Thermal vibrations will allow partial dissociation, which results in a two-step process of defect formation followed by ion migration. The activation energy for this process depends heavily on the distance between the hopping cation and the anion located at the next nearest neighbor, as well as the height of the intervening barrier. (A discussion of the relationship between coulombic and activation energies is provided in Ref. [14] but, in addition to having low coulombic energies, the Ge-chalcogenides also have relatively low activation energies for ion transport.) In this respect, the existence of channels due to the structure of the electrolyte is critical in the ion transport process. As an example of this effect, the Ag^+ conductivity in glassy $AgAsS_2$ is 100-fold larger than that in the crystalline counterpart due to the more "open" structure of the non-crystalline material [15].

The conductivity and activation energy for ion conduction of the ternary glasses is a strong function of the mobile ion concentration. For example, in the Ag concentration range between 0.01 and 3 atom%, the room temperature conductivity of Ag-Ge-S glass changes from 10^{-14} to about $5 \times 10^{-10} \, \Omega^{-1} \, cm^{-1}$, accompanied by a decline in activation energy from 0.9 to 0.65 eV. However, above a small atomic percent, both conductivity and activation energy change more rapidly as a function of Ag concentration (see Figures 16.4a and b, respectively [16]). This change in the slopes of the conductivity and activation energy curves with Ag content in both Ag-Ge-S and

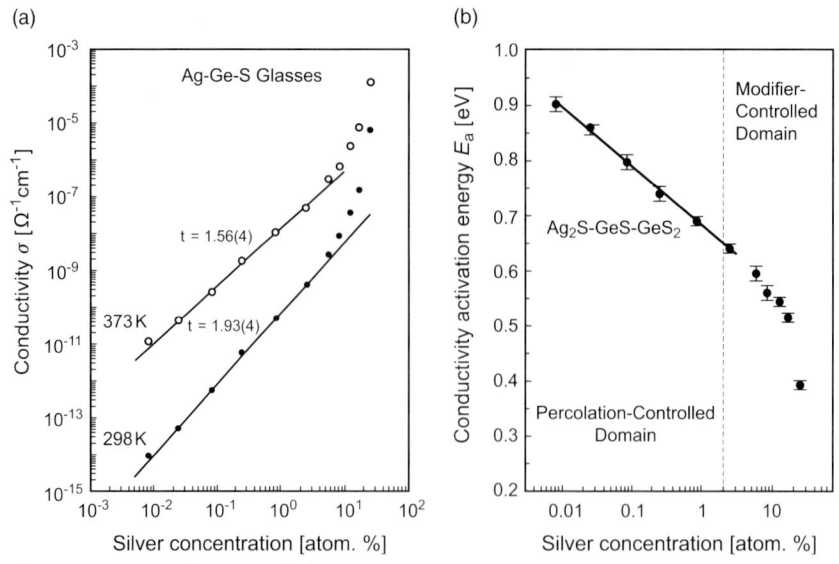

Figure 16.4 (a) Conductivity and (b) activation energy as a function of Ag concentration in Ag-Ge-S ternaries. (From Ref. [16].).

Ag-Ge-Se is a result of a transformation of the ternary material itself caused by the presence of so much silver.

16.2.4
The Nanostructure of Ternary Electrolytes

The transformation that occurs in ternary electrolytes at over a small atomic percent of metal is, by no means, subtle. Indeed, the material undergoes considerable changes in its nanostructure that have a profound effect on its macroscopic characteristics. These changes are a result of *phase separation* caused by the reaction of silver with the available chalcogen in the host to form Ag_2Se in Ag-Ge-Se and Ag_2S in Ag-Ge-S ternaries. For example, if it is assumed that the Ag has a mean coordination of 3, the composition of ternary Ag-Ge-Se glasses may be represented as

$$(Ge_xSe_{1-x})_{1-y}Ag_y = (3y/2)(Ag_2Se) + (1 - 3y/2)(Ge_tSe_{1-t}) \qquad (16.4)$$

where t is the amount of Ge in the Ge-Se backbone $= x(1 - y)/(1 - 3y/2)$ [17]. For a Se-rich glass such as $Ge_{0.30}Se_{0.70}$, $x = 0.30$, and $y = 0.333$ at saturation in bulk glass; hence, $t = 0.40$. This means that the material consists of Ag_2Se and $Ge_{0.40}Se_{0.60}$ (Ge_2Se_3) in the combination

$$16.7\, Ag_2Se + 10\, Ge_2Se_3 = Ag_{0.33}Ge_{0.20}Se_{0.47} \qquad (16.5)$$

This electrolyte has a Ag_2Se molar fraction of 0.63 (16.7/26.7) and a Ag concentration of 33 atom% [18]. It has been determined that the dissolution of Ag into a

Se-rich base glass produces a ternary that is a combination of *separate* dispersed crystalline Ag_2Se and continuous glassy Ge-rich phases [19]. The spacing, s, between the Ag_2Se phase regions (and therefore the thickness of Ge_2Se_3 material between them) can be estimated by assuming that the crystalline regions are spherical and uniform in size and dispersion, so that

$$s = d_c(F_v^{-1/3} - 1) \tag{16.6}$$

where d_c is the average measured diameter of the crystalline Ag-rich phase and F_v is the volume fraction of this phase [20]. The volume fraction in the case of Ag_2Se in $Ag_{0.33}Ge_{0.20}Se_{0.47}$ is 0.57 (for a molar fraction of 0.63), so the average spacing between the Ag-rich regions is approximately 0.2 times their diameter. The average diameter of the Ag_2Se crystallites in Ag-diffused $Ge_{0.30}Se_{0.70}$ thin films was determined, using X-ray diffraction (XRD) techniques, to be 7.5 nm [20], which means that by Equation 16.6, they should be separated by approximately 1.5 nm of glassy Ge-rich material. This general structure has been confirmed using high-resolution transmission electron microscopy (TEM). XRD analysis was also performed on a sulfide-based ternary thin film with similar stoichiometry, $Ag_{0.31}Ge_{0.21}S_{0.48}$, with much the same results. In this case, the Ag_2S crystallites are in the order of 6.0 nm in diameter [21]. Even though the detected Ag-rich phases mainly correspond to the room-temperature polymorphs, which are not particularly good ion conductors at room temperature, the ternary is superionic at room temperature. This is not surprising as defects, interfaces, and surfaces play a considerable role in ion transport and the large surface-to-volume ratio of the crystallites within the ternary is likely to greatly enhance ion transport. In addition, it has been noted that the Ag_2Se phases that form following the solid-state diffusion of Ag into Ge-Se may be "distorted" by the effective pressure of the medium to produce high ion mobility phases [19]. The nano-inhomogeneous ternary is ideal for devices such as resistance change memory cells, as the relatively high resistivity leads to a high off-resistance in small diameter devices, although the availability of mobile ions via the dispersed Ag-rich phases means that the effective ion mobility is high.

The addition of Ag (or Cu) to the chalcogenide base glass can be achieved by diffusing the mobile metal from a thin surface film via *photodissolution*. This process utilizes light energy greater than the optical gap of the chalcogenide glass to create charged defects near the interface between the reacted and unreacted chalcogenide layers [22]. The holes created are trapped by the metal, while the electrons move into the chalcogenide film. The electric field formed by the negatively charged chalcogen atoms and positively charged metal ions is sufficient to allow the ions to overcome the energy barrier at the interface, and so the metal moves into the chalcogenide [23]. Prior to introduction of the metal, the glass consists of GeS_4 ($GeSe_4$) tetrahedra and, in the case of chalcogen-rich material, S (Se) chains. The introduced metal will readily react with the chain chalcogen and some of the tetrahedral material to form the ternary. This Ag–chalcogen reaction, which essentially nucleates on the chalcogen-rich regions within the base glass, results in the nanoscale phase-separated ternary described above [24, 25].

16.3
Electrochemistry and Mass Transport

16.3.1
Electrochemical Cells for Mass Transport

In order to move mass, it is clear that an ion current must be generated. Regardless of ion mobility in the electrolyte, a sustainable ion current will only flow if there is a source of ions at one point and a sink of ions at another; otherwise, the movement of ions away from their oppositely charged fixed partners would create an internal field (polarization) which would prevent current flow. The process of *electrodeposition*, in which cations in the electrolyte are reduced by electrons from a negative electrode (cathode), is essentially an ion sink as ions are removed from the electrolyte to become atoms. However, in the absence of an ion source, the reduction of ions at the cathode will occur at the expense of the electrolyte. The concentration of ions in the solid electrolyte will therefore decrease during electrodeposition until the electrode potential equals the applied potential and reduction will cease. Further reduction requires greater applied voltage (governed by the Nernst equation), so that the deposition process is effectively self-limiting for a moderate applied potential. It should be noted also that a depleted electrolyte could allow the subsequent thermal dissolution of an electrodeposit, which would not occur if the glass was maintained at the chemical saturation point. This has important consequences for the stability of any electrodeposit formed. It is therefore necessary to have an *oxidizable* positive electrode (anode) – one which can supply ions into the electrolyte to maintain ion concentration and overall charge neutrality. In the case of a silver ion-containing electrolyte, this oxidizable anode is merely silver or a compound or alloy containing free silver. So, the most basic mass-transport device consists of a solid electrolyte between an electron-supplying cathode and an oxidizable anode (see Figure 16.5). These devices can have both electrodes in a coplanar configuration (as in Figure 16.5a), or on opposite faces of the electrolyte (Figure 16.5b).

In such a device, the anode will oxidize when a bias is applied if the oxidation potential of the metal is greater than that of the solution. Under steady-state conditions, as current flows in the cell, the metal ions will be reduced at the cathode.

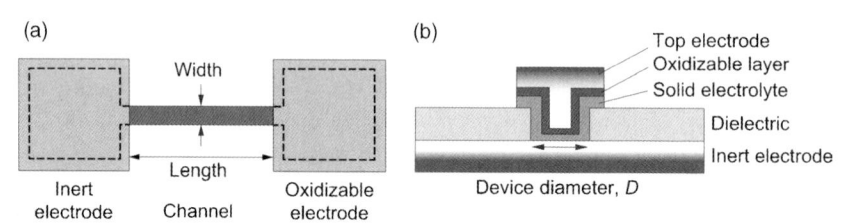

Figure 16.5 Schematic descriptions of two mass transport devices. (a) Coplanar structure that has the two electrodes on the same surface of a solid electrolyte layer. (b) Vertical structure with the solid electrolyte sandwiched between an inert electrode and an oxidizable layer which is covered by the top electrode.

For the case of silver, the reactions are:

$$\text{Anode}: \text{Ag} \rightarrow \text{Ag}^+ + e^- \tag{16.7}$$

$$\text{Cathode}: \text{Ag}^+ + e^- \rightarrow \text{Ag} \tag{16.8}$$

with the electrons being supplied by the external power source. The deposition of Ag metal at the cathode and partial dissolution of the Ag at the anode indicates that device operation is analogous to the reduction–oxidation electrolysis of metal from an aqueous solution and much the same rules apply, except that in this case the anions are fixed. When a bias is applied across the electrodes, silver ions migrate by the coordinated hopping mechanism (as described above) towards the cathode, under the driving force of the applied field and the concentration gradient. At the boundary layer between the electrolyte and the electrodes, a potential difference exists due to the transfer of charge and change of state associated with the electrode reactions. This potential difference leads to polarization in the region close to the phase boundary, known as the *double layer* [26]. The inner part of the double layer, consisting of ions absorbed on the electrode, is referred to as a *Helmholtz layer*, while the outer part, which extends into the electrolyte and is known to have a steep concentration gradient (over a few tens of nanometers in these systems), is called the *diffuse layer*. Electrically, the double layer very much resembles a charged capacitor, with a capacitance in the order of 10^{-14} F μm^{-2} and resistance around 10^{10} Ω μm^2 for a typical solid electrolyte under small applied bias [27]. An important consequence of the electric double layer is that, for the reduction–oxidation reaction to proceed, the applied potential must overcome the potential associated with the double layer. This means that no ion current will flow and no sustained electrodeposition will occur until the *concentration overpotential* is exceeded. Below this threshold voltage, the small observed steady-state current is essentially electron leakage by tunneling through the narrow double layer. Above the threshold, the ion current flows and the ions are reduced and join the cathode, effectively becoming part of its structure, both mechanically and electrically. The nature of the electrodeposits will be discussed in greater detail later in the chapter.

The intrinsic threshold is typically in the order of a few hundred millivolts. As the overpotential is governed by the ease of transfer of electrons from the cathode to the ions in the electrolyte, its precise value depends on factors such as the barrier height between the cathode material (including surface/interface states) and the electrolyte, and the bandgap/dielectric constant of the electrolyte. For example, the threshold voltage of a Ni/Ag-Ge-Se/Ag structure is in the order of 0.18 V, whereas the threshold of W/Ag-Ge-Se/Ag is around 0.25 V. Switching to a larger bandgap material, the threshold of a W/Ag-Ge-S/Ag device becomes closer to 0.45 V. The threshold has an Arrhenius dependence on temperature, with an activation energy that is in the order of a few tenths of an electron volt or less, which means that for the W/Ag-Ge-S/Ag structure, the threshold is still 0.25 V even at an operating temperature of 135 °C. Once a silver electrodeposit has formed on the cathode, the Ag metal becomes the new cathode and the threshold for further deposition of Ag is much less, typically less than half the original threshold. This reduced threshold for electrodeposition following initiation has a profound effect on device operation, and is discussed in more detail below. Of course, long structures that have a high series resistance will require higher voltages to initiate electrodeposition, as most of the applied voltage will be dropped across the electrolyte. For example, the polarization-resistance of a 10 μm^2 electrode will be $10^9\,\Omega$, but if a 50 nm-thick $100\,\Omega\,\text{cm}$ Ag-Ge-Se electrolyte between anode and cathode is 10 μm wide and 100 μm long, then the series resistance will be twice this value and so at least 0.75 V will be needed to drop 0.25 V at the cathode and cause electrodeposition.

Just as in any "plating" operation, the ions nearest the electron-supplying cathode will be reduced first. However, in devices with closely spaced interfaces in which the nanoscale roughness of the electrodes is significant and the fields are relatively high, statistical non-uniformities in the ion concentration and in the electrode topography will tend to promote localized deposition or nano-nucleation rather than blanket plating. Even if multiple nuclei are formed, the one with the highest field and best ion supply will be favored for subsequent growth, extending out from the cathode as a single metallic feature. The nature of this somewhat one-dimensional growth will be discussed later in the chapter. The electrodeposition of metal on the cathode does not mean that ions entering from the oxidizable anode must travel the entire length of the structure to replace those that are reduced. As noted earlier, the ions move through the electrolyte by a *coordinated motion*; the ion closest to the reduced ion will move to the vacated negative site on the hosting material, and those upstream will do likewise, each filling the vacated site of the one downstream, until the last vacated space closest to the anode is filled by the incoming ion. So, in the initial stages, the electrodeposit is actually made up of reduced ions from the electrolyte itself; however, as each ion deposited on the growing electrodeposit corresponds to one that has been removed from the metal source, the net effect is a shift of mass from the anode toward the cathode. It should be noted that the growth process in these structures is more complex than a simple plating operation as the deposition interface is moving toward the source of the ions. As the electrodeposit is physically connected to the cathode, it can supply electrons for subsequent ion reduction; hence, the growing electrodeposit will harvest ions from the electrolyte, plating them onto its surface to extend itself outwards from the cathode. This has two consequences: (i) the growth interface continually moves out to meet the ions; and (ii) the growth closes the gap between the electrodes, thereby increasing the field. Both of these outcomes help to speed the overall growth rate of the deposit. The growth rate and electrodeposit morphology are discussed in the following section.

It should be noted that if the cathode is electrochemically inert (not oxidizable), then the electrodeposition process is reversible by switching the polarity of the applied bias. When the electrodeposit is made positive with respect to the original oxidizable electrode, it becomes the new anode and dissolves via oxidation. During dissolution of the electrodeposit, the balance is maintained by deposition of metal back onto the place where the excess metal for the electrodeposition originated. Once the electrodeposit has been completely dissolved, the process self-terminates. It is important to note that it is the *asymmetry* of the device structure that allows the deposition/dissolution process to be cycled repeatedly.

16.3.2
Electrodeposit Morphology

It is clear that the reduction of the ions results in the formation of neutral metal atoms. However, what is not so obvious is the form that the electrodeposits take, as the process depends on a number of factors and involves not only the basic principles of electrochemistry but also transport phenomena, surface science, and metallurgy [28–30]. In this section, some of the more important issues of electrodeposition and deposit morphology with the ternary solid electrolytes will be considered. This process is best illustrated in structures which support electrodeposit growth between coplanar electrodes on an electrolyte layer [31], although growth through an electrolyte film will also be considered.

In the most general case, the process of deposit formation starts with the nucleation of the new metal atom phase on the cathode, and the deposits develop with a structure that generally follows a Volmer–Weber 3-D island growth mechanism [32]. The addition of new atoms to the growing deposit occurs due to a *diffusion-limited aggregation* (DLA) mechanism [33, 34]. In this growth process, an immobile "seed" is fixed on a plane in which particles are randomly moving around. Those particles that move close enough to the seed in order to be attracted to it attach and form the aggregate. When the aggregate consists of multiple particles, growth proceeds outwards and with greater speed as the new deposits extend to capture more moving particles. Thus, the branches of the core clusters grow faster than the interior regions. The precise morphology of these elongated features depends on parameters such as the potential difference and the concentration of ions in the electrolyte [35]. At low ion concentrations and low fields, the deposition process is determined by the (non-directional) diffusion of metal ions in the electrolyte and the resulting pattern is fractal in nature; that is, it exhibits the same structure at all magnifications. For high ion concentrations and high fields, conditions common in the solid electrolyte devices, the moving ions have a stronger directional component, and *dendrite* formation occurs. Dendrites have a branched nature but tend to be more ordered than fractal aggregates and grow in a preferred axis that is largely defined by the electric field. An example of dendritic growth is shown in Figure 16.6 for a Ag electrodeposit on a Ag-saturated $Ge_{0.30}Se_{0.70}$ electrolyte between Ag electrodes. Figure 16.6a is an optical micrograph of the electrodeposit, showing its dendritic character; while Figure 16.6b is an electron micrograph of a similar deposit, showing the extreme roughness of its surface at the nanoscale.

The above model for electrodeposit evolution assumes a homogeneous electrolyte. However, since electrodeposit growth is obviously related to the presence of available Ag ions in the electrolyte surface, the content and consistency of the electrolyte will have a profound effect on electrodeposit morphology. In the case of electrolyte based on a $Ge_{0.30}Se_{0.70}$ glass, the growth of low (about 20 nm high) continuous dendritic deposits are observed on surface of the films (see Figure 16.7a). In the case of the Ge-rich glasses ($Ge_{0.40}Se_{0.60}$), the growth of isolated, tall (>100 nm) electrodeposits can be seen (Figure 16.7b) [36]. The $Ge_{0.30}Se_{0.70}$ material has the higher chalcogen content of the two, and therefore will possess greater and more uniform quantities of

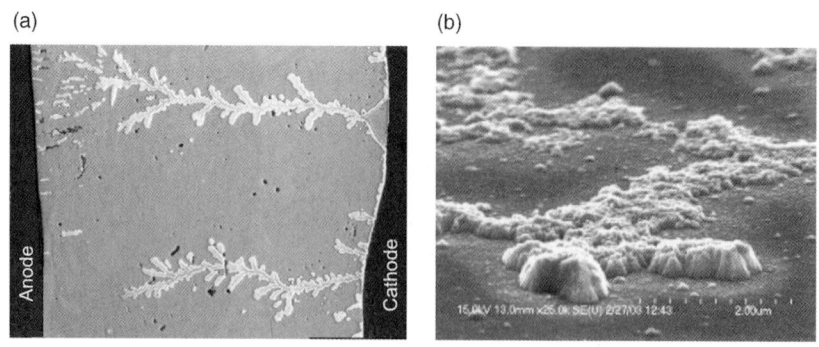

Figure 16.6 Surface electrodeposit of Ag on a Ge-Se-Ag solid electrolyte. (a) Optical micrograph of Ag dendrites. (b) Scanning electron microscopy image of the 3-D structure of the dendrites, showing the nano-roughness and large surface area. (From Ref. [31].).

ion-supplying Ag_2Se following the addition of Ag. This leads to dendritic growth that is closer to that expected with a homogeneous material. The isolated growth on the $Ge_{0.40}Se_{0.60}$ electrolyte is a direct consequence of the greater degree of separation of the Ag-containing phases.

The alternative device configuration has the electrodes on opposite sides of a thin electrolyte film, so that the growth of the electrodeposit is forced to occur *through* rather than *on* the electrolyte. Even though the capture and reduction of ions will essentially be by the same mechanism, it is unlikely that growth inside an electrolyte film will follow the same type of evolution as surface electrodeposition. At this point in time, although our understanding of the exact mechanism of growth within these electrolytes is incomplete, it is clear that the role of the nanoscale morphology should be considered. The confining nature of the medium, with its somewhat flexible channels and voids, will distort the shape of the electrodeposit, and its nano-inhomogeneity (as discussed above) will have a profound effect on local potential and ion supply. The net result is that the electrodeposit will not necessarily appear to be fractal or dendritic in nature, instead taking a form that is governed by the shape of the glassy voids and crystalline regions in the electrolyte. An example of such distorted morphology is shown in Figure 16.8; this is an electron micrograph of an electrodeposit within a 60 nm-thick $Ag_{0.33}Ge_{0.20}Se_{0.47}$ electrolyte between a tungsten bottom electrode and a silver top electrode. This was captured by overwriting a large (5 × 5 μm) device to produce multiple internal electrodeposits and then using a focused ion beam (FIB) system to ion mill a hole through the electrolyte [37]. The filament appears to be around 20 nm across, but this is misleading as the feature continues to grow through ion reduction by the electron beam.

As a final comment on morphology, reversing the bias dissolves the electrodeposit as it becomes the oxidizable element in the electrochemical cell. Macroscopically, this appears to be the reverse of the growth process, with the electrodeposit dissolving backwards from its tip (or tips in the case of a more two-dimensional dendrite). On closer inspection, the deposit actually dissolves near the tip region into a string of

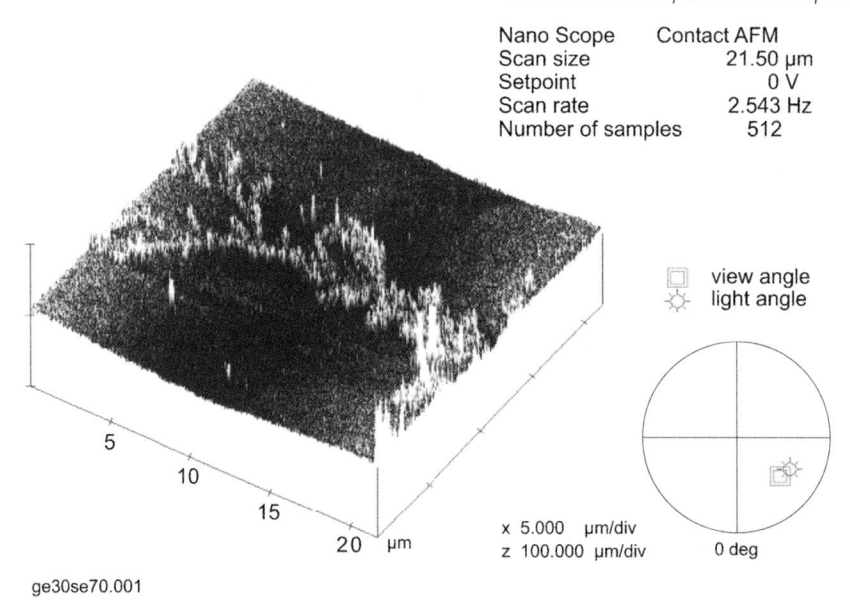

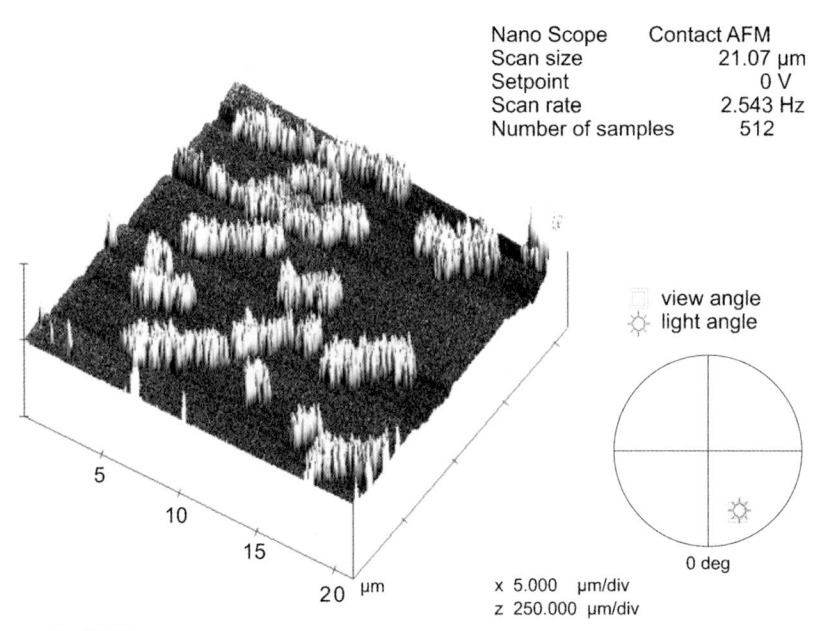

Figure 16.7 Atomic force microscopy image (3-D topographical scan) of (a) Ag electrodeposit grown on Ag-saturated $Ge_{0.30}Se_{0.70}$; the growth is continuous and the maximum electrodeposit height is a few tens of nanometers. (b) Ag electrodeposit grown on Ag-saturated $Ge_{0.40}Se_{0.60}$; the growth appears discontinuous and the maximum electrodeposit height is in the order of 100 nm. (From Ref. [36].).

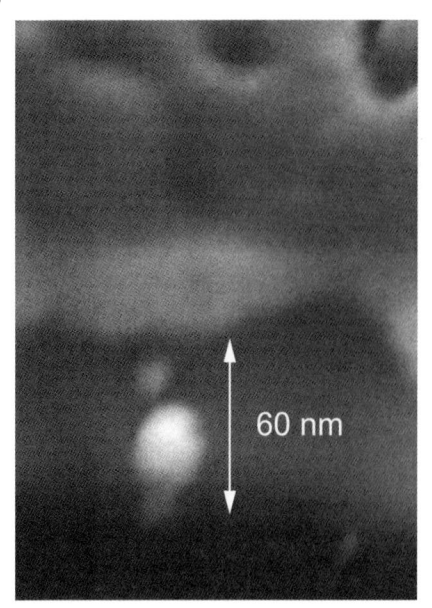

Figure 16.8 Electron micrograph of electrodeposit formation within a 60 nm-thick Ag-Ge-Se solid electrolyte observed by scanning electron microscopy following device sectioning by focused ion beam. (From Ref. [37].).

metal islands which then disappear into the electrolyte. This is a consequence of the uneven nature of the electrodeposit, created in part by the nano-morphology of the electrolyte, which allows some regions (perhaps associated with grain boundaries in the metal) to dissolve slightly faster than others.

16.3.3
Growth Rate

As in any deposition process, the growth rate of the electrodeposit, V, will depend on the ion flux per unit area, F, which corresponds to the current density, J, and the atomic density of the material being deposited, N, by

$$V = F/N = J/qN \tag{16.9}$$

The current density is given as

$$J = \sigma E \tag{16.10}$$

where E is the electric field. In large devices, the field will be relatively low, as will the current density. For example, in a 100 μm-long lateral (coplanar electrode) structure with 10 V applied, the field is 10^3 V cm^{-1}. Ag-saturated ternary electrolytes such as Ag-Ge-Se have a conductivity around 10^{-2} S cm^{-1}, and so the ion current density for this field will be 10 A cm^{-2}. Dividing current density by the charge on each ion (1.60×10^{-19} for Ag$^+$) gives the ion flux density, in this case 6.25×10^{19} ions cm^{-2} s^{-1}.

Using Equation 16.9 above with the atomic density of Ag (5.86×10^{22} atoms cm^{-3}) gives a growth rate of approximately 10^{-3} cm s^{-1}, or $10\,\mu$m s^{-1}. This is a gross simplification, as the complex morphology of the electrodeposits and the moving boundary condition of the advancing electrodeposit will complicate the deposition process. However, this is the approximate average velocity that is measured in a real device for the above conditions.

The electrodeposit growth rates are much more difficult to model in devices that have a thin electrolyte layer sandwiched between two electrodes, and at this point any knowledge of the nano-morphology of the material must be invoked. The average fields in this case range from 10^5 to 10^6 V cm^{-1}, for applied voltages of a few hundred millivolts to a few volts across an electrolyte that is a few tens of nanometers thick. Local fields may be higher still, as most of the applied bias will be dropped across the high-resistance glassy areas between the lower resistivity, metal-rich nanoclusters. Taking a field of 10^6 V cm^{-1} with the conductivity given above suggests that the growth rate will be in the order of 1 cm s^{-1} or 10 nm μs^{-1}. This is much slower than the rate suggested by measured switching speeds observed in actual devices (as will be shown in the next section), so the simple approach that was appropriate for macroscale devices apparently fails at the nanoscale. This may be due to a number of factors. For example, at fields of 10^6 V cm^{-1} or more, the linear conduction equation no longer holds [38] and the mobility will be higher than in the macroscopic case. It is also likely that n_i is larger than the previously assumed 10^{19} cm^{-3}, as the overall silver concentration can be as high as 10^{22} cm^{-3} in Ag-saturated chalcogenide glass electrolytes, and more of this is likely to be mobile due to barrier lowering in materials subjected to such high fields. The overall effect is that the current densities could be sufficiently high to make the electrodeposit growth rates several orders of magnitude higher in nanoscale devices.

16.3.4
Charge, Mass, Volume, and Resistance

The atomic weight of Ag is 107.9 g mol^{-1}, and the metal has a bulk density of 10.5 g cm^{-3}. This means that each atom weighs approximately 1.79×10^{-22} g (107.9 divided by Avogadro's number, 6.022×10^{23} atoms per mol) and this, of course, is the smallest amount of mass that can be transferred using silver as the mobile ion. Each cm^3 of Ag contains 5.86×10^{22} atoms, but a more useful unit in these nanoscale systems in the nm^3; such a volume contains 58.6 atoms on average, and will weigh approximately 10^{-20} g. If this is the electrodeposited mass, then each Ag$^+$ ion requires one electron from the external circuit to become reduced to form the deposited atom. So, each nm^3 of Ag will require 58.6 times the charge on each electron (1.60×10^{-19} C) which is 9.37×10^{-18} C of Faradaic charge (we can also use Faraday's constant, 9.65×10^4 C mol^{-1}, to perform this calculation). This charge is merely the integral of the current over time, and so a constant current of 1 μA would supply sufficient charge in 1 μs to deposit 10^5 nm^3 or around 1 fg (10^{-15} g) of Ag. So, in this mass-transfer scheme, current and time are the control parameters for the amount of mass deposited.

The amount of charge supplied will also determine the volume of the electrodeposit. The increase in metal volume at a point on the surface of the electrolyte (or decrease in volume at the anode) could be useful in a variety of microelectromechanical applications but it is the electrical resistance of this volume that is perhaps of most interest. The resistance, R, of an electrodeposit is given by

$$R = L/\sigma_m A \tag{16.11}$$

where σ_m is the conductivity of the metal, and L and A are its length and cross-sectional area, respectively (volume is $L \times A$). If the electrodeposited material is silver, σ_m will range from a value close to 5×10^5 S cm^{-1} (slightly higher than the bulk resistance) for features with thickness and width that are greater than a few tens of nanometers to much larger values for sub-10 nm features where surface scattering will play a considerable role. For a silver electrodeposit that is 20 μm long, 2 μm wide, and 20 nm thick (not unlike the example shown in Figure 16.7a), the resistance is about 10 Ω. This volume would take a total of 7.50×10^{-9} C to form. Note that if the underlying Ag-Ge-Se electrolyte (with conductivity 10^{-2} S cm^{-1}) was 20 μm long, 20 μm wide, and 50 nm thick, its resistance would be 2×10^7 Ω (or closer to 4×10^8 Ω for Ag-Ge-S), which is many orders of magnitude higher than that of the electrodeposit. Hence, the overall resistance of the newly formed structure is dominated by the electrodeposit. Figure 16.9 shows, graphically, the measured "on" state data from 50 nm-thick silver-doped arsenic disulfide electrolyte on a thick oxide layer on silicon substrates, patterned into channels with large silver contacts (100 × 100 μm) at the ends [39]. The "off" resistance, R_{off}, is a geometric function of the channel dimensions, following $R_{off} = L/\sigma dW + R_c$, where σ is the conductivity

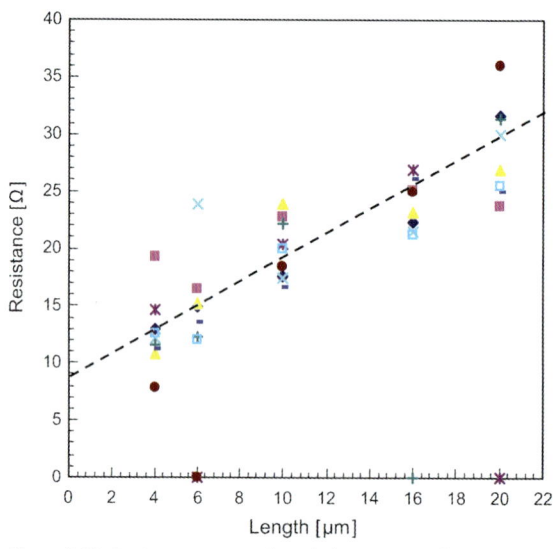

Figure 16.9 Resistance versus length for 10 μm-wide coplanar structures for a 25 mA current limit. The different symbols indicate results from different devices. (From Ref. [39].).

of the electrolyte layer (in the 10^{-3} S cm^{-1} range), d is its thickness, and W and L are width and length, respectively. R_c is the contact resistance (at zero channel length), mainly due to electrode polarization and tunneling at the measurement voltage through the polarization barrier. R_c is in the range of 10^8 to low 10^9 Ω for the electrode configuration used. A $10\times10\,\mu$m ($W\times L$) device therefore exhibits an R_{off} around 1.5 GΩ. The figure shows the results from a number of 10 μm- wide devices for programming using a 5-s voltage sweep from 0.5 to 1.8 V with a 25 mA current limit. This produces a substantial surface electrodeposit with a resistance of around 1 Ω μm^{-1} of device length. The average electrodeposit contact resistance in this case is around 9 Ω.

The resistance of the electrodeposit that forms within the nanostructured electrolyte is also determined by its volume, but in this case the influence of the different phases present at the nanoscale must also be considered. As discussed above, electrodeposition in its early stages is likely to occur on the metal-rich clusters, through the glassy high resistance regions between them. This means that the initial connection through the electrolyte will essentially consist of metallic bridges between the relatively low-resistivity clusters. In the case of a link that is dominated by the conductivity of the clusters rather than that of the metal, an on-resistance in the order of 20 kΩ in a 50 nm-thick Ag-Ge-Se electrolyte would require a conducting region less than 10 nm in diameter (assuming that the conductivity of the Ag$_2$Se material is close to the bulk value of 10^3 S cm^{-1} [40]). In the case where the electrodeposit dominates the pathway – that is, when the electrodeposited metal volume is greater than that of the superionic crystallites in the pathway – the electrodeposit resistivity will determine the on-resistance. In this case, a 10 nm-diameter pathway will have a resistance in the order of 100 Ω. This means that the diameter of the conducting pathway will not exceed 10 nm for typical programming conditions which require on-state resistances in the order of a few kΩ to a few tens of kΩ. The small size of the conducting pathway in comparison to the device area explains why on-resistance has been observed to be independent of device diameter, whereas off-resistance increases with decreasing area [41]. An electrodeposit this small means that the entire device can be shrunk to nanoscale dimensions without compromising its operating characteristics. This has been demonstrated by the fabrication of nanoscale devices as small as 20 nm that behave much like their larger counterparts [20, 42]. The other benefit of forming a small-volume electrodeposit is that it takes little charge to do so; in an extreme case, if half the volume of a sub-10 nm-diameter, 50 nm-long conducting region was pure Ag, only a few fC of Faradaic charge would be required to form a low-resistance pathway.

As discussed above, reversing the applied bias reverses the electrodeposition process so that the electrodeposit itself becomes the oxidizable anode and is thereby dissolved. The amount of charge necessary to do this is essentially the same as that required to grow the link in the first place. However, it is not just the oxidation of the electrodeposited metal that is responsible for breaking of the link, especially in the case of a connection between the electrodes that are mostly metal rather than a chain of metallic and Ag-rich electrolytic clusters. The very narrow (and uneven) link is susceptible to being weakened by electromigration by the current flowing through the metal. This means that the time to failure (initial breaking) of a metallic link depends partly on the current density; it can take years to break the link at a reverse current several orders of magnitude below the critical current density (around 10^7–10^8 A cm^{-2}), but the same link can be broken in less than 1 μs for a reverse current density in excess of this. In practice, for vertical structures, the critical current is typically less than the current limit used to form the link, but not less than 20% of this current. It should be noted that a forward current cannot easily be used to break the link by electromigration if the applied bias is above the threshold for electrodeposition, as any weakness (high-resistance region) in the electrodeposit will be "healed" by electrodeposition in the area due to the elevated voltage drop there. As the resistivity of the electrolyte is many orders of magnitude higher than that of the electrodeposited metal, the overall resistance of the structure rises dramatically as soon as the link is broken. The remainder of the electrodeposited metal in the now-incomplete link is dissolved electrochemically.

16.4
Memory Devices

16.4.1
Device Layout and Operation

As shown in Figure 16.5, the basic elements of a resistance change device – the solid electrolyte, the oxidizable electrode, and the inert electrode – may be configured either laterally or vertically. Whereas lateral devices may have utility in a variety of applications (e.g., microelectromechanical systems; MEMS), it is the vertical configuration that is of most interest in the context of memory devices. Vertical structures occupy the smallest possible area, which is critical for high-density memory arrays. In addition, the distance that the electrodeposit must bridge in order to switch a vertical device to its low-resistance state – a key factor in determining switching speed – is defined by the electrolyte thickness rather than by a lithographically defined gap. As the film thickness can typically be made much smaller than a lateral gap using conventional manufacturing technology, vertical structures switch faster than their lateral counterparts.

A schematic representation of how vertical solid electrolyte memory devices may be integrated in a complementary metal oxide-semiconductor (CMOS) circuit is shown in Figure 16.10. In this case, the inert electrodes are the tungsten plugs that are normally used to connect one layer of interconnect metal to another. The solid electrolyte layer is placed on top of these individual tungsten electrodes, and a common oxidizable electrode (or a bilayer of oxidizable metal and another electrode material) caps the device structures. The individual devices are defined by each tungsten plug. It should be noted that the storage elements are built in the interconnect layers above the silicon devices in a "back-end-of-line" (BEOL) process, which means that the CMOS fabrication scheme need not be changed. A further

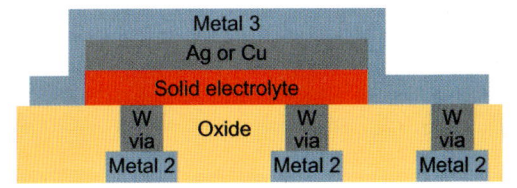

Figure 16.10 Schematic illustration of solid electrolyte device integration between two levels of metal (in this case "metal 2" and "metal 3" in a standard CMOS process. The individual devices are defined by each W via plug under the continuous electrolyte layer. One extra mask step is used to define which tungsten via plugs will be covered with the solid electrolyte and oxidizable metal, and which will be through connections.

advantage is that only one extra mask is required to define which tungsten plugs are covered with the device stack, and which are through-connections to the upper layers of the interconnect. This helps to reduce the cost of integration and also facilitates embedding the memory with standard logic. In order to obtain the maximum performance from the devices, each storage cell is connected through the underlying interconnect to a "select" transistor in a "one transistor-one resistor" (1T1R) cell array (Figure 16.11a). In this scheme, the transistor is used to select the cell, and an appropriate programming voltage is then applied across the device. Passive arrays, in which sneak current paths through the cells are avoided using diode elements in the array itself rather than transistors, are also possible using row and column electrodes with device structures at their intersections (Figure 16.11b). This latter approach does not allow high-speed operation but does lead to the densest array possible as there are no transistors to enlarge the total cell area.

The programming of the solid electrolyte memory devices is relatively straightforward. A forward bias (oxidizable electrode positive with respect to the inert electrode) in excess of the threshold required to initiate electrodeposition is used to write the

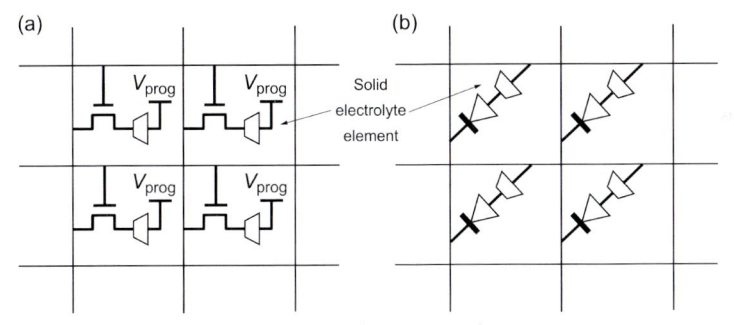

Figure 16.11 (a) Active "1T1R" array configuration. Each memory location consists of a solid electrolyte device and a transistor. V_{prog} is used to set the appropriate voltage across the device for programming. (b) Passive array configuration. Each cell has a diode to prevent "sneak paths" between rows and columns through the cells.

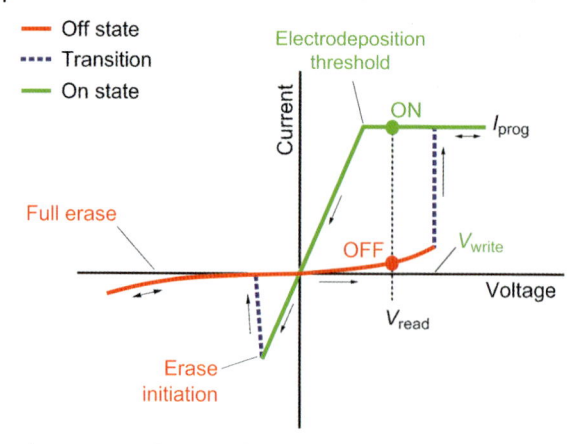

Figure 16.12 Schematic of a current–voltage plot of a solid electrolyte memory device programmed with current limit I_{prog}. The conducting pathway forms at the write voltage (V_{write}) and breaks at the erase initiation voltage/current. Further negative bias is required to fully remove the electrodeposited material and return the device to its original off state. The state of the device is read using a positive voltage below V_{write} to avoid disturbing the state.

device. A negative bias is used to erase the device. Reading the state of the device involves the application of a bias that will not "disturb" or destroy the current state. This typically means that the devices are read using a forward bias that is below the minimum required to write under normal operating conditions. This is shown schematically in Figure 16.12, which shows a current–voltage plot of a solid electrolyte memory device. Only leakage current flows in the off state, but when the conducting pathway forms at the write voltage (V_{write}), the current quickly rises to the programming current limit (I_{prog}). It should be noted that the electrodeposition continues after switching, albeit more slowly than the initial transition, until the voltage across the device reaches the minimum threshold for electrodeposition. The lower on-state resistance is preserved until the erase initiation voltage is reached, at which point the conducting pathway breaks and the device resistance goes high. Further negative bias is required to fully remove the electrodeposited material and return the device to its original off state. The device is read using a positive voltage below V_{write} and the current measured to determine the state. Note that in Figure 16.12, V_{read} has been chosen to be between the minimum voltage for electrodeposition and V_{write} (the consequences of this are discussed in Section 16.4.3). The following section provides results from a variety of fabricated devices.

16.4.2
Device Examples

To date, electrochemical switches have been fabricated using thin Cu_2S [43] and Ag_2S [44] binary chalcogenide electrolytes. The Cu_2S devices have been demonstrated in small memory arrays [45] and reconfigurable logic [46], and although the applications show promise, there is room for improvement in device performance factors such as retention and endurance with this particular electrolyte. The studies on Ag_2S devices have concentrated on switching by the deposition and removal of small numbers of silver atoms in a nanoscale gap between electrodes. This is of major significance as it demonstrates that the electrochemical switching technique has the potential to be scaled to single atom dimensions. Various oxide-based devices have also been demonstrated [47, 48], and these show great promise as easily integrated elements. However, the lower ion mobility in these materials tends to make the devices slower than their chalcogenide counterparts. Devices based on ternary chalcogenide electrolytes, including Ag-Ge-Se, Ag-Ge-S, and Cu-Ge-S have been the most successful to date, with the silver-doped variants having been applied in sophisticated high-density memory arrays [49] and post-CMOS logic devices [50]. Ag-Ge-Te devices have also been explored [51], but these materials have a tendency to crystallize at low temperatures and so may not be the best choice for devices that must be integrated with CMOS using elevated processing temperatures.

To illustrate the operation of devices based on ternary electrolytes, the discussion will be confined here to those utilizing Ag-Ge-S and Ag-Ge-Se materials. Of these, the Ag-Ge-S electrolyte is the most compatible with BEOL processing in CMOS fabrication as it can withstand thermal steps in excess of 400 °C without any degradation of device characteristics. The sulfides possess better thermal stability as there is less change in the nanostructure at elevated temperature than in the case of selenide electrolytes [21, 52]. A typical device operation is shown in Figure 16.13a and b, which provide current–voltage and resistance–voltage plots respectively for a 240 nm-diameter W/Ag-Ge-S/Ag device with a 60 nm-thick electrolyte [53]. The voltage sweep runs from -1.0 to $+1.0$ to -1.0 V, and the current limit is 10 µA. As mentioned above, the write threshold for this material combination is 0.45 V, at which voltage the device

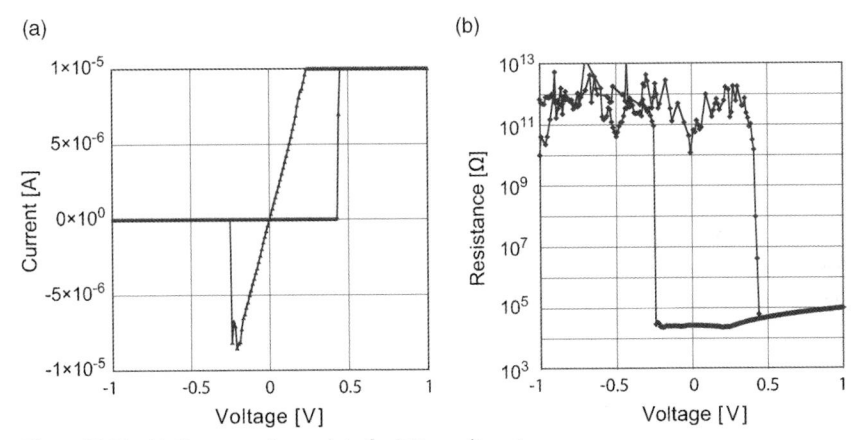

Figure 16.13 (a) Current–voltage plot of a 240 nm-diameter device with a 60 nm-thick Ag-Ge-S electrolyte using a 10 µA current limit. The device has been annealed at 300 °C. The voltage sweep is -1.0 to $+1.0$ to -1.0 V. (b) Resistance–voltage plot of the same device. The voltage sweep is -1.0 to $+1.0$ to -1.0 V. (From Ref. [52].).

switches from an off-state resistance, R_{off}, above $10^{11}\,\Omega$ to an on-state resistance, $R_{on} = 22\,\text{k}\Omega$, more than six orders of magnitude lower for the 10 µA programming current. This apparent rise in resistance following switching is caused by the current limit control in the measurement instrument. Once electrodeposition is initiated, the threshold for further electrodeposition is decreased, as indicated by the presence of a lower voltage (0.22 V) at which the current drops below the current limit on the negative-going sweep. R_{on} is determined by this voltage divided by the current limit (see below). The device transitions to a high-resistance state at -0.25 V, this being due to an initial breaking of the electrodeposited pathway by the reverse current flow. Continuing the negative sweep, the off-resistance remains above $10^{11}\,\Omega$ as the voltage is swept out to -1.0 V with a leakage current of less than 10 pA at maximum reverse bias. When considering the above characteristics, the device may be written using a voltage in excess of 0.45 V, read by applying a positive voltage that is less than 0.45 V, and erased by a bias greater than -0.25 V. These voltages are compatible with devices at the 22-nm node of the International Technology Roadmap for Semiconductors (ITRS).

Figure 16.14 illustrates the dependence of R_{on} on the programming current limit, I_{prog}, in the range 1–10 µA for a W/Ag-Ge-Se/Ag device with a 50 nm-thick electrolyte [54]. For this electrolyte/electrode combination, the write threshold is 240 mV and the electrodeposition threshold 140 mV. R_{on} is related to I_{prog} by $R_{on} = 0.14/I_{prog}$ (the solid line in the figure). This relationship between electrodeposition voltage, programming current, and on-resistance is common to all material combinations. It is explained by the fact that as long as sufficient potential difference is maintained for the situation where electrodeposition is already underway (in this case 140 mV), the reduction of silver ions will continue and the decrease in resistance of the conducting bridge will be maintained, even after it has formed. If the external current source is limited, when the resistance falls to a point where the voltage drop is no longer sufficient to support electrodeposition, the resistance lowering process ceases. The resulting resistance in ohms is therefore given approximately by the minimum potential to sustain electrodeposition in volts divided by the current limit of the external supply in amperes.

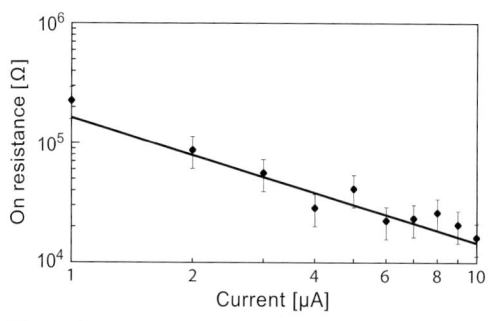

Figure 16.14 On-state resistance versus programming current for a 1 µm-diameter W/Ag-Ge-Se/Ag device. The resistance value (in Ω) is approximately 0.14 divided by the current in A (solid line). (From Ref. [54].).

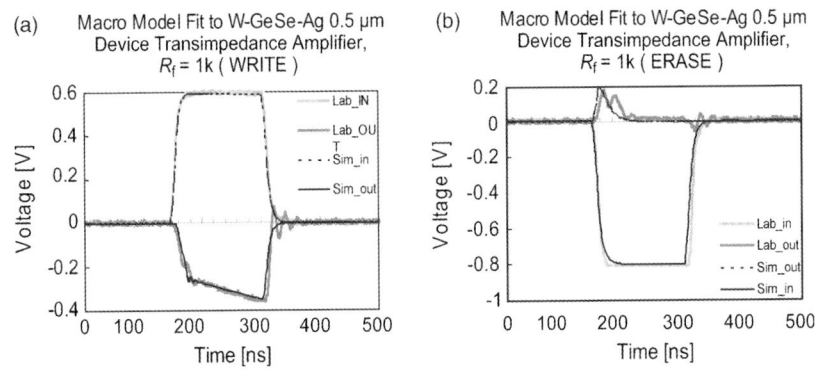

Figure 16.15 (a) Result of a 150-ns write pulse of 600 mV applied to a 500 nm W/Ag-Ge-Se/Ag device showing the output of the transimpedance measurement amplifier. The final on-resistance is 1.7 kΩ. (b) Result for a 150-ns erase pulse of −800 mV on the same device, showing the output of the transimpedance measurement amplifier. Lab_in and Lab_out are the measured input and output signals; Sim_in and Sim_out are the corresponding model generated curves. (From Ref. [55].).

The switching speed of a 500 nm-diameter W/Ag-Ge-Se/Ag device with a 50 nm-thick electrolyte is illustrated in Figure 16.15, which shows both measured and simulated device results for write (Figure 16.15a) and erase (Figure 16.15b) operations [55]. For the write, a 150-ns pulse of 600 mV was applied to the device, and the output of the transimpedance measurement amplifier shows that the device initially switches in less than 20 ns, while the resistance continues to fall more slowly, ultimately reaching an on-resistance of 1.7 kΩ at the end of the write pulse. For the erase, a 150-ns pulse of −800 mV was applied, whereupon the output of the transimpedance measurement amplifier shows that the device transitions to a high-resistance state (the start of the voltage decay in the output signal) in around 20 ns. The electrodeposit is essentially metal that has been added to a chemically saturated electrolyte, and this local supersaturation leads to high stability of the electrodeposit and excellent device retention characteristics. The results of a retention assessment test on a 2.5 μm-diameter W/Ag-Ge-S/Ag device with a 60 nm-thick electrolyte annealed at 300 °C, and programmed using a 0 to +1.0 V sweep is shown in Figure 16.16. The plot shows the off and on resistances measured using a 200 mV read voltage. The off state was in excess of 10^{11} Ω (above the limit of the measurement instrument) and remained undisturbed by the read voltage at this level for the duration of the test. The on resistance remained below 30 kΩ during the test. Following this, the device was erased using a 0 to −1.0 V sweep, and the off-state resistance measured using a 200 mV sensing voltage as before. The device remained above 10^{10} Ω beyond 10^5 s, demonstrating that the erased state was also stable. Other studies have show that both on and off states are also stable at elevated temperature, with a margin of several orders of magnitude being maintained even after 10 years at 70 °C [42]. Figure 16.17 provides an example of cycling for a 75-nm Ag-Ge-Se electrolyte device [20]. Trains of positive (write) pulses of 1.2 V in magnitude and 1.6 μs duration followed by −1.3 V negative (erase) pulses of 8.7 μs duration were

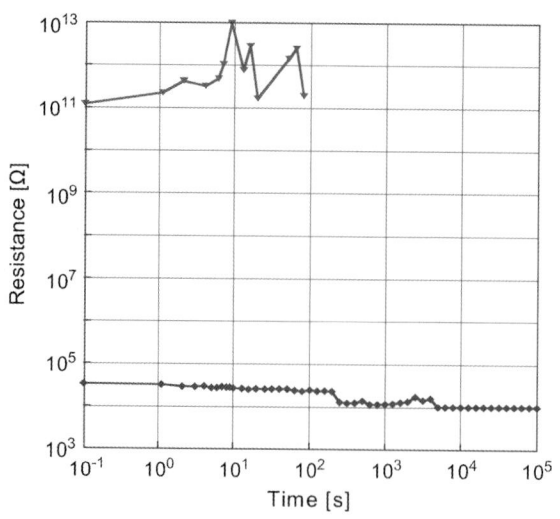

Figure 16.16 Off- (upper plot) and on-state (lower plot) resistance versus time measured using a 200 mV read voltage for a 2.5 μm W/Ag-Ge-S/Ag device programmed using a 10 μA current limit. The off-state resistance remained above $10^{11}\,\Omega$ for the duration of the test. (From Ref. [53].).

used to cycle the devices. A 10 kΩ series resistor was used to limit current flow in the on state. The results are shown in 10^9 and 10^{11} cycle ranges. The data in Figure 16.17 show that there might be a slight decrease in on current, but this is gradual enough to allow the devices to be taken well beyond 10^{11} write–erase cycles (if this decrease is maintained, there will only be a 20% decrease in on current at 10^{16} cycles).

As mentioned above in this section, oxide-based electrolytes may also be used in memory devices. Of these, Cu-WO$_3$ [47] and Cu-SiO$_2$ [48] are of particular interest as they utilize materials that are already in common use in the semiconductor industry,

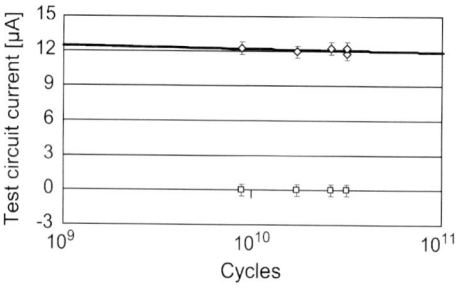

Figure 16.17 Current in the on (upper plot) and off (lower plot) state at various numbers of cycles for a 75-nm Ag-Ge-Se device. The device was cycled using trains of positive (write) pulses of 1.2 V in magnitude and 1.6 μs duration, followed by −1.3 V negative (erase) pulses of 8.7 μs duration. The solid line is a logarithmic fit to the on current data. (From Ref. [20].).

namely Cu and W for metallization and SiO_2 as a dielectric, and this will help to reduce the costs of integration. In general, the switching characteristics for both systems are very similar to those observed in metal-doped chalcogenide glasses, and that is why the same switching mechanism is assumed for the oxide-based cells, even though the material nanostructure is quite different from that found in the ternary chalcogenide electrolytes. For example, in the case of $Cu-WO_3$, the Cu must exist within oxide in unbound form for successful device operation [47]. For $Cu-SiO_2$, the best results are attained via the use of porous oxide, formed by physical vapor deposition, into which the metallic copper is introduced by thermal diffusion so that it exists in "free" form in the nano-voids in the base glass. In the case of $W-(Cu/SiO_2)-Cu$ devices with a 12 nm-thick electrolyte, both unipolar (positive voltage for both write and erase) as well as bipolar switching has been observed [48]. Unipolar switching requires high programming currents (several hundred μA to several mA) to thermally break the electrodeposited copper connection in forward bias. Bipolar switching with a resistance ratio of 10^3 is achieved with switching voltages below 1 V and currents down to the sub-μA range. Highly stable retention characteristics beyond 10^5 s and switching speeds in the microsecond regime have been demonstrated, and the possibility of multi-bit storage exists due to the relationship between on-state resistance and programming current. These results, combined with the initial endurance testing which showed that more than 10^7 cycles were possible with these structures, indicate that this technology shows promise as a low-cost, low-energy Flash memory replacement technology.

16.4.3
Technological Challenges and Future Directions

The above results indicate that memory devices based on electrodeposition in solid electrolytes show great promise. However, although several substantial development efforts are under way, many questions remain unanswered with regards to the physics and long-term operation of this technology. The most pressing issues relate to the reliability of such devices. In any memory technology, the storage array is only as good as its weakest cell. Reduced endurance (cycling between written and erased states), poor retention, and "stuck" bits plague even the most mature memory technologies. It may be many years before the issues concerning the solid electrolyte approach are fully understood, but considerable optimism exists regarding reliability which may set this technology apart from others. For example, many technologies suffer from reduced endurance due to changes in the material system with time. In this respect, solid electrolyte devices can exhibit diminishing off-to-on-resistance ratio with cycle number if incorrect programming (overwriting and/or incomplete erase) leads to a build up of electrodeposited metal within the device structure. The convergence of the off and on states eventually leads to an inability to discriminate between them. However, it is possible electrically to "reset" the solid electrolyte using an extended or "hard" erase; this will then plate the excess material back on to the oxidizable electrode and return the electrolyte to its original composition. This ability to change material properties using electrical signals allows such corrections to be performed in the field, and this may have a profound effect on device reliability. Another issue that can occur in written devices is the upward drift in programmed on-state resistance with time at elevated temperature. This is thought to be due to thermal diffusion of the electrodeposited metal, but it may also be a consequence of electromigration during repeated read operations. However, a read voltage that lies between the write voltage and the minimum voltage for electrodeposition will essentially repair or "refresh" a high-resistance/open on state. To illustrate this, the device characteristics shown in Figure 16.13 are revisited. If a read voltage between 0.22 and 0.45 V is used, an off/erased device will not be written, but a device that has been previously programmed will actually have its on state strengthened. This "auto-refresh" above the minimum threshold for electrodeposition is unique to electrochemical devices. It should be noted that although this effect is extremely useful, it can also lead to problems in incorrectly erased devices (those which are open circuit but still have electrodeposited material on the cathode), as these can also be written at read voltages. Clearly, under-erasing must be avoided in order to maintain high device reliability.

Attention is now turned to the scaling of solid electrolyte memory devices. This involves two points of consideration: physical scaling; and electrical scaling. Physical scaling of the types of device described in the previous section has already been demonstrated to below 22 nm, with good operational characteristics [42]. In addition, studies on the bridging of nanometer-sized gaps between a solid electrolyte and a top electrode seem to suggest that *atomic-scale* electrodeposits could be used to change the resistance of the device, and this may represent the ultimate scaling of the technology [44]. What is not known is how the "high-performance" phase-separated chalcogenide electrolytes will scale, as these contain crystallites that approach 10 nm in diameter. Clearly, further investigations are required in this area, although some are already under way. The other aspect of scaling is electrical scalability. For example, the supply voltage for highly scaled systems around the 22 nm node of the ITRS is on the order of 0.4–0.6 V. This means that, in order to avoid the use of area-, speed-, and energy-sapping charge pumps, the memory cells must be able to operate at the very low voltages at which solid electrolyte devices can function. In addition, the critical current density for 22 nm interconnect is only a few tens of μA, and the devices must also be able to operate at these current levels which, once again, is achievable by solid electrolyte devices.

The final consideration for the future relates to memory density in the Tb (10^{12} bits)/chip regime. Such high storage densities will eventually be required for high-end consumer and business electronics to replace mechanical hard drives in small-form factor, portable systems. If it is assumed that a $20 \times 20 \text{ mm}^2$ chip has an extremely compact periphery such that most of the area is storage array, and a compact cell at $4 F^2$ (where F is the half-pitch), then Tb storage would require F to be 10 nm at most. Such small wires cannot be produced using standard semiconductor fabrication technologies without significant variations, and their current carrying capacity is very small. Backing-off to $F = 22$ nm means that multi-level cell (MLC) storage – more than one bit per physical storage cell – will be necessary to achieve Tb storage. The ability in solid electrolyte devices to control the on-resistance using the programming current allows multiple resistance levels to be stored in each cell. For example, four discrete resistance levels leads to 2 bits of information in each cell (00, 01, 10, 11). Such MLC storage has already been demonstrated in a solid electrolyte memory array that was integrated with CMOS circuitry [56]. A combination of the above characteristics demonstrated physical scalability with low-voltage/-current/-power and MLC operation, and it would appear that solid electrolyte memory devices are a strong contender for future solid-state memory and storage.

16.5
Conclusions

Considering the characteristics described in the previous sections, devices based on mass transport in solid electrolytes appear to be appropriate for use as scalable, low-power memory elements. A reduction in resistance of several orders of magnitude is attainable in vertical devices for a write power below 1 µW, and since the on-resistance is a function of programming current, then MLC operation with simple sensing schemes is possible. The elements are non-volatile, with extrapolated retention results suggesting that the reduced resistance, with a large off-to-on ratio, will persist for well over 10 years. Even sub-100 nm devices show excellent endurance with no significant degradation to over 10^{10} cycles, and with stable operation indicated well beyond this. Both, simulated and measured data show that the devices write and erase within 20 ns, and further scaling – especially in the vertical dimension – is likely to result in even greater programming speeds. It should be noted that write times in the order of a few tens of nanoseconds mean that the write energy is less than 100 fJ, which makes solid electrolyte cells memory one of the lowest energy non-volatile technologies. The low resistivity and small size of the electrodeposits mean that the entire device can be shrunk to nanoscale dimensions, without compromising operating characteristics. This physical scalability, combined with low-voltage and low-current operation, suggests that extremely high storage densities will be possible. The other benefit of forming a small-volume electrodeposit is that it takes little charge to do so – in the order of a few fC to create an extremely stable low-resistance link. The charge required to switch a solid electrolyte element to a non-volatile state is therefore comparable to that required to program a typical dynamic random access memory cell, with the potential to further reduce this charge in future devices.

References

1 Maier, J. (2005) *Nature Materials*, **4**, 805.
2 Faraday, M. (1838) *Philosophical transactions of the Royal Society of London*,
3 Kummer, J.T. and Weber, N. (1966) US Patent 3,458,356.
4 Kirby, P.L. (1950) *British Journal of Applied Physics*, **1**, 193.
5 Bychkov, E., Tsegelnik, V., Vlasov, Yu., Pradel, A. and Ribes, M. (1996) *Journal of Non-Crystalline Solids*, **208**, 1.
6 Miyamoto, Y., Itoh, M. and Tanaka, K. (1994) *Solid State Communications*, **92**, 895.
7 Goodenough, J.B. (2003) *Annual Review of Materials Research*, **33**, 91.

8 Lazure, S., Vernochet, Ch., Vannier, R.N., Nowogrocki, G. and Mairesse, G. (1996) *Solid State Ionics*, **90**, 117.

9 Kreuer, K.-D., Kohler, H. and Maier, J. (1989) *High Conductivity Solid Ionic Conductors* (ed. T. Takahashi), World Scientific, Singapore, p. 242.

10 Kartini, E., Kennedy, S.J., Sakuma, T., Itoh, K., Fukunaga, T., Collins, M.F., Kamiyama, T., Suminta, S., Sugiharto, A., Musyafaah, E. and Bawono, P. (2002) *Journal of Non-Crystalline Solids*, **312–314**, 628.

11 Ogawa, H. and Kobayashi, M. (2002) *Solid State Ionics*, **148**, 211.

12 Shahi, K. (1977) *Physica Status SolidI A-Applied Research*, **41**, 11.

13 Tanaka, K. (2000) Chalcogenide glasses, in *Encyclopedia of Materials: Science and Technology*, Elsevier.

14 Tanaka, K., Miamoto, Y., Itoh, M. and Bychkov, E. (1999) *Physica Status SolidI A-Applied Research*, **173**, 317.

15 Elliott, S.R. (1991) Chalcogenide glasses, in *Materials Science and Technology* (ed. J. Zarzycki), VCH, New York.

16 Bychkov, E. (2000) *Solid State Ionics*, **1111**, 136–137.

17 Mitkova, M., Wang, Y. and Boolchand, P. (1999) *Physical Review Letters*, **83**, 3848.

18 Mitkova, M. and Kozicki, M.N. (2002) *Journal of Non-Crystalline Solids*, **1023**, 299–302.

19 Kozicki, M.N., Mitkova, M., Zhu, J. and Park, M. (2002) *Microelectronic Engineering*, **63**, 155.

20 Kozicki, M.N., Park, M. and Mitkova, M. (2005) *IEEE Transactions on Nanotechnology*, **4**, 331.

21 Balakrishnan, M., Kozicki, M.N., Poweleit, C.D., Bhagat, S., Alford, T.L. and Mitkova, M. (2007) *Journal of Non-Crystalline Solids*, **353**, 1454–1459.

22 Rennie, J.H.S. and Elliott, S.R. (1987) *Journal of Non-Crystalline Solids*, **1239**, 97–98.

23 Kolobov, A.V., Elliott, S.R. and Taguirdzhanov, M.A. (1990) *Philosophical Magazine B*, **61**, 857.

24 Mitkova, M., Kozicki, M.N., Kim, H. and Alford, T. (2004) *Journal of Non-Crystalline Solids*, 338–340. 552.

25 Mitkova, M., Kozicki, M.N., Kim, H.C. and Alford, T.L. (2004) *Thin Solid Films*, **449**, 248.

26 Kotzeniewski, C. in (1997) *The Electrochemical Double Layer* (ed. B.E. Conway), The Electrochemical Society Inc.

27 West, W.C., Sieradzki, K., Kardynal, B. and Kozicki, M.N. (1998) *Journal of the Electrochemical Society*, **145**, 2971.

28 Dini, J.W. (ed.) (1992) *Electrodeposition: the Materials Science of Coatings and Substrates*, Noyes Publications NJ, USA.

29 Budevski, E., Staikov, G. and Lorenz, W.J. (1996) *Electrochemical Phase Formation and Growth*, VCH Publishers, NY, USA.

30 Watanabe, T. (2004) *Nano-Plating Microstructure Control Theory of Plated Film and Data Base of Plated Film Microstructure*, Elsevier.

31 Kozicki, M.N., Maroufkhani, P. and Mitkova, M. (2004) *Superlattices and Microstructures*, **34**, 467.

32 See for example: Henrich, V.E. and Cox, P.A. (1994) *The Surface Science of Metal Oxides*, Cambridge University Press, Cambridge. Chapter 5.

33 Witten, T.A. and Sander, L.M. (1981) *Physical Review Letters*, **47**, 1400.

34 Meakin, P. (1983) *Physical Review A*, **27**, 1495.

35 Sawada, Y., Dougherty, A. and Gollub, J.P. (1986) *Physical Review Letters*, **56**, 1260.

36 Mitkova, M., Kozicki, M.N. and Aberouette, J.P. (2003) *Journal of Non-Crystalline Solids*, **425**, 326–327.

37 Ratnakumar, C., Mitkova, M. and Kozicki, M.N. (November 2006) Proceedings of the 2006 Non-Volatile Memory Technology Symposium, San Mateo, California p. 111.

38 Dignam, M.J. (1968) *Journal of Physics and Chemistry of SolidS*, **29**, 249.

39 Kozicki, M.N., Yun, M., Hilt, L. and Singh, A. (1999) *Electrochemical Society Proceedings*, **13**, 298.

40 Miyatani, S.-Y. (1960) *Journal of the Physical Society of Japan*, **15**, 1586.
41 Symanczyk, R., Balakrishnan, M., Gopalan, C., Grüning, U., Happ, T., Kozicki, M., Kund, M., Mikolajick, T., Mitkova, M., Park, M., Pinnow, C., Robertson, J. and Ufert, K. (November 2003) Proceedings of the 2003 Non-Volatile Memory Technology Symposium, San Diego, California, p. 17-1.
42 Kund, M., Beitel, G., Pinnow, C., Röhr, T., Schumann, J., Symanczyk, R., Ufert, K. and Müller, G. (2005) *IEDM Technical Digest*, 31.5.
43 Sakamoto, T., Sunamura, H., Kawaura, H., Hasegawa, T., Nakayama, T. and Aono, M. (2003) *Applied Physics Letters*, **82**, 3032.
44 Terabe, K., Hasegawa, T., Nakayama, T. and Aono, M. (2005) *Nature*, **433**, 47.
45 Kaeriyama, S., Sakamoto, T., Sunamura, H., Mizuno, M., Kawaura, H., Hasegawa, T., Terabe, K., Nakayama, T. and Aono, M. (2005) *IEEE Journal of Solid State Circuits*, **40**, 168.
46 Sakamoto, T., Banno, N., Iguchi, N., Kawaura, H., Kaeriyama, S., Mizuno, M., Terabe, K., Hasegawa, T. and Aono, M. (2005) *IEDM Technical Digest*, 19.5.
47 Kozicki, M.N., Gopalan, C., Balakrishnan, M. and Mitkova, M. (2006) *IEEE Transactions on Nanotechnology*, **5**, 535.
48 Schindler, C., Thermadam, S.C.P., Waser, R. and Kozicki, M.N.(2007) *IEEE Trans. Electron Devices*, **54**, 2762.
49 Hönigschmid, H., Angerbauer, M., Dietrich, S., Dimitrova, M., Gogl, D., Liaw, C., Markert, M., Symanczyk, R., Altimime, L., Bournat, S. and Müller, G. (June 2006) IEEE VLSI Circuits Symposium, Honolulu, Hawaii, 132.
50 Fujita, S., Fujita, S., Abe, K. and Lee, T.H. (May 2005) *NSTI-Nanotech*, Anaheim, California. 31.04.
51 Kim, C.-J., Yoon, S.-G., Choi, K.-J., Ryu, S.-O., Yoon, S.-M., Lee, N.-Y. and Yu, B.-G. (2006) *Journal of Vacuum Science & Technology B: Microelectronics and Nanometer Structures*, **24**, 721.
52 Mitkova, M., Kozicki, M.N., Kim, H.C. and Alford, T.L. (2006) *Journal of Non-Crystalline Solids*, **352**, 1986.
53 Kozicki, M.N., Balakrishnan, M., Gopalan, C., Ratnakumar, C. and Mitkova, M. (2005) IEEE Non-Volatile Memory Technology Symposium, D5, p. 1.
54 Kozicki, M.N., Gopalan, C., Balakrishnan, M., Park, M., Mitkova, M. (November, 2004) Proceedings of the 2004 Non-Volatile Memory Technology Symposium, Orlando, Florida, USA, p. 10.
55 Gilbert, N.E., Gopalan, C. and Kozicki, M.N. (2005) *Solid-State Electronics*, **49**, 1813.
56 Gilbert, N.E. and Kozicki, M.N. (2007) *IEEE Journal of Solid-State Circuits*, **42**, 1383.

Index

a
aberration 145ff.
– chromatic 146
– correction 147, 197
– lens 146
– optimizer 146
– spherical 146
– theory 146
absorber 196, 198
acceleration 184
– value 184
– voltage 152, 161, 167
activation energy 491f., 495
– curve 491
adatom 307f.
adhesion 221, 321, 429
adsorption 182, 321ff.
– brick-wall 329
– co- 323f.
– optical 53, 130
– process 318
– properties 318
– SAM 321
– site-selective 321
– sites 318
Aharonov-Bohm
– effect 3, 23ff.
– oscillations 26f., 33
– period 27
– rings 28, 223
AlGaAs/GaAs
– heterostructure 10, 13, 15, 25
– interface 9f.
– layer system 9
– ring structure 24, 26
– split-gate point contact 13
amplification 99, 101
amplitude 243ff.

– complex 14f.
– constant 244
– -distance curve 246
– fluctuation 27, 32f.
– free 245f.
– oscillation 25, 29
– probability 16
– total 15, 22
– vibration 244
annealing
– correct 300
– incorrect 300
antibodies 277, 285, 288f.
– antihuman secondary 289
– libraries 289
– molecules 288ff.
– Ronit1 scFv phage library 290
– secondary 293
– soluble 289
antiferromagnet 424ff.
– (AF)-pinned reference structure 425
– (AF)-pinned structure 424
– synthetic (SAF) 424
antiferromagnetic 56
– exchange 60
antiparallel
– configuration 71f.
– moments 66ff.
aperture
– plane 146
– plate system 167f.
– shaped 148ff.
array quality factor (AQF) 431f., 442
Arrhenius
– behavior 469
– law 307
– plot 470
aspect ratio 228, 284

assembly
- bio- 277
- dimer-monomer 327
- free-running 277
- ligase 294
- modular 276f.
- molecular 298
- -self 276f.
- sequence-specific molecular 277
atomic force microscopy (AFM) 239ff.
- Auto-Probe 264
- interface 261
- MFP-3D 246
- multi-tip array 265
- nanomanipulation 255, 262
- non-contact 254
- operation 241, 246f.
- piezo 263
- -SEM approach 255
- software-compensated 263
- tip 240f., 258, 263
- topography 263
autonomous binary p-shift register 294ff.

b

Back-End-Of-Line (BEOL) processing 436, 440f., 466, 504, 507
backscattering 12, 16
ballistic
- channels 74
- leads 7
- regime 6
- system 32
- transport 3, 6, 10, 98
- wires 7
band
- amorphous 460
- conduction 87, 89, 97, 102
- diagram 67, 102, 460f.
- edge smearing 81
- gap 81
- index 89
- offset 367
- paramagnetic 82
- semiconductor conduction 89
- structure 67, 77, 90, 459
- tunneling 357
- velocity 90
bandwidth 189, 195
barrier
- amorphous oxide 80f., 84
- composite 82
- crested 367f.
- effective 357, 367

- energy 113f., 119f.
- epitaxial 67, 84
- epitaxial oxide 77
- height 78, 87, 114f., 120f.
- interface 82
- interface potential 87
- oxidation 402, 410
- region 120f.
- Schottky 87, 90
- /spacer oxides 82
- δ-spike 89
- symmetric 123
- thermionic 87
- thickness 89
- trapezoidal 357
- tunnel 426ff.
- tunnel δ- 94
- tunneling 77, 367f., 375
- width 78, 84, 87, 115
beam
- electron 142ff.
- high-intensity laser 187
- multi-electron- 167ff.
- propagation 183
- pulsed laser 186
- rectangular-shaped 149f.
- superimposed 194
- variable-shaped (VSB) 148ff.
Bennet equilibrium 188
bias
- source-drain 61
- voltage 10, 41, 45f.
- window 43ff.
binary 109ff.
- digit (bit) 109
- information-processing system 110
- logic 112
- state variables 111
- states 109, 111ff.
- transition 116, 123
binary switch 109ff.
- charge-based 123, 125, 130f.
- communicativity 111
- controllability 111
- distinguishability 111ff.
- electron-based 113
- energy barriers 113f.
- multi-electron 122
- optical 130f.
- particle-based 111
- spin-based 124ff.
binary transition 110f.
- per unit area 110
- per unit time 110

Index

- throughput 110f.
binding 289ff.
- differential 289, 293
- high-affinity selective 289
- non-specific 290ff.
- selective 289f., 292
- structured 289
bioassembly of electronic materials 276
biomolecules
- electronic components 276
- electronic conductivity 275
biotechnological tools 277
bit
- anomalous (SILC) 366f.
- erratic 366f.
- multi 372
bit line (BL) 361ff.
- contact 385
Bloch waves 41
Bohr
- magneton 74
- radii 38
Boltzmann
- constant 186
- distribution 114
- equation 69
- exponent 92
- limit 117, 119
bonds
- covalent 321
- dangling 80f.
- non-covalent 321
bottom-up
- approach 305
- method 305
- strategy 275, 305
- technique 320
bound state energies 41
boundary condition 90f.
Bragg
- equation 195
- reflectors 139
buffer gas 193
buried contact (BC) 410

c

cantilever 243ff.
capacitance 38, 387
- constant 393
- storage 384ff.
capacitor 383ff.
- area 385, 387, 311
- cell 385ff.
- 3-D structure 413ff.
- ferroelectric 397, 401ff.
- formation process 386
- over bit-line (COB) structure 402, 408ff.
- planar 413, 415
- planar storage 385
- poly Si 384
- Pt/PZT/Pt 402
- PZT 407, 412ff.
- stacked 384ff.
- storage 383f.
- technologies 385
- trench 384ff.
- -type FRAM 398
- under bit line (CUB) 408ff.
carbon nanotube (CNT) 38f., 52, 56, 277, 284ff.
- addition spectra 56
- conducting 286
- semiconducting 286
carboxylategroup 326f.
carrier
- band velocities 78
- majority 80
- minority 80, 82
- polarization of 94
- recombination 461
catalyst 200
- heterogeneous chiral 321
chalcogenide glasses 490ff.
- $AgAsS_2$ 491
- Ag-Ge-S 492, 507
- Ag-Ge-Se 492, 498, 507
- bandgap 491
- binary 491, 506
- conductivity 491
- Cu-Ge-S 507
- Ge- 491
- ternary 491ff.
chalcogenide material 453ff.
- alloy 459
- amorphous state 455f.
- crystalline state 455f.
- doped 455f.
- eutectic compound 453f.
- long-range order 455
- pseudo-binary 453f.
- short-range order 455
- structure 455
- phase transition 455f.
channel 10ff.
- doping 389
- hot electron injection (CHEI), *see* flash memories
- one-dimensional 10ff.

charge
- distribution 37, 42
- drain 37
- electron 38
- elementary 4
- flow 37
- fractional 37
- gate 37
- image 53
- neutrality 494
- non-switching 398
- quantization 37ff.
- source 37
- state 42
- storage 372, 383
- switching 398
- total 360
- transfer 42
- transport measurements 53
charge trapping 389
- layer 372
- multi-bit 372f.
charging energy 38f., 42, 49, 51, 56
checkerboard pattern 59
chemical
- developer 137, 153, 155, 160
- force microscopy 342
chemical mechanical polishing (CMP) 413f., 416, 428
- process 413
chemical potential 40ff.
- drain 49ff.
- gate 49
- source 49ff.
- synthesis 277
chemical saturation point 494
chemical solution deposition (CSD) 406f., 412
- PZT 407, 412
chemical vapor deposition (CVD) 256, 306, 322, 406
- low-pressure (LPCVD) 378
- metalorganic (MOCVD) 405ff.
- process 406
circuit
- architectures 281
- DNA-templated 277ff.
- near-perfect 277
coarse-grain model 226
coarsening 310
coating 190ff.
- anti-adhesive 229
- metal 282
- multilayer 199

- optical 193
- technology 192
coercive
- field 401, 423ff.
- voltage 401
coherent wave propagation 3
collector 185, 188, 191
- chamber 194
- efficiency 185
- eight-shell 192
- manufacturers 192
- multi-shell 191, 193
- normal-incidence 192
- shell 192
- supplier 191
- surface 193
- Wolter-type nested shell 183, 191
collinear moments 78
collision time 86
complementarity-determining region (CDR) 289f., 293
- loops 290
complementary metal oxide-semiconductor (CMOS) 352, 378, 402, 434f.
- front-end-of-line (FEOL) processing 434, 436
- technology 466
computational elements 109f.
computing system 111
- optical digital 130
conductance 12f., 79
- bulk-to-bulk 84
- bulk-to-surface 84
- differential 48, 50f., 58, 60f.
- Drude 29
- fluctuations 27ff.
- junction 78, 85
- maps 54
- pattern 28
- quantum 46, 48
conduction band 9
- offset 9
- profile 10
conduction measurements 55
conductivity 4, 490ff.
- curves 491
- Drude 16
- mismatch 88, 96
conductor 3f., 7f.
- ballistic 7
- diffuse 3, 6, 14, 17, 29
- metallic 21
- mixed 490
- one-dimensional 11f.

– ring-shaped   3, 21ff.
– superionic   490
conjugated
– cores   333
– OPEs   331
– π-orbitals   331
constriction   12f.
conversion
– coefficient   101, 185
– efficiency   188, 190
Coulomb
– blockade   39, 43, 49, 61, 377
– diamonds   51ff.
– effects   38
– energy   38, 50ff.
– interaction   40f., 49, 51, 53
– peak height   48
coupling
– alkanes   321
– antiparallel   431
– asymmetric   55
– constant   73, 75, 124
– covalent   321
– electron-phonon   58f.
– electron-vibron   57f.
– electronic   321
– ferromagnetic exchange   56
– gate   48, 50f.
– magnetostatic   426
– Néel   426ff.
– non-zero   58
– parallel   431
– strong   42, 47
– structural   321
– weak   42, 44f., 50f.
critical
– density   187
– dimension (CD)   137, 139, 223f., 229
crosslinking
– process   214
– thermal-induced   214
crossover thickness   81
cross-point cell (XPC)
– bit   433
– structure   433
crystalline facets   289ff.
crystallization
– perovskite phase   406
– temperature   406
Curie law for paramagnetism   128
Curie temperature   401, 404
curing   214f.
– UV   230
current   8f.

– contributions   8
– emitter   101
– Hall   74
– injected   66
– in-plane-geometry (CIP)   68ff.
– -perpendicular-to-planes (CPP)   68ff.
– polarized   71, 94
– pulsed   188
– source-drain   123
– status   189, 199
– threshold   96
– tunnel   66f.
– -voltage characteristic   45f., 58f.
– voltage curve   80, 333f.

d

Daata-Das   77
– ballistic spintronic modulator/switch   73
– interference device   66
debris   185f.
– mitigation system   185, 188, 190f., 193, 196
– transmission   185
Debye
– integral   86
– length   120
decoherence   44
deep ultraviolet (DUV) range   185
defect
– charged   493
– density   199
– printable   199
– size   199
– state   80
deflection   241
– cantilever   243, 251
– field   147
– signals   252, 255
deflector   148
– electrostatic   148
– integrated   149
degeneracy points   52, 61
degrees of freedom   59, 99
– nuclear   59
delta-function   46f.
demagnification   148f., 170
demodulation   244
demolding   212, 220f.
– process   227
– temperature   213f.
dendrite   497f.
– formation   497
– growth   497f.
– two-dimensional   498
Density Functional Theory (DFT)   40

density of states (DOS) 8, 12, 41f., 46, 48, 79ff.
– effective 89
– partial 68
– poarized 81
– spin 102
– spin-polarized 87
– total 68
depletion zone 10
deposition 193, 315, 322, 406f.
– /dissolution process 496
– electroless 259, 322
– gas-phase 323
– rate 307
– solution 322f.
– temperature 307
– two-step 324
– vapor phase 322
depth of focus (DOF) 181f., 185
devices
– application-specific integrated circuits (ASIC) 158, 162
– carbon nanotube (CNT) 284, 330f.
– charge-trapping 370
– chemical field effect transistor (CFET) 333f.
– chemical force microscopy 321
– chemical sensors 321
– complementary metal oxide-semiconductor (CMOS) 352, 378, 402, 434f.
– computer-aided design (CAD) 162
– degree-of-freedom haptic 261
– DNA-templated devices 277ff.
– DVD-RAM optical disk 453
– DVD-R/W optical disk 453
– dynamic random access memory (DRAM) 351, 383ff.
– electrical erasable programmable memory (EEPROM) 351f., 370
– electrical programmable memory (EPROM) 351f., 370
– electron interference semiconductor 72
– electron spin resonance (ESR) 128
– exoteric 76
– faulty 277
– ferroelectric random access memory (FRAM) 397ff.
– ferromagnetic terminals 78, 128
– field effect transistors (FET) 121, 162, 433ff.
– FinFET 375
– flash-type memories 351ff.
– floating gate transistor 359, 370
– functional 277
– heterodyne detector 101
– high density memories 349
– interconnecting 278
– island 42
– lab-on-chip 219
– liquid crystal spatial light modulators 130
– magnetic random access memory (MRAM) 70, 128
– magnetic tunnel junctions (MTJ) 421ff.
– magnetoresistive (MRAM) 397, 419ff.
– mass transport 486, 494
– metal oxide semiconductor field effect transistor (MOSFET) 137, 162, 402
– metal oxide silicon (MOS) 359, 377
– micro-electromechanical systems (MEMS) 155, 168, 210, 242
– micro fluidics 210
– molecular 278, 331
– molecular transistor 39
– multifunctional 99
– nanobiotechnology 210
– nano-CMOS 209
– nano-electro-mechanical systems (NEMS) 174
– nano fluidics 210
– nanophotonics 209
– nanopillar 71
– nanoscale planar double gate transistor 163
– nanotube 41
– non-biological 277
– non-linear interference filter 130
– organic nano-electronics 210
– organic opto-electronics 210
– OUM-PCM memory element 449f.
– Ovonic Unified Memory (OUM) 449f.
– phase change memory (PCM) 397, 447ff.
– piezoelectric actuators 242
– quantum interference 73
– read-mostly-memory 448
– read-only memory (ROM) 351, 404
– resistive (RRAM) 397
– rewritable CD 130
– semiconducting single-wall carbon nanotube (SWNT) 284ff.
– semiconductor memory cards (FLASH) 70
– spin-injection 99f.
– spin-LED 94
– spin-orbital (SO) 73
– spin transistor 86, 88
– spintronic 99
– spin-valve transistor 83
– split-gate transistor 356
– static random access memory (SRAM) 351, 435, 441f.
– SWNT-FET 285
– Teramac machine 277
– three-terminal spin injection 86

– three-terminal spintronic   73, 86
– three-terminal switches   37ff.
– weakly coupled   38f.
dielectric   367ff.
– capacitor   387ff.
– constant   367, 369
– gate   388f.
– high-k   387ff.
– nitride-oxide (NO)   387
– ONO   368
– peripheral transistor gate   389
– permittivity   89, 120f., 387
– reliability   389
diffuse regime, see quantum regime
diffusion   6, 195, 307
– adatoms   307
– coefficient   75
– constant   16, 91, 200
– length   201f., 307
– -limited aggregation (DLA) mechanism   497
– sphere   201
– terrace   307
– process   200f.
diffusive motion   16
digamma function   17
dipole
– -pinning effect   403
– -pinning electrons   403
– switching cycles   403
disentanglement   213
dislocations   4, 314f.
– network   315
dissipation
– energy   119
– power   111, 119
DNA   276ff.
– aldehyde-derivatized   286
– -binding proteins   282
– bridge   279f.
– computing   277, 294, 298
– conductivity   279
– -DNA interaction   282
– double-stranded (dsDNA)   278, 282, 285f.
– hybridization   279f.
– junctions   282
– ion-exchanged   279
– λ-   279ff.
– metallization   283f.
– molecules   277ff.
– network   278f.
– non-recurring sequences   293
– scaffold   278, 282, 286
– segment   286
– sequence-specific   282

– sequences   282
– single-starnded (ssDNA)   282, 285f.
– skeleton   279ff.
– synthesis   294
– templated electronics   277ff.
– templates   277
domain
– orientation   70
– pinning   72
domain wall switching
– (pseudo)spin-torque   70
donut problem   165
doping
– acceptor   89, 120, 460
– donor   89, 96, 120, 460
DRAM (dynamic random acess memory)   383ff.
– advanced technology requirements   384f.
– basic operation   383
– capacitorless   393
– cell size   385
– high-density   384
– high-performance embedded   394
– integration   393
– one-transistor (1T-1C)   383
– 1T-1C cell   383
– three-dimensional (3-D) structures   384
drift   247, 252
– compensator   263
– -diffusion expression   75, 91
– thermal   248
– velocity   4, 247
Dresselhaus effect   19, 21, 66
Drude
– conductivity   4
– model   4
D'yakonov-Perel mechanism   19

e
edge
– roughness   184
– spin accumulation   74
Einstein relation   75, 96
elastic
– relaxation   312
– scattering time   5
– strain   311
– strain energy   312
– stress   311
– theory   310
electric
– activity   210
– field   4f.
electrical discharge   186, 188

– pulsed   186
electrochemical cell   494
– mass transport   494
electrochemical potential   9, 11f., 43, 69, 88
electrochemistry   485, 494, 497
– deposition   486
– solid-state   486
electrode
– gate   388f.
– metal gate   389
electrodeposit   494ff.
– charge   501ff.
– formation   500
– growth   497, 499, 501
– growth rate   500
– height   499
– mass   501ff.
– morphology   497
– resistance   501ff.
– stability   494
– surface   498
– volume   501ff.
electrodeposition   485, 494ff.
– process   496
– surface   498
– voltage   506, 508
electrolyte   485, 490ff.
– alkali metal ion   489
– bcc structure   490
– chalcogenides   489
– coordinated motion   496
– depleted   494
– device   506ff.
– electron-supplying cathode   494, 496
– layer   497
– liquid   485
– oxidizable anode   494, 496
– saturated   509
– sublattice   490
– ternary   492
electromigration   439
– issues   439f.
– performance   440
– reliability   440
– resistance   439
electron
– addition energy   52f.
– affinity   96
– beam lithography   9
– charge flux   74
– concentration   4, 32
– density distribution   94
– density of state   85
– discreteness   38

– drift-diffusive   68
– effective mass   78, 84, 90, 101
– energy   4, 151
– gas   5
– hot   82f.
– lenses   143
– mass   4, 38
– mobility   4, 9, 75
– momentum   4, 72
– -nucleus interaction   40
electron optical
– aberration   145
– columns   145ff.
– elements   146f.
– non-ideal elements   145f.
electron optics   142f., 146
electron scattering
– -back   151ff.
– forward   151ff.
– modes   151
electron
– sources   141f., 148
– spin-polarized   65, 101
– spin projection   66
– transfer theory   45
– transmission   332ff.
– transport   4
– unpaired   60
– velocity   8
– -vibron coupling constant   57
– waves   20, 24, 32
– wavelength   3
electronic transport   7
electroplating   485
electrostatic
– energy   49
– potential   40, 43
– repulsion   40
elementary quantum mechanics   3
Elliot-Yafet mechanism   19
ellipsometry   223
embossed structures   227
emission
– efficiency   233
– laser-induced   187
– spectrum   186f.
– spontaneous   233
emitter   190
– non-thermal   186
– thermal   186
enantiomeric form   328f.
energy
– activation   307
– barrier framework   116

– binding   307
– correlation   32f.
– diagram   122f.
– diffusion   307
– discrete   41
– dispersion   11
– eigenvalues   11
– electrostatic   37
– excitation   53
– level   38, 44
– relaxation   313
– separation   11
– transfer   5
Engineering Test Stand (ETS)   196
– projection optics system   196
epitaxial growth   306f., 318, 334
– semiconductor   306
– techniques   306
equilibrium
– configuration   309
– process   309
– spin polarization   75
error
– buttering   147, 150
– classical   114, 116f.
– correction   298
– elongation   299
– phase   199
– probability   116
– quantum   114, 116f.
– rate   299f.
– soft   438
– suppression codes   277, 298, 300
– suppression scheme   277
– synthesis   300
etching   158, 160, 411f.
– beam-assisted   176
– conditions   411f.
– dry   212
– isotropic wet   228
– reactive ion   223
exchange splitting   77, 82
excitation   55f., 210
– charge-neutral   55
– electronic   55, 57
– gapless   66
– lines   53
– subtle   53
– vibrational   57, 59
excited state (ES)   54ff.
expansion coefficient   247
exposure process   161f.
extreme ultraviolet (EUV)
– radiation   182, 185f.

– radiator   185
– range   182
– scanner   182f.

f

fatigue   402f.
– behavior   403
– non-   402
– properties   403f.
feedback
– circuitry   246
– control   247
– electro-bio   277
– loop   254, 277
– off   251f.
– on   252
– system   245
Fermi
– distribution   13, 32
– energy   3f., 11, 29, 32, 41, 45ff.
– factors   80
– functions   40, 47
– level   60, 68, 78, 82, 461
– potential   360
– principle   56
– surface   68
– velocity   5, 38, 68
– wavelength   6, 10, 26, 28
– wavenumber   29
Fermi-Dirac
– distribution   45, 47, 78
– function   43
Fermi's golden rule   58, 67f.
ferrite core memory   420
ferroelectric oxides   401ff.
– Bi-layer structured   401, 403
– BLT   401ff.
– BTO   401
– perovskite-structured   401
– PZT   401f.
– reliability   403
– SBT   401ff.
ferroelectric random access memory
   (FRAM)   397ff.
– capacitor-type   398ff.
– cell operation   399f.
– cell size   410f.
– cell structure   408
– 3-D structure   413ff.
– degradation   412
– high-density   402, 410ff.
– non-volatile   397
– one-transistor-one-capacitor (1T1C) -type
   398ff.

– sense amplifier types   400
– sensing scheme   399f.
– two-transistor-two-capacitor (2T2C) - type   398ff.
ferromagnet-insulator-ferromagnet (FM-I-FM) structures   65, 67, 77
ferromagnet-semiconductor
– boundary   90
– contacts   95
– heterostructure   87, 97
– n$^+$-n'-p heterostructure   101f.
– interface   88f., 92, 95
– junction   87ff.
ferromagnetic   56
– cladding   440
– electrodes   67
– liner field   440
– materials   66
– shell   100
Feynman path   30
field emission   141f.
film
– adhesion   441
– BLT   404
– chalcogenide   493
– delamination   441
– enhanced permeability dielectric (EPD) encapsulating   442
– fatigue-free   403
– ferroelectric   404, 406, 408, 420
– planar   406
– PZT   403, 406, 412ff.
– PZT crystallinity   406
– SBT   403f.
– structure reversible   448
flash memories   351ff.
– cell concepts   355
– charge-trapping   370
– erase mechanisms   353ff.
– floating gate   359ff.
– hot carrier injection mode   353ff.
– NAND   447
– NOR   447
– programming   353ff.
– reliability aspects   365, 367
– scaling   366f.
– thickness   366
– tunneling mode   353, 355
flip-chip bonder   216
floating body cell (FBC)   393ff.
– depleted   393
fluid bed   226
focal point   183, 191
fogging effect   160ff.

foil
– manipulation   248
– trap concept   193
force
– -displacement curve   225
– gradient   246
– interaction   245
– interatomic   241
– long-range electrostatic   241
– setpoint   243
– surface adhesion   252
– tip-sample   246
– van der Waals   256
forward-biased   94f., 102
Frank-Condon factor   59
frequency
– conversion   9
– cyclotron frequency   75
– high   97, 99, 101
– multiplier   99, 101
– operating   99
Front-End-Of-Line (FEOL) processing   434, 436, 466
full-width half maxmum (FWHM)   48, 61, 195
functional
– electronics   279
– film   222
functionality
– electronic   278
– non-biological   278

g

gate   390ff.
– controllability   392
– dual   476
– electrodes   10, 26, 43, 392
– field   42
– fingers   9f.
– length   137, 390ff.
– sidewall   390f.
– splitting   476
– TIS structure   390f.
– voltage   10ff.
Gaussian
– beam strategy   162
– beam writing   159f.
– electron beam   147
– intensity distribution   159
glass-transition temperature   211, 213, 456, 491
gracing incidence angle   191, 194
ground-state (GS)   56ff.
– configuration   53

growth 306ff.
– chamber 306
– CVD 306
– gases 306
– grain 406
– heteroepitaxial 311f.
– homoepitaxial 310
– kinetics 307f., 313
– kinetically limited 307
– MBE 306
– rate 307f.
– self-assembled 307, 314
– self-organized 317
– step-flow 315ff.
– Stranski-Krastanov 312f.
gyromagnetic
– factors 77, 124
– ratio 99, 101

h

half-metallic
– behavior 82f.
– CrO_2/RuO_2 82
– CrO_2/TiO_2 82
– ferromagnet 82
– ferromagnetic junctions 82
– gap 82
– Heussler 83
– multilayer 83
Hall
– effect 75f.
– measurements 4
Hamilton
– formalism 145
– operator 10f.
Hamiltonian 40, 44, 52, 57
– electron 72
– one-electron 40
Hamilton-Jacobi theory 145
Hanle effect 74, 76
harmonic spectrum 58
Hartree 38
Hartree-Fock (HF) theory 40
Heisenberg
– distinguishability length 115ff.
– uncertainty principle 60, 115
hemispherical grains 387f.
heteroepitaxy 310f.
hierarchical assembly 259
holes 101
– heavy 73
homoepitaxy 307
HOMO-LUMO 332
– gap 52f.

hopping
– electron 44
– inelastic 81
– ion 488, 491
hybrid methods 317
hybridization
– gap 82f.
– state 83
hysteresis 249, 423ff.
– curve 399, 423f.
– effects 249
– loop 424f.

i

IBM 109f.
illuminator 183, 191
imaging
– full-field 184
– errors 197
– specifications 197
– system 196
imprint
– single-step 233
impurity 4f., 19, 27, 74, 80ff.
– -assisted conductance 80f.
– -assisted tunneling 81, 86
– channels 81
– collisions 30
– configuration 27f.
– effective 80
– level 81
– magnetic 60
– scattering 74, 77
– states 82
– supression 80f.
inband
– emission 185
– radiation 185
inelastic
– effects 84
– processes 84, 86
information
– carrier 109, 113
– -defining particle 113f.
information-processing
– applications 126f.
– element 109
– systems 109ff.
information
– processor 110
– throughput 111
injection molding 210
integrated circuit 137, 139
– fabrication 173f.

interactions
– atom   241
– dipol-dipol   325
– electron-electron   44
– electron-phonon   44, 85
– electron-photon   44
– electrostatic   325
– exchange   56
– fluid-solid   226
– intermolecular   318, 321, 327
– metal-ligand   321, 329
– non-covalent   321
– van der Waals   325
interface
– conductances   88
– functional   293
– metal-oxide   84
– metal-semiconductor   84
– oxide   83
– roughness   426
– silicon/insulator   360
– silicon-molybdenum   196
– stamp/polymer   227
– user   260f.
interfacial
– adhesion   227
– energy   220f.
– separation   227
interference   3, 20, 23
– constructive   15, 17, 20, 26
– effects   13
– electron   3, 20, 23
– pattern   25, 26
interlayer dielectric (ILD)   229
– UV   229
International Technology Roadmap for Semiconductors (ITRS)   139f., 166, 199, 508
intramolecular confirmation changes   325
intrinsic
– angular momentum   124
– carrier concentration   120
inversion
– asymmetry effect   74
– symmetry   19, 75
– time   75
ion
– beam focused (FIB)   173, 176
– conduction   491
– conductors   493
– current   494
– flux density   500
– 10-fold ionized   186
– mobility   493f.
– milling   174
– optics   173
– projection direct structuring (IPDS)   174f.
– sources   173
– transport   485, 491
ionization
– avalanche impact   461f.
– levels   186
island   41f., 44, 46, 49, 54ff.
– chemical potential   309f.
– coarsening   310
– coupled   42
– density   307f.
– Kranski-Krastanov   312
– mode   312
– morphology   312f.
– nucleation   307f.
– size distribution   309
– sizes   308f.
– thermodynamically stable   309, 311
– three-dimensional (3-D)   311ff.
– two-dimensional (2-D)   307ff.

k

Kalman
– equation   263
– filtering   263
Kane Hamiltonian   73
kinetic
– energy   113f.
– equation   97
Kondo
– effect   59ff.
– peaks   61
– physics   60
– resonance   60f.
– temperature   61
kp method   72

l

Lagrangian function   15f.
Lance-like structure   449
Landau-Lifshitz (LL) equation   72
Landauer-Büttiker formalism   3, 7, 12, 28f.
Landauer
– expression   71
– formula   47f.
Landé gyromagnetic factor   124
laser
– frequency   187
– high-power pulsed   188
– intensity   187
– power   188
– pulse   190

– pulsed CO_2 188
– solid state diode pumped   188
– -triggered vacuum arc   189f.
lattice   4
– constants   310ff.
– mismatch   311, 315f.
– regular triangular   319
– vibrations   5
layer
– AlGaAs spacer   9
– anti-adhesive   221
– antiferromagnet-pinned reference   424
– antiferromagnetic oxide   72
– bi-   426
– boundary   194
– BOX   393f.
– cap   429
– charge-storage   354, 359
– charge-trapping   354, 372, 376
– dead   429
– dielectric   102
– diffuse   495
– δ-doped   97, 101ff.
– double   495
– drain   70
– ferromagnetic   68, 99f.
– free   423f., 426ff.
– GST   449f.
– Helmholtz   495
– insulating   100, 137f.
– interconnect   229f.
– magnetic   72, 99
– multiple   426
– n-semiconductor   96, 98
– n-type δ-doped   9f., 88f.
– oxide-nitride-oxide (ONO)   366, 373, 376
– oxidizable   494
– permalloy free   72
– pinned   423ff.
– protective   198
– reference   427
– residual   222, 224, 229
– seed   428, 432
– semiconductor   88, 97
– SOI   393f.
– space   68
– spacer   432
– strained 2-D   313
– storage   354, 357, 359
– thickness   100, 224
– trapping   370f.
leads   7f., 41ff.
lens
– condenser   148
– electromagnetic   139
– electron   143
– electrostatic   143, 170
– magnetic   144
– non-rotational-symmetric   145f.
– optical   137ff.
– real   146
– rotational-symmetric   146
level
– discrete   44
– separation   39
– spacing   39
– spectroscopy   56
– splitting   56
lifetime   189ff.
– collector   192
– electrode   190
line edge roughness (LER)   199f., 201f.
lithography   137ff.
– atom   210
– bio-inspired   210
– charged particle   137, 139ff.
– chromeless phase (CPL)   159f.
– commercial   155
– deep ultraviolet (DUVL)   138f., 156, 158ff.
– direct-write   151
– electron beam (EBL)   140ff.
– electron beam direct-write (EBDWL)   158, 162ff.
– electron projection (EPL)   139, 164f.
– extended ultraviolet (EUVL)   139, 158, 182ff.
– extreme ultraviolet   181f.
– Gaussian beam   147, 159
– gray-scale   228
– hybrid   158
– ion beam (IBL)   140, 172f.
– ion direct-structuring (IDS)   174
– ion projection (IPL)   139
– low-energy electron beam proximity (LEEPL)   165f.
– maskless (ML2)   139, 164, 170
– mix-and match   158, 172
– mold-assisted   211
– molecular   282f.
– multiple-electron-beam   139
– nanoimprint (NIL)   139f., 158, 170, 209ff.
– non-optical   209f.
– optical   137, 170, 222
– photoactive nanoimprint (PNIL)   170
– process   139
– projection electron beam (PEL)   164f.
– projection maskless (PML2)   167ff.
– proof-of- (POL)   169
– proximity electron (PEL)   139

– proximity maskless   169, 170
– reversed contact UV-NIL (RUVNIL)   228f.
– reversed nanoimprint (RNIL)   228
– thermal nanoimprint   211
– SCattering with Angular Limitation PEL (SCALPEL)   164
– sequence-specific molecular   281ff.
– shaped beam   148
– single-step NIL   215
– soft   210
– step-and-flash imprint (SSIL)   216f., 223
– step-and-stamp imprint (SFIL)   216ff.
– technique   139
– thermal nanoimprint (NIL)   212ff.
– ultraviolet nanoimprint (UV-NIL)   211f., 222
LMR (Laboratory of Molecular Robotics)   250ff.
– nanomanipulation   250
– software   260
loading effect   161f.
localization   16f.
– weak   13ff.
Lorentzian   47
– density   46
– peaks   42

m

macromolecular network   214
macromolecule   214
magnetic   419ff.
– configuration   67
– field   4, 12, 16ff.
– field constant   188
– film stack deposition   437
– flux   14, 19, 22, 24f., 425
– flux quantum   17, 24f.
– memory   70
– moments   67, 70, 78
– multilayers   65, 67
– nanopillars   66
– permeability   128
– relaxation time   17f.
– semiconductor electrode   86
– sensing   99
– storage   420
– transition metals   70
magnetic tunnel junction (MTJ)   66, 421ff.
– encapsulation   438
magnetization   420ff.
– anti-parallel   422f.
– ferromagnet   420
– moment   66
– parallel   422f.

magneto-conductivity   21
magnetoresistance (MR)   19, 23, 25f., 65, 77ff.
– disorder sppresses   82
– Dyakonov's   74, 76
– giant (GMR)   66ff.
– negative   81f.
– positive   74
– ratio   421
– tunnel (TMR)   66ff.
magnetoresistive random access memory (MRAM)   419ff.
– back-end-of-line (BEOL) processing   436, 440f.
– basic   420
– cell structure   429
– circuit design   429ff.
– future   441f.
– hinges   436
– integration   436
– magnetic tunnel junction MTJ-MRAM   421ff.
– process steps   437ff.
– processing technology   436
– reference cell defects   436
– reliability   438
– resistance   423
– structure   421, 423
– toggle   431f.
– toggle-mode   432
magnetostatic demagnetization   426
magnification   146, 183, 191, 196
magnon   84ff.
– density of state   85
– emission   86
– frequency   85
manipulation   257ff.
– distance   253
– DNA   282
– enzymatic   278
– interactive   260
– LMR   257
– of nanoparticles   253
– operation   251
mask   183, 198f.
– blank   198f.
– hard   411, 429, 437f.
– hybrid   228
– length   199
– reflective   139
– soft photoresist   429
– stencil   164
– transmission   164, 174
material

– cap   429
– ferroelctric   397, 401ff.
– magnetic   397
– phase-change   397
– resistive switching   397
matrix-isolation
– 2-D   331
– method   324
maximum binary throughput (BIT)   110
mean free path
– elastic   4f., 9f., 14, 19, 68
– inelastic   4f.
memory cell   352f., 358f.
– AND   358
– array architecture   358
– charge-trapping flash   370, 375f.
– disturbs   361f.
– embedded   420
– EPROM tunnel oxide (ETOX)   361ff.
– ferroelectric random access memory (FRAM)   397ff.
– flash   447
– floating gate memory   361
– magnetoresistive (MRAM)   397
– multi-bit   370f., 377
– multi-bit charge trapping   372ff.
– NAND string   370
– NAND-type   358f., 363ff.
– non-volatile   351f.
– NOR-type   358f., 361ff.
– phase change (PCM)   397, 448ff.
– resistive (RRAM)   397
– silicon nanocrystal flash   377
– SONOS   370f., 375, 377
– stacked gate   361
– TANOS   372
– three-dimensional stacking   370
– virtual ground NOR   358
metal
– lift-off step   223, 227f.
– -ligand 2-D network   329
metal nanoparticle   334ff.
– assembly   337
– one-dimensional assembly   341f.
– preparation   334ff.
– self-assembly   334ff.
– three-dimensional assembly   337f.
– two-dimensional assembly   339ff.
metal
– organothiols   325
– spacer   68
– strap   434f.
metallic
– dots   51

– ferromagnetic electrode   86
– ferromagnetic nanowire   99
– -half   79
– heterostructures   65, 69
– island   39
– nanostructures   33
– nanotube   38
– quantum dot   51, 56
– spacer   72
– spin valves   72
metallization   278, 281ff.
– biomolecules   281
– DNA   283ff.
– gap   283f.
– gold   283f.
– PZT   407
metalorganic chemical vapor deposition (MOCVD)   405ff.
– PZT   407, 412
metrology   233
– non-destructive   223
microfluidics   214, 230
mirror   195ff.
– diameter   197
– real   194
– multilayer   185, 194f.
– reflecting multilayer   182, 196, 198
– spherical multilayer   183
misfit dislocation network   313
mode
– AC   244
– contact   243, 250, 255
– dynamic   243f., 246, 251
– intermitting-contact   244
– magnetic AC (MAC)   244
– magnon   84f.
– non-contact   244
– oscillatory   244
– phonon   84f.
– tapping   244
model
– capacitance   49
– constant interaction   49f.
– defect chemistry   403
– effective circuit   70
– energy barrier   123
– equivalent circuit   69
– honeycomb-chain-trimer   319
– independing particle   40
– Kane's   72
– kp   73
– Luttinger-Kohn's   72
– microscopic   77
– Mott's two-fluid   66, 69

– throughput   185
modulation transfer function (MTF)   200f.
molecular
– beam epitaxy (MBE)   306, 313
– biology   275
– building blocks   320
– electronics   278
– hexagonal network   319
– island   59
– layers   318
– mobility   214
– monolayer   322
– recognition   276
– -scale building blocks   275
– -scale electronics   275, 277
– self-assembly   320
– shift-registers   293
– systems   57
– weight   213
molecule
– chiral organic   327
– information-carrying   278
– non-recurring sequences   293
– oligophenylene ethynylenes (OPE)   331f.
– periodic   293
– reorganization   318
– self-assembly (SAM)   306
– symmetrical   332
monolayer   320
– -high surface steps   315
– thick wires   315
Monte-Carlo-based simulation   152
MOSFET's   77
Mott and Davis's picture   460
multilayer
– coated components   196, 198
– optics   194
– systems   194f.
multi-pass
– gray writing   160
– writing   150
multispin system   66

n

nanoelectronic systems   4f.
nanofabrication   209ff.
– bottom-up approaches   209
– emerging approaches   209
– methods   305
– top-down approaches   209
nanoimprint   210, 217
– process   210, 212
– roll-to-roll   217, 222
– stamp   217

– thermal   212
nanoionics   485f.
nanoislands   313f.
– lateral positioning   314
– semiconductors   313
nanomagnets   70
nanomanipulation   239, 247, 256
– LMR   250
nanoparticle   257, 264, 334ff.
– linking   259f.
– metal   334
– optical properties   334
– pattern   257
nanopatterning   209f., 257
– bottom-up emerging   210
nanostructure
– fabrication   306
– formation   305, 311
– functional   305
– growth   306
– inorganic   318
– self-assembled   306
– semiconductor   311
– supramolecular   327
– templates   318
– thermodynamically stable   309
nanowire   99f., 257f., 315ff.
– conductive   100
– semiconductors   313
– silicide   315
narrow-gap
– n'-region   102
– semiconductor   103
Navier-Stokes equation   224
near-Gaussian intensity   147f.
Newtonian fluid
noise
– environment   400
– immunity   400
non-equilibrium
– electrons   92, 99
– Fermi levels   89
– spin polarization   86
non-recurring sequences   293
non-relativistic action   14
non-volatile memory (NVM)
– market   447
– technologies   447
nucleation   314, 406
– ordered   315
– process   406
– regime   308
nucleoprotein filament   283, 285ff.
nucleus

– critical   307
– subcritical   307
numerical aperture (NA)   137ff.
Nyquist contribution   5
Nyquist-Johnson noise   117

o

oligonucleotide   278f.
– 12-base   279
– sequence   279
– single-stranded   279
– solution   280
– synthetic   294
one-dimensional
– density   8
– electron gases (1DEGs)   19
– leads   7f.
– structure   4
– system   3
– wire   8
Onsager relation   75
operation
– CHANGE "0-1" mode   112f., 122
– continious   189
– elementary switching   112
– pick-and-place   255
– READ   111
– short-term   189
– STORE "0" mode   112f.
– STORE "1" mode   112f.
– switch   118
– TALK   111
– WRITE   111
optical
– imaging   224
– lens   137ff.
– reflection   223
– system   196ff.
orbitals   42, 52f.
– electronic   53, 58
– energies   50
– one-particle   43
– single-electron   40
oscillation
– Al'tshuler-Aronov-Spivak   21f., 24ff.
– cycle   246
– magnetic vortex   72
– pattern   26
– resistance   21f.
oscillator   11, 57ff.
Ostwald ripening   309f., 323
overpotential concentration   495
Ovshinsky's picture   460
oxygen vacancies   403

p

parallel
– configuration   71f.
– moments   66ff.
pattern transfer   215, 230
– technology   233
patterning   151, 153, 173, 317, 429
– ability   228
– 3-D   227f.
– double   181
– electron beam   153
– information   282
– ion-beam-induced   175
– metallization   283
– multiple ion beam projection maskless (PMLP)   175
– process   233
– tunnel junction   437
Pauli's
– matrices   77
– principle   39, 42, 45
penetration depth   182, 199f.
permanent magnetic momentum   124
permittivity of free space   187
perovskite
– capacitor   407
– crystal structure   407
– oxide electrode   407f.
phage   290ff.
– helper   292
phase
– accumulation   5, 14, 16
– breaking   5, 17
– -breaking time   6, 16
phase change memory (PCM)   397, 447ff.
– amorphization processes   458f.
– back-end-of-line (BEOL) processing   436, 440f.
– bipolar p-n-p select device   465
– cell components   464
– cell formation   466
– cell optimization   467
– cell size scaling   476f.
– cell structures   464f., 467f.
– chalcogenide   453ff.
– crystallization processes   458f.
– cycling   470
– design optimization   458f.
– dissipated power   473f.
– failure modes   470
– figure of merit   478
– integration   464ff.
– Joule heating   467, 474
– material   397, 453, 458ff.

– material parameter   457
– material phase transitions   458
– modelling   458
– optimization direction   457
– physics   458f.
– programming characteristics   452
– properties   457
– read and program disturbs   472
– recovery   463
– reliability   469
– reset current   473ff.
– RESET state   450ff.
– retention   469f.
– scaling   472f., 475ff.
– SET resistance   478
– SET state   450, 452, 458f., 469f.
– switching   397, 458f.
– thermal resistance   467
– thermal stability   469
– thickness   468
– 1T1R cell structure   465
– technology   469, 474
– temperature profile distributions   472
– transient behavior   463f.
– voltage scaling   474f.
phase-coherence length   4ff.
phase-coherent pieces   31
phase-coherent transport   3, 5, 13, 34
phase
– change   448, 450
– factor   17
– memory   4, 14
– separation   492f.
– shift   5, 14f., 17, 26f.
– transformation   448f., 455
PHINES (programming by hot-hole injection nitride electron storage)   374
phonon   84ff.
– blockade   58
photodetector   241
photodissolution   493
photolithography   137ff.
– immersion   138f.
– process   137f.
photomask   158ff.
– binary   159f.
– blank   158, 170
– chromeless phase lithography (CPL)   159f.
– fabrication   170ff.
– high-resolution   159
– patterning   159
– phase-shift (PSM)   159f.
– transmission   137ff.
photon   203

– -based solution   182
– statistics   201, 203
– wavelength   137ff.
photonic
– applications   230
– crystals   227f., 233
photoresist   137, 153ff.
– bi-layer process (BLR)   158
– chemically amplified (CA)   156ff.
– diazonaphtoquinone (DNQ)   155f.
– hydrogen silsesquioxane (HSQ)   157f.
photosensitive components   233
physical vapor deposition (PVD)   192, 322
physiosorbed   53
pinned   423ff.
– antiferromagnet-   424f.
– ferromagnet   426
– layer   423ff.
pinning   423ff.
– strength   426, 429
pitch
– bitline (BL)   361, 363
– wordline (WL)   361ff.
pixel values   243
Planck's
– constant   186
– law of radiation   186
– limit   186
plasma
– column   188
– cylinder   188
– discharge-produced   186ff.
– EUV-emitting   185
– EUV-emitting pinch   193
– focus   189
– frequency   187
– gas-discharge   188f.
– hollow-cathode-triggered pinch   189
– hot   186, 188
– ignition   188
– laser-induced   186f., 189
– lithium-based   182
– pressure   188
– source   182
– state   186
plasmonics   233
point contact opening   13
– geometrical shape   13
Poison equation   120
polarity state
– opposite-   404
– stored   404
polarization   65f., 68, 399f., 411
– backswitching   404

– behavior   402
– degradation   411
– effective   78
– field   404
– reduction   403
– remanent   401f.
– retention   403
– reversals   402
– tunnel   68
– tunnel current   78
polarized
– light   66
– radiation   101
– radiation recombination   103
polycyclic aromatics compounds   319
polymer   211ff.
– amorphous   225
– biocompatible   230
– chain   213
– conducting   218, 230
– crosslinked   214, 228
– film   211f., 220, 229
– flow   211f.
– layer   212, 222, 227
– light-emitting   230
– loaded with nanoparticles   230
– molding   210
– nanoimprinted   223
– optics   214
– overflow   229
– pattern   228
– pre-   214
– surface structured method   211
– technology   210
– thermoplastic   212f., 228
– thermosetting   212
– UV-curable   229
– viscosity   213
polymerization   214, 283
– RecA   285
polymethyl methacrylate (PMMA)   155, 202f., 225
polymorph   489
Poole-Frenkel mechanisms   462
positioning paths   264f.
potential
– barrier   92
– 2-D profile   11
– electrochemical   43, 88
– external   49ff.
– parabolic   11
– profile   10f., 19
– saddle-shaped   11
– vector   16, 25
– well   10f.
power
– consumption   127
– spectral density (PSD) function   197
printing   211ff.
– cycle   211
– direct-   230
– micro-contact   210, 331
– nano-contact   331
– pressure   213
– roll-to-roll   217
– techniques   217, 230
– temperature   212ff.
Probe Control Software (PCS)   252
process flow   137f.
projection optics   137ff.
– mirror   139
projection
– Exposure with Variable Axis Immersion Lenses (PREVAIL)   164
– performance   152
– space charge effects   166
– system   145
proof-of-concept (POC) tool   168f.
propagation   8
– electron   9f., 14
– free   10, 15
protein machine   276
proximity
– correction   153f., 159
– effect   152f., 160
– function   153
pushing
– interactive   253
– nanoparticle   253
– operating   252, 254
– paths   264
– perfect   253

q

quantization axis   66, 70f., 77
quantum channels   29
quantum chemical calculations   59
quantum computing   77
Quantum-Confined Stark Effect   131
quantum confinement   49, 53
quantum dot   4, 50, 59f.
– nanotube   38f., 60f.
– self-assembled   39
– semiconducting   38f., 60f., 233
– single molecule   39, 60f.
– spin-polarized   76
– systems   39
– well   65, 131

quantum
- confinement 377
- effects 6
- energy 38, 50
- level splitting 39
quantum mechanical
- behavior 37
- correction 21
- wave function 37
quantum
- mechanics 37
- point contact 3, 9, 13
- quantization 39
- regime 6
- splitting 53
- states 41
- well 9, 20, 21, 66, 102
quantum wire 27f.
- networks 22
quasi steady-state solution 224

r
radiation
- background 113
- blackbody 186
- EUV 194, 199, 203
- polarization 103f.
- power 189
- recombination 103
Radical Innovation MAskless NAnolithogrphy (RIMANA) 169
Rashba effect 20, 21
raster scanning 147
RecA 282ff.
- monomer 283
- nucleoprotein filament 284ff.
- protein 282, 285f.
- servings 285
recognition 288f.
- facet 289
- process 293
- sequence-specific 282
recombination 278
- homologous genetic 282f., 285
- reaction 282
reflection 7, 30, 130
- back- 30
- index 195
- probability 8f., 29f.
reflectionless contacts 47
reflectivity 191f., 193ff.
- angular-dependent 191
- EUV 198
- finite 191

- gracing incidence 192
- incidence 182
- mirror 182
refraction
- electron-optical 143
- index 130, 181, 192, 194, 215
regime
- ballistic 6
- diffuse 6
- quantum 6
n-region 101ff.
n'-region 101
p-region 101ff.
remanence 423
repetition rate 189
replication 219, 278
reservoir 7f., 11f., 40ff.
- drain 41
- source 41
resist 153ff.
- amplified 201ff.
- bi-layer process (BLR) 158, 167
- chemically amplified (CA) 156ff.
- commercial 155
- dispenser 217
- electron 139f., 147, 151, 153ff.
- exposure 153
- hydrogen silsesquioxane (HSQ) 157f.
- inorganic 228
- liquid 214
- monomer 170
- negative 154f.
- plane 137f.
- positive 153ff.
- processing 140
- sensitivity 184f., 199, 201, 203
- sequence-specific 283, 286
- single-layer 167
- thickness 199ff.
resistance 4, 7, 9, 13ff.
- corrosion 321
- fluctuations 27, 31
- four terminal 8
- macroscopic 7
- modulations 25
- off-state 508ff.
- on-state 508ff.
- oxidation 410
resolution 138, 149, 154, 181, 199f.
- capability 155
- enhancement technique (RET) 138, 159
- high 141
- near-atomic scale 199
- optical 151

resonance 44, 48
– energy 45
– frequency 243f.
– line width 48
resonant
– conductance 80
– energy 45
– quantum 48
– state 45
– transmission 81
– tunnel diode-type 83
– tunneling 81f.
retention 365ff.
– characteristics 366
– data 385
retention failure
– mechanism 405
– problem 405
retention
– loss 365, 367, 404
– opposite-state 404, 407
– properties 365, 371, 407
– pulse sequence 405
– result 406
– same-state 404
return probability 16
– total 16
reversed-biased 93
rheology 211f., 224
Richardson's factor 90
root mean square (RMS) figure error 197ff.
rotation matrix 20

s
scan
– line 243
– slow 243
scanner
– DUV 183
– EUVL 195f.
– response 249
scanning electron microscopy (SEM) 141, 163, 215, 223, 255
– field emission 260
scanning probe microscopy (SPM) 239, 324
– imaging 240
– nanolithographic 240
– tip 239
scanning tunneling microscopy (STM) 239, 316, 324
– tip 239f.
scattering
– center 4, 14, 16, 19f., 27
– diffuse boundary 19

– elastic 4ff.
– electron 5
– electron-electron 5
– electron-photon 5
– event 4
– inelastic 3ff.
– intersubband 12
– ionized impurity 9
– length 5
– Mott skew 74
– small angle 199
– small-energy-transfer electron-electron 5
– spin 20, 22
– spin-flip 5
– target 5
scatterometry 223
– spectrum 224
scFv 289ff.
– antibody 291
– fragments 289ff.
– molecule 292
Schottky
– barrier 74, 77, 87ff.
– contact 87, 92, 97
– δ-doped barrier 88ff.
– junctions 66, 78, 87ff.
Schrödinger equation 77
second-order process 44, 59
seeding 405
self-assembled monolayer (SAM) 321ff.
– adsorption of nanocomponents 330
– applications 330
– densely packed 322
– head group 325
– highly ordered 326
– mixed 323
– preparation 322f.
– spacer 325
– substrate 325
– surface modification 330
– terminal functional group 325
self-assembly 276, 283, 287, 305
– complex 276
– epitaxial growth 306
– formation of nanostructures by 306
– of atoms 305
– of molecules 306
– process 326
self-organization 317, 320
semiconductor 94ff.
– conductivity 96
– degenerate 66, 93
– heterostructures 65, 96
– high-resistance 96

– interface   96
– layers   9, 97ff.
– nanostructures   33
– narrower gap   101
– non-degenerate   91, 96
– non-magnetic (NS)   86, 94f.
– n-type   96f., 99ff.
– n^+-type   96
– parameter   96
– quantum dot   51, 56f.
– ring structures   25
– structures   18, 32
SET/RESET programming cycles   469ff.
shallow trench isolation (STI)   361
shear stress   225
shift registers (SRs)   277
Shockley-Hall-Read (SHR) recombination   461
short-chain segments   213
shot noise   201ff.
silicon
– band-gap   354
– hydrogen-passivated   240
– surface   240
single-electron effects   37, 40
single-particle
– energy level   41f., 49f.
– orbitals   41
– Schrödinger equation   49
singlet states   57
Slater determinat   40
solid electrolytes   485ff.
– alkali metal ion   489
– chalcogenides   489
– device application   489
– device layout   504, 506
– future directions   511f.
– inorganic   488
– lithium batteries   488
– mass transport   485f.
– memory devices   504f.
– operation   504
– surface   486
– technological challenges   511f.
– ternary   497
solid metal electrodeposits   488
solid-state memory technology   449
source
– collector module   183
– concept   189
– high power   192
– incoherent plasma   182, 185
– plasma   183
– power specifications   185, 189
– spatially extended plasma   183
– synchrotron radiation   186
– thermal   192
– tin-based gas discharge   193
spacer   194
– materials   194
– oxides   82
spatial frequency range
– high (HSFR)   198
– mid (MSFR)   198f.
– roughness   198
spatial uncertainties   247f., 252
spectrotopic tool   53
spin
– accumulation   69ff.
– angular moment   70
– angular momentum transfer rate   72
– -coated film   211, 228
– coefficient   90
– -coherence time   96
– conductances   71
– current   73
– factor   90
– filter   422
– -flip   56, 68, 71
– density   66
spin-diffusion   88
– depth   93
– length   91, 101
spin
– down states   19, 66f., 71
– drift   88
– -drift length   91
– -effects   19
– ensemble-based quantum computing   66
– extraction   86ff.
– factor   90
– field effect transistor   21
– -flip scattering length   427
– -Hall effect (SHE)   66, 73f., 75f.
– injection   77, 86ff.
– injection/accumulation   86
– injection efficiency   65f., 68, 87f., 90f., 94f.
– injection-extraction processes   65f., 86
– LED   66
– logic circuits   76
– logic devices   66
– majority   78, 80, 82, 102
– minority   70, 78, 80, 82, 102
– -mixing   86
– momentum   70f.
– non-equilibrium   66
– -orbit coupling   3, 19, 21, 66

– -orbital (SO) coupling   72ff.
– -orbital (SO) interaction   74f.
– orientation   19f., 76
– penetration depth   91, 93
– penetration length   95
– polarization   65f., 71, 88, 93ff.
– polarization flux   75
– polarization vector   75
– polarizer   422
– precession   3, 19ff.
– precession frequency   75
– -precession wave vector   73
– projection   67, 70, 80
– relaxation   69
– relaxation time   65, 75, 86, 96
– rotation angle   99ff.
– -selective properties   88, 101
– state   19f.
– subband   78
– -torque (ST) measurements   71
– -torque (ST) switching   66
– transport   65f., 88
– tunneling   66, 89
– unpaired   60
– up states   19, 66f., 71
– -valve   72, 421
– -valve effect   97
– wave function   71
spintronic
– devices   65, 421f.
– effects   65, 67
– mechanism   99
split-gate   9ff.
– electrodes   9ff.
– point   3, 7
– quantum point contact   9ff.
spring constant   244ff.
sputtering   193
stability
– diagram   52ff.
– mechanical   226
stamp   211ff.
– anisotropy   226
– cavities   213, 224ff.
– design   213, 225
– deterioration   223
– imprint   222
– quartz   221
– removal   227
– roll-to-roll   223
– silicon   220
– three-dimensional   230
– transparent   211
– two-dimensional   221

– UV-transparent   221
standard spin algebra   80
statistical mechanics   42
Stefan equation   224
step edge   315f.
– energy   310f.
Stokes equation   226
Stoner spin-splitting   82
storage
– data   397
– element   397
– -node   387
– -node contact   385f.
subbands   11f.
– 1-D   11
substrate   138, 211f., 428f.
– deformation   224
– inorganic   318
– planar   322
– preparation   322
– SAM
– semiconductor   292f.
– single-crystal metal   322
– support   212
– surface   318f.
– p-type   287
– ultra-smooth   428
superexchange process   45
superlattice unit cell   326
superposition   28, 77
superstructure   325f.
supramolecular   320ff.
– assembly   330
– binding   327
– chemistry   321, 327
– 2-D honeycomb network   328
– nanostructures   321
– structure   320, 328f.
– systems   321
surface   319
– effects   77
– energies   221, 312f.
– flat   255
– imperfect   84
– inorganic   293
– lattice   320
– metal layer   223
– model   322
– nanostructured   222
– non-oxidizing   321
– reconstruction   293, 308, 311, 319
– roughness   192, 195, 197, 322
– semiconductor   321
– silicon   320

– specifications   197
– state   84
– states assisted TMR   84
– -stress domain   310
– states assisted tunneling   86
– topology   197, 211
switching   430ff.
– bidirectional   439
– bipolar   511
– characteristics   430, 433, 449
– critical threshold   451
– energy   130f.
– field   440, 451
– memory   448
– modelling of RESET   462f.
– modelling of SET   462f.
– Stoner-Wohlfarth (S-W)   430ff.
– threshold   448, 451
symmetry
– points   72
– translational   41

t

Tamm states   84
template   259f.
– affinity   331
– fabrication   176
– imprint   170, 176
– stripping   322
thermal
– activated process   225
– evaporation   322
– expansion   198
– expansion coefficient (CTE)   198
– fluctuations   57
– stability   195f., 214f.
– temporal   215
– velocity   90
third-order perturbation theory   73
thouless energy   32
time-dependend
– dielectric breakdown (TDDB)   440
– resistance drift (TDRD)   440
time-depending perturbation theory   44
time-energy uncertainty   45
top-down
– approach   305
– method   305
– strategy   275, 305
topography   243
– non-null   250
– signal   250ff.
trajectory   27f.
– closed   19f.

– electron   23, 28
– electron propagation   14
– equation   145
– paraxial   144
– time-reversed   22
transimpedance   509
transistor   383ff.
– array   383ff.
– array field-effect transistor (FET)   390
– fin-array-FET   390ff.
– formation process   385
– performance   384f., 390
– peripheral   392
– recess-channel array (RCAT)   389f.
– silicon-on-insulator (SOI)   390
– surrounding gate (SGT)   391ff.
– trench isolated (TIS)   390ff.
– vertical   391
transit time   98f.
transition   44, 186f.
– classic   114ff.
– 2D to 3D   313
– direct   44
– final state   44
– initial state   44
– intermediate state   44
– over-barrier   116f.
– quantum   114ff.
– spontaneous, see error
– thermal   117
– through-barrier tunneling   116f.
transmission   7, 47
– amplitudes   71
– coefficient   71, 79
– electron microscopy (TEM)   194, 255
– ideal   12
– mask   139, 164
– optical   193
– probability   9, 12, 28, 30f., 71, 78f.
transport   42f.
– ballistic   6f.
– classical   6
– coherent-incoherent   43
– diffuse   7
– elastic-inelastic   44
– linear   52
– non-linear   54
– non-phase coherent   15
– phase coherent   15
– phenomena   6
– properties   9, 27
– quantum ballistic   6
– regimes   6
– resonant   43, 45, 48

– resonant-off-resonant 44
triplet states 57
tunnel
– barrier 42, 67, 94
– barrier dielectrics 440
– current density 78
– exchange vertex 85
– Hamiltonian 84f.
– junctions 49, 66, 79, 82
– magnetoresistance (MR) 65
– spin junctions 65
– width 80
tunneling 59ff.
– band-to-band 373
– co- 44, 59ff.
– condition 115
– conductance 84
– contact resistance 96
– direct 45, 358
– elastic 44, 59f., 85
– elastic coherent 89
– electron 65
– final 67
– Fowler-Nordheim (FN) 357f., 361ff.
– inelastic 59, 61f., 85
– initial 67
– matrix element 67
– out 55
– probability 115f.
– process 42, 62
– quantum mechanical 114
– sequential 44, 59
– spectroscopy 61
– surface-to-surface 84
– thermoemission current 92
– transparency 87
– velocity 78
two-dimensional
– electron gases (2DEGs) 5, 9f., 12f., 15f., 18f., 73
– structure 4

u

UHV 240f.
– methods 322
– system 323

v

vacuum condition 183
Vasko-Rashba
– coupling constant 73
– effects 66
– Hamiltonian 73f.
– spin splitting 73f.
– term 73
vector
– multi-pass strategy 161
– scan exposure strategy 149
– scanning 137f.
– -shaped beam 137, 150
– single-pass strategy 161
– strategy 137
vibrational
– frequencies 59
– modes 57ff.
– molecular states 57, 59
vibrons 57
viscosity 212, 214, 224f.
– inherent shear-rate-dependent 225
voltage
– operation 412
– scaling 412
Volmer-Weber 3-D island growth mechanism 497

w

wafer 140, 155, 162f., 166, 169, 181
– full-patterned 216
– silicon 137, 215, 322
– throughput 182, 184, 199
wavefunction 40, 78
– transverse 41
wavelength 181f.
– centroid 199
– De Broglie 210
– laser 187
wavevektor 32
weak
– antilocation 19ff.
– coupling 42, 45, 49
– localization effect 3, 15f., 18ff.
welding 258, 260
Wenzel-Kramers-Brillouin (WKB) approximation 115
Wetting control 321
Wien's law 186
wire
– boundaries 19
– fluctuations 31
– InGaAs/InP 21
– structures 19
wordline (WL) 361ff.

z

Zeeman
– effect 57
– splitting 77, 125
zero-dimensional structure 4